BASIC PHYSIOLOGY
AND
ANATOMY

BASIC PHYSIOLOGY
AND
ANATOMY

THIRD EDITION

Ellen E. Chaffee, R.N., M.N., M.Litt.

Science Coordinator and *Associate Professor of Nursing*
University of Pittsburgh School of Nursing

Esther M. Greisheimer, Ph.D., M.D.

Professor Emeritus of Physiology
Temple University Medical Center, Philadelphia
Formerly Professor of Physiology
Woman's Medical College of Pennsylvania, Philadelphia
Formerly Associate Professor of Physiology
The University of Minnesota, Minneapolis

J. B. LIPPINCOTT COMPANY
Philadelphia   New York   Toronto

Third Edition

Distributed in Great Britain by
Blackwell Scientific Publications
London   Oxford   Edinburgh

ISBN 0-397-47299-4

Library of Congress Catalog Card Number 73-20126

Printed in the United States of America

1 3 5 7 9 8 6 4 2

Library of Congress Cataloging in Publication Data

Chaffee, Ellen E.
  Basic physiology and anatomy.

  Includes bibliographies.
  1. Human physiology. 2. Anatomy, Human.
I. Greisheimer, Esther Maud, 1891-   joint author.
II. Title. [DNLM: 1. Anatomy. 2. Physiology.
QS4 C433b 1974]
QP34.5.C48 1974   612   73-20126
ISBN 0-397-47299-4

To our students—
past, present, and future

Preface to the Third Edition

As investigators in laboratories all over the world continue to unravel the mysteries of the human body, students and teachers of physiology and anatomy are hard pressed to keep up with the flood of discoveries. This state of affairs is complicated by the fact that scientists often propose more than one theory to explain the significance of their laboratory work; also there frequently are differences of opinion. Such details can be included in the more advanced textbooks, but for the beginning student we feel that there should be a judicious paring of such information. Hopefully, we have accomplished this task.

However, when one attempts to describe the awesome wonders of the human body, it is extremely difficult to decide just what must be omitted for the sake of brevity in an introductory textbook and just what must be included for the equally important sake of clarity. We realize that it would be impossible to write a textbook that is just "perfect" for everyone, but we have endeavored to meet the expressed needs of students and teachers in many schools across the country. However, in the final analysis, the educational process depends to a large extent upon teachers and students who are willing to go beyond the confines of any single textbook. To facilitate such activity we have included a list of supplementary readings for each chapter.

An outstanding feature of this edition is the incorporation of approximately 220 illustrations executed by Neil O. Hardy, the eminent medical illustrator. Mr. Hardy's drawings reflect not only his artistic competence, but also his imaginative use of the third dimension.

In response to many requests for expanded coverage of intermediary metabolism, we sought the assistance of Virginia G. Braley, R.N., Ph.D. Dr. Braley, an Associate Professor of Nursing at the University of Pittsburgh, has had many years of experience in teaching this difficult subject matter; thus we feel fortunate in having her collaborate with us. Chapter 17 has been extensively revised and much of it has been completely rewritten; there is expanded coverage of the role of enzyme systems and of the citric acid cycle. The electron transport system has been added, and the section on heat production and temperature regulation has been expanded. Dr. Braley also has written an entirely new chapter dealing with Body Fluids and Electrolytes, including brief discussions of the effects of fluid and electrolyte imbalances.

In addition to updating content in all chapters, some areas of major additions, revision, or expansion include the following: exchange of fluids through capillary membranes; excitability of muscle tissue; transmission across synapses; spinal reflexes; functions of the cerebrum; the limbic system; sleep; the effects of drug abuse on the central nervous system; the nature of sound; the thymus; blood pressure; regulation of the circulation; circulatory shock; pulmonary volumes; the work of breathing; exchange of gases; the vitamins; cyclic AMP; the prostaglandins; stress concepts; and effects of nondisjunction of the sex chromosomes (the latter by Dr. Braley).

We hope that this textbook will be helpful to students and teachers alike as they work together in studying the human body. As in previous editions, we shall welcome all evaluations and suggestions for improvement.

In conclusion, I want to extend all good wishes to Dr. Esther M. Greisheimer, my friend and coauthor of this textbook. Dr. Greisheimer has retired for a well-deserved rest from the rigors of teaching, writing, and research. All of us who have been privileged to know this remarkable woman wish her much happiness in the years ahead. Personally, I want to acknowledge once again my very deep appreciation and gratitude for her wise counsel and guidance. The following quotation from *The Education of Henry Adams* seems very appropriate when one considers Dr. Greisheimer's many contributions through the years:

"A teacher affects eternity—he never can tell where his influence stops."

ELLEN E. CHAFFEE

CONTENTS

Introduction to Human Anatomy and Physiology

"*Another book was opened, which is the book of life.*"

Rev. 20:12

From the simplest of microscopic creatures to the complex body of man runs a protoplasmic thread that distinguishes living matter from the inanimate stones and clods. Each one of us knows how it feels to be alive, but even our greatest scientists as yet cannot explain the mystery of life itself.

However, centuries of observation and experimentation have led to the establishment of certain facts and theories which help in our general understanding of all living matter, whether it be plant or animal. This highly organized body of knowledge is referred to as **biology,** *the study of life and living things. Such a vast subject would be impossible to study were it not broken down into plant and animal divisions, plus an infinite variety of subdivisions, two of which are human anatomy and physiology. These biologic sciences are of basic importance to those who would not only recognize and preserve good health, but give intelligent care to the sick.*

ORIENTATION

The significance of anatomy

Human anatomy is primarily concerned with **structure,** with the way in which our bodies are built. The word *anatomy* is of Greek origin; it is derived from the prefix *ana-* meaning *apart* and the root *tome* meaning *a cutting,* so it literally means *a cutting apart.* The Latin word *dissection* is derived from the prefix *dis-* meaning *apart* and the root *secare* meaning *to cut,* so it likewise means *to cut apart.* Although both terms, anatomy and dissection, mean a cutting apart, the former has acquired a much broader application and now includes many subdivisions. The terms indicate that the anatomy of early days consisted of that learned by dissecting human bodies.

Our study of anatomy will be taken up according to the systems of the body (**systemic anatomy**) and will be devoted primarily to **gross anatomy,** the study of structures which can be seen with the naked eye. The study of **microscopic anatomy** will be limited to relatively brief studies of cells (**cytology**) and tissues (**histology**) as seen with the aid of a microscope. There are many other subdivisions of anatomy, but they are beyond the scope of this textbook.

The significance of physiology

Human physiology is concerned with **function,** with the study of all activities which are characteristic of living matter. It is perhaps the most fascinating of all the sciences and is essential for total understanding of the body, but a knowledge of basic anatomy is prerequisite to the comprehension of physiology. In other words, anatomy and physiology are inseparable companions when one considers the body as an integrated whole. For example, cells are not only the structural units of the body, but they are the functional units as well.

All living organisms are made up of cells, and each cell, in turn, is composed of **protoplasm.** T. H. Huxley described protoplasm as "the physical basis of life," and this semiliquid material is indeed the essential substance in every living cell. Whether an organism exists as a single cell, such as the lowly amoeba, or as the fantastic human complex of trillions of cells, there are certain **basic life processes** which are essential for its survival in a hostile world. These processes are as follows:

Excitability, or **irritability,** is the capacity to respond to a stimulus (any change in the environment). We constantly use this fundamental characteristic of excitability as a test of aliveness or awareness in another person or in an animal. Children pinch each other, call each other names, poke animals, and make loud noises to see if they can call forth responses in any living organism in their vicinity.

Conductivity refers to the capacity of a cell to transmit a wave of excitation from the point of stimulation to other parts. In higher animals, irritability and conductivity are developed most highly in nerve cells.

Contractility refers to the ability of a cell to undergo shortening and to change its form. This property is highly developed in muscle cells, and makes possible all muscular activity.

Metabolism in its broadest sense includes all physical and chemical processes involved in the activities of life. However, in common use it is limited to those processes by which food substances are converted into energy or into complex substances such as protoplasm; in this respect, there are two phases of metabolism: (1) *anabolism,* or the *building-up* processes which involve the formation of complex substances from simpler ones, and (2) *catabolism,* or the *breaking-down* processes, particularly the breakdown of food with a resultant liberation of energy.

The detailed processes of food-getting and food-utilization, which are such an important part of the metabolic picture, include ingestion, digestion, absorption, assimilation, respiration, secretion, excretion, and egestion.

The **ingestion,** or taking in, of food is followed by **digestion,** a process whereby complex foods are broken down to simple substances which can be absorbed, or taken into the protoplasm of a cell. **Assimilation** refers to the processes within cells by which the absorbed material is used to build protoplasm. **Respiration** usually involves the use of oxygen and the production of carbon dioxide (CO_2) with the release of energy.

Excretion refers to the ability of a cell to eliminate waste products which it has produced during its metabolic activities (e.g., water, urea, CO_2). Specialized cells also can remove raw materials from the blood and build useful substances, such as digestive juices and hormones; the process whereby such substances leave the cell is called **secretion. Egestion** refers to the elimination of indigestible or unusable material which has not taken part in metabolic activities.

Growth is the increase in the amount of protoplasm from sources within an organism. It is due to an increase in the number of cells by cell division, rather than to an increase in the size of cells. It includes the process of repair by which damaged parts are replaced. When one reaches the adult stage the body no longer increases in size, and repair becomes predominant as growth ceases. **Reproduction** involves not only the formation of new cells within the growing body, but also the formation of entirely new individuals. This type of reproduction keeps life going from one generation to the next.

Physiology is a young science in comparison with anatomy, and it is very dependent on several other branches of knowledge. For example, a knowledge of chemistry is indispensable in unraveling the secrets of digestion, muscle contraction, and metabolism as a whole. A knowledge of physics aids our understanding of the mechanics of respiration, vision, and hearing. The science of pathology (study of the causes and the effects of disease) has contributed much information to physiology regarding the function of the endocrine glands and the importance of vitamins. The pathologic changes in the body which follow disturbances in the endocrine system or occur when one is deprived of certain vitamins disclose facts of great importance to the physiologist. The functions of various parts of the nervous system have been made more understandable by the contributions of psychology and psychiatry. Surgery has extended the boundaries of physiologic knowledge by noting the effects which follow removal of various organs or parts of organs. The science of electronics is making it possible for investigators to study many functions of the body by entirely new methods and is responsible for the revision of some of the ideas of function.

As you study human anatomy and correlate it with physiology, you soon will begin to appreciate the beauty and the wonder of the various systems. Most fascinating of all are the amazing interrelationships between all of the systems. You will see how various reflexes tend to keep heart rate, blood pressure, respiration, body temperature, and composition of the blood relatively constant. It is only when the body fails to maintain this delicate balance that sickness results. The acquisition of a working knowledge of anatomy and physiology is like laying a foundation for a house; it should be done well so that the base for advance study is sound.

DESCRIPTIVE TERMINOLOGY

Throughout the study of anatomy and physiology you will be confronted constantly with new terminology. The process of adding these words to your scientific vocabulary will be greatly simplified if you try to link them with their origins or meanings. Anatomic terms are as essential to the student of anatomy as tools are to a carpenter, and without a growing vocabulary this course will become more bewildering every day.

Many of the terms in use today have been derived from Greek or Latin sources; they consist of prefixes, suffixes, and combining forms. For example, the word *pericarditis* tells us its meaning when we look at its construction. *Peri-* is a commonly encountered prefix meaning *immediately around, cardi-* is a combining form used in reference to the heart, and the suffix *-itis* means *inflammation.* Therefore, pericarditis refers to an inflammation of the membrane which surrounds the heart. Another common suffix is *-al,* which means *pertaining to; radial* means *pertaining to the radius,* and *femoral* means *pertaining to the femur.* The root *myo-* means *muscle; myocardium* means the *muscle of the heart,* and *myoneural* junction means the junction between *muscle and nerve.* Additional prefixes, combining forms, and suffixes used in the formation of anatomic and medical words may be found on pages 527 to 529.

You will encounter many of the same terms again and again throughout the course and will soon learn to recognize them as old friends. The use of a good medical dictionary, in addition to the glossary in the back of this book, will help you to understand and remember new terms as they appear. Study with the intent to remember; write the difficult words, spell them aloud, use them in conversation. Since you need to learn countless terms, you may as well make a game of it and have fun in the process. If you really put forth effort you will be gratified with your progress.

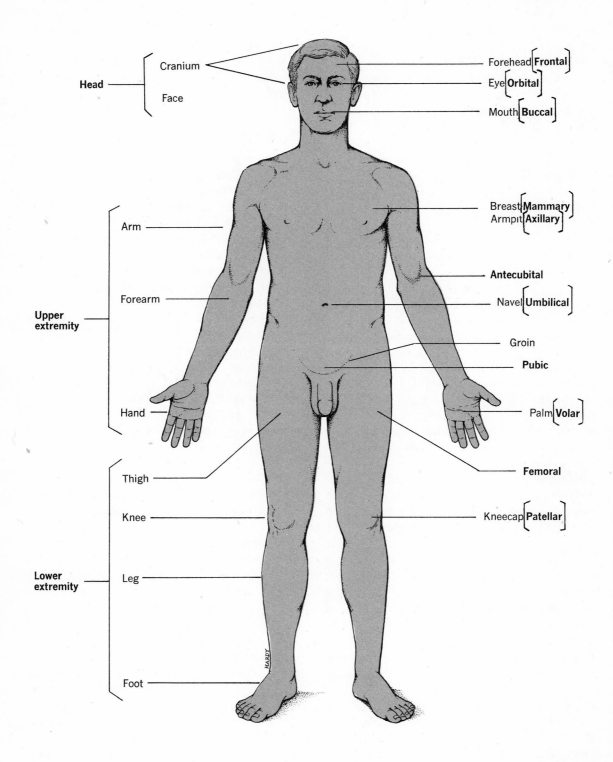

Fig. 1-1. The anatomic position, anterior view. Common terms used in reference to body parts are shown in roman type, anatomical terms for regions are in **boldface type.**

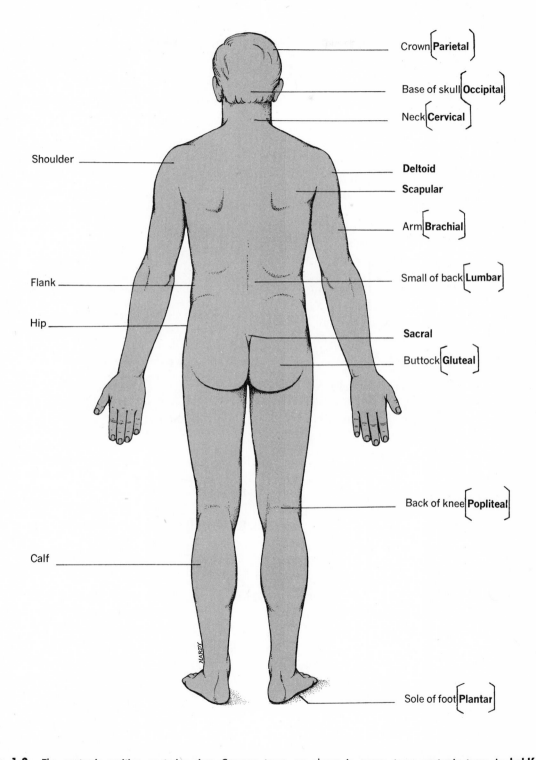

Crown [Parietal]

Base of skull [Occipital]

Neck [Cervical]

Shoulder

Deltoid

Scapular

Arm [Brachial]

Flank

Small of back [Lumbar]

Hip

Sacral

Buttock [Gluteal]

Back of knee [Popliteal]

Calf

Sole of foot [Plantar]

Fig. 1-2. The anatomic position, posterior view. Common terms are shown in roman type, anatomic terms in **boldface type.**

Probably never again in your educational journey will you be able to see daily progress so clearly as in anatomy. But if you neglect your study for a few days and get behind you will find a mountainous task awaiting you. Your attitude toward the course may well determine your level of achievement.

Body parts and regions

Familiarity with terminology used in reference to the surface anatomy of your body should precede any study of underlying structures. Figures 1-1 and 1-2 illustrate the various parts and regions of the body as a whole; common words appear in roman, or regular type, while the anatomic terms which should be added to your vocabulary appear in boldface type. For example, the forehead is the **frontal** region; the neck is the **cervical** region. If you wish to avoid confusion during the study of bones and muscles, you should no longer refer to your **upper extremities** as arms or to your **lower extremities** as legs; pay particular attention to these parts and their subdivisions as illustrated in Figure 1-1.

For convenience of description and reference, the abdomen is arbitrarily divided into nine regions by imaginary planes, two horizontal and two vertical, as shown by lines drawn on the body surface (Fig. 1-3). The location of the various abdominal viscera is noted in Table 1-1.

Terms of location and position

Of fundamental importance are the terms used in describing the location and the position of various body parts. When using these terms we must always consider the body to be in the **anatomic position,** or standing erect, facing forward, with the upper extremities at the sides and palms turned forward (Fig. 1-1).

superior, or *cranial:* above, or toward the head end of the body (e.g., the chest is superior to the abdomen).

inferior, or *caudal:* below, or toward the tail end of the body (e.g., the neck is inferior to the head).

anterior, or *ventral:* nearer the front or belly side of the body (e.g., the breasts are on the anterior, or ventral, aspect of the chest).

Table 1-1   Abdominal Viscera

Viscus	Location
Right hypochondriac	Greater part of right lobe of the liver; hepatic flexure of the colon; and part of the right kidney
Right lateral, or lumbar	Ascending colon; part of right kidney; and, sometimes, part of ileum
Right inguinal, or iliac	Cecum and appendix; termination of ileum
Epigastric	Greater part of left lobe and part of right lobe of liver with gallbladder; part of stomach; part of duodenum, pancreas, upper end of spleen; parts of kidneys and suprarenal glands
Umbilical	Greater part of transverse colon; part of duodenum; part of jejunum and ileum; part of mesentery and greater omentum; part of kidneys and ureters
Hypogastric, or pubic	Ileum; bladder in children; bladder, if distended, in adults; uterus when pregnant; sigmoid flexure
Left hypochondriac	Part of stomach; part of spleen; tail of pancreas; splenic flexure of colon; part of left kidney; and, sometimes, part of left lobe of liver
Left lateral, or lumbar	Descending colon; part of jejunum; part of left kidney
Left inguinal, or iliac	Sigmoid colon; jejunum and ileum

posterior, or *dorsal:* nearer the back side of the body (e.g., the shoulder blades are on the posterior, or dorsal, aspect of the chest).

medial: nearer the midline of the body (e.g., the nose is medial to the eyes).

lateral: farther from the midline of the body (e.g., the eyes are lateral to the nose).

internal: deeper within the body (e.g., the internal surface of the nose is lined with mucous membrane).

external: nearer the outer surface of the body (e.g., the external surface of the nose is covered with skin).

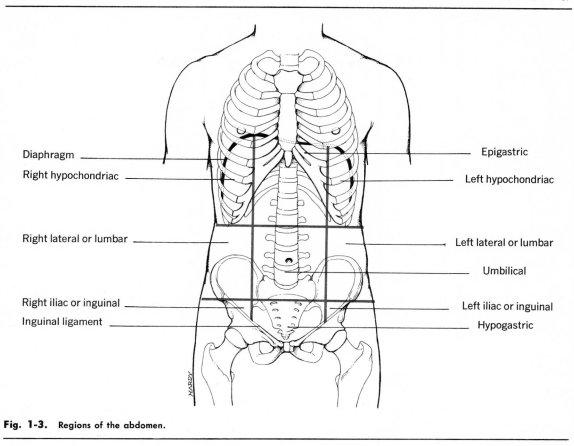

Fig. 1-3. Regions of the abdomen.

Diaphragm

Right hypochondriac

Right lateral or lumbar

Right iliac or inguinal

Inguinal ligament

Epigastric

Left hypochondriac

Left lateral or lumbar

Umbilical

Left iliac or inguinal

Hypogastric

NOTE: Internal and external may also be used in reference to nearness to or distance from the midline. For example, the medial rectus muscle of the eye also may be called the internal rectus, and the lateral rectus muscle of the eye may be called the external rectus.

proximal: nearer to the body or to the origin of a part (e.g., with respect to the upper extremity, the arm is proximal to the forearm).

distal: farther from the body or from the origin of a part (e.g., the hand is distal to the elbow).

central: the principal part; situated at, or related to, a center (e.g., the brain and the spinal cord comprise the central nervous system).

peripheral: toward the surface of the body, or extensions from, the principal part or center (e.g., the peripheral nervous system refers to nerves which go out to or come in from all parts of the body).

parietal: pertaining to the walls of a cavity (e.g.,

parietal blood vessels are those which supply structures in the walls of a cavity, such as the chest or the abdomen).

visceral: pertaining to the organs within a cavity (e.g., visceral blood vessels supply the viscera or organs within a cavity, such as the chest or the abdomen).

The preceding terms will be used with great frequency during this course, and after a bit of practice they should be an integral part of your anatomic vocabulary. Figure 1-4 illustrates comparative terminology in quadruped and man.

The fundamental planes of the body

In order to show the various internal structures of the body, sections are cut in different planes as shown in Figure 1-5.

The **midsagittal plane** passes through the midline of the body lengthwise, from front to back, and divides it into exactly even right and left

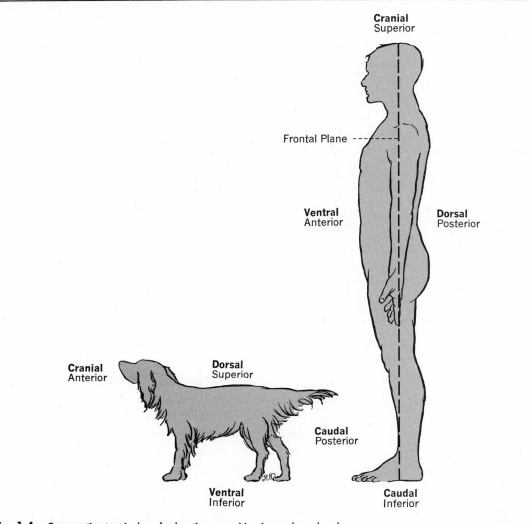

Fig. 1-4. Comparative terminology for location or position in quadruped and man.

halves, excluding such internal organs as the heart and the liver, which do not lie exactly in the midline. This plane is so named because it passes through the sagittal suture of the skull (p. 68). Any other plane which divides the body into right and left portions but does not pass through the midline of the body is referred to as a **sagittal** plane.

The **coronal,** or **frontal plane,** takes its name from the coronal suture of the skull (p. 68). It passes through the body from top to bottom and divides it into front and back portions.

The **transverse,** or **horizontal plane,** passes through the body dividing it into upper and lower portions.

STRUCTURAL FEATURES OF THE BODY

Just as a city, a factory, a hospital, or even your wristwatch has a plan of organization, so does the human body. In each case, a large unit is made up of smaller units, all of which work together in close harmony for the welfare of the unit as a whole. Should one or more of the small units fail to carry out its function, the end result could be a breakdown of the entire complex; for example, just as a broken mainspring stops the action of your watch, so does a failing heart stop the activity of your body. Throughout your study of anatomy and physiology, always remember that the human body is an integrated whole.

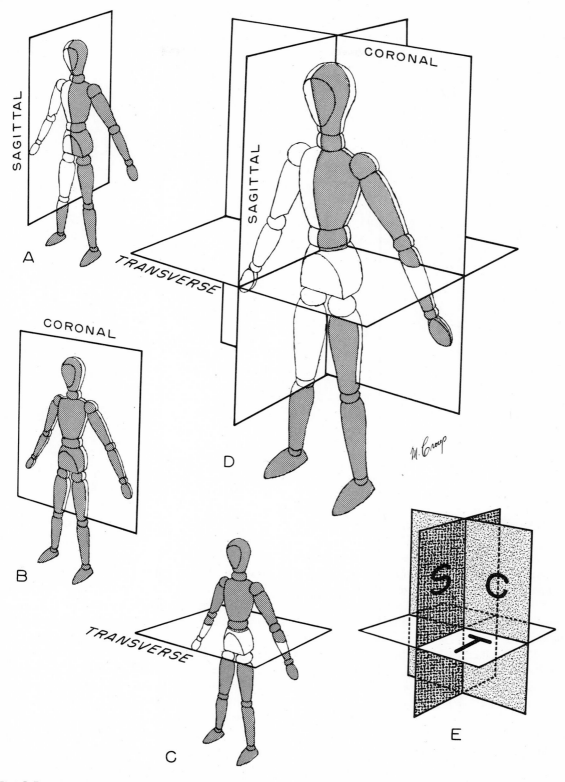

Fig. 1-5. The fundamental planes of the body.

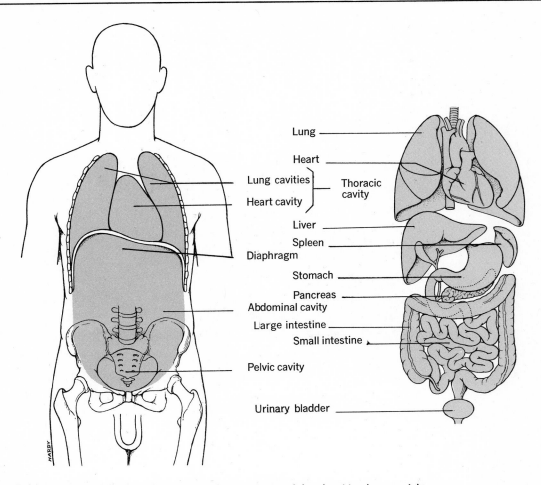

Fig. 1-6A.   Diagram of the vertebrate structure of man. Contents of dorsal cavities shown at right.

The vertebrate structure of man

Man is a member of the vertebrate group of animals. In addition to possessing a vertebral column (backbone), vertebrates are characterized by the presence of dorsal and ventral cavities. Figure 1-6A and B illustrates the vertebrate structure of man.

The **dorsal cavity** lies within the skull and the backbone. It consists of two portions: (1) the *cranial portion*, which is occupied by the brain, and (2) the *vertebral portion*, which is occupied by the spinal cord.

The **ventral cavity** is divided into thoracic and abdominal portions by a muscular partition called the diaphragm. Each of these portions may be subdivided further. The *thoracic cavity* comprises two pleural compartments, each of which is occupied by a lung, and the pericardial compartment, which is occupied by the heart. In addition to the heart, the space between the lungs contains the following: trachea, esophagus, thymus, large blood vessels, lymphatic vessels, and nerves.

The *abdominal cavity* lies inferior to the diaphragm. It is occupied by the stomach, the small intestine, most of the large intestine, the liver, the gallbladder, the pancreas, the spleen, the kidneys, the adrenal glands, and the ureters. The lowermost portion of the abdominal cavity is called the *pelvic cavity*; it is occupied by the urinary bladder, the end of the large intestine (sigmoid colon and rectum), and certain reproductive organs (uterus, uterine tubes, and ovaries in the female; prostate gland and seminal vesicles in the male).

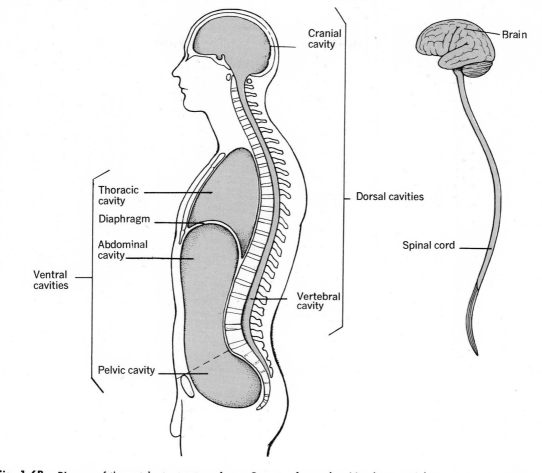

Fig. 1-6B. Diagram of the vertebrate structure of man. Contents of ventral cavities shown at right.

Figures 1-7 and 1-8 illustrate the anterior and the posterior surface projections of the various organs to help with your orientation to their location.

Components

Detailed information relating to the structural and the functional units of the body will be found in subsequent chapters. However, it usually is helpful if one has a general idea of the total organizational picture before studying the components.

Let us compare the organization of the body with that of a large hospital. When you first walk into a hospital corridor you see a large number of persons, each walking briskly about, apparently busy with his own tasks.

Each person may be thought of as a unit and is comparable with a **cell** in the body. There is a great difference in appearance among the persons whom you see and also in the work that each does. There is an even greater difference among the various cells of the body and the functions that they perform.

As you become a little better acquainted with the hospital, you realize that the persons who do a certain type of work, such as nurses, constitute one large group; and each large group carries out certain definite tasks. The nurses give nursing care to patients; the physicians take histories, do physical examinations, and prescribe certain tests, medications, and treatments for patients. Each of these groups, the nurses and the physicians, is made up of persons who are performing specific

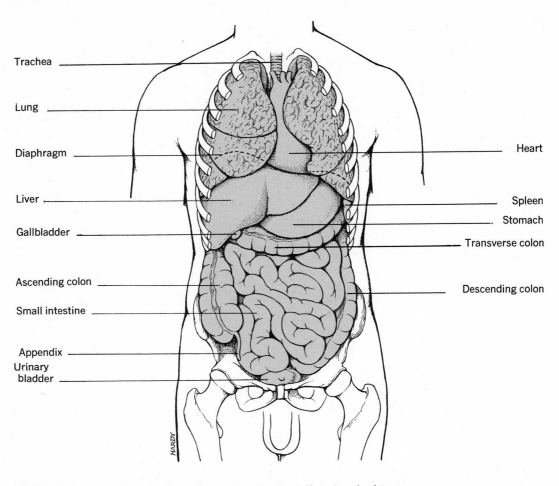

Fig. 1-7. The anterior surface projection of some of the thoracic, abdominal, and pelvic viscera.

functions. Similarly, in the body there are groups of cells which perform specific functions. Such groups of cells are called **tissues.**

In a larger sense, nurses and physicians work together to improve the health of patients; in other words, two groups with different individual functions work together for one purpose. Likewise, in the body, tissues of various kinds are grouped together to form **organs,** which perform a surprising number of related functions.

No matter how efficient nurses and physicians may be, hospitals would not exist long without supervisors, maintenance men, elevator operators, telephone operators, cooks, maids, and a host of other workers to carry out innumerable duties. In the body, groups of organs work together to accomplish a particular function, such as to supply the body with oxygen; these groups of organs are called **systems.** It is as impossible for the body to function properly without the cooperation of each individual cell as it is for a hospital to run efficiently without the cooperation of each person, no matter whether the work is giving nursing care or cooking meals.

Without carrying the analogy further, we shall list the systems of the body. They are: the skeletal, muscular, nervous, digestive, respiratory, circulatory, excretory, reproductive, and endocrine systems. Our plan is to have you study cells, then tissues, and finally systems.

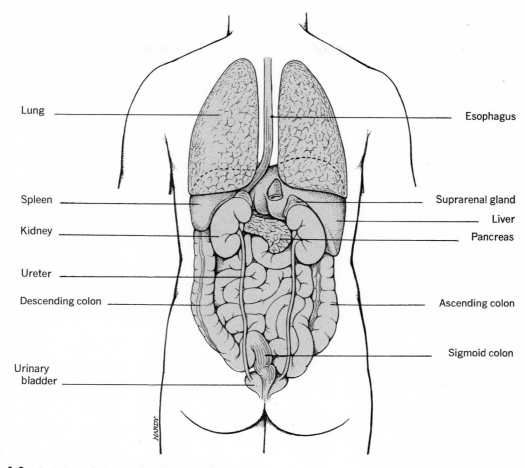

Lung

Esophagus

Spleen

Suprarenal gland

Liver

Kidney

Pancreas

Ureter

Descending colon

Ascending colon

Sigmoid colon

Urinary
bladder

Fig. 1-8. Posterior surface projection of some of the viscera.

SUMMARY

1. **Biology—organized body of knowledge dealing with life and living things**
A. Anatomy concerned with structure of body
  a. Systemic: studied according to body systems
  b. Gross: seen with naked eye
  c. Microscopic: seen only with microscope
B. Physiology concerned with function of body
  a. Cells are functional and structural units
  b. Basic life processes essential to survival
    (1) Excitability or irritability
    (2) Conductivity
    (3) Contractility
    (4) Metabolism
    (5) Growth and reproduction
  c. Dependent on other branches of knowledge
2. **Descriptive terminology**
A. Body parts and regions (reviews Figs. 1-1 and 1-2)
B. Terms of location and position (body always considered to be in anatomic position)

a. Superior (cranial): above; inferior (caudal): below
b. Anterior (ventral): in front; posterior (dorsal): in back
c. Medial: nearer midline; lateral: farther from midline
d. Internal: deeper within body; external: nearer outer surface of body
e. Proximal: nearer the origin; distal: farther from origin
f. Central: principal part; peripheral: toward body surface, extensions from the center
g. Parietal: pertaining to walls of cavity; visceral: pertaining to organs within a cavity
C. Fundamental planes
a. Midsagittal: exactly even right and left halves of body
b. Sagittal: right and left halves, but not necessarily even

    c. Coronal or frontal: front and back portions
    d. Transverse or horizontal: upper and lower portions

3. **Structural features of the human body**
A. Cavities
    a. Dorsal: cranial portion; vertebral portion
    b. Ventral
      (1) Thoracic cavity
      (2) Abdominal and pelvic cavities
B. Component units: cells, tissues, organs, and systems; all functioning as an integrated whole

QUESTIONS FOR REVIEW

1. Now that you have begun the study of physiology and anatomy, how would you explain the following terms to an interested friend?
Excitability, excretion, growth, anterior, inferior, medial, proximal, visceral, transverse plane.
2. What anatomic term would apply to each of the following regions of the body?
Forehead, armpit, navel, thigh, base of skull, small of back, buttock.
3. What is the difference between a sagittal plane and a coronal plane?
4. Distinguish between the thoracic and abdominal cavities.

REFERENCES AND SUPPLEMENTAL READINGS

Goss, C. M. (ed.): Gray's Anatomy of the Human Body, ed. 28. Philadelphia, Lea & Febiger, 1966.
Guyton, A. C.: Textbook of Medical Physiology, ed. 4. Philadelphia, W. B. Saunders, 1971.

Organization of the Living Body

2

"Thou from primeval nothingness did call First chaos, then existence."
"Ode to God" • Gabriel Derzhavin

Perhaps the most striking property of living matter is its organization. A cell is the smallest unit of life, a microscopic bit of protoplasm; yet each of its parts, as well as the cell as a whole, is so organized that it performs special functions. For example, muscle cells are so constructed that they can shorten and so move part of the body; nerve cells can transmit impulses to and from the brain and the spinal cord; gland cells produce secretions that are essential to the health of the body.

Although cells vary in function, size, shape, and composition, they have much in common. In this chapter we shall consider not only the structural and functional similarities between cells, but also the way in which they combine to form tissues and membranes.

PROTOPLASM

Protoplasm is the only known form of matter in which life is evident. It is a viscid, translucent, colloidal substance that makes up the essential material of every cell.

Composition

Even though the chemical composition of protoplasm has been determined with accuracy, scientists cannot yet make synthetic protoplasm in the laboratory. It is the *organization* of protoplasm that is unique, and apparently cannot be duplicated. None of the chemical elements composing protoplasm are any different from those which are found in nonliving material, but their organization with relation to each other spells the difference between life and death for any given cell. Table 2-1 merely indicates those elements and compounds which are found in protoplasm; the way in which these substances are put together to form living protoplasm is not yet known.

While examining Table 2-1 you should keep in mind that only the gases (oxygen and nitrogen) are present in the body in their elemental state. The other elements are present in the form of inorganic compounds (those lacking carbon) and organic compounds (those containing carbon). The percentage distributions as shown are for the adult human body as a whole.

Importance of inorganic compounds

Water is by far the most abundant of the compounds in all living tissue, and life can continue for only a few days without it. Some of the functions of water are as follows:

1. Water serves as a solvent. It holds the various components of protoplasm in solution so that they are free to move about within the cell or from one cell to another. Most chemical reactions take place only in solution.

2. The high chemical stability of water makes it possible for chemical reactions to occur in water without involving the water itself.

3. Many substances ionize in water; that is, their molecules separate into ions which bear electrical charges. For example, sodium chloride (NaCl) separates into a positive sodium ion (Na^+) and a negative chloride ion (Cl^-). Such

Table 2-1   Composition of Protoplasm

Atomic Structure	Percentage
Elements	
Oxygen	65.00
Carbon	18.00
Hydrogen	10.00
Nitrogen	3.00
Calcium	2.00
Phosphorus	1.00
Potassium	0.35
Sulfur	0.25
Chlorine	0.15
Sodium	0.15
Magnesium	0.05
Iron	0.004
Other elements	0.046
Compounds	
INORGANIC	
Water	50 to 60
Inorganic salts	4.4
ORGANIC	
Proteins	16.0
Lipids	13.0
Carbohydrates	0.6

substances in solution are called *electrolytes* (capable of conducting an electric current). The properties and the classes of electrolytes are determined by the kinds of ions they yield when they dissolve in water. These classes are acids, alkalies or bases, and salts. All of these are of great importance in living matter.

The maintenance of a proper balance between acid and base in the body is essential to life itself. Hydrogen (H^+) and hydroxyl (OH^-) ions are involved in all chemical processes taking place in living cells, and their concentration determines the degree of acidity or alkalinity of body fluids (i.e., the pH). If there are more hydrogen ions than hydroxyl ions in a fluid, it is acid in reaction. On the other hand, an alkaline fluid would have more hydroxyl ions than hydrogen ions. Table 2-2 will refresh your memory in regard to the determination of pH.

Acids are formed constantly in the body as a result of metabolic activity. For example, carbonic acid (or its precursor, carbon dioxide) is produced by the oxidation of all types of food. Substances which produce *bases*, such as sodium, potassium,

Table 2-2 Determination of pH

Normality			Concentration H ions	Concentration OH ions	pH
Normal HCl	I N C R E A S I N G	A C I D I T Y	1	10^{-14}	0
0.1 N HCl			10^{-1}	10^{-13}	1*
0.01 N HCl			10^{-2}	10^{-12}	2
0.001 N HCl			10^{-3}	10^{-11}	3
0.0001 N HCl			10^{-4}	10^{-10}	4
0.00001 N HCl			10^{-5}	10^{-9}	5
0.000001 N HCl			10^{-6}	10^{-8}	6
Pure water-neutral			10^{-7}	10^{-7}	7
0.000001 N NaOH	I N C R E A S I N G	A L K A L I N I T Y	10^{-8}	10^{-6}	8
0.00001 N NaOH			10^{-9}	10^{-5}	9
0.0001 N NaOH			10^{-10}	10^{-4}	10
0.001 N NaOH			10^{-11}	10^{-3}	11
0.01 N NaOH			10^{-12}	10^{-2}	12
0.1 N NaOH			10^{-13}	10^{-1}	13
Normal NaOH			10^{-14}	1	14

* It is evident that every rise of 1 in pH means lowering of the concentration of H ions to 1/10 of its previous value.

calcium, and magnesium, are taken into the body in our food. Since living cells are very sensitive to changes in reaction, toward either acidity or alkalinity, there are various means of keeping the pH relatively constant in the body. For example, buffer salts in the blood (p. 465) tend to keep the pH of that fluid within the very narrow range of 7.35 to 7.45. A slight shift in the pH of the blood, which indicates the condition of the entire body, can seriously alter the function of many cells. Acid-base balance is considered in greater detail on pages 464 to 468.

4. Another role played by water is in the regulation of body temperature. Evaporation of the water in perspiration cools the skin and helps to prevent a rise in body temperature. Human cells function most efficiently at optimum body temperature, which is 98.6° F. (37° C.). Elevation of the body temperature (fever) may interfere with the activity of many cells.

Inorganic salts are found in the cells and the fluids (blood, interstitial fluid, and lymph) of the body. They occur primarily as chlorides, carbonates, sulfates, and phosphates of sodium, potassium, calcium, and magnesium. The principal *extracellular ions* (outside of the cell) are sodium and chloride; the principal *intracellular ions*

(within the cell) are potassium and phosphate. Inorganic salts are needed for such things as:

1. Maintenance of proper osmotic conditions;

2. Maintenance of proper acid-base balance (e.g., the buffer salts in the blood);

3. Clotting of blood (calcium);

4. Development of bones and teeth (calcium and phosphorus);

5. Formation of hemoglobin in red blood cells (iron);

6. Normal functioning of muscle and nerve (sodium, potassium, and calcium).

The proper concentration and distribution of various salts is referred to as **electrolyte balance.** Alteration of this balance can lead to serious disturbances. Chapter 19 is devoted to the study of body fluids and electrolytes; it describes the highly complicated process of fluid and electrolyte balance, which involves not only the intake of adequate food and water, but also the activity of various organs such as the kidneys, the heart, the lungs, and certain endocrine glands.

Importance of organic compounds

Proteins form the framework of protoplasm and are the most abundant of the organic constituents. Growth of new tissue and repair of old tissue depend on the availability of protein. Thus growing children, elderly persons, and those who have undergone surgical operations or other stressful experiences need more protein in their diet than do healthy young adults.

Proteins are large molecules consisting of many small structural units called *amino acids;* they are the building blocks of the body. Twenty different amino acids commonly are found in protein molecules, but the number and the arrangement of these building blocks vary with the kind of protein that is being constructed. For example, the amino acid arrangements in a muscle cell are quite different from those found in a nerve cell, inasmuch as each type of tissue is characterized by its own specific kinds of protein. The importance of the chromosomes and genes in protein synthesis is described on page 28.

In addition to the role they play in building and repairing tissues, proteins are also extremely important in maintaining the colloid osmotic pressure of the blood (p. 34), and in the formation

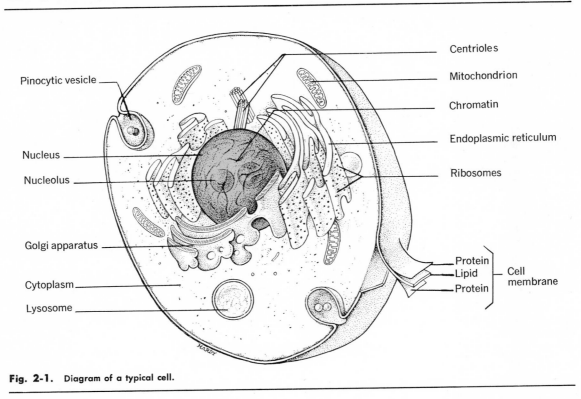

Fig. 2-1. Diagram of a typical cell.

of enzymes and certain hormones. Under certain conditions protein is oxidized to supply energy for the body.

Carbohydrates include the starches and the sugars. These large units are broken down by digestive processes to simple sugars such as *glucose*, which is the principal sugar present in the blood. Carbohydrates are the most abundant and economical source of energy. However, if one's diet includes more carbohydrate than is required to meet the immediate energy needs of the body, some of the excess is converted to *glycogen* (animal starch) and some to adipose tissue (fat) for storage purposes. When one goes without food (as in starvation) the glycogen and fat deposits are used up first; then the protein of protoplasm itself is broken down and oxidized, because energy needs take precedence over all others. This gradual destruction of protoplasm accounts for the extreme emaciation and ultimate death of starvation victims.

Lipids include both fats and fat-related substances. They are sources of energy and are stored as reserve food in adipose tissue. These fatty deposits of adipose tissue serve also as shock absorb-

ers in the support and the protection of various organs and as an insulator, preventing excessive loss of body heat. Lipids are present in high concentration in cell membranes, and they play an important role in maintaining normal permeability (p. 22).

Therefore, each foodstuff has its own contribution to make to the welfare of the body as a whole, and a balanced diet contributes greatly to the maintenance of good health.

ORGANIZATION OF THE CELL

At one time cells were considered to be merely microscopic droplets of protoplasm, or little bags of water containing assorted chemicals and enzymes. Now electron microscopy, coupled with sophisticated biochemical research, indicates that each cell is a highly organized molecular factory. Inasmuch as entire books have been written about the detailed structure and function of these microscopic masterpieces, it is obvious that our discussion must be limited to broad concepts and generalizations regarding the living cell. Figure 2-1 illustrates the structure of a so-called typical cell, but it must be kept in mind that there are many variations in cell structure and function. For example, a brain cell does not look or act like a liver cell, yet both cells have a nucleus, a cytoplasm that contains various organelles (little organs), and an external cell membrane. Although these parts have specialized functions, they are so highly organized and interrelated that the cell functions as a unified whole.

The nucleus

The nucleus may quite accurately be termed the control center of the cell. Not only does it exert a controlling influence over the chemical reactions that occur in these factories, but it is also of fundamental importance in reproduction of the cell as a whole. These vital functions depend upon the fact that the nucleus is an amazing storehouse of genetic information.

In a resting cell (i.e., one not ready to divide) the nucleus shows darkly staining granules that are referred to as *chromatin material*. Actually this chromatin material is arranged in loosely coiled strands that are better known as **chromosomes,** but they are not readily visible as such until they become more tightly coiled just prior to division of the cell. Every normal somatic (body) cell has exactly 46 chromosomes in its nucleus, and each chromosome in turn may have thousands of genes distributed along its thread-like structure. The genes are the ultimate units that determine what a cell will be and what activities it will carry out.

Chemically, each chromosome consists of **deoxyribonucleic acid** (*DNA*) combined with a protein, but it is the DNA that constitutes the primary hereditary material of each cell. For an understanding of how genetic information can be stored in a chemical compound, and how such a compound can be precisely duplicated, it is necessary to know the basic structure of DNA.

In 1953, Dr. James Watson and Dr. Francis Crick first introduced the concept that a DNA molecule resembles a spiral ladder. The sides of the ladder are formed by alternating units of deoxyribose sugar and phosphate, while the rungs are composed of pairs of nitrogenous bases. One end of each nitrogenous base is attached to a sugar-phosphate unit, and the other end is linked to its partner on the opposite side of the ladder to complete the rung. These nitrogen compounds are only four in number: adenine, thymine, guanine, and cytosine, or, as they are referred to in scientific shorthand, A, T, G, and C. However, the pairing of these bases is highly specific, as A is so constructed that it will fit precisely only with T; and G, only with C. Thus the sequence of bases along one side of the ladder must necessarily determine the sequence along the other. This fact is of fundamental importance in understanding the ability of DNA to duplicate itself prior to cell division. Figures 2-2 and 2-3 illustrate the basic structure of DNA.

Although the sides of the ladder are an important part of the DNA molecule, it is the order or sequence of bases in the rungs of the ladder that constitutes the genetic code. In other words, the order in which A, T, G, and C appear apparently spells out the genetic instructions that control the activities of the cell in general, and the construction of complex protein molecules in particular. Because most of the molecules in every cell are proteins, their production is the key to life itself. For example, enzymes (the essential organic catalysts of the body) are proteins, and every one of the *thousands* of different chemical reactions that take place in the body is expedited by a specific enzyme. Thus DNA has the monumental task of directing the formation of thousands of different proteins just to meet the enzyme requirements of the body. Added to this task is the formation of many other proteins that serve as the major structural components of skin, muscle, blood vessels, and all internal organs.

How does DNA accomplish this task? First let us think of a protein molecule as resembling a line of alphabet blocks, each lettered block representing one of 20 amino acids; the sequence of these amino acid blocks determines the protein "word" that will be spelled out. Inasmuch as the fact that DNA carries the genetic code letters A, T, G, and C has already been established, it now remains to see how this chemical alphabet of only four letters can be used to write "protein words" that may contain anywhere from a mere 50 letters to tens of thousands of letters. The secret lies in the fact that this four-letter alphabet is used to write only *three-letter code words*. Because a total of only 20 code words (one for each of the amino acids) is needed to construct any one of the thou-

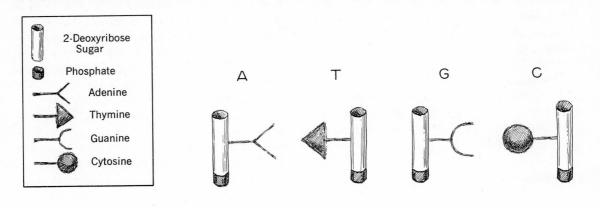

Fig. 2-2. Structural units (nucleotides) of a DNA molecule. Each unit consists of a phosphate group and a sugar group to which is attached a nitrogenous base (adenine or thymine, or guanine or cytosine).

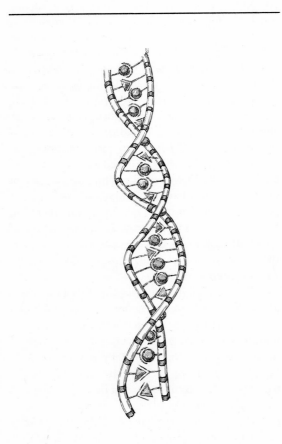

Fig. 2-3. Schematic representation of the spiral ladder arrangement of repeating nucleotide units found in the DNA molecule. It is thought that anywhere from 500 to 1,000 of these rungs make up a single gene, and that there are over 1,000 genes in a single chromosome.

sands of body proteins, the four letters are more than adequate. For example, when one does not have to worry about pronouncing the words, any four letters can be so arranged as to form 64 different three-letter combinations or code words. Then, if one thinks of the 20 amino acid code words as being analogous to 20 letters of the English alphabet, it does not require too much imagination to visualize the almost infinite number of "protein words" of varying lengths that could be constructed. As a matter of fact, it has been estimated that if all of the coded DNA instructions found in one single cell were translated into English they would more than fill 1,000 volumes of an encyclopedia. Figure 2-3 illustrates the spiral ladder appearance of only a small fraction of a DNA molecule—if all of the DNA in a single human cell nucleus were stretched out in a thin thread it would be about 3 feet long!

In addition to the chromosomes, the nucleus also contains one or more small rounded bodies known as **nucleoli** (sing. *nucleolus*). A nucleolus does not have a limiting membrane; it is a collection of loosely bound granules that are composed primarily of protein and ribonucleic acid (RNA). RNA is formed from DNA and is part of an elaborate messenger system that enables DNA to control the manufacture of protein (p. 28). It is believed that the nucleolus may be the place where the actual protein factories (ribosomes) are assembled and whence they migrate into the cytoplasm, where the bulk of protein synthesis occurs.

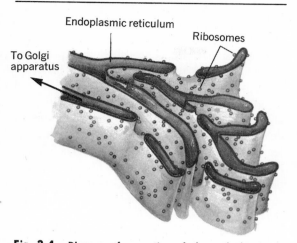

Fig. 2-4. Diagram of a portion of the endoplasmic reticulum and related ribosomes.

The cytoplasm

Although the cell nucleus is the control center, most of the actual work is done in the surrounding cytoplasm, in which are found organelles that have highly specific functions. These organelles include the ribosomes, endoplasmic reticulum, lysosomes, centrioles, and mitochondria.

Ribosomes are tiny rounded bodies that play an important part in the manufacture of protein molecules. They are rich in RNA, and serve as the site where the genetic code is read and then acted upon. Some ribosomes float freely in the cytoplasm, and are concerned with the production of proteins that are required for the cell's own use. Other ribosomes are attached to the membranes of the endoplasmic reticulum, described below, and are generally concerned with the synthesis of protein that is to be exported from the cell and used elsewhere in the body.

The **endoplasmic reticulum** (ER) is a system of delicate membranes so interconnected and arranged as to form a network of tiny canals. The most common type of ER has a granular, rough-surfaced appearance, because the outer surface of the membrane is studded with ribosomes (Fig. 2-4). Communicating with the endoplasmic reticulum is the **Golgi apparatus,** an array of parallel membranes arranged as flattened sacs. It is believed that this apparatus is concerned with the synthesis of large carbohydrate molecules which

are then combined with protein molecules arriving from the endoplasmic reticulum. As these glycoprotein compounds accumulate, the Golgi sacs become fatter. Little vesicles break away and travel to the cell membrane for delivery to the extracellular environment (Fig. 2-5).

Other membranous structures in the cytoplasm are the **lysosomes.** These organelles are tiny closed sacs containing powerful digestive enzymes that are capable of breaking down worn-out parts of the cell itself, as well as molecules of protein, glycogen, and any other substances that may enter the cell by the process of phagocytosis or pinocytosis (p. 25). These enzymes are so potent that if they are released into the cytoplasm through rupture of their confining membrane the entire cell (and possibly some of the surrounding cells) will be destroyed; for this reason lysosomes are sometimes called "suicide bags."

Close to the nucleus, in all cells capable of reproducing themselves, lie one or two cylindrical organelles called *centrioles.* During cell division centrioles are involved in the formation of the mitotic spindle, as described on page 26.

Every living cell requires a constant supply of energy to carry on its many activities, and the powerhouse that produces the greatest amount of this vital commodity is the **mitochondrion.** The number of mitochondria found in any given cell varies from hundreds to thousands, depending upon the amount of energy required by that cell to carry out its assigned functions. The average mitochondrion roughly resembles a bean pod and consists of two surrounding membranes, as shown in Figure 2-6. Various enzymes within the fluid interior make possible the complex oxidative reactions whereby the food that we eat is burned

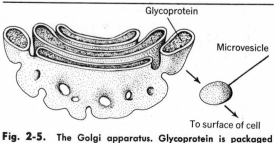

Fig. 2-5. The Golgi apparatus. Glycoprotein is packaged in microvesicles and dispatched toward the cell membrane for secretion into the extracellular environment.

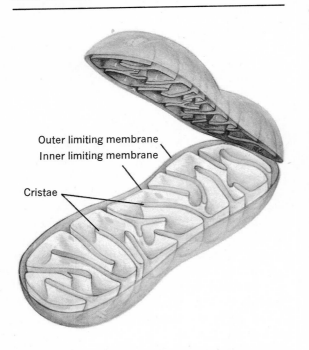

Outer limiting membrane
Inner limiting membrane

Cristae

Fig. 2-6. Diagram of a mitochondrion. The inner limiting membrane is thrown into transverse folds (cristae) that project into the fluid-containing interior.

to release its energy. Other enzymes, located along the folds of the inner membrane, help to trap this food energy and to store it temporarily in the chemical bonds of *adenosine triphosphate* (*ATP*). This energy-rich compound then diffuses out of the mitochondrion and is ready to power the various metabolic processes that take place throughout the cell. Although some ATP is produced by anaerobic reactions in the cytoplasm, 90 to 95 per cent of all the ATP produced by a cell is formed by aerobic reactions within the mitochondria. Energy metabolism is discussed in more detail in Chapter 17.

Other substances that may be present in the cytoplasm include stored foods, pigments, and the like. These may have resulted from activities carried on by the cell, but they are not part of the metabolic machinery.

The cell membrane

The cytoplasm and its contents are separated from the surrounding liquid environment by a thin elastic membrane that is fairly resistant to damage, minor injuries being rapidly repaired by cytoplasmic constituents. The cell membrane consists of two layers of protein molecules enclosing a middle layer of lipid (fat) molecules. Tiny openings or *pores* pass through these layers from one side of the membrane to the other. This membrane does much more than merely hold the cell together. Its physical and chemical properties enable it to play an important regulatory role in determining just what substances may enter or leave the cell and in what quantity at any given time; that is, the cell membrane has the property of **selective permeability**. Generally speaking, there are three main processes whereby substances may pass through the cell membrane: diffusion, active transport, and ingestion (i.e., pinocytosis or phagocytosis).

Diffusion is a physical process that involves the constant scattering or spreading out of molecules of gases or liquids until they tend to become equally concentrated in all parts of the container. This process occurs because molecules are constantly moving, bumping into one another, and bouncing away. Thus they tend to scatter from areas in which they are more highly concentrated to areas in which they are less concentrated. For example, when one enters a room in which there are flowers, such as lilies of the valley, one immediately notes the fragrance. Molecules of the perfume diffuse from the flowers to all parts of the room, and within a short time the fragrance is uniform throughout the area. Diffusion of substances through the cell membrane generally follows this same pattern of movement; that is, from areas of greater concentration to those of lesser concentration. However, certain structural and chemical features of the cell membrane enable it to be highly selective of the substances that may pass through it, regardless of their concentration.

First, let us consider the lipid or fatty layer of the cell membrane and the related factor of *lipid-solubility*. Substances that are lipid-insoluble are prevented from entering or leaving the cell, but molecules such as oxygen (O_2) and carbon dioxide (CO_2) that dissolve easily in fats cross the membrane without difficulty. This exchange of gases (O_2 moving into the cell and CO_2 moving out of the cell) is of fundamental importance. Other lipid-soluble substances such as ether, chloroform, and alcohol also can enter cells, but they may interfere with normal activities. For example, when ether enters nerve cells in the brain,

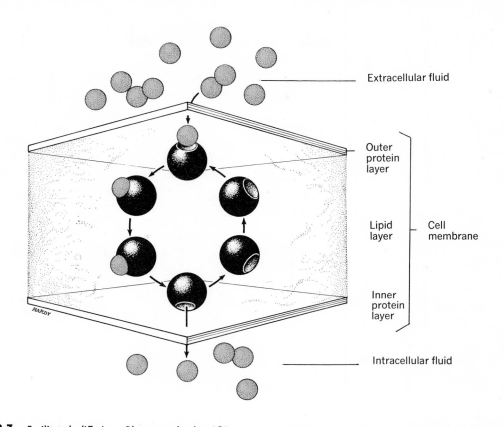

Fig. 2-7. Facilitated diffusion. Glucose molecules (◯) are most concentrated in the extracellular fluid. After passively diffusing through the outer protein layer of the cell membrane, they are picked up by a glucose carrier, represented by the circular baskets, which makes them temporarily lipid-soluble. Thus they can pass through the middle lipid layer to the inner protein layer, where they are released by the carrier and can enter the cell.

their function is so altered that the person becomes temporarily unconscious.

Another factor that influences the permeability of cell membranes is the *size of the pores*. Substances such as water molecules, urea molecules (waste products), and chloride ions are much smaller than the pores; thus they pass through with ease. Water is by far the most abundant substance to diffuse through the cell membrane, but ordinarily it diffuses out of the cell in amounts that are exactly equal to those that enter the cell; when this balance exists there has been no *net* diffusion of water. However, this precise balance may be upset if a *concentration gradient for water* should develop (i.e., if there should be more water molecules on one side of the membrane than on the other). When this state of affairs exists, a net diffusion of water oc-

curs, and the cell either shrinks or swells, depending on the direction of the net diffusion. This process of a net diffusion of water due to a concentration gradient is referred to as osmosis; it is discussed further on page 32, and illustrated in Figure 2-16.

Particles that are too large to squeeze through the tiny pores in a cell membrane must enter by other means. For example, glucose and other simple sugars may enter the cell by a process known as *facilitated diffusion*. In facilitated diffusion a special carrier substance in the cell membrane combines with the sugar, making it temporarily lipid-soluble, so that it can diffuse from an area of greater concentration outside of the cell to an area of lesser concentration within the cell (Fig. 2-7). This process is somewhat similar to that of active transport, which will be described

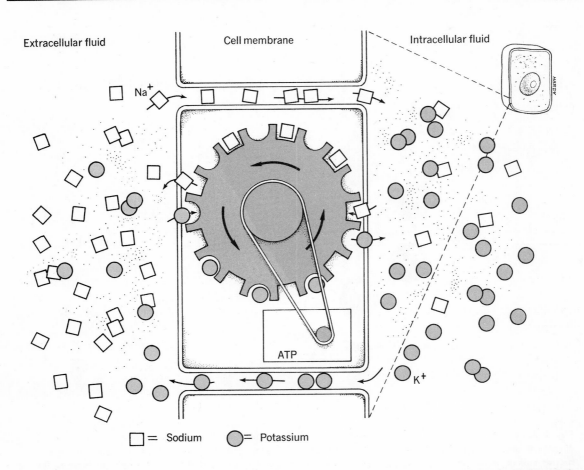

Extracellular fluid   Cell membrane   Intracellular fluid

Na$^+$

K$^+$

ATP

□ = Sodium   ⬤ = Potassium

Fig. 2-8. Active transport. Sodium (□) that diffuses into the cell through a pore in the cell membrane is actively pumped out of the cell by a carrier system, represented by the wheel. A similar fate awaits potassium (○) that diffuses out of the cell as it is actively pumped back into the cell. The energy for this transport is obtained from ATP.

shortly. Other large molecules, such as protein or bacteria, may enter a cell by the process of pinocytosis or phagocytosis (p. 25).

Still another factor that influences permeability is the *electrical charge* on the cell membrane and on the ion which is seeking entrance. For example, if both cell and ion carry a positive charge, the substance is repelled; on the other hand, a negatively charged ion may freely enter or leave a positively charged cell.

Active transport is a physiologic process whereby a cell releases energy for the movement of certain substances through its membrane. In contrast to facilitated diffusion, substances involved in active transport are moved from areas of lower concentration to areas of higher concentration. Thus it is said that they are *moving*

against a concentration gradient or being pumped uphill rather than coasting downhill, as is the case in diffusion. As might be expected, this pumping action requires the expenditure of a considerable amount of energy, and the mitochondria are kept busy producing the necessary ATP power source.

The basic mechanism of active transport is thought to involve certain *carrier systems* that exist within cell membranes. Such carriers are highly specific in that each one transports only certain substances. This specificity probably is determined by the chemical nature of the carrier or by the nature of the enzymes that catalyze the specific chemical reactions. A few substances, such as sodium ions, potassium ions, and amino acids, are thought to have active transport systems in all cells of the body. On the other hand, glucose

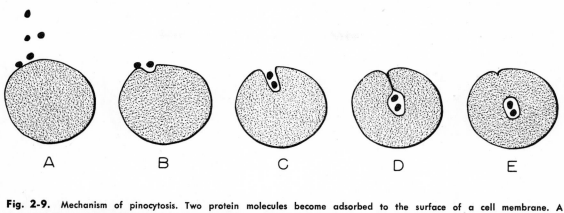

Fig. 2-9. Mechanism of pinocytosis. Two protein molecules become adsorbed to the surface of a cell membrane. A portion of the membrane then invaginates, carrying the protein with it.

and certain other sugars undergo active transport in only a few places, such as the intestine and the kidney. For example, glucose is continually pumped from the intestinal tract into the bloodstream, even though its concentration in the intestine may be very low; this prevents fecal loss of a valuable source of energy.

The sodium and potassium transport mechanism plays an important role in maintaining the proper concentration of ions within the cell and in the surrounding extracellular fluid. Proper cell function is related to the existence of a greater concentration of potassium ions (K^+) within the cell than there is in the surrounding interstitial fluid, while the opposite is true of sodium ions (Na^+); that is, Na^+ is the chief extracellular ion, only a relatively small amount being found within the cell. Inasmuch as a certain amount of K^+ tends to diffuse *out* of the cell (where it is more concentrated) and a certain amount of Na^+ tends to diffuse *into* the cell (where it is less concentrated), it is absolutely essential that an active transport mechanism pump these wandering ions against their concentration gradients, that is, into the cell in the case of K^+ and out of the cell in the case of Na^+. Figure 2-8 illustrates the carrier system theory of active transport of sodium and potassium.

Substances also may enter a cell by a process of ingestion known as **pinocytosis** or **phagocytosis**, terms derived from Greek roots meaning *drinking by cells* and *eating by cells*. Figure 2-9 illustrates the mechanism of pinocytosis as two protein molecules become adsorbed to the outer surface of a cell membrane, then are engulfed by the cell as the membrane invaginates or infolds, carrying with it the protein and a variable amount of fluid from the extracellular environment. The invaginated portion of the membrane then breaks away from the cell surface and forms a little vesicle that moves off into the cytoplasm. The lysosomes then play a part in the breakdown of the vesicle and in the digestion of its contents. Phagocytosis occurs in essentially the same way as pinocytosis, but the former term is used when larger particles such as bacteria and cell fragments are ingested.

CELL DIVISION

Growth of the body, replacement of cells that have a short life span, and the repair of various injuries all depend upon a process of cell division known as *mitosis*. We do not yet know what starts mitosis, or all of the factors that control it, but we do know that it follows a distinct pattern requiring a variable length of time.

When cells are not undergoing division they are said to be in a resting stage, or **interphase,** during which time they carry on their specialized tasks. Also during interphase, a very important preliminary step must be taken prior to division of the cell as a whole. This preliminary activity involves a duplication of the chromosomes, the gene-carrying threads that contain the all-important deoxyribonucleic acid, or DNA. Because the chromosomes with their genetic material deter-

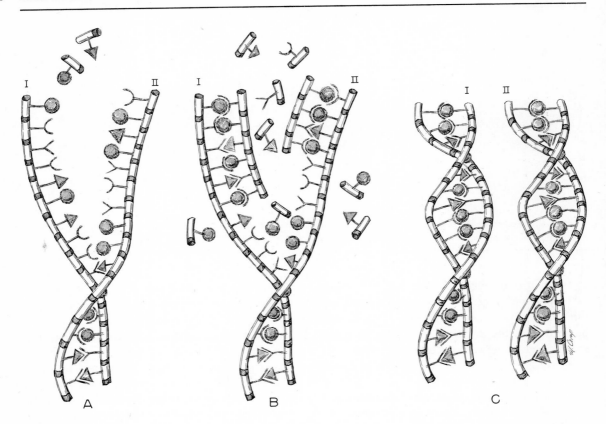

Fig. 2-10. Schematic representation of the replication of DNA: A, prior to cell division, the bonds between the nitrogenous bases are broken, the two strands separate, and each strand takes with it the bases attached to its side. B, the bases attached to each single strand attract free-floating nucleotide units and pair off in the usual way: adenine with thymine, guanine with cytosine. C, the end result is two exact replicas of the original DNA molecule, and the cell is ready to undergo division.

mine what kind of cell a cell will be and what activities it will carry out, it is essential that they be duplicated before the mother cell divides and forms two new daughter cells.

Duplication of the chromosomes occurs within the nucleus a few hours before mitosis takes place. Each one of the 46 chromosomes splits longitudinally and forms two exact replicas of itself; these replicas are called *chromatids.* The net result of this duplication is 92 chromatids, which ultimately will become the 46 chromosomes in each of the two daughter cells formed during mitosis. Figure 2-10 illustrates the way in which the DNA component is able to duplicate itself.

Although mitosis is a continuous process requiring a variable amount of time, it usually is divided into four stages for convenience of description (Fig. 2-11).

1. Prophase. The centrioles separate and begin to move toward opposite sides of the cell. The chromosome threads become more tightly coiled, and each one can be seen to consist of two chromatids. The nuclear membrane and the nucleolus disappear.

2. Metaphase. By the time that the nuclear membrane has dissolved, fine tubules can be seen radiating from the centrioles through the nuclear region to the midline of the cell. The divergence of these tubules gives the appearance of a spindle. The paired chromatids of each chromosome have become aligned in orderly fashion in the middle of the cell, and are attached to the tubules at that point.

3. Anaphase. The two chromatids of each chromosome become completely separated from each other, and each one is now considered to be

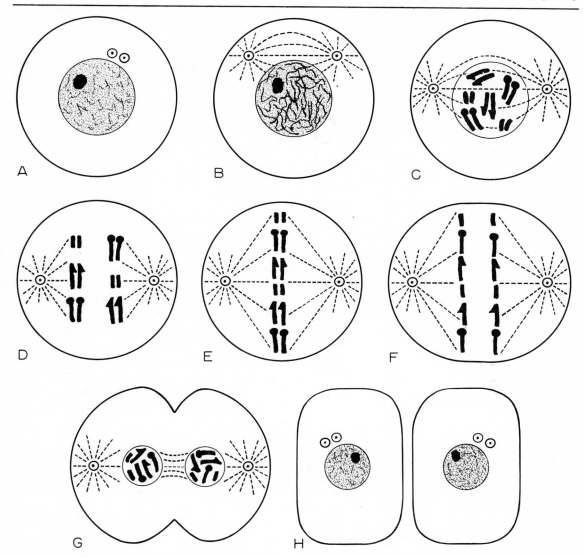

Fig. 2-11. Schematic representation of mitosis: *A,* interphase (resting); *B, C,* and *D,* prophase; *E,* metaphase; *F,* anaphase; *G,* telophase; *H,* daughter cells (resting). See text for description.

a chromosome in its own right. The spindle tubules then pull the separated chromatids to their respective ends of the cell.

4. Telophase. The chromosomes uncoil and begin to assume the threadlike form characteristic of interphase. The nuclear membrane and nucleoli of the daughter cells are formed. Each centriole has produced a daughter centriole in preparation for the next division. The cell membrane constricts at the midpoint of the elongated cell, and two new cells are formed, each with 46 chromosomes. The miracle of mitosis may then

be repeated; the two cells becoming four, and so on.

Some cells in the body undergo mitosis throughout life; for example, cells in the growth layer of the skin are constantly dividing and pushing upward to replace those that are worn away by friction as one works and plays. Other cells are unable to reproduce themselves after their initial work of organ formation has been completed; for example, cells of the central nervous system can never be replaced if they are destroyed through injury or disease. Normally cell division proceeds at an orderly pace, just rapidly enough to ensure proper growth or repair of tissues. Occasionally, however, certain cells may begin to divide with abnormal rapidity, with no regard for the needs of the body. These abnormal cells crowd out the surrounding normal cells, robbing them of their nourishment and resulting in a malignant growth which is called **carcinoma** or **cancer.** Despite the tremendous amount of research in this area, we do not yet know what starts cells on this mad course of lawless, uncontrolled reproduction. However, investigators have found that the genetic makeup of certain cancer cells is different from that of normal cells. Thus it may be that cancer results from a mutation in some part of the genetic system. Another theory undergoing intensive study concerns the possibility that the genetic system may be altered by the presence of a virus in the cell. If even a single cell is so altered, it will grow and proliferate indefinitely since the mechanism that normally limits its reproductive capacity has been eliminated.

PROTEIN SYNTHESIS

The synthesis of protein, a vital function of every living cell, takes place along the ribosomes. However, these protein factories must be provided with specific instructions and with the proper raw materials for construction of such complex molecules. As described on page 19, deoxyribonucleic acid (DNA) is the source of genetic information that determines protein structure; thus it must provide the specific blueprints. The DNA in the nucleus forms vast numbers of molecules called *ribonucleic acid* (RNA). The RNA molecules then travel to the cytoplasm and complete the various tasks according to DNA's master plan.

Although RNA is formed from DNA, it differs from DNA in that it contains ribose sugar groups instead of deoxyribose, and uracil instead of thymine. In forming RNA, the DNA strands separate, and RNA nucleotides are attracted to the appropriate bases (uracil with adenine, guanine with cytosine); because this happens along only one half of the DNA ladder, RNA molecules occur as single strands.

There are three functional types of RNA, each equipped to carry out specific tasks in the assembly of a protein molecule. One type is called *messenger* RNA (m-RNA), and on this molecule DNA prints the genetic instructions needed to make one specific kind of protein; thus there must be as many different m-RNA molecules as there are different proteins. Another type of RNA, called *transfer* RNA (t-RNA), is equipped to recognize a specific amino acid and then attach itself to that amino acid; thus there must be as many different t-RNA molecules as there are amino acids. Prior to their pickup by t-RNA, the amino acids are activated by being coupled with energy-rich ATP. A third type of RNA, *ribosomal* RNA (r-RNA), plays a part in translating the genetic code brought from DNA by m-RNA.

As shown in Figure 2-12, m-RNA carries the blueprint from DNA to the ribosomes. As soon as m-RNA is attached to the ribosomes it is ready to direct the alignment of amino acids that are carried by t-RNA. In a manner that resembles an automobile assembly line, amino acids are dropped into position by t-RNA as dictated by the coded message on m-RNA. The amino acids are bonded together; after the chain is completed, the finished protein molecule moves away from the ribosome. This molecule may be used by the cell that produced it, or it may be exported from the cell for use elsewhere in the body.

It is believed that m-RNA carries only one message; thus, it probably is broken down to its basic nucleotide units following the synthesis of a protein molecule. On the other hand, it is believed that the ribosomes and the t-RNA molecules are free to work again.

MAINTAINING HOMEOSTASIS IN THE LIQUID CELLULAR ENVIRONMENT

Over a billion years ago life apparently began in the sea. Although this organization of protoplasm was probably in the form of simple microscopic plants, it was life. The sea water constantly bathed these cells, delivering their food and serving as a receptacle for wastes. Now, one thousand million years later, the same basic pattern prevails in our own bodies. Even though we live on dry land and breathe air, we literally have a sea within, as water bathes every living cell in a manner quite similar to that of the oceanic waters

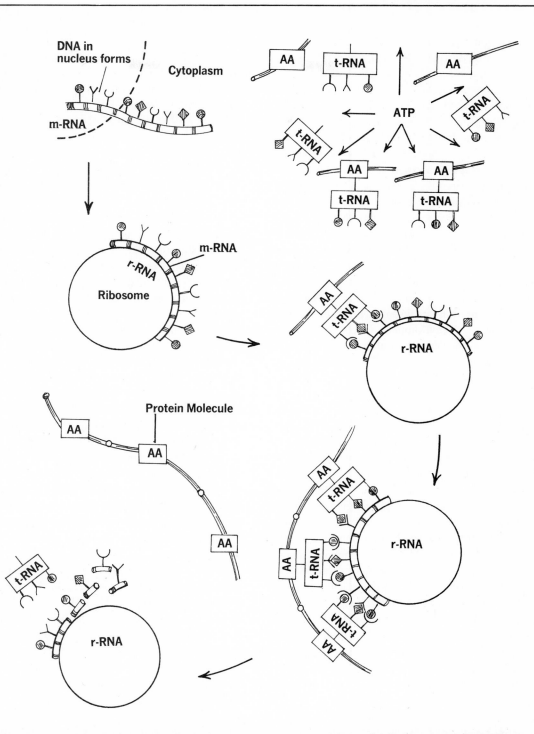

Fig. 2-12. Diagram of protein synthesis. *(Upper left)* A small section of single strand messenger RNA (m-RNA), seen leaving nucleus, attaches to ribosome containing ribosomal RNA (r-RNA). *(Upper right)* Amino acids (AA) in cytoplasm are activated by ATP and coupled to transfer RNA (t-RNA). *(Center right)* Three-letter code word of t-RNA then is attracted to appropriate site on m-RNA, ensuring proper sequence of amino acids. *(Lower right)* Amino acids are bonded together forming protein molecules of varying sizes. *(Lower left)* Protein molecule moves away from ribosome.

of long ago. This intercellular water is called **interstitial fluid** or *tissue fluid* (meaning in the spaces between cells); it serves as the "middleman" between the bloodstream and the cells, which also are composed primarily of water. In fact, the human body as a whole is approximately 50 to 60 per cent water: about two thirds of this water is found within the cells and the remainder, in the tissue spaces and in the bloodstream.

Sea water and body fluid are remarkably similar in chemical composition in that both contain *water* and *electrolytes*. The principal ions in both fluids are sodium, potassium, magnesium, calcium, chloride, sulfate, and phosphate, but the total concentration is much greater in sea water than in our body fluid. In other words, creatures in the sea live in a saltier environment than do cells in the human body. However, whether we are dealing with sea water or with body fluid, the important thing to keep in mind is the fact that each liquid environment meets the particular needs of those cells which it serves.

Just as the ocean is a dynamic, moving mass of water, so is the human body. From the great spurting of blood in our arteries, to the microscopic seeping of water through a cell membrane, our body fluids are constantly on the move, constantly striving to maintain a fairly uniform environment for the working cells of the body. For example, as a cell uses molecules of food and oxygen, other such molecules must be moving in to take their place; at the same time, secretions and excretions resulting from cellular activity must be removed from the immediate vicinity and carried to other cells for use or for elimination from the body. Thus we can see that the internal environment is truly in a state of *dynamic equilibrium*, or **homeostasis;** it is constantly changing, yet it remains essentially unchanged insofar as volume, pH, and chemical composition are concerned. Every system in the body plays some part in maintaining this constancy of environment, and our very survival depends on their continued ability to do so.

The circulatory "subway" system

Much as a subway train serves a city by picking up passengers at one point and discharging them at another, so the circulatory system serves the human body. Various substances are carried as passengers in the bloodstream as it is pumped through a series of tubes or blood vessels. Each beat of the heart sends a spurt of blood into thick-walled vessels called *arteries*. These vessels branch out to all parts of the body, gradually becoming so small that they are no longer visible to the naked eye. With the help of a microscope, however, one can see that these tiny arteries, or arterioles, open into a network of still smaller tubes which are called *capillaries*.

The capillaries are microscopic vessels whose walls are only one cell thick. Between these cells are the tiny openings or intercellular spaces through which the blood picks up and discharges its various passengers. However, capillary membranes, unlike regular cell membranes, *do not exhibit selective permeability*, and any particle smaller than the intercellular spaces may pass through. Large particles, such as most of the colloidal blood proteins, are unable to pass through the openings in this microscopic sieve, so they normally remain in the bloodstream. On the other hand, water, oxygen, carbon dioxide, crystalloids, and other tiny molecules can pass freely through the capillary membrane. For example, as blood passes through the lung capillaries it picks up some oxygen and gets rid of some carbon dioxide; in muscle capillaries it delivers food and oxygen needed by the muscle cells, and, at the same time, it picks up carbon dioxide and other waste products resulting from the metabolic activity of these hardworking cells. Similar patterns of pickup and delivery may be observed in other tissues of the body.

When blood reaches the end of the microscopic capillary network, it flows into a series of slightly larger tubes called venules, then into still larger vessels, the *veins*, and finally back into the heart where it is again pumped out into the arteries.

Thus we can see that the circulatory system is a completely closed circuit of tubes, all parts being of equal importance in the total functioning of the system. However, we must not lose sight of the fact that *the capillary is the focal point of the entire system,* for only in this thinwalled portion of the circuit can the necessary exchanges be made between the blood and its surrounding environment.

Movement of substances through membranes

Maintenance of homeostasis in our liquid cellular environment depends on the movement of substances through the membranes which separate our fluid compartments. For example, materials needed by the cell must pass from the bloodstream through the capillary membrane into the interstitial fluid, then through the cell membrane and into the cell; materials produced by the cell must travel in the opposite direction, first passing through the cell membrane into the interstitial fluid, then through the capillary membrane and into the blood.

Movement of substances through a cell membrane by the processes of diffusion, active transport, pinocytosis, and phagocytosis are described on pages 22 to 25. We shall now consider the processes of filtration, diffusion, and osmosis, with emphasis on capillary membranes in addition to cell membranes.

Filtration. Filtration is the process whereby water and dissolved substances are literally pushed through a permeable membrane from an area of higher pressure to an area of lower pressure. This mechanical force is referred to as **hydrostatic pressure,** or *pushing pressure.* Figure 2-13 illustrates a laboratory demonstration of filtration in which water and dissolved substances are pushed through filter paper. Due to the weight of the fluid in the funnel, together with atmospheric pressure, movement of fluid occurs from the area of greater hydrostatic pressure in the funnel to the area of lower hydrostatic pressure in the empty beaker below. Since the filter paper is not permeable to the large sand particles, they are held back.

The formation of interstitial fluid is an excellent example of filtration in the human body. Due to the pumping action of the heart, the hydrostatic pressure exerted by the blood against the walls of the capillaries is greater than that of the surrounding interstitial fluid; therefore, water and dissolved substances filter through the capillary membrane into the interstitial fluid. Large particles such as red blood cells and *most* of the proteins remain in the bloodstream. However, a number of small protein molecules, such as albumin, also are pushed out into the interstitial fluid; normally these molecules are returned to the bloodstream

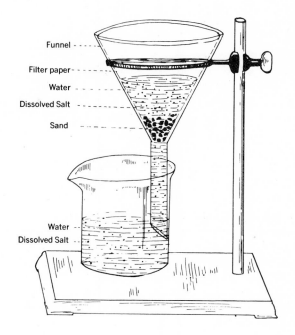

Fig. 2-13. Diagram representing filtration.

by way of lymphatic vessels (see Fig. 2-17) and are of little significance. On the other hand, in certain instances the appearance of extravascular (outside of the blood vessels) protein may be a sign of trouble. This is particularly true when protein is found in the urine. Normally only water and crystalloid solutes are filtered from the blood as it passes through the kidney capillaries (Fig. 2-14), but in certain cases of kidney infection the capillary membranes may be injured to such an extent that larger particles, such as proteins and even red blood cells, also may filter out and thus appear in the urine; these symptoms of kidney damage are referred to as *proteinuria* (protein in the urine) and *hematuria* (blood in the urine).

Diffusion. As described on page 22, diffusion is a physical process that involves the constant scattering or spreading out of molecules of gases or liquids until they are equally concentrated in all parts of the container. In the body there is diffusion of gases between the air sacs or alveoli of the lungs and the blood. Molecules of oxygen diffuse from the air sacs in the lungs into the blood, and carbon dioxide molecules diffuse from the blood

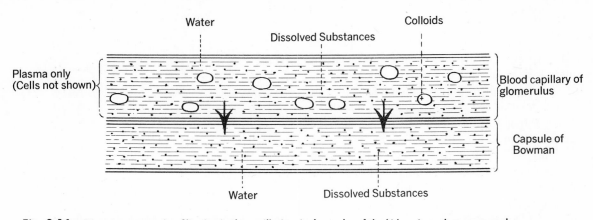

Fig. 2-14.   Diagram representing filtration in the capillaries. A glomerulus of the kidney is used as an example.

into the air sacs in the lungs. When the blood reaches the capillaries in the tissues, oxygen diffuses from the blood into the interstitial fluid and on into tissue cells, while carbon dioxide diffuses from the tissue cells into the interstitial fluid and on into the blood. In every instance the molecules diffuse from the region in which they are highly concentrated to the region in which they are less highly concentrated. A diagram of diffusion is shown in Figure 2-15.

Molecules diffuse easily through membranes which are permeable to them. For example, crystalloid waste products produced by a cell have no difficulty in passing from the cell (where their concentration is great) through the cell membrane and into the interstitial fluid (where their concentration is less), then through the capillary membrane into the bloodstream; such substances are said to be *diffusible*. On the other hand, colloidal substances such as proteins, which are too large to pass through normal membranes, are said to be *nondiffusible*. Protoplasm, which is essentially protein in nature, is formed within the cell by various combinations of animo acids that are actively transported into the cell.

Osmosis. Osmosis is a physical process involving the passage of water through a semipermeable membrane that separates two solutions of different concentrations. A semipermeable membrane is one which is *partly permeable*, in that it is permeable to water but either impermeable or only slightly permeable to the solute. In osmosis, water moves from the weaker solution, where there are more water molecules, to the more concen-

trated solution, where there are fewer water molecules; thus water moves according to a *concentration gradient*. However, students frequently find it helpful to think of osmosis in terms of water being *pulled* from the weaker solution into the more concentrated solution in an attempt to equalize concentrations on both sides of the membrane, or of water being drawn into the concentrated solution in order to dilute it.

Theoretically the osmotic process will stop when the concentration of dissolved substance is the same on both sides of the membrane. However, as water molecules move into the more concentrated solution there is an increase in the volume and in the pressure of that solution. This process continues until the pressure reaches a height that will prevent any more water molecules from passing through the membrane. This pressure, resulting from the influx of water into the more concentrated solution, is referred to as the **osmotic pressure,** or *pulling pressure*, of that solution.

The more solute particles in any given solution, the higher the osmotic pressure; the fewer solute particles, the lower the osmotic pressure. For example, a 10 per cent NaCl solution contains more sodium and chloride ions than a 1 per cent solution and therefore would have the greater osmotic pressure. A simple way of describing the process of osmosis is to say that water goes where the most salt is to be found.

Osmosis can be demonstrated in the laboratory by the use of red blood cells and salt solutions of various strengths. Normally red blood cells are

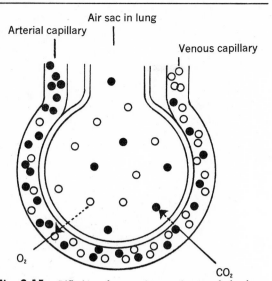

Fig. 2-15. Diffusion of gases in an air sac of the lung. The circles represent oxygen and the black dots represent carbon dioxide. The net diffusion of each gas is shown by changes in the number of dots and circles in the arterial and venous ends of the capillary.

suspended in the plasma (liquid portion of the blood), the concentration of dissolved substances, such as salts, inside the red cell being the same as the concentration of dissolved substances in the plasma. When two solutions have the same concentration of dissolved substances they are said to be *isotonic*; that is, to have the same osmotic pressure. Since plasma is isotonic with the red blood cells, water moves freely in equal amounts in both directions across the cell membrane, and there is no change in the shape of the cell. However, if red cells are removed from the plasma and placed in a concentrated or **hypertonic** salt solution, there will be a net diffusion of water *out* of the cells into the salt solution. As a result, the cells become shrunken and appear to be knobby or *crenated*. On the other hand, if the cells are placed in a dilute or **hypotonic** solution, there will be a net diffusion of water *into* the cells causing them to swell until they burst and lose their oxygen carrier, hemoglobin. This destruction of red blood cells with the loss of hemoglobin is called *hemolysis* (meaning *blood breakdown*). Crenation and hemolysis are illustrated in Figure 2-16.

Theoretically only isotonic solutions should be injected into the bloodstream (intravenously) in order to prevent undesirable shrinking or swelling of the red blood cells. However, hypertonic solutions may be used on certain occasions; if these solutions are administered slowly and in relatively small amounts, they will be safely mixed with the plasma. The two solutions used most frequently for intravenous injection (exclusive of whole blood and plasma) are 0.9 per cent sodium chloride, which is called physiologic or normal saline, and 5.0 per cent glucose. Both of these solutions are isotonic with the body fluids.

Since osmotic pressure is directly related to the number of particles in solution, it is logical to assume that the size of the solute particles would also play a part in the determination of this pressure. For example, a solution which contained 10 Gm. of a colloid would have a lower osmotic pressure than a solution which contained 10 Gm. of an electrolyte (such as NaCl), for the simple reason that colloids are large particles and there could be proportionately fewer of them in solution. However, the colloid osmotic pressure of the blood is very important in the return of fluid from the intercellular spaces. Since colloids are not diffusible, they are much more concentrated in the bloodstream than in the interstitial fluid. For this reason they generate a much higher osmotic or pulling pressure in the bloodstream than in the interstitial fluid. Crystalloids, on the other hand, are diffusible; their concentration is practically the same on both sides of the capillary membrane, and they generate little or no effective

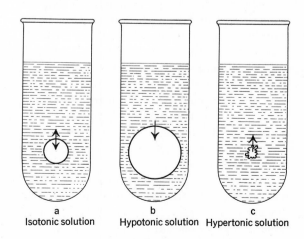

a
Isotonic solution

b
Hypotonic solution

c
Hypertonic solution

Fig. 2-16. Osmosis: A, red blood cells undergo no change in size in isotonic solutions. B, they increase in size in hypotonic solutions. C, they decrease in size in hypertonic solutions.

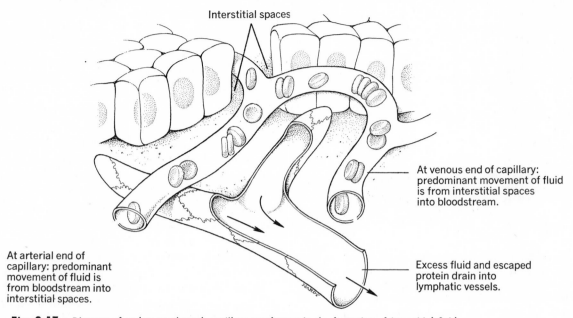

Interstitial spaces

At venous end of capillary: predominant movement of fluid is from interstitial spaces into bloodstream.

At arterial end of capillary: predominant movement of fluid is from bloodstream into interstitial spaces.

Excess fluid and escaped protein drain into lymphatic vessels.

Fig. 2-17.   Diagram of exchanges through capillary membranes in the formation of interstitial fluid.

osmotic pressure. Thus, for all practical purposes, when considering the interchange of fluids through capillary membranes we shall be most concerned with the colloid osmotic pressure (COP).

Exchanges through capillary membranes

As blood moves through the capillaries, there is a constant interchange of water and dissolved substances. This fascinating process involves the mechanisms previously described. According to the famous hypothesis advanced by E. H. Starling in 1896, hydrostatic and osmotic pressures within and without the capillaries are responsible for the transfer of fluids across capillary membranes. These pressures vary considerably in different tissues and in different species; so for our purpose we shall arbitrarily select a capillary in the skin of a man.

The **capillary hydrostatic pressure** (Cap-HP) is primarily responsible for the formation of the interstitial fluid. As blood flows away from its pressure source (the heart), its pressure becomes progressively lower; consequently, the Cap-HP is normally higher at the arterial end of the capillary (average 32 mm. Hg) than it is at the venous

end (average 15 mm. Hg). This means that there is proportionately more filtration occurring at the arterial end of the capillary than at the venous end. The **interstitial fluid hydrostatic pressure** (ISF-HP) is difficult to measure, and figures may vary from −7 mm. Hg to 6.5 mm. Hg, depending on the technique used by the investigator. However, small positive pressures most often are registered, and an ISF-HP of 2 mm. Hg will be used for the purpose of this discussion. This pressure would tend to push fluid back into the bloodstream in opposition to the Cap-HP which is pushing fluid out.

The **colloid osmotic pressure of capillary blood** (Cap-COP) plays an important role in the return of interstitial fluid to the bloodstream (Fig. 2-17). Since most of the colloidal plasma proteins (e.g., albumin, globulin) are unable to pass through the capillary membrane, their osmotic pressure remains essentially the same (about 25 mm. Hg) throughout the capillary bed. This means that water is being pulled *in* at approximately the same rate in all parts of the capillary bed. However, as previously noted (p. 31), a certain amount of protein escapes from the bloodstream into the interstitial fluid where it tends to draw fluid *out* of the capillaries; this **interstitial fluid colloid osmotic pressure** (ISF-COP) averages about 4 mm. Hg.

With the preceding figures at hand, it is now possible to compare the inward and outward forces that are responsible for the transfer of fluids across capillary membranes.

At Arterial End of Capillary

Outward Forces

Cap-HP: 32 mm. Hg
ISF-COP: 4 mm. Hg
Total: 36 mm. Hg

Inward Forces

Cap-COP: 25 mm. Hg
ISF-HP: 2 mm. Hg
Total: 27 mm. Hg

Thus there is a *net outward force* of 9 mm. Hg, and the predominant movement of fluid is from the blood into the interstitial spaces.

At Venous End of Capillary

Outward Forces

Cap-HP: 15 mm. Hg
ISF-COP: 4 mm. Hg
Total: 19 mm. Hg

Inward Forces

Cap-COP: 25 mm. Hg
ISF-HP: 2 mm. Hg
Total: 27 mm. Hg

Now we have a *net inward force* of 8 mm. Hg, and the predominant movement of fluid is from the interstitial spaces into the bloodstream.

As you can see, there is a 1 mm. Hg difference between the net inward and outward forces (i.e., 9 mm. Hg outward and 8 mm. Hg inward). This means that more fluid leaves the bloodstream than can be returned. The excess fluid, plus protein molecules that escape from the capillary, enter neighboring lymphatic vessels and are drained away, ultimately returning to the bloodstream.

When this dynamic fluid balance is upset during the course of certain disease processes, there may be an accumulation of interstitial fluid that is referred to as *edema*. On the other hand, if ex-cessive quantities of interstitial fluid move into the bloodstream, the tissues become *dehydrated*. Fluid and electrolyte balance are presented in more detail on pages 451 to 464.

THE PRIMARY TISSUES

A tissue consists of a group of cells that are similar in structure and in intercellular substance, arranged in a characteristic pattern. In the human body there are four basic or primary tissues: epithelial, connective, muscular, and nervous tissue. The primary tissues, in turn, may be subdivided into various types, each type being specialized for the performance of a specific task or tasks.

Epithelial tissue

Epithelial tissue covers the body surfaces and forms a lining for its cavities, including the tubular structures. In certain parts of the body epithelial tissue grows into the body substance to form structures called glands.

General Functions of Epithelial Tissue. The epithelial tissues of the body are concerned with protection, secretion, absorption, and filtration. For example, the outer layer of the skin (epidermis) offers *protection* from such things as bacterial invasion; cells in the respiratory tract have fast-moving, hairlike projections (cilia) which are constantly at work sweeping dirt and dust particles out of the lungs. *Secretion* is a specialty of epithelial cells in the various glands of the body. Epithelial cells highly specialized for *absorption* are found in the small intestine and in the tubules of the kidneys. The formation of interstitial fluid throughout the body depends very much upon the process of *filtration*, which occurs through the delicate epithelial tissue in the capillaries of the circulatory system.

Characteristics of Epithelial Tissue. The cells of epithelial tissues fit together so closely that there is no true intercellular substance other than certain structural components which hold the cells together. Epithelial tissues occur in the form of sheets or membranes that are not very strong. The important source of support comes from the underlying connective tissues to which the epithelial tissues are firmly attached by a permeable

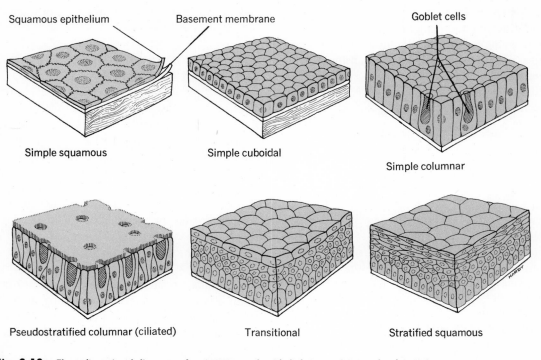

Fig. 2-18. Three-dimensional diagrams of various types of epithelial tissues. See text for description.

but *adhesive basement membrane* (Fig. 2-18).

Epithelial tissues have no blood vessels of their own; thus they depend upon capillaries in the underlying connective tissue for supplies of food and oxygen and for the removal of waste products. Well-nourished epithelial cells have an excellent capacity for regeneration, so that cells lost during the wear and tear of everyday living are readily replaced by mitosis (p. 25).

Types of Epithelial Tissue. The covering and lining division of epithelial tissue consists of two general classes: *simple* epithelium, which consists of one layer of cells; and *stratified* epithelium, which consists of more than one layer of cells. The terms squamous, cuboidal, and columnar indicate the general shape of the cells (in stratified tissue, these terms refer to the surface cells).

The glandular division of epithelial tissue forms the exocrine and some endocrine glands of the body. Endocrine glands pour their secretion into the bloodstream, while the exocrine glands pour their secretion onto the surface from which they originated.

1. **Simple squamous epithelium** consists of one layer of thin, flat cells fitted together in such a way that they resemble a mosaic (Fig. 2-18). This type of epithelium is adapted for filtration and diffusion, rather than for withstanding wear and tear. Consequently, simple squamous epithelium lines the blood and lymph vessels—here it is called *endothelium*; in the microscopic capillaries this single layer of cells makes possible the formation of interstitial fluid. The air spaces of the lungs also are lined with simple squamous epithelium, through which oxygen and carbon dioxide diffuse without difficulty.

2. **Simple cuboidal epithelium** consists of a single layer of cells that somewhat resemble tiny cubes when seen in a side view (Fig. 2-18). It is not found in very many places in the body; a few examples are the thyroid gland, the surface of the ovary, and a portion of the kidney tubules.

3. **Simple columnar epithelium** is composed of cells that are taller than they are wide (Fig. 2-18). These cells not only protect underlying tissues but also are specialized for additional functions. For example, consider the simple columnar epithelial lining of the small intestine, in which

two types of cells are interspersed with one another. One type is specialized for the *selective absorption* of the products of digestion. A second type is specialized for the secretion of a slippery, protective substance called *mucus;* such a cell is called a *goblet cell* because the nucleus lies near the basement membrane, and the remainder of the cell is shaped like a goblet (Fig. 2-18). The mucus secreted by goblet cells keeps the epithelial surface moist.

Simple columnar epithelium is also in the uterus, in the uterine tubes, and in some of the respiratory passages. In addition to goblet cells, the epithelium in the aforementioned sites presents still another type of specialized cell, one that has hairlike processes on its free surface (Fig. 2-18); these processes are called *cilia*, and they act together to move particles and secretions, such as mucus, across the surface of the tissue. Repeatedly they bend down rapidly with a forceful stroke in one direction and then return to their original position. The movement passes over the ciliated surface in waves, as the cilia are rigid when bending down and limp when returning to the erect position.

4. Pseudostratified columnar epithelium is so named because it appears at first glance to be composed of several layers of cells (Fig. 2-18). However, closer inspection shows that although some of the cells in contact with the basement membrane do not reach the free surface of the tissue, most of the cells are tall, and do reach the surface. Both ciliated cells and goblet cells are seen in this epithelium, which lines most of the respiratory passages.

5. Stratified epithelium consists of more than one layer of cells and is named according to the shape of the cells in the top layer. This tissue serves to protect body surfaces that are subject to much wear and tear in daily living. As the surface cells are brushed off by friction, new ones are pushed up from below to take their place; this cell replacement is made possible by cell division taking place in the deepest layer of the tissue, in which cells are columnar in shape.

(a) *Stratified squamous keratinized* epithelium makes up the epidermis of the skin (see Fig. 2-26). The superficial cells in this tissue are flattened, dead cells that have become fused together, forming a tough, waterproof layer of keratin material. The epidermis protects the underlying living tissues from a variety of injuries; it is described in more detail on page 43.

(b) *Stratified squamous nonkeratinized* epithelium is also protective in nature, but it differs from the keratinized tissue in that all of its cells are alive. Such epithelium is found on wet surfaces that are subjected to wear and tear, such as the inside of the mouth, where it provides protection from coarse foods.

(c) *Transitional epithelium* is a peculiar type of stratified epithelial tissue found in structures which are subjected to periodic distention, such as the urinary bladder. It is like stratified squamous nonkeratinizing epithelium except that the superficial cells are rounded when the bladder is empty and flattened when it is distended. In the latter case they are stretched and drawn out into squamouslike cells without breaking apart (see Figs. 2-18 and 18-7).

6. Glandular epithelium gives rise to secretory units which are called glands. The mucus-secreting goblet cells previously described cannot produce more than a small fraction of the various secretions needed in the body. To meet the needs for additional secretions, epithelial cells in many regions have turned inward from the covering or lining surfaces and grown into or invaded the underlying supportive connective tissue. The manner in which such masses of secreting cells (called glands) develop is shown in Figure 2-19. These cells are highly specialized to secrete; this means that they remove materials from the blood and manufacture new substances which they then extrude from their cytoplasm.

(a) *Endocrine glands.* Some glands lose their connection with the epithelial surface from which they developed; consequently, they possess no ducts through which they can extrude their secretions. The secretions which are produced by endocrine glands are extruded into blood or lymph. The endocrine glands comprise the endocrine system to which Chapter 20 is devoted.

(b) *Exocrine glands.* The exocrine glands retain their connection with the epithelial surface from which they originated. The connections are called the ducts of the glands, and they convey the secretions from the glands to the surface. The secretory units are clusters of cells located at the ends of the ducts, and the shape of the unit determines the type of gland; Figure 2-20 illustrates various types of exocrine glands.

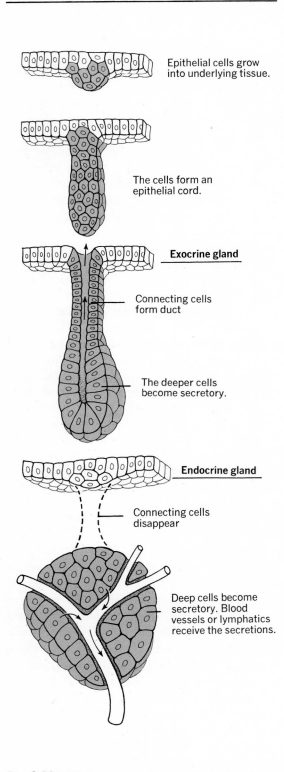

Epithelial cells grow into underlying tissue.

The cells form an epithelial cord.

Exocrine gland

Connecting cells form duct

The deeper cells become secretory.

Endocrine gland

Connecting cells disappear

Deep cells become secretory. Blood vessels or lymphatics receive the secretions.

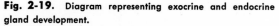

Fig. 2-19. Diagram representing exocrine and endocrine gland development.

Connective tissue

Connective tissue is the most abundant and the most widely distributed of all the primary tissues. It is found in all parts of the body.

General Functions of Connective Tissue. The connective tissues perform a variety of functions, ranging from broad areas of support, protection, and the binding together of other tissues, to the production of certain components of the blood, storage of fat, and protection of the body against bacterial invasion and disease. Some of these functions depend primarily upon the activities of special connective tissue cells, while others depend more upon the intercellular substance produced by the cells.

Characteristics of Connective Tissues. In addition to their widespread distribution in the body, connective tissues are characterized by vascularity; with a few exceptions, such as cartilage and tendons, connective tissues are abundantly supplied with blood vessels. These tissues also are characterized by the quantity and the quality of their intercellular substance. At one extreme are the blood-forming (hemopoietic) tissues, which consist almost entirely of cells; at the other extreme are connective tissues, such as bone and cartilage, which consist chiefly of intercellular substance. Between the two extremes are those tissues that are a representative mixture of cells and intercellular substance.

The **intercellular substance,** or matrix, is the nonliving material between the cells of connective tissue. It is produced by the cells and consists of two basic types: fibrous and amorphous (meaning without form).

Fibrous intercellular substance includes three types of fibers: collagenic, elastic, and reticular; all three are protein in nature. Collagen is the most common protein in the body, and it is extremely tough. Therefore, *collagenic fibers*, also called white fibers, are strong and tough, yet flexible and resistant to a pulling force. *Elastic fibers* consist of the protein elastin; they permit connective tissue to be stretched and then return to the original form when the stretching force is removed. *Reticular fibers* are delicate fibers arranged in networks; they are found wherever connective tissue joins the other tissues, such as in or near basement membranes.

Amorphous intercellular substance is the background substance in which fibers and cells are embedded. It varies in consistency from a liquid (in blood), through a soft, jellylike material (in loose connective tissues), to the hard substance of bone.

The **cells of connective tissue** produce the intercellular substance, store fat, make new blood cells, eat bacteria and cell debris (phagocytosis), produce antibodies that give us immunity to certain diseases, and make heparin, which prevents the clotting of blood. Some of the specialized cells of connective tissue are described below.

1. *Fibroblasts* are the most numerous cells in connective tissue. They contain an abundance of rough-surfaced endoplasmic reticulum (p. 21) which indicates that they are active in the synthesis of protein for use in the extracellular environment; for example, they form the fibrous intercellular substance needed for growth and repair of tissues.

2. *Macrophages,* or histiocytes, are large ameboid cells that engage in phagocytosis, or the ingestion of bacteria, other cells, and any foreign material which enters the tissues. They are found in all loose connective tissue and are an important part of the *reticuloendothelial system*; this term is used to designate certain modified endothelial cells which occur in various parts of the body and have the ability to get rid of foreign matter which comes in contact with them. In addition to the macrophages in loose connective tissue, the reticuloendothelial system includes the cells of the spleen, the lymph nodes and the bone marrow, as well as the Kupffer cells in the liver. The main function of all these cells is phagocytosis, which means that they protect the body from injury by foreign substances, and they rid the body of worn-out red blood cells and other cellular debris.

3. *Mast cells* are connective tissue cells that are concerned with the production of heparin, an anticoagulant (p. 269). They also are believed to contain much of the histamine (an inflammatory substance) in the body. Inflammation is described on pages 264 to 265.

4. *Fat cells* are fibroblasts which are specialized to store fat. Under the microscope these cells resemble a signet ring as the large fat droplet within each cell flattens the nucleus and the cytoplasm to one side.

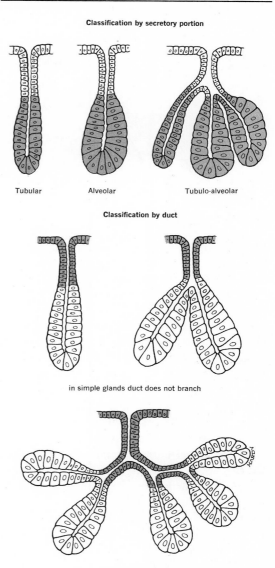

Classification by secretory portion

Tubular   Alveolar   Tubulo-alveolar

Classification by duct

in simple glands duct does not branch

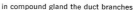

in compound gland the duct branches

Fig. 2-20. Diagram showing the classification of exocrine glands.

5. *Plasma cells* are small rounded cells that are the primary producers of circulating antibodies (gamma globulins, page 267). They develop most commonly in the loose connective tissues that support the wet epithelial membranes lining the respiratory and intestinal tracts. However, they also may develop in lymphatic tissue (p. 324) in any part of the body.

Types of Connective Tissue. The following tissues are found in the human body: ordinary

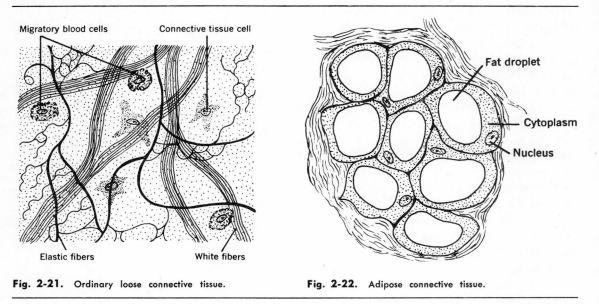

Fig. 2-21. Ordinary loose connective tissue.

Fig. 2-22. Adipose connective tissue.

loose connective tissue, adipose tissue, dense fibrous tissue, cartilage, bone, hemopoietic tissue, blood, and lymph.

1. **Ordinary loose connective tissue** (*areolar*) contains all types of fibers loosely interwoven and embedded in a soft, jellylike background substance (Fig. 2-21). As its name suggests, loose connective tissue has a certain degree of flexibility; its intercellular substance is so arranged that the tissue can be stretched to some extent in various directions without being injured. This tissue is called ordinary because it is so abundant and so widely distributed throughout the body. It is found immediately beneath most epithelial membranes, and it serves as a delicate packing along the course of blood vessels, nerves, and the ducts of glands. In addition, it extends into and between various organs of the body. Loose connective tissue plays an important nutritive role in that it carries many capillaries, and its soft, jellylike substance is ideally suited for the diffusion of materials back and forth between the bloodstream and neighboring tissue cells. This tissue, with its many macrophages and plasma cells, also serves as the body's second line of defense against bacteria that may penetrate the skin or mucous membranes.

2. **Adipose tissue** is a specialized variation of loose connective tissue. It consists primarily of closely packed fat cells with elastic and collagenic fibers running in all directions. Each specialized cell contains such a large droplet of fat that the nucleus and the cytoplasm are barely visible (Fig. 2-22). Adipose tissue is found under the skin, around the kidneys and the eyeballs, and as soft elastic pads between organs; it also forms padding around joints and is found in the marrow of our long bones. This fatty tissue of the body is a reserve food supply, an insulator against heat loss, and a support and protection for various organs.

3. **Dense fibrous connective tissue** differs from the loose variety in that its major function is performed by its intercellular substance; the cells that it contains are primarily concerned with producing that substance (i.e., collagenic and elastic fibers). In most dense connective tissue, collagenic fibers are more abundant than the elastic type; however, elastic fibers are prominent in such places as the walls of the trachea and the bronchial tubes of the respiratory system, and the walls of blood vessels, particularly the large arteries.

Tendons (attaching muscles to bones) and *ligaments* (holding bones together at joints) are both constructed primarily of collagenic fibers that are regularly arranged; that is, they run more or less in the same plane and in the same direction. This arrangement provides great strength and the ability to withstand tremendous pulling when force is applied in one general direction. On the other hand, when the fibers are irregularly arranged (i.e., run in different planes and directions), the tissue can withstand stretching in any direction

(Fig. 2-23); such tissue is found in the deep layer of the skin, in the capsules of various organs, and as an outer wrapping (fascia) for muscles.

4. **Cartilage**, or *gristle*, is a special kind of dense connective tissue in which the shapeless intercellular substance exists in the form of a very firm jelly or gel, and the fibers form a dense network. Cartilage cells, called *chondrocytes*, lie in little cavities called lacunae, usually in groups of two or four. There are no blood vessels in cartilage; its cellular nourishment and waste removal are accomplished by the diffusion of substances through the intercellular substance between the chondrocytes and capillaries located outside of the cartilaginous structure. The fibrous connective tissue membrane that covers cartilage is called the *perichondrium*.

Three types of cartilage occur in the body: hyaline, fibrous, and elastic. *Hyaline cartilage* (Fig. 2-24) is the most common type; it is pearly white and glassy or translucent in appearance. It forms the skeleton in the embryo and provides models in which most of the bones develop. It covers the articulating ends of bones in movable joints, and makes up the rib cartilages and the cartilages of the nose and the larynx.

Fibrous cartilage (fibrocartilage) is less firm than hyaline cartilage, but it has great strength owing to the regular arrangement of its collagenic fibers (Fig. 2-25). Probably the best known examples of fibrocartilage are the invertebral disks, springy shock absorbers between the bodies of the vertebrae. The center of each disk is a soft mass (nucleus pulposus) which occasionally ruptures into the vertebral canal (p. 73). Fibrous cartilage also is found in the form of semilunar disks in the knee joint.

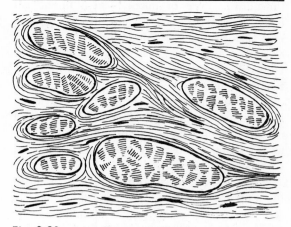

Fig. 2-23. Dense fibrous connective tissue.

Elastic cartilage is more flexible than the other types due to the presence of a network of elastic fibers in its intercellular substance. It enters into the formation of several cartilaginous structures, such as the external ear and the epiglottis. Elastic cartilage allows some degree of change in these structures, yet strengthens them and maintains their shape.

5. **Osseous connective tissue** is most frequently referred to as *bone tissue*. Its origin, structure, and function are described in Chapter 3.

6. **Hemopoietic tissue, blood,** and **lymph** are described in detail in Chapter 11.

Muscle tissue

The third primary tissue is muscle tissue, an amazing collection of highly specialized cells which have the capacity to shorten and thus produce movement of some body part. There are

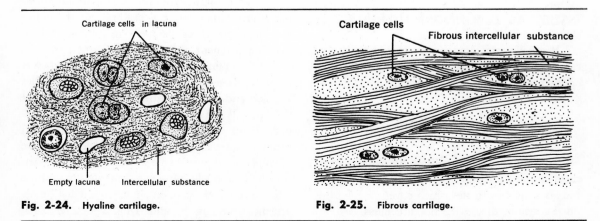

Fig. 2-24. Hyaline cartilage.

Fig. 2-25. Fibrous cartilage.

three distinct types of muscle tissue in the human body: skeletal, visceral or smooth, and cardiac muscle. A complete description of these tissues is given in Chapter 4. Only a brief outline will be presented at this time as we survey body organization as a whole.

1. **Skeletal muscle,** attached to the skeleton, is a voluntary type of muscle tissue (under control of the conscious will). Contraction of skeletal muscle results in movement of external body parts, such as the extremities.

2. **Visceral,** or **smooth muscle,** is found in the walls of the internal organs and the blood vessels. It is an involuntary type of muscle tissue (not under conscious control). Contraction of visceral muscle results in such movement as the peristaltic waves in the stomach and the intestine during the digestion of food.

3. **Cardiac muscle** is found only in the walls of the heart. This involuntary type of muscle compresses the heart chambers and forces blood into the arteries.

Nerve tissue

Nerve tissue is the fourth primary tissue. It is composed of specialized cells which not only receive stimuli but also conduct impulses to and from all parts of the body. *Nerve cells,* or *neurons,* vary markedly from other cells in their appearance. For example, their cytoplasm is drawn out into long stringlike processes which may be several feet in length, although microscopic in diameter. Chapter 6 presents a complete description of this tissue, which is essential for the coordination and the correlation of all body activities.

TISSUES AS BUILDING MATERIALS

We have seen how groups of cells band together to form tissues; now we shall see how various tissues are combined to form membranes, skin, and internal organs.

Membranes

The simplest combination of tissues is observed in the formation of membranes. A membrane is a sheet of tissues which covers or lines the body surface or divides organs into lobes.

Mucous membranes line body cavities and passageways which open to the outside world, such as the digestive tract, the respiratory passages, and the urinary and the reproductive tracts. These membranes have epithelium on their free surface and a layer of connective tissue beneath. All mucous membranes contain a certain number of goblet cells which secrete the slippery, protective mucus, but the other types of cells may vary. For example, many of the cells in the respiratory passages are ciliated for the purpose of sweeping out mucus and particles of dirt, while the columnar cells of the intestine are specialized for the absorption of food.

Serous membranes line the closed cavities of the body (those that do not open to the outside world) and cover the various organs which are located within those cavities. These thin glistening membranes are composed of epithelium on the free surface, with connective tissue below. Cells in the surface layer secrete a slippery serous fluid which protects against friction when organs rub against the body wall or glide over one another. For example, the walls of both lung compartments of the thorax are lined with serous membrane (the parietal pleura), and each lung is covered with serous membrane (the visceral pleural); as the lungs expand with air, the slippery visceral layer glides against the equally slippery parietal layer with a minimum of friction. A similar arrangement of serous membranes, the pericardium and the peritoneum, protect the heart and the abdominal viscera, respectively.

Synovial membranes are composed entirely of connective tissue. Cells in the surface layer secrete a slippery fluid that has the consistency of raw egg white; this lubricant is called synovial fluid. Synovial membranes are associated with bones and joints (see Chap. 3). For example, joint cavities are lined with these slippery membranes which permit movement of bones without friction.

The **superficial fascia,** or *subcutaneous tissue,* is an extensive fibrous membrane which underlies the skin and forms a continuous sheet throughout the body. It is a combination of adipose and ordinary loose connective tissues and is attached firmly to the deep layer of the skin (dermis). The amount of adipose tissue varies in different persons but those who possess a generous amount are said to be well padded. The superficial fascia attaches the skin to underlying structures, such as the deep fascia, the periosteum of bones, and the perichondrium of cartilage. However, the very continuity of these tissues may contribute to a serious spread of infection. For example, when bacteria enter the subcutaneous tissue through a break in the skin they are immediately attacked by the defensive macrophages in the areolar tissue. Should the bacteria win this preliminary battle, they may multiply and spread in all directions through the subcutaneous tissue, resulting in a diffuse and widespread inflammation called *cellulitis*. It is chiefly in the superficial fascia that abnormal collections of fluid (*edema*) occur.

The **deep fascia** lies under the superficial fascia and is composed of dense connective tissue. It contains no fat and is in close relation to bones, ligaments, and muscles. Sheets of deep fascia are wrapped around muscles, supporting them and forming partitions between them. Deep fascia encloses glands and viscera, and it also forms sheaths for nerves and blood vessels.

Miscellaneous **fibrous membranes,** composed entirely of connective tissue, include such specialized structures as the *periosteum* (covering bone), the *dura mater* (covering the brain), and the *sclera* (outer covering of the eye); these will be described at a later time in connection with their related organs.

The skin

The skin, or integument, is much more than a simple hide which covers the body. It serves as an example of a more complex combination of tissues than that found in the membranes previously described. Skin is subjected to many more insults than are the internal membranes, and its structure varies accordingly.

Structure of the Skin. The skin covers the entire surface of the body. It consists of two main layers which are quite different in character; these layers are the *epidermis* and the *dermis* which are firmly cemented together. However, excessive rubbing of the skin (as with poorly fitted shoes) may cause the epidermis to be forcibly separated from the dermis; interstitial fluid accumulates in this area, further separating the two layers and resulting in a blister. Immediately under the dermis, but not part of the skin, is the subcutaneous tissue or superficial fascia, described previously as a fibrous membrane.

The **epidermis** is the thin surface layer of skin. It is composed of stratified squamous keratinizing epithelium which, when viewed under the microscope, is seen to consist of several distinct layers or strata of cells. Since epithelial tissues have no blood vessels, the only living cells are in the deepest layer which is nourished by interstitial fluid from capillaries in the dermis. These cells are constantly undergoing mitosis, and the daughter cells are pushed upward toward the surface, away from their source of nourishment; as these cells die (from lack of interstitial fluid) they undergo a chemical transformation and their soft protoplasm becomes keratinized, or cornified. Since the epidermis is exposed to the air, these horny waterproof cells prevent dehydration of the deeper layers.

The epidermis varies in thickness in different parts of the body. For example, it is thinnest on the eyelids and thickest on the palms and the soles. A person who spends many hours on his feet may develop calluses on his soles, as one who does difficult manual work may have calluses on his hands; this extra thickening of the epidermis offers additional protection to underlying tissues.

Figure 2-26 illustrates the four layers of epidermis. These layers are (1) *the stratum corneum,* or horny layer, which consists of dead, keratinized cells; (2) *the stratum lucidum* and (3) *the stratum granulosum* which are middle layers where the epithelial cells gradually die and become keratinized; and (4) *the stratum germinativum* which is adjacent to the dermis and contains the only epidermal cells which can reproduce themselves. The cells in this deep layer also produce *melanin,* the pigment which is present in the skin of nonwhites, and in the area of the nipple in white races; however, exposure of white skin to ultraviolet light will result in a production of melanin and the much sought after sun tan. Freckles are merely irregular patches of melanin.

The **dermis,** also called the *corium* or true skin, is the inner layer of the skin. It is composed of connective tissues. The surface of the dermis is uneven due to the presence of conelike elevations, or papillae, which serve to attach it to the epidermis. If you look at your fingers and palms you can see ridges in the skin; these are the papillae of the dermis which project into the epidermis. The ridges develop during the third and fourth fetal months, and the pattern never changes except to enlarge. The unchanging pattern formed

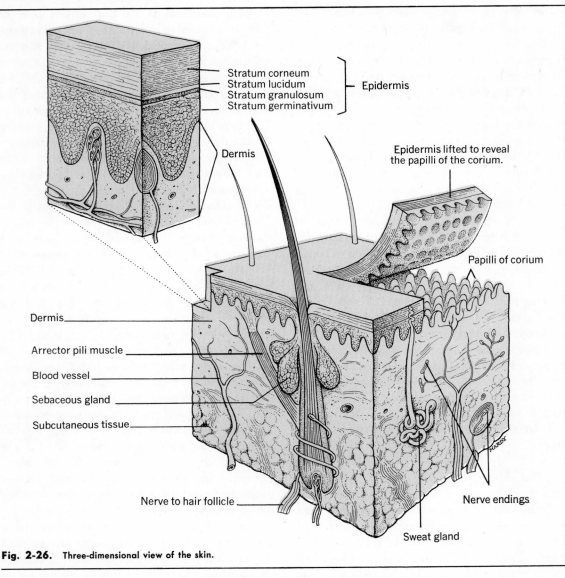

Fig. 2-26. Three-dimensional view of the skin.

by these ridges is peculiar to the individual and is an excellent means of identification.

Both the papillary and the reticular layers of the dermis contain bundles of collagenic fibers; networks of elastic fibers are located between the bundles. The cells in this connective tissue are fibroblasts, fat cells, and some macrophages. The fibers give strength, extensibility, and elasticity to the skin. The extensibility of the skin is evident when one has swollen extremities; it is even more evident when the skin over the abdomen is stretched during pregnancy, in extreme obesity, or when there is an accumulation of a large amount of fluid in the abdomen. The stretched skin is smooth and glistening, and can be injured by excessive stretching. The little tears which may occur during stretching remain visible afterward as silvery white streaks called *striae*.

In youth the skin is extensible and elastic, but as one grows older certain changes occur. The skin becomes thinner, there are fewer elastic fibers, and fat disappears from the subcutaneous tissue; this combination of events results in skin which is wrinkled in appearance.

The **blood supply** to skin is of great importance in the regulation of body temperature as well as for nourishment of the tissue cells. The arteries of the skin are arranged in the form of a network in the subcutaneous tissues just below the dermis. Branches from this network lead into capillary beds which permeate the dermis and ultimately drain into veins. The blood vessels of the skin are so numerous that they can hold a large proportion of the total blood supply of the body. When the skin is exposed to heat, or when the internal temperature of the body begins to rise, the small arteries of the skin dilate and allow more blood to flow through them, thus favoring the loss of heat from the body. On the other hand, when the skin is exposed to cold, or when the internal body temperature begins to drop, the vessels constrict and less blood flows through them; this means that less heat is lost from the body.

Interference with the blood supply to the skin will cause the death of cells and result in areas of ulceration. *Decubitus ulcers* (bedsores) may occur when bedridden persons spend prolonged periods of time in one position. The weight of the body exerts pressure on the skin, particularly where it lies over bony prominences, interfering with its blood supply and resulting in the death of tissue. Good nursing care, which includes frequent turning of the patient's body, massaging, and keeping the skin clean and dry, helps to prevent this distressing condition.

The **nerve supply** of the skin consists of fibers which carry impulses to and from the central nervous system. The nerves which reach the skin from the central nervous system are distributed to the smooth muscle in the walls of the small arteries in the dermis, to the smooth muscle around the roots of the hairs, and to the sweat glands of the skin. The nerves which carry messages from the skin to the central nervous system show specialized endings (receptors, p. 231) of many types. They are affected by changes in the environment of the individual, and the messages which they transmit to the central nervous system arouse sensations in the brain. By this means we acquire information about the world in the immediate vicinity of the body.

Appendages of the Skin. The cutaneous glands, the hair, and the nails comprise the appendages of the skin. All are specializations or differentiations of the epidermis and are protective in nature.

The **cutaneous glands** are of three types: sebaceous, sweat, and ceruminous glands. The *sebaceous glands* are found almost everywhere on the surface of the body, two prominent exceptions being the palms and the soles. These glands lie in the dermis, and their excretory ducts usually open into the necks of hair folicles, as shown in Figure 2-26. Sebaceous glands produce an oily secretion (called *sebum*) which prevents hair from becoming dry and brittle; it also forms an oily layer on the skin surface keeping it soft and helping to waterproof it. Persons with lazy sebaceous glands tend to have dry hair and skin, while those with active glands have hair and skin that are more oily. *Blackheads* are discolored accumulations of dried sebum. *Acne* (pimples) may be due to infection of sebaceous glands.

Sweat glands are distributed over the entire skin; they are most numerous in the axillae, and on the palms, the soles, and the forehead. It is estimated that there are about 2 million sweat glands in the skin of an adult. These glands are located in the dermis, and their ducts pass through the epidermis, in a spiral manner, to the surface of the skin where they open in pores.

Sweat glands are supplied by nerves which carry messages to promote perspiration when the outside temperature is high or when there is an elevation of body temperature. The main purpose of sweat glands is to act as an emergency mechanism when it is necessary for the body to lose heat by the evaporation of water from its surface. The pores of the sweat glands are rather prominent, and many persons use astringents on the skin to close the pores and make them appear smaller.

Ceruminous glands are modified sweat glands; they are found only in the skin of the passage leading into the ear (external auditory meatus). They secrete a yellow waxy substance (called *cerumen*) which helps to protect the tympanic membrane. However, if this wax accumulates in large amounts, it may interfere with hearing.

The *mammary glands,* or breasts, are specialized cutaneous glands which produce milk for the nourishment of the young. They are described on page 509.

Hairs and hair follicles (Fig. 2-26) are epithelial structures. The follicles are formed by a down-growth of cells of the epidermis into the dermis or into the subcutaneous tissue. Hair is formed by well-nourished germinal cells in the deepest part of the follicle. As these cells divide, the daughter cells are pushed upward, getting farther and farther away from their source of nourishment; eventually the cells die and become keratinized.

A hair usually consists of a central core, or medulla, surrounded by a cortex. Hair color depends upon the quality and quantity of pigment (melanin) in the cortex. White hairs have no pigment; when they are mixed with pigmented hairs, the mixture that results is called *gray hair* (true gray hair is rare).

The root of a hair is that part embedded in the follicle (usually slanting from the perpendicular), while the shaft projects above the surface of the skin. Two or more sebaceous glands are associated with each follicle; they lie in a little triangle formed by the slanted follicle and a tiny bundle of smooth muscle fibers, the *arrector pili* or *pilomotor muscle*. One end of this involuntary muscle bundle is attached to the sheath of the follicle; the muscle then slants upward to the point at which its other end is attached to the papillary layer of the dermis. Contraction of this muscle results in two pronounced effects: first, because the muscle pulls on the deep part of the follicle, the hair is made to stand on end as the follicle becomes more perpendicular; second, as the muscle pulls on its place of attachment to the dermis it tends to produce a dimple in the skin and give the well-known appearance of goose pimples. In addition, contraction of the muscle squeezes the sebaceous glands, forcing sebum into the follicle and onto the skin. Cold is an important stimulus that leads to the contraction of the pilomotor muscles.

Hairs are present over most of the body surface, with certain exceptions such as the palms and the soles, but are most abundant on the scalp. Straight or curly hair is related to the degree of curvature of the follicle and the form of the hair. If the follicle is unbent and the shaft is cylindrical, the hair is straight. If the follicle is bent and the shaft is flattened, the hair is curly.

Throughout life, one constantly loses hairs, but they are replaced by the continued division of epidermal cells in the germinal area. Cutting or shaving hair has no effect on its growth, the maximal rate being about ½ inch per month.

The **nails** are the horny scales of epidermis that overlie the dorsal surfaces of the fingers and the toes. The dermis, which the nail covers, is modified to form the nail bed. New formation of nail takes place in the epithelium at the proximal portion of the nail bed. As new formation takes place the nail moves forward and thus grows in length. The crescent at the base of the nail is called the lunula; it may be slightly or completely overlapped by the cuticle or *eponychium*.

Functions of the Skin. The skin performs many essential functions, such as protection, temperature regulation, and those of a sense organ. Relatively minor functions include the excretion of water and salts and the formation of vitamin D.

Protection of underlying tissues is an outstanding function of the intact skin. It acts as a mechanical barrier to prevent the invasion of bacteria and other foreign matter, and to prevent physical injury to the delicate tissues below. The skin is essentially a waterproof covering which makes it possible for the body to maintain a high content of water even in dry air. The importance of the skin in this respect is dramatically illustrated in the case of a person who suffers extensive burns over the entire body. Destruction of the skin results in such a rapid loss of fluid from the underlying tissues that the water balance of the entire body is upset; the victim may suffer complete circulatory collapse and die. It might also be added that without this waterproof exterior we would be unable to swim in fresh water without becoming swollen, or in salt water without becoming shrunken.

Temperature regulation depends on several factors, but the skin plays an important role. Its many sweat glands constitute one of the mechanisms for cooling the body; the *evaporation* of sweat from the surface of the body facilitates a loss of heat when the environment becomes too warm. An even greater amount of body heat is lost through *radiation*; thus the amount of blood brought to the skin is an important factor in temperature regulation. When blood vessels are dilated, more heat is lost; when the vessels are constricted, less heat is lost.

The skin is a very efficient **sense organ**, containing nerve endings which are responsive to touch, pain, and changes in temperature. It is an important source of information which enables us to withdraw from harmful stimuli (such as a hot stove) and to make other adjustments to the ever-changing world in which we live.

Other functions of the skin include the excretion of small mounts of nitrogenous waste products and NaCl in the sweat, the formation of vitamin D on exposure to the ultraviolet rays in sunlight, and the storage of water, sodium chloride, and glucose.

Organs

As stated previously, tissues are combined to form organs. We have considered the structure of some of the tissues of the body and have had examples of simple combinations of tissues to form membranes, and a complex combination of tissues to form the skin. The next step in considering the organization of the body is to think of the components of organs.

An organ is a part of the body which performs a definite function. Organs have connective tissue frameworks, with special blood, lymph, and nerve supplies, in addition to highly specialized cells which are characteristic for each organ. The specialized cells form the *parenchyma* of the organ, which is the important and predominant tissue. The accessory supporting tissues form the *stroma* of the organ. The stomach, the kidney, and the liver are examples of organs.

THE SYSTEMS

The final units of organization in the body are called systems. A system is a group of organs, each of which contributes its share to the function of the whole. The systems of the body are skeletal, muscular, nervous, circulatory, respiratory, digestive, excretory, endocrine, and reproductive.

The **skeletal system** includes not only the bones and the joints, but the strong ligaments that hold them together. This system provides a protective and supporting framework for the body as well as supplying a series of bony levers on which the skeletal muscles can act. Each bone is an organ, and connective tissue predominates in this system.

The **muscular system** is composed of three types of muscle plus the connective tissue necessary to harness the pulling power. Many movements of the body are due to the contraction of skeletal muscles which pull on the bones. Movement of food through the digestive tract depends on the contraction of visceral muscle, and blood is pumped around the body by cardiac muscle. Each muscle is an organ, and muscular tissue predominates in this system.

The **nervous system** is made up of the brain, the spinal cord, and the peripheral nerves. The organs of special sense (e.g., the eye and ear) are included with the nervous system. This is the great correlating and controlling system of the body. Messages flash with great rapidity over all parts of this system, as over telephone and telegraph wires. The nervous system is intimately connected with all other systems, and, by means of the organs of special sense, it is the source of information about the outside world and about one's own interior. Each part of the central nervous system is an organ. Nervous tissue predominates, and the cells are highly specialized for the conduction of messages.

The **circulatory system** comprises the heart, blood vessels, lymph vessels, and nodes. The heart serves as the pump, while the blood vessels carry food and oxygen to all parts of the body and return wastes to the organs of excretion. The heart and each vessel are organs. Many types of tissue are present in the circulatory system, and many types of specialized cells are found.

The **respiratory system** comprises the respiratory passages and the lungs. It makes possible the oxygenation of the blood and the elimination of carbon dioxide. The lungs and each part of the passageways are organs; many types of cells and tissues are found in this system.

The **digestive system** comprises the alimentary tract, its associated glands, the tongue, and the teeth. It converts food into substances which can be absorbed and utilized by the various tissues of the body. Each part of the system, such as the stomach, the pancreas, and the liver, is an organ. Many types of tissues and an even greater number of specialized cells are found in this system.

The **excretory system** proper includes the kidneys and the excretory ducts. Each kidney and ureter and the urinary bladder are organs. Nitrogenous waste, excess inorganic salts, and water are eliminated by this system. Many types of tissues and cells are found in the excretory system.

The **endocrine system** comprises the glands of internal secretion. The pituitary and the adrenal glands are examples of the organs in this system, and the cells of each are highly specialized to produce their characteristic secretions (hormones). The function of this system is to provide chemical correlation which is just as important as nervous correlation, but much slower. In comparing the two, one might say that the speed of nervous correlation is like the telephone, while that of endocrine correlation is like the postal service.

The **reproductive system** is concerned with the perpetuation of the species. The organs of the male reproductive system include the testes (which produce sex cells called spermatozoa), a series of excretory ducts, and various accessory structures. The female reproductive system consists of the ovaries (which produce sex cells called ova), uterine tubes, uterus, vagina, and associated external structures. Many tissues are found in this system, and the cells in each are specialized to carry out their particular functions.

All of the systems are closely interrelated and coordinated for the sole purpose of maintaining homeostasis. As a successful baseball team requires the close cooperation of nine different men, so the healthy body depends on the coordinated teamwork of its nine systems.

Whenever something happens in one part of the body to disturb the existing condition, compensatory responses occur in other parts of the body to restore the original condition. This constant give and take between systems serves to maintain that dynamic equilibrium which we call homeostasis. For example, suppose that you suddenly look at your watch and see that it is time for class. You run up the steps, and your muscles require more oxygen than they had been using when you were sauntering along toward school. To supply extra oxygen, you breathe more rapidly, the heart beats faster, and the blood pressure rises in order to bring a more abundant supply of blood to the active muscles. You may feel warm and begin to perspire by the time you reach the classroom. This is a brief list of the changes which occur during a bout of exercise.

As you would anticipate, damage to any part of the body has widespread effects on other parts. The whole field of medicine is concerned with the changes that occur in the human body when there is damage to such organs as the heart, the lungs, or the liver. It is interesting to know that there are some organs which function for a time and then undergo involution, or atrophy. For example, the ovaries in women cease to function after the menopause. Normal adjustments are made by the body when such organs undergo involution.

The beauty of the organization of the body should make a lasting impression on you. It is truly remarkable how well the body works if properly nourished and given adequate rest. It is equally remarkable how few difficulties are due to defective parts present at birth or to natural breakdowns of parts of the body before one reaches old age. Intricate integration and correlation of functions enable us to maintain life with a feeling of well-being and to perpetuate the species by reproduction.

SUMMARY

1. Cells—smallest units of life
2. Protoplasm
A. Composition
   a. Inorganic compounds
   b. Organic compounds
B. Importance of inorganic compounds
   a. Water: a solvent; high chemical stability; substances ionize in water; $H+$ and $OH-$ involved in all chemical processes in cell; aids in regulation of body temperature
   b. Inorganic salts: maintenance of proper osmotic conditions; acid-base balance; clotting of blood; formation of bones, teeth, hemoglobin; normal functioning of muscle and nerve; electrolyte balance. Principal extracellular ions: sodium and chloride; intracellular ions: chiefly potassium and phosphate
C. Importance of organic compounds
   a. Proteins: framework of protoplasm; growth and repair of tissue
   b. Carbohydrates: primary source of energy; glucose is simple sugar in blood
   c. Lipids: source of energy; stored as adipose tissue (reserve food, protect and support organs, insulate); role in permeability of cell membrane
3. Cell structure
A. Nucleus
   a. Control center
   b. DNA plays key role
B. Cytoplasm
   a. Surrounds nucleus; concerned with functional activities of cell
   b. Contains centrioles, mitochondria, Golgi apparatus, endoplasmic reticulum, ribosomes, lysosomes
   c. RNA important in synthesis of protein
C. Cell membrane
   a. Consists of lipoproteins; pores between molecules

allow passage of certain substances

b. Selective permeability influenced by size of pores, lipid-solubility, electric charge on membrane and ion seeking entrance, electrolyte balance; processes of passive and active transport

4. *Cell division (mitosis)*

A. Provides for growth and repair

B. DNA replicates prior to division

C. Prophase, metaphase, anaphase, telophase

5. *Protein synthesis*

A. m-RNA from DNA to ribosomes

B. t-RNA picks up activated amino acids

C. Assembly on ribosomes (r-RNA)

6. *Maintaining homeostasis in liquid cellular environment*

A. Interstitial fluid serves as "middleman" between bloodstream and cell

B. Homeostasis: state of dynamic equilibrium in internal environment

C. Circulatory "subway" system

a. Transportation of materials throughout the body

b. Closed circuit of tubes

c. Capillaries: focal point of system

D. Movement through membranes

a. Filtration

b. Diffusion

c. Osmosis

d. Exchange of fluid through capillary membranes dependent on hydrostatic and osmotic pressures in blood and interstitial fluid; dynamic balance prevents accumulation of fluid

7. *Primary tissues*

A. Epithelial

a. Covers body and lines cavities; protects, secretes, and absorbs; no blood vessels

b. Types: simple squamous, cuboidal, columnar; pseudostratified columnar; stratified squamous; transitional; glandular

B. Connective

a. Supports, protects, and binds together other body structures; stores fat and produces new blood cells; most abundant of all primary tissues; most have an excellent blood supply; fibrous and amorphous intercellular substance; fibroblasts, macrophages, mast, fat, plasma cells

b. Types: ordinary loose, white fibrous, cartilage, osseous

C. Muscle

a. Cells have capacity to shorten; produce movement of body parts

b. Types: skeletal, visceral or smooth, cardiac

D. Nerve

a. Nerve cells receive stimuli, conduct impulses to and from all parts of body

8. *Tissues as building materials*

A. Membranes: sheets of tissues that cover or line cavities

a. Mucous

b. Serous

c. Synovial

d. Superficial fascia

e. Deep fascia

f. Periosteum, dura mater, sclera

B. Skin

a. Epidermis: thin surface layer, consisting of stratified squamous epithelium; surface cells horny and keratinized; deep cells are living, undergo mitosis constantly; melanin is pigment. Dermis, or true skin: inner layer; consists of connective tissues, subdivided into papillary and reticular layers

b. Blood supply abundant; important in temperature regulation

c. Nerve supply to skin keeps us aware of environment

d. Appendages: cutaneous glands, hair, and nails

e. Functions: protection, temperature regulation, sense organ, excretion of water and salts, formation of vitamin D, storage of water, sodium chloride, and glucose

C. Organ

a. A part of the body which performs a definite function; connective tissue framework

b. Specialized cells form parenchyma; accessory supporting structures form stroma

9. *Systems: skeletal, muscular, nervous, circulatory, respiratory, digestive, excretory, endocrine, reproductive*

A. All are closely interrelated and coordinated for purpose of maintaining homeostasis

QUESTIONS FOR REVIEW

1. Destruction of the skin involves what primary tissues?

2. What pressures are involved in filtration?

3. What happens to red blood cells if they are placed in a hypotonic solution?

4. What food material is the most essential for repair of tissue?

5. What is the importance of the ribosomes, endoplasmic reticulum, and Golgi apparatus?

6. Why are lysosomes called "suicide bags"?

7. What is ATP and where is most of it produced?

8. Distinguish between active and passive transport through cell membranes.

9. What is the importance of mitosis?

10. Why is DNA of fundamental importance in protein synthesis?

11. Distinguish between m-RNA, t-RNA, and r-RNA in terms of their general functions.

12. What does the term homeostasis imply?

13. What is the predominant movement of fluid at the arterial end of the capillary? Why?

14. Distinguish between epithelial and connective tissues in terms of their *general* functions.

15. Distinguish between mucous and serous membranes in terms of their location and function.

16. What is the role of the skin with regard to regulation of body temperature?

REFERENCES
AND SUPPLEMENTAL READINGS

Allison, A.: Lysosomes and disease. Sci. Am., 217:62, Nov. 1967.

Best, C. H., and Taylor, N. B.: The Physiological Basis of Medical Practice, ed. 8. Baltimore, Williams & Wilkins, 1966.

Green, D. E.: The mitochondrion. Sci. Am., 210:63, Jan. 1964.

Guyton, A. C.: Textbook of Medical Physiology, ed. 4. Philadelphia, W. B. Saunders, 1971.

Ham, A. W.: Histology, ed. 6. Philadelphia, J. B. Lippincott, 1969.

Metheny, N. M., and Snively, W. D.: Nurses' Handbook of Fluid Balance, ed. 1. Philadelphia, J. B. Lippincott, 1967.

Montagna, W.: The skin. Sci. Am., 212:56, Feb. 1965.

Neutra, M., and LeBlond, C. P.: The Golgi apparatus. Sci. Am., 220:100, Feb. 1969.

Nomura, M.: Ribosomes. Sci. Am., 221:28, Oct. 1969.

Ross, R., and Bornstein, P.: Elastic fibers in the body. Sci. Am., 224:44, June 1971.

Ruch, T. C., and Patton, H. D.: Physiology and Biophysics, ed. 19. Philadelphia, W. B. Saunders, 1965.

Starling, E. H.: J. Physiol., 19:312–326, 1896.

Temin, H. M.: RNA-directed synthesis. Sci. Am., 226:24, Jan. 1972.

The Skeletal System

"Oh them bones, them dry bones . . ."
Negro Spiritual

All too often the words of the old song, "Oh them bones, them dry bones," are taken quite literally by beginning students of anatomy and physiology. However, in the next few pages you will see that those 206 bones are not solid, inert masses. The latter impression is one that is all too frequently gained following examination of the typical skeleton in the classroom closet. Nothing could be further from the truth! Rather, these vital structures are dynamic factories where millions of cells are engaged in work on a production line that would be the envy of any modern assembly plant.

For example, cells in the red bone marrow division manufacture blood corpuscles *in fantastic quantities*. It has been estimated that an average of one million wornout red blood corpuscles are destroyed every second of every day in the body; thus it is essential that an equal number of new ones be formed to replace them. Perhaps this will give some idea of the bustling activity that goes on within the bones.

Less dramatic, but equally important, are the other functions of the skeleton. In combination with cartilaginous tissue it forms a solid supporting framework for the softer tissues and provides protection for them. Some parts of the skeletal system connect with each other in such a way that they form structures like cages or boxes in which internal organs are lodged. An example of this is the thorax, or chest, in which the lungs and the heart occupy a protected position. The skeleton also furnishes surfaces for the attachment of muscles, tendons, and ligaments, which in turn pull on the individual bones and make it possible for us to move about. Thus one can see that bones and muscles are inseparable companions in the erect and moving body. It is only for ease of study that we consider them as individual units.

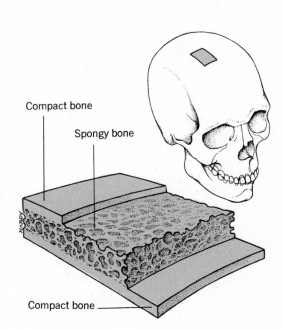

Fig. 3-2. Diagram of the arrangement of compact and spongy bone tissue in a flat bone of the skull.

CLASSIFICATION OF BONES

Bones are classified on the basis of their shape rather than according to size. For example, the relatively small bones in your fingers are still referred to as *long bones*. More attention will be given to structural details later on, but first let us consider the four classes as follows:

1. **Long bones** are found in the upper and the lower extremities; the only exceptions being the bones of the wrist, the ankle, and the knee-cap. A typical long bone consists of a shaft, or *diaphysis*, and two ends, or *epiphyses* (Fig. 3-1). The diaphysis is essentially a strong tube of compact bone with a roughly hollowed out space down the center. This space is referred to as the *medullary cavity*, and it is filled with yellow bone marrow. Each epiphysis consists of an outer compact shell with a spongy interior. In children, red bone marrow is found in all epiphyses, but it is eventually replaced by yellow marrow except in the proximal epiphyses of the adult arm and thigh bones.

2. **Short bones** are located in the wrist and the ankle. They might be described as cubical

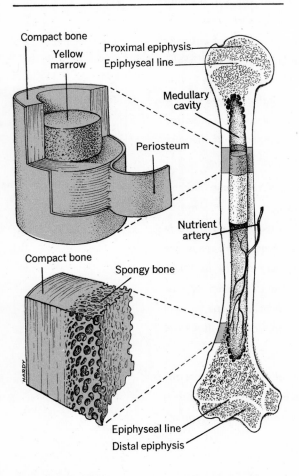

Fig. 3-1. Diagram of a long bone shown in longitudinal section.

in shape and consist of spongy bone covered with a shell of compact bone.

3. **Flat bones** are relatively thin, each being composed of two plates of compact bone which enclose between them a layer of spongy bone (Fig. 3-2). The ribs and several of the skull bones are good examples of this type.

4. **Irregular bones** appear in various shapes and include all that are not in one of the preceding classes. For example, bones of the spinal column and the jaw are irregular bones. Bulky portions of these bones consist of spongy bone surrounded by a layer of compact bone, while the thinner parts may be composed of two plates of compact bone with a small amount of spongy bone between them.

Sesamoid bones, with the exception of the patella, or kneecap, are very small, rounded bones. They develop in the capsules of joints or in tendons, and their function is to reduce friction. The number of sesamoid bones is variable.

BONE FORMATION

To better understand the structural features of bone tissue, let us start at the very beginning of its formation when an embryo has been developing in its mother's uterus for approximately 2 months.

At the end of this 8-week period there is a complete little skeleton, but it is composed entirely of *hyaline cartilage*, except for some of the face bones and the skull bones which are temporarily constructed of *fibrous membrane*. Obviously, cartilage and membrane cannot carry out the functions which we have ascribed to the bony skeleton; therefore, we shall now see what great changes are brought about in a relatively short period of time.

Intramembranous bone formation

Bones of the skull are formed by the intramembranous method and will be considered first. Try to visualize little pieces of membrane, which are rough patterns for the bone-to-be. In the center of each individual membrane special bone-building cells, or **osteoblasts,** produce an intercellular substance in which calcium salts are deposited.

From this primary center, **ossification,** or bone formation, advances in radiating columns toward the edges of the membrane. However, the membrane also continues to grow for some time, and ossification of the skull bones is not yet complete when the baby is born. This is most noticeable at the *corners* of adjoining membranes and accounts for the *soft spots*, or **fontanels,** which may be felt on an infant's head. Mothers often worry needlessly about injuring these areas when washing the baby's head or brushing the hair. They should be reassured that, with normal handling, the chance of injury is remote.

Eventually, ossification is completed. The cranial bones then consist of hard outer and inner plates of compact bone with spongy bone between. The latter layer is referred to as *diploë*.

Endochondral, or cartilaginous, bone formation

Most of the bones of our body are preformed in hyaline cartilage, and the endochondral or cartilaginous method of bone formation is responsible for converting these cartilage models into true bones. This process might be compared with the building of a bridge, in which a temporary wooden frame is erected first, and then gradually is replaced by a permanent steel structure.

Ossification begins soon after the second month of uterine life. The first change is an increase in the size of cartilage cells in certain areas. These areas, so far as long bones are concerned, are in the center of the shaft, or diaphysis, and in each end, or epiphysis. After the cartilage cells enlarge they begin to degenerate, and, eventually, they leave spaces into which the osteoblasts, or bone-forming cells move. The two processes go on simultaneously; that is, destruction of cartilage and formation of bone take place in adjacent areas, as illustrated in Figure 3-3. Bone formation spreads out from the centers of ossification in the diaphysis and the epiphyses until only two thin strips of cartilage remain, one at each end of the bone between the diaphysis and the epiphyses. These strips persist as **epiphyseal cartilages** until growth of the bone is completed, at which time these cartilages also become transformed into bone, and the epiphyses are said to be *closed*. The final stages in the replacement of cartilage occur long after birth in some bones.

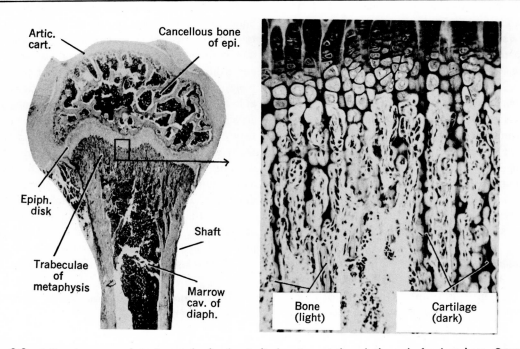

Fig. 3-3. *Left,* a low-power photomicrograph of a longitudinal section cut through the end of a long bone. Osteogenesis has spread out from the epiphyseal center of ossification so that only the articular cartilage above and the epiphyseal disk below remain cartilaginous. *Right,* under high power, on the shaft side of the disk are trabeculae, which consist of cartilage cores on which bone has been deposited.

GROWTH OF BONE

Bones grow **in circumference** by the deposit of bone beneath the periosteum by the apposi-tional method or side to side application of suc-cessive layers. The deeper layer of the periosteum is composed of osteoblasts, which are responsible for the new bone that is formed (Fig. 3-4). As

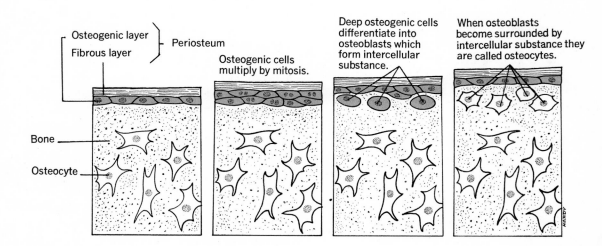

Fig. 3-4. Diagram of the process whereby bone increases in circumference. New layers of bone are produced by cells of the osteogenic layer of periosteum in appositional growth.

new bone is formed on the external surface, other cells called *osteoclasts* dissolve the bony tissue adjacent to the medullary cavity and in this way enlarge the marrow cavity to keep pace with the increase in circumference of the shaft.

Bones grow **in length** due to the activity of the cartilage cells in the epiphyseal cartilages and the replacement of these by osteoblasts. As ossification occurs in the cartilage adjacent to the shaft of the bone, new cartilage cells are formed at the distal end. The length of the shaft is increased in this manner, and growth in length continues until the epiphyses are closed, which means that all the cartilage has been transformed into bone. Cessation of growth in bones occurs at about 18 years in females and soon after 20 in males. An immature long bone is illustrated in Figure 3-5.

Certain *hormones*, chemical messengers produced by the endocrine glands, play a very prominent role in promoting or in inhibiting the growth of bone. For example, if there is an excessive amount of the pituitary growth hormone, bones

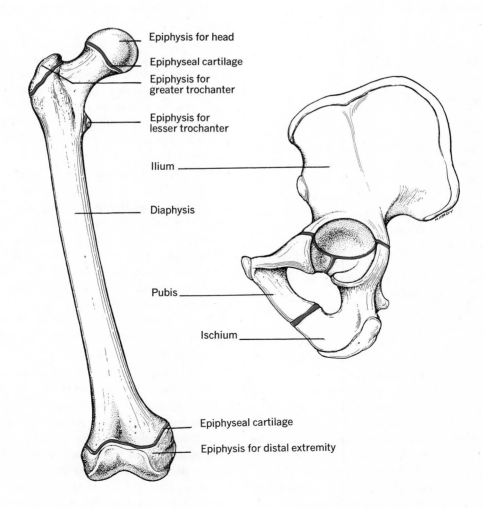

Fig. 3-5. Immature bone: *right,* femur, showing epiphyseal cartilages for head, trochanters and distal end of bone. *Left,* hip bone, showing the cartilage at which growth is taking place between the ilium, the ischium and the pubis.

may continue to grow for a prolonged period of time. The resulting condition is referred to as *gigantism.*

Bone-age studies are often an important diagnostic tool in the evaluation of endocrine disturbances. X-ray pictures of the long bones of a child will show the presence of epiphyseal cartilages; they appear as dark lines on the relatively white bone, since cartilage is of a lesser density. Normally these cartilage bands become smaller and smaller as ossification proceeds, and an expert can tell the age of a child by just looking at x-ray pictures of the bones (see Fig. 3-39). In certain abnormal conditions, however, the cartilage bands may continue to grow, and a youngster with a chronologic age of 12 years, for example, may have a bone age of only 8 years.

STRUCTURAL DETAILS

Composition of bone

Bone, or osseous connective tissue, is no different from other tissues insofar as it is maintained by living cells. However, we all know from personal experience that bone is hard and cannot be cut with a knife, while other tissues are soft and pliable. Therefore, the essential difference must lie in the composition of the *intercellular substance,* or *matrix.*

Osteocytes, or *bone cells,* are responsible for the maintenance of this matrix which is composed of various organic and inorganic constituents. The chief organic constituent is *collagen,* and the chief inorganic constituent is *calcium phosphate.* However, calcium is also present as carbonate and fluoride; magnesium and sodium are other inorganic components of bone.

These salts are constantly being deposited and constantly being withdrawn as the bone plays a part in maintaining the normal level of calcium in the bloodstream. Various hormones, which we shall study at a later time, exert great influence on this continuous interchange of inorganic salts. For example, if the parathyroid glands are too active, the amount of calcium in the bloodstream rises sharply. To meet this demand there may be a massive withdrawal of calcium from the bones, leaving them weak and subject to spontaneous

fractures. **Fracture** is a term that refers to any break in a bone. It may be complete or only partial.

While the inorganic constituents make up approximately two thirds of the solid material in bone, the organic matter is no less important. If a fresh bone is placed in dilute nitric acid, the inorganic salts will be removed. The remaining material still resembles a bone, but it is so flexible or rubbery that it can be tied in a knot. On the other hand, if a bone is placed over a hot flame, the organic matter will be removed, and the remaining bone is so brittle that it crumbles at the slightest touch. The conclusion to be drawn is obvious: there must be appropriate amounts of both organic and inorganic constituents if one is to have functional bone tissue for support and protection.

In *rickets* there is defective calcification of bones due to a disturbance in mineral metabolism. The bones remain soft, a condition which may result in bowlegs and malformations of the head, the chest, and the pelvis. The cure and the prevention of rickets hinge on the intake of liberal amounts of calcium, phosphorus, and vitamin D in the diet.

Thinking in terms of early bone formation and growth it is easy to understand why the bones of children normally contain more organic matter and less inorganic matter than the bones of adults. Consequently, since young bones are so much more flexible and less brittle, it stands to reason that children can better tolerate a tumble to the sidewalk. If a child's bone should happen to be fractured it would probably resemble the splitting that one encounters when trying to remove a branch from a growing bush; thus the term *green stick fracture* is often used in such cases.

At the other end of the scale, one is more apt to see a generalized loss of cortical bone substance; this condition is known as **osteoporosis.** Recent studies indicate that the primary defect is excessive bone resorption, most probably due to an inadequate intake of dietary calcium. Although osteoporosis almost invariably accompanies the aging process, it also is seen in cases of malnutrition, vitamin C deficiency, and various endocrine deficiencies. Individuals suffering from osteoporosis frequently experience bone pain and pathologic fractures.

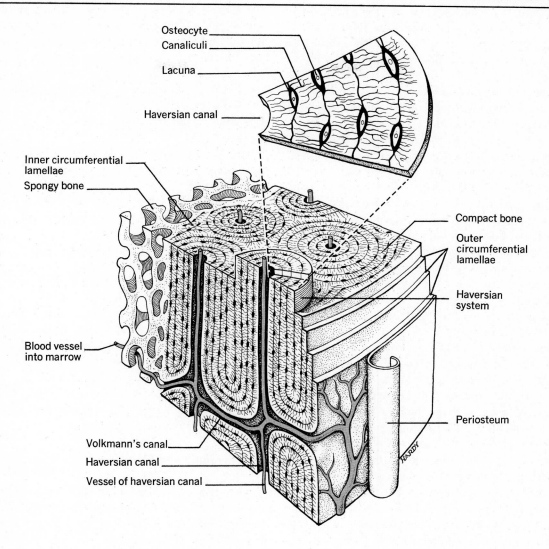

Osteocyte
Canaliculi
Lacuna

Haversian canal

Inner circumferential lamellae
Spongy bone

Blood vessel into marrow

Compact bone

Outer circumferential lamellae

Haversian system

Periosteum

Volkmann's canal
Haversian canal
Vessel of haversian canal

Fig. 3-6. Diagram of haversian systems as seen in a wedge of compact bone tissue. The periosteum has been peeled back to show a blood vessel entering one of Volkmann's canals. *Upper right,* sketch showing osteocytes lying within lacunae; canaliculi permit interstitial fluid to reach each lacuna.

The haversian systems

One of the basic principles of physiology is that all living cells must constantly receive nourishment and eliminate waste products resulting from their metabolic processes. Osteocytes, the living bone cells, are no exception to this rule even though they are surrounded by calcified intercellular substance. Fantastic systems of tiny canals permeate bone tissue and convey materials to and from every single osteocyte. These microscopic,

functional units are known as haversian systems (Fig. 3-6).

When bone formation gets under way, the osteoblasts proceed to supervise the deposition of calcium salts on all sides, thus tending to wall themselves off in little lakes, or **lacunae.** To prevent complete isolation they send out little extensions of their cytoplasm, (somewhat resembling the tentacles of an octopus). As the inorganic salts continue to be deposited, the presence of these cytoplasmic processes keeps microscopic channels, or **canaliculi,** open for the movement of interstitial fluid. When bone formation is complete, the osteoblasts withdraw their little processes, but the bony tunnels remain. Unable to leave the lacunae, the osteoblasts, now referred to as **osteocytes,** assume the role of *maintenance cells* and supervise the constant interchange of inorganic salts in healthy adult bone.

The canaliculi extending from one lake, or lacuna, communicate with those from neighboring lacunae and form a network of permeating channels throughout the matrix. This network has final connections with the central or **haversian canal** which contains the all-important blood vessels essential for the delivery of nutrients and the pickup of wastes which move back and forth through the canaliculi according to the varying needs of the osteocytes (Fig. 3-6).

Further examination of bone tissue will show that osteocytes are arranged in layers; one such layer of cells, together with the matrix around them, forms a sheet called a **lamella.** The lamella may then be arranged in one of two general designs, each of which gives rise to a characteristic type of bone structure. The first type is **cancellous** or *spongy bone,* in which the lamellae are arranged as a scaffolding or latticework of needle-shaped spicules of bone, with marrow-filled spaces between (Fig. 3-2). The second type is **compact bone,** in which the lamellae are arranged to form an apparently solid mass, as shown in Figure 3-6. Compact bone appears along lines of stress in the shaft of long bones and forms a protective outer shell in others. It is extremely strong and provides sturdy support as well as protection. Cancellous bone, on the other hand, is lighter in weight. It is found under the compact shell in short, flat, and irregular bones, and at the epiphyses of long bones where a relatively large area is needed for articulations. Were it not for this rather light but strong type of osseous tissue our bones would be unduly heavy.

Bone marrow

Bone marrow arises from early embryonic structures and invades the cartilage as it is being replaced by bone. In adult bone, two types of marrow are present: red and yellow. *Yellow marrow* is found, especially, in the medullary cavities of long bones and consists largely of adipose tissue.

Red marrow is found in the bodies of the vertebrae, diploë of the cranial bones, the sternum, the ribs, and the proximal epiphyses of femur and humerus. The red marrow consists of a large number of cells contained in the meshes of a delicate connective tissue which is abundantly supplied with blood vessels.

Some of the *functions* of red bone marrow are:

1. Formation of the red blood cells which are so important in the transport of oxygen and carbon dioxide;

2. Formation of some of the white blood cells (granular leukocytes), the most active of our infection fighters;

3. Formation of thrombocytes, or platelets, which aid in the clotting of blood;

4. Phagocytosis, the destruction of old, worn-out red blood cells and foreign particles.

Red marrow is the only source of red blood cells and granular leukocytes after birth; consequently, improper functioning may lead to abnormalities such as certain types of anemia. In some areas, red marrow is changed to yellow marrow by the invasion of fat cells. This begins soon after birth, and, in the adult, red marrow is found only in those areas mentioned previously.

Coverings and blood supply

Our bones are almost completely covered with a dense fibrous membrane called the **periosteum,** the only exception being the joint, or articular, surface. Those areas where bones come into contact with each other (in freely movable joints) are covered with a cushion of hyaline cartilage referred to as the **articular cartilage** (Fig. 3-51). The periosteum is firmly attached by small thread-like fibers called Sharpey's fibers which actually penetrate the bone. Ligaments and tendons are also anchored securely to bone as their fibers interlace with those of the periosteum. The inner layer of the periosteum gives rise to the osteoblasts which are so important in the early growth of bone, as well as in the repair of fractures which may occur when our bones are subjected to undue stress.

In addition to its aforementioned functions, the periosteum also plays an important role in the nourishment of bone tissue as a whole. Numerous **periosteal blood vessels** enter the bone through tiny openings called Volkmann's canals which in turn connect with the central canals of the microscopic haversian systems (Fig. 3-6). The medullary cavity of long bones is supplied with blood by way of the **nutrient** or **medullary artery,** which enters the shaft at an oblique angle (Fig. 3-1). It then sends branches upward, downward, and into the adjacent bony tissue to *anastomose*, or join, with the haversian vessels. Similar arteries supply the central parts of short, flat, and irregular bones. Lymphatic vessels and nerves also have been traced into the substance of bone, offering even further proof that we are dealing with vital, dynamic tissue! And, like any other tissue of the body, a bone may become infected. The resulting inflammation of the bone and its marrow is called *osteomyelitis*, a very serious condition.

DESCRIPTIVE TERMS

When one examines the various bones of the skeleton, it will be noted that many of them have marked projections, or processes. Some have depressions, or even a hole, which further helps in their identification. Familiarity with the following terms should prove to be helpful when studying the bones.

head: a rounded end which extends from a constricted portion, or neck, and fits into a joint. For example, the *head of the femur*, or thigh bone (see Fig. 3-44).

condyle: a slightly rounded projection for articulation (at a joint) with another bone. For example, the *condyles of the femur* which articulate with the tibia, or shin bone (see Fig. 3-44).

crest: a ridge of bone to which muscles are attached. For example, the flaring upper border of the hipbone, the *iliac crest* (see Fig. 3-40).

spine or spinous process: a relatively sharp projection for muscle attachments. For example, one can feel the *spinous processes* of the vertebrae just under the skin of the back. (see Figs. 3-25 and 3-26).

trochanter: a very large process to which muscles are attached. This is used to identify two large projections near the upper end of the femur, the *greater and the lesser trochanters* (see Fig. 3-44).

tubercle: a small rounded projection for muscle attachment. For example, at the upper end of the humerus, or arm bone, one can find the *greater and lesser tubercles* (see Fig. 3-35).

tuberosity: a large, roughened projection to which muscles attach. For example, the *tuberosities of the ischium* (lower hipbone) help to bear the weight of your body as you sit in a chair while reading these lines (see Fig. 3-40).

foramen: a hole in a bone. For example, the *foramen magnum* in the occipital bone of the skull (see Fig. 3-11), through which the spinal cord passes.

fossa: a shallow or hollow place in a bone. For example, the *olecranon fossa* at the lower end of the humerus (see Fig. 3-35).

sinus: an air-filled cavity in a bone. For example, the *maxillary sinus* in the upper jaw (see Fig. 3-15).

meatus: a tube-shaped opening in a bone. For example, the external *auditory*, or *acoustic*, *meatus* which is the canal leading to the tympanic membrane (see Fig. 3-9).

Other examples of projections and depressions will be encountered as you study the individual bones of the body.

THE SKELETON AS A WHOLE

The adult human skeleton is composed of 206 bones (excluding the small sesamoid bones which are variable). A greater number of bones may be seen in the skeleton of a newborn baby, but during normal growth and development some of these bones fuse, or unite, to form just one bone where previously there had been several.

For purposes of study, the skeleton is divided into two main parts. Logically, these divisions are as follows:

1. The *axial skeleton*, which is composed of the bones of the skull, the thorax, and the vertebral column. These bones, 80 in number, form the *axis* of the body.

2. The *appendicular skeleton*, which consists of the bones of the shoulder, the upper extremities, the hips, and the lower extremities. These bones, 126 in number, are attached to the axial skeleton as *appendages*.

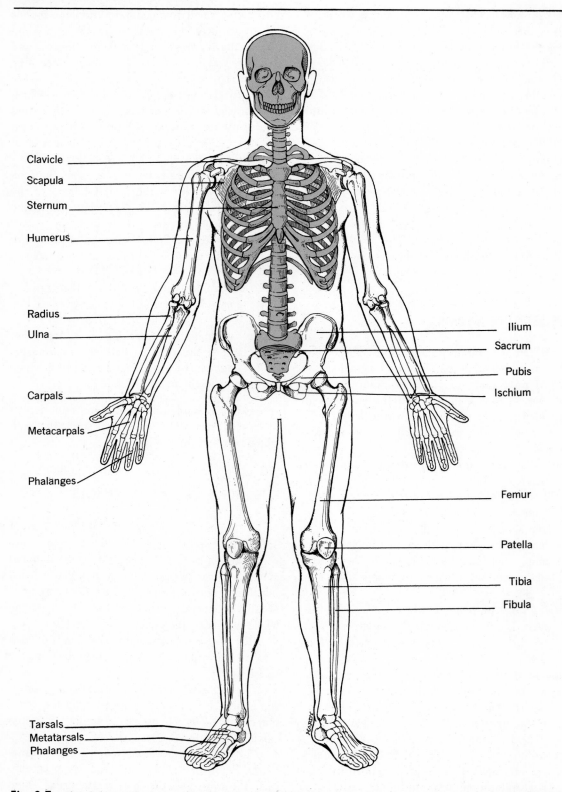

Clavicle

Scapula

Sternum

Humerus

Radius

Ulna

Carpals

Metacarpals

Phalanges

Ilium

Sacrum

Pubis

Ischium

Femur

Patella

Tibia

Fibula

Tarsals
Metatarsals
Phalanges

Fig. 3-7. The skeleton. The bones of the head and the trunk forming the axial skeleton are shown in red, and those of the extremities forming the appendicular skeleton are shown in black.

The divisions of the skeleton are presented in color in Figure 3-7—the axial portion in blue, the appendicular in yellow. Details of individual bones will be shown at appropriate intervals as they are presented for consideration.

Study of the bones is much more interesting if one can locate the bones in his own body and, insofar as possible, feel their general outline. Examination of an actual skeleton is also quite helpful. *Osteology*, or the study of bones, is a distinct science in itself, so the following descriptions will necessarily be limited to those features which are most outstanding.

THE AXIAL SKELETON

Bones of the skull

A close examination of the skull will reveal 28 distinct bones. This number includes the six small bones of the middle ears which are concerned with the transmission of sound waves; they will be described in Chapter 10. The remaining bones form the framework of the head and provide excellent protection for such important organs as the brain, the eyes, and the ears. With the exception of the lower jaw bone, which is freely movable, the skull bones are joined together and immovably fixed in position. These joints, or *sutures*, are described on page 89. To facilitate study, the bones of the skull will be considered in the following subdivisions: *the cranium, superficial bones of the face, and deeper bones of the face.*

1. The cranial bones: one frontal; one occipital; two parietal; one sphenoid; two temporal; one ethmoid.

The **frontal bone** (Figs. 3-8 to 3-10) is shaped somewhat like a scoop; it forms the anterior roof of the skull, the forehead, and assists in the formation of the roof of the orbital and the nasal cavities. The *supraorbital margins* are arched rims of bone above the orbit where they afford

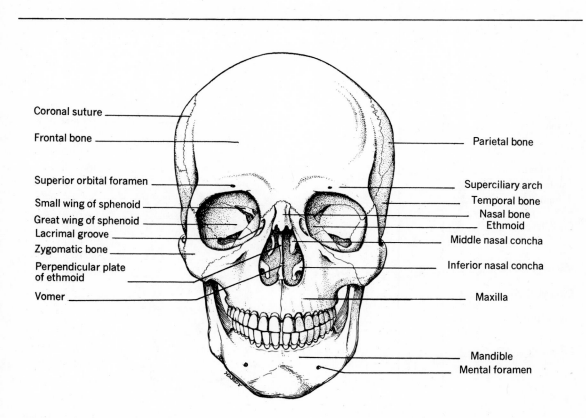

Coronal suture

Frontal bone

Superior orbital foramen

Small wing of sphenoid

Great wing of sphenoid

Lacrimal groove

Zygomatic bone

Perpendicular plate of ethmoid

Vomer

Parietal bone

Superciliary arch

Temporal bone

Nasal bone

Ethmoid

Middle nasal concha

Inferior nasal concha

Maxilla

Mandible

Mental foramen

Fig. 3-8. Anterior view of the skull.

considerable protection to the eye. They can be felt through the skin in the region of the eyebrows. The *superciliary ridges* are elevations above the orbital margins. Beneath these, within the bone, are the *frontal sinuses*, or air-filled cavities, which communicate with the nasal cavities.

The **parietal bones,** somewhat saucer-shaped, form much of the top and the sides of the cranium (Figs. 3-8 and 3-9).

The **temporal bones** are relatively complicated bones which form part of the sides and part of the base of the skull (Figs. 3-9 and 3-10). They articulate with the lower jaw, or mandible. For ease of description, each bone shall be divided into four parts: squamous, tympanic, mastoid, and petrous.

1. *Squamous portion:* a thin, flat plate of bone at the side of the skull. The *zygomatic process* projects forward from the lower lateral surface and articulates with the temporal process of the zygomatic, or cheek bone, to form the *zygomatic*

arch. The inferior surface is occupied largely by the *mandibular fossa* which receives the articular condyle of the mandible. This freely movable joint enables one to move the lower jaw when eating and talking.

2. *Tympanic portion:* a curved plate, inferior to the squamous portion. It makes up part of the wall of the *external acoustic meatus*, or bony canal of the external ear. This canal is closed at its medial end by the tympanic membrane

3. *Mastoid portion:* posterior to the squamous and tympanic areas. The *mastoid process* projects downward immediately behind the external meatus; it contains many air spaces, referred to as *mastoid air cells*, which communicate with the middle ear. Infection of the latter (otitis media) may extend into the air cells causing *mastoiditis.* Only a thin plate of bone separates these air cells from the coverings of the brain, and infection may result in serious complications.

4. *Petrous portion:* a wedge-shaped mass of

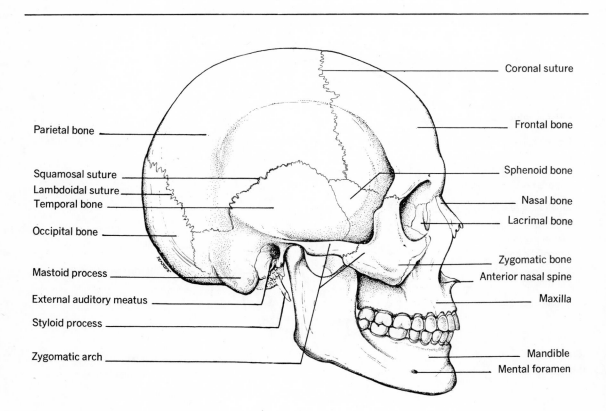

Fig. 3-9. Lateral view of the skull.

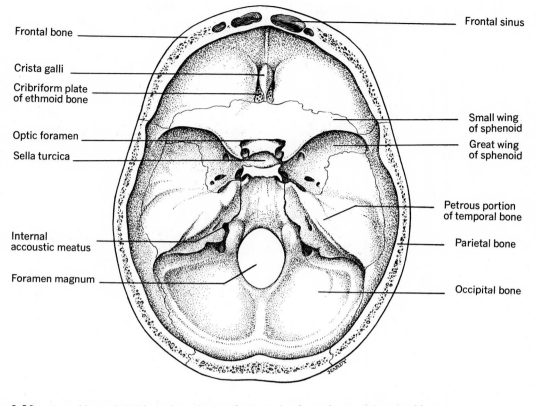

Frontal bone

Crista galli

Cribriform plate
of ethmoid bone

Optic foramen

Sella turcica

Internal
accoustic meatus

Foramen magnum

Frontal sinus

Small wing
of sphenoid

Great wing
of sphenoid

Petrous portion
of temporal bone

Parietal bone

Occipital bone

Fig. 3-10.   View of base of skull from above showing the internal surfaces of some of the cranial bones.

bone constituting part of the base of the skull (Fig. 3-10). It contains the essential organs of hearing and position sense. On the medial side can be seen the *internal acoustic meatus,* a foramen which allows passage of the facial and the acoustic nerves on their way to the brain. The *styloid process* projects downward from the inferior surface of the bone (Fig. 3-9).

The **occipital bone** forms the back and a large part of the base of the skull (Figs. 3-9 and 3-10). The *foramen magnum* is a large opening through which the spinal cord communicates with the brain (Fig. 3-10). On each side of the foramen magnum are the smooth *condyles* for articulation with the *atlas,* or first bone in the vertebral column. Nodding of the head, forward and backward, will demonstrate the action of this joint. The *external occipital protuberance* is a prominent elevation on the posterior aspect of the bone, in the midline. It can be felt as a definite bump

just under the skin. The *superior nuchal line* is a curved ridge of bone passing laterally from the protuberance to the mastoid portion of the temporal bone.

The **sphenoid bone** is sometimes referred to as the *keystone* due to its central location and to its articulations with all of the other cranial bones (Figs. 3-10 and 3-11). When seen from above, the sphenoid fills the space between the orbital plates of the frontal bone anteriorly, the temporal and the occipital bones posteriorly. It bears some resemblance to a bat with its wings extended (Fig. 3-12). The middle portion, or *body,* contains the *sphenoid sinuses,* or air cells, which open into the nasal cavity. The upper surface presents a marked depression in the form of a saddle called the *sella turcica;* the pituitary gland lies in this depression. Extending laterally from the body are the two *great wings* and, above them, the two *small wings.* Projecting downward from the undersurface of the

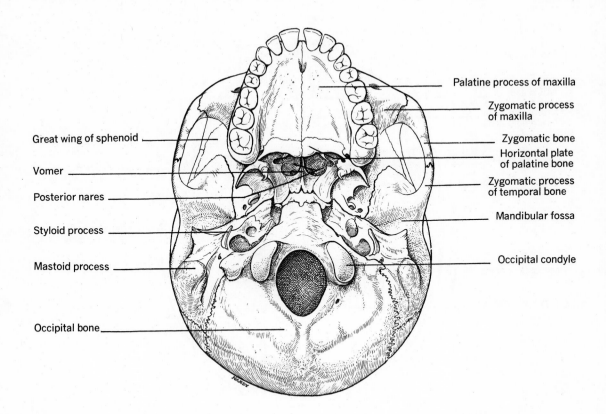

Fig. 3-11. View of base of the skull from below, with the lower jaw removed, showing some of the cranial bones and the bony roof of the mouth.

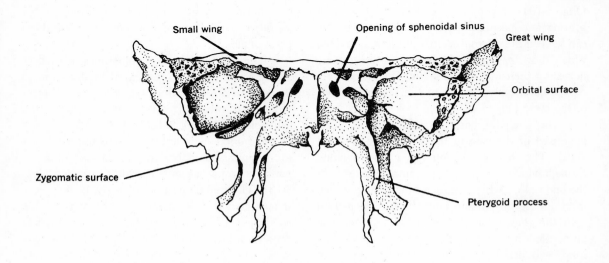

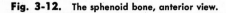

Fig. 3-12. The sphenoid bone, anterior view.

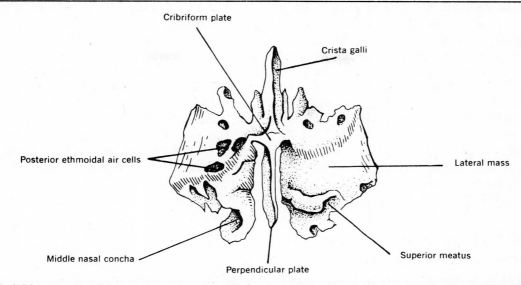

Fig. 3-13. The ethmoid bone, posterior view. The left lateral mass has been opened to show the posterior ethmoidal air cells, or sinuses.

body and forming part of the lateral walls of the nasal cavity are the two *pterygoid processes*. These would correspond to the feet of this bat-shaped bone.

The **ethmoid bone** is a light, cancellous bone located between the orbital cavities. Looking at the base of the skull from above (Fig. 3-10) one

can see the *horizontal*, or *cribriform, plate* which forms the roof of the nasal cavities and part of the base of the cranium. From the upper surface of this plate extends the *crista galli* (cock's comb), a triangular projection which provides a place for the attachment of the covering of the brain. Closer inspection of this plate will reveal many small

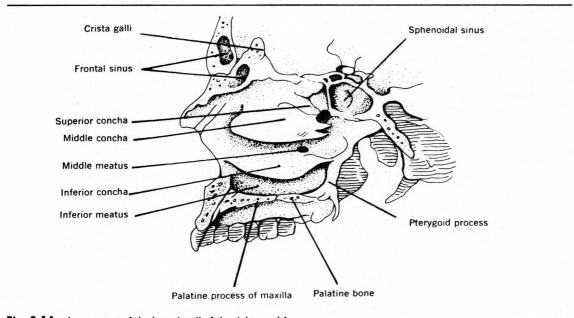

Fig. 3-14. Inner aspect of the lateral wall of the right nasal fossa.

openings, the *olfactory foramina*, through which pass the nerves of smell on their way to the brain. An anterior view of the skull (Fig. 3-8) will reveal the *perpendicular plate* as it forms the upper part of the nasal *septum*, or partition, which separates the nasal cavities. On each side, the *lateral masses* (Fig. 3-13) enter into the formation of the orbital cavity and lateral wall of the nose. They contain the *ethmoid sinuses*, or air cells, which open into the nasal cavities. Two scroll-like ridges of bone, the *superior* and the *middle conchae*, or *turbinates*, project medially from the lateral masses and serve to increase the total surface area of the nasal cavities (Fig. 3-14).

2. **Superficial bones of the face:** two nasal; two maxillae; two zygomatic; one mandible.

The **nasal bones** are small oblong bones, which lie side by side, forming the so-called bridge of the nose (Fig. 3-8). The anterior portion of the nose, with which we are most familiar, is composed of cartilage and skin.

The **zygomatic bone** is the cheek bone, located at the upper and lateral part of the face (Fig. 3-8). In addition to forming the prominence of the cheek, it enters into the formation of part of the lateral wall and the floor of the orbital cavity. It articulates posteriorly with the zygomatic process of the temporal bone. This union forms the prominent *zygomatic arch* which can be felt on the side of the face in front of the ear.

Maxilla is the term used in reference to the upper jaw as a whole. However, there are actually two *maxillae* which fuse in the midline (Fig. 3-8). They not only form part of the floor of the orbital cavities, but also the *horizontal*, or *palatine*, *processes* form the anterior three fourths of the roof of the mouth, or *hard palate*. Faulty union of these bones may result in a congenital deformity known as *cleft palate* in which there is an opening in the roof of the mouth. Infants born with this defect have difficulty in nursing, since their mouths communicate with the nasal cavities above, and air, rather than milk, is sucked in. The anterior and lateral aspects of the central portion, or *body*, form part of the face below the eye and above the teeth. On the medial aspect, the body forms part of the lateral wall of the nasal cavity. The *maxillary sinus*, or *antrum of Highmore*, is a large air space inside the body (Fig. 3-15); it opens into the nasal cavity. The *alveolar process* contains sockets for the upper teeth.

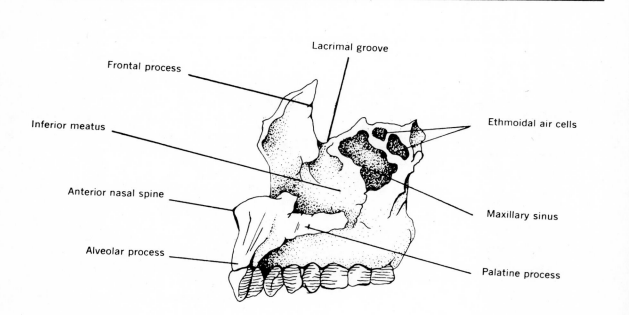

Lacrimal groove

Frontal process

Inferior meatus

Anterior nasal spine

Alveolar process

Ethmoidal air cells

Maxillary sinus

Palatine process

Fig. 3-15.   Right maxilla, showing some of the ethmoidal cells and the maxillary sinus.

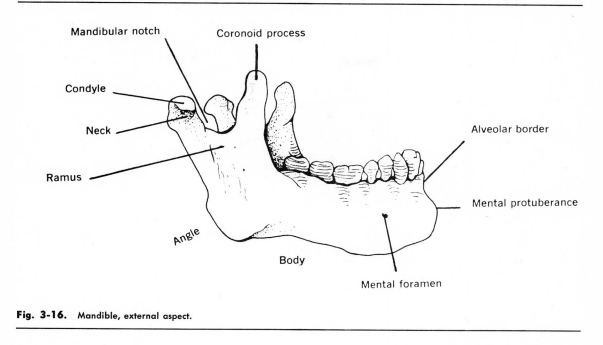

Fig. 3-16. Mandible, external aspect.

The **mandible,** or lower jaw, is the largest bone of the face (Figs. 3-9 and 3-16). Its horseshoe-shaped central portion is referred to as the *body,* and the upper margin or alveolar process contains sockets for the lower teeth. The *mental foramen* permits the passage of nerves and blood vessels to the chin. Extending upward from the posterior portion of the body are two *rami* (singular, *ramus*). Two prominent processes can be noted on top of each ramus. The posterior, or *condyloid process,* articulates with the temporal bone to form the *temporomandibular joint,* the only movable joint of the skull. If a finger is placed just in front of the opening of the external ear, the condyle of the mandible can be felt sliding back and forth during movements of the lower jaw. The anterior, or *coronoid process,* provides a point of attachment for some of the muscles that move the jaw. At birth, the mandible consists of two parts, but these fuse into one bone during the first year. It undergoes several changes in shape as teeth erupt for the first and then the second time. In old age, after the teeth are lost, the alveolar process is absorbed, and the chin appears more prominent.

3. Deeper bones of the face: two lacrimal; two palatine; one vomer; two inferior conchae.

The **lacrimal bones** are thin, fragile bones, about the size of a fingernail. They lie in the front part of the medial wall of the orbital cavity, just lateral and posterior to the nasal bones. Each lacrimal bone contains a groove which is part of the *lacrimal canal* from the orbit to the nasal cavity; the tear duct passes through the canal.

The **vomer,** shaped roughly like the blade of a plow, forms the lower back part of the nasal septum. It lies between the perpendicular plate of the ethmoid above and the maxillary and palatine bones below (Figs. 3-8 and 3-17).

The **palatine bones** are generally described as being L-shaped, but it must be kept in mind that the one on the left is reversed (i.e., ⌐). These bones are located at the back part of the nasal cavities between the maxillae and the pterygoid processes of the sphenoid. The upper end of each bone forms part of the floor of the orbital cavity; the upright portion forms part of the lateral wall of the nasal cavity; the horizontal section forms part of the floor of the nasal cavity and the posterior fourth of the hard palate (Fig. 3-11).

The **inferior conchae,** or *turbinates,* extend horizontally along the lateral wall of the nasal cavities (Fig. 3-14). Each is composed of a thin plate of bone which is slightly curled upon itself, hence the expression *scroll-like* in shape.

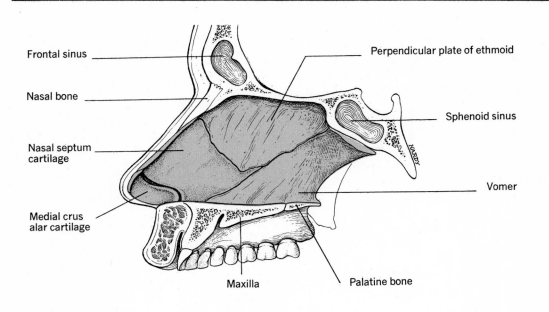

Frontal sinus

Nasal bone

Nasal septum
cartilage

Medial crus
alar cartilage

Perpendicular plate of ethmoid

Sphenoid sinus

Vomer

Maxilla

Palatine bone

Fig. 3-17. Lateral view of the nasal septum showing relationship of bony and cartilaginous structures.

The cranial sutures

Immovable articulations between the various cranial bones are referred to as sutures. The following may be seen either in Figure 3-9 or in Figure 3-19: (1) *coronal*, between the frontal and the parietal bones; (2) *lambdoidal*, between the parietal and the occipital bones; (3) *sagittal*, between the two parietal bones; and (4) *squamosal*, between part of the temporal and the parietal

Fig. 3-18. Six children with microcephaly from the same family. (From M. Bartalos: Genetics in Medical Practice. Philadelphia, J. B. Lippincott, 1968.)

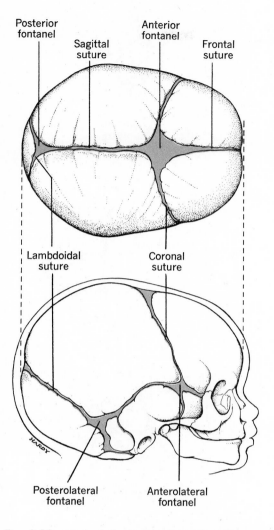

Fig. 3-19. Fetal skull showing fontanels and sutures: *above,* superior aspect; *below,* lateral aspect.

bones. These are the major sutures of the skull. In the newborn infant there is a frontal suture between the two halves of the frontal bone; however, these halves unite in the second year, and the frontal suture is mentioned only to facilitate understanding of the anterior fontanel which is described in the next paragraph. Premature ossification of the sutures results in an abnormal smallness of the head, or *microcephaly* (Fig. 3-18). Such individuals are idiots as they have an arrested growth of the brain as well as an arrested growth of the skull.

The fontanels

As mentioned previously (p. 53), ossification of the cranial bones is not yet complete at birth. The unossified membranous areas are referred to as fontanels. There are six such fontanels at the angles of the parietal bones (Fig. 3-19). The largest is the *anterior* fontanel, a diamond-shaped area at the junction of the sagittal, coronal, and frontal sutures; it usually closes at about 18 months of age. The *posterior* fontanel is a triangular space at the junction of the sagittal and lambdoidal sutures; it closes in about 6 to 8 weeks. The *anterolateral* fontanels, on each side, are at the junction of the frontal, the parietal, the sphenoid, and the temporal bones. Similarly, the *posterolateral* fontanels are located at the junction of the temporal, the parietal, and the occipital bones. The lateral fontanels close within a few months after birth. The unossified cranial sutures and fontanels are of great importance during the birth of a baby. They allow for a certain amount of overriding of the cranial bones and a temporary molding of the baby's head as it passes through the narrow birth canal.

The air sinuses

You will recall that during the description of the bones of the skull it was noted that four of the cranial bones and one of the face bones contain air spaces, or *sinuses*. These spaces tend to decrease the weight of the skull and are important as resonance chambers in the production of voice. They may be divided into two groups:

1. The **paranasal sinuses** (see Fig. 15-2), which communicate with the nasal cavity, include the *frontal,* the *maxillary,* the *sphenoid,* and the *ethmoid* sinuses. They are lined with ciliated mucous membrane which is continuous with that of the nose. Inflammation of this lining is referred to as sinusitis.

2. The **mastoid sinus** in each temporal bone, opens into the middle ear. The mucous membrane lining of these structures is continuous with that of the nasopharynx, or superior portion of the throat. Inflammation of this sinus is referred to as *mastoiditis.*

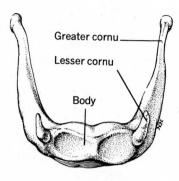

Greater cornu

Lesser cornu

Body

Fig. 3-20. The hyoid bone, anterior view.

The hyoid bone

The hyoid is an isolated, U-shaped bone lying in the anterior part of the neck just below the chin. It consists of a central portion, or *body*, and two projections on either side called the *greater* and the *lesser cornua* (Fig. 3-20). The hyoid is suspended from the styloid processes of the temporal bones by the *stylohyoid ligaments*. It serves for the attachment of certain muscles which move the tongue and aid in speaking and swallowing.

Bones of the vertebral column

The vertebral column, or backbone, of an adult consists of 26 separate bony segments arranged in series to form a strong, flexible pillar. Children have 33 or 34 vertebrae, but some of these eventually fuse to form single segments as noted in the next paragraph. The bones of the vertebral column are named and numbered, from above downward, on the basis of their location (Fig. 3-21). There are seven cervical vertebrae in the neck, twelve thoracic vertebrae in the thorax, five lumbar vertebrae in the small of the back, one sacrum (resulting from the fusion of five separate vertebrae), and one coccyx (resulting from the fusion of four or five vertebrae).

Fig. 3-21. Lateral view of adult vertebral column showing curves.

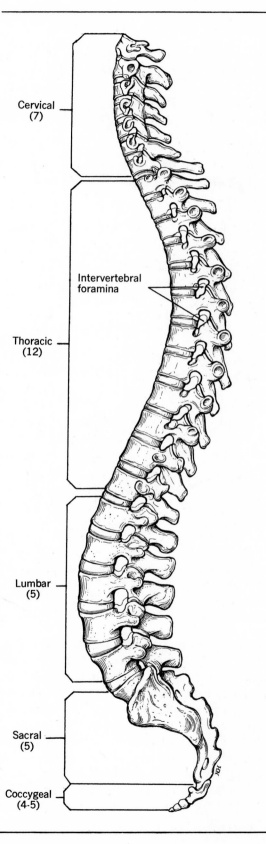

Cervical (7)

Intervertebral foramina

Thoracic (12)

Lumbar (5)

Sacral (5)

Coccygeal (4-5)

General structure of vertebrae

Although there are characteristic differences between bones in each of the aforementioned groups, there are marked similarities in general structure. Each vertebra, with the exception of the first two cervical, has a drum-shaped *body* located anteriorly where it serves as the weight-bearing part. An *intervertebral disk* of compressible fibrous cartilage acts as a cushion and shock absorber between each pair of vertebrae as their bodies are piled one on the other. Extending posteriorly from the body are two short thick projections, the *pedicles*. Arising from the pedicles are a pair of broad bony plates, the *laminae*; they meet and fuse in the midline posteriorly. If the laminae fail to unite, the resulting condition is referred to as *spina bifida*; the contents of the spinal canal, either the spinal cord or its coverings, or both, may protrude. From the junction of laminae the *spinous process* arises and projects backward and downward where it can usually be felt just under the skin of the back. *Transverse processes* extend laterally from the junction of pedicle and lamina on either side. These processes, as well as the spinous process, provide places for the attachment of muscles and ligaments. The large opening in the center of each vertebra is referred to as the *vertebral*, or *spinal foramen*, and when all of the vertebrae are arranged in series we have provided a hollow bony cylinder which surrounds and protects the spinal cord. Facilitating movement of this bony column are the smooth articular surfaces on each vertebra. Two *superior articular processes* and two *inferior articular processes* project vertically near the junction of pedicles and laminae. In action, the inferior processes of one vertebra articulate with the superior processes of the vertebra immediately below, and so on down the column. In front of each of these joints can be seen an *intervertebral foramen*, or opening, through which emerges one of the spinal nerves as it leaves the vertebral cavity. These foramina can be noted between each two vertebrae in the column (Fig. 3-21).

Regional variations of vertebrae

The Cervical Vertebrae. This part of the vertebral column is extremely flexible, permitting a wide range of movement of the head and the neck. The seven vertebrae are smaller than those lower in the column, but the main distinguishing characteristic is the presence of three foramina instead of only one. In addition to the usual *vertebral foramen*, there is an opening in each transverse process called the *transverse foramen* (Fig. 3-22) through which pass the vertebral arteries on their way to the brain.

The first two cervical vertebrae have additional distinctive features and have been given specific names. The **atlas**, or first cervical (Fig. 3-23), supports the head. (It was named for the Greek god who supposedly carried the world on his shoulders.) Having no body or spinous process

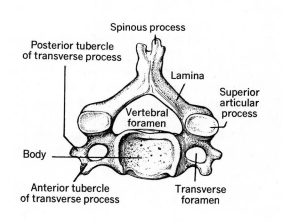

Fig. 3-22. Fourth cervical vertebra, superior aspect.

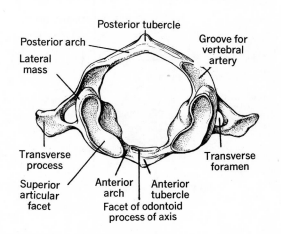

Fig. 3-23. The atlas, superior aspect.

it is little more than a bony ring which receives, on its superior articular surface, the condyles of the occipital bone. By means of this joint you are able to rock your skull back and forth, as in nodding agreement. The second cervical vertebra is called the **axis** (Fig. 3-24), and it is easily identified by the presence of a large toothlike projection, the *odontoid process*, which arises from the body. This process fits up into the anterior part of the atlas and forms the axis of rotation. As you turn your head from side to side, the atlas moves with the skull, and the two turn on the axis with the odontoid process serving as a pivot. The seventh cervical vertebra has a very prominent spinous process which can be felt at the nape of the neck.

The Thoracic Vertebrae. These vertebrae, 12 in number, become progressively larger from above downward. Their main distinguishing characteristic is the presence of smooth areas, or *facets*, on the body and transverse processes for articulation with the ribs (Fig. 3-25).

The Lumbar Vertebrae. Nearing the end of the column are the larger, heavier lumbar vertebrae which are well equipped for weight-bearing (Fig. 3-26). Short spinous processes, gross size, plus the absence of transverse foramina and rib facets facilitate identification of these vertebrae.

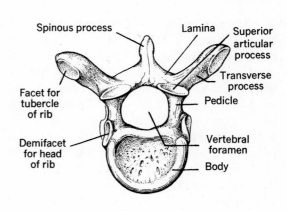

Fig. 3-25.   Sixth thoracic vertebra, superior aspect.

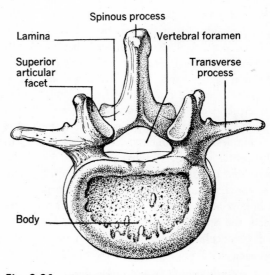

Fig. 3-26.   Third lumbar vertebra, superior aspect.

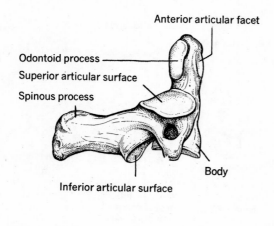

Fig. 3-24.   The axis, lateral view.

The Sacrum. Formed by the fusion of five sacral vertebrae, this bone resembles a somewhat concave triangle, the base articulating with the fifth lumbar vertebra, the apex extending downward, and its concave surface facing forward. Wedged between the two hipbones, the sacrum fills in the back part of the bony pelvis (see Fig. 3-41). The *sacral promontory* is a prominent lip at the upper anterior margin of the *body*; it is one of the important obstetric landmarks in determining the size of a pelvis for childbirth.

The Coccyx. Attached to the apex of the sacrum, the coccyx, or tailbone, is formed by the fusion of four rudimentary vertebrae. This triangular bone is the last segment of the vertebral column (see Fig. 3-21).

The vertebral column as a whole

The human vertebral column is a versatile arrangement of bony segments which are linked in series by ligaments. Not only does it provide the strength and rigidity necessary for *support* of the head, the trunk, and the upper extremities, but it is also flexible to an amazing degree. One has merely to watch an acrobatic dancer for full appreciation of the mobility of this column. In addition to its supporting function, this hollow cylinder of bones effectively surrounds and *protects* the soft spinal cord as well as the origins of the 31 pairs of spinal nerves. Therefore, fractures of the vertebrae (*broken neck* or *broken back*) may result in paralysis or even death, depending on the location and the severity of the fracture. Less serious, but extremely painful, is the so-called slipped or *ruptured disk* (technically called herniation of the nucleus pulposus). In these cases, the soft central portion of an intervertebral disk is forcibly pushed into the spinal canal where it presses on the spinal cord or on spinal nerve roots (Fig. 3-27). This condition may be caused by a fall or by improper body mechanics when lifting heavy objects.

The **normal curves** of the vertebral column provide some of the resilience and spring so essential in walking and jumping. In a newborn infant the whole vertebral column is concave forward, a primary curve. When the infant begins to assume the erect posture, secondary curves, or those which are concave to the rear, may be noted. The vertebral column normally presents four curves (see Fig. 3-21). The *thoracic* and the *sacral curves* are primary curves and are present at birth. The *cervical curve* appears when the infant begins to hold up his head at about 3 months, and the *lumbar curve* appears when the infant begins to walk.

Exaggerated curves of the vertebral column may be due to disease, injury or, in some cases, to poor posture. An increase in the thoracic curve is called *kyphosis*, or *hunchback*. An increase in the lumbar curve is referred to as *lordosis*, or *sway-*

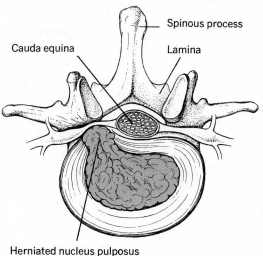

Cauda equina

Spinous process

Lamina

Herniated nucleus pulposus compresses nerve root

Fig. 3-27. Diagram of a ruptured intervertebral disk. The soft, central portion of the disk is protruding into the vertebral canal where it exerts pressure on a spinal nerve root.

back. A lateral curvature, called *scoliosis,* is present to some degree in most persons; however, severe cases may require corrective surgery.

Ligaments of the vertebral column help to hold the bony segments in proper position as shown in Figure 3-28. These strong bands of connective tissue are somewhat extensible, and the intervertebral disks are compressible; thus we are able to bend and to straighten our vertebral column within certain limits.

The *anterior longitudinal* ligament and the *posterior longitudinal* ligament connect the bodies of the vertebrae. They extend from the axis to the sacrum and might be compared with two long strips of adhesive tape, one extending along the anterior surface and one along the posterior surface of the vertebral bodies and disks.

Ligamenta flava are rather broad, thin ligaments which connect lamina to lamina from the axis down to the sacrum.

The *ligamentum nuchae* extends from the external occipital protuberance downward over the spinous processes to the seventh cervical vertebra. This ligament helps to support the weight of the head in four-legged animals such as cows and

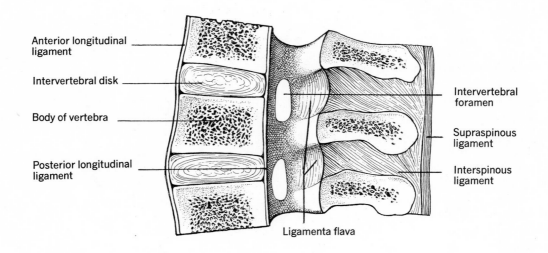

Anterior longitudinal ligament

Intervertebral disk

Body of vertebra

Posterior longitudinal ligament

Intervertebral foramen

Supraspinous ligament

Interspinous ligament

Ligamenta flava

Fig. 3-28. Sagittal section through part of the vertebral column, showing relationships between bony and soft tissues.

horses, but it is of less importance in man.

The *supraspinous ligament* connects the tips of the spinous processes from the seventh cervical vertebra downward to the sacrum.

Interspinous ligaments connect adjacent spinous processes between the tips and the lamina.

The bones of the thorax

Generally speaking, the thorax is a cone-shaped bony cage, narrow at the top and broad at its base. Twelve pairs of ribs form the bars of this cage, assisted by the sternum, or breastbone, anteriorly, and the 12 thoracic vertebrae posteriorly. It is covered with muscles and skin, and the floor is formed by the dome-shaped diaphragm. In fact, it is the upward curving of this muscular floor that makes it possible for the thorax to protect not only the heart and the lungs, but also the liver and the stomach which lie immediately below the diaphragm in the upper part of the abdominal cavity. In addition to its protective function, the thorax plays an important role in the breathing process and helps to support the bones of the shoulder girdle.

The **sternum** lies in the midline in the front of the thorax; it is a flat, narrow bone about 6 inches (15 cm.) long. It somewhat resembles a dagger in shape and is composed of three parts:

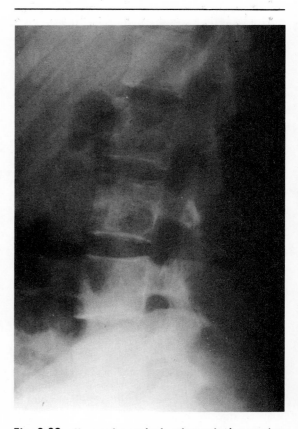

Fig. 3-29. X-ray of vertebral column, lumbar region, lateral view. Dark areas between bodies of vertebrae indicate position of intervertebral disks.

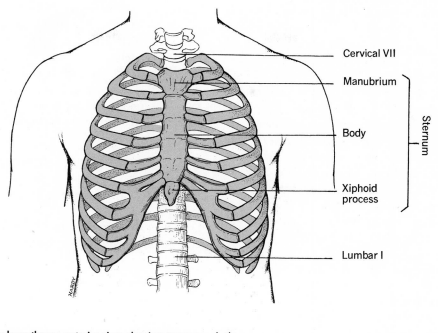

Fig. 3-30. The bony thorax, anterior view, showing sternum and ribs.

the *handle* or *manubrium* above, the *body*, or *gladiolus*, in the middle, and the *xiphoid process* below (Fig. 3-30). The clavicle and the first rib cartilage articulate with the manubrium, while the next nine rib cartilages join the body, either directly or indirectly as will be described later. The point of union between the manubrium and the body forms the *sternal angle*, a relatively prominent ridge which can be felt just under the skin; it is an important landmark, since the second rib articulates with the sternum at this point. The superficial location of the sternum, as well as its

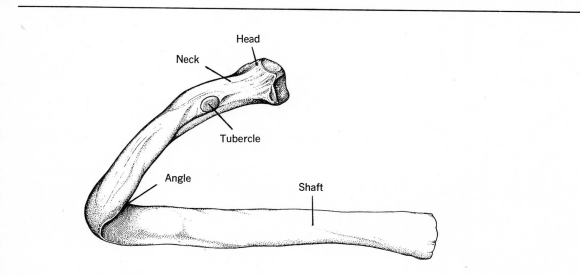

Fig. 3-31. A central rib, left side of body, posterior view.

structure, make it a site for biopsy (examination of tissue from living subjects) of red bone marrow. This procedure is referred to as *sternal puncture* and is often of great value in the diagnosis of blood disorders such as the various types of anemia.

The **ribs,** 24 in number, are located 12 on each side of the thoracic cavity. They are slender, curved, and relatively flexible bones. All of the ribs are attached posteriorly to the thoracic vertebrae, but the anterior attachment varies. None of the ribs actually join the sternum directly; instead there are resilient bars of hyaline cartilage, the *costal cartilages,* which extend forward from the ends of the ribs, with the exception of the last two pairs, toward the sternum. The first seven pairs of ribs are referred to as true, or *vertebrosternal,* ribs since their costal cartilages actually reach the sternum. The remaining five pairs are false ribs; of these, the first three are called *vertebrochondral* since their costal cartilages turn upward to join the cartilage of the rib above. The last two pairs have no anterior attachment at all and are known as the *floating* or *vertebral* ribs

(Fig. 3-30). The spaces between the ribs are referred to as the *intercostal spaces;* muscles, blood vessels, and nerves are located here.

A typical rib (Fig. 3-31) has a *head* and a tubercle which articulate with the facets on a thoracic vertebra. The *shaft* curves around the side of the chest, slanting downward and forward, making the sternal end lower than the vertebral end. The upper ribs are shorter than the lower, with the exception of the last two.

THE APPENDICULAR SKELETON

Bones of the upper extremity

The 32 bones in each of the upper extremities may be divided into four groups for ease of study.

1. Shoulder girdle
   one clavicle
   one scapula
2. Arm
   one humerus

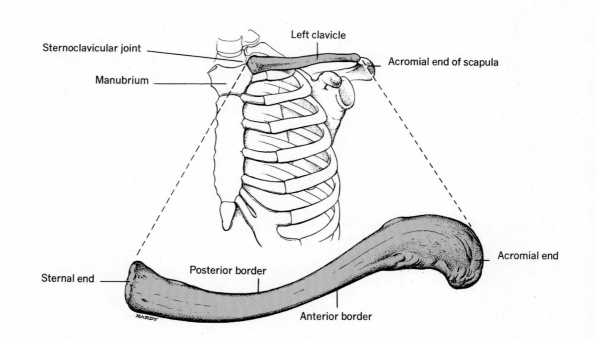

Fig. 3-32. Left clavicle.

3. Forearm
 one ulna
 one radius
4. Wrist and hand
 eight carpal bones
 five metacarpal bones
 fourteen phalanges

1. The Shoulder Girdle. Despite its name, the shoulder *girdle* does not form a complete ring of bone but presents an especially large gap posteriorly between the medial border of the scapula and the vertebral column. This space is occupied by some of the muscles which move the scapula and contribute to the great mobility of the arm. The shoulder girdle serves to attach the bones of the upper extremity to the axial division of the skeleton and to provide places of attachment for many muscles.

The **clavicle,** or collar bone, is a slender bone with a double curvature (Fig. 3-32). It lies horizontally in the root of the neck, just above the first rib. Medially, the clavicle articulates with

the sternum forming the *sternoclavicular joint,* the only bony joint between the shoulder girdle and the trunk; laterally it articulates with the acromial end of the scapula. Functionally, the clavicle serves to brace the shoulder. Fractures of this slender bone are fairly common when undue stress is applied along its shaft, as in falling heavily to one side and landing on the shoulder.

The **scapula,** or shoulder blade, is located on the upper back part of the thoracic cage. It is a rather flat, triangular-shaped bone which can be felt readily by reaching over the right shoulder with the left hand, placing it on the back, and then moving the right shoulder in various directions. One also may feel the prominent *spine* of the scapula (Fig. 3-33) as it passes across the upper posterior surface and ends in the large, flat *acromiol process* which forms the tip of the shoulder and articulates with the clavicle. The acromion, as well as the clavicle, may be fractured when one falls on the shoulder. Figure 3-34 best shows the *glenoid cavity,* a smooth shallow socket that receives the head of the humerus. Just above

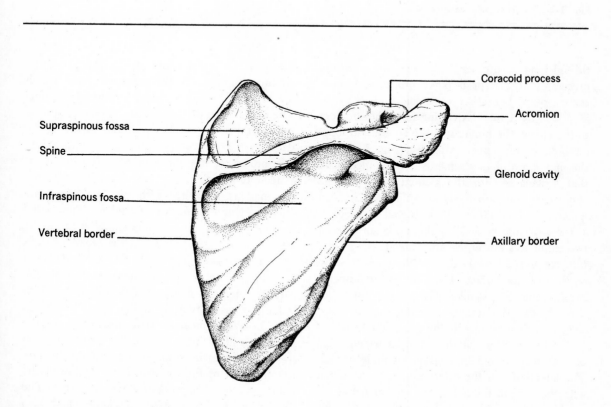

Fig. 3-33. Right scapula, posterior view.

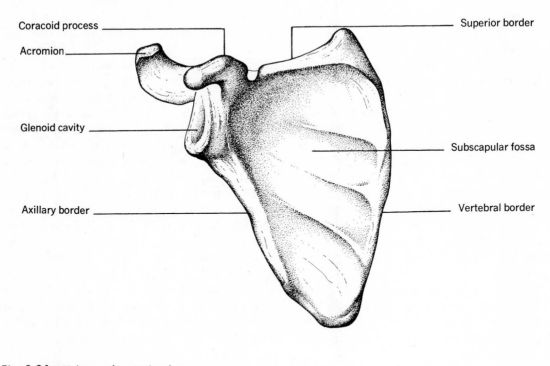

Coracoid process

Acromion

Glenoid cavity

Axillary border

Superior border

Subscapular fossa

Vertebral border

Fig. 3-34.   Right scapula, anterior view.

the glenoid cavity can be seen the *coracoid process,* a hooked or beaklike projection, which curls forward beneath the clavicle.

2. *The Arm.* Extending from the shoulder to the elbow, the **humerus** is the largest and longest bone of the upper extremity (Fig. 3-35). At the upper end of the humerus is a rounded *head* that articulates with the glenoid cavity of the scapula, forming a ball-and-socket joint. The head is joined to the shaft by a constricted area known as the *anatomic neck.* Near this neck can be seen two eminences for muscle attachment: the *greater tubercle* on the lateral side, and the *lesser tubercle* on the anterior surface. The shaft of the humerus is rather long and slender. The portion just below the tubercles is referred to as the *surgical neck,* since it is most frequently the site of fracture.

The lower end of the humerus widens, and one can see two smooth areas on the undersurface. The lateral area is the *capitulum,* for articulation with the head of the radius; the medial area is the *trochlea,* for articulation with the ulna. Two depressions may also be noted at the distal end of

the bone. On the anterior aspect, just above the articular surfaces, is the *coronoid fossa* which makes room for the coronoid process of the ulna when the forearm is flexed, or bent upward toward the arm. On the posterior aspect is the deep *olecranon fossa,* which receives the olecranon process of the ulna when the forearm is extended, or straightened. The two projections from the margins, on a line with the depressions, are the *medial* and the *lateral epicondyles.* The medial epicondyle is the more prominent and can easily be felt on the inner aspect of the elbow joint. A large nerve passes just below this projection and may be pinched when you bump your elbow; the strange tingling sensation which follows has been referred to as hitting the funny bone.

3. *The Forearm.* The **ulna** lies on the medial or little finger side, of the forearm. It can be felt most easily through the skin on the posterior aspect of the forearm, from the tip of the elbow to the small projection in the wrist. The ulna (Fig. 3-36) consists of a slender shaft with an enlarged, hook-shaped upper end that is very

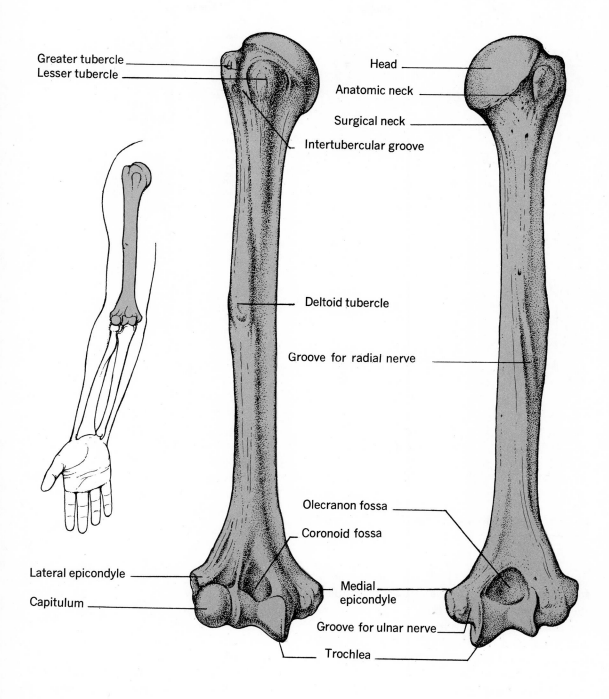

Greater tubercle
Lesser tubercle
Head
Anatomic neck
Surgical neck
Intertubercular groove
Deltoid tubercle
Groove for radial nerve
Olecranon fossa
Coronoid fossa
Lateral epicondyle
Capitulum
Medial epicondyle
Groove for ulnar nerve
Trochlea

Fig. 3-35. Right humerus: *left,* anterior view; *right,* posterior view.

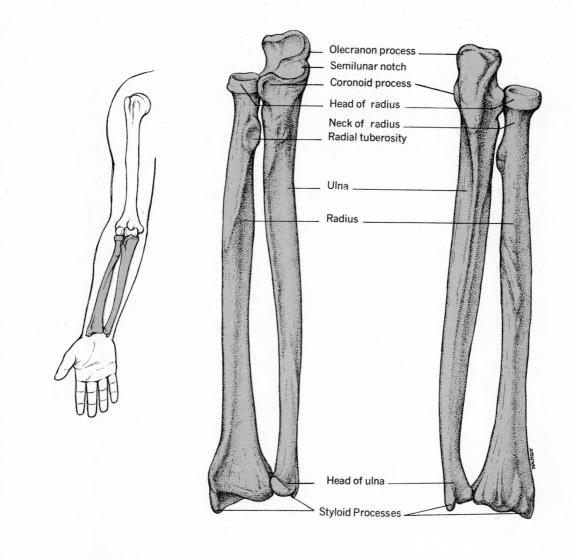

Fig. 3-36. Right radius and ulna: *left,* anterior aspect; *right,* posterior aspect.

distinctive. The upper, back part of the hook forms the prominence of the elbow and is called the *olecranon process.* In front of the olecranon is a concave surface, the *semilunar notch,* which articulates with the trochlea of the humerus. Curving upward and forward from the notch is the *coronoid process.*

The distal end of the ulna is small; it presents a knobbed portion, the *head,* and a downward projection on the medial side, the *styloid process.* On the lateral surface at both the upper and lower ends are small, smooth areas for articula-

tion with the radius. The ulna does not articulate directly with the wrist bones, but with a fibro-cartilaginous disk instead.

The **radius** lies on the lateral, or thumb side, of the forearm. It is more difficult to feel than the ulna since it is better padded with muscles, except at the wrist. The radius consists of a shaft, with a small upper end and a large lower end (Fig. 3-36). Proximally, a disklike *head,* supported by a *neck,* articulates with the capitulum of the humerus and the side of the ulna. The prominent *radial tuberosity* (for attachment of the biceps muscle) can be seen on the shaft just below the head. On the undersurface of the distal end is a smooth surface which articulates with two of the wrist bones and with the head of the ulna. The *styloid process* is lateral to the articular surface.

When the palm is up, or forward, the radius and the ulna are parallel; when the palm is turned down, or back, the lower end of the radius moves medially around the head of the ulna so that the shafts of the two bones are crossed. A break in the lower third of the radius is referred to as *Colles' fracture*. It is one of the most common fractures and is caused by a fall on the palm of the hand, at which time the weight of the body is quite sud-

denly transmitted through the radius to the ground. The styloid process of the ulna also may be broken off due to the pull on its ligaments.

4. The Wrist and the Hand. The wrist contains eight small **carpal bones,** arranged in two rows of four each (Fig. 3-37). Beginning on the lateral, or thumb side, in each row they are as follows:

Proximal Row	Distal Row
navicular	greater multangular
lunate	lesser multangular
triquetral	capitate
pisiform	hamate

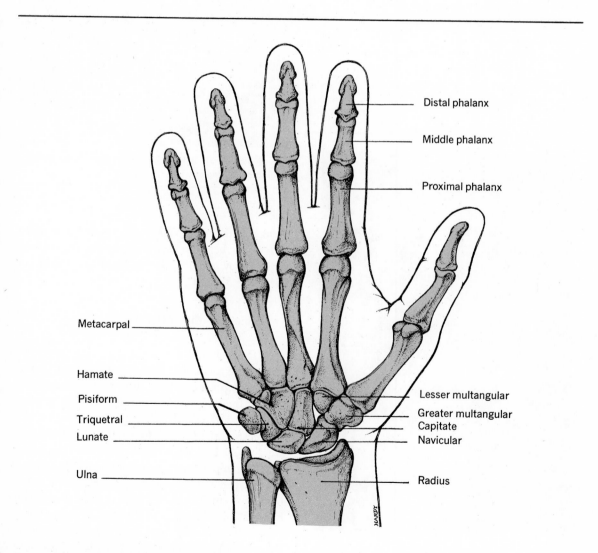

Fig. 3-37. Bones of right hand, anterior view.

The carpal bones are bound firmly and closely together by ligaments.

The five **metacarpal bones** form the bony framework of the body of the hand; they are numbered from the lateral side (Fig. 3-37). These bones are relatively long and cylindrical, and their rounded distal ends form the knuckles. Proximally, the metacarpals articulate with the carpal bones and with each other; distally, they articulate with the first row of phalanges.

There are 14 **phalanges,** or finger bones, in each hand, two for the thumb, and three for each finger (Fig. 3-37). They are referred to as the first, or proximal; the second, or middle; and the third, or distal phalanx. Obviously, the thumb has only a proximal and a distal phalanx.

The carpals, the metacarpals and the phalanges (a total of 27 bones in each hand) are joined in such a way that each contributes to the dexterity and the mobility of the human hand. Of particular importance is the ability to place the

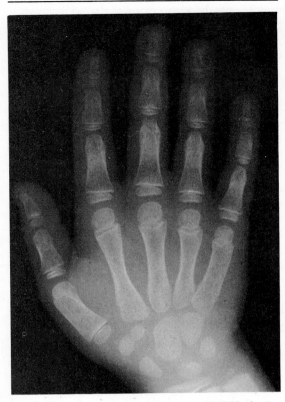

Fig. 3-39. X-ray of hand of three-year-old child, showing ossification in progress.

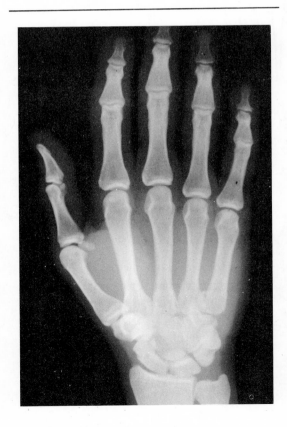

Fig. 3-38. X-ray of right hand, posterior view.

thumb against the various fingers—one of many differences between man and the four-legged animals. X-ray studies of the hand are shown in Figures 3-38 and 3-39.

Bones of the pelvis

The pelvis is a strong, bony ring somewhat resembling a basin. It is composed of four bones: the two hipbones, which form the front and the sides of the ring; and the sacrum (wedged between the hipbones) and the coccyx which complete the ring posteriorly. (Note: the sacrum and the coccyx are described on page 72.)

The **hipbone,** or *os coxae,* is a large, irregularly shaped bone. Each bone begins its development as three separate parts: the ilium, the ischium, and the pubis. Eventually these segments fuse to form a solid bone, but the original names are retained for purposes of description. The lines of fusion are in the **acetabulum,** a prominent cup-

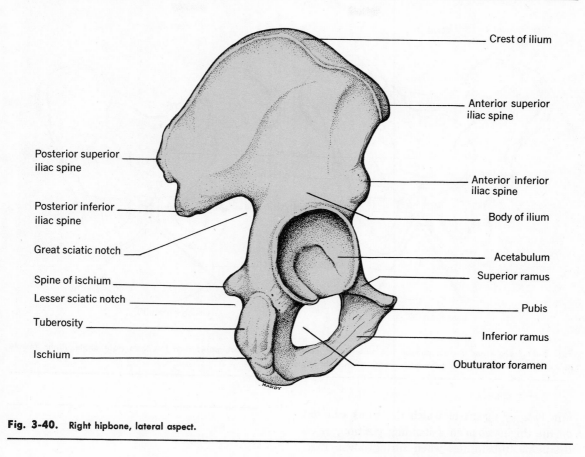

Crest of ilium

Anterior superior
iliac spine

Posterior superior
iliac spine

Anterior inferior
iliac spine

Posterior inferior
iliac spine

Body of ilium

Great sciatic notch

Acetabulum

Spine of ischium

Superior ramus

Lesser sciatic notch

Pubis

Tuberosity

Inferior ramus

Ischium

Obuturator foramen

Fig. 3-40. Right hipbone, lateral aspect.

shaped socket on the lateral aspect (Fig. 3-40); the head of the femur articulates with the hipbone in this socket.

The upper, flared portion of the hipbone is the **ilium** which forms the prominence of the hip. The curved rim of the ilium, along the upper border, is called the *iliac crest*; this part can be felt easily near the level of the waist and it is a good site for biopsy of red bone marrow. At either end of the crest there are bony projections, but the more prominent one is the *anterior superior iliac spine* which serves as an important anatomic landmark. The inner surface of the ilium is somewhat concave and contributes to the basinlike appearance of the pelvis. The articulation of the ilium with the sacrum, posteriorly, forms the *sacroiliac joint*. The use of improper body mechanics when lifting heavy objects may result in torn ligaments around this joint, a very painful injury.

The **ischium** (Fig. 3-40) is the lowest and strongest portion of the hipbone. It extends down-

ward from the acetabulum and expands into the large *ischial tuberosity* which helps to support the weight of the trunk when one sits down. Just above the tuberosity is a pointed projection, the *ischial spine*, which is an important landmark in obstetrics. A ramus, or branch, curves forward from the body of the ischium to join the pubis.

The anterior portion of the hipbone is called the **pubis**. It consists of two bars of bone, the *superior* and *inferior rami*, which meet in a flat *body*. Anteriorly, the body units with its fellow in the midline, forming a joint known as the *symphysis pubis*. The *pubic arch* is the angle formed by the two inferior rami at this point. The superior ramus ascends to the ilium, and the inferior ramus descends to the ischium. The *obturator foramen* is a large hole in the hipbone, bounded by the pubis and the ischium.

The **pelvis as a whole** protects the urinary bladder, some of the reproductive organs, and the distal end of the large intestine; it serves also as a

Male   Female

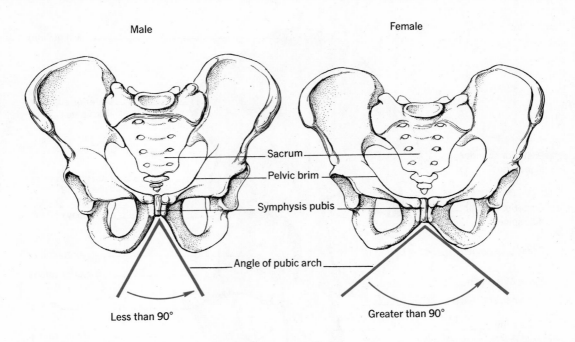

Sacrum

Pelvic brim

Symphysis pubis

Angle of pubic arch

Less than 90°   Greater than 90°

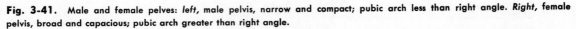

Fig. 3-41. Male and female pelves: *left,* male pelvis, narrow and compact; pubic arch less than right angle. *Right,* female pelvis, broad and capacious; pubic arch greater than right angle.

firm base by means of which the trunk can rest on the thighs when in a standing position, or on the ischial tuberosities when sitting down. Fractures of the pelvis are serious and painful injuries which often necessitate several weeks of rest in bed.

A curved line drawn from the sacral promontory, downward and forward on each side (along the rounded ridges of the ilia), to the upper margin of the symphysis pubis helps to identify the boundaries of the *pelvic brim* (Fig. 3-41), which separates the greater from the lesser pelvis. The greater, or false, pelvis is the expanded portion above the brim, bounded by the flared portion of each ilium and by the muscular walls of the abdomen. The lesser, or true, pelvis lies below the brim; it is bounded in front and on the side by the pubis, the lower part of each ilium, and the ischium, and behind by the sacrum and the coccyx.

The **true pelvis** consists of an inlet, a cavity and an outlet. The *inlet,* or pelvic brim (Figs. 3-41 and 3-42), is of special importance during the birth of a baby; if any of its diameters are too small it may be impossible for the infant's skull to enter the true pelvis for a normal or natural de-

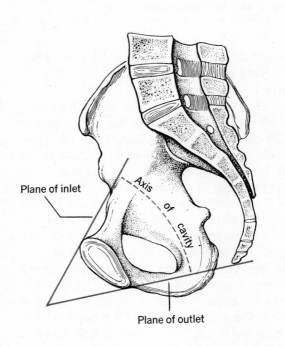

Plane of inlet

Plane of outlet

Fig. 3-42. Midsagittal section of pelvis.

livery. The cavity is a short, curved canal, considerably deeper in the region of the sacrum and coccyx than in the region of the symphysis pubis (Fig. 3-42). The urinary bladder lies behind the symphysis pubis, and the rectum lies in the curve along the sacrum and coccyx. In women, the vagina lies between the urinary bladder and the rectum. The *outlet* is the space between the tip of the coccyx, behind, the ischial tuberosities, at the sides, and the pubic arch, in front. An x-ray of the pelvis is shown in Figure 3-43.

one femur

one patella (included here for convenience)

2. Leg
one tibia
one fibula

3. Foot
seven tarsal bones
five metatarsal bones
fourteen phalanges

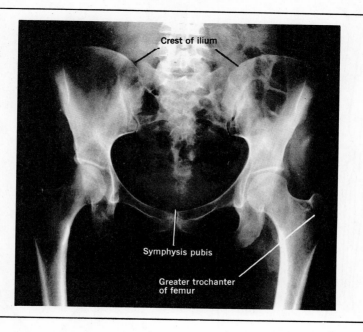

Fig. 3-43. X-ray of pelvis.

Differences between the Male and the Female Pelvis. In women, the pelvis is adapted for pregnancy and childbirth, the bones being lighter and smoother, the cavity of the true pelvis roomier, the inlet and the outlet larger, and the coccyx more movable than in men. The female pelvis is broad and shallow with a wide pubic arch (greater than a right angle), while the male pelvis is narrow and deep with a pubic arch that is more pointed (less than a right angle). Figure 3-41 shows the male and the female pelves.

Bones of the lower extremity

The 30 bones of the lower extremity may be divided into three groups for ease of study.

1. Thigh

1. Bones of the Thigh. The femur is the heaviest, longest, and strongest bone in the body (Fig. 3-44). It transmits the entire weight of the trunk from hip to tibia and is covered so thickly with muscles it can be felt only near its ends. At the upper end of the femur is a rounded *head* supported by a constricted portion, or *neck*, which joins the cylindrical shaft at an angle. The head fits into the acetabulum of the hipbone, forming a deep articulation designed for the task of transmitting weight to the femur. The relatively long neck enables the femur to be moved freely even though the head is more or less buried. However, the angular junction with the shaft increases the risk of fracture, particularly in the aged person whose bones tend to become more brittle. A fracture through the neck of the femur is commonly

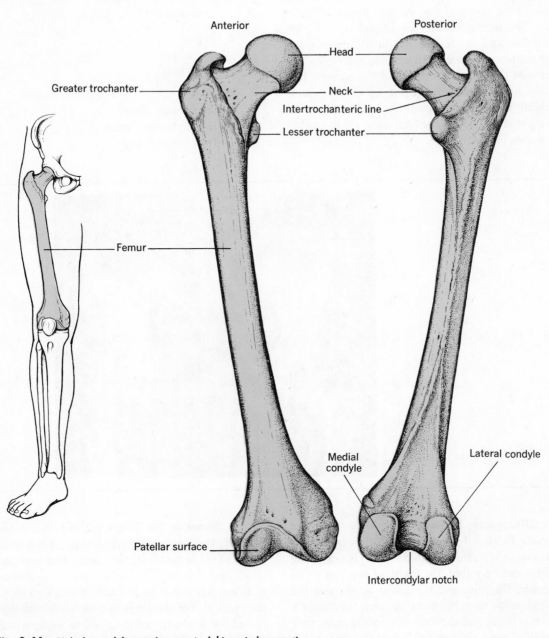

Fig. 3-44. **Right femur:** *left,* anterior aspect; *right,* posterior aspect.

referred to as a *broken hip.* At the top of the shaft, on the lateral side, is a very large protuberance, the *greater trochanter,* which can be felt under the skin on the outer aspect of the upper thigh. The *lesser trochanter,* a smaller elevation, is on the medial side, below the neck.

The *linea aspera,* a prominent ridge for the attachment of muscles, can be seen extending lengthwise on the posterior side of the shaft. At the lower end of the femur are two large, smooth bulges, the *medial* and the *lateral condyles,* which articulate with the tibia. They can be felt at the

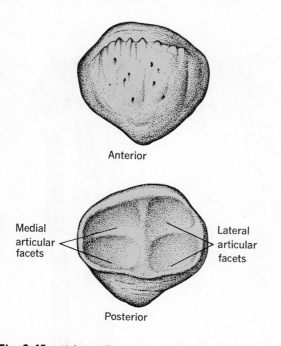

Anterior

Medial
articular
facets

Lateral
articular
facets

Posterior

Fig. 3-45. Right patella, anterior and posterior views.

sides of the knee. The *intercondyloid fossa* is a deep depression between the condyles posteriorly. On the anterior aspect is a smooth surface for articulation with the patella.

The **patella,** or kneecap (Fig. 3-45), forms the prominence in front of the knee. It is the largest sesamoid bone in the body and is embedded in the tendon of the quadriceps femoris (the principal anterior thigh muscle) where it crosses the knee to insert on the tibia. Its posterior surface is smooth for articulation with the femur. Functionally, the patella serves to protect the front of the knee joint, but, because of its position, it is frequently dislocated or even fractured.

2. Bones of the Leg. The **tibia,** or shin bone, is the medial and larger bone of the leg. It is second to the femur in length and transmits the weight of the trunk to the foot. The upper end of the tibia is quite broad and presents two masses of bone on either side, the *medial* and the *lateral condyles* (Fig. 3-46), the smooth upper surfaces of which articulate with the condyles of the femur. Between these surfaces is a projection called the *intercondylar eminence.* On the anterior surface, just below the condyles, there is a

knoblike projection, called the tibial *tuberosity,* that can be felt just under the skin. The anterior border of the shaft is also quite superficial, and the overlying skin may be painfully bruised when we bump our shins.

The lower end of the tibia widens somewhat and has a smooth surface for articulation with the talus (a tarsal bone) below, and the fibula, laterally. A downward projection, the *medial malleolus,* forms the prominence on the inner aspect of the ankle; it can be identified easily as it lies just under the skin.

The **fibula** is a long, slender bone on the lateral side of the leg. The upper end is expanded slightly to form a *head* (Fig. 3-46) which articulates only with the tibia, not with the femur. The fibula does not enter into formation of the knee joint; it is not a weight-bearing bone. At the lower end, a downward projection, the *lateral malleolus,* forms the prominence on the outer aspect of the ankle joint by articulating with the tibia to form the side of the socket for the talus; the latter fits between the medial and the lateral malleoli and receives the weight of the body from the tibia.

Fracture of the fibula a few inches above the ankle and a splitting off of the medial malleolus of the tibia is referred to as *Pott's fracture;* it is usually caused by a sudden, sharp turn of the ankle.

3. Bones of the Foot. The structure of the foot is basically similar to that of the hand. However, the foot serves to support the weight of the body and to provide a strong lever for the muscles of the calf in walking and leaping; thus we would expect the foot to be stronger and less mobile than the hand. Since arches provide great supporting strength in any type of construction, it is not surprising to note that the foot possesses two such curves: the *longitudinal arch,* which extends from heel to toe, and the *transverse arch,* which extends crosswise in the foot. These springy arches are maintained by powerful ligaments, muscles, and tendons which hold the bones in proper alignment (p. 144).

Of the seven **tarsal bones,** the uppermost is the *talus* (Fig. 3-47) which articulates with the tibia and the fibula as described previously. The largest and strongest of the tarsal bones is the *calcaneus,* or heel bone. It lies below and behind the talus and is shaped like the handle of a revolver. The calcaneus forms the base of the heel

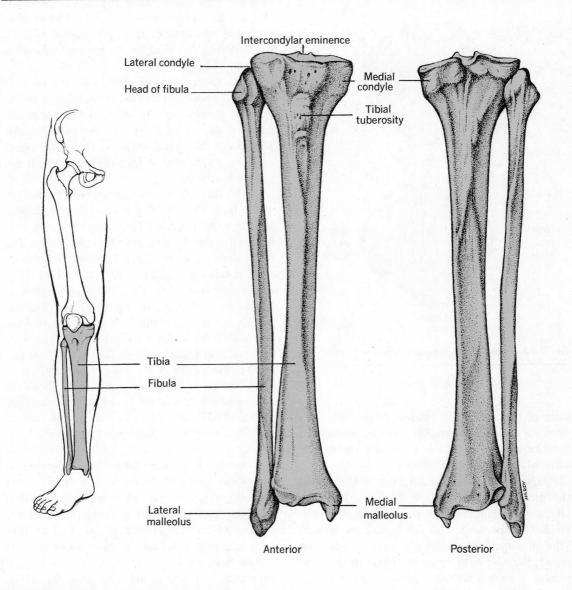

Fig. 3-46.   Right tibia and fibula, anterior and posterior aspects.

and transmits the weight of the body from the talus to the ground. The *cuboid*, which lies on the lateral side of the foot, articulates with the anterior end of the calcaneus, while the *navicular*, on the medial side, articulates with the anterior end of the talus. In front of the navicular are the three small *cuneiform bones*.

The five **metatarsal bones**, lying anterior to the tarsus, are relatively cylindrical and slightly curved (Fig. 3-47). The bases of the first three

articulate with the cuneiform bones, and those of the outer two articulate with the cuboid. The heads of the five metatarsals form the ball of the foot.

There are fourteen **phalanges** in each foot. Of these, two are in the great toe, and three in each of the other toes. In marked contrast to the phalanges of the hand, these phalanges are relatively unimportant, due to the limited mobility of the great toe.

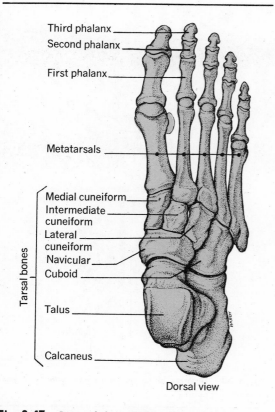

Third phalanx

Second phalanx

First phalanx

Metatarsals

Tarsal bones
Medial cuneiform
Intermediate cuneiform
Lateral cuneiform
Navicular
Cuboid

Talus

Calcaneus

Dorsal view

Fig. 3-47. Bones of the right foot, dorsal aspect.

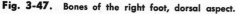

THE ARTICULATIONS, OR JOINTS

Those areas where bones come into close contact with one another are referred to as **articulations,** or joints. *Arthro-* is a combining form which is frequently used when referring to a joint; for example, arthrology is the branch of anatomy which deals specifically with the joints, and arthritis refers to inflammation of a joint.

The bones are efficiently bound together by connective tissues which vary according to the amount of movement that is required at each type of joint. The body can make thousands of movements because of the mobility of most of its joints. On the other hand, immobility may be desirable in some areas, such as between the cranial bones which protect the brain.

For our purpose, joints are best classified according to the amount of movement which they permit. On this basis there are three classes: *synarthroses,* or immovable joints; *amphiarthroses,* or slightly movable joints; and *diarthroses,* or freely movable joints. Structural types will be included in the discussion of each class.

The synarthroses, or immovable joints

This class includes all joints in which the surfaces of the bones are in almost direct contact and in which there is no appreciable movement, as in the joints between the bones of the skull, excepting the mandible. The bones are fastened together by fibrous tissue, by cartilage, or by bone itself as described below.

A *suture,* or sutura, is that type of articulation in which the adjoining margins of the bones are united by a thin layer of fibrous tissue. Some sutures consist of jagged, interlocking indentations and projections of the bone margins; others involve an overlapping of the margins; in either case the bones are slightly separated by a thin layer of fibrous tissue (Fig. 3-49).

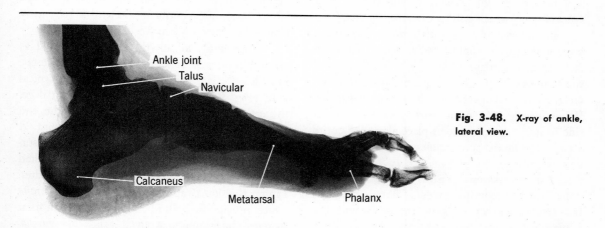

Ankle joint
Talus
Navicular

Calcaneus

Metatarsal

Phalanx

Fig. 3-48. X-ray of ankle, lateral view.

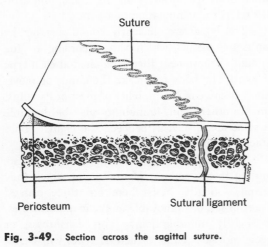

Fig. 3-49. Section across the sagittal suture.

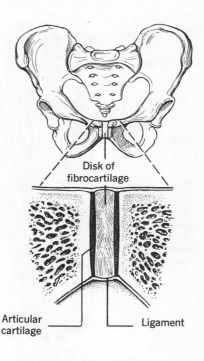

Fig. 3-50. Diagram of an amphiarthrosis, symphysis type: the symphysis pubis.

A *synchondrosis* (meaning *together* and *cartilage*) is a joint in which the connecting tissue is cartilage. This is a temporary form of joint, since the cartilage is eventually converted into bone. Examples are the joints between the epiphyses and the shafts of immature bones (see Fig. 3-5). When growth is complete, the connecting link of cartilage undergoes ossification; the resulting bony joint is referred to as a *synostosis* (meaning *together* and *bone*).

The amphiarthroses, or slightly movable joints

There are two types of amphiarthrotic joints, the symphysis and the syndesmosis. Functionally, the sacroiliac joint has only limited movement and belongs in this group; structurally, it is a synovial joint which will be described later.

In a *symphysis* articulation the adjoining bony surfaces are connected by broad, flattened disks of fibrocartilage (Fig. 3-50). Examples are: the symphysis pubis in the pelvis, and the intervertebral disks between the bodies of vertebrae. While sitting in a chair, turn your trunk from side to side; the action takes place along the entire vertebral column, since each vertebra can move only slightly.

The *syndesmosis* joint is one in which the bone margins are united by dense fibrous tissue, as between the lower ends of the tibia and the fibula.

The diarthroses, or freely movable joints

Most of the joints in the body belong to this class and are so constructed as to provide great freedom of movement, the hip and the shoulder being the most movable of all. For example, you can swing your upper extremity around in great circles, and, with practice, you can do almost the same thing with your lower extremity by swinging it from the hip. Other joints, such as the elbow, may not permit movement in so many planes, but they are still freely movable.

Structurally, all diarthroses are classified as *synovial joints*, or those in which fluid is present in a closed cavity between two bones. In a typical joint (Fig. 3-51) a thin layer of hyaline cartilage, the *articular cartilage*, serves to cushion the end of each bone as contact is made in the joint. A strong *capsule*, or envelope, of fibrous tissue surrounds the joint as a whole and is firmly attached to both bones, thus helping to hold them together. Lining this joint capsule is a *synovial membrane* which provides a built-in lubricating system as it secretes the slippery synovial fluid, or *synovia*,

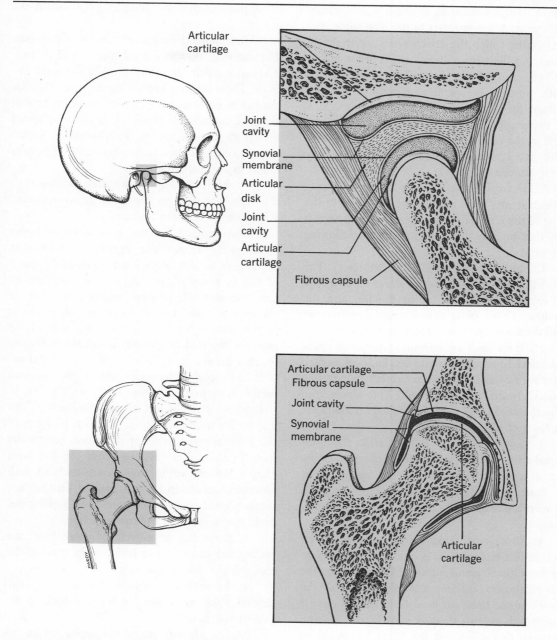

Articular cartilage
Joint cavity
Synovial membrane
Articular disk
Joint cavity
Articular cartilage
Fibrous capsule

Articular cartilage
Fibrous capsule
Joint cavity
Synovial membrane
Articular cartilage

Fig. 3-51. Structure of synovial joints. Joint cavities have been enlarged. *Top,* temperomandibular joint. *Bottom,* hip joint.

(meaning *with egg,* since it resembles egg white in consistency). Just as the moving parts of an automobile must be kept lubricated if they are to run smoothly and with a minimum of wear, so our joints must be similarly protected.

Where additional strength is to be desired, such as in the hip and the knee joints, ligaments are found within the joint itself, binding one bone firmly to the other; these are referred to as *interosseous ligaments.* Even though joints are further strengthened by overlying muscles and tendons, it is possible for the bones to become completely separated and displaced; this injury is referred to as a *dislocation,* or *luxation.* An incomplete or partial dislocation is called a *subluxation* of the joint. A *sprain* refers to the wrenching of a joint, during which the attachments are stretched or torn, but there is no displacement of the bones. The knee joint, in addition to its interosseous ligaments, contains two semilunar disks of articular cartilage, the medial and the lateral *menisci*. Sudden twisting or wrenching of the knee joint may result in the detachment of one or both of these cartilages, a fairly common injury, especially among athletes.

While freely movable joints are basically similar in structure, there are distinct variations in the shape of the articulating surfaces, and it is on this basis that we may divide them into the following subgroups:

1. *Ball - and - socket joint* (enarthrosis). The rounded head of one bone is received into a cuplike cavity in the other, as at the hip and the shoulder. These joints permit angular movement in all directions, and axial rotation as well.

2. *Hinge joint* (ginglymus). A convex cylindrical surface meets a concave cylindrical surface in such a way as to permit movement only in one plane, forward and backward. The elbow, the knee, the ankle, and the interphalangeal joints are examples of hinge joints.

3. *Condyloid joint* (ellipsoidal). An oval-shaped articular surface fits into an elliptical cavity in such a way as to permit angular movement in all planes, but no axial rotation. The wrist joint, between the radius and the carpal bones, is a good example.

4. *Pivot joint* (trochoid or rotary). Movement in this type of joint is limited to rotation, as one bone turns about another. For example, the atlas rotates around the odontoid process of the axis when one turns his head from side to side. If it were not for this type of joint in the forearm (between the radius and the ulna) we would be unable to turn a doorknob or to use a screwdriver.

5. *Saddle joint*. The end of one bone is convex in one direction and concave in the other, while the end of the other bone is exactly the opposite. Observe the great range of movement possible at the carpometacarpal joint at the base of your thumb; this is a saddle joint.

6. *Gliding joint* (arthrodia). This type is best seen between the various carpal bones at the wrist and the tarsal bones of the foot where the articular surfaces permit one bone to glide against the other.

Movements Possible at Diarthrotic Joints. The movements of the head, trunk, and extremities result chiefly from the revolving of joint surfaces. These surfaces revolve around axes that are located in or near the joint and produce distinct kinds of movement. However, it is to be noted that these movements are often combined in the various joints. Seldom is only one type of motion found in any particular joint.

As the name implies, **angular movement** is the type that changes the angle between bones. It may be divided into four kinds:

1. *Flexion* decreases the angle between bones or brings the bones closer together, as in bending the fingers to close your hand. One might think of this in terms of moving a body part forward from the anatomic position, except in the case of the knee and the foot joints where movement is to the rear. When we are standing in the erect position, the foot is normally in a state of flexion. Further flexion of the foot is called *dorsi flexion*.

2. *Extension* increases the angle between bones or separates them more widely, as in straightening the fingers to open the hand. When one stands in the anatomic position, all parts of the body are in a state of extension, except the foot. In the latter case we often use the term *plantar flexion*, rather than extension, and such movement results in our standing on tip toes. Extreme extension of a body part, as in arching the back, is referred to as *hyperextension*.

3. *Abduction* is movement away from the midsagittal plane of the body, as in moving the arm straight out to the side.

4. *Adduction* is movement toward the midsagittal plane, as in bringing the arm back to the side of the body.

NOTE: Movements of the digits are exceptions, as the point of reference is the median longitudinal axis of the hand or the foot rather than the midline of the body; therefore, abduction would be a spreading of the fingers or toes, while adduction would refer to pulling them back together.

When all of the angular movements are combined in succession so that the distal end of an extremity describes a circle and the shaft of the extremity describes the surface of a cone, the resulting movement is referred to as **circumduction.**

For example, if one stands with arm outstretched at the blackboard and draws a large circle, the upper extremity must go through flexion, abduction, extension, and adduction successively.

In **rotation** there is a revolving or twisting of a part of the body around the longitudinal axis of that part with no change in its position. For example, when one turns his head from one side to the other (as when indicating disagreement), the atlas is rotating around the axis. Rotation also is possible at the shoulder and the hip joints. *Internal*, or *medial*, rotation involves the twisting or rotating of a part toward the midsagittal plane of the body, as when standing with the toes of each foot pointed inward (pigeon-toed) instead of forward. *External*, or *lateral*, rotation involves the twisting or rotating of a part away from the midsagittal plane of the body, as happens when one is standing with the toes directed outward instead of forward.

There are **special movements** of the forearm, the ankle, the lower jaw, and the clavicle.

1. *Supination* is the movement that brings the forearm into the anatomic position, with the palm forward, and the radius and the ulna parallel.

2. *Pronation* is the act of turning the palm down, or backward; the radius is crossed over the ulna.

Demonstrate these movements on yourself by holding one forearm with the opposite hand, then turning your palm downward and upward several times; you can feel the radius change its position with respect to the ulna.

3. *Inversion* is the act of turning the sole of the foot inward.

4. *Eversion* is the act of turning the sole of the foot outward.

Most of us have a moderate amount of inversion or eversion at all times. Examine the heels of your shoes. If they are worn down on the outer edge, it indicates that you walk with your feet slightly inverted, but if the wear is on the inner edge, then the tendency is toward eversion. *Clubfoot*, or *talipes*, is a term used in reference to various types of foot deformities; for example, a radical amount of inversion is referred to as talipes varus, while radical eversion is called talipes valgus.

5. *Protraction* is the act of moving a part forward, as in sticking out your tongue or your jaw.

6. *Retraction* refers to the act of drawing a part back.

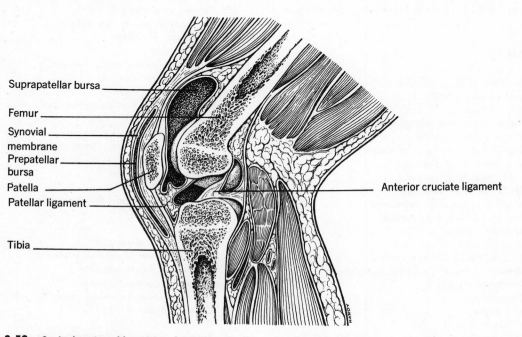

Suprapatellar bursa

Femur

Synovial membrane

Prepatellar bursa

Patella

Patellar ligament

Tibia

Anterior cruciate ligament

Fig. 3-52. Sagittal section of knee joint, showing prepatellar and suprapatellar bursae.

The bursae

Structure. The *bursae* (singular, *bursa*, meaning *a purse*) are small clefts, or fissures, lined with synovial membrane which thus converts them into closed sacs containing synovial fluid.

Location and Function. Bursae are found in those areas where pressure is exerted during the movement of body parts; for example, between bones and their overlying skin, muscles, or tendons (Fig. 3-52). These little sacs act as slippery cushions which tend to reduce friction between the moving parts, and they are named according to their location. A few of the more important bursae are: *acromial, subdeltoid, subscapular, olecranon,* and *preptellar.*

Injury to bursae may result in inflammation, or *bursitis,* and one finds it extremely painful to move the affected part. The prepatellar bursa, due to its location between the skin and the patella, is frequently injured by those who work for prolonged periods in the kneeling position; the resulting irritation and enlargement of this bursa is popularly called housemaid's knee. A *bunion* is a swelling of a bursa of the foot, particularly at the metatarso-phalangeal joint of the great toe, which is then forced outward, making the joint unduly prominent. It may develop as a result of pressure or friction from improperly fitted shoes.

SUMMARY

1. Function of the skeleton
A. Produce blood cells
B. Support body
C. Attachment of muscles and tendons
D. Protect internal organs
2. Classification on basis of shape
A. Long bones (e.g., humerus)
B. Short bones (e.g., carpals)
C. Flat bones (e.g., sternum)
D. Irregular bones (e.g., vertebrae)
E. Sesamoid bones (e.g., patella)
3. Bone formation
A. Intramembranous
B. Endochondral or cartilaginous
4. Growth of bone: circumference, length
5. Structural details
A. Composition
    a. Inorganic
    b. Organic
B. The haversian system
C. Bone marrow: red and yellow

D. Coverings and blood supply
    a. Periosteum
    b. Articular cartilage
    c. Periosteal blood vessels
    d. Nutrient or medullary artery
6. Descriptive terms
A. Projections: head, condyle, crest, spine, trochanter, tubercle, tuberosity
B. Depressions and openings, foramen, fossa, sinus, meatus
7. Skeleton as a whole—206 bones in adult
A. Divisions: axial and appendicular
8. Axial skeleton
A. Skull
    a. Cranium: frontal, parietal, temporal, occipital, sphenoid, ethmoid
    b. Superficial bones of face: nasal, zygomatic, maxillae, mandible
    c. Deeper bones of face: lacrimal, vomer, palatine, inferior conchae
B. Cranial sutures
    a. Immovable joints
    b. Coronal, lambdoidal, sagittal, squamosal
C. Fontanels
    a. Incomplete ossification of skull bones at birth
    b. Anterior, posterior, anterolateral, posterolateral
D. Air sinuses
    a. Paranasal: frontal, maxillary, sphenoid, ethmoid
    b. Mastoid of temporal bone
E. Hyoid bone
F. Vertebral column
    a. Adult and child
    b. General structure of vertebrae
    c. Regional variations
    d. Column as a whole
G. Thorax: bony cage, sternum, ribs
9. Appendicular skeleton
A. Upper extremity
    a. Shoulder girdle: clavicle, scapula
    b. Arm: humerus
    c. Forearm: ulna, radius
    d. Wrist and hand: carpals, metacarpals, phalanges
B. Pelvis
    a. Hipbone (ilium, ischium, pubis)
    b. As a whole: hipbones, sacrum, coccyx
C. Lower extremity
    a. Thigh: femur, patella
    b. Leg: tibia, fibula
    c. Foot: tarsals, metatarsals, phalanges
10. Articulations (joints)
A. Classification
    a. Synarthroses: immovable
    b. Amphiarthroses: slightly movable
    c. Diarthroses: freely movable
B. Movements at diarthrotic joints
    a. Angular: flexion, extension, abduction, adduction
    b. Circumduction
    c. Rotation
    d. Special: supination, pronation, inversion, eversion, protraction, retraction

C. Bursae
   a. Clefts lined with synovial membrane; reduce friction between moving parts
   b. Named for location under muscles, tendons, bones; e.g., acromial, subdeltoid, olecranon, prepatellar.

QUESTIONS FOR REVIEW

1. What is the function of the intervertebral disks?
2. In performing a laminectomy for the removal of a herniated disk, what structures would the physician remove?
3. Why are the bones of elderly individuals more easily fractured than those of children?
4. What kinds of motion are possible in a ball-and-socket joint?
5. Your friend asks you to explain his fracture of the head of the femur to him. What do you tell him?
6. In surgery, the doctor removes the fibula because of bone disease. Will this individual be able to walk without support after healing? Why?
7. What is the difference between the male and the female pelvis?
8. What are fontanels?
9. At what age do you expect the growth of long bones to be complete? How can you tell this has occurred?
10. Describe briefly, the functions of the skeleton and the periosteum.
11. Distinguish between osteoblasts and osteoclasts.
12. In terms of the haversian systems, how do osteocytes receive nourishment and eliminate wastes?

REFERENCES AND SUPPLEMENTAL READINGS

Goss, C. M. (ed.): Gray's Anatomy of the Human Body, ed 28. Philadelphia, Lea & Febiger, 1966.

Ham, A. W.: Histology, ed 6. Philadelphia, J. B. Lippincott, 1969.

Robbins, S. L., and Angell, M.: Basic Pathology, ed 1. Philadelphia, W. B. Saunders, 1971.

Muscle Tissue

Were it not for muscle tissue, our bones and joints would be quite useless, food could not move through the digestive tract, and blood could not circulate through the vessels of the body. In short, movement of any type would be impossible, and life as we know it could not exist.

Muscle tissue is most famous for its ability to contract, or to shorten, thus producing movement of internal and external body parts. For example, one very important activity begins long before you are born and continues until your last minute of life; this vital movement is the contraction of the heart muscle which results in the pumping of blood to all parts of your body. A fetus also can move its extremities several months before birth; when this begins, the mother says that she "feels life" and knows that the fetus is alive within her body. Other important movements are those related to breathing, or getting oxygen into and carbon dioxide out of the lungs. The first gasp or cry of the newborn infant indicates that the muscles of respiration have begun to function.

Soon after birth we find other important muscular activity getting under way, activity over which we have no voluntary control: the movement of food material along the gastrointestinal tract, the emptying of the bowel, the transport of urine from the kidneys to the urinary bladder, and then the emptying of this bladder. These activities are made possible by muscle tissue which is located in the walls of various organs.

The developing child's mastery of the muscles that move his extremities is fascinating to watch.

Within a few months after birth, the infant learns to hold his spoon and then his cup. The maintenance of posture is another important accomplishment; first the child learns to control those muscles that enable him to sit and then to stand erect. A little later he learns to walk and to run. But it takes several years before movements can be so coordinated as to permit the child to button his clothes, tie his shoe strings, and write his own name. Still more years are necessary before muscular control permits such things as playing the piano, fingering a violin, or using a typewriter. In fact, some persons who are all thumbs may never acquire skill in using their hands for delicate work such as threading needles, sewing a fine seam, or using small tools of any kind.

In addition to the work accomplished by its contraction, another important function of muscle tissue is the production of heat. The chemical changes which occur during muscle contraction produce much of the heat that is needed to maintain our normal body temperature. For example, when you shiver in cold weather, you are experiencing involuntary contractions of certain muscles; this is the body's attempt to produce more heat and keep you warm.

The above examples will give you an idea of the variety of activities carried out by muscle tissue, and help you to appreciate more fully muscles that are functioning normally. Consider what a difference it would make in our lives if we could not walk, talk, feed ourselves, or perform the hundreds of tasks that make up a day's work.

TYPES OF MUSCLE TISSUE

Muscles serve as the engines of our body and are so constructed as to provide varying degrees of speed and power. There are three types of muscle tissue (skeletal, smooth, and cardiac), each being uniquely designed for the performance of a certain type of work.

Skeletal muscle

Skeletal muscles are attached to the bones, cover the skeleton and give shape to the body, as well as make it possible for us to move about during our work and play. This type of muscle also is referred to as **striated,** due to the striped appearance of its individual fibers when seen under a microscope. A third name, **voluntary,** is sometimes used to indicate that this type of muscle is under the control of our conscious will. Skeletal muscle tissue is like a highspeed hot rod engine; it can develop great power but runs for relatively short periods of time.

Nonstriated, or smooth muscle

Smooth muscle is very prominent in the walls of viscera such as the stomach, the intestine, and the urinary bladder. For this reason it is sometimes called **visceral** muscle. Smooth muscle also occurs in the walls of blood vessels, in glands, and in the skin. When seen under the microscope, it appears to be smooth, or nonstriated, since it lacks the characteristic striped appearance of other muscle tissue. The term **involuntary** is used to indicate that this type of muscle is not subject to our conscious control. Smooth muscle might be compared with the engine of a freight elevator: geared for slow speed and heavy duty, it continues to function day after day in a slow but steady manner.

Cardiac muscle

This highly specialized tissue is found only in the walls of the heart, hence the name cardiac muscle. It too is **involuntary,** as we have no conscious control of its activity. When seen under the microscope, cardiac muscle is a branching network of fibers which are striated. Our cardiac engine works steadily day after day, and yet it is capable of vigorous response for short periods, such as when we make a dash for the bus or run up a flight of stairs.

STRUCTURE OF MUSCLE TISSUE

Microscopic structure

Skeletal muscle cells are drawn out into long threadlike cylinders which vary in length from 1 to 40 mm. (Fig. 4-1). Since their length is relatively so much greater than their diameter, these cells are often referred to as muscle *fibers.* Each fiber is wrapped in a delicate connective tissue sheath called the *endomysium* (*within muscle*), which contains a network of capillaries for nourishment of the fiber. The cell membrane is called *sarcolemma* (*sarco*, meaning *flesh*, and *lemma*, meaning *husk*). Skeletal muscle fibers are multinucleated, their many nuclei lying between the sarcolemma and the *sarcoplasm* (*muscle cytoplasm*). Embedded in the sarcoplasm and running the entire length of the cell are many fine fibers which are called myofibrils. Each myofibril, in turn, consists of *actin* and *myosin* filaments, the large protein molecules that are ultimately responsible for muscle contraction. Figure 4-8 illustrates one contractile unit (called a *sarcomere*), of a myofibril. The arrangement of actin and myosin contributes to the alternating light and dark sections that are responsible for the characteristic striated appearance of skeletal muscle tissue. A network of tubules, the *sarcoplasmic reticulum,* extends between the myofibrils; this network provides the plumbing needed for the distribution of substances that are required for contraction of

Fig. 4-1. Longitudinal section of striated muscle showing cross-striations and the nuclei at the periphery. ×275.

the myofibrils. Numerous mitochondria also are found in the sarcoplasm.

Smooth muscle cells are spindle-shaped, which means that they are long and tapering (Fig. 4-2). Each cell, or fiber, has a single nucleus which is centrally located. The length of smooth muscle fibers varies markedly in different parts of the body. In the walls of tiny blood vessels we find the shortest fibers—only 0.02 mm. long, or 1/50 as long as the shortest skeletal muscle fiber. The longest fibers are found in the walls of a pregnant uterus, in which they may be 0.5 mm. long, but they still are shorter than the shortest skeletal fiber. Myofibrils are present in the sarcoplasm, but they are nonstriated in appearance. Smooth muscle has an abundant blood supply.

Cardiac muscle cells are roughly quadrangular in shape with a single central nucleus (Fig. 4-3). The cells are arranged end to end and end to side, forming a network of branching fibers. The dark-stained bands noted at intervals are called *intercalated disks*, and studies with the electron microscope indicate that they represent cell boundaries. This latticework arrangement of cardiac cells forms a *functional syncytium*, so that impulses transmitted over the membrane of any single fiber spread to all of the fibers (i.e., either of the atria or of the ventricles, as will be described in more detail in Chapter 12). Myofibrils are present, as are the transverse striations mentioned previously. Mitochondria are particularly abundant in the sarcoplasm of cardiac muscle. This important muscle tissue also has an excellent blood supply, which is essential to life itself. Any disease process that interferes with this blood supply may result in a so-called heart attack.

Gross structure

All muscles contain connective tissues that not only provide for the support and the nourishment of working fibers, but also help to direct their "pull" into the performance of specific tasks, such as moving an arm, or emptying the stomach.

Each **skeletal muscle** has an outer sheath or overcoat of connective tissue called the *epimysium* (*upon muscle*). Extending inward from the epimysium are little partitions of connective tissue which divide the muscle into bundles, or *fasciculi*; these partitions are called *perimysium* (*around*

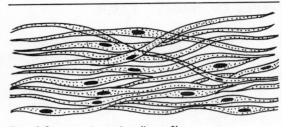

Fig. 4-2. Smooth muscle cells, or fibers.

muscle), since they surround bundles of muscle fibers. Endomysium surrounds individual fibers as previously noted in the discussion of microscopic features. A diagrammatic three dimensional view of skeletal muscle is shown in Figure 4-4. The epimysium, the perimysium, and the endomysium are continuous with the tendon, which merges with the periosteum of the bone to which a muscle is attached and on which it exerts its pull when contracting (see Fig. 5-2). In fact, the connective tissue component of skeletal muscle has been compared with a harness, by means of which the pull exerted by a horse can be brought to bear on a wagon.

Smooth muscle tissue is arranged in sheets or bundles which form contractile layers in the walls of blood vessels and viscera. The cells of one layer lie in a circular plane while those of another layer lie in a longitudinal plane. These layers in turn are bound firmly to other components of the wall by connective tissue. Con-

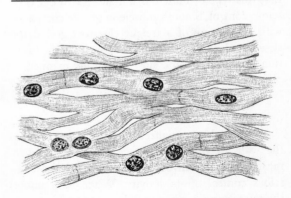

Fig. 4-3. Cardiac muscle. Branching fibers, centrally placed nuclei, and intercalated disks are evident.

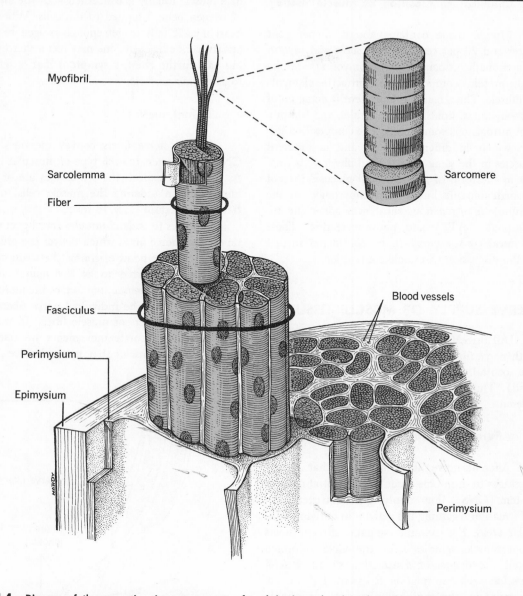

Myofibril

Sarcolemma

Fiber

Fasciculus

Perimysium

Epimysium

Sarcomere

Blood vessels

Perimysium

Fig. 4-4. Diagram of the connective tissue components of a skeletal muscle. The relationships between a muscle bundle (fasciculus), a single muscle cell (fiber), and a myofibril also are indicated. The structure of a sarcomere is shown in Figure 4-8.

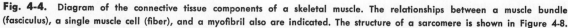

traction of muscle fibers of the layers results in various changes in the shape and size of organs as needed to carry out body functions. For example, when one looks at a bright light, it is smooth muscle that automatically decreases the size of the pupil and regulates the amount of light that can enter the eye.

Cardiac muscle tissue is composed of bundles

which are arranged concentrically as spirals or figures-of-eight so that when contraction occurs, the cavities of the heart are temporarily obliterated and blood is forced out. The muscle tissue of the upper chambers (atria) is not continuous with that of the lower chambers (ventricles), but the two groups work in close harmony, as described in Chapter 12.

Chemical composition of muscle tissue

Muscle tissue contains about 75 per cent water and 20 per cent protein; *actin* and *myosin* (functionally associated as actomyosin) are the main proteins composing the contractile elements of muscle. The remaining 5 per cent is made up of carbohydrates, lipids, inorganic salts, and nonprotein nitrogenous compounds. The composition differs widely in different species and in different muscles in the same animal. Carbohydrate is present in the form of glygogen and glucose. Present in small amounts, but of great importance, are the nonprotein nitrogenous substances adenosine triphosphate (ATP) and phosphocreatine. These chemical constituents will be considered further in the discussion of muscle contraction.

NERVE SUPPLY OF MUSCLE TISSUE

All three types of muscle tissue are supplied with nerve fibers that carry messages to and from the central nervous system (brain and spinal cord). These fibers are classified as afferent and efferent.

Afferent, or sensory nerves

Afferent nerve fibers are those that convey messages from muscle tissue to the central nervous system (CNS), thus supplying headquarters with information regarding the activity in outlying muscular areas. For example, **somatic afferent** fibers from muscle spindles keep the CNS informed about the degree of contraction of all *skeletal muscles;* such information is essential if we are to maintain our equilibrium and to coordinate all muscular movements. **Visceral afferent** fibers keep the CNS informed about the condition of cells in *smooth muscle* and *cardiac muscle.* For example, excessive contraction (spasm-cramp) or excessive stretching (distention) of smooth muscle will arouse sensations of pain. If the intestine is distended with gas that cannot escape by the rectum, pain results; as you may know, gas pains can be very severe. When the gas escapes, either spontaneously or after the insertion of a rectal tube, the distention is relieved and the pain disappears. Information sent to the brain from cardiac muscle usually is concerned with the amount of oxygen being supplied to the cells. When the heart muscle fails to get enough oxygen to carry out its work efficiently, one may feel a sharp warning pain in the chest, a symptom that is referred to as *angina pectoris.*

Efferent nerves

Efferent nerve fibers convey messages from CNS headquarters to each type of muscular tissue. As you might expect, these messages are of great importance in ordering the muscle cells to contract or, in certain cases, to relax.

Messages to *skeletal muscles* travel over **somatic efferent** nerve fibers which deliver the order for contraction. The point of contact between muscle and nerve is referred to as the *neuromuscular junction,* the *myoneural junction,* or the *motor end plate* (Fig. 4-5). The individual nerve fibers supply varying numbers of muscle fibers. In muscles in which fine, accurate movements are essential, such as in the muscles that move the eyes, each

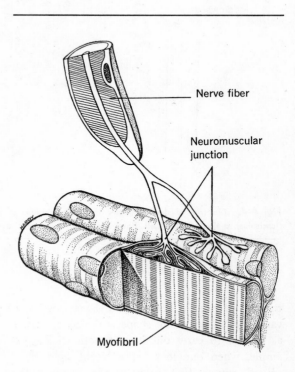

Nerve fiber

Neuromuscular junction

Myofibril

Fig. 4-5. Neuromuscular junction (motor end plates) in striated muscle.

nerve fiber supplies only three to six muscle fibers. On the other hand, in muscles in which such fine control is unnecessary, as in the muscles of the leg, each nerve fiber divides extensively and supplies as many as 150 muscle fibers. A nerve cell, its outgoing fiber, and all the muscle fibers supplied constitute a **motor unit.** Each skeletal muscle is made up of an enormous number of motor units.

Skeletal muscle is completely dependent on its nerve supply, and, should these efferent fibers be destroyed through injury or disease, the end-result will be *paralysis.* Since paralyzed muscles cannot be used voluntarily, they undergo *atrophy* (wasting or decrease in size) unless properly treated. Much of the training and the treatment of paralysis victims is directed toward the prevention of atrophy of the involved muscles. Atrophy will occur, likewise, in muscles rendered immobile by splints or casts, or if their tendons are cut. When muscles cannot be contracted voluntarily, they should be massaged, exercised by passive movements, or stimulated to contract by electrical methods. This is a part of the field of physical medicine or physiotherapy.

Messages to *smooth muscle* and to *cardiac muscle* travel over **visceral efferent,** or *autonomic* nerves (p. 220). In contrast to skeletal muscles which contract only when they receive orders from the CNS, smooth muscle and cardiac muscle may contract quite independently of such messages. However, there are times when the activity of these tissues must be temporarily increased or decreased, and messages from the CNS are essential on such occasions. To meet this dual need, there is a double nerve supply to smooth muscle and to cardiac muscle; one nerve supply carries orders for an increase in the activity that is already going on, and the other for a decrease in such activity. This regulation of activity is entirely unconscious, and we have no voluntary control over it.

PHYSIOLOGIC PROPERTIES OF MUSCLE TISSUE

All three types of muscle tissue have the same fundamental properties:

1. Excitability (irritability), or the capacity to respond to a stimulus;

2. Contractility, or the capacity to shorten;

3. Tonus, or the capacity to maintain a state of steady, partial contraction;

4. Extensibility, or the capacity to stretch when force is applied;

5. Elasticity, or the capacity to regain original form when the force is removed.

Excitability

The capacity to respond to a change in the environment (i.e., a stimulus) is referred to as excitability. This property is highly developed in muscle fibers and in nerve fibers (p. 163), as both are capable of transmitting electrochemical impulses along their membranes. The details of electrical membrane potentials are not within the scope of this textbook, but a few broad, general statements will lay the groundwork for an understanding of how muscle and nerve fibers respond to stimuli.

Resting Membrane Potential. The outer surface of a resting fiber or cell membrane bears a positive electrical charge, and the inner surface bears a negative charge (Fig. 4-6). Such electrical potentials exist across most of the cell membranes in the body and undoubtedly play a part in cellular functions. However, these membrane potentials are of particular importance in excitable cells such as muscle and nerve cells.

The origin of membrane potentials involves several factors which will be summarized briefly as follows:

First, as described on page 25, the sodium-potassium pump plays an important role in maintaining the proper concentration of ions within and without the cell. Moving against concentration gradients, sodium is pumped *out* of the cell and potassium is pumped *into* the cell; the energy for this active metabolic process is derived from adenosine triphosphate (ATP).

Second, the normal resting muscle or nerve membrane is much more permeable to potassium than it is to sodium; this means that sodium diffuses into the cell only with great difficulty, but potassium can move out with relative ease. Calcium ions play an important role in determining membrane permeability as described on page 462.

Third, within the cell, there are large numbers of *negative* ions that cannot diffuse through the membrane at all, or that do so very poorly (e.g., organic sulfate ions and organic phosphate ions).

Thus there will be a tendency for positive charges to line up *outside* the membrane while an equal number of negative charges line up *inside* the membrane; this results in the development of a resting membrane potential.

Action Potential. If a resting muscle or nerve cell membrane is stimulated chemically, electrically, or mechanically, its permeability to sodium ions is suddenly increased. Large numbers of these positively charged ions rush into the cell causing a reversal of the electrical potential (i.e., the interior becomes positive and the exterior becomes negative); this stage of the action potential is called *depolarization*. Within a tiny fraction of a second after depolarization takes place, the membrane again becomes relatively impermeable to sodium ions, there is a rapid diffusion of potassium ions to the exterior of the cell, and the normal resting potential is restored; this stage is called *repolarization*. The sodium-potassium pump operates far too slowly to restore the resting potential in a tiny fraction of a second; thus it is the outward diffusion of potassium that does the job. However, there is a limit to the number of times that an action potential can be elicited without reestablishing the ionic concentration gradients of sodium and potassium; this active process of recharging the fiber is accomplished by the sodium-potassium pump.

The **propagation of action potentials** has been compared with lighting a trail of gun powder —the flame in one section serving to ignite the next. In an excitable fiber, depolarization begins at the point of stimulation and immediately excites adjacent portions of the membrane; these newly depolarized areas excite areas further along the membrane, and so on over the course of the fiber (Fig. 4-6). This transmission of the depolarization process is called an impulse, either a *nerve impulse* or a *muscle impulse*, depending on the type of fiber involved. While impulses usually start at one end of a fiber, there is no *single* direction of propagation; if a fiber is stimulated in its midportion, the impulse will travel in all directions from the point of initial stimulation.

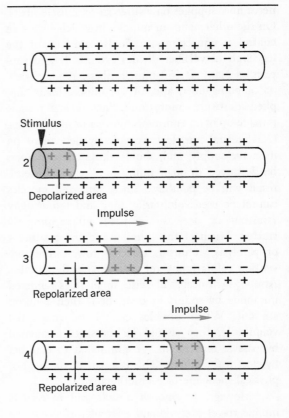

Fig. 4-6. Diagram of membrane potentials. 1. *Resting membrane potential:* outer surface bears a positive charge and inner surface bears a negative charge. 2. *Action potential, first stage:* stimulation of fiber results in depolarization, i.e., outer surface becomes negative and inner surface becomes positive. 3. *Action potential, second stage:* repolarization occurs as the resting potential is restored. 4. Propagation of impulse continues in direction of arrow.

Rhythmicity refers to a repetitive discharge of action potentials. Automatic depolarization followed by repolarization can occur in all excitable tissues if the threshold for stimulation is sufficiently lowered. Such discharges occur normally in cardiac muscle and in most smooth muscle fibers; however, in skeletal muscle and in nerve fibers these discharges usually involve some abnormality (e.g., changes in calcium concentration).

Stimuli. We shall now consider the various disturbances, or stimuli, that serve to excite the three types of muscle tissue. In the intact body, skeletal muscles are excited exclusively by nerve impulses coming from the brain and the spinal cord. If such impulses fail to reach them, they will be completely relaxed and unable to respond.

Smooth muscle responds to a great variety of stimuli, such as distention, temperature changes, and nerve impulses. In addition, smooth muscle is sensitive to changes in the composition of the blood or in the interstitial fluid that bathes it. Thus hormones, drugs, salts, acids, and bases may affect the activity of smooth muscle. As previously noted, smooth muscle in certain parts of the body displays the property of rhythmicity. This self-excitatory process plays an important part in promoting peristalsis (p. 401) along various tubular structures.

Cardiac muscle is duly famous for its rhythmicity (see above) which develops as soon as the heart is formed in the embryo and long before any nerves have grown out to it. However, when the nerve supply is established, the heart will respond to nerve impulses which order an increase or a decrease in its rate and force of contraction. Cardiac muscle also may be stimulated by mechanical, electrical, thermal, and chemical agents. Doctors frequently make a life-saving practical application of these facts when a patient's heart suddenly stops beating for no apparent reason (*cardiac arrest*). For example, the heart may be stimulated by firm, rhythmic pressure on the sternum (external cardiac massage). If the chest is open, the doctor may rhythmically squeeze the heart with his hand, or he may use a special machine that sends an electrical current through the walls of the heart. In any case, he is depending on the property of excitability of cardiac muscle as he endeavors to re-establish the heart beat.

Neuromuscular Transmission. As noted in the preceding discussion, all three types of muscle tissue can be excited by nerve impulses. Because this is of fundamental importance insofar as skeletal muscle is concerned, we shall discuss briefly the neuromuscular transmission of impulses. Although we do not yet fully understand all of the details, we do know that when nerve impulses reach a neuromuscular junction in skeletal muscle, a chemical substance, acetylcholine, is released by the end plate of the nerve fiber (Fig. 4-7). **Acetylcholine** excites the muscle fiber and initiates the process of contraction. Within a fraction of a second, acetylcholine is destroyed by the

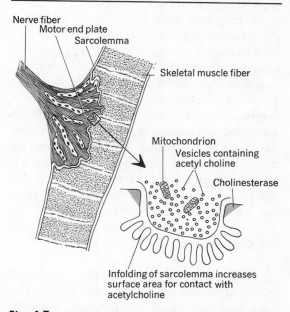

Fig. 4-7. Diagram of a single end plate in striated muscle. The general appearance of the junction as a whole is shown on the left. The arrow indicates a small section of the end plate and sarcolemma that has been even more greatly enlarged. (Modified from A. C. Guyton: Textbook of Medical Physiology, ed. 4. Philadelphia, W. B. Saunders, 1971.)

enzyme *cholinesterase*; this prevents re-excitation of the muscle until the arrival of another nerve impulse. The chemical excitation of smooth muscle and cardiac muscle follows a similar pattern, but it is best described after study of the autonomic division of the nervous system (p. 227).

Certain drugs, such as curare and similar muscle relaxants, can block the passage of nerve impulses at neuromuscular junctions, thus temporarily preventing the contraction of the muscles. Used in conjunction with various types of anesthetics, these drugs produce excellent muscular relaxation; this is very important to the surgeon who has made an incision through the strong muscles of the abdominal wall and is trying to examine the internal organs. Perhaps you are wondering about the effect of these drugs on the muscles of respiration. For some unknown reason, the muscles of the abdominal wall and the extremities respond to smaller doses of curarelike drugs than do the muscles of respiration. However, if larger doses are administered, the respiratory muscles also will be paralyzed; in such cases artificial respiration must be given until the effects of the drug wear off, usually within 5 minutes.

Myasthenia gravis is a condition characterized by weakness of skeletal muscles due to blocking of nerve impulses at the neuromuscular junction. It usually begins first in muscles of the face, then it spreads to other muscles until the patient is eventually incapacitated. Death by asphyxia follows severe involvement of the respiratory muscles. The cause of myasthenia gravis is uncertain, but it may be due to an inability of the end plates to secrete adequate quantities of acetylcholine; another possible cause may be the presence of excessive quantities of cholinesterase.

Contractility

When appropriately stimulated, all muscle cells or fibers have the capacity to shorten, or contract. However, before one can fully appreciate the overall sequence of events involved in muscle contraction, there must be some understanding of the basic molecular changes that take place, as well as the all-important sources of energy for those changes.

Molecular Changes. In the earlier discussion of the microscopic structure of muscle tissue (p. 97), it was noted that each muscle fiber contains many myofibrils embedded in the sarcoplasm. Each myofibril, in turn, consists of myosin and actin filaments, the large protein molecules that are ultimately responsible for muscle contraction. These filaments are more precisely arranged in skeletal and cardiac muscle than they are in smooth muscle, but the contractile process appears to be essentially the same in all three types of muscle tissue. The precise arrangement of the actin and myosin filaments in skeletal muscle is shown diagrammatically in Figure 4-8. Note that the thick filaments (myosin) are arranged longitudinally in the middle of the relaxed sarcomere, but do not reach the ends of the sarcomere. On the other hand, the thin filaments (actin) extend from the ends (Z bands) of the sarcomere, but do not reach the middle. Contraction of the sarcomere takes place when there is an interaction between these filaments, causing the thin actin filaments to move inward among the myosin filaments and pull with them the Z bands to which they are attached. The chemical changes involved in this process are extremely complex and not yet

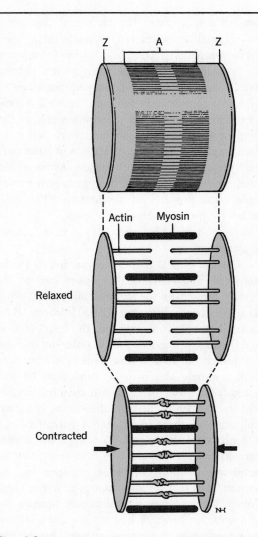

Fig. 4-8. Diagram of actin and myosin filaments in one sarcomere, or contractile unit, of a myofibril in skeletal muscle. Contraction occurs when the thin actin filaments move inward among the myosin filaments and pull with them the Z bands to which they are attached.

completely understood. However, this actomyosin team is so strong that it can move loads many times its own weight; pound for pound, it packs more power than any engine yet designed by man.

Energy Sources. The immediate energy source for the contractile process is **adenosine triphosphate** (ATP). This energy-rich phosphate compound sometimes is referred to as the spark plug of the muscular engine. When a muscle cell is stimulated, an enzyme causes one of ATP's three phosphate radicals to split off, leaving *adenosine diphosphate* or ADP; during this splitting

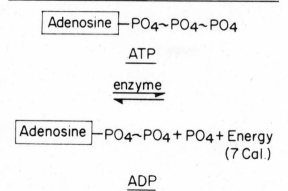

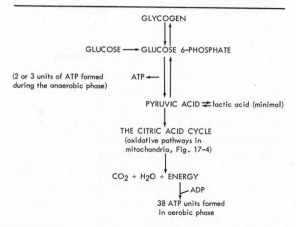

Fig. 4-9. Schematic representation of the reversible action between adenosine triphosphate (ATP) and adenosine diphosphate (ADP). The symbol $\sim$ represents high energy bonds between phosphate radicals. Splitting of a high energy bond releases 7 calories of energy, as when the above reaction proceeds to the right. When ADP is reconverted to ATP, as the reaction proceeds to the left, 7 calories of energy are required for the PO_4 bonding process.

Fig. 4-10. A summary of the highlights of anaerobic and aerobic pathways in the breakdown of carbohydrate leading to the formation of ATP. This figure also illustrates the steady state (i.e., during moderate exercise sufficient oxygen is available to oxidize pyruvic acid and prevent the formation of any significant amount of lactic acid).

process a burst of energy is released (Fig. 4-9). The ultramicroscopic sparks of energy from a variable number of ATP units power the interaction between actin and myosin, causing the fiber to shorten.

Within a fraction of a second after the contractile process ends, the ADP is converted back to ATP. Most of this immediate reconversion occurs when ADP reacts with **phosphocreatine,** another high-energy compound found in the sarcoplasm. Phosphocreatine is present in muscle in far greater concentration than is ATP, but its energy must be funneled through ATP before it can be utilized by the muscle fiber.

ADP + Phosphocreatine → ATP + Creatine

Some ADP also diffuses out of the myofibril and into the mitochondria, where foods are being oxidized in the citric acid cycle, a complex series of chemical reactions that not only require a constant supply of oxygen, but also liberate large quantities of energy for the formation of ATP (p. 418). However, these oxidative processes do not produce ATP in the quantity needed during prolonged muscular activity, and the available ATP would soon be used up were it not immediately resynthesized by the interaction of ADP and phosphocreatine. During periods of rest, when energy requirements are reduced, enough ATP is formed by the aforementioned oxidative processes in the mitochondria to convert the creatine

back to phosphocreatine, thus recharging the chemical battery of the muscular engine.

Another source of energy for muscular contraction is found in the *anaerobic breakdown of carbohydrates.* Through a complex series of chemical reactions, not requiring the simultaneous use of oxygen, glycogen and glucose are broken down to *pyruvic acid;* during this process a relatively small amount of ATP is formed (Fig. 4-10). However, during strenuous exercise even this small amount of ATP is of considerable importance.

The course of events following the formation of pyruvic acid depends on whether or not oxygen is available in sufficient quantities for the amount of work being done. For example, during moderate activity one is able to breathe in enough oxygen so that some of the pyruvic acid is oxidized to CO_2 and H_2O with the liberation of sufficient energy to convert the remaining pyruvic acid back to glycogen. Exercise physiologists sometimes refer to this as the *steady state* (Fig. 4-10). On the other hand, during bursts of strenuous activity, such as running up a flight of stairs, a person cannot breathe in oxygen rapidly enough to handle the increasing amounts of pyruvic acid being produced in the hard-working muscular engines. Under these conditions pyruvic acid is converted to *lactic acid,* which begins to accumulate in the body. As lactic acid accumulates the person is said to be running up an *oxygen debt.* (Fig. 4-11). However, nature has decreed that one must repay this obligation, so a person will be out of breath until the debt is paid in full (Fig. 4-12). During this period of rapid breathing the lactic acid is changed back to pyruvic acid, some of which is then oxidized to provide energy for the conversion of the remainder to glycogen, as well as for the restoration of phosphocreatine and ATP, as previously noted. Thus the oxidative processes in the mitochondria serve not only to recharge the cellular batteries, but also to prevent the accumulation of lactic acid and help prevent the depletion of one's glycogen fuel supply.

If for any reason a person is unable to repay his oxygen debt and prevent the accumulation of lactic acid, *muscle fatigue* will result. Excessive amounts of lactic acid tend to depress the activity in muscle cells, leading to decreased irritability and interfering with their ability to contract. Therefore, if the recovery phase lags too far behind the contraction phase, the muscles will be physically unable to respond in their usual efficient manner, thus forcing one to stop his activity and rest until conditions return to normal.

Heat production is an important consideration in the functioning of any engine, and muscles are no exception. Most of the energy liberated during muscle contraction appears as heat, but only a small part (20 to 30 per cent) can be converted to work; muscular efficiency is roughly the same as an ordinary steam engine in that respect. However, at this point the comparison breaks down, since most of the remaining heat is used for the very important and useful purpose of maintaining body temperature.

In extremely hot weather we often tend to reduce our muscular activity, wear lightweight clothing to facilitate heat dissipation, and take things easy. On the other hand, when cold weather arrives we put on heavier clothing to prevent heat loss, and move briskly in an effort to generate even more body heat. If we cannot do this, nature will take a hand in our affairs and increase heat production by making us shiver.

Heat production does not stop when we go to bed at night, but it is greatly reduced. While we are asleep, a certain amount of heat is produced by the constant activity of respiratory muscles, the heart muscle and smooth muscle in the walls of blood vessels, and various organs.

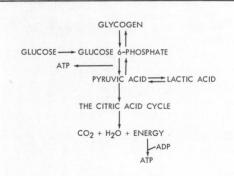

Fig. 4-11. Oxygen debt. During severe exercise the aerobic phase lags behind the more rapid anaerobic phase and lactic acid begins to accumulate. The oxygen debt must be repaid by rapid breathing, or muscle fatigue will result.

In concluding this discussion of the property of contractility, we present a brief description of the overall sequence of events that follows excitation of a muscle fiber. The various types of contractions and the strength of contractions will be considered later in the chapter.

Contraction of Skeletal Muscle. As described on page 102, stimulation of a muscle fiber results in an action potential that spreads throughout the entire fiber. The electrical current associated with an action potential flows through the sarcoplasmic reticulum to reach all myofibrils, even those deep within the muscle fiber. It is believed that this flow of current causes the release of *calcium ions* that are bound in the tubules of the sarcoplasmic reticulum of a resting fiber.

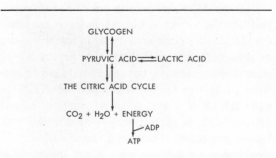

Fig. 4-12. Repayment of oxygen debt. During the period of rest and rapid breathing, lactic acid is oxidized to pyruvic acid. About 20 per cent of this pyruvic acid then is oxidized to CO_2 and H_2O with the production of energy for conversion of the remaining pyruvic acid to glycogen.

These calcium ions diffuse to the interior of each myofibril where they enable myosin to act like an enzyme, causing the breakdown of ATP to ADP. The burst of energy liberated during this breakdown powers the interaction between actin and myosin, thus causing the fiber to contract. Within a tiny fraction of a second after the action current has passed, calcium returns to its binding sites in the sarcoplasmic reticulum, and the muscle fiber relaxes.

The term *all-or-nothing response* refers to the fact that when a stimulus is applied to a muscle fiber, an action potential travels over the entire fiber or else fails to stimulate it at all. Thus either the entire fiber contracts or none of it contracts. However, the *force* of the resulting contraction can and does vary with the contractile state of the fiber when it is stimulated. For example, the contraction may be very weak if the nutritional status of the fiber is poor. On the other hand, a well-nourished fiber that has been warmed will contract even more strongly than usual.

Contraction of Smooth Muscle. The actual contractile process in smooth muscle is apparently the same as in skeletal muscle, the only significant difference being in the timing of the process. For example, the development of a contraction in smooth muscle is much slower than in skeletal muscle, and relaxation at the end of contraction is equally slow, possibly owing to the development of much stronger bonds between the contractile elements. This feature may help to explain why smooth muscle can maintain a considerable amount of tension over long periods of time without using much ATP energy, which is quite unlike skeletal muscle.

Contraction of Cardiac Muscle. Because cardiac muscle is striated like skeletal muscle, and has the same arrangement of actin, myosin, and sarcoplasmic reticulum, the basic mechanism of contraction is essentially the same. However, there are certain important differences in the manner of excitation and in the duration of depolarization. These will be described in Chapter 12, which deals exclusively with the heart.

Muscle tone, or tonus

All healthy muscles exhibit the property of tonus which is a state of steady, partial contraction, varying in degree but present at all times.

Tonus in **skeletal muscle** is due to the stretch reflex, an involuntary response to the stretching of muscle fibers; this means that it is completely dependent on the central nervous system for maintenance. Our skeletal muscles are so arranged in the body that they are held in slight stretch over the joints; this is particularly true of the *extensor* muscles that enable us to stand erect. Many sensitive nerve endings (receptors) are located in each fiber, and these are stimulated by any stretching of the muscle across a joint. As a result of this stimulation, nerve impulses are sent in to the CNS which, in turn, causes impulses to be sent back to the part of the muscle that is being stretched, ordering contraction of the fibers.

During the maintenance of tonus, groups of motor units work in relays or shifts. This means that one group of motor units is active for a short period of time; when this group relaxes, another group becomes active.

Muscle tone is often reduced when one is in poor health, and particularly so when prolonged bed rest is necessary. Such persons should have assistance when getting out of bed for the first time; otherwise they may suffer a fall as their weakened muscles fail to support the body weight. Early ambulation of hospital patients markedly reduces this hazard.

Tonus in **smooth muscle** differs widely from that in skeletal muscle and is entirely independent of the nervous system. However, smooth muscle is extremely sensitive to the chemical composition of its environment. The response to any particular stimulus may be rhythmic contraction, an increase in tonus, or a decrease in tonus (relaxation). In fact, a given stimulus which produces contraction under some circumstances may produce relaxation under other circumstances, depending on the condition of the muscle itself and its environment. Even though smooth muscle has the same fundamental properties as skeletal and cardiac muscle, it is not as uniform or as constant in its reactions because it is influenced by such a large number of extrinsic factors.

Smooth muscle tonus is influenced markedly by the presence of drugs in the interstitial fluid bathing the muscle fibers. Figure 4-13 illustrates the effect of epinephrine (Adrenaline) and pituitrin on the tonus of a strip of smooth muscle taken from a rat. The curarelike drugs, in addition to their paralyzing effect on skeletal muscle, lead

Fig. 4-13. Smooth muscle from a strip of rat uterus. Three normal, spontaneous contractions are shown at the left. When epinephrine was added, complete inhibition resulted. With the addition of pituitrin, the contractions began again with an increase in tonus. When this was washed out (right), the normal contractions began.

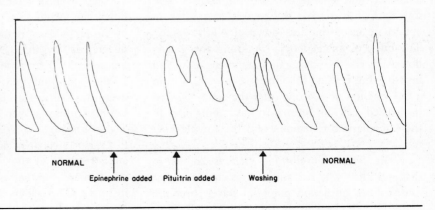

NORMAL   Epinephrine added   Pituitrin added   Washing   NORMAL

to a temporary loss of tone and activity in the intestinal tract, the blood vessels, and the smooth muscle generally.

Tonus in **cardiac muscle** is independent of nerve impulses and is synonymous with its general physiologic condition. When the heart is in good condition it has a high tonus and empties itself almost completely with each beat. A heart in poor condition is dilated at the end of each beat and contains a large amount of residual blood.

Extensibility and elasticity

All types of muscle tissue have the capacity to stretch when force is applied and to regain their original form when the force is removed. From a practical standpoint these properties are extremely important in that they lessen the danger of muscle rupture during times of great stress. The fact also has been established that muscle contracts more forcibly when it is slightly stretched.

The extensibility of smooth muscle, in particular, is truly remarkable. For example, this property makes possible the increase in size of the urinary bladder as urine accumulates. If the bladder could not distend, there would be a constant dribbling of urine instead of a periodic emptying of the bladder. Figure 4-14 illustrates the bladder in an empty and in a markedly distended condition. Distention of such a degree as illustrated is serious and occurs only under abnormal circumstances. Normally as the bladder is moderately distended the sense of fullness in the bladder merges into pain, and the bladder is emptied long before it is stretched to a dangerous extent.

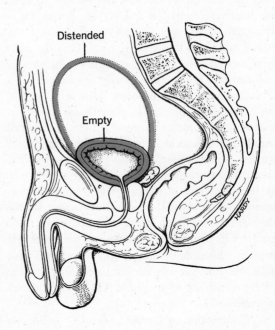

Distended

Empty

Fig. 4-14. Urinary bladder, empty and distended.

MECHANICAL PHENOMENA OF CONTRACTION

Much of our knowledge of muscular contraction has been gained from laboratory studies of an isolated muscle (gastrocnemius) taken from the leg of a frog. The muscle is suspended between a clamp and a writing lever tipped with a stylus. The stylus rests against a drum, and, as

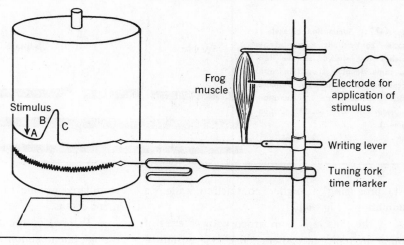

Fig. 4-15. Simple twitch of frog muscle following a single electrical stimulus: *A,* latent period, *B,* period of contraction, and *C,* period of relaxation.

the drum turns, any movement of the muscle is graphically recorded (Fig. 4-15). Such muscles will respond to many types of stimulation, such as the application of a hot wire, a pinch, a crystal of salt, and a strong salt solution, but the stimulus of preference is an electric shock. Electrical stimuli are least harmful, most easily regulated, and most like nerve impulses which stimulate muscles in the intact body. Consequently, the discussion that follows is based on the results of electrical stimulation.

Types of contractions

If a single electric shock is sent through the muscle, it responds by giving one contraction which is called a **simple twitch.** A possible example of this in our own bodies might be the involuntary blinking movement of the eye, but the rest of our movements are far more complex. A record of the simple twitch is shown on the drum in Figure 4-15. Note that a short period of time elapses after the stimulus is applied and before the muscle contracts; then the lever describes a curve. The period of time between the application of the stimulus and the beginning of the contraction is called the *latent period.* The time during which the muscle is shortening is called the *contraction period* and that during which the muscle is lengthening is called the *relaxation period.* The duration of each of these periods varies with the muscle studied. For example, Figure 4-16 illustrates the comparative times of three different muscles; a muscle which moves the eyeball, and

two calf muscles which move the foot. As one would expect, the eye muscle is most rapid in its response while the calf muscles are proportionately slower, in accordance with the function which each performs in the body.

We shall now leave the simple twitch and consider the results that follow the application of two stimuli in quick succession. If the second stimulus follows the first too quickly, the muscle will not respond to the second one. We know from this fact that there is a short *refractory period* following stimulation of skeletal muscle, during which it will not respond to a second stimulus, no matter how strong that stimulus is.

If a little more time is allowed to elapse between the first and the second stimulus (i.e., if time for recovery of excitability is allowed, or if we wait until the refractory period is over), the muscle responds to both the first and the second stimulus, and the second response is added to the

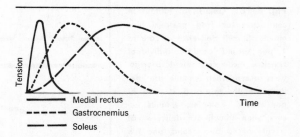

Fig. 4-16. Relative durations of simple twitches of different types of skeletal muscles. The medial rectus of the eye is rapid in its responses. The soleus twitch is of long duration. The gastrocnemius is intermediate.

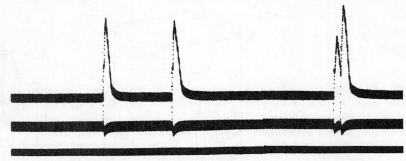

Fig. 4-17. Summation of skeletal muscle contractions. Two single twitches are shown on the left. The same stimulus is used in rapid succession for the next two twitches, which demonstrate summation. (Courtesy of Dr. M. J. Oppenheimer.)

first, producing a higher contraction called a **summation** (Fig. 4-17).

If the muscle is bombarded with a series of stimuli in rapid succession, there is a sustained contraction with no apparent relaxation. This response is called a *complete* **tetanic contraction,** or *tetanus* (Fig. 4-18). (NOTE: This should not be confused with the disease entity known as *tetanus,* or *lockjaw.* Victims of this dread condition do have sustained tetanic contractions, but the problem lies in their inability to end them voluntarily. Many muscles are involved, but those which cause the jaws to be clamped shut are responsible for the common expression *lockjaw.*) The rate at which stimuli need to be applied to produce complete tetanus varies in different muscles. In the soleus muscle of a mammal, complete tetanic contractions are produced when the rate of stimulation is about 30 per second; in the extrinsic eye muscles about 350 stimuli per second are required to produce complete tetanic contractions. In any event, our normal body movements depend on these smooth contractions rather than simple twitches. Tetanus is closely related to muscle tone, but in the latter case only a few motor units are

stimulated at one time so that no actual movement occurs.

Treppe, or the staircase effect, is an interesting phenomenon involving changes in the strength of muscle contraction. When one begins to use a muscle that has been resting, the initial strength of contraction may be as little as one-half of its strength some 30 or 50 contractions later. In other words, from the onset of activity the strength of contraction gradually increases until a plateau is reached. This staircase effect probably is caused by electrolyte changes that occur during a series of contractions. For example, it has been suggested that the loss of potassium from a fiber may increase the rate of liberation of calcium ions from the sarcoplasmic reticulum. Another factor may be the rise in temperature that accompanies muscle contraction.

Not all muscular contractions result in actual shortening of fibers and movement of some body part. For example, if you push against a stone wall, there will be no appreciable movement; yet your muscles are expending a great deal of energy. At such times there is a marked increase in tonus but little shortening; such contractions are called **iso-**

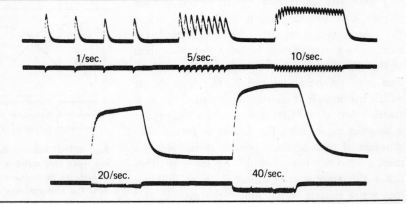

Fig. 4-18. Genesis of tetanus in contraction of frog gastrocnemius muscle. Stimuli delivered at rate of 1 per second produce individual twitches. Incomplete tetanic contractions occur when stimuli are delivered at the rate of 5, 10, and 20 per second. Complete tetanus occurs when stimuli are delivered at a rate of 40 per second, and the height of contraction is markedly increased over the height for single twitches. (Courtesy of Dr. M. J. Oppenheimer.)

1/sec.   5/sec.   10/sec.

20/sec.   40/sec.

metric. When a reasonable load is to be moved, tonus is relatively unchanged, but the muscle shortens. The latter is referred to as an **isotonic** contraction.

Strength of contractions

Various factors may alter the strength of contraction of any muscle. Some of these will be mentioned briefly.

Satisfactory metabolic conditions tend to strengthen contractions. Such conditions exist when adequate food and oxygen are available, when lactic acid is efficiently oxidized, and when waste products are removed promptly.

The **initial length of the muscle fibers** is of great importance in determining the strength of contraction of any given muscle. As illustrated in Figure 4-19, a muscle that is moderately stretched before contraction begins will be able to contract much more forcibly than one that is unstretched.

The **strength of the stimulus** can alter the strength of contraction. A minimal stimulus is the weakest one that will initiate the contraction of a muscle, and that contraction also will be weak. As the strength of the stimulus is increased, the height of contraction becomes greater and greater until the maximal response is obtained.

This change in the height of contraction of skeletal muscle with increasing strength of stimulation is due to the fact that more and more motor units are brought into action as the stimulus is increased. Keep in mind that this is true only for skeletal muscle. If the same experiment is carried out on an isolated quiescent ventricle of the heart (one that is no longer beating spontaneously), it will be noted that the height of contractions is the same whether minimal or maximal stimuli are used. Since cardiac muscle fibers form a functional syncytium, and since there are no motor units as in skeletal muscle, the entire muscle mass behaves as a unit. The responses of skeletal and cardiac muscle obtained by increasing strengths of stimuli are shown in Figure 4-20. In the intact body we can execute contractions of different strengths by using varying numbers of motor

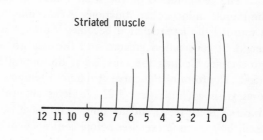

Striated muscle

12 11 10 9 8 7 6 5 4 3 2 1 0

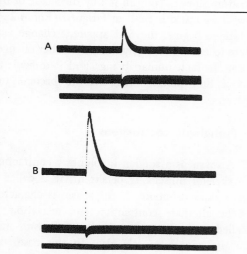

Fig. 4-19. A : A frog gastrocnemius that is unstretched develops a small amount of tension when stimulated. B : When the muscle is stretched, indicated by the raised baseline, the tension developed is much greater. (Courtesy of Dr. M. J. Oppenheimer.)

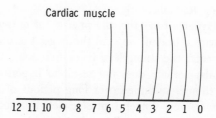

Cardiac muscle

12 11 10 9 8 7 6 5 4 3 2 1 0

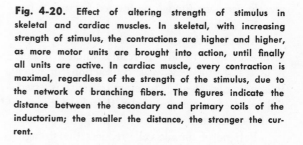

Fig. 4-20. Effect of altering strength of stimulus in skeletal and cardiac muscles. In skeletal, with increasing strength of stimulus, the contractions are higher and higher, as more motor units are brought into action, until finally all units are active. In cardiac muscle, every contraction is maximal, regardless of the strength of the stimulus, due to the network of branching fibers. The figures indicate the distance between the secondary and primary coils of the inductorium; the smaller the distance, the stronger the current.

units. In other words, we can grade our muscle contractions; we use few units in weak contractions and more units in strong contractions.

The **rate of stimulation** (as measured by action currents in a single motor nerve fiber) increases from 5 to 50 per second as voluntary contractions progress from weak to strong. Some motor units respond at only 8 stimuli per second, others at 12, and so on, up to 50 per second. This results in *asynchronism*, the net result of which is a smooth contraction. If all of the motor units in use contracted and relaxed at the same rate, the result would be a series of jerks, instead of a smooth contraction. In pathologic cases this condition does exist and is called *clonus*.

The **weight of the load** imposed on a muscle, when increased to optimal limits, will stretch the fibers and cause the muscle to contract more forcibly. However, as the load is gradually increased, the latent period becomes longer, the height of the contraction lower, and the relaxation faster.

The **temperature of the fluid** which bathes the muscle influences contractions. As the muscle is warmed the latent period becomes shorter, the height of contraction greater, and the relaxation more rapid. Cooling the muscle has the opposite effect. The form of the curve at different temperatures is shown in Figure 4-21. Athletes are well aware of these facts, and no doubt you have noticed their warmup exercises before contests which require muscular effort. Another familiar sight is the baseball pitcher who wears a jacket between innings, even though the weather is warm, to prevent his arm muscles from cooling out and contracting less efficiently.

Hypertrophy (*increase in size*) of muscle occurs with repeated use, and the larger the muscle, the greater the strength of its contractions. Weak muscular activity does not result in hypertrophy, even though sustained over long periods of time. However, very forceful muscle activity, such as isometric exercises, will result in an increase in muscle size even though performed for only a few minutes each day.

Atrophy (*decrease in size*) of muscle, is the reverse of hypertrophy. It results when a muscle is used only for very weak contractions or when it is not used at all. Therefore, exercise is essential in strengthening muscles that have become weakened through dissuse while in splints or casts.

Fatigue of muscle cells reduces the strength

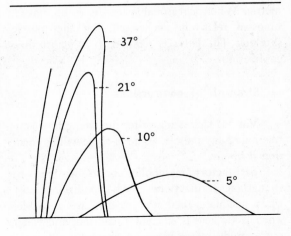

Fig. 4-21. Effect of altering the temperature on simple muscle twitch. Temperature expressed in degrees of centigrade.

of contraction. Accumulation of excessive amounts of lactic acid is probably a key factor in the intact body (p. 106). However, if you continue to stimulate a frog's muscle after removing it from the body, fatigue will occur, since it is no longer being supplied with oxygen and food, and its waste products are not being removed.

As fatigue progresses, the latent period becomes longer, the height of contraction less, and relaxation slower. You can tell by the shape of the curve if a muscle is fresh or fatigued (Fig. 4-22). As fatigue begins, the first apparent change is a failure to relax completely; later the other changes appear. In fact, human beings find it difficult to relax if they permit themselves to become too tired.

Pathologic contractions

A **spasm** is a sudden involuntary contraction of a skeletal muscle. If persistent, it is called a *tonic spasm*, or *cramp*. If the spasm is characterized by alternate contraction and relaxation, it is called a *clonic spasm*, or *clonus*. Spasms may occur in smooth muscle also. In this case, the spasm is evident as a transitory constriction of a passage, a duct, or an orifice, an example of this being the spasm during the passage of a gallstone or a kidney stone. Spasmodic contractions are accompanied by pain.

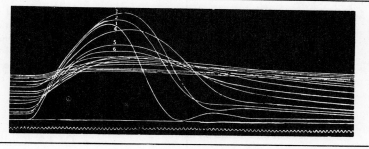

Fig. 4-22. Muscle twitches showing fatigue due to repeated stimulation. The first six contractions are numbered.

Fasciculation is the contraction of some motor units whose nerve cells are undergoing pathologic changes; it resembles localized shivering in some respects and is evident in superficial muscles just under the skin.

Fibrillation is the uncoordinated contraction of individual fibers; therefore, the whole muscle does not contract. This cannot be seen readily in muscles under the skin, but if it happens to involve the tongue muscles it can be seen as a quivering of parts of the protruded tongue. When fibrillation occurs in cardiac muscle, the heart seems to tremble or to be in a state of continuous convulsive movement; if this involves the lower chambers (ventricles) of the heart, death is usually quite sudden.

Rigor mortis is a state of contracture of all the muscles of the body after death. There are no action potentials, but the muscles become stiff and rigid. It is believed that this rigidity is due to the absence of ATP which normally prevents the bonding of actin and myosin in the resting state. Within 15 to 25 hours after death, the muscle proteins are destroyed by intracellular and bacterial enzymes and the rigor ends.

SUMMARY

1. Muscle function
A. Movement: voluntary; involuntary
B. Production of heat (body temperature)

2. Types of muscle tissue
A. Striated or skeletal: attached to bones; give shape to body; voluntary
B. Nonstriated or smooth: in walls of viscera and skin: involuntary
C. Cardiac: in walls of heart; involuntary; striated

3. Structure of muscle tissue
A. Microscopic
  a. Skeletal muscle: cells (fibers) threadlike cylinders; multinucleated; sarcolemma; sarcoplasm; myofibrils
  b. Smooth muscle: spindle-shaped cells single nucleus; nonstriated myofibrils
  c. Cardiac muscle: cells quadrangular; form network of branching fibers; intercalated disks; functional syncytium; blood supply essential to life
B. Gross structure
  a. Connective tissues support and nourish
  b. Skeletal muscle: covered by epimysium; partitioned into bundles by perimysium; individual fibers surrounded by endomysium; attached to tendons or bones by connective tissue harness
  c. Smooth muscle: arranged in sheets or bundles; in circular or longitudinal plane; produces changes in size and shape of organs
  d. Cardiac muscle: composed of bundles; arranged in spirals or figures-of-eight; contraction obliterates cavities and empties them; muscle of atria not continuous with ventricles
C. Chemical composition: 75 per cent water; 20 per cent protein; carbohydrate as glycogen and glucose; ATP and phosphocreatine chief NPN substances

4. Nerve supply of muscle tissue
A. Afferent, or sensory fibers
  a. Somatic afferent: inform CNS about degree of contraction of skeletal muscles
  b. Visceral afferent: inform CNS about condition of cells in smooth and cardiac muscle
B. Efferent fibers
  a. Messages to skeletal muscles over somatic efferent fibers; motor unit is nerve cell, its fiber and all muscle fibers supplied by it; nerve injury results in paralysis
  b. Messages to smooth and cardiac muscle over visceral efferent fibers; most organs have double nerve supply; may contract independently of nerve supply

5. Physiologic properties of muscle tissue
A. Excitability, or irritability
  a. Capacity to respond to stimulus
  b. Resting membrane potential
  c. Action potential: depolarization; repolarization
  d. Propagation of action potentials; muscle and nerve impulses
  e. Rhythmicity
  f. Stimuli: nerve impulses; distention; temperature changes; composition of blood and interstitial fluid

B. Contractility
  a. Capacity to shorten
  b. Molecular changes: interaction between actin and myosin
  c. Energy sources: immediate source is ATP; phosphocreatine aids restoration of ATP; oxidative processes of citric acid cycle; anaerobic breakdown of carbohydrate; steady state; oxygen debt; repayment of oxygen debt; fatigue
  d. Heat production
  e. Calcium ions turn muscle on and off
  f. All-or-nothing response
  g. Skeletal muscle contracts more rapidly than smooth muscle
C. Muscle tone, or tonus
  a. State of steady, partial contraction present in muscle at all times
  b. In skeletal muscle: due to stretch reflex; dependent on nerve supply
  c. In smooth muscle: independent of nerve supply; widely fluctuating; influenced by chemical composition of environment
  d. In cardiac muscle: independent of nerve supply; tonus is synonymous with general physiologic condition
D. Extensibility and elasticity
  a. Capacity to stretch when force applied and to regain original form when force removed
  b. Prevents muscle rupture under stress; most marked in smooth muscle

6. *Mechanical phenomena of contraction*
A. Types of contraction
  a. Simple twitch: single contraction followed by relaxation
  b. Refractory period
  c. Summation
  d. Tetanic contraction: smooth; sustained; makes possible all normal body movements
  e. Treppe: the staircase effect of increasing strength of contractions
  f. Isometric contraction: increase in tonus; length unchanged
  g. Isotonic contraction: muscle shortens; tonus unchanged
B. Strength of contraction affected by
  a. General metabolic conditions
  b. Initial length of fibers
  c. Strength of stimulus
  d. Rate at which stimulus is applied
  e. Weight of the load to be moved
  f. Temperature of fluid around muscle fibers
  g. Size of muscle
  h. Fatigue
C. Pathologic contractions: spasm; fasciculation; fibrillation; rigor mortis

QUESTIONS FOR REVIEW

1. What substances tend to *decrease* in quantity during strenuous spurts of exercise?
2. What causes the feeling of weakness you experience when you make a "mad dash" to catch your bus?
3. What is muscle tone? What controls muscle tone in the skeletal muscles?
4. Distinguish between skeletal, smooth, and cardiac muscle in terms of location and function.
5. Briefly, in terms of membrane potentials, what occurs when a muscle is stimulated?
6. How are impulses transmitted at neuromuscular junctions?
7. What are the sources of energy for muscle contraction?
8. Why are curarelike drugs of value in surgery?
9. Why are calcium ions essential for the contractile process?
10. Describe some of the factors that may alter the strength of contraction of a muscle.

REFERENCES AND SUPPLEMENTAL READINGS

Guyton, A. C.: Textbook of Medical Physiology, ed. 4. Philadelphia, W. B. Saunders, 1971.

Ham, A. W.: Histology, ed. 6. Philadelphia, J. B. Lippincott, 1969.

Hoyle, G.: How is muscle turned on and off? Sci. Am., 222:845, April 1970.

Porter, K. R., and Franzini-Armstrong, C.: The sarcoplasmic reticulum. Sci. Am., 212:73, March 1965.

The Skeletal Muscles

"A hard task and the muscle to achieve it . . ."
The Teamster • Henry H. Bashford

Over 600 skeletal muscles are under our command as we move about at work and play. Varying in size, shape, power, and method of attachment, these muscular engines enable us to execute all manner of voluntary movements. They also make up the so-called red meat of the body, comprising about 36 per cent of the body weight in women and 42 per cent in men. The combined weight of all the skeletal muscles in the body is about three times as great as that of all the bones.

Since much of the body is composed of skeletal muscles, most of our body heat is produced by the activity of these contractile organs. However, this chapter deals primarily with the study of individual muscles and the way in which they move various parts of the body. Body temperature is discussed in Chapter 17.

GENERAL PRINCIPLES

Naming the skeletal muscles

Each skeletal muscle in the body has a name; many of these are now in English, others are still in Latin. Learning the names of some of the more important skeletal muscles will be easier if you understand that names have been assigned in a logical manner. The following list of ways in which muscles are named includes examples of each.

1. *Location:* intercostal muscles (between the ribs); brachii (region of the arm); tibialis posterior (behind the tibia)

2. *Direction of fibers:* rectus (straight); transversus (across); obliquus (in an oblique direction)

3. *Action:* adductor, flexor, and levator (lifter)

4. *Shape or size:* deltoid (from Greek letter *delta*); trapezius (like a trapezoid); maximus (largest); minimus (smallest); longus (long); brevis (short)

5. *Number of heads of origin:* biceps (2 heads); triceps (3 heads); quadriceps (4 heads)

6. *Points of attachment:* sternocleidomastoid (sternum, clavicle, mastoid process of temporal bone); sacrospinalis (sacrum and vertebrae of spinal column)

As you progress in your study of the muscles you will discover that some names are combinations of two of the above, such as adductor magnus, biceps brachii, and rectus abdominis.

Parts of a skeletal muscle

The end of the skeletal muscle that is the relatively more fixed point of attachment is called the **origin.** The end that is freely movable is called the **insertion.** When the muscle contracts, the insertion is pulled toward the origin. However, you will note later that the origin is fixed absolutely in only a small number of muscles, such as those of the face. In many other muscles, the origin and the insertion are *functionally* interchangeable, which means that these can *act from either end.* For example, muscles which pull the thigh up toward the trunk also may pull the trunk down toward the thigh, as in bending forward. That portion of the muscle between the origin and the insertion is called the *body,* or *belly,* of the muscle.

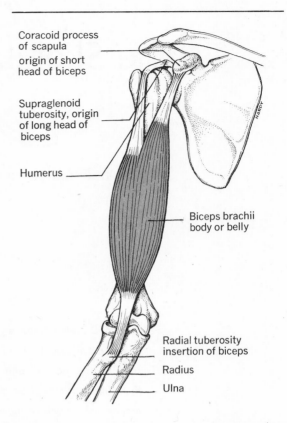

Coracoid process of scapula
origin of short head of biceps

Supraglenoid tuberosity, origin of long head of biceps

Humerus

Biceps brachii body or belly

Radial tuberosity insertion of biceps

Radius

Ulna

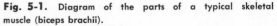

Fig. 5-1. Diagram of the parts of a typical skeletal muscle (biceps brachii).

The parts of a typical muscle (biceps brachii) are shown in Figure 5-1.

Attachments of skeletal muscles

Extensions of the connective tissue of muscle attach the muscles to bone, cartilage, skin, mucous membrane, or fasciae. The attachments to periosteum of bone or perichondrium of cartilage are broad. This type of attachment is found in the intercostal muscles and in some of the muscles attached to the shoulder and the hip girdles.

The connective tissue may be prolonged in the form of tendons, as shown in Figure 5-2. Tendons attach to bone, cartilage, or other structures. They vary greatly in appearance; many of them resemble glistening white cords, some are like flat ribbons, and others are broad sheets called *aponeuroses.* In each case the type of attachment is best suited to the particular work which that muscle performs. Tendons are extremely strong,

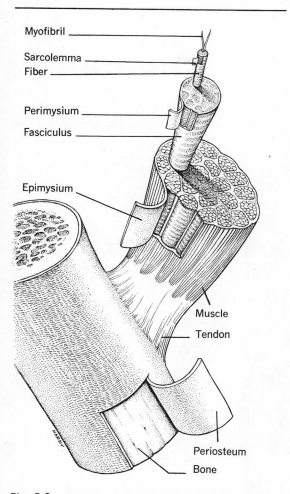

Myofibril

Sarcolemma

Fiber

Perimysium

Fasciculus

Epimysium

Muscle

Tendon

Periosteum

Bone

Fig. 5-2. Diagram of the connective tissue harness (tendon) that attaches skeletal muscle to bone.

and severe stress will usually result in a broken bone rather than a torn tendon.

Muscles are separated from one another and held in position by sheets of tough fibrous connective tissue (deep fascia). *Bursae,* the closed sacs that contain a small amount of fluid (see p. 94), reduce friction in areas where there are gliding movements between muscles and their underlying parts. Important bursae are found at the elbow, the shoulder, the hip, and the knee.

Skeletal muscle action

Skeletal muscles are responsible for the movements of parts of the body or the body as a whole and are used in both voluntary and reflex move-

ments. When muscles contract, they become shorter and thus exert a pull on their attachments. Most skeletal muscles are attached to bones (which function as levers) and move those bones at the joints (which serve as fulcrums). Muscles usually do not lie directly over the part that is to be moved; instead they are located above or below it, in front of or behind it, depending on the desired movement. For example, the muscle bodies which move the forearm are located in the region of the arm, their tendons crossing the elbow joint to insert on the bones of the forearm.

Many skeletal muscles are arranged in pairs; when one of a pair contracts, the other muscle of the pair relaxes. The first is called the **prime mover,** and the second the **antagonist.** The relaxation of the antagonist makes possible coordinated movement. A diagram of coordinated movement is illustrated in Figure 5-3. When the opposite movement is performed, the prime mover becomes the antagonist, and vice versa.

When you wish to perform a particular act, such as rubbing your eye, the act as a whole is thought of, and you do not analyze it into the individual muscles that you are going to use. Even our simplest acts require the coordinated use of many muscles in different functional capacities; this is essential to the precision of the act.

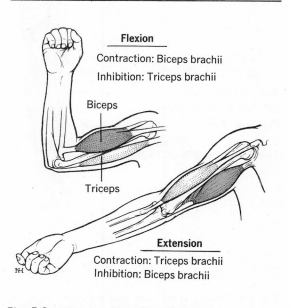

Flexion

Contraction: Biceps brachii
Inhibition: Triceps brachii

Biceps

Triceps

Extension

Contraction: Triceps brachii
Inhibition: Biceps brachii

Fig. 5-3. Diagram of coordinated movement.

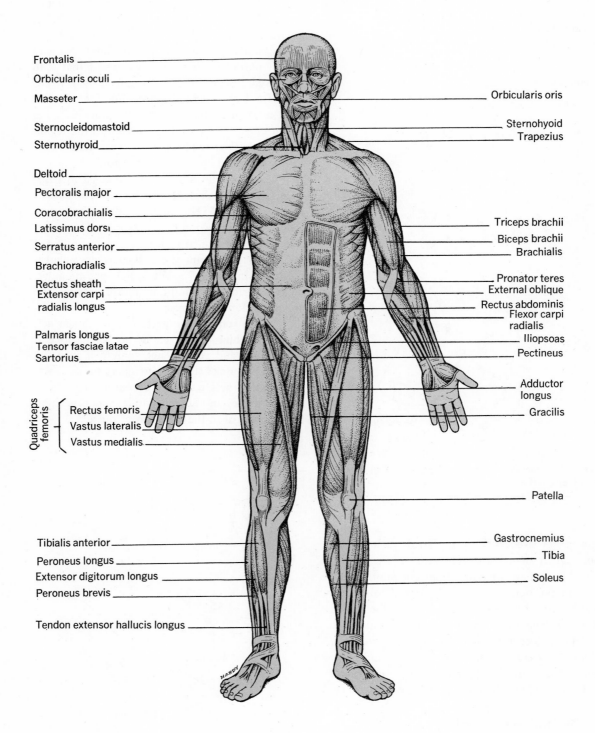

Frontalis

Orbicularis oculi

Masseter

Orbicularis oris

Sternocleidomastoid

Sternohyoid

Trapezius

Sternothyroid

Deltoid

Pectoralis major

Coracobrachialis

Latissimus dorsi

Triceps brachii

Biceps brachii

Brachialis

Serratus anterior

Brachioradialis

Rectus sheath

Extensor carpi
radialis longus

Pronator teres

External oblique

Rectus abdominis

Flexor carpi
radialis

Palmaris longus

Tensor fasciae latae

Sartorius

Iliopsoas

Pectineus

Adductor
longus

Gracilis

Quadriceps
femoris

Rectus femoris

Vastus lateralis

Vastus medialis

Patella

Tibialis anterior

Gastrocnemius

Peroneus longus

Tibia

Extensor digitorum longus

Soleus

Peroneus brevis

Tendon extensor hallucis longus

Fig. 5-4. Muscles of the body, anterior view.

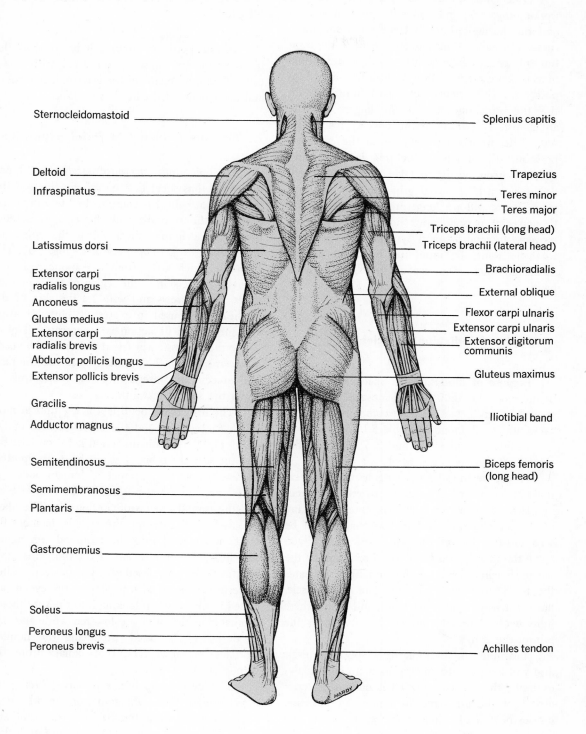

Sternocleidomastoid

Deltoid

Infraspinatus

Latissimus dorsi

Extensor carpi radialis longus

Anconeus

Gluteus medius

Extensor carpi radialis brevis

Abductor pollicis longus

Extensor pollicis brevis

Gracilis

Adductor magnus

Semitendinosus

Semimembranosus

Plantaris

Gastrocnemius

Soleus

Peroneus longus

Peroneus brevis

Splenius capitis

Trapezius

Teres minor

Teres major

Triceps brachii (long head)

Triceps brachii (lateral head)

Brachioradialis

External oblique

Flexor carpi ulnaris

Extensor carpi ulnaris

Extensor digitorum communis

Gluteus maximus

Iliotibial band

Biceps femoris (long head)

Achilles tendon

Fig. 5-5. Muscles of the body, posterior view.

Other muscles act as **synergists** (meaning *working together*) and enable the prime movers to perform the desired act efficiently and smoothly. In some instances, the synergists control the position of intermediate joints, so that the prime mover may exert its action on a distal joint. Suppose you wish to clench your hand; in order to do this the prime movers flex the fingers and the antagonists (extensors of the fingers) relax. The synergistic muscles fix the wrist in the extended position. If you have something in your hand which someone wants to take from you, he will flex your wrist, and your grip will be lost, since your fingers will no longer remain flexed. Even though that person knows nothing about prime movers, antagonists, and synergists, he can get what he wants by interfering with the coordinated movements that accompany clenching your hand.

Study of individual muscles

Because of limitations of space, only a few of the more than 600 skeletal muscles can be described in any detail. A more complete listing of these muscles is found in Table 5-1 at the end of this chapter. Figures 5-4 and 5-5 illustrate the muscles of the body as a whole.

As you begin to learn the names and the locations of prominent muscles, try to visualize them in relation to your own body. Movements of many of the muscles described in this chapter can be seen and felt through the skin, particularly if you make the muscles *work* for you. For example, the biceps brachii flexes the forearm and supinates the hand; to increase the work of this muscle, have a friend hold your hand and resist these movements. Note the muscle activity under the skin of the arm.

Learning the location of a specific muscle also will help you to analyze its main action. For example, the triceps brachii lies on the posterior aspect of the arm, and its tendon of insertion crosses *behind* the elbow joint on its way to the ulna. Since *muscles can only pull* (not push), the action of the triceps brachii would necessarily be extension of the forearm. Understanding this basic principle is essential for learning the action of any muscle.

MUSCLES OF THE HEAD AND THE NECK

Included in this group will be some of the more prominent muscles concerned with facial expression, movements of the eyeball, mastication, and movements of the head as a whole.

Movement related to facial expression

The **epicranius** (*occipitofrontalis*) is a broad musculofibrous sheet that covers the top and the sides of the skull; it is composed of two muscular portions, the frontalis and the occipitalis, connected by a fibrous sheet called the *epicranial aponeurosis*, or *galea aponeurotica* (Fig. 5-6). The *frontalis*, which arises from the aponeurosis and inserts into the skin in the region of the eyebrows, raises the eyebrows and wrinkles the skin of the forehead. It is used in expressing both attention and surprise; if its action is exaggerated it gives rise to the expression of fright or horror. The *occipitalis* arises from the occipital bone and the mastoid process and inserts into the aponeurosis; it draws the scalp backward.

The **orbicularis oculi** is a flat, elliptical muscle that sweeps around the circumference of the orbit (Fig. 5-6). Its main action is to constrict the opening into the orbit, shut out light, and protect the eye against the entrance of foreign bodies. It arises from the nasal portion of the frontal bone and the frontal process of the maxilla. The fibers extend laterally, occupy the eyelids, encircle the orbit, and extend into the temporal region and the cheek. When only part of the muscle is working, the eyelids are closed gently as in sleep. When more fibers are called into action, the eyelids are closed firmly, as when looking into the sun; tissues overhanging the orbit are lowered while the skin of the cheek is raised, producing the familiar wrinkles, or *crows feet*, that radiate from the corner of the eye.

The **levator palpebrae superioris** (*lifter of the upper eyelid*) is the antagonist of the orbicularis oculi. It arises from the back of the orbit and passes forward to be inserted into the tarsal plate (dense connective tissue in the upper lid) as shown in Figure 5-7. This muscle opens the eye by drawing the upper eyelid upward and backward.

The **orbicularis oris** consists of numerous layers of muscle fibers that surround the opening of the mouth and extend in different directions (Fig. 5-6). It is composed partly of fibers derived from other facial muscles, which are inserted into the lips, and partly of fibers distinctive of the lips. This muscle closes the lips and compresses them against the teeth; it also may cause protrusion or pouting of the lips.

The **buccinator** (Fig. 5-6) is the muscle coat of the cheeks. It arises from the maxilla and the mandible and extends forward to be inserted into the sides of the mouth. This muscle draws the corner of the mouth laterally, pulls the lips against the teeth, and flattens the cheek. It aids in mastication, swallowing, whistling, and blowing wind instruments. It prevents food from being pocketed between the teeth and the cheek. When the cheeks have been distended with air, the buc-

cinator muscles expel it between the lips, as in blowing the trumpet, from which the muscle derived its name.

The **zygomatic muscle** arises from the zygomatic bone and inserts into the orbicularis oris (Fig. 5-6). It raises the corner of the mouth and draws it laterally, as in smiling and laughing.

The **platysma** is a broad sheet of muscle that lies just under the skin in the anterolateral region of the neck (Fig. 5-6). Its fibers arise from fascia covering the pectoral and the deltoid muscles and extend obliquely upward across the clavicle to insert on the mandible and into the skin around the mouth. The platysma is a muscle of facial expression which acts particularly on the skin of the lower lip and the neck. It wrinkles the skin of the neck, depresses the corner of the mouth, and expresses melancholy, sadness, fright, and suffering. It also relieves the pressure on the veins that lie

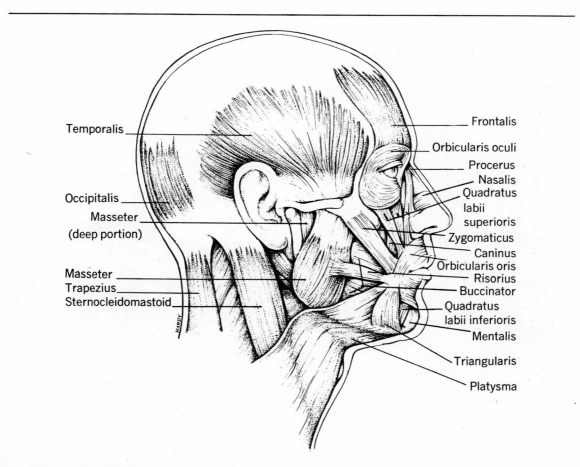

Fig. 5-6. Muscles of facial expression.

under it and thus aids the circulation of the blood.

Many other facial muscles help in the expression of various emotions and in emphasizing spoken words. If you watch the faces of your classmates, you will see how frequently their expressions change due to the varying activities of these muscles. Look in the mirror and assume various facial expressions, such as smiling, frowning, and determination; note which muscles are called into action.

Movement of the eyeball

There are six muscles that control the movement of the eyeball and make it possible for us to glance in all directions without moving the head (Fig. 5-7). The four rectus muscles arise from the margin of the optic foramen at the apex of the orbital cavity. Each muscle extends forward to in-

sert into the sclera, or outer coat, of the eyeball on four different sides as indicated by their names (i.e., superior, inferior, medial, and lateral recti).

The **superior rectus** directs the eyeball upward as in looking at the ceiling, and the **inferior rectus** turns the eyeball downward, as in looking at the floor. The **medial rectus** directs the eyeball inward, and, working with its fellow on the opposite side, enables one to gaze at near objects. The **lateral rectus** directs the eyeball outward; it normally works with the medial rectus of the other eye, as in glancing to one side.

There are two oblique muscles (superior and inferior) which work with a high degree of coordination as they tend to rotate the eyeball on its axis. The **superior oblique** arises at the apex of the orbit, extends forward, and passes through a ring of cartilage at the upper medial angle of the orbit. This determines the direction of its pull,

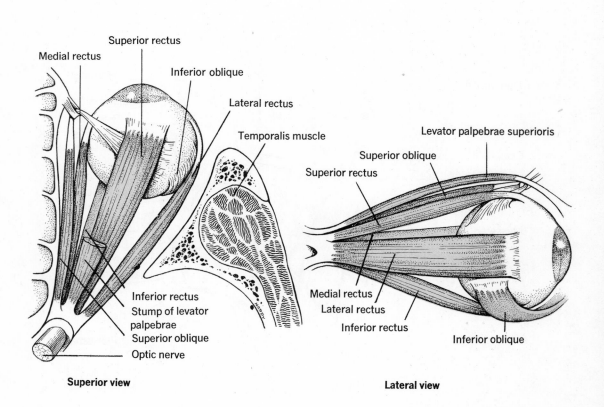

Superior view

Lateral view

Fig. 5-7. Extrinsic muscles of the eye: *left,* within the orbital cavity, viewed from above; *right,* lateral view.

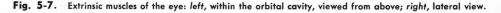

which is downward and laterally, with a medial rotation. The **inferior oblique** arises from the medial part of the orbital floor and curves around the under portion of the eyeball to insert on the lateral aspect of the sclera; it turns the eyeball upward and laterally, with a lateral rotation.

Strabismus refers to a deviation of one of the eyes from its proper direction; it may be due to a muscle defect or to a lack of coordination in the movement of the two eyes. For example, if one or both eyes are turned inward (*crosseyed*), the defect is referred to as *convergent strabismus*, and if one or both eyes are turned outward (*walleyed*), it is referred to as *divergent strabismus*.

Movement of the mandible and the tongue

Chewing and swallowing of food are made possible by the coordinated action of many muscles.

The **masseter** (Fig. 5-6) arises from the zygomatic arch and is inserted on the lateral surface of the ramus and the angle of the mandible. If you close your jaws tightly, you can feel this muscle while it is in action; it appears as a bulging mass near the angle of the jaw.

The **temporal** muscle (Fig. 5-6) is a fan-shaped muscle that arises from the temporal fossa, extends downward beneath the zygomatic arch, and inserts on the coronoid process of the mandible. If the jaws are closed firmly, the temporal muscle bulges out in the temple.

The **pterygoids** (Fig. 5-8) are deeply placed and extend from the under aspect of the skull to the medial surface of the ramus to the mandible. The *external pterygoid* muscle assists in opening the mouth. The side-to-side movements, such as occur in grinding the teeth, are produced by the pterygoid muscles of the two sides acting in alternation. If the external pterygoids of the two sides act together, they protrude the jaw.

The muscles of mastication enable one to close the mouth and clench the teeth. Normally, the mouth is closed; it is remarkable how the expression can change by merely opening the mouth. When the muscles are relaxed, the mandible drops, due to the pull of gravity.

Movements of the tongue are important in speaking, in keeping food between the teeth during mastication, and in the process of swallowing.

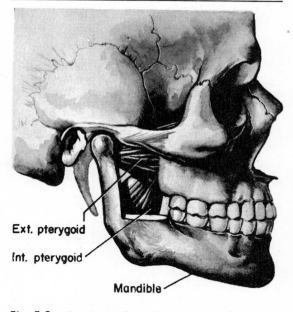

Ext. pterygoid

Int. pterygoid

Mandible

Fig. 5-8. The pterygoid muscles; a portion of the ramus of the mandible has been cut away.

The most characteristic feature of the tongue is its mobility. It is composed of a compact mass of interwoven muscle fibers covered with mucous membrane. There are two groups of muscles in the tongue, the intrinsic and the extrinsic.

The *intrinsic* muscles lie entirely within the tongue and are responsible for its mobility and changes in general shape. The intrinsic muscles are important in speaking, mastication, and swallowing.

The *extrinsic* muscles arise outside the tongue and insert into it as described below.

The **genioglossus** forms part of the body of the tongue. It arises from the mandible, behind the point of the chin, and spreads out in a fanlike manner to be inserted along the whole length of the undersurface of the tongue (Fig. 5-9). It performs various functions according to the part of it that contracts. The anterior fibers withdraw the tongue into the mouth and depress its tip; the middle fibers draw the base of the tongue forward and protrude it from the mouth. If the two genioglossus muscles act along their entire length, they draw the tongue downward. When this is done, the superior surface of the tongue becomes concave, like a channel. Fluids readily pass along this channel toward the pharynx during the act of sucking.

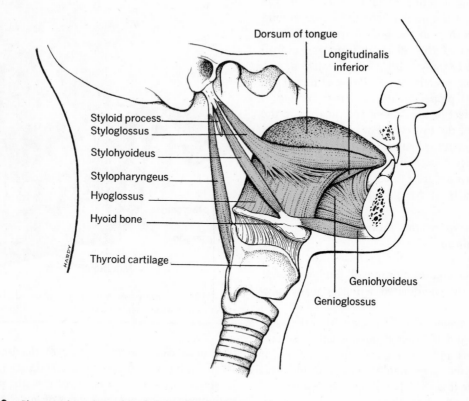

Fig. 5-9. Pharyngeal muscles and extrinsic muscles of tongue.

The **styloglossus** (Fig. 5-9) arises from the styloid process of the temporal bone and is inserted into the whole length of the side and the underpart of the tongue. The styloglossus muscles retract the tongue, elevate the margins, and raise the base of the tongue.

Movement of the head and the neck

Many muscles are involved in movement of the head as a whole. The sternocleidomastoid and the semispinalis capitis are typical examples.

The **sternocleidomastoid** (Figs. 5-10 and 5-12) is a strong, bandlike muscle that lies superficially across the side of the neck. It can be palpated in its entire length if you resist the pressure of one hand under your chin. It arises from the upper border of the sternum and the clavicle and extends obliquely upward and backward to be inserted on the mastoid process of the temporal bone. The sternocleidomastoid draws the head

toward the shoulder of the same side and rotates the head so as to carry the face toward the opposite side. When the head is abnormally fixed in this position one is said to have *torticollis*, or *wry neck*; this condition may be due to injury or disease of the sternocleidomastoid muscle, either alone or in association with a similar involvement of the trapezius.

If the two sternocleidomastoid muscles act together, they flex the cervical part of the vertebral column and bring the head forward. If the head is fixed, these muscles elevate the thorax and, in this way, aid in forced inspiration.

The **semispinalis capitis** (Figs. 5-11 and 5-13) arises from the last five cervical and upper five thoracic vertebrae; it extends upward to insert on the occipital bone. When these two muscles act together they extend the head, but if one acts alone, it bends the head to that side and rotates it so that the face is turned to the opposite side.

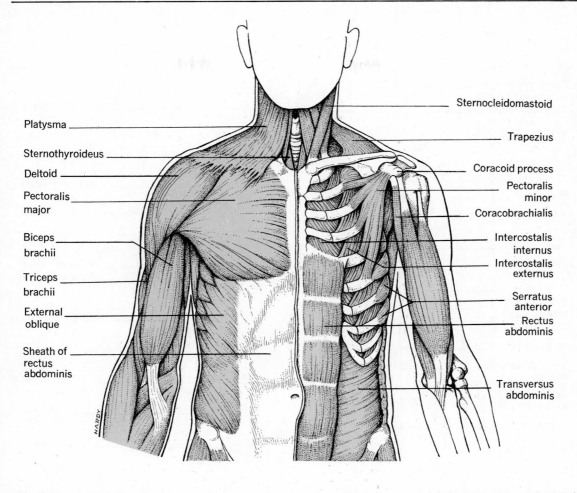

Platysma

Sternothyroideus

Deltoid

Pectoralis major

Biceps brachii

Triceps brachii

External oblique

Sheath of rectus abdominis

HARDY

Sternocleidomastoid

Trapezius

Coracoid process

Pectoralis minor

Coracobrachialis

Intercostalis internus

Intercostalis externus

Serratus anterior

Rectus abdominis

Transversus abdominis

Fig. 5-10. Muscles of neck, thorax, and arm, anterior view. Superficial muscles removed from left side so that deep muscles may be visualized better.

MUSCLES OF THE UPPER EXTREMITY

The muscles of the upper extremities are used in writing, eating, manipulating tools and machines, climbing, defending, attacking, and innumerable other activities. They even play a part in expressing one's emotions.

Movement of the shoulder and the arm

The shoulder joint has the widest range of movement of all the joints due to its large humeral head, shallow socket, and loose capsule. In addition, muscles which change the position of the scapula further increase this range of movement. The scapula can be moved forward, backward, upward, downward, and rotated as it follows the movements of the arm. The head of the humerus is kept in the glenoid cavity during these movements. The arm can be abducted, adducted, flexed, extended, and rotated at the shoulder joint.

Perform the following acts, and in each case try to determine the direction of the movements of the arms and the shoulders: fold your arms across your chest; make yourself round-shouldered; throw out your chest and throw your shoulders back; tie your shoestrings; scratch your back; place a book on your head and move your arms in various directions without dropping the book. If you have been a careful observer during these acts, you noted that one muscle takes part in several acts, and that different parts of one muscle may take part in opposite acts, according to variations in the relative fixation of the attachments.

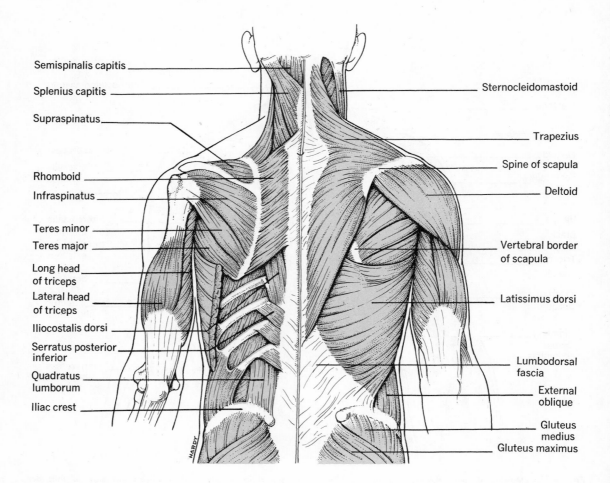

Semispinalis capitis

Splenius capitis

Supraspinatus

Rhomboid

Infraspinatus

Teres minor

Teres major

Long head
of triceps

Lateral head
of triceps

Iliocostalis dorsi

Serratus posterior
inferior

Quadratus
lumborum

Iliac crest

Sternocleidomastoid

Trapezius

Spine of scapula

Deltoid

Vertebral border
of scapula

Latissimus dorsi

Lumbodorsal
fascia

External
oblique

Gluteus
medius

Gluteus maximus

Fig. 5-11. Muscles of neck, thorax, and arm, posterior view. Superficial muscles removed from left side.

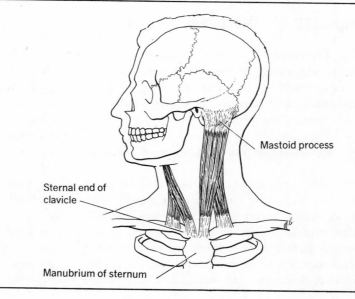

Fig. 5-12. Sternocleidomastoid muscles.

Mastoid process

Sternal end of
clavicle

Manubrium of sternum

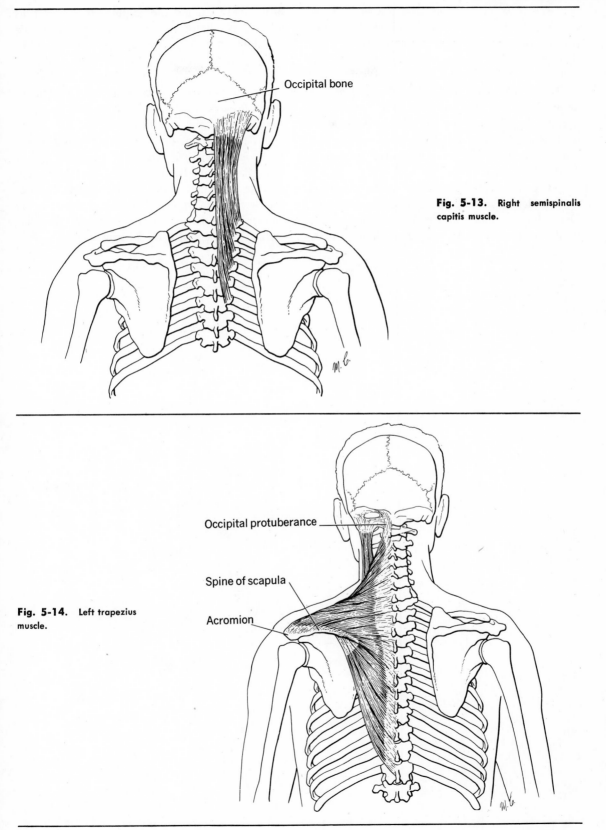

Occipital bone

Fig. 5-13. Right semispinalis capitis muscle.

Occipital protuberance

Spine of scapula

Acromion

Fig. 5-14. Left trapezius muscle.

The **trapezius** is a very large muscle that lies superficially at the back of the neck and the upper part of the thorax (Figs. 5-11 and 5-14). The right and the left trapezius muscles together form a trapezoid, which is the reason the muscle was so named. Each trapezius arises from, or has its origin from, the occipital bone, the ligamentum nuchae, the spines of the seventh cervical, and all of the thoracic vertebrae. Each is inserted on the clavicle, the acromion, and the spine of the scapula on its own side.

The degree of contraction of the trapezius muscles determines the position of the shoulders. Depending on which part of the muscle is more fixed, the trapezius can take part in many acts, such as (1) pulling the scapulae back and thus bracing the shoulders, (2) raising and shrugging the shoulders, (3) extending the head, and (4) turning the head from side to side. When all parts act together, they rotate the scapulae so that we can raise our arms as in reaching for objects on an overhead shelf. *Stiff neck* may be due to an inflammation of the trapezius.

The **latissimus dorsi** (meaning *widest of the back*) is shown in Figures 5-11 and 5-15. It is a large, flat, triangular muscle that covers the loin and the lower half of the thoracic region. It arises mainly from the spines of the lower thoracic, the lumbar, and the sacral vertebrae and the iliac crest. The fibers of each latissimus dorsi muscle

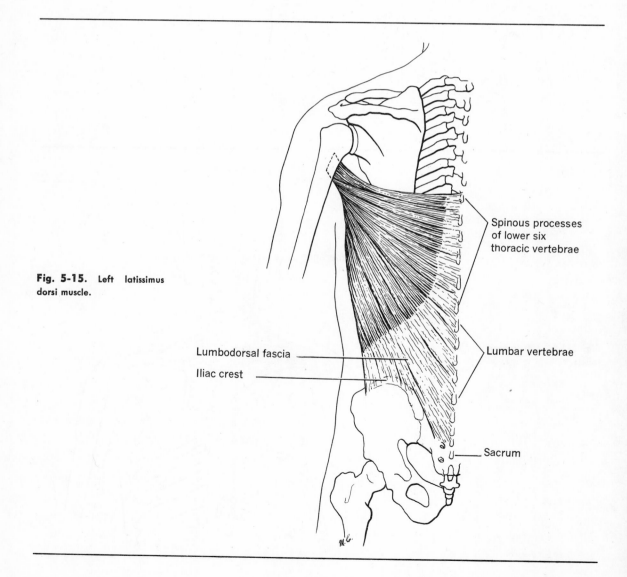

Fig. 5-15. Left latissimus dorsi muscle.

Spinous processes of lower six thoracic vertebrae

Lumbar vertebrae

Lumbodorsal fascia

Iliac crest

Sacrum

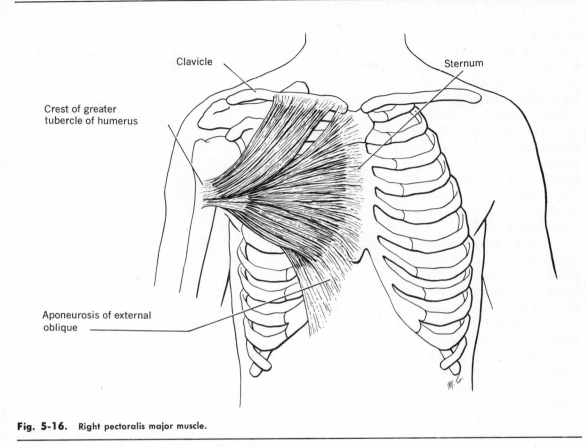

Clavicle

Sternum

Crest of greater
tubercle of humerus

Aponeurosis of external
oblique

Fig. 5-16. Right pectoralis major muscle.

converge to a narrow tendon and insert on the anterior surface of the upper part of the humerus on its side of the body.

With the trunk fixed, the latissimus dorsi powerfully extends the arm, drawing it downward and backward while simultaneously rotating it inward, as in the well-known swimming stroke. This muscle also adducts the arm behind the back. The latissimus dorsi is the chief muscle used in giving a downward blow with the upper extremity. When both arms are fixed, the latissimus dorsi muscles assist the abdominal and the pectoral muscles in pulling the trunk forward, as in climbing. These muscles also raise the lower ribs, and, in this way, can act as muscles of respiration when a person is breathing forcibly.

The **pectoralis major** (Figs. 5-10 and 5-16) is a large, thick, fan-shaped muscle on the upper, anterior part of the chest. The fibers of the two pectoralis major muscles form the anterior axillary folds in front of the armpits. Each muscle arises from the clavicle, the sternum, and the upper six costal cartilages and is inserted on the greater tubercle of the humerus of its side of the body.

The pectoralis major adducts, flexes, and rotates the humerus inward. It draws the shoulder girdle forward and depresses it. If the arm has been raised, the pectoralis major helps to bring it back to the side. These are the chief muscles of flight in birds.

The **pectoralis minor** (Fig. 5-10) is a thin, triangular muscle that lies under the pectoralis major. It pulls the scapula forward and depresses the tip of the shoulder.

Both pectoralis major and minor muscles are removed during *radical mastectomy*, an extensive operation for cancer of the breast. Postoperative exercises are very important for such persons, since muscles which formerly acted primarily as synergists now must assume the role of prime movers.

The **serratus anterior** is a large muscle that occupies the side of the chest and the medial wall of the axilla or armpit (Figs. 5-10 and 5-17). Its fibers arise like the teeth of a saw or in a jagged fashion, suggesting the name serratus. It arises from the upper eight or nine ribs along the front and the sides of the thorax and is inserted on the vertebral border of the scapula.

The serratus anterior muscles are the fixation muscles for the scapulae. They are important in pushing, during which motion they carry the scapulae forward. These muscles assist the trapezius muscles in supporting weights on the shoulders, and they aid in raising the arms above the horizontal level.

The **coracobrachialis** is a bandlike muscle that extends from the coracoid process of the scapula to the humerus (Fig. 5-18). It assists in flexion and adduction of the arm and helps to keep the head of the humerus in the glenoid cavity.

The **deltoid** is a thick, powerful, shield-shaped muscle that covers the shoulder joint and gives roundness to the upper part of the arm just below the shoulder (Figs. 5-10 and 5-19). It arises from the clavicle, the acromion, and the spine of the scapula. Its fibers unite to form a thick tendon that inserts on the deltoid tubercle on the lateral aspect of the humerus.

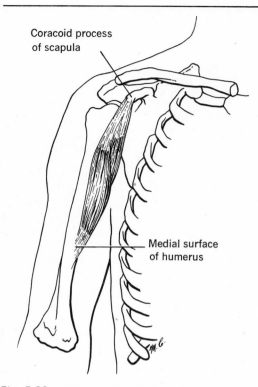

Coracoid process
of scapula

Medial surface
of humerus

Fig. 5-18. Right coracobrachialis muscle.

When the whole deltoid contracts, it abducts the arm and raises it laterally to the horizontal position. The anterior fibers aid in flexion of the arm, and the posterior fibers aid in extension of the arm.

The **supraspinatus,** the **infraspinatus,** and the **teres minor** (see Fig. 5-11) pass over the capsule of the shoulder joint, to which their tendons adhere, and insert on the greater tubercle of the humerus. They help to protect the shoulder joint, as well as to assist in lateral rotation of the arm. The supraspinatus, in addition, aids the deltoid in abducting the arm and helps to fix the head of the humerus in the glenoid cavity.

The **teres major** (Fig. 5-11) arises from the axillary border of the scapula and ascends to the upper, anterior surface of the humerus. This thick, flat muscle extends the arm and adducts it behind the back.

Movement of the forearm and the hand

The **biceps brachii** (Figs. 5-1 and 5-10) is a large, conspicuous muscle which small boys like

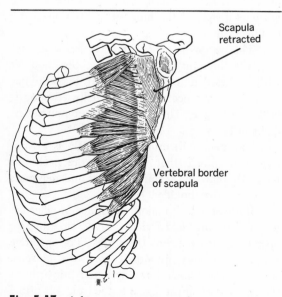

Scapula
retracted

Vertebral border
of scapula

Fig. 5-17. Left serratus anterior muscle.

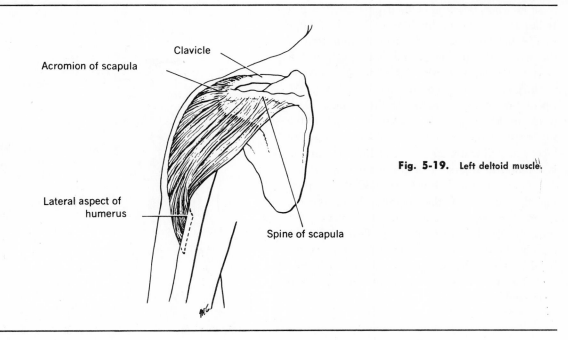

Acromion of scapula

Clavicle

Lateral aspect of humerus

Spine of scapula

Fig. 5-19.   Left deltoid muscle.

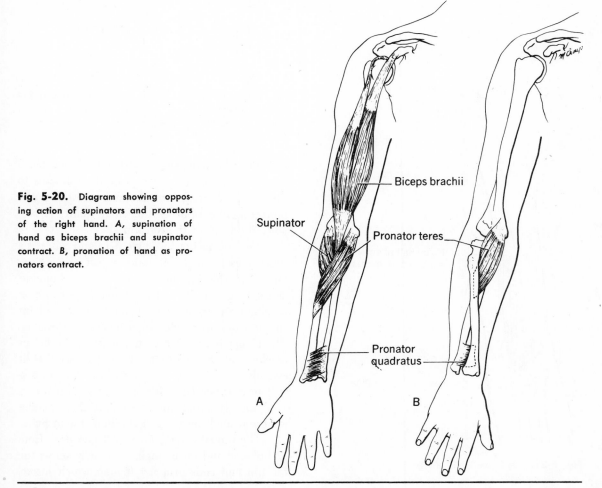

Fig. 5-20. Diagram showing opposing action of supinators and pronators of the right hand. *A,* supination of hand as biceps brachii and supinator contract. *B,* pronation of hand as pronators contract.

Biceps brachii

Supinator

Pronator teres

Pronator quadratus

A

B

to show when boasting of their physical strength. It lies anterior to the humerus and forms a large part of the substance of the arm. As indicated by the name, the biceps brachii has two heads of origin; the long head arises from the upper margin of the glenoid cavity, and the short head from the tip of the coracoid process of the scapula. Both heads insert by way of a single tendon on the tuberosity at the proximal end of the radius. The biceps brachii assists in flexion of the forearm and is a powerful supinator of the hand (Fig. 5-20). Action of this muscle can be studied best by placing the fingers of the opposite hand on the anterior surface of the arm and then flexing the forearm. While maintaining the flexed position, rapidly alternate supination and pronation of the hand and feel the activity of this muscle. Note that flexion is most powerful when the hand is supinated.

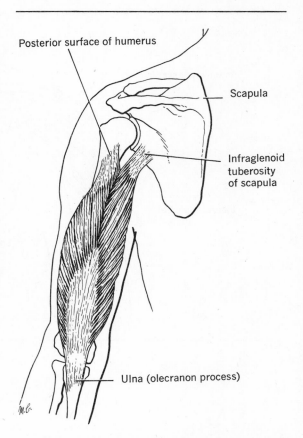

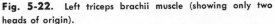

Fig. 5-22. Left triceps brachii muscle (showing only two heads of origin).

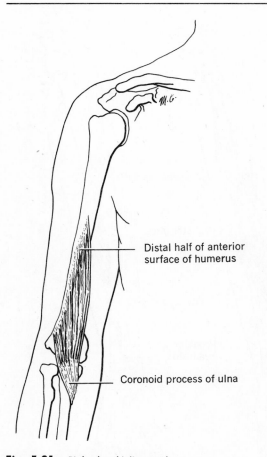

Fig. 5-21. Right brachialis muscle.

The **brachialis** (Figs. 5-10 and 5-21) extends from the distal, anterior half of the humerus to the coronoid process of the ulna, covering and protecting the front of the elbow joint. It is a powerful flexor of the forearm.

The **triceps brachii** (Figs. 5-11 and 5-22) is a single muscle on the posterior aspect of the arm, but it arises by three heads: two from the humerus and a long one from the scapula. The common tendon of insertion crosses behind the elbow joint and attaches to the olecranon process of the ulna. The triceps brachii is the antagonist of the biceps brachii and the brachialis, which means that it extends the forearm. This muscle is very important in pushing, since it converts the arm into a solid rod. It also helps to support the shoulder joint and hold the head of the humerus in place.

The **pronator teres** (Fig. 5-20) extends from its humeral and ulnar origins obliquely across the forearm and ends in a flat tendon which inserts on the body of the radius. This muscle forms the medial boundary of the antecubital fossa in front of the elbow joint. It pronates the hand.

Movement of the hand and the fingers

For a detailed listing of the muscles involved in movement of the hand and the fingers, see Table 5-1 at the end of this chapter. Contraction of muscles which originate in the forearm and pull across the wrist joint, together with those in the hand, makes possible an amazing variety of movements of the hand and the fingers. For purposes of general description, the muscles of the forearm may be divided into two subgroups: anteromedial and posterolateral.

The **anteromedial muscles** of the forearm, crossing in front of the wrist joint, are pronators and flexors of the hand and the fingers. The **posterolateral muscles,** crossing the posterior aspect of the wrist joint, are supinators and extensors of the hand and the fingers. Feel the movement of

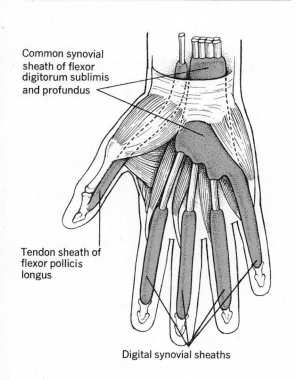

Common synovial sheath of flexor digitorum sublimis and profundus

Tendon sheath of flexor pollicis longus

Digital synovial sheaths

Fig. 5-23. Diagram of flexor tendon sheaths in the hand.

these muscles under the skin of your forearm while opening and closing the hand.

While flexing and extending the fingers, note the numerous tendons which are usually visible as well as palpable at the hand and the wrist. These glistening white cords are encased in slippery synovial tendon sheaths which reduce friction between the moving tendons and the stationary bones. Figure 5-23 shows the artificially distended sheaths of the flexor tendons of the right hand.

Injuries of the hand may easily sever the tendons of the muscles of the forearm. Repair after such lacerations must be exact in order for normal movements to be restored. When such repairs are attempted in outpatient clinics, sometimes wrong tendons have been sutured together. Generally it is realized that it is imperative to do such repairs of tendons of the wrist and the hand in the operating room where better facilities are available. Infection in the hand may travel along the tendon sheaths producing a *tenosynovitis*, or inflammation of a tendon and its sheath. Another complication, one that frequently follows fracture of the distal end of the radius, is stiffness of the wrist and the fingers due to adhesions (sticking together) of the extensor tendons and their sheaths to surrounding structures.

The **thenar eminence** (thumb side of the palm) is made up of muscles that arise from the carpus and the metacarpus and insert on the metacarpal and first phalanx of the thumb. A similar group about the little finger makes up the **hypothenar eminence.** These muscle groups can be seen in Figure 5-23, but they are not labeled. These muscles of the hand act on the first and fifth metacarpal bones and bring the thumb and the little finger into contact with each other. Observe all the intricate movements that you can perform with your thumb and fingers. The free movements of the thumb are of great importance in the human hand, and many are distinctly characteristic of man.

MUSCLES OF THE TRUNK

The muscles of the trunk will be considered in the following groups: (1) muscles of respiration, (2) muscles of the abdominal wall, (3) muscles of the pelvic floor, and (4) deep muscles of the back.

Muscles of respiration

The diaphragm and the external intercostals are the chief muscles of respiration. Other muscles of the thorax, the neck, and the abdomen play a relatively minor role in normal breathing, but they may become very important when breathing is labored or forced in any way (p. 372).

The **external intercostals** fill in the spaces between ribs (see Fig. 5-27). They arise from the lower portions of the first 11 ribs, and the fibers extend downward and forward to be inserted on the upper borders of the ribs below. These muscles play an important part in respiration by enlarging the thoracic cavity from side to side and from front to back as they elevate the ribs.

The **internal intercostals** (Fig. 5-27) lie under cover of the external intercostals and are similar to them, except in the direction of their fibers, which pass downward and backward. These muscles probably depress the ribs to some extent, and may decrease the size of the thorax.

The **diaphragm** is a dome-shaped muscular partition which completely separates the thoracic and the abdominal cavities (see Fig. 1-5). The heart and the lungs rest on the upper, convex surface of this muscle as it forms the floor of the thoracic cavity. The lower, concave surface forms the roof of the abdominal cavity (Fig. 5-28), lying directly over the liver on the right side and over the spleen and the stomach on the left side. The central portion of the diaphragm is composed of fibrous tissue and is called the *central tendon*. The peripheral muscular portion of the diaphragm arises from the xiphoid process of the sternum, the costal margins of the lower six ribs and the vertebral column. The fibers converge to be inserted into the central tendon. There are three large openings in the diaphragm; one transmits the vagus nerves and the esophagus; the second transmits the aorta, azygos vein, and the thoracic duct, and the third transmits the inferior vena cava.

When the muscular fibers of the diaphragm contract, the central portion is pulled downward, and thus the thoracic cavity above is enlarged from top to bottom. Because of this, the diaphragm is the principal muscle of inspiration. The level of the dome is changing constantly during respiratory movements. Also, its level varies with the amount of distention of the stomach and the intestines. When one is standing or sitting, the diaphragm descends to a lower level than when one is lying down, the abdominal organs tending to pull the diaphragm down in the first case and to push it up in the second. Therefore, patients who are in respiratory distress will breathe more easily if the head of the bed is elevated to support them in a sitting position.

Muscles of the abdominal wall

The anterolateral abdominal wall is composed of four pairs of flat muscles. Each muscle is enclosed by fascia which continues beyond the muscle to form aponeuroses. Fusion of right and left aponeuroses occurs in the midline from sternum to pubis and forms a sort of seam called the *linea alba* (meaning *white line*) (Fig. 5-24).

The **external oblique** is the strongest and most superficial muscle of the lateral abdominal wall. It is a broad, thin, muscular sheet, as shown in Figure 5-24. It is composed of muscular tissue on the lateral wall and an aponeurosis on the anterior wall of the abdomen. It arises from the sides of the lower eight ribs by slips that interlace with those of the serratus anterior. The fibers extend downward and forward to be inserted on the iliac crest, the anterior superior iliac spine, and the pubic tubercle. Between the iliac spine and the pubic tubercle, the lower border of the muscle has no bony attachment but folds back on itself, like the hem on a dress, to form a thick band called the *inguinal ligament* (see Figs. 5-26 and 5-28), an important anatomic landmark.

The large aponeurosis of the external oblique muscle forms part of the anterior abdominal wall and inserts into the linea alba. The **subcutaneous,** or **superficial inguinal ring,** is an opening in the aponeurosis just above and lateral to the pubic tubercle (Fig. 5-24); it is one of the weak places in the abdominal wall.

The posterior portions of the external oblique muscles are overlapped by the latissimus dorsi muscles.

The **internal oblique** muscle (Fig. 5-24) also is a broad, thin sheet of muscular tissue. It is located between the external oblique and the transversus abdominis muscles. It arises from the lumbar fascia, part of the iliac crest, and lateral two thirds of the inguinal ligament. The fibers extend

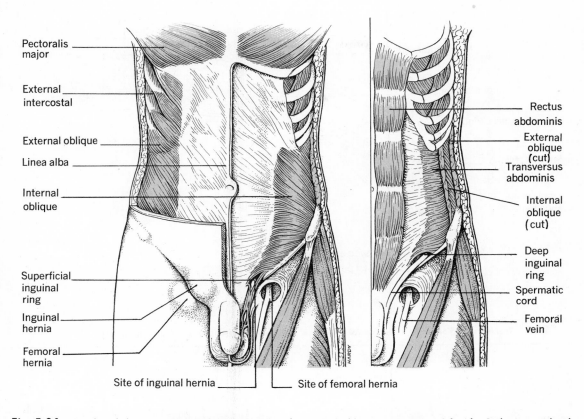

Labels, left figure: Pectoralis major; External intercostal; External oblique; Linea alba; Internal oblique; Superficial inguinal ring; Inguinal hernia; Femoral hernia

Labels, right figure: Rectus abdominis; External oblique (cut); Transversus abdominis; Internal oblique (cut); Deep inguinal ring; Spermatic cord; Femoral vein

Site of inguinal hernia   Site of femoral hernia

Fig. 5-24. Muscles of the anterolateral abdominal wall. *Left,* external oblique cut away on left side. *Right,* rectus sheath, external and internal oblique cut away.

upward and forward. Some are inserted on the cartilages of the seventh, the eighth, and the ninth ribs; the remainder of the fibers spread out like a fan and become an aponeurosis, which inserts into the linea alba. The lower part of the aponeurosis inserts on the pubic bone; the upper part splits to enclose the rectus abdominis before it reaches the linea alba.

The **transversus abdominis** (Fig. 5-24) is deeply placed in the lateral abdominal wall. It arises from the iliac crest, the inguinal ligament, the deep fascia of the back, and the lower ribs. The fibers pass directly forward around the abdominal wall to form what has been called a *living girdle.* They end in an aponeurosis which helps to form the sheath of the rectus abdominis before inserting into the xiphoid process, the linea alba and the pubic bone. There is an opening in the aponeurosis of the transversus abdominis half-

way between the anterior superior iliac spine and the pubis. This opening is called the **abdominal,** or **deep inguinal ring** (Fig. 5-26); it is another of the weak places in the abdominal wall. The fibrous sheet internal to the transversus abdominis muscle is called the transversus fascia, and internal to this fascia is the serous membrane known as the peritoneum (p. 385).

The external oblique, the internal oblique, and the transversus muscles, with their aponeuroses, form a triple wall for the abdomen (Fig. 5-25). The muscle fibers run in three directions and greatly strengthen the wall.

The **rectus abdominis** muscle (Fig. 5-24) is a long, straplike muscle that arises from the pubic bone and its ligaments. It widens as it extends upward to be inserted on the anterior surface of the xiphoid process and the costal cartilages of the fifth, the sixth, and the seventh ribs. It is en-

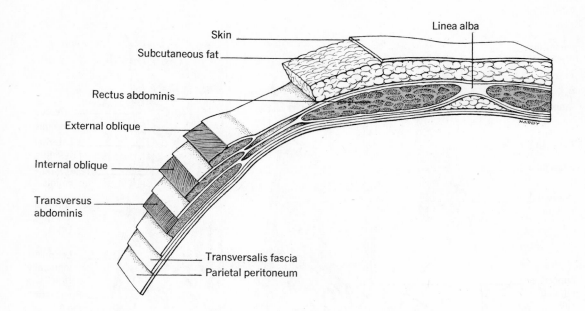

Fig. 5-25. Transverse section of the anterolateral abdominal wall showing arrangement of muscular layers.

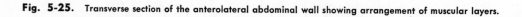

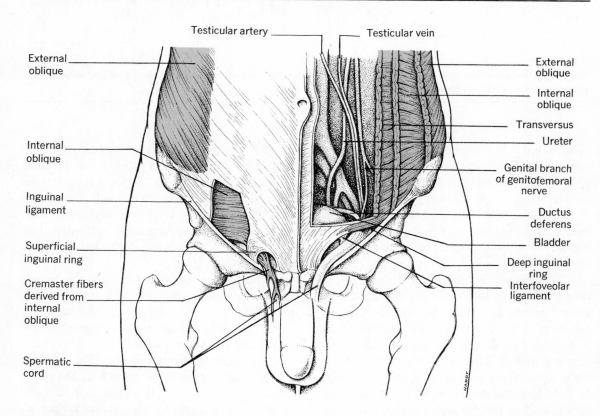

Fig. 5-26. Inguinal canal and contents. Greater part of abdominal wall removed on left side to expose these structures. Superficial muscles of abdomen shown on right side of the body.

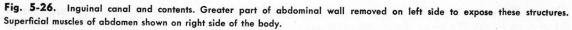

tirely enclosed in the sheath formed by the apo-neuroses of the muscles of the lateral wall of the abdomen.

Functions of the anterolateral abdominal wall muscles include: (1) protection of the underlying viscera, (2) providing aid in expelling substances from the body by compressing the abdominal cavity and increasing the intra-abdominal pressure (e.g., coughing, vomiting, defecation, and childbirth), and (3) assistance in bending the body forward or flexing the vertebral column. These muscles are antagonists of the diaphragm, relaxing as it contracts and contracting as it relaxes.

The **inguinal canal** is an oblique tunnel about 4 cm. long which passes through the lower abdominal wall parallel with, and a little above, the inguinal ligament. The inguinal canal extends from the abdominal inguinal ring to the subcutaneous inguinal ring, as shown in Figure 5-26. It transmits the spermatic cord (p. 497) in the male, and the round ligament of the uterus (p. 504) in the female.

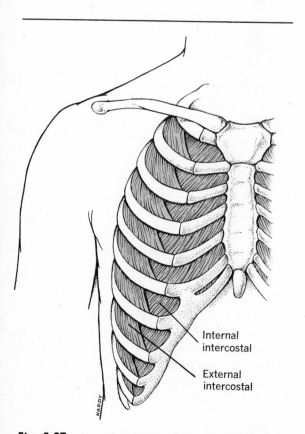

Fig. 5-27. Internal and external intercostal muscles.

Weak places in the abdominal wall include: (1) the subcutaneous and the abdominal inguinal rings, as described above; (2) the femoral ring, the oval upper end of the femoral canal, lying behind the inguinal ligament; and (3) the umbilicus, a puckered scar that marks the closure of the umbilical opening of the fetus.

Owing to the upright position of man, the pressure of the weight of the abdominal viscera falls on the ventral part of the abdominal wall. As one stands, any weak places in this portion of the wall will be put under added strain.

Hernia, or rupture, refers to the abnormal protrusion of an organ, or part of an organ through the wall of the cavity that normally contains it. This term is most frequently applied to the protrusion of an organ, usually the small intestine, through one of the weak places in the abdominal wall. If such herniation occurs at the umbilicus, it is referred to as an umbilical hernia; if in the area of the inguinal rings, an inguinal hernia, or in the femoral ring, a femoral hernia (Fig. 5-24).

The basic underlying cause of any hernia is a sudden or prolonged increase in the intra-abdominal pressure. Decreased muscle tone in the abdominal wall which may follow an illness or occur naturally in older persons can be a contributing factor. Activities which promote the occurrence of a hernia are those that increase the pressure of abdominal viscera against the ventral body wall, e.g., coughing, heavy lifting, or straining.

Muscles of the posterior abdominal wall assist in movements of the vertebral column and the thigh. The **quadratus lumborum** (Fig. 5-28) lies lateral to the vertebral column and extends from the last rib to the iliac crest; the kidneys are in close relationship to this muscle. In addition to its respiratory function of fixing and depressing the last rib and helping to anchor the diaphragm, the quadratus lumborum also aids in flexing the vertebral column. The **iliopsoas muscle** is described with those of the lower extremity (p. 142).

Muscles of the pelvic floor

The outlet of the true pelvis (p. 84) is closed by a shallow hammock of muscles and fascia which form the pelvic floor (Fig. 5-29). They are an extremely important group of muscles, for on

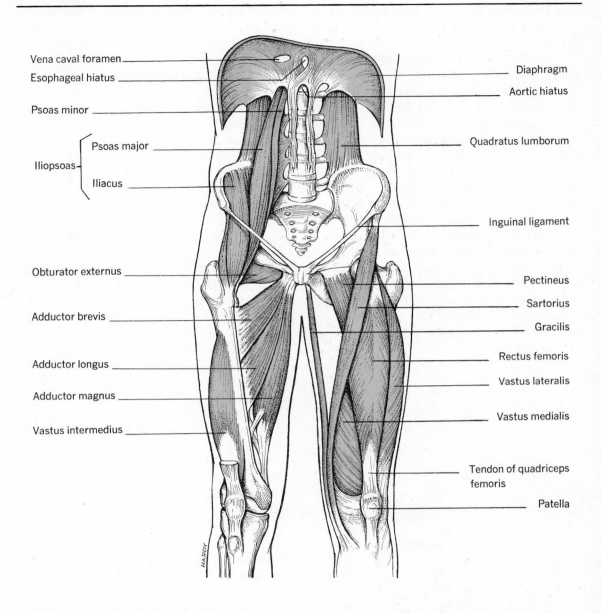

Vena caval foramen
Esophageal hiatus
Psoas minor
Iliopsoas
Psoas major
Iliacus
Obturator externus
Adductor brevis
Adductor longus
Adductor magnus
Vastus intermedius

Diaphragm
Aortic hiatus
Quadratus lumborum
Inguinal ligament
Pectineus
Sartorius
Gracilis
Rectus femoris
Vastus lateralis
Vastus medialis
Tendon of quadriceps femoris
Patella

HARDY

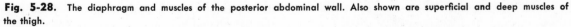

Fig. 5-28. The diaphragm and muscles of the posterior abdominal wall. Also shown are superficial and deep muscles of the thigh.

them rests part of the weight of the abdominal and the pelvic viscera.

The **levator ani** muscle arises from the pubis and the spine of the ischium; its fibers extend downward, backward, and toward the midline. The posterior fibers are inserted on the coccyx; the anterior fibers unite with those from the opposite side forming a seam in the midline. These muscles constrict the lower part of the rectum

and pull it forward, thus aiding in defecation.

The **coccygeus** muscle arises from the spine of the ischium and inserts on the sacrum and the coccyx. The coccygeus and the levator ani provide support for the pelvic viscera and resist increases in intra-abdominal pressure. These muscles are occasionally stretched and torn during childbirth.

The **external anal sphincter** is a funnel-shaped

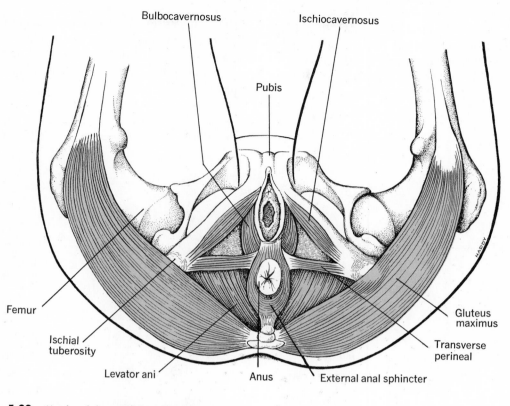

Fig. 5-29. Muscles of the pelvic floor (female perineum).

muscle which surrounds the anal canal. It is attached to the coccyx and to the skin surrounding the anus. This important sphincter muscle keeps the anal canal and the anus closed; like all skeletal muscle, it is under control of the conscious will.

Deep muscles of the back

The deep muscles of the back lie on either side of the vertebral column and form a complex group. They are attached to the sacrum, to the ilium, to the spines, the transverse and the articular processes, and the laminae of the lumbar, the thoracic, and the cervical vertebrae, and to the posterior aspects of the ribs and the base of the skull. They are the splenius capitis and cervicis, the sacrospinalis, the semispinalis, the multifidus, the rotatores, the interspinales, and the intertransversarii muscles. Only the sacrospinalis will be described.

The **sacropinalis** is the longest muscle in the body. It is an elongated muscular mass, consisting of separate slips, which extends from the sacrum to the skull (Fig. 5-11). It arises from the lower and posterior part of the sacrum, the posterior portion of the iliac crests, and the spines of the lumbar and the lower two thoracic vertebrae. It climbs the back in a series of columns and extends as high as the cervical vertebrae and the temporal bone. The lateral muscular column is called the *iliocostalis* muscle; and the medial is called the *longissimus* muscle. The sacrospinalis muscle is attached on the ribs and the vertebrae all the way up the back to the mastoid process of the temporal bone.

The sacrospinalis helps to maintain the vertebral column in the erect posture. At certain times these muscles bend the trunk backward. This occurs when it is necessary to counterbalance the effect of weight at the front of the body, such as in pregnancy. Since the vertebral column is drawn backward in pregnancy, the gait is peculiar and very characteristic.

If the sacrospinalis of one side acts alone, it bends the vertebral column to that side. You can easily observe the great flexibility of the vertebral column if you extend, flex, abduct, and adduct it. Both the cervical region and the lumbar region possess the greatest freedom of movement.

MUSCLES OF THE LOWER EXTREMITY

The muscles of the lower extremity are used chiefly to support the body in the standing position and are specialized for locomotion. Movements of the pelvic girdle as a whole are greatly restricted.

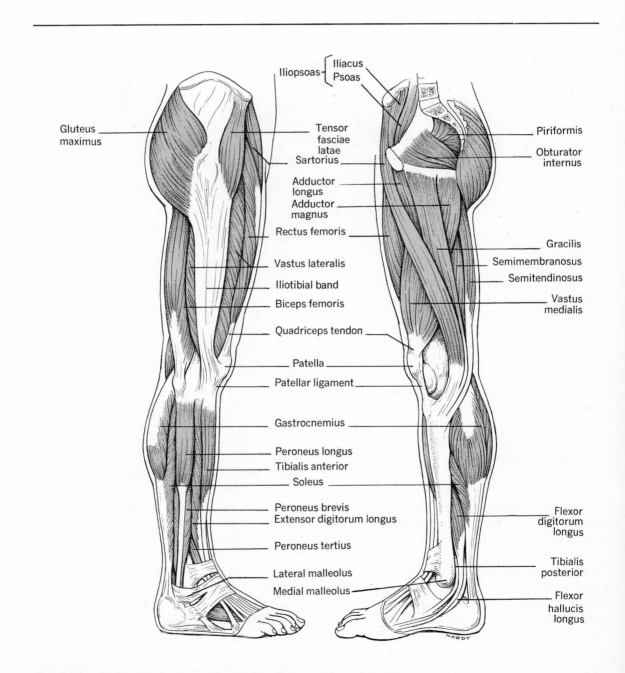

Fig. 5-30. Muscles of the lower extremity: *left*, lateral aspect; *right*, medial aspect.

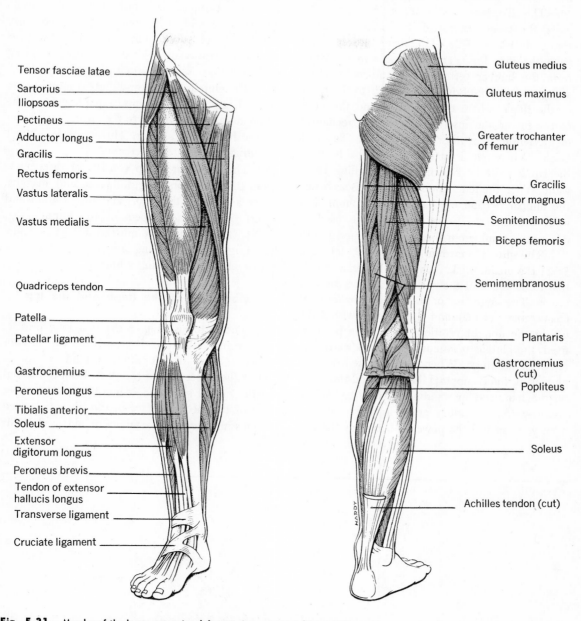

Tensor fasciae latae
Sartorius
Iliopsoas
Pectineus
Adductor longus
Gracilis
Rectus femoris
Vastus lateralis
Vastus medialis
Quadriceps tendon
Patella
Patellar ligament
Gastrocnemius
Peroneus longus
Tibialis anterior
Soleus
Extensor digitorum longus
Peroneus brevis
Tendon of extensor hallucis longus
Transverse ligament
Cruciate ligament

Gluteus medius
Gluteus maximus
Greater trochanter of femur
Gracilis
Adductor magnus
Semitendinosus
Biceps femoris
Semimembranosus
Plantaris
Gastrocnemius (cut)
Popliteus
Soleus
Achilles tendon (cut)

Fig. 5-31. Muscles of the lower extremity: *left*, anterior aspect; *right*, posterior aspect.

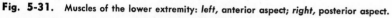

When one is sitting, the pelvis is flexed, and the weight of the body is transmitted through the sacrum and the hipbones to the ischial tuberosities. When one is standing, the pelvis is extended, and the weight of the body is transmitted to the femoral bones through the acetabula. When one is walking, the pelvis is rotated forward toward the limb that is being moved forward.

Movement of the thigh

The hip joint is a ball-and-socket joint that permits free movement in all directions, except as limited by muscles around it. Flexion and extension make possible such acts as walking and running. Abduction, adduction, circumduction, and rotation also occur and are important in balancing the body during all types of activities.

The **iliopsoas** is a compound muscle. Its parts, the *psoas major* and the *iliacus*, are illustrated in Figure 5-28. The iliac portion arises from the iliac fossa, and the psoas major arises from the lumbar vertebrae. The fibers combine and pass along the brim of the pelvis to the front of the thigh, where they wind around the neck of the femur and insert on the lesser trochanter of the femur. This muscle is a powerful flexor of thigh. When one is standing, it helps to hold the trunk erect by keeping it from falling backward. When the femur is fixed, the iliopsoas bends the lumbar portion of the vertebral column forward and tilts the pelvis forward. It is used in raising the trunk from the recumbent position.

The **gluteus maximus** (Figs. 5-30, *left* and 5-32) is a broad, thick, fleshy, quadrilateral muscle that forms much of the buttock. Its fibers are very coarse. The large size of the gluteus maximus is characteristic of man and is associated with walking in the upright position. When the pelvis is fixed, the gluteus maximus is a very powerful extensor of the thigh. When the femur is fixed, the gluteus maximus supports the pelvis and the trunk on the femur, as in standing on one leg. When you rise to a standing position, after stooping over, you can feel the powerful action of the glu-

teal muscles. The gluteus maximus helps to rotate the lower extremity outward. It is used in walking up stairs and in climbing hills, and you may notice soreness after undue climbing. This large muscle is occasionally selected as the site for intramuscular injections of various drugs.

The **gluteus medius** and **gluteus minimus** are located under the maximus, extending from the outer surface of the ilium to the greater trochanter of the femur (Fig. 5-31). These muscles abduct the thigh and rotate it medially.

The **adductor group** consists of three muscles: the adductors longus, brevis, and magnus. They arise from the pubis and the ischium and insert on the linea aspera of the femur (Fig. 5-28). These important muscles adduct and flex the thigh as in grasping a saddle between the knees when one is riding a horse.

Movement of the thigh and the leg

The **sartorius** (Fig. 5-28) is a long muscle, narrow and ribbonlike. It arises from the anterior superior spine of the ilium, extends obliquely across the upper part of the thigh from the lateral to the medial side, and is inserted on the upper part of the medial surface of the tibia. It is used in crossing the legs, as tailors of old used to sit

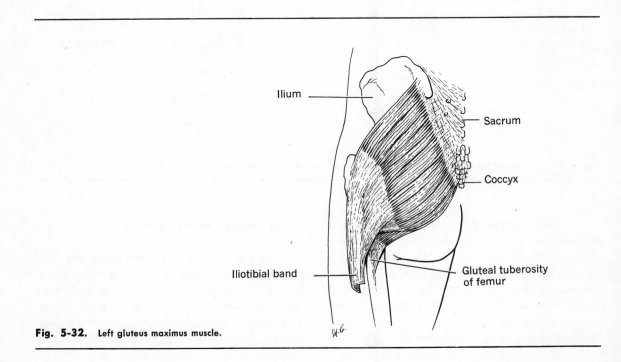

Fig. 5-32. Left gluteus maximus muscle.

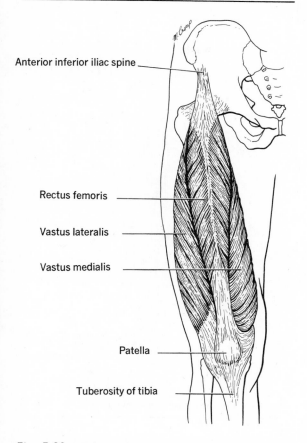

Anterior inferior iliac spine

Rectus femoris

Vastus lateralis

Vastus medialis

Patella

Tuberosity of tibia

Fig. 5-33. Right quadriceps femoris muscle, showing only three heads of origin (vastus intermedius is not visible).

the semitendinosus and the semimembranosus. These muscles extend from the ischial tuberosity to the proximal ends of the fibula and the tibia (Fig. 5-31). You can feel the tendons of the hamstring muscles as ridges behind the knee where they help to form the boundaries of the popliteal space (the diamond-shaped area back of the knee). The main action of the hamstrings is to flex the leg on the thigh. Extreme relaxation of these muscles is required for such exercise as touching the toes without bending the knees and for doing the high kick.

Movement of the leg and the foot

Since the muscles of the leg are responsible for the movements of the foot and the toes, it is important to have the movements clearly in mind before describing the muscles. In the foot, upward movement at the ankle joint is called *dorsal flexion* of the foot, and downward movement at the ankle joint is called *plantar flexion* (extension) of the foot. Upward movement of the toes is called *extension* and downward movement is called *flexion* of the toes. In other words, upward and downward movements at the ankle and the metatarsophalangeal joints are described by different terms. Turning the sole of the foot inward is called *inversion* and turning it outward is called *eversion.*

The **tibialis anterior** (Figs. 5-4 and 5-34) originates from the lateral condyle and the shaft of the tibia; its tendon of insertion crosses over the front of the tibia and attaches to the first metatarsal and the first cuneiform bones. This muscle dorsally flexes and inverts the foot. Injury to the nerve supply of the tibialis anterior (p. 183) results in a condition known as *foot drop* (inability to pull the foot up toward the shin). Such persons have difficulty in walking due to failure of the foot to clear the ground on the forward step; to compensate for this defect they walk with a characteristic high step and flop of the affected foot.

The **tibialis posterior** (Figs. 5-30 and 5-31) arises from the posterior surfaces of the tibia and the fibula; its tendon passes behind the medial malleolus to insert on several of the tarsal bones (navicular, calcaneus, cuneiforms, and cuboid) and on the base of the second, the third and the fourth metatarsal bones. It plantar flexes, extends,

cross-legged on the floor or the table (which explains the origin of the name). It flexes and rotates the thigh laterally, and also flexes the leg.

The **quadriceps femoris** is the large fleshy mass covering the front and the sides of the femur. It is composed of four parts: the rectus femoris, which arises from the ilium; and the vasti (lateralis, medialis, and intermedius), which arise from the femur, as shown in Figures 5-28 and 5-33. There is a common tendon of insertion for the four parts, and in this tendon, which is closely applied to the capsule of the knee joint, is the patella. The tendon is inserted on the tubercle of the tibia. The chief action is extension of the leg, but it also helps to keep the femur erect on the tibia; this means that the quadriceps is important in both walking and standing. The rectus femoris portion of the muscle flexes the thigh.

The **hamstring muscles** are the biceps femoris,

and inverts the foot. The tendon of the tibialis posterior is an important factor in the maintenance of the longitudinal arch of the foot (p. 87).

The **peroneus longus** (Figs. 5-30 and 5-35), on the lateral aspect of the leg, originates from the head and the lateral surface of the fibula; it ends in a long tendon which passes behind the lateral malleolus, crosses the sole of the foot obliquely, and inserts on the first cuneiform and the base of the first metatarsal bone. This muscle plantar flexes and everts the foot; it is an antagonist of the tibialis anterior. The peroneus longus helps to maintain the transverse arch of the foot (p. 87).

The **gastrocnemius** and the **soleus** are superficial muscles that form the powerful muscular mass referred to as the calf of the leg. The two heads of the gastrocnemius arise from the condyles of the femur (Figs. 5-30 and 5-36), while the

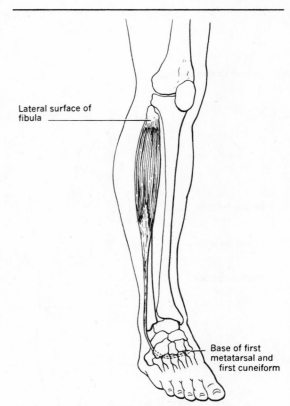

Fig. 5-35. Left peroneus longus muscle.

soleus arises from the upper ends of the tibia and the fibula (Fig. 5-31) underneath the gastrocnemius. Both muscles insert on the calcaneus by way of the *tendon of Achilles,* the thickest and strongest tendon in the body, and they are the chief plantar flexors of the foot at the ankle joint. These powerful muscles are used in standing, walking, and leaping. In walking, they raise the heel from the ground; the body is supported on that foot, and the opposite limb is free to be carried forward. These muscles enable one to stand on the tips of the toes; consequently, they are very well developed in toe dancers.

Movement of the toes

Deep in the sole of the foot, there are numerous muscles that act on the toes. Flexion, extension, abduction, and adduction occur, but these movements are quite limited in comparison with those of the fingers. For a listing of muscles in this group consult Table 5-1 at the end of the chapter.

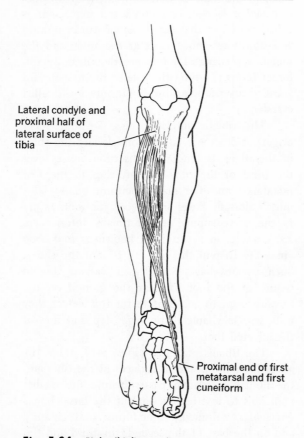

Fig. 5-34. Right tibialis anterior.

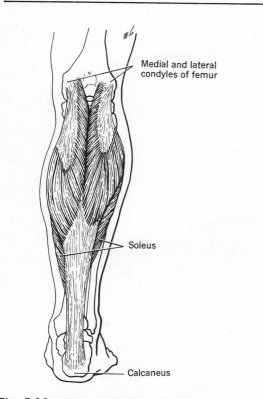

Fig. 5-36. Right gastrocnemius and soleus muscles.

The arches of the foot

The human foot is unique among all feet in the animal kingdom in that it is arched. Some of the advantages of this arched state are as follows: it distributes body weight over the foot, provides strength and rigidity for use of the foot as a lever and also a protected space for the soft structures of the sole (blood vessels and nerves), provides a resilient spring, in harmony with the muscles, to absorb jarring shocks.

The bones of the arch are supported by a posterior base (the calcaneus) and an anterior base (the metatarsal heads). On the medial aspect of the foot, the calcaneus extends upward and forward, carrying the talus above and in front of it. The talus forms the highest point of the arch and is the keystone. From the talus, moving downward and forward in order, are the navicular, the three cuneiforms, and the three medial metatarsals whose heads meet the ground (Fig. 5-37). Laterally, the arch is much lower. The calcaneus

rises to meet the cuboid which in turn is directed downward and forward to the lateral two metatarsals, their heads meeting the ground. This placement of bones between the calcaneus and the metatarsal bones is called the **longitudinal arch.** The foot also shows a **transverse** arch, formed by the metatarsal bones in front and the distal row of tarsals behind.

Support of the arches depends on various ligaments and muscles, the latter being of paramount importance. For example, the tibialis anterior braces the keystone of the arch and supports it from above; the tibialis posterior braces the tarsal bones and supports the arch from below; the peroneus longus crosses under the sole of the foot and forms a sling that helps to maintain the transverse arch.

The strength of the muscles supporting the arches can be increased by exercise. This is particularly important for those who must spend long hours on their feet. Sales clerks, teachers, policemen, and others who do a great deal of standing often find that the supporting muscles and ligaments of the arches are unable to meet the demands placed on them. When this occurs, the arch sags downward and may disappear entirely, resulting in *flatfoot* (Fig. 5-38). Figure 5-39 shows prints of the right foot with high, normal and low arches, and flatfoot.

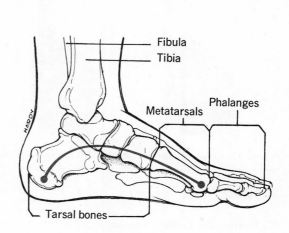

Fig. 5-37. Bones of the foot and ankle. Longitudinal arch indicated by curved line passing through tarsal bones.

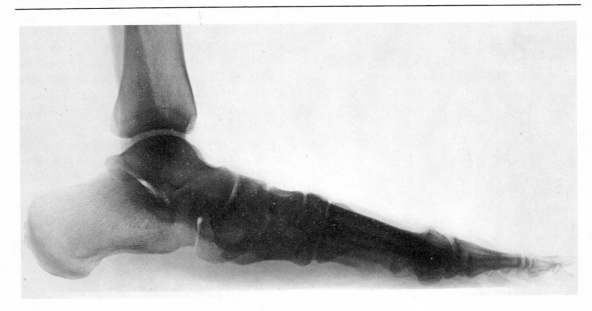

Fig. 5-38. X-ray of bones of left foot; flatfoot. (Courtesy of M. R. Marino, D.P.M.)

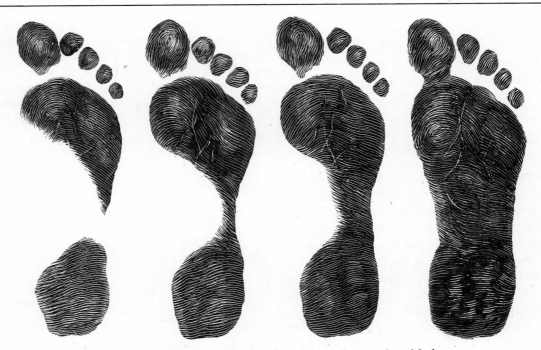

Fig. 5-39. Prints of right foot, showing (left to right) high arch, moderate arch, low arch, and flatfoot.

SUMMARY

1. Naming the skeletal muscles
A. Location: e.g., intercostals
B. Direction of fibers: e.g., transversus
C. Action: e.g., flexor
D. Shape or size: e.g., deltoid, maximus, brevis
E. Number of heads of origin: e.g., biceps

F. Points of attachment: e.g., sternocleidomastoid
2. Parts of a skeletal muscle
A. Origin: more fixed point of attachment
B. Insertion: more movable point of attachment
C. Body or belly: portion of muscle between origin and insertion

3. *Skeletal muscle action*

A. Muscles become shorter when they contract and exert pull on attachments
   a. Most muscles attached to bones which act as levers
   b. Bones moved at joints which serve as fulcrums
   c. Muscle bodies usually lie above or below, in front of or behind part to be moved
B. Many muscles arranged in pairs
   a. Prime mover contracts
   b. Antagonist relaxes
C. Synergists work with prime movers

4. *Study of individual muscles*

A. Consult Table 5-1 at end of chapter
B. Visualize muscles in relation to your own body
   a. See and feel them under the skin when possible
   b. Make them work for you by resisting their movements

QUESTIONS FOR REVIEW

1. Assume the various positions shown below, and in each case identify (A) the type of movement, and (B) the names of one or two muscles whose contraction makes possible that particular action.

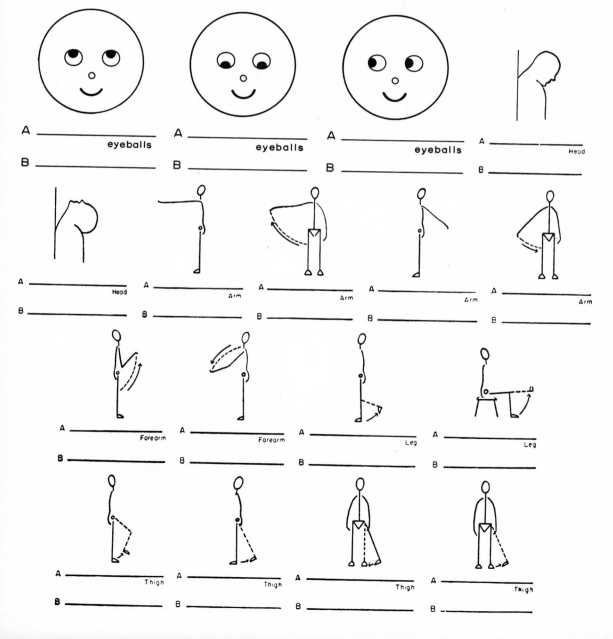

2. Distinguish between prime mover, synergist, and antagonist in skeletal muscle action.
3. In radical surgery for breast cancer, the pectoralis major and minor muscles are removed; what other muscles can assume the functions of the pectoral muscles?
4. Briefly, what are the functions of the anterolateral abdominal wall muscles?
5. What is a hernia? Why may coughing or straining promote herniation?
6. Describe the weak areas in the abdominal wall.
7. Which muscle is particularly characteristic of man and associated with walking in the upright position? What is the action of this muscle?

REFERENCES AND SUPPLEMENTAL READINGS

Anson, B. J. (ed.): Morris' Human Anatomy, ed. 12. New York, Blakiston Div. McGraw-Hill, 1966.
Goss, C. M. (ed.): Gray's Anatomy of the Human Body, ed. 28. Philadelphia, Lea & Febiger, 1966.

Table 5-1   The Skeletal Muscles

The name, origin, insertion, nerve supply, and action of certain additional muscles are included in the following table. Those muscles which have been described in the text are printed in **boldface** type.

Muscle	Origin	Insertion	Innervation and Action
Muscles of the Head and the Neck			
MOVEMENTS RELATED TO FACIAL EXPRESSION			
The Epicranius			
Occipitalis	Superior nuchal line of occipital bone	Epicranial aponeurosis	Facial nerve. Tenses epicranial aponeurosis and draws scalp backward.
Frontalis	Epicranial aponeurosis	Skin of eyebrow and of root of nose	Facial nerve. Elevates eyebrow; wrinkles forehead.
Levator palpebrae superioris	Margin of optic foramen	Upper eyelid	Oculomotor nerve. Raises upper lid.
Orbicularis oculi	Medial palpebral ligament, frontal bone, and maxilla	Lateral and medial palpebral ligaments; some fibers encircle eye	Facial nerve. Closes eyelids; stretches skin of forehead.
Corrugator	Frontal bone	Skin of eyebrow	Facial nerve. Draws skin of brow downward and medially, as in frowning.
Procerus	Cartilages of nose	Skin over root of nose	Facial nerve. Draws skin of forehead down.
Nasalis	Maxilla above incisor and canine teeth; dorsum of nose	Skin of nasolabial groove and margin of nostril	Facial nerve. Draws wings of nose laterally and upward; constricts nostrils.
Orbicularis oris	Various muscles running into lip	Fibers surround oral opening, forming a sphincter	Facial nerve. Draws lips together.
Quadratus labii superioris			
Angular head	Root of nose	Alar cartilage; upper lip	Facial nerve. The whole muscle raises the upper lip. The angular head also lifts the wings of the nose.
Infra-orbital head	Maxilla below orbit	Upper lip	

Table 5-1   (Continued)

Muscle	Origin	Insertion	Innervation and Action
Zygomatic head	Zygomatic bone	Upper lip	
Quadratus labii inferioris	Mandible below canine and premolar teeth	Lower lip	Facial nerve. Draws lower lip downward.
Incisivus labii (inferior and superior)	Maxilla and mandible, near canine and lateral incisor teeth	Orbicularis oris muscle	Facial nerve. Draws corners of lips medially.
Zygomatic	Zygomatic bone	Orbicularis oris muscle	Facial nerve. Raises corner of mouth and draws it laterally.
Canine	Canine fossa of maxilla	Orbicularis oris muscle	Facial nerve. Raises corner of mouth and draws it medially.
Risorius	Subcutaneous tissue over parotid gland	Skin and mucous membrane at corner of mouth	Facial nerve. Draws corner of mouth laterally.
Triangularis	Mandible below canine, premolar, and first molar teeth	Orbicularis oris muscle	Facial nerve. Draws corner of mouth downward.
Buccinator	Maxilla, mandible, and pterygomandibular raphe	Orbicularis oris muscle and skin of lips	Facial nerve. Draws corner of mouth laterally; pulls lips and cheek against teeth.
Mentalis	Mandible, below lower lateral incisor	Skin of chin	Facial nerve. Draws up skin of chin.
Platysma	Fascia of upper pectoral and deltoid muscles	Lower border of mandible; skin of lower part of cheek and corner of mouth	Facial nerve. Wrinkles skin of neck; depresses corner of mouth.
MOVEMENT OF THE EYEBALL			
Superior rectus	Margin of optic foramen	Eyeball, above pupil	Oculomotor nerve. Turns eyeball upward.
Medial rectus	Margin of optic foramen	Eyeball, medial to pupil	Oculomotor nerve. Turns eyeball medially.
Inferior rectus	Margin of optic foramen	Eyeball, below pupil	Oculomotor nerve. Turns eyeball downward.
Lateral rectus	Margin of optic foramen	Eyeball, lateral to pupil	Abducens nerve. Turns eyeball laterally.
Superior oblique	Margin of optic foramen	Lateral side of eyeball	Trochlear nerve. Turns eyeball downward and laterally; rotates medially.
Inferior oblique	Medial part of floor of orbit	Lateral side of eyeball	Oculomotor nerve. Turns eyeball upward and laterally; rotates laterally.
MOVEMENT OF THE MANDIBLE, THE TONGUE, AND THE PHARYNX			
Masseter	Zygomatic arch	Lateral surface of ramus of mandible	Masticator or motor root of trigeminal nerve. Raises mandible and draws it forward.
Temporal	Temporal fossa of temporal bone	Coronoid process of mandible	Masticator or motor root of trigeminal nerve. Raises mandible.

Table 5-1   (Continued)

Muscle	Origin	Insertion	Innervation and Action
External pterygoid	Lateral pterygoid plate, sphenoid, and palatine bones	Neck of condyle of mandible	Masticator or motor root of trigeminal nerve. Draws mandible forward and sideward; aids in opening mouth.
Internal pterygoid	Maxilla and palatine bone	Medial surface of ramus of mandible	Masticator or motor root of trigeminal nerve. Draws mandible upward and sideward.
Genioglossus	Spine of mandible near midline	Fascia of tongue; hyoid bone	Hypoglossal nerve. Anterior fibers retract tongue; remainder draws it forward and depresses its tip. Draws hyoid bone upward and forward.
Hyoglossus	Body and great cornu of hyoid bone	Fascia of tongue	Hypoglossal nerve. Depresses side of tongue and retracts it.
Styloglossus	Styloid process of temporal bone	Side of tongue	Hypoglossal nerve. Draws tongue backward.
Glossopalatinus	Aponeurosis of soft palate	Side and under surface of tongue	Pharyngeal plexus (vagus nerve). Draws side of tongue upward and soft palate downward. Constricts faucial isthmus.
Inferior constrictor of pharynx	Lateral surfaces of thyroid and cricoid cartilages	Dorsal part of pharynx	Pharyngeal plexus. Constricts pharynx; aids in swallowing.
Middle constrictor of pharynx	Greater and lesser cornua of hyoid bone; stylohyoid ligament	Dorsal part of pharynx; occipital bone	Pharyngeal plexus. Constricts pharynx; aids in swallowing.
Superior constrictor of pharynx	Pterygoid process, pterygomandibular raphe, and mylohyoid ridge of mandible	Dorsal part of pharynx; occipital bone	Pharyngeal plexus. Constricts pharynx; aids in swallowing.
Stylopharyngeus	Styloid process of temporal bone	Thyroid cartilage and lateral wall of pharynx	Glossopharyngeal nerve. Lifts pharynx in act of swallowing.
Pharyngo-palatine	Aponeurosis of soft palate and cartilage of auditory tube	Thyroid cartilage and lateral wall of pharynx	Pharyngeal plexus (vagus nerve). Closes opening between nasal and oral pharynx; depresses soft palate.
Levator veli palatini	Under surface of temporal bone and cartilage of auditory tube	Aponeurosis of soft palate	Pharyngeal plexus (vagus nerve). Raises soft palate; narrows pharyngeal opening of auditory tube.
MOVEMENT OF THE HEAD AND NECK			
Sternocleido-mastoid	Manubrium of sternum and medial third of clavicle	Mastoid process and occipital bone	Accessory, second and third cervical nerves. Bends head and neck toward shoulder; rotates head toward opposite side. Both sides acting together flex the neck and bring the head forward.
Semispinalis capitis	Last five cervical and upper five thoracic vertebrae	Occipital bone, between inferior and superior nuchal lines	Posterior rami of upper five cervical nerves. Extends head and bends it laterally.
Longus coli	Lower five cervical and first three thoracic vertebrae	Upper six cervical vertebrae	Second to sixth cervical nerves. Flexes neck.

Table 5-1 (Continued)

Muscle	Origin	Insertion	Innervation and Action
Longus capitis	Transverse processes of third to sixth cervical vertebrae	Basilar part of occipital bone	First four cervical nerves. Flexes head and rotates it toward same side.
Rectus capitis anterior	Lateral part of atlas	Basilar part of occipital bone	First cervical nerve. Flexes head and rotates it toward same side.
Rectus capitis lateralis	Transverse process of atlas	Lateral part of occipital bone	First cervical nerve. Bends head to side.
Rectus capitis posterior major	Spine of axis	Inferior nuchal line of occipital bone	Suboccipital nerve. Extends and rotates head.
Rectus capitis posterior minor	Posterior tubercle of atlas	Inferior nuchal line of occipital bone	Suboccipital nerve. Extends head.
Obliquus capitis inferior	Spine of axis	Transverse process of atlas	Suboccipital nerve. Rotates head.
Obliquus capitis superior	Transverse process of atlas	Inferior nuchal line of occipital bone	Suboccipital nerve. Extends head.

Muscles of the Upper Extremity

MOVEMENT OF THE SHOULDER AND THE ARM

Muscle	Origin	Insertion	Innervation and Action
Trapezius	Occipital bone, nuchal ligament, seventh cervical, and all thoracic vertebrae	Lateral third of clavicle, acromion, and lateral part of spine of scapula	Accessory, second, third, and fourth cervical nerves. Upper part raises shoulder, extends head, bends head toward shoulder, and turns head toward opposite side. Lower part draws scapula downward and inward, rotating the inferior angle laterally. The whole muscle rotates the scapula so that the lateral angle points upward.
Latissimus dorsi	Spines of last six thoracic and upper lumbar vertebrae, lumbodorsal fascia, and crest of ilium	Anterior surface of upper part of humerus	Thoracodorsal nerve. Adducts, extends, and rotates arm medially. Depresses shoulder.
Levator scapulae	Transverse processes of first four cervical vertebrae	Upper part of vertebral border of scapula	Third and fourth cervical nerves. Draws scapula upward.
Minor rhomboid	Nuchal ligament, last cervical, and first thoracic vertebrae	Vertebral border of scapula	Dorsal scapular nerve. Draw scapula upward and medially; rotate it so as to depress the tip of the shoulder.
Major rhomboid	Upper four thoracic vertebrae	Vertebral border of scapula	
Pectoralis major	Cartilages of second to sixth ribs, sternum, and medial half of clavicle	Humerus, on a line extending downward from the greater tubercle	Anterior thoracic nerves (medial and lateral). Adducts, flexes, and rotates arm medially.
Pectoralis minor	Second to fifth ribs, near their cartilages	Coracoid process of scapula	Medial anterior thoracic nerve. Pulls scapula forward.
Subclavius	First rib near its cartilage	Lower surface of clavicle	Fifth cervical nerve. Depresses scapula and shoulder.

Table 5-1 (Continued)

Muscle	Origin	Insertion	Innervation and Action
Serratus anterior	Lateral surfaces of upper eight or nine ribs	Vertebral border of scapula	Long thoracic nerve. Draws scapula forward and laterally; rotates it so as to raise the tip of the shoulder.
Coraco-brachialis	Carocoid process of scapula	Middle third of humerus	Musculocutaneous nerve. Adducts and flexes arm; aids in medial rotation of the arm.
Deltoid	Spine of scapula, acromion, and lateral part of clavicle	Deltoid tubercle of humerus	Axillary nerve. Abducts arm; anterior part flexes and rotates arm medially; posterior part extends and rotates arm laterally.
Supraspinatus	Supraspinous fossa of scapula	Greater tubercle of humerus	Suprascapular nerve. Abducts arm.
Infraspinatus	Infraspinous fossa of scapula	Greater tubercle of humerus	Suprascapular nerve. Rotates arm laterally.
Teres major	Axillary border of scapula	Anterior surface of upper part of humerus	Lower subscapular nerve. Adducts, extends, and rotates arm medially.
Teres minor	Axillary border of scapula	Greater tubercle of humerus	Axillary nerve. Rotates arm laterally
Subscapularis	Subscapular fossa of scapula	Lesser tubercle of humerus	Subscapular nerves. Rotates arm medially; holds humerus in glenoid cavity.
MOVEMENT OF THE FOREARM AND THE HAND			
Biceps brachii	Short head from coracoid process; long head from scapula above glenoid fossa	Tubercle on proximal end of radius	Musculocutaneous nerve. Flexes and supinates forearm; flexes and rotates arm medially; long head abducts, short head adducts the arm.
Brachialis	Ventral surface of distal half of humerus	Tubercle on proximal end of ulna	Musculocutaneous nerve. Flexes forearm.
Triceps brachii	Long head from infraglenoid tubercle of scapula. Lateral and medial heads from posterior surface of humerus	Olecranon of ulna	Radial nerve. Extends forearm. Long head also extends and adducts the arm.
Pronator teres	Medial epicondyle of humerus and coronoid process of ulna	Middle third of lateral surface of radius	Median nerve. Pronates and flexes forearm.
Supinator	Lateral epicondyle of humerus; proximal fifth of ulna	Proximal third of radius	Radial nerve. Supinates forearm.
Anconeus	Lateral epicondyle of humerus	Proximal part of ulna	Radial nerve. Extends forearm.
Brachioradialis	Ridge above lateral epicondyle of humerus	Proximal part of styloid process of radius	Radial nerve. Flexes forearm.
Pronator quadratus	Distal fourth of volar surface of ulna	Distal fourth of volar surface of radius	Median nerve. Pronates the forearm.

Table 5-1   (Continued)

Muscle	Origin	Insertion	Innervation and Action
MOVEMENT OF THE HAND AND THE FINGERS			
Flexor carpi radialis	Medial epicondyle of humerus	Base of second metacarpal	Median nerve. Flexes and abducts the hand.
Palmaris longus	Medial epicondyle of humerus	Fascia of palm	Median nerve. Flexes the hand.
Flexor carpi ulnaris	Medial epicondyle of humerus and proximal two thirds of ulna	Pisiform, hamate, and fifth metacarpal	Ulnar nerve. Flexes and adducts the hand.
Flexor digitorum sublimis	Medial epicondyle of humerus, coronoid process of ulna, and proximal part of radius	Volar surfaces of second phalanges of fingers	Median nerve. Flexes fingers and hand.
Flexor digitorum profundus	Proximal three fourths of ulna and interosseous membrane	Volar surfaces of third phalanges of fingers	Median and ulnar nerves. Flexes fingers and hand.
Flexor pollicis longus	Volar surface of radius and interosseous membrane	Terminal phalanx of thumb	Median nerve. Flexes thumb.
Extensor carpi radialis longus	Ridge above lateral epicondyle of humerus	Base of second metacarpal	Radial nerve. Extends and abducts hand.
Extensor carpi radialis brevis	Lateral epicondyle of humerus	Bases of second and third metacarpals	Radial nerve. Extends and abducts hand.
Extensor digitorum communis	Lateral epicondyle of humerus	By four tendons into phalanges of of fingers	Radial nerve. Extends wrist and fingers.
Extensor digiti quinti proprius	Septum separating it from extensor digitorum communis	Phalanges of little finger	Radial nerve. Extends little finger.
Extensor carpi ulnaris	Lateral epicondyle of humerus; proximal three fourths of ulna	Base of fifth metacarpal	Radial nerve. Extends and adducts hand and fifth metacarpal.
Abductor pollicis longus	Dorsal surface of ulna, radius, and interosseous membrane	Base of first metacarpal	Radial nerve. Abducts the thumb.
Extensor pollicis brevis	Middle part of dorsal surface of radius and interosseous membrane	Base of first phalanx of thumb	Radial nerve. Extends thumb and abducts first metacarpal.
Extensor pollicis longus	Middle third of dorsal surface of ulna and interosseous membrane	Base of second phalanx of thumb	Radial nerve. Extends second phalanx of thumb.

Table 5-1   (Continued)

Muscle	Origin	Insertion	Innervation and Action
Extensor indicis proprius	Distal third of dorsal surface of ulna and interosseous membrane	Aponeurosis on dorsal surface of index finger	Radial nerve. Extends index finger.

The Thenar Muscles

Muscle	Origin	Insertion	Innervation and Action
Abductor pollicis brevis	Transverse carpal ligament and lateral part of carpus	Base of first phalanx of thumb	Median nerve. Abducts thumb; flexes first and extends second phalanx.
Flexor pollicis brevis	Transverse carpal ligament; distal row of carpal bones	Base of first phalanx of thumb	Median nerve. Flexes and adducts the thumb.
Opponens pollicis	Transverse carpal ligament and greater multangular bone	Lateral border of metacarpal of thumb	Median nerve. Flexes, adducts, and rotates thumb medially.
Adductor pollicis	Second and third metacarpals, capitate, and ligaments	Base of first phalanx of thumb	Ulnar nerve. Adducts and flexes the thumb.

The Hypothenar Muscles

Muscle	Origin	Insertion	Innervation and Action
Palmaris brevis	Palmar aponeurosis	Skin of medial side of palm	Ulnar nerve. Draws skin of medial side of hand toward center.
Abductor digiti quinti	Pisiform bone and ligaments	First phalanx of little finger	Ulnar nerve. Abducts little finger; flexes first phalanx of little finger.
Flexor digiti quinti brevis	Transverse carpal ligament and hamate bone	First phalanx of little finger	Ulnar nerve. Flexes first phalanx of little finger.
Opponens digiti quinti	Transverse carpal ligament and hamate bone	Fifth metacarpal	Ulnar nerve. Flexes and adducts little finger.

The Intermediate Muscles

Muscle	Origin	Insertion	Innervation and Action
Lumbricales (4)	Tendons of flexor digitorum profundus	Tendons of extensor digitorum communis	Ulnar nerve, except the lateral two lumbricales, which are supplied by the median nerve. All flex first and extend second and third phalanges. Lumbricales draw fingers toward thumb. Volar interossei draw fingers toward middle finger, and dorsal interossei draw fingers away from middle finger.
Interossei volares (4)	First, second, fourth, and fifth metacarpals	First phalanx of thumb, index, ring, and little fingers	
Interossei dorsales (4)	Adjacent sides of metacarpal bones	First phalanx of middle three fingers, two muscles inserting on middle finger	

Muscles of the Trunk

THE MUSCLES OF RESPIRATION

Muscle	Origin	Insertion	Innervation and Action
External intercostal	Lower border of each rib	Upper border of next rib	Intercostal nerves. Elevate ribs and enlarge thorax.

Table 5-1 (Continued)

Muscle	Origin	Insertion	Innervation and Action
Internal intercostal	Lower border of each rib	Upper border of next rib	Intercostal nerves. Contract thorax (probably).
Diaphragm	Xiphoid process, lower six ribs and their cartilages, lumbar vertebrae	Central tendon of diaphragm	Phrenic nerve. Expands thorax; compresses contents of abdominal cavity.
Levatores costarum	Transverse processes of thoracic vertebrae	Next lower rib	Intercostal nerves. Bend vertebral column laterally; extend and rotate it.
Transversus thoracis	Dorsal surface of lower half of sternum	Cartilages of second to sixth ribs	Second to sixth intercostal nerves. Depresses ribs (in expiration).
Serratus posterior superior	Nuchal ligament and upper three thoracic vertebrae	Second to fifth ribs	Upper four intercostal nerves. Raises ribs; in this way enlarges thorax.
Serratus posterior inferior	Last three thoracic and first two lumbar vertebrae	Last four ribs	Ninth to eleventh intercostal nerves. Draws lower ribs outward; enlarges thorax.
Anterior scalene	Transverse processes of fourth to sixth cervical vertebrae	First rib	Fifth, sixth, and seventh cervical nerves. Raises first rib and bends neck to same side.
Middle scalene	Transverse processes of third to seventh cervical vertebrae	First and second ribs	Fourth to eighth cervical nerves. Raises first and second ribs; bends neck to same side.
Posterior scalene	Transverse processes of fifth and sixth cervical vertebrae	Second rib	Seventh or eighth cervical nerves. Action is the same as that of the scalenus medius. All scalene muscles aid in forced inspiration.
MUSCLES OF THE ABDOMINAL WALL			
External oblique	Lower eight ribs	Iliac crest, inguinal ligament, and linea alba	Lower seven intercostal nerves and iliohypogastric nerve. Compresses abdomen; flexes and rotates vertebral column.
Internal oblique	Inguinal ligament, iliac crest, and lumbodorsal fascia	Lower three ribs, linea alba, and pubic bone	Iliohypogastric, ilio-inguinal, and last three intercostal nerves. Compresses abdomen; flexes and rotates vertebral column.
Cremaster	Upper border of inguinal ligament	Pubic tubercle	Genitofemoral nerve. Lifts testis toward subcutaneous inguinal ring.
Transversus abdominis	Lower six ribs, lumbodorsal fascia, iliac crest, and inguinal ligament	Linea alba and pubic tubercle	Iliohypogastric, ilio-inguinal, genitofemoral, and last five intercostal nerves. Compresses abdomen.
Rectus abdominis	Symphysis pubis and body of pubis	Fifth to seventh costal cartilages; xiphoid process	Lower six intercostal nerves. Depresses thorax; flexes vertebral column and pelvis.
Pyramidalis	Body of pubis	Linea alba	Last thoracic nerve. Tenses linea alba.
Quadratus lumborum	Iliac crest, lower three lumbar vertebrae	Upper three lumbar vertebrae, twelfth rib	First three or four lumbar nerves. Flexes vertebral column laterally; extends vertebral column.

Table 5-1 (Continued)

Muscle	Origin	Insertion	Innervation and Action
MUSCLES OF THE PELVIC FLOOR AND THE PERINEUM			
Levator ani	Body of pubis, ischial spine, and obturator fascia	Coccyx and raphe joining coccyx to rectum	Fourth sacral nerve. Flexes coccyx; raises anus; resists downward pressure of abdominal viscera.
Coccygeus	Ischial spine	Fourth and fifth sacral vertebrae and coccyx	Third and fourth sacral nerves. Flexes and abducts coccyx.
External anal sphincter	Fibers of this muscle surround the anus and are attached to the skin and to the coccyx.		Inferior hemorrhoidal nerves. Keeps anus closed.
Transversus perinei profundus	Inferior ischial ramus	Fibers interdigitate in midline	Perineal nerve. Draws back central tendon of perineum.
Sphincter urogenitalis	Pubic ramus	Fibers interdigitate in midline, some encircle urethra	Perineal nerve. Compresses the uretha; compresses vagina or Cowper's glands.
Ischiocavernosus	Ischial tuberosity and inferior ischial ramus	Crus penis (in male) Crus clitoridis (in female)	Perineal nerve. Constricts crus penis or clitoridis.
Bulbocavernosus	Median raphe on ventral side of bulb; central tendon of perineum	Dense tissue covering root of penis	Perineal nerve. Compresses bulb of urethra.
Transversus perinei superficialis	Ischial tuberosity	Central tendon of perineum	Perineal nerve. Fixes central tendon of perineum.
DEEP MUSCLES OF THE BACK			
Splenius cervicis	Third to sixth thoracic vertebrae	Upper two or three cervical vertebrae	Posterior rami of second, third, and fourth cervical nerves. Bend and rotate head toward the side of the muscle which is acting.
Splenius capitis	Nuchal ligament and upper thoracic vertebrae	Mastoid process and occipital bone	
Sacrospinalis (iliocostalis, longissimus, spinalis)	Ilium, sacrum, lumbar, thoracic, and last four cervical vertebrae; posterior parts of ribs	Lumbar, thoracic, and cervical vertebrae; posterior parts of all ribs; mastoid process	Posterior rami of spinal nerves from first cervical to fifth lumbar. Bends head and vertebral column to side; extends head and vertebral column.
Semispinalis cervicis and dorsi	All thoracic vertebrae	Lower six cervical and upper five thoracic vertebrae	Posterior rami of third to sixth cervical and third to sixth thoracic nerves. Extend and rotate vertebral column.
Multifidus	Iliac crest, sacrum, and all vertebrae below fourth cervical	Spines of vertebrae up to second cervical	Posterior rami of spinal nerves. Extends and rotates the vertebral column.
Rotatores	Transverse processes of vertebrae	Next vertebrae above	Posterior rami of spinal nerves. Extend and rotate vertebral column.
Interspinales	Spines of vertebrae	Spines of next vertebrae above	Posterior rami of spinal nerves. Extend vertebral column.

Table 5-1 (Continued)

Muscle	Origin	Insertion	Innervation and Action
Intertransversarii	Transverse processes of vertebrae	Transverse processes of next vertebrae	Posterior rami of spinal nerves. Bend vertebral column to side.

Muscles of the Lower Extremity

MOVEMENT OF THE THIGH

Muscle	Origin	Insertion	Innervation and Action
Gluteus maximus	Iliac crest, lateral surface of ilium, sacrum, coccyx, and sacrotuberous ligament	Iliotibial band and gluteal tuberosity of femur	Inferior gluteal nerve. Extends thigh; rotates thigh laterally.
Gluteus medius	Lateral surface of ilium	Great trochanter of femur	Superior gluteal nerve. Abducts thigh; aids in rotation.
Gluteus minimus	Lateral surface of ilium; capsule of hip joint	Great trochanter of femur	Superior gluteal nerve. Abducts and rotates thigh medially.
Tensor fasciae latae	Anterior part of iliac crest	Iliotibial band	Superior gluteal nerve. Tenses fascia lata; flexes, abducts, and rotates thigh medially.
Piriformis	Ventral surface of sacrum and sacro-tuberous ligament	Great trochanter of femur	First or second sacral nerve, or both. Extends, abducts, and rotates thigh laterally.
Obturator internus	Pelvic surface of pubis, ischium, and obturator membrane	Trochanteric fossa of femur	Fourth lumbar to second sacral nerves. Rotates thigh laterally.
Quadratus femoris	Ischial tuberosity	Line below great trochanter of femur	Fourth lumbar to first sacral nerves. Rotates thigh laterally.
Gemellus superior	Ischial spine	With obturator internus	Nerve to obturator internus ⎫ Rotate thigh laterally.
Gemellus inferior	Ischial tuberosity	With obturator internus	Nerve to quadratus femoris ⎭
Psoas major	Bodies of twelfth thoracic to fifth lumbar vertebrae and invertebral disks	Small trochanter of femur	Second to fourth lumbar nerves. Flexes thigh; adducts and rotates it medially.
Iliacus	Iliac fossa of ilium	Small trochanter of femur	Femoral nerve. Flexes thigh; adducts and rotates it medially.
Iliopsoas	The iliacus and psoas major muscles are often considered together as the iliopsoas.		
Psoas minor	Twelfth thoracic and first lumbar vertebrae	Iliopectineal eminence of os coxae	First and second lumbar nerves. Flexes the pelvis.
Pectineus	Crest of pubis	Pectineal line of femur, below small trochanter	Femoral nerve. Flexes and adducts thigh.
Adductor longus	Superior ramus of pubis	Middle third of linea aspera of femur	Obturator nerve. Adducts, flexes, and rotates thigh laterally.
Adductor brevis	Inferior ramus of pubis	Upper third of linea aspera of femur	Obturator nerve. Adducts thigh.

Table 5-1   (Continued)

Muscle	Origin	Insertion	Innervation and Action
Adductor magnus	Rami of pubis and ischium, and ischial tuberosity	Linea aspera of femur; turbercle above medial condyle	Obturator and sciatic nerves. Adducts thigh; aids in flexion, extension, and lateral rotation.
Obturator externus	Lateral surface of pubis, ischium, and obturator membrane	Trochanteric fossa of femur	Obturator nerve. Rotates thigh laterally; aids in adduction.
MOVEMENT OF THE LEG AND THE THIGH			
Gracilis	Rami of pubis and ischium	Tibia, below medial condyle	Obturator nerve. Adducts and rotates thigh medially; flexes leg.
Sartorius	Anterior superior iliac spine	Medial surface of proximal end of tibia	Femoral nerve. Flexes thigh and leg; rotates thigh laterally.
Quadriceps femoris	Arises by four heads from ilium and femur	Tuberosity of tibia, through a common tendon	
Rectus femoris	Anterior inferior iliac spine	Common tendon	Femoral nerve. All parts extend leg. Rectus femoris flexes thigh.
Vastus medialis	Medial lip of linea aspera	Common tendon	
Vastus lateralis	Lateral lip of linea aspera	Common tendon	
Vastus intermedius	Anterior surface of femur	Common tendon	

The Hamstrings

Muscle	Origin	Insertion	Innervation and Action
Biceps femoris	Long head from ischial tuberosity; short head from middle third of linea aspera	Head of fibula; lateral condyle of tibia	Sciatic nerve. Flexes leg and rotates it laterally; long head also extends and adducts thigh.
Semitendinosus	Ischial tuberosity	Proximal part of medial surface of tibia	Sciatic nerve. Flexes leg and rotates it medially; extends and adducts thigh and rotates it medially.
Semimem-branosus	Ischial tuberosity	Medial condyle of tibia	Sciatic nerve. Flexes leg and rotates it medially; extends and adducts thigh and rotates it medially.
MOVEMENT OF THE LEG AND THE FOOT			
Gastrocnemius	Medial and lateral condyles of femur	Calcaneus, through tendon of Achilles	Tibial nerve. Flexes leg; plantar flexes, adducts, and inverts foot.
Soleus	Head and proximal third of fibula, and middle third of tibia	Calcaneus, through tendon of Achilles	Tibial nerve. Plantar flexes, adducts, and inverts foot.
Plantaris	Line above lateral condyle of femur	Through a slender tendon terminating in the fibrous tissue of the heel	Tibial nerve. Flexes leg and plantar flexes foot. Its action is very weak.

Table 5-1 (Continued)

Muscle	Origin	Insertion	Innervation and Action
Popliteus	Lateral condyle of femur	Proximal fourth of tibia	Tibial nerve. Flexes leg and rotates it medially.
Flexor digitorum longus	Dorsal surface of tibia and fascia covering tibialis posterior	Phalanges of lateral four toes	Tibial nerve. Flexes digits; plantar flexes and inverts foot.
Flexor hallucis longus	Distal two thirds of fibula and fascia covering tibialis posterior	Terminal phalanx of big toe	Tibial nerve. Flexes big toe; plantar flexes and inverts foot.
Tibialis posterior	Posterior surfaces of tibia, fibula, and interosseous membrane	Navicular, cuboid, and all cuneiform bones; second to fourth metatarsals	Tibial nerve. Adducts, plantar flexes, and inverts foot.
Tibialis anterior	Lateral condyle and proximal part of tibia and interosseous membrane	First cuneiform and base of first metatarsal	Deep peroneal nerve. Dorsal flexes and inverts foot.
Extensor digitorum longus	Lateral condyle of tibia, interosseous membrane, and shaft of fibula	Phalanges of four lateral toes	Deep peroneal nerve. Dorsal flexes foot; everts foot and extends toes.
Peroneus tertius	Distal third of fibula and interosseous membrane	Bases of fourth and fifth metarsal	Deep peroneal nerve. Dorsal flexes foot and everts it.
Extensor hallucis longus	Middle half of fibula; distal half of interosseous membrane	Second phalanx of big toe	Deep peroneal nerve. Extends big toe; dorsal flexes foot and inverts sole.
Peroneus longus	Lateral condyle of tibia, head and shaft of fibula, and intermuscular septum	First cuneiform and base of first metatarsal	Common peroneal nerve. Plantar flexes, abducts, and everts foot. Supports arch of foot.
Peroneus brevis	Middle third of fibula, and intermuscular septa	Tuberosity of fifth metatarsal	Superficial peroneal nerve. Everts foot; aids in extension. (Plantar flexion.)
MOVEMENT OF THE TOES			
Extensor digitorum brevis	Calcaneus and cruciate ligament	Phalanges of medial four toes	Deep peroneal nerve. Extends medial four toes.
Quadratus plantae	Calcaneus and ligaments	Tendon of flexor digitorum longus	Lateral plantar nerve. Aids flexor digitorum longus.
Flexor digitorum brevis	Calcaneus, plantar aponeurosis, and septa	Phalanges of lateral four toes	Medial plantar nerve. Flexes toes.
Abductor hallucis	Calcaneus and ligaments	First phalanx of big toe	Medial plantar nerve. Abducts big toe.
Flexor hallucis brevis	Cuneiform bones and ligaments	First phalanx of big toe	Medial plantar nerve. Flexes first phalanx of big toe.

Table 5-1   (Continued)

Muscle	Origin	Insertion	Innervation and Action
Adductor hallucis			
Oblique head	Cuboid, third cunei-form, second and third metatarsals, and ligaments	First phalanx of big toe	Lateral plantar nerve. Adduct big toe; aid in flexion of big toe. Transverse head holds heads of metatarsal bones together.
Transverse head	Capsules of metatar-sophalangeal joints	Sheath of tendon of flexor hallucis longus	
Abductor digiti quinti	Calcaneus and fas-cia of sole	First phalanx of little toe	Lateral plantar nerve. Abducts and flexes little toe.
Flexor digiti quinti brevis	Cuboid and fifth metatarsal	First phalanx of little toe	Lateral plantar nerve. Abducts and flexes little toe.
Opponens digiti quinti	Cuboid	Fifth metatarsal	Lateral plantar nerve. Draws little toe medially and plantarward.
Lumbricales (4)	Tendons of flexor digitorum longus	Phalanges of lat-eral four toes	Medial plantar nerve to most medial lumbrical. Others by lateral plantar nerve. These muscles flex the first phalanx. Dorsal interossei abduct from, plantar interossei adduct toward, second toe. Lumbricales extend second phalanges.
Interossei dorsales (4)	Metatarsal bones and ligaments	Phalanges of mid-dle three toes	
Interossei plantares (4)	Metatarsal bones and ligaments	Medial side of pha-langes of digits	

Organization of the Nervous System

6

"Sow a Thought, and you reap an Act;
Sow an Act, and you reap a Habit;
Sow a Habit, and you reap a Character;
Sow a Character, and you reap a Destiny."

of Unknown Authorship

At this very moment your body is being bombarded with stimuli of many kinds, and you are making constant adjustments (either consciously or unconsciously) to these environmental changes. For example, if the light rays reflected from this page are too bright, you may unknowingly begin to squint, that is, to partly close your eyes; as the chair seems to become harder, you may shift and squirm around to relieve the pressure on various parts of your anatomy. Microscopic sentries (nerve endings) in all parts of the body are constantly reporting on conditions in their area, constantly informing headquarters (the brain and the spinal cord) of any changes that may require some type of positive action on the part of your body.

The following sequence of events will serve to illustrate the comprehensive way in which our nervous system functions. While chewing a piece of candy, you become aware of a severe toothache. Since this is the fourth time that sweets have caused such a pain, you realize that the possibility of a cavity must no longer be ignored. A telephone call to the dentist results in an immediate appoint-

ment. As you hurry toward his office, worrying about the expenses of filling the tooth and the possibility of losing it, you step off the curb to cross the street, without your usual vigilance.

There is an immediate screech of brakes, and you are back on the sidewalk in a flash, trembling, perspiring, your knees shaking, and your heart pounding like a trip-hammer. You feel so weak that a few minutes elapse before you can continue on your way.

In terms of functions of the nervous system, the sweet saliva penetrated a crack in the enamel of a tooth and reached the nerve endings. This constituted a change in the environment of those sentries, and they swiftly notified the brain of that fact.

When the impulses reached the brain they produced a sensation of pain. The recognition of pain, together with the memory of the three previous similar experiences and your knowledge of the consequences of neglected cavities, led to the decision to attend to the situation without further delay. However, while crossing the street, you were so busy thinking about the consequences of the toothache that you were heedless of what you were doing. The screech of brakes was a startling change in the external environment, and impulses flashed to your brain which instantaneously cancelled the original order to cross the street. At the same time a new set of impulses started the emergency response which enabled you to jump back to the safety of the sidewalk before you were really aware of what had happened.

From this experience we might summarize some of the basic functions of the nervous system as follows:

1. The conversion of stimuli to nerve impulses which are then conducted to the central nervous system (CNS);

2. The interpretation of incoming impulses and the making of decisions about them on the basis of memory;

3. The sorting out of incoming impulses, unimportant ones being ignored, and important ones receiving immediate attention;

4. The dispatching of nerve impulses from the CNS to effectors (muscle tissue and glands) which carry out orders for appropriate response to a stimulus. In this case, your responses to the toothache consisted of telephoning the dentist and walking toward his office, which involved chiefly the skeletal muscles. However, your responses to the screech of brakes involved not only skeletal muscles, but also other effectors, as evidenced by the pounding of your heart and an increased flow of perspiration.

The final scene in this drama of response to a stimulus consists of a full awareness of what has happened. Impulses are transmitted from the active effectors back to the brain, informing it that assignments have been or are being carried out. You know that your skeletal muscles are quivering, that your heart is pounding, and that you have narrowly escaped serious injury or even death. While you continue to tremble, this experience is vividly stored in your memory, and it will be some time before you again step off the curb without giving it more thought.

GENERAL PLAN OF THE NERVOUS SYSTEM

Mechanisms of correlation

Just as a large city relies heavily on the telephone system to help in the correlation of its many activities, so does the human body rely on the nervous system. A complex network of peripheral nerve fibers, similar to telephone wires, connects all parts of the body with the CNS which serves as a switchboard. Nerve impulses travel over these living wires at a tremendous rate of speed and enable us to make rapid adjustments in both the external and the internal environment of the body.

The endocrine system, in contrast to the nervous system, resembles the slower moving postal service of a city, but it also plays an important role in the field of communication and correlation.

Chemical messengers (hormones) secreted by the endocrine glands are carried in the bloodstream to all parts of the body where they influence various activities. These specialized glands are described in Chapter 20.

Divisions of the nervous system

The nervous system is the most highly organized system of the body, and all of its parts function as an inseparable unit. However, for purposes of study we must first consider individual structures if we are to understand the unified whole.

Organs of this system may be classified according to their location or according to their function.

1. Anatomic divisions:
   a. The central nervous system (CNS) consists of the brain within the skull, and the spinal cord within the vertebral canal.

b. The peripheral nervous system consists of those nerves that connect the CNS with the body wall and with the viscera; these are the cranial and the spinal nerves.

2. Physiologic divisions:

a. The cerebrospinal, or voluntary, nervous system consists of those nerve fibers that connect the CNS with structures of the body wall (e.g., skeletal muscles and the skin); therefore, this division also is known as the *somatic* system.

b. The autonomic, or involuntary, nervous system consists of those fibers that connect the CNS with smooth muscle, cardiac muscle, and glands. It also is referred to as the *visceral* system, as it supplies the internal organs.

Regardless of the classification which may be used, one must never forget the basic unity of the nervous system. For example, one part of our nervous system coordinates the activities involved in external processes related to the seeking, preparing, and eating of food, while another part is concerned with the coordination of internal processes such as digestion, respiration, and circulation of blood, all of which are essential for the nourishment of individual cells. The ultimate goal of all parts of the nervous system is the same as that of every other system—the maintenance of homeostasis!

STRUCTURE OF NERVOUS TISSUE

The structural features of the nervous system include the nerve cells, or neurons, and a variety of supporting cells referred to as **neuroglia**. These glial cells are named after the Greek word meaning *glue* since some of them cluster around neurons within the CNS and offer structural support. Although neuroglia serves the same general function as connective tissue in other organs, one should be aware that neuroglia is extremely delicate, inasmuch as it does not have any tough intercellular substance such as collagen or elastin. Thus, when one examines fresh brain and spinal cord tissues, they are soft and jellylike.

Neuroglial cells also play a protective role within the CNS. Presumably, some of them (*astrocytes*) serve as part of the blood-brain barrier which protects the neurons from harmful substances in the bloodstream. Other glial cells (*microglia*) have phagocytic properties; they increase in number during inflammatory processes and can engulf bacteria and cellular debris when necessary. *Oligodendrocytes* are glial cells that are believed to be of importance in the formation of myelin sheaths around fibers within the CNS, a process that will be described on page 164.

The neuron

Neurons, or nerve cells, are microscopic structural units that are uniquely designed for the transmission of impulses to and from all parts of the body. The length of these cells varies from just a fraction of an inch to as much as 3 or 4 feet in some parts of the body. Although neurons vary in shape as well as size, they have certain basic similarities in that each consists of a cell body and elongated processes (Figs. 6-1 and 6-2).

The **cell body** is that part of a neuron which contains the nucleus. The nucleus, in turn, contains chromatin material and usually only one prominent nucleolus. Fine fibrils, referred to as *neurofibrils*, form a network in the cytoplasm of the cell body and extend into the cell processes. Scattered throughout the cytoplasm are *Nissl bodies*, clumps of flat membranous vesicles with many ribosomes between them. It is believed that they are concerned with the continuous formation of protein molecules needed to replace those that are used by the neuron in various metabolic activities. The synthesized protein that streams out into the axon is called *axoplasm*. Mitochondria and the Golgi apparatus also are present in the cytoplasm.

An intact, well-nourished cell body is essential to the life of the neuron as a whole. If the cell body dies, the remainder of the neuron also dies and can never be replaced, because these specialized cells cannot reproduce themselves. Therefore, this vital portion of each neuron is located in a relatively protected area; cell bodies are found only within the gray matter of the brain, in the spinal cord, and in ganglia (small nodules of nervous tissue lying close to the CNS).

The **cell processes** are elongated extensions of the cell body, and it is this anatomic feature that gives the neuron its unique appearance; no other cells in the body have such a drawn out shape.

These processes also are referred to as **nerve fibers.** *Dendrites* (meaning *treelike*) are those fibers that carry impulses toward the cell body, and they vary greatly in different types of neurons. For example, some dendrites are very long and branch only at the end, while others are short and branch freely. An *axon* is the fiber that carries impulses away from the cell body; it frequently has one or more side branches (collaterals) that come off at right angles. Each neuron has only one axon, and this fiber may be long or short depending on the location of its cell body.

Both dendrites and axons may develop *sheaths* which completely envelop them, or they may remain naked (i.e., consist only of a conducting core or axis cylinder). The sheaths which envelop certain fibers are of two types: myelin and neurolemma (also spelled neurilemma).

The **myelin sheath** is a white, lipid, and protein substance that surrounds many of the nerve fibers of the body. However, it is not a continuous sheath, as it is interrupted at intervals by constrictions called the *nodes of Ranvier.* Myelin is an excellent insulator that prevents almost all flow of sodium and potassium ions. Thus the uninsulated areas at the nodes of Ranvier play an important role in the conduction of nerve impulses, as will be described on page 170.

Cell processes supplied with myelin sheaths are referred to as *myelinated*, or white fibers, and those that lack such a sheath are called *unmyelinated*, or gray fibers (Fig. 6-2). Both myelinated and unmyelinated fibers are found in the peripheral nerve trunks. Within the CNS itself, bundles of myelinated fibers form the so-called *white matter* of the brain and the spinal cord. Conversely, the *gray matter* of the CNS consists of unmyelinated fibers and cell bodies of certain neurons.

The **neurolemma sheath** is a delicate sheet of protoplasm formed by special cells called Schwann cells that wrap themselves around peripheral nerve fibers; it ends at the point at which a fiber enters or leaves the CNS. The neurolemma of unmyelinated fibers is wrapped directly around the axis cylinder; on myelinated fibers it surrounds the myelin and comes into contact with the axis cylinder only at the nodes of Ranvier. The neurolemma sheath cells play an important part in the myelination of peripheral nerve fibers (it is believed that neuroglia cells perform a similar function with

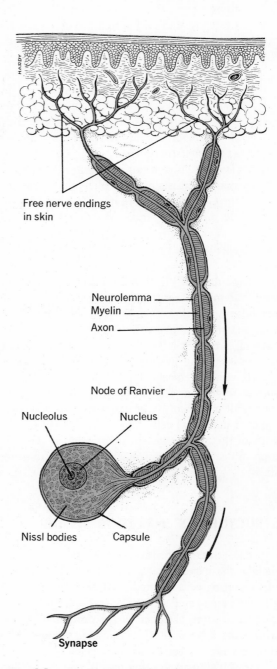

Fig. 6-1. Typical afferent (sensory) neuron.

Free nerve endings in skin

Neurolemma
Myelin
Axon

Node of Ranvier

Nucleolus   Nucleus

Nissl bodies   Capsule

Synapse

regard to the myelination of fibers within the CNS, in which neurolemma is absent).

The neurolemma sheath probably plays a protective role and aids in maintaining the integrity of normally functioning nerve fibers. However,

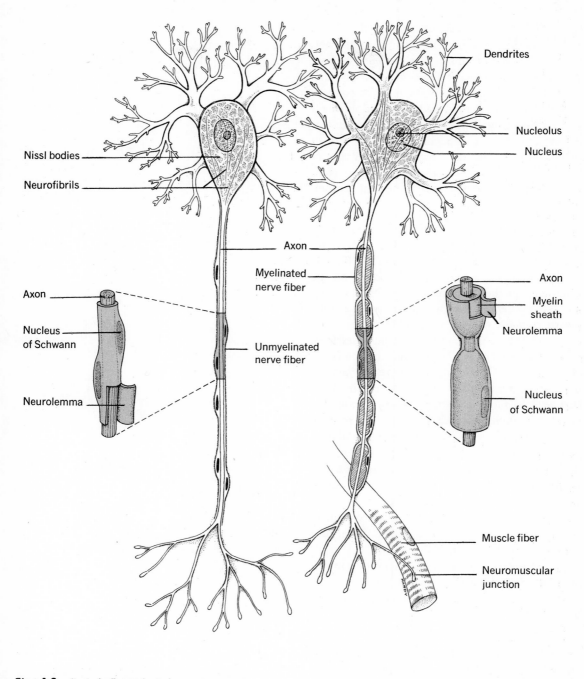

Fig. 6-2. Typical efferent (motor) neuron.

following injury to a fiber, the neurolemma assumes added importance because it is essential for regeneration of the fiber. Figure 6-7 illustrates in diagrammatic form the changes that occur in a nerve when it is cut and when it regenerates.

Briefly, if a nerve fiber is crushed or cut so that it is no longer connected to its cell body, the cytoplasm and the myelin in the distal stump degenerate and are engulfed by phagocytic cells. Meanwhile, if the cut ends are not too widely separated, neurolemma sheath cells grow into the cut and unite the stumps. The nerve fibers in the proximal stump that are still connected to their cell bodies then send buds or sprouts into the network of sheath cells; many of these buds become obstructed, but some of them do grow through the old area of injury and down into the cords of neurolemma sheath cells waiting to enfold them and complete the repair process. Regeneration is a slow process and, depending upon the length of the injured fibers, may require many weeks or months. Within the brain and the spinal cord, where there is no neurolemma, there is no regeneration of injured fibers.

Functional Classification of Neurons. 1. **Afferent** (*sensory*) neurons (Fig. 6-1) transmit impulses from the periphery to the CNS. Specialized *receptors* or sense organs (located on the distal ends of the dendrites) are excited by stimuli; this *excitatory state*, or *nerve impulse*, then sweeps over the neuron to the CNS.

2. **Efferent** neurons transmit impulses from the central nervous system to some effector (muscle or gland). There are four types of efferent neurons: *motor neurons* (Fig. 6-2) carry impulses that lead to the contraction of skeletal muscles; *secretory neurons* transmit impulses that provoke the secretion of glands; *accelerator* and *inhibitory neurons* supply smooth muscle and cardiac muscle with impulses that speed or slow activities respectively. The point of contact between nerve fiber and effector is called the *neuroeffector junction*. (NOTE: the terms neuromuscular and myoneural junctions are limited to muscle tissue; they are not applicable to glands).

3. **Central**, or *internuncial*, neurons transmit impulses from one part of the brain or the spinal cord to another. Consequently, their processes do not leave the CNS. These neurons also may be called *intermediate*, or *association neurons*.

Structural Classification of Neurons. 1. **Unipolar** neurons have only one process extending from the cell body, but this process divides into a central and a peripheral branch (Fig. 6-1); the central branch is the axon, and the peripheral branch functions as a dendrite.

2. **Bipolar** neurons have two processes, a single dendrite and a single axon (Fig. 6-3).

3. **Multipolar** neurons have several dendrites and one axon (Fig. 6-4).

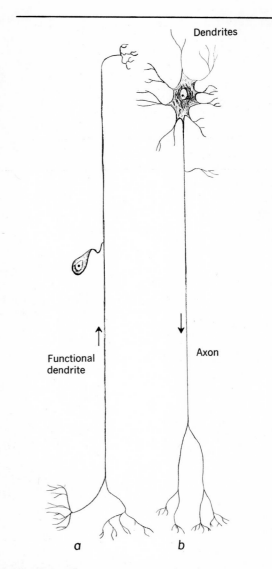

Fig. 6-3. Bipolar neurons: *A,* from olfactory mucous membrane; *B,* from retina.

Synapses

The entire nervous system is composed of chains of neurons, but there is no anatomic continuity between them. This means that neurofibrils do not extend from one neuron into the next. Each neuron is a separate unit, but it is in contact with other neurons. The points of contact are called *synapses*. The term synapse was derived from a Greek word that means *to clasp*, as one would clasp hands with a friend. Although the

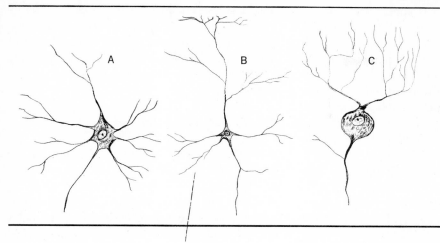

Fig. 6-4. Multipolar neurons: A, from spinal cord; B, from cerebral cortex; C, from cerebellar cortex.

hands would be in contact, there would be no continuity of the flesh.

Since impulses are transmitted away from cell bodies over axons, a synapse is necessarily a point of junction between the axon of one neuron and a dendrite, or the cell body, of another neuron; the tiny space that separates the presynaptic axon terminal from the membrane of the postsynaptic neuron is called the synaptic cleft. Although synaptic junctions show many structural variations, most axon terminals end in small knoblike expansions that contain mitochondria and many synaptic vesicles. The mitochondria provide the energy required for synthesis of chemical transmitter substances that are stored in the vesicles. The transmission of impulses across synapses is described on page 171.

A single axon may have extensive terminal arborizations and thus convey impulses to a number of neurons. Conversely, a single neuron may receive impulses from the axons of a great many other neurons—all of which points up the fantastic complexity of the nervous system (Fig. 6-5).

Nerves and tracts

Peripheral nerves, such as the sciatic, are composed of an infinite number of nerve fibers; these fibers are arranged in related bundles, much like the wires in a large telephone cable. Individual nerve fibers (axons and dendrites of various neurons) are bound to one another by delicate connective tissue called *endoneurium*. Groups of these individual fibers are bundled together and surrounded by stronger connective tissue which is called *perineurium*. When several nerve fiber bundles are grouped and bound together by *epineurium*, the resulting organ is called a **nerve** (Fig. 6-6).

Within the CNS, bundles of nerve fibers are referred to as **tracts** rather than as nerves. How-

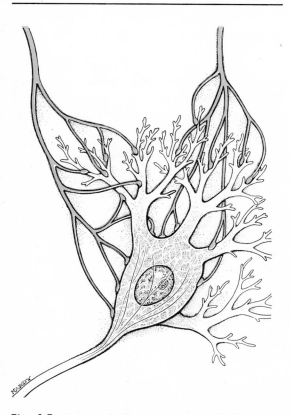

Fig. 6-5. Synaptic knobs on dendrites and cell body of a neuron.

ever, both nerves and tracts appear white to the naked eye, since most of their individual fibers are myelinated.

With respect to function, peripheral nerves are divided into four types:

Somatic afferent nerves carry impulses from the skin, the skeletal muscles, tendons, and joints to the CNS.

Somatic efferent nerves carry impulses from the CNS to the skeletal muscles.

Visceral afferent fibers carry impulses from the viscera to the CNS.

Visceral efferent fibers carry impulses from the CNS to smooth muscle, cardiac muscle, and glands.

Neuritis is a term which refers to either inflammatory or degenerative lesions of a nerve. It may result in pain, hypersensitivity, loss of sensation, muscular paralysis, and atrophy.

Neuralgia is the condition in which there are severe paroxysmal pains along the course of a nerve; there is no structural change in the nerve, as in neuritis. The pain is sharp and stabbing. A ruptured intervertebral disk pressing on nerve fibers may cause neuralgia.

PHYSIOLOGY OF NERVOUS TISSUE

Total functioning of the nervous system depends not only on the structural integrity of neu-rons, but also on their highly developed physiologic properties of excitability and conductivity. From receptor to effector, the transmission of nerve impulses is a complicated process, and only a brief summary is within the scope of this textbook.

The reflex arc and act

Reflex arcs are the *functional units* of the nervous system as they provide pathways over which nerve impulses travel from receptors to effectors. The activity that results from the passage of impulses over a reflex arc is called a *reflex act*, or simply, a *reflex*. Generally speaking, a reflex is a response to a stimulus, but the term usually means only our involuntary responses. We may or may not be aware of reflex activities, but they occur without any conscious assistance on our part.

The basic components of a typical reflex arc consist of:

1. The *receptor* (sense organ) that is excited by an appropriate stimulus;

2. The *afferent neuron* which transmits the impulse (excitatory state) from the receptor to the CNS;

3. The *central*, or *internuncial neuron*, which transmits the impulse from afferent to the efferent neuron; (NOTE: the great majority of reflex arcs involve many neurons; there are rela-

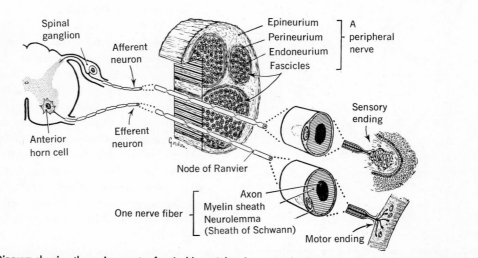

Fig. 6-6. Diagram showing the various parts of a sizable peripheral nerve, individual nerve fibers, and nerve endings.

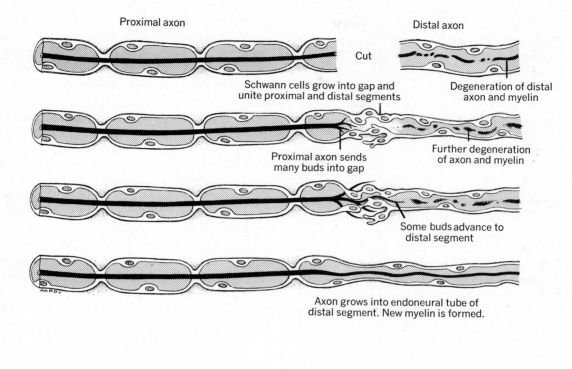

Proximal axon

Distal axon

Cut

Schwann cells grow into gap and
unite proximal and distal segments

Degeneration of distal
axon and myelin

Proximal axon sends
many buds into gap

Further degeneration
of axon and myelin

Some buds advance to
distal segment

Axon grows into endoneural tube of
distal segment. New myelin is formed.

Fig. 6-7. Diagram of changes that occur in a nerve fiber that has been cut and then regenerates.

tively few two-neuron arcs in which afferent and efferent neurons synapse directly.)

4. The *efferent neuron* which transmits the impulse from the CNS to the effector;

5. The *synapses*, or points of contact between related neurons (i.e., between 2 and 3, and 3 and 4 above);

6. The *neuroeffector junction*, or point of contact between the efferent fiber and the effector that brings about a response.

For example, when your hand touches a hot stove, it is withdrawn swiftly. In terms of the reflex arc, the following sequence of events occurs in just a fraction of a second. Special heat receptors in the skin are stimulated by the hot stove, and the resulting nerve impulse flashes over the dendrite, the cell body, and the axon of an afferent neuron into the dorsal horn of gray matter in the spinal cord. At this point, the nerve impulse crosses the synapse between the axon of the afferent neuron and the dendrite of a central neuron; continuing along the dendrite, the cell body, and

the axon of this central neuron, the impulse then crosses another synapse to the dendrite of an efferent neuron in the ventral horn of spinal cord gray matter. Speedily moving along the dendrite, the cell body and the axon of this motor neuron, the nerve impulse reaches the motor end plate in a skeletal muscle, activating the contractile protoplasm and enabling you to withdraw from the harmful stimulus (see Fig. 7-11).

Characteristics of nerve tissue

The two outstanding characteristics of nerve tissue are excitability and conductivity. Both are inseparably related to depolarization and repolarization of cell membranes, as previously described on page 102, you may find it helpful to review this material before proceeding.

Excitability is the capacity to be affected by a stimulus (i.e., a change in the environment), and neurons may be stimulated at any point along their entire length. Many persons discover this fact for themselves when they accidentally strike an elbow, or "funny bone", in such a way that the ulnar nerve is pinched. The resulting waves of depolarization, or nerve impulses, spread in both directions from the point of stimulation; within a fraction of a second there may be a twitch of certain forearm muscles, and one is aware of a sensation that is sometimes described as pins and needles in the little finger and the ring finger of the hand. However, under normal conditions nerve impulses of the reflex arc originate in the receptors. The physiologic property of excitability is most highly developed in these afferent nerve endings, and the *threshold* or *minimal stimulus* (weakest stimulus that will excite the receptor) is very low. Each receptor, except those for pain, is specialized for the reception of a particular type of stimulus which is called its *adequate stimulus*. For example, light rays are the adequate stimuli for the receptors of vision, and sound waves for those of hearing.

There is no completely satisfactory classification of receptors, but the following one is useful. Those receptors which are affected by changes in the external environment are called *exteroceptors* and are concerned with touch, cutaneous pain, heat, cold, smell, vision, and hearing. Those receptors which respond to changes in skeletal muscles, tendons, joints, and the labyrinth of the inner ear are called *proprioceptors* and are concerned with muscle sense, sense of position, and movement of the body in space. Those receptors which respond to changes in the viscera are called *interoceptors* and are concerned with such sensations as visceral pain, hunger, and thirst as well as reflex activities.

Receptors not only initiate the involuntary reflex acts, but they also help to keep the brain informed of environmental changes that require some voluntary response in making adjustments. If the weather suddenly turns cold, your receptors may start the reflex that results in shivering, but you will *consciously* go home to secure heavier clothing.

Conductivity, the ability to transmit nerve impulses, is a highly developed property of nerve tissue. Because of this property the CNS is informed of all changes in the environment, and then may dispatch orders to various effectors to make necessary adjustments. There is no visible change in a nerve fiber as it conducts an impulse,

but various types of investigation have led to the establishment of certain facts and theories regarding conduction. A few of these will be summarized briefly.

1. *A nerve impulse is self-propagating.* This means that as one portion of a neuron becomes depolarized it serves as a stimulus for the depolarization of adjacent portions. Thus, once an action current is started, the wave of depolarization will sweep over the entire membrane (the all-or-nothing law).

2. *The electrical changes* related to depolarization and repolarization were previously described on page 102. At this point it should be added that the action currents or impulses that sweep over nerve fibers may be recorded and studied. Records of the so-called brain waves are really records of electrical changes in the brain. Such records are called **electroencephalograms.** They are of great diagnostic importance in the study of brain disorders. However, it should be noted that whether nerve impulses are traveling over central nerve fibers or peripheral nerve fibers, the *electrochemical disturbances* involved are identical. We have a sensation of pain from an impulse in one fiber and the contraction of a muscle in another because of the location of the endings of the two fibers. It is possible to have sensations of pain only in the brain, and to have contraction only in a muscle. Impulses may be compared with electrical currents flowing through wires. In one case the current will produce heat or toast bread; in a second case it will ring a doorbell. The electrical current is the same in each wire; the difference is in the ending to which it is conducted.

3. *A nerve impulse travels at a definite speed* that depends upon the size of the fiber. Large, myelinated fibers conduct more rapidly than do small, naked fibers. As previously noted, myelin is an insulator that interferes with the flow of sodium and potassium ions; thus depolarization occurs only at the uninsulated areas or nodes of Ranvier. It is believed that the high velocity of conduction in myelinated fibers exists because the impulse jumps from node to node rather than passing over the entire fiber; this is referred to as a *saltatory* type of impulse conduction. Studies have shown that very large fibers can conduct impulses as rapidly as 120 meters per second, a distance longer than a football field. This is a slower rate than that at which sound waves travel, but

even in a man who is 6 feet tall, the impulses could travel in large afferent nerve fibers from toes to brain and back to the toes over large efferent fibers in less than 1/20 of a second. On the other hand, the smallest fibers conduct impulses as slowly as 0.5 meter per second, about the distance from the foot to the knee.

4. *During the time that a nerve fiber membrane is depolarized, it will not respond to another stimulus*, no matter how strong that stimulus may be; this brief interval of time (about 1/2500 second) is called the *absolute refractory period*. Thus such a fiber theoretically could transmit a maximum of about 2500 impulses each second. Immediately following the absolute refractory period, during the time that repolarization is taking place, a stimulus must be stronger than normal to excite the fiber once again; this interval of time is called the *relative refractory period*.

5. The *chemical changes* that occur in nerve fibers during the conduction of impulses are not completely understood. There is a greater use of oxygen and increased production of carbon dioxide; glucose is used during conduction, and some ammonia is produced. However, explanations of the chemical changes that accompany activity in nerve fibers are not nearly as well worked out as they are for skeletal muscle activity. The important thing to keep in mind is that the all-important brain cells depend as much on glucose as on oxygen; deprivation of either one or both can result in serious brain damage.

6. *Thermal changes* accompany conduction of nerve impulses, but the amount of heat produced is only 1/10,000 of that which results from the contraction of an equal weight of muscle. In other words, you never will become as overheated from thinking as you might from vigorous exercise.

7. The *transmission of impulses across synapses* apparently depends upon the liberation of chemical transmitter substances at the axon terminals of presynaptic neurons. As previously described in relation to the neuromuscular transmission of impulses (p. 103), it is believed that transmitter substance is stored in little synaptic vesicles in the terminal knobs of the axons (see Fig. 4-7). When a nerve impulse sweeps over an axon, the vesicles release a little squirt of transmitter substance into the synaptic cleft; this substance, in turn, brings about a transient change

in the membrane permeability of the postsynaptic neuron. Depending on the chemical identity of the transmitter substance and the way it interacts with the membrane of the postsynaptic neuron, the permeability change may result in excitation or in inhibition of the postsynaptic neuron.

At *excitatory synapses*, there is an increase in membrane permeability to sodium ions, and if sufficient numbers of sodium ions diffuse inward, the neuron will discharge. A given neuron may fire an impulse when stimulated by only a few incoming fibers or it may not fire until stimulated by many incoming fibers; this threshold of response can be raised or lowered by many factors. Acetylcholine is the excitatory transmitter substance liberated at synapses in the autonomic ganglia (p. 220) and probably at many of the synapses in the central nervous system. Cholinesterase is the enzyme that inactivates acetylcholine and stops synaptic transmission of impulses. Norepinephrine also can excite some of the neurons in the CNS; this transmitter is discussed further on page 228.

Special neurons are involved at *inhibitory synapses*. For example, one type of neuron liberates a chemical substance that increases membrane permeability to potassium ions but not to sodium ions. As the positive potassium ions leave their place of greatest concentration (within the cell), the interior of the neuron becomes even more negative than it is in the resting state; thus for a fraction of a second the neuron is in a state of hyperpolarization. These are the synapses that can inhibit the discharge of a neuron even though it may be receiving a volley of excitatory impulses. The chemical identity of inhibitory transmitter substance is not yet known.

All parts of the CNS are continually bombarded by a barrage of incoming impulses from the peripheral nerves. Thus inhibition is just as important as excitation in determining the overall function of the nervous system. Not only does inhibition refine the activity of motor neurons which control the muscles, but it also protects the brain from being swamped by insignificant information. Without inhibition, the brain would be in such a state of continual excitement that it would be useless.

Since reflex arcs are the functional units of the nervous system, and since each of these units involves a variable number of synapses, we shall summarize the special characteristics of synaptic transmission. Of primary importance is the fact that impulses can pass only in one direction across a synapse, from the axon of one neuron to the dendrites, or the cell body, of another neuron. If an individual nerve fiber is activated in its midportion, the impulse is conducted equally well in either direction. Thus synapses ensure that nerve fibers normally are used only for one-way transmission of signals. Some of the other ways in which synaptic transmission differs from that in a nerve fiber will be mentioned briefly. Over a reflex arc (1) conduction is slower; (2) a single stimulus may not produce a response, but repetition of that stimulus may do so (summation); (3) there is much greater fatigability; (4) there is more dependence on oxygen supply; and (5) there is more susceptibility to anesthetics and other drugs. For example, strychnine blocks the action of inhibitory transmitter substance, and, in the absence of normal inhibition, the effects of the excitatory transmitter become overwhelming; as a result, neurons become so excited that they fire repetitively, causing uncoordinated muscular contractions known as convulsions.

Reaction time

The time which elapses between the application of a stimulus and the start of a response is referred to as *reaction time*. This time will vary in different individuals under different conditions. Some of the factors which will cause variations in reaction time are:

1. The strength of the stimulus;
2. The nature of the stimulus (e.g., sound, pain);
3. The number of synapses through which the impulse must travel;
4. The condition of those synapses (e.g., fatigue, drugs, alcohol may delay transmission of impulses and increase the reaction time).

SUMMARY

1. Basic function of the nervous system
A. Convert stimuli to nerve impulses and conduct to CNS
B. Interpret incoming impulses and decision making
C. Sort out impulses
D. Dispatch impulses to effectors
E. Awareness of body activities

2. General plan of the nervous system
A. Mechanisms of correlation
  a. Nervous system
  b. Endocrine system
B. Divisions of the nervous system
  a. Anatomic
    (1) Central: brain and spinal cord
    (2) Peripheral: cranial and spinal nerves
  b. Physiologic
    (1) Cerebrospinal or voluntary: somatic
    (2) Autonomic or involuntary: visceral
C. Ultimate goal of nervous system: maintenance of homeostasis

3. Structure of nervous tissue
A. Neuroglia
  a. Delicate supporting cells
  b. Protective functions
  c. Formation of myelin sheaths in CNS
B. Neuron
  a. Cell body: contains nucleus; essential to life of neuron
  b. Cell processes: extensions of cell body; dendrites and axons
  c. Myelin sheath: is insulator; fibers with sheath, myelinated; those without, unmyelinated

  d. Neurolemma sheath: essential for regeneration of injured fibers
C. Functional classification of neurons
  a. Afferent or sensory: conduct from **periphery to** CNS
  b. Efferent: conduct from CNS to effectors
  c. Central or association: transmit impulses from one part of brain and cord to another
D. Structural classification of neurons
  a. Unipolar
  b. Bipolar
  c. Multipolar
E. Synapses
  a. Points of contact between neurons
  b. No anatomic continuity
  c. Impulse passes only from axon to dendrite or to cell body of next neuron
F. Nerves and tracts
  a. Nerves composed of peripheral fibers bundled together; endoneurium, perineurium, and epineurium
  b. Tracts are fiber bundles within the CNS
  c. Functional types of peripheral nerves
    (1) Somatic afferent and efferent
    (2) Visceral afferent and efferent

4. Physiology of nervous tissue
A. Reflex arcs
  a. Functional units of nervous system
  b. Activity resulting from passage of impulses over reflex arc called reflex act
  c. Basic components

(1) Receptor and afferent neuron
(2) Central or association neuron
(3) Efferent neuron and neuroeffector junction
(4) Synapses between related neurons

B. Excitability of nerve tissue
   a. Nerve fiber may be stimulated at any point, but receptor usually starts the impulses involved in reflex
   b. Property of excitability most highly developed in receptor
   c. Must have adequate stimulus
   d. Classification of receptors: exteroceptors; proprioceptors; interoceptors

C. Conductivity of nerve tissue
   a. Capable of transmitting nerve impulses
   b. Impulse is self-propagating
   c. Electrical changes accompany passage of impulse
   d. Impulse travels at a definite speed; larger fibers conduct more rapidly
   e. Refractory period: absolute; relative
   f. Chemical changes occur in conducting fibers
   g. Thermal changes
   h. Transmission at synapses depends upon excitatory and inhibitory transmitter substances; synapse more susceptible than nerve fibers to oxygen lack, fatigue, and drugs; conduction is slower

D. Reaction time
   a. That between application of stimulus and start of response

QUESTIONS FOR REVIEW

1. Which part of the neuron is essential to its life?
2. What factors are essential for the regeneration of a severed peripheral nerve fiber?
3. What is the outer covering of a peripheral nerve?
4. Summarize basic functions of the nervous system.
5. How does the nervous system compare with the endocrine system as a means of communication?
6. Distinguish between central and peripheral components of the nervous system.
7. How do neuroglial cells differ from neurons?
8. What is the neuroeffector junction?
9. Describe the components of a typical reflex arc.
10. What is the saltatory type of impulse conduction?
11. How are impulses transmitted across synapses?
12. Why are inhibitory synapses just as important as excitatory ones?

REFERENCES AND SUPPLEMENTAL READINGS

Chusid, J. G.: Correlative Neuroanatomy and Functional Neurology, ed. 14. Los Altos, Lange Medical Publications, 1970.

Eccles, J.: The synapse. Sci. Am., 212:56, Jan. 1965.

Guyton, A. C.: Textbook of Medical Physiology, ed. 4. Philadelphia, W. B. Saunders, 1971.

Stent, G. S.: Cellular communication. Sci. Am., 227:43, Sept. 1972.

Wilson, V. J.: Inhibition in the central nervous system. Sci. Am., 214:102, May 1966.

The Spinal Cord
and Spinal Nerves

"Oh the nerves, the nerves;
the mysteries of this machine called man!"
The Chimes • Charles Dickens

The spinal cord and its 31 pairs of nerves might be roughly compared with a telephone switchboard and its miles of cables. As you dial the telephone number of a friend, electrical impulses flash over wires to the central switchboard; from here they are routed over appropriate circuits to still other wires that lead to the bell in your friend's telephone. For more complicated calls, the services of special operators and supervisors are required. In a similar manner the spinal cord serves as a switchboard for the nerve impulses which are transmitted over the millions of fibers in our spinal nerves. Many of our daily activities are carried out more or less automatically without any special thought processes; however, they are always being monitored and influenced through a more complicated switchboard, the brain.

THE SPINAL CORD

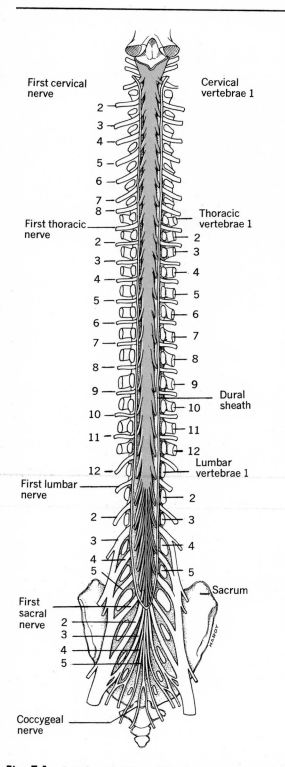

First cervical nerve

Cervical vertebrae 1

2
3
4
5
6
7
8

First thoracic nerve

Thoracic vertebrae 1

2
3
4
5
6
7
8
9

Dural sheath

10
11
12

Lumbar vertebrae 1

First lumbar nerve

2
3
4
5

Sacrum

First sacral nerve

2
3
4
5

Coccygeal nerve

Fig. 7-1. Spinal cord lying within the vertebral canal, spinous processes and laminae have been removed; dura and arachnoid have been opened. Spinal nerves are numbered on the left side; vertebrae are numbered on the right side.

The spinal cord (continuous with the brain stem) begins at the level of the foramen magnum and extends downward through the vertebral canal to the level of the disk between the first and the second lumbar vertebrae (Fig. 7-1).

This elongated cylindrical mass of nervous tissue consists of **31 segments,** each of which gives rise to a pair of spinal nerves. The spinal cord is conspicuously enlarged in two regions: the **cervical enlargement** gives origin to nerves which supply the upper extremities, and the **lumbar enlargement** gives origin to nerves which supply the lower extremities.

Extending the full length of the spinal cord are two clefts that partially divide the spinal cord into right and left halves. The *ventral median fissure* is a relatively wide cleft that effectively separates the anterior portions of the cord. Posteriorly, separation is achieved by a partition which is called the *dorsal median septum;* the position of this septum is marked by a shallow groove, the *dorsal median sulcus.* In the central midportion of the cord, a narrow bridge (commissure) of gray and white fibers, surrounding the central canal, serves to join the two halves (Fig. 7-2). The central canal contains cerebrospinal fluid (p. 209).

Coverings of the spinal cord

In addition to the bony vertebral column, the spinal cord is covered and protected by three **meninges,** or membranes (Fig. 7-3) which are downward extensions of those which cover the brain.

The **dura mater** (*hard mother*) is the outermost membrane, and it is composed of dense fibrous connective tissue. This tough, tubular sheath extends downward to the level of the second sacral segment where it ends as a blind sac. The dura mater also covers the spinal nerve roots as they leave the spinal cord and extend to the intervertebral foramina.

The space between the walls of the vertebral canal and the outer surface of the dura mater is called the *epidural* or *extradural space;* it contains a network of blood vessels and adipose and areolar connective tissue. The *subdural space* is that between the inner surface of the dura and the underlying arachnoid membrane; it is quite

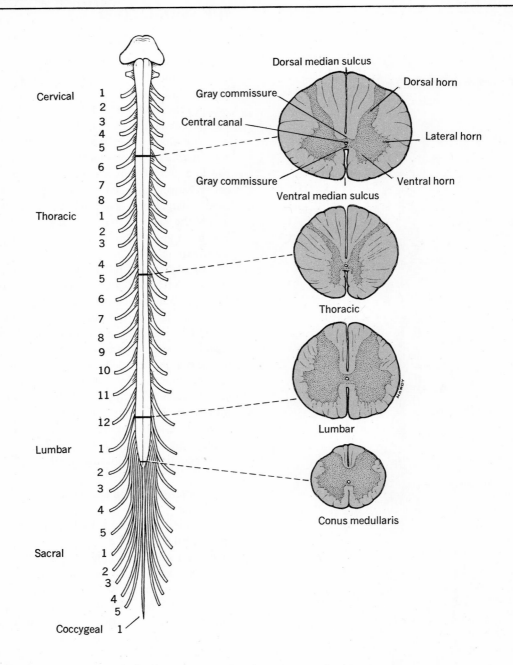

Fig. 7-2.   Cross-sectional views of the spinal cord showing regional variations in the gray matter.

limited and contains only a small amount of fluid.

The **arachnoid membrane** (*like a spider web*) is a thin, transparent sheath which lies immediately under the dura and follows it to the end of the dural sac. The innermost layer of the meninges, called the **pia mater** (*gentle mother*) is very

closely applied to the spinal cord. This delicate membrane contains many blood vessels and plays an important part in the nourishment of cells within the spinal cord.

The **subarachnoid space** is a relatively roomy area between the arachnoid and the pia mater. It

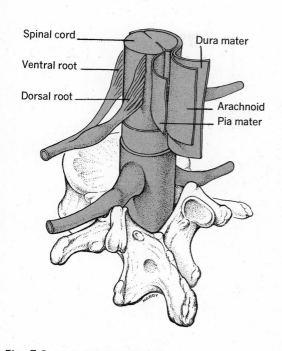

Fig. 7-3. Spinal cord and meninges.

is filled with a liquid commonly called the *spinal fluid*; however, a more accurate term is **cerebrospinal fluid**, since it also is found in the cranial subarachnoid spaces. Analysis of this fluid is a valuable diagnostic aid in many cases, and samples may be obtained by spinal tap, or lumbar puncture. This procedure requires the insertion of a long needle between the third and the fourth (sometimes the fourth and the fifth) lumbar vertebrae and into the subarachnoid space. Since the spinal cord ends at the level of the disk between the first and the second lumbar vertebrae, there is little danger of striking it with the needle. The formation, the circulation, and the functions of cerebrospinal fluid are described on pages 209 to 211.

Inflammation of the meninges is called **meningitis**. It may involve the dura only, or it may involve the arachnoid and the pia mater. Meningitis may be caused by a variety of infectious agents, such as the tubercle bacillus, meningococcus, streptococcus, or staphylococcus.

Internal structure

In cross section the spinal cord presents a somewhat oval-shaped appearance. It is seen to contain an H-shaped mass of gray matter which is surrounded by white matter.

The **gray matter** of the spinal cord is composed of cell bodies and unmyelinated nerve fibers. For descriptive purposes, the columns of the H are referred to as the right and left ventral, or anterior horns, the lateral horns, and the dorsal or posterior horns (see Fig. 7-2).

The **white matter** of the spinal cord consists primarily of myelinated fibers in a network of neuroglia. These fibers serve to link various segments of the spinal cord and to connect the spinal cord with the brain. The white matter in each half of the cord is divided into three columns (or funiculi): the ventral, the lateral, and the dorsal columns (Fig. 7-2). Each column is subdivided into *tracts* which are merely large bundles of nerve fibers arranged in functional groups. *Ascending tracts* transmit impulses to the brain, and *descending tracts* transmit impulses from the brain to various levels of the spinal cord.

THE SPINAL NERVES

There are 31 pairs of spinal nerves, and each pair is numbered according to the level of the spinal cord segment from which it originates. There are eight pairs of cervical nerves, twelve pairs of thoracic, five pairs of lumbar, five pairs of sacral, and one pair of coccygeal nerves.

Origin of spinal nerves

Each spinal nerve arises by two roots: a dorsal, or posterior, root and a ventral, or anterior, root (Fig. 7-4).

1. The **dorsal root** of a spinal nerve may also be called the *sensory root*, since it contains afferent nerve fibers that conduct nerve impulses to the spinal cord. On each dorsal root there is a marked swelling called the **dorsal root ganglion**, or *spinal ganglion*, which contains the cell bodies of both somatic and visceral afferent neurons. The axons of these afferent neurons extend into the dorsal gray matter of the spinal cord.

2. The **ventral root** of a spinal nerve may also be referred to as the *motor root*; it contains efferent fibers that conduct impulses away from the spinal cord. The dendrites and the cell bodies of

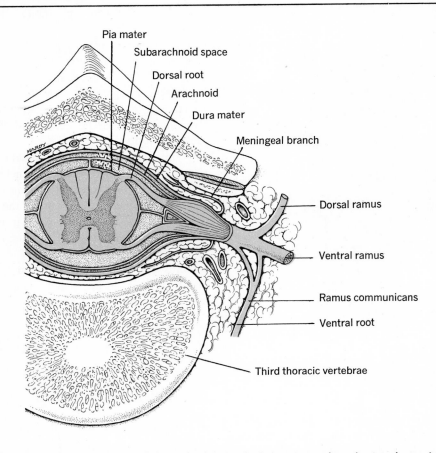

Pia mater
Subarachnoid space
Dorsal root
Arachnoid
Dura mater
Meningeal branch
Dorsal ramus
Ventral ramus
Ramus communicans
Ventral root
Third thoracic vertebrae

Fig. 7-4. Cross-section of vertebral column at the level of the third thoracic vertebra, showing the meninges, the spinal cord, and the origin of a spinal nerve and its branches or rami.

somatic efferent neurons are located in the ventral horns of gray matter in the cord. Figure 7-12 illustrates the synaptic connections between related neurons of the spinal nerve roots. Cell bodies of visceral efferent neurons are located in the lateral horns, as described in Chapter 9.

The dorsal and the ventral roots extend to the intervertebral foramen which corresponds to their spinal cord segment of origin. Early in fetal life, the spinal cord and the vertebral column are the same length, but soon the vertebral column begins to elongate more rapidly than the cord. Because of this difference in growth, the cord segments are displaced upward from their corresponding vertebrae, the discrepancy being greatest at the lower end of the cord. Since the spinal cord ends in the upper lumbar region of the vertebral column, the lower lumbar, sacral, and coccygeal nerve roots must descend further and further to reach their

respective intervertebral foramina (Fig. 7-1). This drooping of lower nerve roots somewhat resembles the tail of a horse, and thus it has been named the *cauda equina* (*tail, horse*).

As a dorsal and a ventral root reach their intervertebral foramen, they unite to form a **spinal nerve,** as shown in Figure 7-4. Therefore, all spinal nerves are *mixed* nerves, since they contain both sensory and motor fibers for the conduction of nerve impulses between the spinal cord and the periphery. Severing of a spinal nerve results in paralysis of muscle and loss of sensation in the areas supplied by the nerve.

Distribution of spinal nerves

As each spinal nerve emerges from the intervertebral foramen, a small *meningeal branch* (Fig. 7-4) is given off; this branch re-enters the vertebral

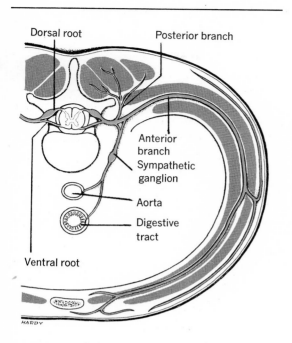

Fig. 7-5. Schematic cross-section through vertebral column and left side of trunk showing anterior and posterior rami of one spinal nerve. Also shown are fibers leading to a sympathetic ganglion and to internal organs.

canal, where it innervates the vertebrae, ligaments, and blood vessels of the spinal cord and the meninges. The two *primary branches,* however, are the anterior and the posterior rami (Fig. 7-5).

The **posterior rami** consist of those motor and sensory fibers that supply the skin and the longitudinal muscles of the back.

The **anterior rami** are larger than the posterior; they consist of motor and sensory fibers which supply all of the structures of the extremities and the lateral and the anterior portions of the trunk. However, the anterior rami (except those of 11 thoracic nerves) do not go directly to skin or muscle; they first combine in such a way as to form complicated networks of nerve fibers which are called **plexuses**. Emerging from each plexus are nerves which are named according to the general region which they supply (e.g., ulnar, femoral). Each of these nerves, in turn, may have numerous branches which are named for the structure which they supply, such as articular, muscular, or cutaneous branches.

The thoracic and upper lumbar spinal nerves also give rise to *white rami* (visceral efferent branches), or preganglionic autonomic nerve fibers; these fibers are described in Chapter 9.

The Cervical Plexus. The anterior rami of the first four cervical spinal nerves combine to form the cervical plexus which lies deep in the neck on each side (Fig. 7-6). Branches from this plexus are distributed to the muscles and the skin of the neck and the posterior part of the scalp. The **phrenic nerve,** composed of fibers from the third, the fourth, and the fifth cervical spinal nerves, is the most important branch of the cervical plexus, as it supplies motor fibers to the muscular *diaphragm.* Crushing of the spinal cord above the level of phrenic nerve origin, as in the case of a broken neck, will result in respiratory paralysis and death since orders from the brain no longer can reach the phrenic nerves, and the diaphragm ceases to contract.

The Brachial Plexus. The anterior rami of the last four cervical and the first thoracic spinal nerves combine to form the brachial plexus (Fig. 7-7) which supplies a number of neck and shoulder muscles in addition to the entire upper extremity. The chief nerves that emerge from this plexus on each side of the body, are (1) the **musculocutaneous,** which supplies motor fibers to the anterior arm muscles (e.g., biceps, brachii) and sensory fibers to the skin of the lateral forearm; (2) the **axillary,** which supplies motor fibers to the deltoid muscle and sensory fibers to the skin over the deltoid; (3) the **median,** which supplies motor fibers to flexor muscles in the anterolateral aspect of the forearm and the hand, and sensory fibers to the skin in the radial half of the palm; (4) the **ulnar,** which supplies motor fibers to the flexor muscles in the anteromedial aspect of the forearm and the hand, and sensory fibers to the skin of the same region; and (5) the **radial** (largest branch of the plexus), which supplies motor fibers to the muscles and sensory fibers to the skin on the posterior aspect of the arm, the forearm, and the hand. The distribution of these nerves is shown in Figure 7-8.

Injuries to the nerves that supply the upper extremities are not uncommon. For example, the improper use of crutches may cause undue pressure on the nerves in the axilla, resulting in a temporary paralysis, mainly of structures supplied by the *radial nerve.* Intramuscular injections aimed at the deltoid muscle, but accidentally given into the groove just inferior to it, may result in

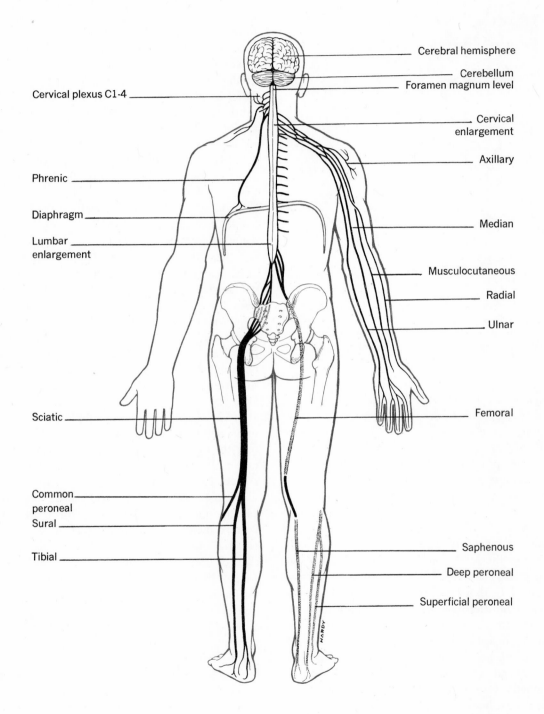

Fig. 7-6. Distribution of certain peripheral nerves, posterior aspect.

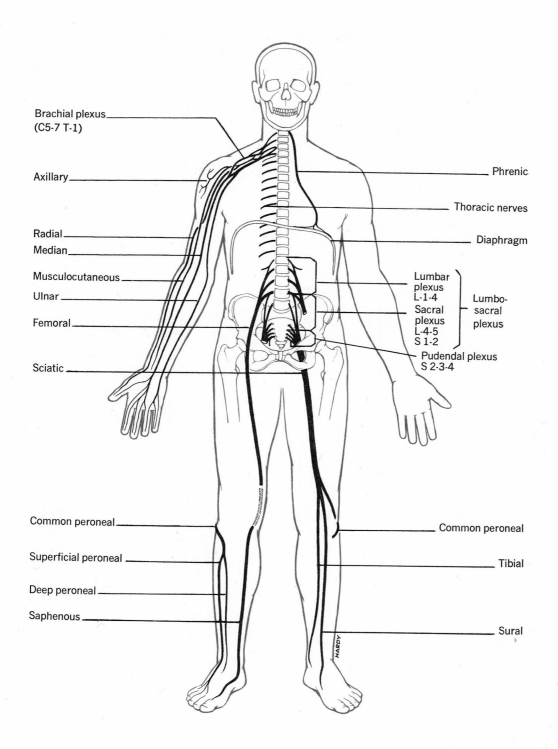

Brachial plexus
(C5-7 T-1)

Axillary

Radial
Median

Musculocutaneous

Ulnar

Femoral

Sciatic

Common peroneal

Superficial peroneal

Deep peroneal

Saphenous

Phrenic

Thoracic nerves

Diaphragm

Lumbar
plexus
L-1-4

Sacral
plexus
L-4-5
S 1-2

Lumbo-
sacral
plexus

Pudendal plexus
S 2-3-4

Common peroneal

Tibial

Sural

HARDY

Fig. 7-7. Distribution of certain peripheral nerves, anterior aspect.

injury to the radial nerve as it winds around the humerus to enter the triceps muscle; this may be avoided by making certain that the needle enters the deltoid about two fingers breadth below the acromion process of the scapula, the bony prominence at the tip of the shoulder. The most characteristic sign of radial nerve damage is an inability to extend the hand at the wrist, a condition that is called *wrist drop.*

When the *median nerve* alone is damaged, there is inability to pronate the forearm or to flex the wrist properly, since the pronators and most of the flexors are supplied by it. The second phalanges of the middle and the index fingers cannot be flexed. The thumb cannot be flexed and abducted, which is one of the most characteristic features of the normal hand. The ability to appose the thumb to any one of the fingers, as is done in picking up very small objects, is lost.

After damage to the *ulnar nerve,* flexion and adduction of the wrist are impaired. There is difficulty in spreading the fingers; the hand is clawed. When both the ulnar and the median nerves are injured at the wrist, as is frequently the case, all small muscles of the hand are paralyzed.

The Intercostal Nerves. With the exception of the first, the anterior rami of the 12 thoracic spinal nerves follow a simple segmental pattern and do not enter into the formation of plexuses. They are referred to as intercostal nerves, since they run through the intercostal spaces in the thorax. These nerves supply motor fibers to the intercostal muscles and to the muscles of the abdominal wall, as well as sensory fibers to the overlying skin.

The Lumbar Plexus. The anterior rami of the first four lumbar spinal nerves combine to form the lumbar plexus (Fig. 7-7). Nerves derived from this plexus supply motor and sensory fibers to the structures of the lower abdominal wall, the external genitalia, and part of the lower extremity. The **femoral nerve,** largest branch of the plexus, supplies motor fibers to the iliopsoas and the quadriceps femoris muscles, and sensory fibers to skin on the anteromedial aspect of the thigh. The *saphenous nerve* is a long cutaneous branch of the femoral which supplies the skin overlying the medial aspect of the leg and the foot. The *obturator nerve* supplies motor fibers to the adductor muscles of the thigh.

Injuries to the femoral nerve may result from

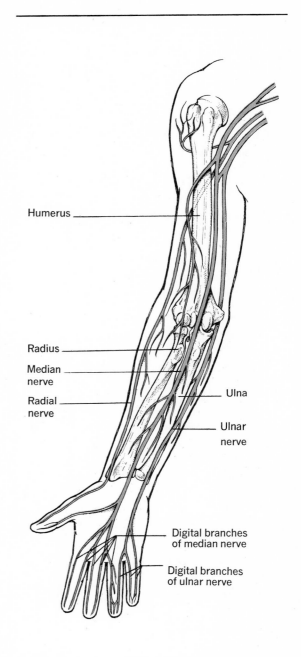

Fig. 7-8. Course of radial, ulnar, and median nerves in right upper extremity.

pelvic tumors or fractures of the pelvis and the upper femur. In such an event, there is inability to extend the leg, and sensation is lost in those skin areas supplied by the nerve. If the injury is high enough to involve the iliopsoas, there is inability to flex the thigh.

sciatic nerve, the largest and longest nerve in the body, is the main branch of the sacral plexus. Covered by the gluteus maximus muscle, this great nerve curves downward through the buttock, resting on the posterior surface of the ischium, between the ischial tuberosity and the greater trochanter of the femur. It then descends through the thigh, supplying branches to the hamstring muscles and the overlying skin before it divides into the tibial (medial popliteal) and the common peroneal (lateral popliteal) nerves (Fig. 7-9).

The *tibial nerve* supplies the posterior calf muscles concerned with plantar flexion and inversion of the foot and with plantar flexion of the toes. The *common peroneal nerve* supplies the lateral and anterior muscles of the leg and the dorsal muscles of the foot, effecting dorsal flexion and eversion of the foot.

Injuries to the sciatic nerve may result from a herniated intervertebral disk, a dislocated hip, compression by the pregnant uterus, or injections of various drugs into or near the nerve. In adults, the upper outer quadrant of the buttocks is one site for intramuscular injections. However, it is essential that the person giving the injection knows the location of the underlying sciatic nerve and how to select the proper site for insertion of the needle. The classic method of dividing the buttocks into four quadrants by intersecting perpendicular lines may be dangerous, because of the absence of well-defined landmarks for locating the vertical line. Daniel J. Hanson, M.D., recommends the following minor modification for locating the injection site: "A line drawn from the posterior superior iliac spine to the greater trochanter of the femur is lateral to and parallel with the course of the sciatic nerve. Any injection lateral and superior to this line will be removed from the course of the sciatic nerve and will be within the region of the greatest gluteal mass. . . ." (Fig. 7-10). Should the sciatic nerve be injured by drug injection, experience has shown that the common peroneal nerve fibers (located on the lateral side of the sciatic trunk) usually suffer the greater damage, resulting in *foot drop* due to paralysis of the dorsiflexor muscles of the foot and the toes.

Sciatica is one type of neuritis. This condition is characterized by sharp, shooting pain along the course of the sciatic nerve; there may be tenderness, numbness, and tingling in addition to the pain. If this inflammatory condition persists, there

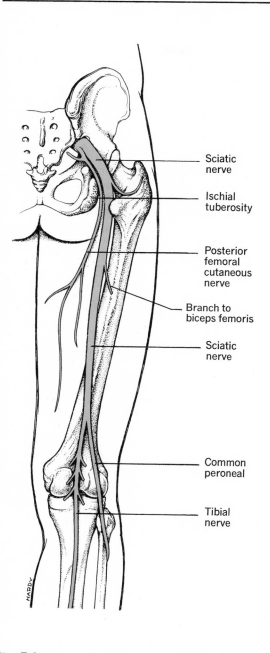

Sciatic nerve

Ischial tuberosity

Posterior femoral cutaneous nerve

Branch to biceps femoris

Sciatic nerve

Common peroneal

Tibial nerve

Fig. 7-9. Course of sciatic nerve in right hip and thigh.

The Sacral Plexus. The anterior rami of the fourth and fifth lumbar and the first three sacral spinal nerves enter into formation of the sacral plexus (Fig. 7-7). Branches from this plexus supply motor and sensory fibers to structures in the buttocks, the perineum (area between pubis and coccyx), and the lower extremities. The may be atrophy (wasting) of those muscles supplied by the nerve.

The Pudenal Plexus. The anterior rami of the second, the third, and the fourth sacral spinal nerves contribute fibers to this plexus. The *pudendal nerve* is the largest branch, and it supplies the levator ani muscle, the skin, and other structures of the perineum. This nerve may be blocked by drugs to facilitate childbirth.

FUNCTIONS OF THE SPINAL CORD

The spinal cord serves as a center for spinal re-

flexes and as a conduction pathway to and from the brain.

Reflex center

The reflex arc (p. 168) depends not only on afferent and efferent fibers, but also on a central structure where these fibers can be connected; the spinal cord serves this function for first level, or spinal cord reflexes. Second and third level reflexes involve centers in the brain and are described in Chapter 8.

Examples of spinal reflexes are the *stretch reflexes* and the *flexor reflexes*. The *knee jerk* is an

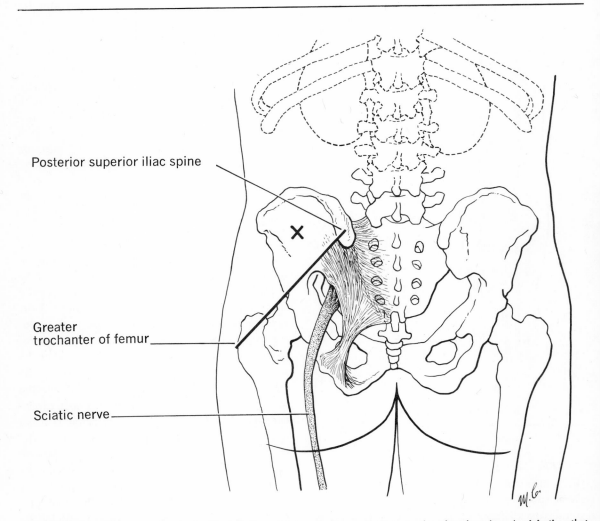

Posterior superior iliac spine

Greater trochanter of femur

Sciatic nerve

Fig. 7-10. Diagram of landmarks for the injection of intramuscular medications in the gluteal region. An injection that is lateral and superior to a line drawn from the posterior superior iliac spine to the greater trochanter of the femur will be removed from the course of the sciatic nerve; the letter X indicates one such site.

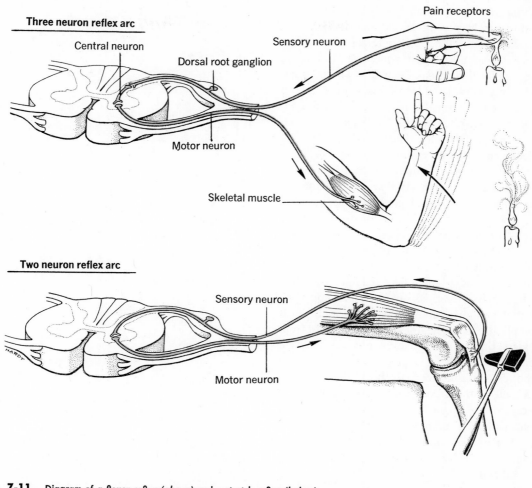

Fig. 7-11. Diagram of a flexor reflex *(above)* and a stretch reflex *(below)*.

example of the stretch reflex, and one that you can demonstrate easily on your own body. Sit with the thighs crossed so that one leg swings free of the floor in a relaxed manner; with the side of your hand, strike the patellar tendon (just below the kneecap) and note the involuntary extension of the leg. When this tendon is struck, the quadriceps muscle of the thigh is stretched slightly, and proprioceptors are stimulated. Impulses travel from these receptors in the muscle over dendrites in the femoral nerve to cell bodies in the dorsal root ganglion and then over the axons to synapses in the ventral horn of gray matter. Since the stretch reflex is a two-neuron circuit, central cells are not required, and impulses travel directly to the dendrites of motor cell bodies in the ventral horn, pass over the dendrites and the cell bodies, and on to the axons and to the muscle fibers of the quadriceps femoris muscle, causing contraction (Fig. 7-11). Probably three segments of the spinal cord in the lumbar region and hundreds of incoming and outgoing fibers are involved in this reflex. It is due to stretch reflexes in our extensor muscles (those that oppose gravity) that we are able to maintain the standing position.

Not all spinal reflexes are as simple as the knee jerk, as will be noted in the following illustration.

Suppose that while walking in your bare feet across a dark room, you stub your toe on a heavy chair. In a fraction of a second, you withdraw the injured foot while shifting your body weight to the other leg. At the same time, your upper extremities reach out to help maintain your balance or to grab something to keep you from falling. Another fraction of a second later, you are aware of what has happened and notice that your heart is beating more rapidly than usual. In this situation, when nerve impulses initiated by receptors in the injured foot reached the spinal cord, they activated a variety of postsynaptic neurons. On the affected side, efferent fibers stimulated the skeletal muscles that enabled you to withdraw your foot from the harmful object; this is a *flexor reflex*. Simultaneously, impulses flashed over fibers that cross to the opposite side of the spinal cord whence impulses were dispatched over efferent fibers to extensor muscles in the unaffected leg; this is called a *crossed extensor reflex* and serves to push the entire body away from the object causing the painful stimulus. Thousands of other neurons also were involved as you reached out with the upper extremities and maintained your balance. In such a complicated response, the entire spinal cord from the sacral through the cervical region was involved. Even visceral efferent neurons were stimulated which caused the heart to beat more rapidly. While all of these events were taking place, impulses arriving at the brain made you conscious of the pain in your foot a fraction of a second after it had been withdrawn from the stimulus. In general, such reflex activity is protective in nature and is illustrated in Figure 7-12.

Testing of various reflexes is an important part of any physical examination. If they are normal, it indicates that the afferent fibers, the reflex center in the CNS, the efferent fibers, and the muscles themselves all are working properly. Figure 7-11 illustrates testing of the knee jerk, or patellar reflex.

Conduction pathway

In its function as a conduction pathway to and from the brain, the spinal cord makes use of *tracts*. Ascending tracts transmit impulses to the brain, and descending tracts transmit impulses from the brain to various levels of the spinal cord. In many cases, the tracts of the spinal cord are given names that indicate the origin and the destination of the constituent fibers.

The **major ascending pathways** in each side

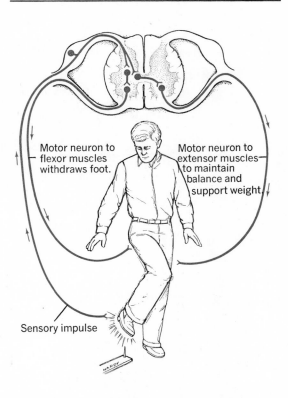

Motor neuron to flexor muscles withdraws foot.

Motor neuron to extensor muscles to maintain balance and support weight.

Sensory impulse

Fig. 7-12. Diagram of flexor and crossed extensor reflexes.

of the spinal cord consist of the dorsal columns, the spinocerebellar, and the spinothalamic tracts (Fig. 7-13).

1. The **dorsal columns** (*fasciculus gracilis* and *fasciculus cuneatus*) consist of axons of cell bodies in the dorsal root ganglia. These axons ascend to the medulla, where they synapse with other neurons which cross to the opposite side and ascend to the thalamus, a relay station in the brain. Eventually nerve impulses reach the highest centers of the brain, and the medullary crossing of fibers means that impulses from the right side of the body are transmitted to the left side of the brain, while impulses from the left side of the body reach the right side of the brain. Some of the impulses which travel over fibers in this tract were originally started by proprioceptors in skeletal muscle, tendons, and joints; these are impulses of *conscious muscle sense*, which means that when they reach the brain they give rise to sensations of body position and movement. Other fibers in the dorsal columns conduct impulses from ex-

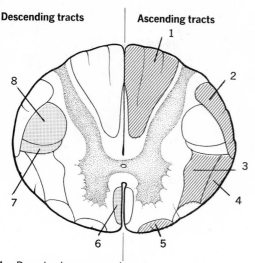

Descending tracts   **Ascending tracts**

1 Dorsal columns conscious
   muscle sense, precise touch

2 Dorsal spinocerebellar tract
   unconscious muscle sense

3 Lateral spinothalamic tract
   pain and temperature

4 Ventral spinocerebellar tract
   unconscious muscle sense

5 Ventral spinothalamic tract
   light touch

6 Ventral pyramidal tract voluntary
   control of skeletal muscle

7 Extrapyramidal tract
   automatic control of skeletal muscle

8 Lateral pyramidal tract
   voluntary control of skeletal muscle

All tracts are bilateral

Fig. 7-13. Cross-section of spinal cord showing principal conduction pathways. These tracts are bilateral, but in this diagram ascending tracts are numbered only on the right, and descending only on the left.

teroceptors for precise touch, vibration, and pressure. When these impulses reach the brain, one is able to recognize the size, shape, and texture of objects; one also is able to localize the exact part of the body that is touched (Fig. 7-14).

Injury to the dorsal columns results in ataxic movements, since proprioceptive impulses cannot reach the brain. This patient cannot tell where his body parts are without looking at them; he walks on a wide base, slapping his feet on the floor, and usually watches his legs so that he will know where they are. In the dark, or with his eyes closed, he may fall. There also is an impairment of vibratory sensation, touch discrimination, and precise localization. Vertebral fractures, tumors, and diseases such as syphilis of the spinal cord (*tabes dorsalis*) may lead to posterior column damage.

2. The dorsal and ventral **spinocerebellar tracts** consist of axons which arise from central cell bodies in the spinal cord and end in the cerebellum; the majority of these fibers are uncrossed. Nerve impulses that travel in these tracts originate in proprioceptors in muscles, tendons, and joints, as do those that travel in the dorsal columns; however, synaptic connections and final destination differ, as spinocerebellar impulses do not reach the conscious centers of the brain. These tracts transmit impulses concerned with *unconscious muscle sense* and play an important part in reflex adjustments of posture and muscle tone.

Injury to these tracts leads to uncoordinated voluntary movements, chiefly because of abnormal tone in the skeletal muscles.

3. The **spinothalamic tracts** consist of axons that arise from central cell bodies in the spinal cord and end in the thalamus. Other neurons connect the thalamus with sensory areas in the brain. The *ventral spinothalamic* tract transmits impulses concerned with crude, or light touch sensations, while fibers in the *lateral spinothalamic tract* transmit impulses concerned with *pain* and *temperature* sensations. These impulses arise from exteroceptors in the skin and travel over afferent neurons which synapse with the spinothalamic neurons in the dorsal gray matter of the cord. The axons of these central neurons cross to the opposite side of the cord before ascending to the thalamus (Fig. 7-14).

The lateral spinothalamic tracts may be intentionally cut by a surgical procedure called *chordotomy*. This operation often makes life bearable for those who suffer from intractable pain (that which cannot be relieved by drugs) as a result of cancer. There is no sensation of pain if the pathways that transmit impulses from the diseased area are cut. However, one must keep in mind the fact that temperature sensations also are abolished, and the patient may be burned by hot water bottles or heating pads unless care is exercised.

The **descending pathways** in each side of the spinal cord consist of two groups of tracts: the pyramidal and the extrapyramidal tracts (see Fig. 7-13).

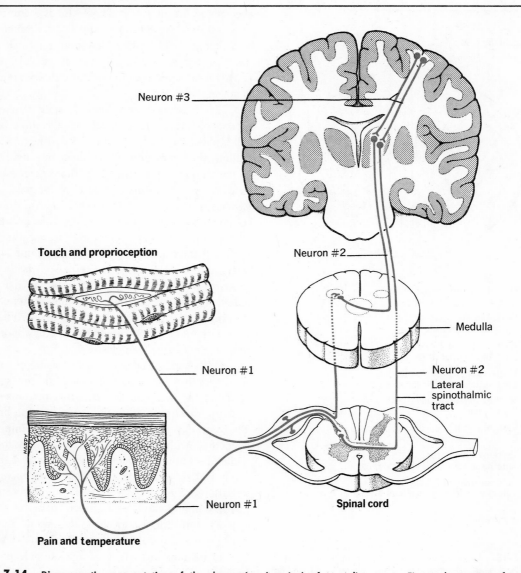

Touch and proprioception

Pain and temperature

Fig. 7-14. Diagrammatic representation of the decussation (crossing) of ascending tracts. First order neurons for touch and proprioception ascend in the dorsal columns to the medulla; here they synapse with second order neurons which cross to the opposite side before ascending to the thalamus. First order neurons for pain and temperature enter the dorsal gray matter of the cord; here they synapse with second order neurons which cross to the opposite side and ascend in the lateral spinothalamic tract to the thalamus. Third order neurons connect the thalamus with the cerebral cortex.

1. The **pyramidal,** or **corticospinal tracts,** are bundles of axons that arise from cell bodies in the motor areas of the frontal lobes of the brain and eventually end in the spinal cord segments. These tracts are the most important and prominent of all the descending bundles, each consisting of over 1 million fibers. Most of the corticospinal fibers cross in the medulla and descend as the large *lateral corticospinal* (crossed pyramidal) tract in each side of the spinal cord; during the descent, each tract gradually diminishes in size as fibers leave to terminate in the ventral gray matter. Those fibers that do not cross in the medulla continue as the smaller ventral corticospinal tracts; some of these fibers cross at each segment of the spinal cord, and a few do not cross at all, but all of them ultimately terminate in the ventral gray matter (see Fig. 8-7).

The corticospinal tracts conduct impulses from the brain to motor neurons in the ventral gray matter of the spinal cord, which in turn conduct the impulses to skeletal muscles and bring about our *fine voluntary movements*, especially those of the fingers, and the hand, which enable us to acquire various manual skills.

To determine the functional status of these tracts, a physician will stroke the lateral side of the sole of the foot, using a blunt object. Normally, there will be a brisk flexion of the toes. The abnormal response is characterized by extension of the great toe and fanning of the other toes; this is known as the *Babinski sign,* or a *positive Babinski reflex,* and usually is indicative of injury to the corticospinal tracts. However, this is not an infallible sign since it can be elicited in the newborn infant, in intoxicated adults, and following a generalized seizure.

2. The **extrapyramidal tracts** include descending pathways (other than the pyramidal tracts) which descend in the spinal cord to synapse with motor neurons in the ventral gray matter. Thus one can see that these busy ventral horn cells truly are the final common path to the skeletal muscles; they are stimulated not only by afferent neurons in segmental reflexes, but also by pyramidal and extrapyramidal fibers from the brain.

The extrapyramidal system (p. 200) is much more intricate than the pyramidal. For example, impulses dispatched from the premotor area of the cerebral cortex will encounter synaptic detours over various subcortical circuits before reaching their ultimate destinations in the spinal cord. Thus the extrapyramidal system is more easily considered as a functional rather than an anatomic unit.

Some of the descending tracts of this system are the *rubrospinal, reticulospinal,* and *vestibulospinal.* The rubrospinal and most of the reticulospinal fibers cross in the midbrain before descending in the spinal cord; most of the vestibulospinal fibers are uncrossed. In general, these tracts convey impulses related to the unconscious or automatic coordination of gross skeletal muscle activity.

The brain cells and their axons in the descending tracts are called *upper motor neurons* in contrast to the ventral horn cells of the spinal cord which are called *lower motor neurons.* Injury to upper motor neurons results in a spastic type of paralysis, the skeletal muscles becoming stiff and resisting passive movements; spasticity occurs when lower motor neurons no longer receive inhibitory or damping impulses from the brain. On the other hand, when lower motor neurons are injured, a flaccid paralysis results; the skeletal muscles become soft and flabby, even reflex movements are absent, and the muscles eventually atrophy.

Injury to the spinal cord

Bullet wounds and fracture dislocations of vertebrae may result in a complete **transection,** or severing of the spinal cord. In such a case, the skeletal muscles below the level of injury are paralyzed, and sensation is abolished as all descending and ascending pathways are interrupted. If the transection occurs below the fifth cervical segment, the phrenic nerves to the diaphragm will not be affected and the patient will continue to breathe. However, paralysis below this point is complete, and even spinal reflexes are temporarily abolished. This severe depression or collapse of cord function is known as *spinal shock* and is believed to be due to loss of stimulation from higher levels.

Limited reflex activity within the isolated spinal cord segments will reappear within a variable period of time; however, muscle tone and the deep reflexes are greatly increased in intensity. The uncontrolled muscle tone leads to alternating spasms of the flexor and the extensor muscles, but eventually one group or the other will predominate. It is believed that the position assumed by the extremities during the early stages of paralysis is an important factor in determining whether flexor or extensor spasms will eventually predominate.

The physical and emotional care of such a patient presents many problems, particularly if the injury is high enough to involve the upper extremities. This patient is completely helpless; he can breathe and talk, his heart beats, and the blood circulates, but he cannot walk, feed himself, or scratch his nose when it itches. Bladder and rectal functions also are greatly disturbed in such transections of the cord as voluntary control is lost. Members of the health team play an important role in proper positioning of the patient, and in the prevention of decubitus ulcers and urinary tract infection. Kidney infection is a serious complication of spinal cord transection. Since a cord bladder does not empty completely, there is residual urine in which bacteria may flourish, and the resulting infection may spread up the ureters to the kidneys. However, antibiotic drugs have changed the picture greatly, and spinal cord damage does not necessarily mean an early death.

Because of the persistent efforts of dedicated members of the medical and allied health professions, many of these patients can be rehabilitated to an amazing degree. This is particularly true of those whose spinal cord injury is below the midthoracic region. A large percentage of these persons can be taught to walk with the aid of braces and crutches, and many of them can become completely self-supporting. One must never lose sight of these facts when one is encouraging and working with those who have suffered a spinal cord injury.

SUMMARY

1. Spinal cord
A. Location and description
  a. From foramen magnum to level of disk between L-1 and L-2
  b. Thirty-one segments; each gives rise to pair of nerves
  c. Cervical enlargement: origin of nerves to upper extremity
  d. Lumbar enlargement: origin of nerves to lower extremity
  e. Ventral median fissure separates anterior portions of cord
  f. Dorsal median septum separates posterior portions of cord
  g. Bridge of gray and white fibers joins two halves of cord in midportion
B. Coverings of the cord
  a. Dura mater: outer covering
  b. Arachnoid membrane: middle covering
  c. Pia mater: innermost covering
  d. Subarachnoid space between arachnoid and pia mater; contains cerebrospinal fluid
C. Internal structure
  a. In cross section: H-shaped mass of gray matter; unmyelinated fibers and cell bodies
  b. White matter surrounds gray; myelinated fibers in network of neuroglia

2. Spinal nerves
A. Origin
  a. Dorsal root contains afferent fibers
    (1) Dorsal root ganglion, contains cell bodies of afferent neurons
    (2) Fibers may synapse with central neurons or ascend to brain
  b. Ventral root contains efferent fibers
  c. The two roots extend to intervertebral foramen; drooping of lower roots called cauda equina
  d. Roots unite to form spinal nerves
B. Distribution of spinal nerves
  a. Each branches after leaving intervertebral foramen
  b. Posterior rami supply skin and longitudinal muscles of back
  c. Anterior rami supply extremities, lateral and anterior portions of trunk; rami enter into formation of plexuses

  d. Cervical plexus: supplies muscles, skin of neck, posterior scalp; phrenic nerve to diaphragm
  e. Brachial plexus: supplies neck and shoulder muscles and upper extremity;
    (1) Musculocutaneous nerve to anterior arm
    (2) Axillary nerve to deltoid
    (3) Median nerve to anterolateral forearm and hand
    (4) Ulnar nerve to anteromedial forearm and hand
    (5) Radial nerve to posterior arm, forearm, and hand
  f. Intercostal nerves: supply motor fibers to intercostal muscles and muscles of abdominal wall
  g. Lumbar plexus: femoral nerve to iliopsoas and quadriceps; obturator nerve to adductor muscles of thigh
  h. Sacral plexus: sciatic nerve supplies hamstrings and muscles of leg and foot; tibial and common peroneal are main branches
  i. Pudendal plexus: pudendal nerve to levator ani muscle and perineum

3. Functions of spinal cord
A. Reflex center
  a. Provides central point for synaptic connections between afferent and efferent neurons
  b. Stretch reflexes
  c. Flexor reflexes
  d. Crossed extensor reflexes
B. Conduction pathway
  a. Ascending pathways: transmit impulses to brain
    (1) Dorsal columns: conscious muscle sense, pressure, and precise touch
    (2) Spinocerebellar tracts: unconscious muscle sense in reflex adjustments of posture and muscle tone
    (3) Spinothalamic tracts: pain and temperature (lateral tract); crude touch and pressure (ventral tract)
  b. Descending pathways: transmit impulses from the brain
    (1) Pyramidal, or corticospinal tracts, concerned with fine, voluntary movements

(2) Extrapyramidal tracts concerned with automatic coordination of skeletal muscle activity

C. Injury to spinal cord
   a. Transection results in complete paralysis of skeletal muscle and loss of sensation below level of injury
   b. Reflex activity will return

QUESTIONS FOR REVIEW

1. What is the result of lower motor neuron destruction such as might occur in poliomyelitis?
2. Where are the cell bodies of the efferent neurons which supply skeletal muscles?
3. At what level is a needle usually introduced in a spinal tap? Why?
4. In advanced syphilis there may be destruction of the dorsal columns of white matter in the spinal cord. What symptoms would you expect to see in an individual with this condition?
5. How do spinal nerve roots differ from the spinal nerve branches or rami?
6. Why does an injury to the spinal cord *above* the C-3 level cause death?
7. What is wrist drop and what may cause it?
8. Where may intramuscular injections be given with relative safety? Why?
9. Describe the components of the knee jerk reflex.
10. What is the significance of the positive Babinski reflex?
11. If the lateral spinothalamic tracts are cut, what would be the neurological deficit?
12. Distinguish between flaccid and spastic paralysis.

REFERENCES AND SUPPLEMENTAL READINGS

Best, C. H., and Taylor, N. B.: The Physiological Basis of Medical Practice, ed. 8. Baltimore, Williams & Wilkins, 1966.

Chusid, J. G.: Correlative Neuroanatomy and Functional Neurology, ed. 14. Los Altos, Lange Medical Publications, 1970.

Elliott, F. A.: Clinical Neurology. Philadelphia, W. B. Saunders, 1964.

Goss, C. M. (ed.): Gray's Anatomy of the Human Body, ed. 28. Philadelphia, Lea & Febiger, 1966.

Guyton, A. C.: Textbook of Medical Physiology, ed. 4. Philadelphia, W. B. Saunders, 1971.

Truex, R. C., and Carpenter, M. B.: Human Neuroanatomy, ed. 6. Baltimore, Williams & Wilkins, 1969.

The Brain and the Cranial Nerves

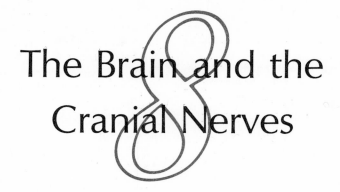

"Our purses shall be proud, our garments poor:
For 'tiz the mind that makes the body rich."
The Taming of the Shrew • William Shakespeare

Just as the executive offices of many large corporations are located on the uppermost floors of the building, so is the human brain positioned in the body. Enclosed in a snug bony box on top of the vertebral column, the brain receives a constant stream of information which has been relayed upward through the spinal cord and the cranial nerves. Junior executives in the lower brain divisions are capable of handling routine problems and dispatching orders to various parts of the body, but command decisions based on associative memory and processes of evaluation are made only by a board of directors located in the topmost executive suites (the cerebral cortex). It is the amazing development of this upper brain division that truly distinguishes man from lower animals. For example, the chimpanzee, the most intelligent of the apes, is unable to carry on a conversation, to read a book, or to dress himself and go to school, no matter how much training he may receive. On the other hand, human beings quickly acquire these skills and then proceed to even more

complicated mental activities. Man, possessing the ability to think, to reason, and to exercise judgment, has the most elaborate brain of any living creature.

The average adult human brain, or encephalon, weighs approximately 3 pounds, but its size varies with the age, the sex, and the size of the individual. For example, men generally have larger brains than women because they have larger heads, but this does not mean that they are necessarily any smarter, since size normally is not indicative of intelligence. Housed in the cranial cavity, the brain is the largest and most complex mass of nervous tissue in the body, and it attains its full growth by the 18th or the 20th year of life. However, it should be noted that although physical growth ceases, there is an unlimited number of neural pathways to be developed as long as we use our brains. In other words, mental exercise is just as important to brain development as physical exercise is to the development of sturdy muscles.

Based on embryologic development, there are three primary divisions of the brain: the *forebrain*, the *midbrain*, and the *hindbrain* (Fig. 8-1). Within these divisions are spaces (ventricles) which are concerned with the formation and the circulation of cerebrospinal fluid (see Fig. 8-12). Each primary division is then subdivided into various parts, a few of which are listed below:

DIVISIONS OF THE BRAIN

1. Forebrain (prosencephalon)
    A. Cerebral hemispheres (telencephalon)
        a. cerebral cortex
        b. basal ganglia (cerebral nuclei)
        c. lateral ventricles
    B. Diencephalon
        a. thalamus
        b. hypothalamus
        c. third ventricle
2. Midbrain (mesencephalon)
    A. Cerebral peduncles
    B. Corpora quadrigemina
    C. Cerebral aqueduct
3. Hindbrain (rhombencephalon)
    A. Cerebellum
    B. Pons
    C. Medulla oblongata
    D. Fourth ventricle

Study of various animal brains shows that, generally speaking, there is a marked correlation between the size of a given part and the importance of that part in the life of the animal. Thus in lower vertebrates, such as the fish and the frog, the cerebrum is extremely small, but in man this all-important part fills most of the cranial cavity.

Much like a large flower, the human **cerebrum** is attached to the lower portions of the brain, referred to as the **brain stem.** Following our pattern of proceeding from lower to higher levels of nervous system activity, we shall first describe the brain stem and the cerebellum, the diencephalon, and finally the cerebral hemispheres. A discussion of the meninges, cerebrospinal fluid, and the cranial nerves completes this chapter.

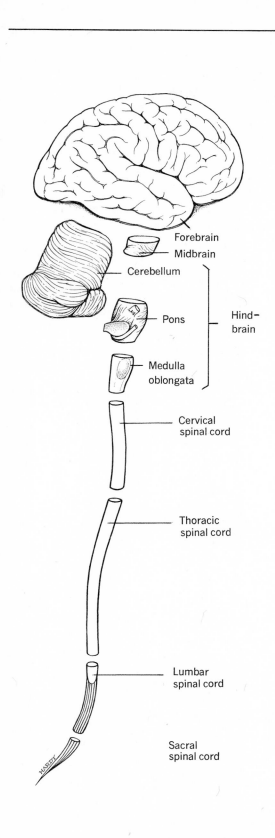

Forebrain
Midbrain
Cerebellum
Pons
Hind-brain
Medulla oblongata
Cervical spinal cord
Thoracic spinal cord
Lumbar spinal cord
Sacral spinal cord

Fig. 8-1. Diagram of the central nervous system. Various parts of the brain and the spinal cord have been separated from one another.

THE BRAIN STEM AND CEREBELLUM

The brain stem serves not only as a conduction pathway to and from the cerebrum, but also as an important reflex center. It consists of the medulla oblongata, the pons, and the midbrain. The cerebellum is attached to the medulla and pons.

Medulla oblongata

The medulla oblongata, or *bulb* of the brain, is an expanded continuation of the spinal cord, extending slightly forward and upward from the level of the foramen magnum to the pons (Fig. 8-2). It is composed of white fiber tracts that conduct impulses to and from the spinal cord and the brain; as previously noted, many of these tracts cross to the opposite side as they pass through the medulla. Scattered throughout the white matter are small nuclei, or controlling centers, of gray matter. Cranial nerves nine, ten,

eleven, and twelve have their nuclei of origin in the medulla.

In addition to being a **conduction pathway** and **decussation** (crossing) **center,** the medulla contains many of the *vital centers,* so called because they are essential to one's very life. These centers are (1) the *cardiac,* to speed or slow the heart rate; (2) the *vasomotor,* to constrict or dilate the blood vessels; and (3) the *respiratory,* to change the rate and the depth of breathing. Afferent impulses from various parts of the body reach these centers and are integrated with impulses from higher levels of the brain; from these centers efferent impulses are then sent to appropriate organs to coordinate their activities for the good of the body as a whole. For this reason, injury to the medulla, as in basal skull fracture or by a disease such as bulbar poliomyelitis, frequently is fatal.

Other reflex centers located in the medulla are those of coughing, sneezing, vomiting, and swallowing. Afferent impulses may be started by various changes in the external environment or by changes within the body itself. For example, the

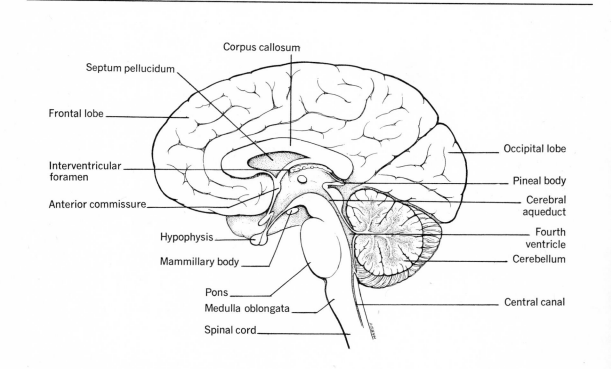

Fig. 8-2. Midsaggital section of the brain.

swallowing reflex is initiated when a mass of food touches the mucous membrane lining of the throat. Afferent impulses are sent to the medulla and shunted to the swallowing center; in a fraction of a second, efferent impulses are dispatched to all effectors concerned with the act of swallowing. This means not only contraction of the throat muscles, but also inhibition of breathing while food passes through the throat to the esophagus; elevation of the soft palate closes the upper respiratory passages, and elevation of the larynx closes the opening into the trachea below. Place your fingers on the upper part of your throat (pharynx) and swallow; note the various movements. These are just a few of the things which occur during swallowing, yet even a newborn baby can perform this complicated reflex act if something is placed in his mouth.

The pons

The term *pons* means *bridge*, a word that is quite descriptive of its functions. The pons lies between the medulla and the midbrain, in front of the fourth ventricle and the cerebellum (Fig. 8-2). It consists of longitudinal and transverse white fiber tracts, the former connecting the medulla with the cerebrum, and the latter connecting the two halves of the cerebellum with the pons. In addition to its role as a **conduction pathway,** the pons also contains the motor and the sensory nuclei of cranial nerves five, six, seven, and eight. Neurons in the lower portion of the pons assist in the regulation of respiration.

The cerebellum

The term cerebellum means *little brain,* in contrast to the larger cerebrum which overlaps it from above. The cerebellum lies in the lower, back part of the cranial cavity, overlapping the pons and the medulla. Its outer portion is composed of a mantle of gray matter, the *cerebellar cortex,* which is thrown into slender folds and furrows; beneath the gray matter are white matter tracts which resemble branches of a tree as they extend to all parts of the cortex. Deep within the white matter are masses of gray matter which are called the *cerebellar nuclei.* In the midsagittal section, this arrangement of gray and white matter in the cerebellum presents such a

treelike appearance that it is called the *arbor vitae,* or *tree of life* (see Fig. 8-2).

The cerebellum is somewhat oval in form, consisting of a constricted central portion and two lateral masses, the *cerebellar hemispheres.* It is attached to the brain stem by three paired bundles of nerve fibers, the *cerebellar peduncles.* The inferior peduncles connect the cerebellum with the medulla and the spinal cord. The middle peduncles, largest of the three paired bundles, transmit impulses between the cerebellar cortex and the cerebral cortex by way of the pons. The superior peduncles connect the cerebellum with the midbrain, from which fibers course upward to the cerebrum and downward to the pons, the medulla, and the spinal cord. Over these various pathways the cerebellum transmits and receives messages to and from other levels of the nervous system.

Functionally, the cerebellum is primarily concerned with the **coordination of muscular activity;** consequently, its connections with proprioceptor endings in skeletal muscles, tendons, joints, and the inner ear (vestibular sense organ, p. 253) are of major importance. However, the cerebellum is not concerned with sensation, nor does it initiate voluntary acts as does the cerebrum. It is primarily involved with reflexes for the regulation of *muscle tone,* the maintenance of *equilibrium* and *posture,* and the *smooth performance* of *voluntary movements.* For example, impulses from the cerebrum initiate our voluntary movements, but impulses from the cerebellum also are needed to coordinate the contraction of prime movers and the relaxation of antagonists if we are to have accurate voluntary movements.

The disturbances that follow injury to the cerebellum vary with the severity of the lesion. If all parts of the cerebellum are involved, there will be a loss of equilibrium, a reeling walk (*cerebellar ataxia*), reduced muscle tone (*hypotonia*), and abnormal postural reflexes. Voluntary movements are inaccurate, and there may be tremors in the muscles when one attempts to perform some activity. However, there is no known sensory loss and no paralysis.

The midbrain

The midbrain is a short, narrow segment of nervous tissue that connects the forebrain with the hindbrain; it extends from the pons to the inferior surface of the cerebrum (Fig. 8-2). A narrow canal, the *cerebral aqueduct*, passes through the central gray matter of the midbrain and serves to connect the third ventricle of the diencephalon with the fourth ventricle of the hindbrain for the circulation of cerebrospinal fluid (p. 209 to 211).

The ventral portion of the midbrain consists of two fiber bundles, the *cerebral peduncles*, which constitute the main conduction pathway between forebrain and hindbrain. Cranial nerves three and four also originate from this ventral area. The dorsal portion, or roof, of the midbrain consists of four rounded eminences, the *corpora quadrigemina* (two superior and two inferior colliculi), each of which has connections with higher brain centers and with the nuclei of those cranial nerves which supply the muscles of the eyeball (p. 212). The corpora quadrigemina are concerned with **visual** and **auditory reflexes,** such as movement of the eyeball in accordance with changes in the position of the head, and turning the head in order to hear sounds which originate at one side of the body.

The midbrain also contains centers for **postural** and **righting reflexes.** Postural reflexes are concerned with the position of the head in relation to the trunk and with adjustments of the extremities and the eyes to the position of the head. Righting reflexes are concerned with the orientation of the head in space; men and animals are alike in wishing to keep the head right side up.

In summary, it is evident that the midbrain is not only a *conduction pathway*, but also a *reflex center* of great importance, as are all the levels of the CNS below it. The more complicated reflexes which occur through activities in the brain stem are sometimes referred to as *second level* reflexes, in contrast to the simpler *first level* reflexes of the spinal cord.

THE DIENCEPHALON

Situated above the midbrain, the diencephalon is almost completely covered by the cerebral hemispheres. It consists of the *thalamus, epithalamus, subthalamus,* and *hypothalamus;* these structures surround the third ventricle, a cleftlike cavity in the diencephalon. At this time we shall limit our discussion to the hypothalamus and the thalamus. The pineal gland, which is part of the epithalamus, is described on page 477.

The hypothalamus

The hypothalamus forms the floor and part of the lateral walls of the third ventricle. Within its substance are several masses of cell bodies (nuclei), and it has extensive connections with other parts of the central nervous system, as well as with the posterior lobe of the pituitary gland. Most of the functions of the hypothalamus are either directly or indirectly related to the fact that it is the chief subcortical center for the **regulation of visceral activities.** Both the sympathetic and the parasympathetic divisions of the autonomic nervous system (Chap. 9) are under the control of the hypothalamus; thus it is possible to coordinate the responses that maintain optimal internal conditions to meet the changing needs of the body. Some outstanding examples are the *control of body temperature, control of the motility* and *secretions of the gastrointestinal tract,* and *control of arterial blood pressure.*

Behavior and emotional responses. Neurons of the hypothalamus have extensive connections not only with autonomic centers, but also with the thalamus and the cerebral cortex; thus it is possible for our emotions to influence visceral effectors on certain occasions. For example, in response to a rising body temperature, the hypothalamus orders the dilatation of superficial blood vessels, and the skin appears flushed as heat is dissipated through radiation. However, this dilatation also may occur when the hypothalamus is stimulated by higher brain centers, as when one blushes with embarrassment. Extreme anxiety, worry, and tension may play a part in bringing about hyperactivity of either the sympathetic or the parasympathetic system (p. 223). The hypothalamus also is believed to play a central role in the emotional expression of rage.

The **regulation of water balance** through the control of water excretion by the kidney is an involuntary activity that does not operate directly through the autonomic nervous system. This important regulatory activity involves special nerve cell bodies within the hypothalamus and thousands of axons that descend into the posterior lobe of the pituitary gland (p. 473). A neurosecretion, called *antidiuretic hormone* or ADH, is produced by the nerve cell bodies and then flows down the axons to the posterior pituitary where it is stored until needed. Other neurons in the hypothalamus, known as *osmoreceptors*, are sensitive to changes in the concentration of the extracellular fluid. For example, if the amount of water in the body decreases, there is a relative increase in the osmotic pressure of the extracellular fluid as it becomes more concentrated. It is believed that the increased osmotic pressure pulls water out of the osmoreceptors, thus stimulating them to discharge impulses at a very rapid rate; these impulses promote the release of ADH from the posterior pituitary. This hormone is carried in the bloodstream to the kidney tubules, where it greatly accelerates the reabsorption of water, thus preventing its loss in the urine. On the other hand, if the amount of water in the body tends to increase, there will be a relative decrease in the osmotic pressure of the extracellular fluid; water will enter the osmoreceptors, thus decreasing their rate of discharge and inhibiting the secretion of ADH. The excess water then will be eliminated in the urine.

Another neurosecretion, *oxytocin* (p. 476), is produced by cells in the hypothalamus and stored in the posterior lobe of the pituitary until needed; oxytocin acts on smooth muscle, particularly the wall of the uterus, where it promotes contraction.

The hypothalamus has control centers for many other involuntary activities of the body that do not operate directly through the autonomic nervous system. These include the regulation of body weight through *appetite control*, an intimate concern with normal *sleeping-waking mechanisms*, and the *regulation of many endocrine glands* through its influence on the secretory activity of the anterior lobe of the pituitary gland (p. 475).

Injury to the hypothalamus may result in a variety of symptoms which include: (1) *diabetes insipidus*, the condition in which a patient passes excessive amounts of dilute, sugar-free urine, (2) loss of temperature control, (3) obesity, and (4) pathologic sleep during which the patient may sleep for abnormally long periods of time.

The thalamus

The right and left thalami are elongated, egg-shaped masses that make up the great bulk of the diencephalon; they form the lateral walls of the third ventricle (Fig. 8-3) and are composed primarily of gray matter which contains many nuclei.

Each thalamus is an important **relay center** for incoming impulses from all parts of the body and a **sensory integrating organ** of great complexity. The relatively simple afferent impulses coming from the periphery are sorted out and rearranged; those having to do with similar functions are grouped together and then relayed to a definite area in the cerebral cortex. It also is believed that the activities of the thalamus are related to consciousness, to a crude sort of awareness, such as the recognition of pain and temperature and of the *affective* quality of sensations (i.e., whether pleasant or unpleasant). But true discrimination (the comparison of stimuli as to intensity or locality) is a function of the cerebral cortex.

In addition to the peripheral afferent fibers, there are also cortical fibers that bring the thalamus under the control of the cerebral cortex. Efferent connections with the hypothalamus and the basal ganglia (p. 200) enable the thalamus to exert its influence on both visceral and somatic effectors.

THE RETICULAR FORMATION

Somewhat like the core of an apple, the reticular formation lies within the brain stem and the basal part of the diencephalon. It extends from the superior portion of the spinal cord upward and forward through the medulla, the pons, the midbrain, and into the hypothalamus. The reticular formation is composed of small islands of gray matter (cell bodies) separated by fine bundles of nerve fibers that run in various directions. It has widespread afferent connections and, in turn, sends efferent impulses over ascending relays into the cerebral cortex and over descending relays into the spinal cord.

The basic importance of the reticular formation involves its regulation of the background activities of the rest of the nervous system by the excitation of certain activities and the inhibition of others. Caudally, it facilitates both reflex and voluntary movements, thus contributing to optimal motor performance. The cephalic influence of the reticular formation is one of activation and regulation of cortical activities that underlie the state of wakefulness or alertness on which the highest functions of the cerebral cortex depend.

THE CEREBRUM

The cerebrum is the largest part of the human brain, and it fills the entire upper portion of the cranial cavity. This mass of nervous tissue consists of literally billions of neurons and synapses which form a fantastic network of neural pathways.

The cerebrum is almost completely separated into right and left halves, or **hemispheres,** each of which is further subdivided into four lobes which are named according to the overlying cranial bones: *frontal, parietal, temporal,* and *occipital* lobes (see Fig. 8-5). Both cerebral hemispheres consist of (1) an external layer of gray matter, the **cerebral cortex,** (2) underlying **white matter tracts,** and (3) deeply situated masses of gray

matter, the **basal ganglia** (*cerebral nuclei*). Within each hemisphere is an elongated cavity, or *lateral ventricle,* which is concerned with the formation of cerebrospinal fluid (p. 209).

The cerebral cortex

The surface layer of each cerebral hemisphere consists of an elaborate mantle of gray matter which is called the cerebral cortex. This is the topmost executive suite in the nervous system organization, and it has been estimated that there are 10 to 14 billion neurons in this area. Generally speaking, the amount of cortical gray matter is one indication of *potential* intelligence; however, there are many other factors which are involved in the *realization* of this potential, and individual variations are great.

To provide a maximal amount of this im-

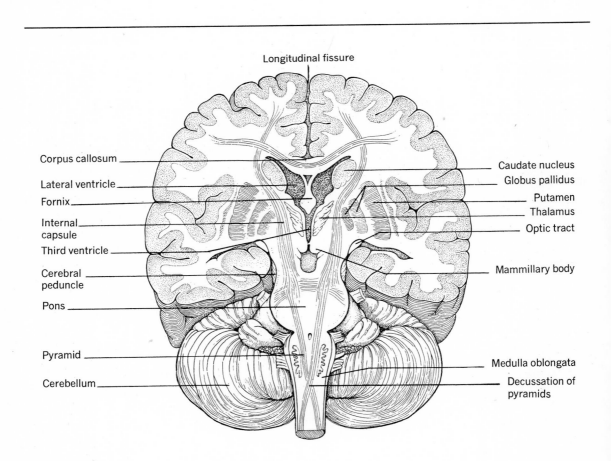

Longitudinal fissure

Corpus callosum

Lateral ventricle

Fornix

Internal capsule

Third ventricle

Cerebral peduncle

Pons

Pyramid

Cerebellum

Caudate nucleus

Globus pallidus

Putamen

Thalamus

Optic tract

Mammillary body

Medulla oblongata

Decussation of pyramids

Fig. 8-3. Oblique coronal section through the cerebrum and brain stem.

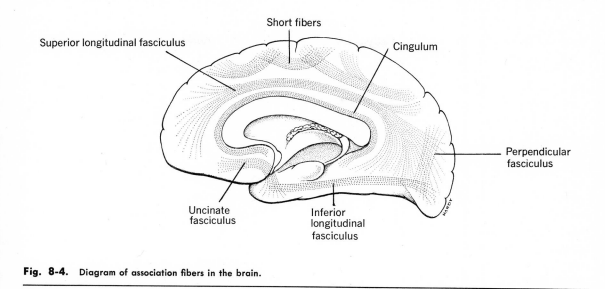

Fig. 8-4. Diagram of association fibers in the brain.

portant gray matter while keeping the size of the brain within reasonable limits, the cortex is arranged in a series of folds or ridges as it dips down into underlying regions. Each fold is called a *convolution*, or *gyrus*. Between these folds are grooves, or furrows, which give the exterior of the brain a wrinkled appearance; a deep groove is called a *fissure*, while a shallow one is referred to as a *sulcus*. Figure 8-3 illustrates the way in which the cortical folds and grooves increase the total surface area of gray matter.

The cerebral hemispheres are separated from each other by the **longitudinal fissure**, a deep groove that extends from front to back in the midline. Anteriorly and posteriorly the separation is complete, but in the middle portion the fissure extends only to the *corpus callosum*, a massive band of white fibers that serves to unite the two hemispheres (see Fig. 8-3). Posteriorly the cerebrum is separated from the cerebellum by the deep **transverse fissure** (see Fig. 8-2).

In each hemisphere there are three fissures which serve as boundary lines between the lobes: (1) the **central fissure** (of Rolando) which separates the frontal and the parietal lobes, (2) the **lateral fissure** (of Sylvius) which lies between the

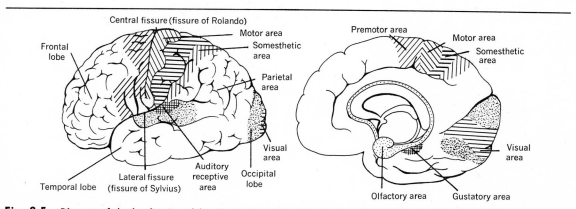

Fig. 8-5. Diagram of the localization of function in the cerebral hemisphere. Various functional areas are shown in relation to the lobes and fissures: *left*, lateral view; *right*, medial view.

frontal and the parietal lobes above and the temporal lobe below, and (3) a small *parieto-occipital* fissure located between the parietal and the occipital lobes (Fig. 8-6). The insula, or island, of Reil is a part of the cortex which lies buried in the lateral fissure and can be seen only when the lips of that fissure are drawn apart.

White matter

Beneath the cerebral cortex there are myelinated nerve fibers which are arranged in related bundles or tracts. These fibers travel in three principal directions and are considered as three groups.

1. **Commissural fibers** are those that transmit impulses from one hemisphere to the other. The largest group of these fibers is the *corpus callosum* as shown in Figure 8-3.

2. **Projection fibers** transmit impulses from one level of the CNS to another. For example, the *internal capsule* (Fig. 8-3) is figuratively a superhighway of ascending and descending fibers.

3. **Association fibers** transmit impulses from one part of the cerebral cortex to another on the same side (Fig. 8-4). These fibers do not cross from one hemisphere to another, nor do they pass to lower levels of the CNS.

The basal ganglia

The basal ganglia, or cerebral nuclei, are four deep-lying masses of gray matter within the white matter of each hemisphere. Two of these are the *caudate nucleus* and the *lentiform nucleus*, the latter being subdivided into the putamen and the globus pallidus (see Fig. 8-3). These nuclei, along with that portion of the internal capsule that lies between them, compose the *corpus striatum*, an important part of the extrapyramidal motor pathway.

Functions of the cerebrum

Because of the fantastic activity of his cerebral cortex, man is the most intelligent creature on earth. In a limited way, lower animals can learn through experience, but only man has the neural mechanism necessary for the more complex discrimination and correlation of sensory impulses, and for greater utilization of past experiences.

Practically every part of the cerebral cortex is connected with subcortical centers and, strictly speaking, there are no perfectly circumscribed areas in the cortex which are purely motor or purely sensory. However, in a very general way there are regions which are primarily concerned with the *expressive* phase of cortical functioning and other regions which are primarily *receptive* in nature. These generalizations facilitate study of the brain, but one must never lose sight of the fact that the cortex as a whole makes possible the *associative memory* reactions which are so characteristic of man.

The **motor area** of each hemisphere is located in the frontal lobe, immediately anterior to the central fissure (Fig. 8-5). Most of the fibers of the pyramidal tract arise from cell bodies in this area. The pyramidal system is primarily concerned with facilitating the fine, versatile movements that enable us to acquire various manual skills.

The **premotor area** lies immediately in front of the motor area; axons descending from this area transmit impulses to the basal ganglia and to certain related nuclei in the brain stem (e.g., the substantia nigra, the red nucleus, and the reticular formation). Collectively, these structures are referred to as the *extrapyramidal system*, a term that serves to emphasize their important functional relationships. In general, this system is concerned with the coordination of gross skeletal muscle activities that are largely reflex in nature, such as postural adjustments, defensive reactions, and feeding.

In man, there are various accessory or associated movements that accompany voluntary activities. For example, the swinging of the arms as one walks, the changes in facial expression, and the gesticulations that one makes while carrying on a conversation are all associated movements. These movements are dependent on impulses from the premotor area and the basal ganglia.

The work of the extrapyramidal system is done on an unconscious level; (that is, it is more or less automatic in that we don't have to *think* about the mechanics of sitting, standing, or walking). However, the functions of this system are subordinate to orders from the cerebral cortex and are brought into balance with the activities of the pyramidal system.

Figure 8-7 illustrates two anatomic facts that are of considerable clinical importance: (1) fibers

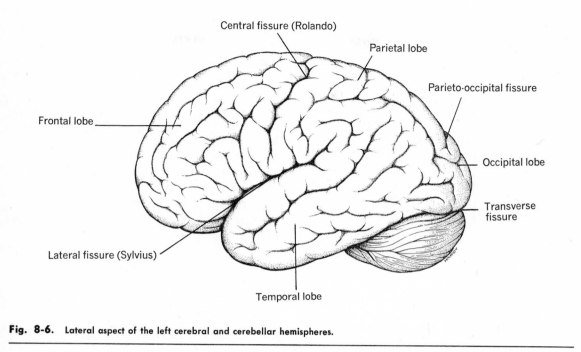

Central fissure (Rolando)

Parietal lobe

Parieto-occipital fissure

Frontal lobe

Occipital lobe

Transverse fissure

Lateral fissure (Sylvius)

Temporal lobe

Fig. 8-6. Lateral aspect of the left cerebral and cerebellar hemispheres.

from both the motor and the premotor areas are funneled through the narrow internal capsule as they descend to lower levels of the CNS, and (2) because of the crossing of fibers, the cortex of one side of the cerebrum controls the skeletal muscles on the opposite side of the body.

The **motor speech area** (*Broca's area*) lies at the base of the motor area and slightly anterior to it. This portion of the cortex is concerned with the ability to form words both in speaking and in writing. The motor speech area usually is located in the left hemisphere of right-handed persons and in the right hemisphere of some left-handed persons. This is due to the fact that although the cerebral hemispheres appear to be exact duplicates of each other, there are a number of functions that are not represented equally at the cortical level. In other words, in certain higher functions, one hemisphere appears to be the leading one; thus it is referred to as the dominant hemisphere. This phenomenon of *cerebral dominance* probably is most prominent in relation to the complex aspects of language and speech. Handedness is related to cerebral dominance, but not in such a clear-cut way as was assumed in the past. For example, left-handedness is less definite

than right-handedness, and it is less regularly associated with dominance in either hemisphere. The degree of cerebral dominance apparently varies from one person to another and with respect to different functions.

The **prefrontal area** is that portion of the frontal lobe which lies anterior to the premotor area. It has extensive connections with cortical areas in other parts of the brain and is thought to play an important part in *complex intellectual activities* such an planning for the future, solving mathematical and philosophical problems, and controlling one's behavior in accord with the standards of society.

The prefrontal area also exerts an influence on certain autonomic functions of the body through the transmission of impulses to the hypothalamus, either directly or indirectly by way of the thalamus. For example, feelings of anxiety, fear, or anger may cause an increase in heart rate, elevation of blood pressure, and interference with the digestion of food. In extreme cases there may even be pathologic changes in the viscera (p. 489).

Much information concerning the functions of the prefrontal area was obtained with the advent of *psychosurgery* and an operation called

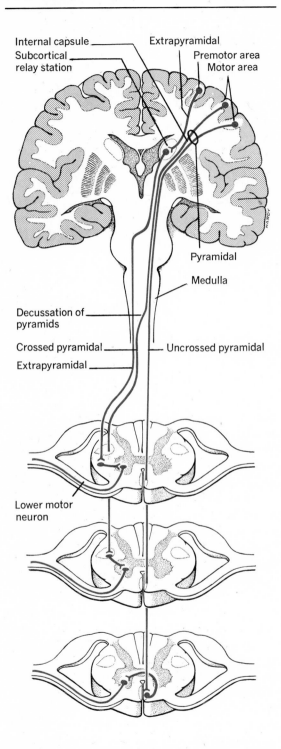

Fig. 8-7. Diagram of motor pathways between the cerebral cortex, one of the subcortical relay stations and lower motor neurons in the spinal cord. Decussation (crossing) of fibers means that each side of the brain controls skeletal muscles on the opposite side of the body.

prefrontal lobotomy. The objective of this procedure was to cut the nerve fibers that linked thought centers in the prefrontal area with emotional centers deep within the brain, thus relieving persons who suffered from certain forms of mental illness (e.g., morbid anxiety, obsessive-compulsive states). However, the surgeon couldn't tell which fibers were specifically involved in emotions; thus he was forced to cut other fibers as well. As a result, many lobotomy patients suffered varying degrees of intellectual and emotional deterioration. They were highly distractible and unable to pursue long and complicated thoughts. They were also unable to analyze situations and their judgment was poor. Even more striking were the alterations in emotional behavior, as the patients were prone to quickly changing moods of hate, joy, and sadness. They became careless in their personal habits and unconcerned with financial or domestic problems.

More recently, as a result of the great strides made in fundamental research on the brain, new psychosurgical procedures have been developed. These new procedures are more precise and can be aimed at a much smaller target, thus avoiding the adverse side effects of lobotomy. One such target is a segment of the cingulum in the limbic system (p. 203 to 204).

The **general sensory area** (*somesthetic area*) is located in the parietal lobe on the postcentral gyrus (Fig. 8-5). While crude sensations of pain, temperature, and touch enter consciousness at the level of the thalamus, *true discrimination* is a function of the cerebral cortex. The position of the body in space, the recognition of size, shape, and texture of objects, the comparison of stimuli as to intensity and locality, all depend on the activities of this area. The left hemisphere receives impulses from the right side of the body, and vice versa.

The **auditory area** lies in the superior part of the temporal lobe, and each hemisphere receives impulses from both ears. The **visual area** is located in the occipital lobe. The right occipital cortex receives impulses from the right half of each eye, and the left cortex receives impulses from the left half of each eye.

The **olfactory** and **gustatory areas** have not yet been determined with certainty. However, it is thought that one area for the sense of smell (olfactory) is located in the hippocampal gyrus on the medial aspect of the temporal lobe (Fig. 8-5), and the area for taste (gustatory) close by at the base of the postcentral gyrus.

The remaining areas of the cerebral cortex are referred to as **association areas,** and they play an important role in the integration of the receptive and the expressive phases of cortical functioning. For example, when you are taking an examination, the occipital lobe serves as the primary receptive area for visualizing the words in each question, and the frontal lobe directs those activities which are involved in expressing your answers. However, impulses must be relayed back and forth through a complex maze of association areas (probably involving the entire cerebral cortex) before you are able to understand the meaning of the question, to sort out and to organize those facts which are related to the question, and to reach a logical conclusion based on your past experience.

Memory and Learning. One of the most striking features of the cerebral cortex is its ability to retain, modify, and reuse information which has been received through the senses; this is the basis of *associative memory* and is the foundation of all knowledge. Such things as abstract reasoning, judgment, and moral sense are made possible by the almost unbelievable interplay between the billions of neurons in the human cerebrum. The brain has been compared to a high-speed computer, but cerebral circuits are infinitely more complex than any machine, and we have only theories to explain human memory banks. Also, little is known about the important process of *forgetting* information that has been acquired through the senses.

Neuropsychologists, who are studying the functional organization of the brain, believe that memory functions in two distinct ways: on a short-term basis and on a long-term basis. *Short-term memory* enables you to retain a telephone number just long enough to dial it, to remember the clothing that you wore yesterday, or to tell a friend about your activities over the weekend. However, such information usually is not retained unless it is repeated or rehearsed. Although forgetting certain information may be annoying on occasion, it does have its advantages in that the brain does not become cluttered with unimportant material.

On the other hand, there is *long-term memory* for special events and for the retention of intellec-

tual skills. This type of memory is, in fact, *learning:* the process whereby information is stored for recall when needed at any time. Some researchers believe that long-term memory involves permanent changes in proteins in the cortical nerve cells and at the synapses. Emotional centers in the limbic system also may play a major role in the learning process; any teacher or parent knows that learning involves reward and punishment.

The retrieval of information from long-term storage may be a complicated process, the major problem being that of finding access to specific units of information. Just as one probes the card catalog to find the location of a particular book in the library, one must also probe the brain to find a particular unit of information that has been stored away until needed. Students frequently come to grips with this problem during examinations. For example, they can *see* the information on a certain page in their notes, but are unable to retrieve it for the practical purpose of answering a question.

Acquired or **conditioned reflexes,** so prominent in all human behavior, are also dependent on associative memory. These reflexes must be *learned* by each individual, since they are not transmitted from parent to child as are the unconditioned or natural reflexes such as the knee jerk or swallowing. Since acquired reflexes involve the cerebral cortex, they are sometimes referred to as *third level* reflexes in contrast to those which involve only subcortical centers in the brain (*second level*) or in the spinal cord (*first level*). Voluntary control of urination is a good example of an acquired reflex. The infant bladder when filled with urine automatically empties itself, and it is only through training and practice that this spinal reflex is brought under control of the cerebral cortex. In fact, most of our daily activities involve behavior patterns which we learned as children. For example, we get dressed every day, giving little or no thought to the shoelace, the button, or the buckle that presented such problems when we were young; and the use of knife, fork, and spoon no longer requires our complete attention at the dinner table.

The **limbic system** plays an important role in controlling the emotions and basic drives of fear, hunger, pleasure, and sex. It includes the limbic cortex and associated subcortical structures, such as the hypothalamus, anterior thalamic nuclei, epithalamus, and amygdaloid complex. The specific functions of all parts of the limbic system are not yet well understood, but the intimate relationship between the limbic cortex and the hypothalamus have led some physiologists to refer to the system as the *visceral brain*.

The *hypothalamus* (p. 196) is the most important part of the limbic system as far as behavior is concerned. This is due to the fact that surrounding parts of the system transmit most of their signals through the hypothalamus.

The *limbic cortex* is an arcuate convolution of gray matter surrounding the central structures of the system. It forms part of the medial frontal and temporal lobes (e.g., cingulate and parahippocampal gyri), and is the oldest part of the cerebral cortex. This part of the brain is thought to be a storehouse for information related to past experiences with pain, pleasure, odors, and sexual activities. Information of this nature apparently is combined with that coming from the prefrontal area and from sensory areas in other parts of the brain. Signals then are dispatched to initiate the appropriate behavioral response (e.g., excitement, rage, sexual activity). Thus it has been said that the limbic cortex occupies an intermediate, associative position between the neocortex (most highly evolved cortex) and the lower, subcortical centers; however, the neocortex has a relatively minor controlling influence on the limbic system. This fact is well demonstrated by the success of the lie detector or polygraph; a normal person is unable to control his autonomic reactions during the emotional stress of telling a lie.

Sleep is a state in which the cerebral cortex is relatively inactive and consciousness is lost. The body is incapable of any purposeful action or thought, yet most of us can be brought back to a state of wakefulness by various stimuli, such as the ringing of an alarm clock. We still do not understand the precise mechanism of sleep, nor do we know exactly why sleep is necessary, but we do know that it is an integral part of life. Severe sleep deprivation results in varying degrees of mental deterioration and hallucinations; however, these symptoms vanish after the person has been permitted to sleep.

There are two different forms of sleep. One, called nondreaming, or slow sleep, is characterized by a progression through various stages of diminished brain activity (i.e., drowsy, light sleep, medium sleep, and deep sleep). The deepest sleep usually occurs about 1½ hours after going to sleep. A second form of sleep is called REM (rapid eye movement) sleep, or paradoxical sleep. After about 1 or 2 hours of deep sleep, the person drifts back into the lighter stages. Now his eyes begin to move very rapidly under his eyelids, much as if he were watching a motion picture. If the person is awakened during REM sleep, he almost always reports that he has been dreaming. A person usually spends a total of about 1½ to 2 hours in REM sleep during an average night. Everyone dreams, but some people do not remember that they do so.

Various physiologic changes accompany sleep. Muscle tone is greatly diminished, heart rate and blood pressure are decreased, the metabolic rate is diminished by about 10 per cent, body temperature falls on an average of about 0.5° F., and respiration is slower and shallower; however, the energy-producing work of the body continues, as food is digested and assimilated.

The **effect of drugs** on the human nervous system is a matter of concern to the general public, as well as to members of the health professions. In recent years, there has been a marked increase in drug abuse and in drug dependence (addiction), particularly in the younger population. The greatest controversy seems to swirl around the use of *marihuana* (cannabis) and whether or not it has any harmful effects. It has been claimed that large numbers of individuals smoke marihuana without any harmful effects, but these claims are unsupported by scientific evidence. On the other hand, a large number of scientifically controlled studies now provide a mounting body of evidence that marihuana is not the harmless drug that many smokers believe it to be. The chief dangers seem to be alterations in mood, judgment, and perception, as well as the precipitation of psychotic states.

Individuals who smoke marihuana on a regular basis will experience a variety of mental changes. For example, there is a limited attention span, decreased ability to concentrate, impaired memory, distorted sense of time, and a loss of interest in personal appearance. The intensity of these symptoms seems to be directly related to the frequency and duration of drug use. Remission of symptoms also is related to the length of time that marihuana is used. Short-term smokers usually have complete recovery if they stop smoking. However, those who use the drug for three years or longer show a poor recovery rate, and heavy doses on a daily basis may lead to atrophy of the brain.

Opiates are drugs that produce both psychic and physical dependence. Two well-known examples are *morphine* and *heroin*, the latter being the main drug in the illegal traffic in the United States. The signs and symptoms of dependence on these drugs varies with the dose and the degree of tolerance. Nausea and vomiting usually follow the first dose, but tolerance develops rapidly and, with repeated use, the individual begins to experience a reduction of anxiety, a feeling of increased ability to perform, and drowsiness. At first, the body temperature falls slightly, there is a decrease in heart rate and blood pressure, the pupils are constricted, and the individual is constipated. With continued use, tolerance develops and most of the effects disappear. Except for constricted pupils and chronic constipation, the user appears to be normal, both mentally and physically.

Withdrawal or abstinence leads to signs of autonomic dysfunction and CNS irritability. These signs range from mild to severe, depending on the quantity of drug and the length of time it has been used. In severe cases, there is restlessness, sweating, dilated pupils, insomnia, muscle cramps, chills, fever, elevated respiratory rate, elevated blood pressure, vomiting, and diarrhea. After 36 to 72 hours, the acute symptoms begin to decline rapidly, and most will have disappeared by the fifth to fourteenth day. During the next 2 to 4 months, there is a gradual decrease in irritability and muscular pains, and the person is able to sleep.

The appeal of heroin is the feeling of escape that it provides, but once the user has become physically dependent (hooked) on the drug, he must have access to a daily supply or else suffer the pangs of withdrawal. Thus drug-seeking behavior is constantly reinforced, and the user will go to great lengths to secure his supply. Viral hepatitis is common in addicts who inject their drugs intravenously. Sudden death is also a frequent occurrence, possibly due to respiratory arrest, convulsions, or both.

Hallucinogenic drugs, such as *LSD* (lysergic acid diethylamide) and *mescaline*, seem to en-

hance the activity of excitatory pathways in the brain. At the same time there is a depression of the inhibitory pathways that normally enable one to suppress most of the barrage of sensory stimuli that continually arrive in the brain. As a result, the individual is flooded with a torrent of sensations that cannot be ignored, and a variety of hallucinations occur. For example, there may be an uninterrupted stream of fantastic images that are constantly changing shape, an intense play of colors and flickering lights, the faces of other people in the room may appear as weird masks, and the individual may feel as if he were out of his own body. Psychiatric complications are common; these states may last for days or weeks, and a few persons have remained permanently psychotic.

Alcohol probably is the dominant drug. It is taken to relieve insomnia, to relieve anxiety, and to facilitate socializing at parties. Alcohol depresses the activity of the CNS and may produce both psychic and physical dependence. Many people believe that alcohol is a stimulant, but this is an illusion due to the fact that one early effect is a loss of inhibitions. For example, a person who tends to be rather shy may become talkative and euphoric after taking a few drinks; as the blood alcohol level continues to rise, he may become argumentative and aggressive. Failing muscular coordination results in slurred speech, in staggering and falling. Finally, there will be sedation and sleep.

When the alcohol is metabolized and its sedative effects are removed, the person may experience a hangover. The severity of this reaction is proportional to the amount and duration of drinking. Some of the symptoms may be headache, tremor, pallor, nausea, vomiting, nervousness, and difficulty in thinking.

Amphetamines are drugs that strongly stimulate the nervous system, and they may produce both psychic and physical dependence. Small oral doses usually cause only slight nervousness, loss of appetite, insomnia, relief of fatigue, and a feeling of increased mental and physical efficiency. However, tolerance develops rapidly and the dose must be increased to obtain the desired effects. Amphetamines are frequently abused by students, by individuals who are trying to lose weight, and by truck drivers who must stay awake to drive long distances. However, the most serious form of abuse involves the self-administration of amphetamines by intravenous injection; individuals who do this are called "speed freaks" by their peers. They may inject the drug as often as every two hours when they are on a "run." During this time, the user is euphoric and in a state of hyperalertness; his pupils are dilated, his blood pressure is elevated, and he neither eats nor sleeps. He may develop paranoid ideas and assault some innocent bystander. Eventually, he becomes exhausted, or depletes his drug supply, and the run stops; this event is referred to as "crashing."

Amphetamine drug abuse may have serious consequences for the individual and for society. A single dose has been known to cause an acute psychotic reaction. Suicide is common. Some users, while experiencing the grand illusion that they can float or fly, have leaped to their death from tall buildings. Infections also contribute to the death toll as viral hepatitis and bacterial endocarditis may be transmitted by the sharing of contaminated hypodermic needles.

Cerebral lesions

A detailed description of tumors, abscesses, depressed skull fractures is obviously not within the scope of this textbook, but a few basic principles related to the more common lesions may facilitate future studies.

The blood supply to the brain (p. 300) is of primary importance, and any lesion that interferes with its normal distribution may cause irreparable damage to nerve cells. For example, a clot may form within a vessel (*cerebral thrombosis*) and block the flow of blood to a particular area, or there may be bleeding into the brain substance (*cerebral hemorrhage*) from a ruptured blood vessel. Such lesions are called **cerebral vascular accidents** (*stroke* or *apoplexy*).

The location of these injuries determines the symptoms that will be presented by the patient. For example, let us consider a small lesion which is localized in part of the motor cortex. Since the pyramidal tract is widely spread out in the frontal area, such cortical lesions involve only part of the fibers, thus producing a contralateral muscular paralysis of a single extremity (*monoplegia*) or possibly paralysis of facial muscles depending on which portion of the cortex is injured. A similar lesion in the sensory cortex would result in a contralateral loss of sensation. On the other hand, if the lesion is in the area of the internal capsule and the basal ganglia, all of the fibers that supply the opposite side of the body may be involved, and a complete *hemiplegia* (one-sided paralysis) will result. If this lesion is in the dominant hemisphere there may be speech defects in addition to the contralateral motor and sensory losses. Headache, dizziness, mental confusion, and coma usually are present during the acute phase. If the patient survives the initial attack, he may be rehabilitated to an amazing degree, but it is a long slow process requiring persistence and patience on the part of all concerned.

Epilepsy is one of the most important of the convulsive states, and it is characterized by sudden, transient alterations of brain function. In many cases the attacks are the result of abnormally active brain tissue caused by injury, infection, or unknown agents. In *grand mal seizures*, the patient usually has a specific warning (aura) before the attack; this may be a visual image, an odor, a sound or a tingling in some region of the body. Loss of consciousness follows, and the patient falls to the floor. Strong contractions of the jaw muscles during the ensuing convulsions may injure the tongue, and there frequently is involuntary emptying of the bladder and the bowel. A variable period of sleep and stupor follows the attack. Drugs that depress the excitability of the nervous system may be administered to epileptic patients in an attempt to prevent attacks.

Parkinson's syndrome (*paralysis agitans*) is linked to a disorder in the basal ganglia. It is characterized by muscular rigidity and tremor which result in disturbances of posture and voluntary movement. Tremor of the hands results in a characteristic rubbing together of the thumb and the fingertips which is called the *pill rolling* movement. These patients have no change in facial expression (masklike appearance), their arms do not swing when walking, and their bodies are bent forward to such an extent that they walk with a peculiar running gait.

Patients with this disease have diminished amounts of dopamine in their basal ganglia. Dopamine, an intermediate product in the synthesis of norepinephrine, is believed to function as an inhibitory transmitter substance that serves to balance the excitatory effects of acetylcholine; thus its absence could account for the aforementioned symptomatology.

PROTECTION OF THE BRAIN

The outer covering of the brain is osseous connective tissue which forms the cranial bones. The inner covering is membranous and consists of the three meninges. Cerebrospinal fluid fills the brain cavities and the surrounding subarachnoid space.

The cranial bones

Housed in its snug bony box, the brain is well protected from ordinary bumps and jolts. As described in Chapter 3, the frontal, the parietal, the occipital, the temporal, the sphenoid, and the ethmoid bones all contribute to the formation of the walls of the cranial cavity in which the brain lies.

The meninges

The membranes that cover the brain are continuous with those of the spinal cord and have the same names, but a few anatomic variations should be noted.

The cranial portion of the *dura mater* consists of two layers. It not only covers the brain, but also lines the interior of the skull and serves as the internal periosteum of the cranial bones. The two layers of the dura mater are in contact with each other in some places, and in other places they are separated as the inner layer dips inward to form protective partitions between parts of the brain. The spaces or channels which are formed by this separation of dural layers (Fig. 8-8) are filled with blood which is flowing out of the brain; thus they are an important part of the circulatory sys-

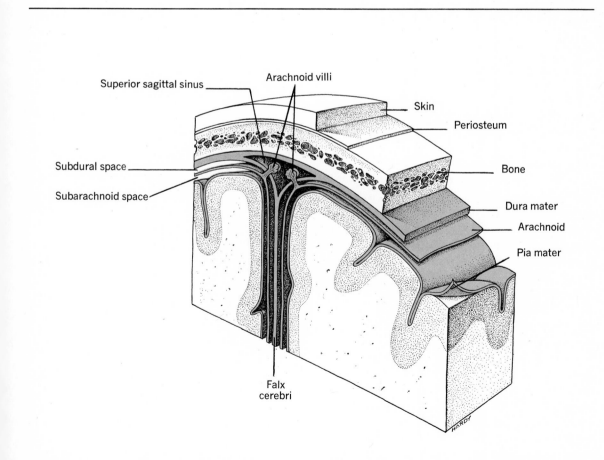

Fig. 8-8. The cranial meninges. Arachnoid villi shown within superior sagittal sinus are one site of passage of cerebrospinal fluid into the blood.

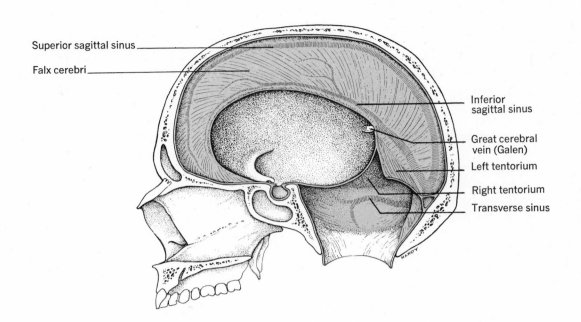

Falx cerebri

Inferior
sagittal sinus

Great cerebral
vein (Galen)

Left tentorium

Right tentorium

Transverse sinus

Fig. 8-9. Cranial dura mater. Skull opened to show the falx cerebri and the right and left portions of the tentorium cerebelli, as well as some of the cranial venous sinuses.

tem. These blood-filled channels are called the *cranial venous sinuses* (not to be confused with the air sinuses of the cranial bones as described on p. 309).

The projection of dura mater which dips into the longitudinal fissure and separates right and left halves of the brain is called the *falx cerebri* (Fig. 8-9); the superior sagittal venous sinus lies in its upper margin (Fig. 8-8). A second projection of dura mater dips into the transverse fissure and separates the cerebrum from the cerebellum; this projection is called the *tentorium cerebelli* (Fig. 8-9). All of the cranial venous sinuses formed by these extensions of the dura are described in Chapter 13.

The cranial portion of the **arachnoid membrane** forms a sort of roof over the pia mater to which it is joined by a network of trabeculae (Fig. 8-10). It also sends little tuftlike extensions through the inner layer of dura mater and into the cranial venous sinuses. These tiny fingers, called *arachnoid villi*, are most numerous along the superior sagittal sinus, but they also are found along the other venous sinuses of the brain. The

arachnoid villi help in the return of cerebrospinal fluid to the bloodstream.

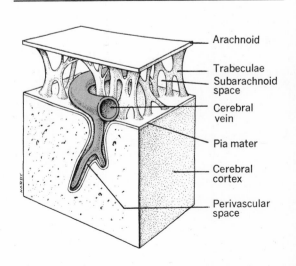

Arachnoid

Trabeculae
Subarachnoid
space

Cerebral
vein

Pia mater

Cerebral
cortex

Perivascular
space

Fig. 8-10. Diagram showing relations of the pia mater, arachnoid, blood vessels, and the brain. The perivascular spaces are in communication with the subarachnoid space.

The inner layer of the meninges, or **pia mater,** covers the brain very closely and follows all of the hills and valleys of the brain surface. The larger blood vessels of the brain lie in the subarachnoid space, and branches from these pass through the pia mater and enter the brain substance, as shown in Figure 8-10. In certain places the pia mater is caught in the folds of the developing brain, and bits of it remain there and become part of the *choroid plexuses* of the ventricles. The pia mater and the arachnoid are connected by a network of fine, threadlike structures (trabeculae); in some areas over the brain it is difficult to separate them.

Inflammation of the meninges is called **meningitis.** It may involve the dura only, or it may involve the arachnoid and the pia mater. It may be due to a variety of infectious agents, such as the tubercle bacillus, meningococcus, streptococcus, or staphylococcus.

The cerebrospinal fluid

The cerebrospinal fluid is a clear, colorless, watery fluid that fills the ventricles of the brain, the subarachnoid spaces around the brain and the spinal cord, and the perivascular spaces related to those blood vessels that penetrate the brain substance (Fig. 8-10). Some of this fluid is formed by filtration through capillary walls, just as is interstitial fluid in other parts of the body, but the major portion is formed by the choroid plexuses (networks of blood vessels) in the ventricles of the brain. In the latter case, as fluid filters through the walls of the choroid plexuses it is modified slightly by the secretory activities of the epithelial cells; for this reason, cerebrospinal fluid is referred to as *modified interstitial fluid.* Some of its constituents are water, sodium chloride, potassium, glucose, traces of protein, and a few white blood cells (3 to 8 per cu. mm.); it has a specific gravity of 1.004 to 1.008. The quantity of cerebrospinal fluid is subject to individual variation, but the average range is from 100 to 200 cc. in the adult.

Functionally, cerebrospinal fluid protects the brain and the spinal cord from mechanical injury by acting as a watery shock absorber. It serves also as a "middleman" in the exchange of nutrients and wastes between the bloodstream and the busy cells of the CNS.

Because most of the cerebrospinal fluid is

formed from the blood that passes through the choroid plexuses in the ventricles of the brain, one should be familiar with the pathway that it follows while seeping downward to the subarachnoid space for distribution to all parts of the CNS. The following structures are of great clinical importance in the circulation of cerebrospinal fluid.

1. The **first** and **second ventricles,** or **lateral ventricles,** are elongated cavities that lie within the cerebral hemispheres, as illustrated in Figure 8-11. Each contains a large choroid plexus.

2. From each lateral ventricle cerebrospinal fluid passes through a narrow, oval opening, the **interventricular foramen** (of Monro) into the third ventricle.

3. The **third ventrical** is merely a slitlike cavity located between the right and the left thalami, but it too contains a choroid plexus.

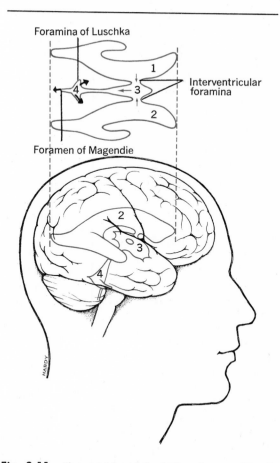

Fig. 8-11. The ventricles of the brain: *above,* outline of ventricles with arrows indicating direction of flow of cerebrospinal fluid; *below,* relation to brain as a whole.

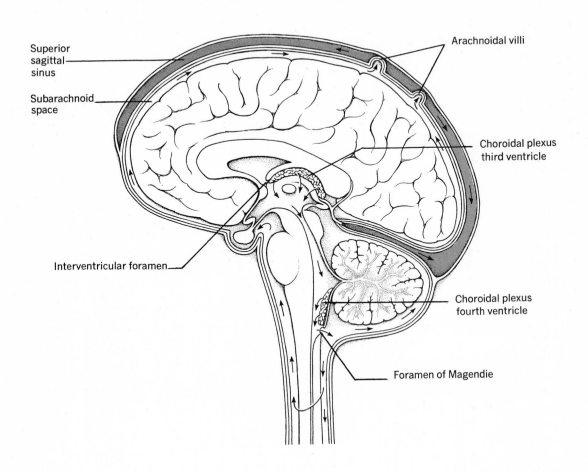

Fig. 8-12. Diagram of the flow of cerebrospinal fluid from the time of its formation from blood in the choroid plexuses until its return to the blood in the superior sagittal sinus. (Note: Plexuses in the lateral ventricles are not illustrated.)

4. The **cerebral aqueduct** (of Sylvius), which passes through the midbrain, is the connecting link between the third and the fourth ventricles.

5. The **fourth ventricle** is a diamond-shaped cavity located between the medulla and the pons anteriorly and the cerebellum posteriorly. The *central canal* of the spinal cord opens into this cavity from below.

6. From the fourth ventricle there are three openings into the cranial subarachnoid space: two lateral openings, called the **foramina of Luschka,** and a medial opening called the **foramen of Magendie.** Figures 8-11 and 8-12 illustrate various features of the ventricular portion of the cerebrospinal fluid pathway.

In the subarachnoid space, cerebrospinal fluid ebbs and flows around the spinal cord and the brain. The cerebrospinal fluid seeps into the sheaths of all cranial and spinal nerves, and into the perivascular spaces (Fig. 8-10) where it mingles with ordinary interstitial fluid. Inasmuch as cerebrospinal fluid is being formed constantly, it must be removed at a comparable rate or a marked increase in intracranial pressure will result. The largest portion of this fluid is absorbed from the cranial subarachnoid space into the bloodstream by way of the arachnoid villi that project into certain of the cranial venous sinuses (Fig. 8-8); it is believed that processes of filtration and osmosis are largely responsible for this absorption of fluid.

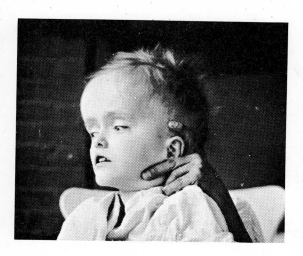

Fig. 8-13. Hydrocephalus. This child was 11 years of age at the time the photograph was taken. (From Minnesota School for the Feeble Minded.)

Because the blood in the venous sinuses has a low hydrostatic pressure and a high osmotic pressure (owing to its protein content), conditions are ideal for the movement of cerebrospinal fluid (with its low protein content) from the arachnoid villi into the bloodstream. The relatively small amount of fluid that passes outward along the sheaths of the cranial and spinal nerves probably enters blood and lymph capillaries as does interstitial fluid in other parts of the body.

Brain lesions which interfere with the passage of cerebrospinal fluid from the ventricles to the subarachnoid space may result in a condition known as **internal hydrocephalus,** or *water in the brain* (Fig. 8-13). For example, if a tumor presses on the midbrain and compresses the cerebral aqueduct, cerebrospinal fluid will accumulate in the lateral and the third ventricles of the brain, greatly distending them and exerting pressure on the brain substance.

Examination of the cerebrospinal fluid is almost a routine procedure in clinical neurology, as both the composition and the pressure of this fluid may provide valuable diagnostic information. Samples of fluid usually are withdrawn through a long spinal needle which taps the large subarachnoid space below the level of the spinal cord (*lumbar puncture* or *spinal tap,* as described on p. 177). Various laboratory tests may then be ordered; for example, cell counts, chemical analysis, presence of bacteria.

Lumbar puncture usually is performed with the patient lying on his side; in this position the cerebrospinal fluid pressure ranges from 70 to 200 mm. of water (average 125 mm. of water). The *Queckenstedt test* is performed by compressing large veins in the neck and temporarily stopping the flow of blood from the cranial venous sinuses. As blood backs up in the brain, its pressure rises, interfering with the drainage of cerebrospinal fluid; as a result there is normally a prompt rise in cerebrospinal fluid pressure which is maintained until the neck veins are released from compression. If the pressure fails to rise and fall promptly, the doctor presumes that there is a block in the system somewhere between the site of puncture and the site of compression. Further definitive studies then must be carried out to locate the lesion.

A **pneumoencephalogram** is an x-ray picture of the skull following replacement of the cerebrospinal fluid with air by means of a lumbar puncture. The ventricles and the subarachnoid spaces usually can be visualized by this method, a valuable tool in localizing intracranial pathology. The air will be absorbed and fluid replaced within 24 to 72 hours.

A more rapid and less traumatic diagnostic study than the **pneumoencephalogram** is the **echoencephalogram** The use of ultrasound echoes to locate the position of midline structures (e.g., falx cerebri, third ventricle) is of particular value in the diagnosis of expanding intracranial lesions that cause distortion of the aforementioned structures.

THE CRANIAL NERVES

There are 12 pairs of cranial nerves originating from the undersurface of the brain (Fig. 8-14) and passing through small foramina in the skull to reach their destination. The cranial nerves are numbered according to the order in which they arise from the brain (front to back), and their names are descriptive of their distribution or of their function. Most are classified as mixed nerves, having both motor and sensory fibers, but three pairs contain only sensory components (olfactory,

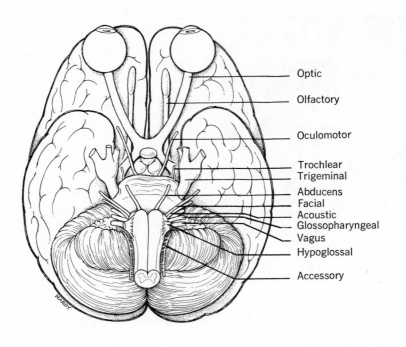

Fig. 8-14. Base of brain showing entrance or exit of cranial nerves. The eyeballs are shown schematically in relation to the optic nerves.

optic, and acoustic). Cell bodies of the primary afferent neurons are located in ganglia which lie outside the brain, while the motor or efferent fibers arise from cell bodies within the brain (nuclei of origin). Some cranial nerves contain only somatic fibers, others contain both somatic and visceral components. Details of the origin and the distribution of visceral fibers are presented in Chapter 9.

1. The **olfactory nerves** are those that conduct impulses related to the sense of smell. They are composed of axons that pass from cell bodies in the olfactory portion of the nasal mucosa upward through the olfactory foramina of the ethmoid bone to reach the brain. Injury results in loss of the sense of smell (*anosmia*).

2. The **optic nerves** are the nerves of vision. Each arises from cell bodies in the retina of the eye and passes through the optic foramen to reach the base of the brain. Injury to the optic nerve results in blindness. (The optic chiasma and the optic tract are described on p. 244).

3. The **oculomotor nerves,** as indicated by the name, are concerned with eye movements. Arising from the midbrain, each nerve supplies motor fi-

bers to the levator palpebrae superioris and to all the extrinsic muscles of the eyeball *except* the superior oblique and the lateral rectus. Visceral efferent fibers supply the constrictor muscle (sphincter pupillae) of the iris and the muscle for accommodation of the lens (ciliaris). Afferent fibers transmit impulses concerned with muscle sense from proprioceptors in the eye muscles to the brain. Signs of oculomotor nerve damage are drooping (ptosis) of the upper eyelid, turning of the eyeball downward and outward, and dilatation of the pupil.

4. The **trochlear nerves** are the smallest of the cranial nerves. Each supplies motor and sensory fibers to the superior oblique muscle of the eye on its own side.

5. The **trigeminal nerves,** arising from the pons, are the largest of the cranial nerves. Although they are often referred to as the great sensory nerves of the head and the face, they actually are mixed nerves. The afferent or sensory portion of each consists of three large divisions: *ophthalmic, maxillary,* and *mandibular branches.* Sensory fibers are supplied to the front part of the

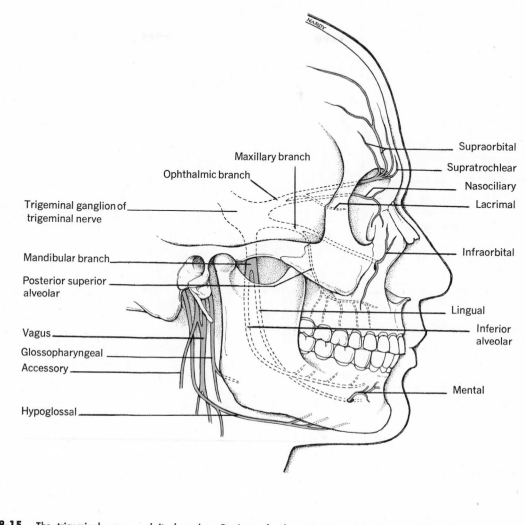

Supraorbital

Supratrochlear

Nasociliary

Lacrimal

Maxillary branch

Ophthalmic branch

Infraorbital

Trigeminal ganglion of trigeminal nerve

Mandibular branch

Posterior superior alveolar

Lingual

Inferior alveolar

Vagus

Glossopharyngeal

Accessory

Mental

Hypoglossal

Fig. 8-15. The trigeminal nerve and its branches. Portions of other cranial nerves also are shown (e.g. vagus and accessory).

head, the orbit, the face, the nose, the mouth, the tongue, the teeth, and the upper part of the throat (nasopharynx). Motor fibers are supplied to the muscles of mastication by way of the mandibular branch. The distribution of the trigeminal nerve is shown in Figures 8-15 and 8-16.

Tic douloureux, a paroxysmal neuralgia of the trigeminal nerve, is an extremely painful condition. Cutting the nerve at its origin, or injecting it with alcohol, may relieve the pain, but the patient also will lose all other sensations in the areas served by the nerve; thus he might burn his tongue with hot coffee, be unaware of a decayed

tooth that should be treated by the dentist, or fail to realize that a cinder is injuring the delicate cornea of his eye. He also would experience difficulty in chewing his food.

6. The **abducens nerves** are small nerves that arise from the pons. Each supplies both motor and sensory fibers to the lateral rectus muscle of the eye on its own side. Injury to the abducens nerve results in the eyeball being turned inward.

7. The **facial nerves,** arising from the pons, are mixed nerves; however, they are often referred to as the great motor nerves of the face, as they supply efferent fibers to all the muscles concerned

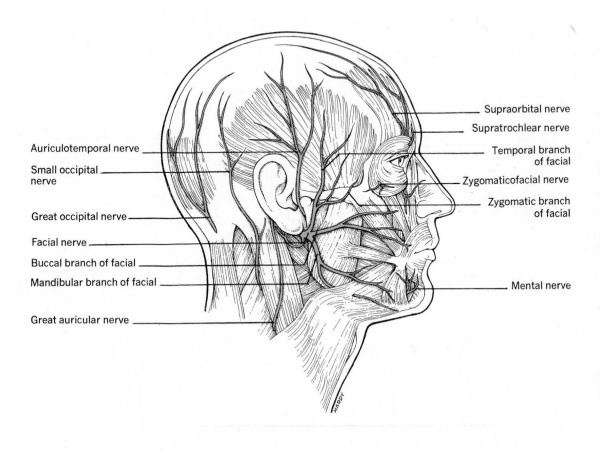

Auriculotemporal nerve

Small occipital nerve

Great occipital nerve

Facial nerve

Buccal branch of facial

Mandibular branch of facial

Great auricular nerve

Supraorbital nerve

Supratrochlear nerve

Temporal branch of facial

Zygomaticofacial nerve

Zygomatic branch of facial

Mental nerve

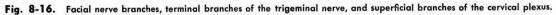

Fig. 8-16. Facial nerve branches, terminal branches of the trigeminal nerve, and superficial branches of the cervical plexus.

with facial expression (Fig. 8-16). Afferent fibers transmit impulses concerned with muscle sense from proprioceptors in those muscles to the brain. The facial nerves also supply sensory fibers for taste on the anterior two thirds of the tongue. Visceral efferent (secretory) fibers are supplied to the submaxillary and the sublingual salivary glands.

Injury to the facial nerve produces a paralysis of the muscles of facial expression (*Bell's palsy*). Figure 8-17 illustrates the resulting masklike expression on the affected side; note the drooping of the corner of the mouth and the inability to close the eye or to wrinkle the forehead. There is also loss of taste in the anterior portion of the tongue and interference with the secretion of saliva.

8. The **acoustic nerves** are concerned with

hearing and with equilibrium. Each consists of two distinct fiber divisions, both of which are purely sensory. The *cochlear* division arises from cell bodies in the cochlea of the internal ear (p. 250) and transmits impulses related to the sense of hearing. The *vestibular* division arises from cell bodies in a ganglion near the vestibule and the semicircular canals of the inner ear (p. 253); it transmits impulses which are initiated by changes in the position of the head and by the movement of the body through space. Injury to the cochlear portion of the acoustic nerve results in deafness; injury to the vestibular portion results in disorientation accompanied by dizziness and nystagmus.

9. The **glossopharyngeal nerves,** as indicated by the name, are distributed to the tongue and the pharynx (Figs. 8-15 and 8-18); they arise from the

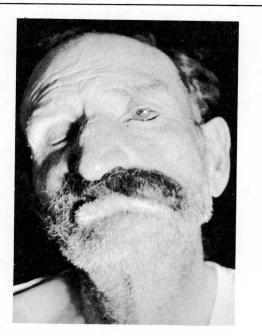

Fig. 8-17. Left-sided facial paralysis. Note drooping mouth and inability to close the eye.

medulla oblongata. These nerves are very important in the act of swallowing, since they supply sensory fibers to the lining of the pharynx and motor fibers to some of the underlying constrictor muscles; they also supply sensory fibers for taste on the posterior third of the tongue. Special visceral afferent fibers conduct impulses from the large artery (internal carotid) in each side of the neck, keeping vital centers in the medulla informed about the pressure and the oxygen tension of the blood. Visceral efferent (secretory) fibers are supplied to the parotid salivary glands. Injury to the glossopharyngeal nerve results in loss of the gag reflex, loss of sensation in the throat and the base of the tongue, difficulty in swallowing, and disturbance of the carotid sinus and carotid body reflexes (p. 342 and p. 344).

10. The **vagus nerves**, after arising from the medulla, live up to their name which means *wanderer*. In other words, these are the only cranial nerves that extend beyond the region of

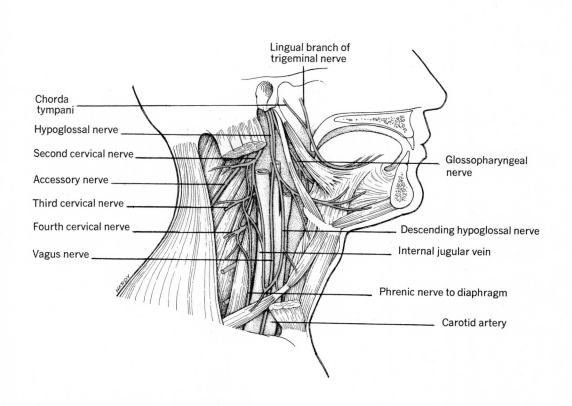

Fig. 8-18. Cranial and spinal nerves related to the neck.

the head and the neck. Each vagus nerve contains both somatic and visceral fibers, but the latter have by far the most extensive distribution as they supply most of the organs in the neck, the thorax, and the abdomen. Figure 9-2 illustrates the fiber pathways of the vagus nerve as it makes an outstanding contribution to the craniosacral division of the autonomic system; functions of this system are described in Chapter 9. In addition to its thoracic and abdominal distribution, the vagus also helps with the action of swallowing by sending motor fibers to skeletal muscle in the soft palate and the pharynx. A very important branch of the vagus (recurrent laryngeal nerve) supplies motor fibers to the muscles in the larynx enabling one to speak. Complete bilateral vagal paralysis is rapidly fatal, but if only one nerve is involved, there will be such things as impairment

of voice, difficulty in swallowing, dilatation of the stomach, and other visceral disturbances.

11. The **accessory nerves** arise from both the medulla and the spinal cord (cervical segments); they were formerly called *spinal accessory* because of the cord component. Each supplies somatic motor and sensory fibers to the sternocleidomastoid and part of the trapezius muscles. They also share some of the functions of the vagus nerves. Following injury to the accessory nerve there is difficulty in rotating the head and in raising the shoulder on the affected side.

12. The **hypoglossal nerves** supply motor and sensory fibers to the muscles of the tongue (Fig. 8-18). Injury results in difficulty in speaking, chewing, and swallowing.

Functions of the cranial nerves are summarized in Table 8-1 on page 218.

SUMMARY

1. Brain, or encephalon, in bony box atop vertebral column
A. Size varies with individual
   a. Physical growth ends by 18th or 20th year
   b. Development depends on mental exercise
B. Divisions of the brain
   a. Forebrain: cerebral hemispheres and diencephalon
   b. Midbrain
   c. Hindbrain
2. Brain stem and cerebellum
A. Medulla oblongata
   a. Bulb of brain; conduction pathway and decussation center
   b. Contains vital centers
   c. Centers for higher reflexes
B. Pons
   a. Between medulla and midbrain
   b. Part of conduction pathway to cerebrum
   c. Connects halves of cerebellum
C. Cerebellum
   a. Attached to brain stem by peduncles
   b. Concerned with coordinating muscular activity
   c. Does not function on conscious level
   d. Reflexes for regulation of muscle tone; maintenance of equilibrium and posture; smooth performance of voluntary movements
D. Midbrain
   a. Cerebral peduncles: main conduction pathway to and from forebrain
   b. Centers for visual and auditory reflexes, postural and righting reflexes
3. The diencephalon
A. Hypothalamus (in floor of third ventricle)
   a. Regulation of visceral activities

   b. Many connections with thalamus and cerebral cortex: emotions may influence visceral effectors
B. Thalamus (in lateral walls of third ventricle)
   a. Relay center of incoming impulses and sensory integrating organ
   b. Crude recognition of unlocalized pain and temperature
4. Reticular formation
A. Core of brain stem
B. Regulates the background activities of the nervous system
C. Concerned with alertness or wakefulness
5. The cerebrum
A. Description
   a. Largest part of brain
   b. Divided into two hemispheres
   c. Each hemisphere consists of four lobes
B. Cerebral cortex
   a. Surface layer of gray matter; highest level of the nervous system
   b. Thrown into folds (convolutions) and grooves (fissures)
   c. Major fissures
     (1) Longitudinal
     (2) Transverse
     (3) Central (Rolando)
     (4) Lateral (Sylvius)
C. White matter tracts
   a. Commissural fibers
   b. Projection fibers
   c. Association fibers
D. Basal ganglia
   a. Four deep masses of gray matter
   b. Links in the extrapyramidal motor pathway
E. Functions of cerebrum

a. Expressive phase
   (1) Motor area
   (2) Premotor area
   (3) Basal ganglia
   (4) Extrapyramidal system
   (5) Motor speech area (Broca's)
   (6) Prefrontal cortex
b. Receptive phase
   (1) General sensory area
   (2) Auditory area
   (3) Visual area
   (4) Olfactory area
   (5) Gustatory area
c. Association areas: role in integrating receptive and expressive phases
d. Memory and learning: ability to retain, modify, and reuse information received through the senses; the basis of all knowledge; short- and long-term
e. Acquired or conditioned reflexes: learned through experience rather than inherited
f. Limbic system: control of emotions and basic drives
g. Sleep: nondreaming; REM
h. Effect of drugs: marihuana; opiates; hallucinogenic; alcohol; amphetamines.

F. Cerebral lesions
   a. Interference with blood supply to brain: cerebral vascular accident
   b. Epilepsy: abnormally active brain tissue
   c. Injury to basal ganglia: paralysis agitans or Parkinson's disease

6. *Protection of the brain*

A. Cranial bones
B. Meninges
   a. Dura mater
   b. Arachnoid
   c. Pia mater
C. Cerebrospinal fluid
   a. Clear, colorless, watery cushion to absorb shock: aids in exchange of nutrients and wastes; formed and drained continuously
   b. Circulation pathway: lateral ventricles; interventricular foramina; third ventricle; cerebral aqueduct; fourth ventricle; foramina of Luschka and Magendie; subarachnoid space; arachnoid villi; cranial venous sinuses
   c. Examination of fluid for constituents and for pressure is valuable diagnostic tool

7. *Cranial nerves*

A. Twelve pairs arising from base of brain; numbered from front to back
B. Table 8-1 is a review of functions

QUESTIONS FOR REVIEW

1. What and where are the vital centers?
2. What part of the CNS is primarily concerned with the coordination of muscular activity?
3. How does the hypothalamus contribute to the maintenance of homeostasis?
4. In terms of sensation, how does the thalamus differ from the general sensory area of the cerebrum?
5. Distinguish between commissural and projection fibers; give an example of each.
6. Briefly, what is the function of the extrapyramidal system?
7. What and where is Broca's area?
8. What is associative memory?
9. What is the general function of the limbic system?
10. What physiologic changes accompany sleep?
11. Is alcohol a stimulant or a depressant? Explain.
12. What will be the end result of a cerebral vascular accident in the region of the internal capsule?
13. List in proper sequence the structures through which a drop of cerebrospinal fluid flows as it travels from the lateral ventricles to the subarachnoid space.
14. What cranial nerve is primarily involved in
    a. an inability to smile or whistle?
    b. loss of sensation in the face?
    c. an inability to shrug the shoulders?
    d. blindness?
    e. drooping of the eyelid, dilated pupil, eyeball turned outward?

REFERENCES AND SUPPLEMENTAL READINGS

Beeson, P. B., and McDermott, W.: Cecil-Loeb Textbook of Medicine, ed. 13. Philadelphia, W. B. Saunders, 1971.

Best, C. H., and Taylor, N. B.: The Physiological Basis of Medical Practice, ed. 8. Baltimore, Williams & Wilkins, 1966.

Chusid, J. G.: Correlative Neuroanatomy and Functional Neurology, ed. 14. Los Altos, Lange Medical Publications, 1970.

Heimer, L.: Pathways in the brain. Sci. Am., 225:48, July 1971.

Kandel, E. R.: Nerve cells and behavior. Sci. Am., 223:57, July 1970.

Kolansky, H., and Moore, W. T.: Toxic effects of chronic marihuana use. JAMA, 222:35, Oct. 2, 1972.

Luria, A. R.: The functional organization of the brain. Sci. Am., 222:66, March 1970.

Peterson, L. R.: Short term memory. Sci. Am., 215:90, July 1966.

Pribam, K. H.: The neurophysiology of remembering, Sci. Am., 220:73, Jan. 1969.

Truex, R. C., and Carpenter, M. B.: Human Neuroanatomy, ed. 6. Baltimore, Williams & Wilkins, 1969.

Table 8-1   Functions of Cranial Nerves

Number	Name	Structures Innervated by Efferent Components	Structures Innervated by Afferent Components	Functions
I	Olfactory	None	Olfactory mucous membrane	Nerve of smell
II	Optic	None	Retina of eye	Nerve of vision
III	Oculomotor	Superior, medial, inferior recti; inferior oblique; levator palpebrae superioris; ciliary; sphincter of iris	Same muscles (muscle sense)	Motor and muscle sense to various muscles listed; accommodation to different distances; regulates the amount of light reaching retina; most important nerve in eye movements
IV	Trochlear	Superior oblique	Same muscle	Motor and muscle sense to superior oblique; eye movements
V	Trigeminal	Muscles of mastication	Skin and mucous membranes in head; teeth; same muscles	Nerve of pain, touch, heat, cold to skin and mucous membranes listed; same for teeth; movements of mastication and muscle sense
VI	Abducens	Lateral rectus	Same muscle	Motor and muscle sense to lateral rectus; eye movements
VII	Facial	Submaxillary and sublingual glands; muscles of face, scalp, and a few others	Same muscles; taste buds of anterior two thirds of tongue	Taste to anterior two thirds of tongue; secretory and vasodilator to two salivary glands; motor and muscle sense to facial and a few other muscles
VIII	Acoustic (cochlear and vestibular portions	None	Cochlear organ of Corti. Vestibular-semicircular canals, utricle, and saccule	Cochlear division is nerve of hearing; vestibular division is concerned with registering movement of the body through space and with the position of the head
IX	Glossopharyngeal	Superior pharyngeal constrictor; stylopharyngeus muscle; parotid gland	Taste buds of posterior one third of tongue; parts of pharynx; carotid sinus and body; stylopharyngeus muscle	Taste to posterior one third of tongue and adjacent regions; secretory and vasodilator to parotid gland; motor and muscle sense to stylopharyngeus; pain, touch, heat, and cold to pharynx; afferent in circulatory and respiratory reflexes
X	Vagus	Muscles of pharynx, larynx, esophagus, thoracic and abdominal viscera; coronary arteries; walls of bronchi; pancreas; gastric glands	Same muscles; skin of external ear; mucous membranes of larynx, trachea, esophagus; thoracic and abdominal viscera; arch of aorta; atria; great veins	Secretory to gastric glands and pancreas; inhibitory to heart; motor to alimentary tract; motor and muscle sense to muscles of larynx and pharynx; constrictor to coronaries; motor to muscle in walls of bronchi; important in respiratory, cardiac, and circulatory reflexes
XI	Accessory	Sternocleidomastoid and trapezius muscles; muscles of larynx	Same muscles	Motor and muscle sense to muscles listed; shares certain function of vagus
XII	Hypoglossal	Muscles of tongue	Same muscles	Motor and muscle sense to muscles of tongue; important in speech, mastication, and deglutition

The Autonomic Division of the Nervous System

"What a piece of work is man!"
Hamlet • William Shakespeare

The autonomic division of the nervous system consists of those neural pathways which are primarily concerned with the regulation of visceral activities, such as the beating of the heart, the secretion of glands, and the movement of food through the digestive tract. As the term **autonomic** implies, this portion of the nervous system has a certain degree of functional independence; it is not under direct voluntary control as is the somatic division which innervates skeletal muscles. Although visceral structures are less dependent on nerve impulses than are the skeletal muscles, their activities are constantly being monitored and regulated by the central nervous system.

This division of the nervous system into visceral and somatic portions is convenient from a physiologic standpoint, but it does not mean that we are dealing with two anatomically distinct systems. For example, the higher centers of the brain regulate both somatic and visceral activities, and peripherally, all spinal nerves and many of the cranial nerves contain both somatic and visceral fibers. Moreover, most body responses to internal and external changes in the environment usually involve skeletal muscle activity on the one hand and simultaneous changes in smooth muscle, cardiac muscle, and glands on the other. Such coordination of somatic and visceral activity is basic to the maintenance of homeostasis. In the light of these facts it is apparent that the autonomic division is only one aspect of an integrated nervous system, but, in accordance with custom, it will be given special consideration in this text.

GENERAL ORGANIZATION

The innervation of visceral structures is fundamentally similar to that of somatic structures in that spinal reflex arcs and higher brain centers are involved. Although *visceral* afferent components are of importance in initiating certain reflexes, major emphasis will be placed on the *visceral efferent* pathways over which impulses are transmitted from the CNS to visceral effectors (smooth muscle, cardiac muscle, and glands).

Visceral afferent fibers

Various visceral and viscerosomatic reflexes are initiated by nerve impulses which originate in the viscera and are transmitted over afferent fibers to the spinal cord and the brain stem. These fibers are contained in cranial and spinal nerves, thereby traveling with somatic afferent fibers but completely independent of them. The cell bodies of visceral afferent neurons are located in the dorsal root ganglia of spinal nerves or in comparable ganglia of cranial nerves. Visceral afferent impulses are primarily concerned with reflex control of the heart, the caliber of blood vessels, respiratory activities, elimination. However, visceral afferent impulses also may give rise to the so-called organic sensations (e.g., hunger, nausea) and to sensations of visceral pain (p. 233 to 234).

Visceral efferent fibers

Visceral efferent fibers are those that transmit impulses from the CNS to smooth muscle, cardiac muscle, and glands. These fibers differ from somatic efferent fibers in that there are two neurons in the pathway from the CNS to the visceral effector. (The somatic efferent pathway to skeletal muscle consists of only one neuron.) *Visceral efferent neuron No. 1* has its cell body in the brain stem or in the spinal cord, from which the axon extends to a ganglion of the autonomic group; such axons are called **preganglionic fibers.**

Within the autonomic ganglion lie the dendrites and the cell body of *visceral efferent neuron No. 2.* The preganglionic fiber (axon No. 1) synapses with the dendrites or the cell body of neuron No. 2; the axon of this second neuron, called a **postganglionic fiber,** transmits impulses from the ganglion to the visceral effector. Figure 9-1 illustrates the general arrangement of these visceral efferent fibers.

Preganglionic fibers have a less extensive origin than somatic efferent fibers. The former are limited to three regions of outflow from the CNS.

The *first* region is cranial. Preganglionic fibers accompany somatic efferent fibers in the oculomotor (third), facial (seventh), glossopharyngeal (ninth) and vagus (tenth) cranial nerves and are called *cranial autonomics.*

The *second* outflow of autonomic fibers is from the thoracic and upper lumbar regions of the spinal cord. These preganglionic fibers, which form part of the anterior roots of the spinal nerves from the eighth cervical to the second lumbar, are called the *thoracolumbar autonomics.* This group of fibers passes through the intervertebral foramina as part of the respective spinal nerves, then leaves each spinal nerve as a white ramus and continues on to an autonomic ganglion, as shown in Figure 9-1. Of particular interest in this group are those preganglionic fibers which enter into the formation of the **splanchnic nerves;** fibers from the fifth to the ninth thoracic spinal nerves form the *greater splanchnic* nerve, and those from the ninth and the tenth thoracic nerves form the *lesser splanchnic* nerve. These important nerves pierce the diaphragm and enter the celiac plexus (p. 224).

The *third* outflow is in the sacral region. Preganglionic fibers, which are present in the anterior roots of the second, the third, and the fourth sacral spinal nerves, are called *sacral autonomics.*

The autonomic ganglia

Synaptic connections between neurons in the visceral efferent pathway occur in autonomic ganglia which are divided into three groups: lateral, collateral, and terminal. The **lateral,** or **vertebral ganglia,** are beadlike structures which are united by threads of nervous tissue (axons and dendrites), forming two distinct chains. Each chain lies close to the bodies of the vertebrae on its own side of the vertebral column, extending downward through the neck, the thorax, and the abdomen to the coccyx (Fig. 9-2). There are usually 22 ganglia in each chain: three cervical, eleven thoracic, four lumbar, and four sacral, but individual variations are common.

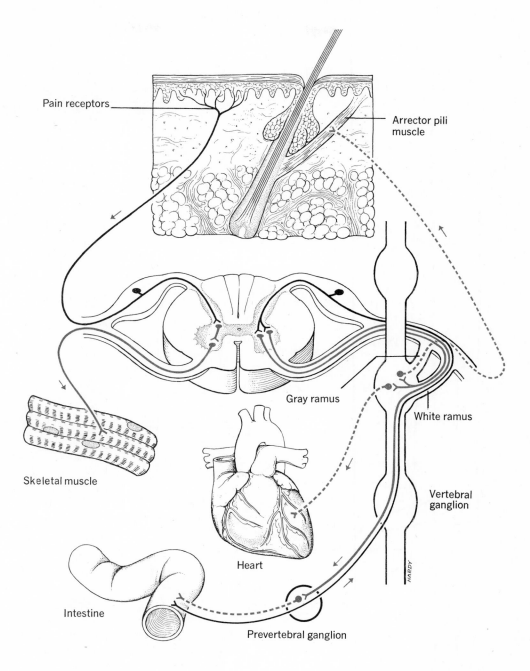

Pain receptors

Arrector pili muscle

Gray ramus

White ramus

Skeletal muscle

Heart

Vertebral ganglion

Intestine

Prevertebral ganglion

Fig. 9-1. Examples of somatic and visceral reflex pathways. Arrows indicate direction of impulse transmission. Afferent neurons are shown in black; somatic central and efferent neurons are shown in color on the left. Visceral efferent neurons, shown in color on the right, are preganglionic (solid lines), and postganglionic (dashed lines); synapses occur either in vertebral ganglia or in prevertebral ganglia.

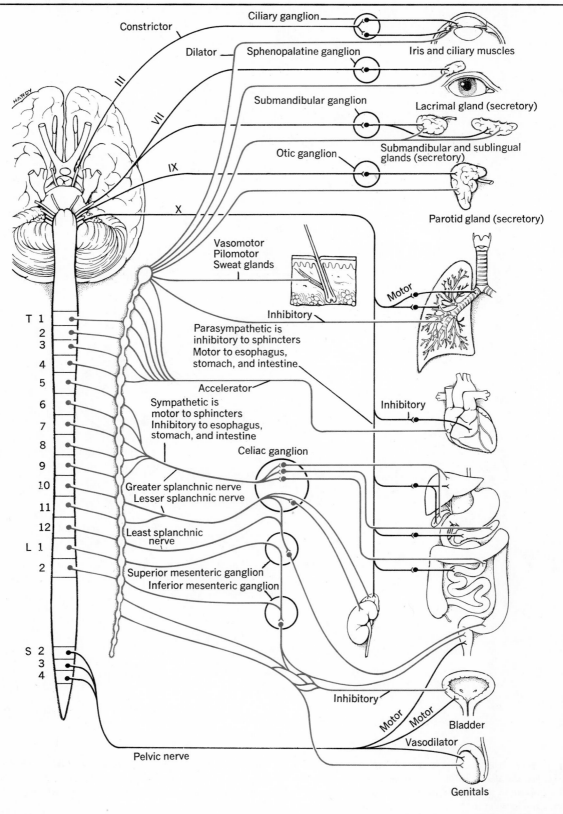

Constrictor

Ciliary ganglion

Iris and ciliary muscles

Dilator

Sphenopalatine ganglion

III

Submandibular ganglion

Lacrimal gland (secretory)

VII

Otic ganglion

Submandibular and sublingual glands (secretory)

IX

X

Parotid gland (secretory)

Vasomotor
Pilomotor
Sweat glands

Motor

Inhibitory

Parasympathetic is
inhibitory to sphincters
Motor to esophagus,
stomach, and intestine

Accelerator

Sympathetic is
motor to sphincters
Inhibitory to esophagus,
stomach, and intestine

Inhibitory

Celiac ganglion

T 1
2
3
4
5
6
7
8
9
10
11
12
L 1
2

Greater splanchnic nerve
Lesser splanchnic nerve

Least splanchnic
nerve

Superior mesenteric ganglion
Inferior mesenteric ganglion

S 2
3
4

Inhibitory

Motor Motor

Bladder

Vasodilator

Pelvic nerve

Genitals

The lateral ganglia receive preganglionic fibers from the thoracolumbar region of the spinal cord; some of these fibers terminate here, either in the ganglion which they entered first, or they may pass up or down the chain and terminate in a ganglion which is above or below their level of entrance. Of the postganglionic fibers leaving the lateral ganglia, some pass directly to visceral structures in the head, the neck, and the chest, while others join the spinal nerves (as a gray ramus) and supply peripheral structures, such as the sweat glands and smooth muscle in the walls of blood vessels and around hair follicles (Fig. 9-1). However, some of the thoracolumbar preganglionic fibers do not terminate in the lateral ganglia but merely pass through them and emerge as the splanchnic nerves which make their synaptic connections in the collateral ganglia of the abdomen.

The **collateral,** or **prevertebral ganglia,** lie in front of the vertebral column, and close to the large abdominal arteries from which their names are derived. For example, the two *celiac* ganglia lie one on either side of the celiac artery as it arises from the abdominal aorta; the *superior* and *inferior mesenteric* ganglia lie near the roots of their respective arteries. The collateral ganglia receive preganglionic fibers from the thoracolumbar region of the spinal cord. Postganglionic fibers follow the course of various arteries as they extend from the ganglia to visceral effectors in the abdomen and pelvis. An exception to this pattern concerns those preganglionic fibers, conveyed by the splanchnic nerves, that are destined for the adrenal medullae (p. 483); these fibers pass through the collateral ganglia, without synapses, to the adrenal medullae, which serve as ganglia in this particular instance, the cells of which correspond to postganglionic fibers.

Terminal ganglia constitute the third group of autonomic ganglia and, as their name implies, are located near the end of the visceral efferent pathway. These ganglia lie close to or within the walls of the visceral effectors which their postganglionic fibers supply; they receive preganglionic fibers from the cranial and the sacral regions of the CNS.

Divisions of the autonomic system

Structurally and functionally, the autonomic portion of the nervous system can be divided into two distinct parts: (1) the thoracolumbar or sympathetic division, and (2) the craniosacral or parasympathetic division.

The **thoracolumbar,** or **sympathetic division,** consists of the *thoracic* and *lumbar* outflow of visceral efferent fibers. The term *sympathetic* pertains to a mutual relation between parts, whereby a change in one has an effect on the other. Preganglionic fibers of this division make their synaptic connections with many neurons in various *lateral,* or *collateral,* ganglia. For example, fibers entering the sympathetic chains may establish connections with as many as nine lateral ganglia, and the branches of a single axon may synapse with 30 or more ganglionic neurons. This arrangement makes possible a widespread radiation of impulses and an extensive distribution of postganglionic fibers.

The **craniosacral,** or **parasympathetic division,** consists of the *cranial* and *sacral* outflow of visceral efferent fibers. The term *parasympathetic* means *beside* or *near* the sympathetic yet having an opposite effect on individual organs. The preganglionic fibers of this division make their synaptic connections only in *terminal* ganglia, and individual fibers synapse with only a few ganglionic neurons (in contrast to the multisynaptic connections of the sympathetic division). Consequently, there is no widespread radiation of impulses and a relatively limited distribution of parasympathetic postganglionic fibers.

The great autonomic plexuses

The autonomic plexuses are complicated networks of fibers found in the thoracic, the abdominal, and the pelvic cavities. They serve as areas of redistribution for both the sympathetic and the parasympathetic fibers which enter into their formation. The cardiac, the celiac, and the hypogastric plexuses are of greatest prominence.

The **cardiac plexus** is located at the base of the heart, overlying the roots of the great blood

---

Fig. 9-2. Diagram of the automatic nervous system. Parasympathetic or craniosacral fibers are shown in black, while the sympathetic or thoracolumbar fibers are shown in red. Note that most organs have a double nerve supply.

vessels. From this plexus, and smaller subplexuses, sympathetic and parasympathetic fibers are routed to the heart, the lungs, and other visceral effectors in the thorax.

The **celiac plexus** (*solar plexus*) is located behind the stomach, overlying the abdominal aorta and the celiac artery. In addition to the sympathetic splanchnic nerves, the celiac ganglia, and related postganglionic fibers, this great plexus also consists of parasympathetic preganglionic fibers which pass through on their way to terminal ganglia. From the celiac plexus (and smaller subplexuses) fibers are routed to most of the abdominal viscera.

The **hypogastric plexus** lies in front of the fifth lumbar vertebra and the sacral promontory. It receives fibers from both of the autonomic divisions and distributes them to the pelvic viscera.

Although it is evident that the autonomic plexuses consist of intermixtures of both sympathetic and parasympathetic fibers, it must be emphasized that the sympathetic fibers may be preganglionic, postganglionic, or both, due to the anatomic location of their ganglia. On the other hand, the parasympathetic fibers are preganglionics which are merely passing through the plexus en route to their terminal ganglia.

PHYSIOLOGIC CONSIDERATIONS

The sympathetic and the parasympathetic divisions of the autonomic nervous system are primarily concerned with the regulation of visceral activities; thus they play an important role in the maintenance of homeostasis. Most of the visceral effectors in the body have a double autonomic nerve supply, one from the sympathetic and one from the parasympathetic division. These divisions are usually antagonistic in their effect on individual organs or structures, one serving to initiate or increase activity, the other to inhibit or decrease the activity. For example, stimulation of sympathetic fibers which supply the heart will speed its action, but stimulation of parasympathetic fibers will slow its rate of beat. There are a few visceral effectors which receive only one nerve supply; the sweat glands, pilomotor muscles, and smooth muscle in the walls of certain blood vessels are supplied with sympathetic fibers only (see Table 9-1 on page 229).

The parasympathetic (craniosacral) division

The parasympathetic division is primarily concerned with those activities which deal with the restoration and the conservation of bodily energy and with the elimination of body wastes. To a large extent, such activities are initiated by internal changes in the viscera themselves, changes which act on the interoceptors of visceral afferent neurons. These activities are almost entirely involuntary and unconscious; the few visceral afferent impulses that do reach the conscious level are poorly localized and vague. However, muscular spasms or the distention of the walls of visceral structures may result in sensations of pain. Because of the limited distribution of efferent fibers, parasympathetic activity may involve only one organ at a time; this means that there is no widespread response as is the case in the sympathetic division.

The functions of the parasympathetic division are illustrated in Figure 9-2. A list of these functions follows:

1. Motor, or Augmentative Functions. This division causes *contraction* of the sphincter muscle of the iris, which makes the pupil of the eye smaller; of the ciliary muscle of the eye, which adjusts the lens system for vision at different distances; of the smooth muscle of the bronchioles, which makes them smaller, causing air to be moved in and out of the lungs with more difficulty; and of the smooth muscle of the walls of the stomach, the intestine, and the urinary bladder, which causes each of these organs to become more active.

2. Inhibitory Functions. The impulses that reach certain effectors by way of this division lead to a decrease in the activity or the *relaxation* of smooth muscle. The smooth muscle in the walls of blood vessels of the salivary glands and the external genitalia is relaxed, which means that these blood vessels dilate. Impulses that reach the heart cause it to beat more slowly and weakly. The sphincters of the gastrointestinal tract and the urinary bladder are relaxed by this division.

3. Secretory Functions. Impulses that reach the stomach and the pancreas stimulate them to produce more secretions. Impulses that reach the salivary glands lead to the production of a thin, watery saliva.

In summary, the functions of the parasympathetic division are *conservative* and *restorative*. For example, by narrowing the pupil of the eye, the retina is shielded from excessive light; by promoting the digestive processes, energy-giving foods can be taken into the body and stored for future use; by slowing the heart rate, cardiac muscle has more opportunity to rest; by emptying the bladder and the bowel, waste products are removed from the body.

The sympathetic (thoracolumbar) division

The sympathetic division is greatly influenced by changes in the external environment, changes that stimulate the exteroceptors of somatic afferent neurons. Although constantly involved in the routine day-to-day maintenance of homeostasis, this division probably is best known for the widespread response which results from stimulation of the unit as a whole. Fear, anger, severe pain, loss of blood, asphyxia, and other forms of physical danger will strongly stimulate the sympathetic system. At such times this division is primarily concerned with the mobilization of bodily reserves. For this reason the sympathetic division might be called the *emergency system*, since it prepares us for the intense muscular activity that may be involved in meeting a stress situation. Whether we stand our ground and fight it out or take to our heels in flight, there are certain visceral changes that take place as we prepare for action (see Fig. 9-3).

The functions of the thoracolumbar division of the autonomic system are illustrated in Figure 9-2. A list of these functions follows:

1. Motor, or Augmentative Functions. This division causes *contraction* of the following: the radial muscle of the iris, which means dilatation of the pupil of the eye; the pilomotor muscles, which make the hair stand on end or cause goose-flesh; and the smooth muscle in the sphincters of the gastrointestinal tract and the urinary bladder for retention of contents. When the sympathetic fibers to the heart are stimulated, they cause it to beat faster and more forcibly. When impulses reach the smooth muscle of the blood vessels in the skin and the viscera, they cause the muscle to contract to a greater extent; this is called *vasoconstriction*, since the blood vessels become constricted or smaller.

2. Inhibitory Functions. This means that when impulses reach certain effectors by way of the sympathetic autonomic fibers, the smooth muscle *relaxes*. The smooth muscle which is so affected is that in the blood vessels of skeletal and cardiac muscle (leading to dilatation of these vessels), the muscle in the walls of the gastrointestinal tract (which stops contracting), the wall of the urinary bladder, and the walls of the bronchioles (which makes it easier for air to get into and out of the lungs).

3. Secretory Functions. Sympathetic fibers carry impulses that increase the activity of sweat glands and promote the secretion of epinephrine and norepinephrine by the adrenal glands. Impulses to the salivary glands result in the production of a thick, viscid saliva, which leads to a feeling of dryness in the mouth.

4. Glycogenolytic Function. When impulses reach the cells of the liver, glycogen is broken down to glucose which then is released to the bloodstream, adding to the available sources of energy.

In summary, the sympathetic division not only counterbalances routine parasympathetic effects, but also prepares the body for *fight* or *flight* in times of physical danger or in emotional crises. For example, imagine yourself walking down a deserted street late at night. The mere *thought* of possible danger may be sufficient to make your palms moist with sweat and your heart begin to beat more rapidly. Then, suddenly, your fears seem to be realized as a figure steps out of a doorway and walks toward you. Impulses flash from the visual cortex and related association areas in the cerebrum to the hypothalamus, whence they are relayed to lower levels which prepare you to meet the emergency (Fig. 9-3).

The entire sympathetic division is strongly stimulated. As a result, there is dilatation of the pupils and an increased secretion of sweat. The heart beats faster, and blood vessels of the viscera and the skin are constricted, thus raising blood pressure and shunting blood into the dilated vessels of skeletal muscle, heart muscle, lungs, and brain. Breathing becomes more rapid, and the bronchioles dilate to facilitate the movement of air into and out of the lungs. The blood sugar level rises as liver glycogen is changed to glucose and put into circulation where it will be available to supply the increasing energy needs of the body. The secretion of epinephrine intensifies and pro-

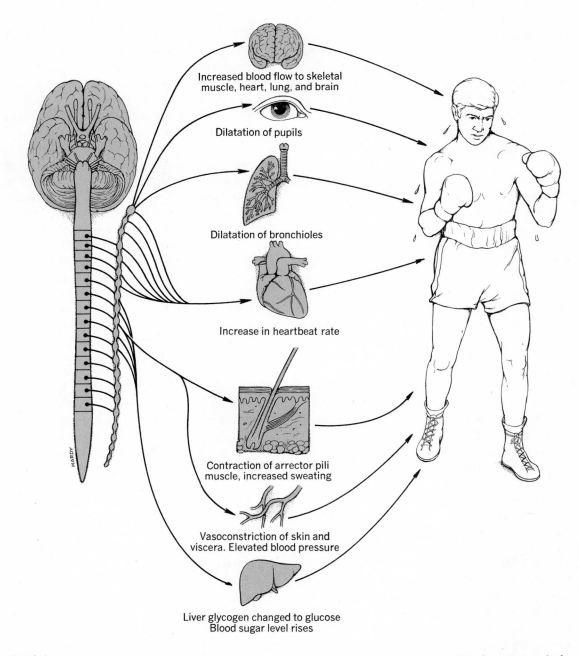

Increased blood flow to skeletal
muscle, heart, lung, and brain

Dilatation of pupils

Dilatation of bronchioles

Increase in heartbeat rate

Contraction of arrector pili
muscle, increased sweating

Vasoconstriction of skin and
viscera. Elevated blood pressure

Liver glycogen changed to glucose
Blood sugar level rises

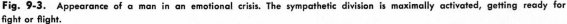

Fig. 9-3. Appearance of a man in an emotional crisis. The sympathetic division is maximally activated, getting ready for fight or flight.

longs the activity of the entire sympathetic division. Since you are going to fight or to take flight, the above changes in the body are primarily concerned with supplying the skeletal muscles and the vital organs with increased amounts of blood that

is rich in oxygen and glucose. Peaceful activities, such as the various digestive processes, are slowed down or even stopped as you prepare for offensive or defensive action. Then the lights of a passing automobile reveal the approaching figure to be a policeman who has been checking building security. With a sigh of relief, you begin to relax as the emergency situation is ended.

Coordination of visceral activity

As stated previously, visceral structures have a certain degree of functional independence in that one does not voluntarily alter their activity. However, they are continuously monitored and regulated by cortical and subcortical centers. Microscopic sentries, the interoceptors (p. 170), are constantly dispatching reports to CNS headquarters, keeping it informed of tissue activities and needs in all parts of the body; consequently, it is imperative that these reports be sorted out and correlated if appropriate visceral responses are to be ordered.

The *cerebral cortex* is apparently the master switchboard for all incoming calls after they have undergone preliminary sorting and grouping in the *thalamus*. However, the cortex shunts most of its calls to the *hypothalamus* which then assumes the responsibility for coordinating the appropriate visceral activities. In this way the cerebral cortex is relieved of the burden of dealing with the constant demands for routine visceral adjustments. Control of sympathetic responses is related to the lateral and posterior regions of the hypothalamus; the anterior and medial hypothalamic regions are concerned with control of parasympathetic activities.

Although the autonomic nervous system has long been considered inferior to and independent of the cerebrospinal system, a number of studies have shown that animals can learn visceral responses in the same way that they learn skeletal responses. For example, rats have been taught to decrease their heart rate and blood pressure and to control the caliber of their blood vessels and the rate of urine formation. Other investigators have begun human studies in this area, and, if they are successful, it could lead to a radically new approach to the treatment of such diseases as essential hypertension, cardiac arrhythmias, and psychosomatic symptoms.

Chemical mediators

Chemical substances liberated by nerve fibers are of fundamental importance in the transmission of impulses at synapses and at neuroeffector junctions (i.e., neuromuscular or neuroglandular junctions). Two extremely important chemical mediators are acetylcholine and norepinephrine.

Acetylcholine is released by all preganglionic fibers, sympathetic and parasympathetic, at their endings in the autonomic ganglia; this mediator then serves to excite the postganglionic fibers (Fig. 9-4). Those postganglionic fibers related to the parasympathetic division in turn release acetylcholine at their endings at neuroeffector junctions, and the transmission process is completed. All fibers that liberate acetylcholine are referred to as *cholinergic fibers.* As *previously* noted in the de-

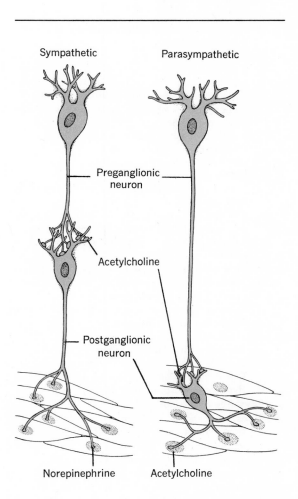

Sympathetic   Parasympathetic

Preganglionic neuron

Acetylcholine

Postganglionic neuron

Norepinephrine   Acetylcholine

Fig. 9-4. The chemical mediators. Acetylcholine is liberated by preganglionic fibers at synapses in all autonomic ganglia, and by parasympathetic postganglionic fiber endings. Norepinephrine is liberated at sympathetic postganglionic fiber endings.

scription of neuromuscular transmission in skeletal muscle (p. 103) acetylcholine is destroyed very rapidly by the enzyme cholinesterase; thus the duration of its activity is limited.

Norepinephrine (sympathin) is the chemical mediator that is released at most, but not all, postganglionic fiber endings of the sympathetic division (Fig. 9-4). One such exception is at the neuroeffector junctions in sweat glands; these fibers liberate acetylcholine. Because norepinephrine closely resembles epinephrine in its effects, the fibers that release it are called *adrenergic fibers*. In general, the effect of these substances is more widespread, because they are augmented by the secretions of the adrenal medullae (p. 483) and are removed somewhat more slowly from the tissues. Norepinephrine is inactivated by the enzyme catechol-o-methyl transferase (COMT).

SUMMARY

1. *Autonomics regulate visceral activity; only one aspect of nervous system*
2. *General organization*
   A. Visceral afferent fibers
      a. Reflex control of activities concerned with maintenance of life
   B. Visceral efferent fibers
      a. Preganglionic fibers from CNS to autonomic ganglia
      b. Postganglionic fibers from ganglia to effectors
      c. Preganglionic limited to three regions of outflow from CNS
         (1) Cranial autonomics
         (2) Thoracolumbar autonomics
         (3) Sacral autonomics
   C. Autonomic ganglia
      a. Lateral or vertebral: beaded chains along vertebral column
      b. Collateral or prevertebral: lie in front of vertebral column in relation to abdominal arteries
      c. Terminal ganglia: lie close to or within walls of effector
   D. Divisions of autonomic system
      a. Thoracolumnar or sympathetic synaptic connections in lateral and collateral ganglia
      b. Craniosacral or parasympathetic: synaptic conections in terminal ganglia
   E. Autonomic plexuses
      a. Areas of redistribution of fibers
      b. Cardiac: visceral effectors in thorax
      c. Celiac: visceral effectors in abdomen
      d. Hypogastric: visceral effectors in pelvis
3. *Physiologic considerations*
   A. Parasympathetic or craniosacral division
      a. Conservation and restoration of bodily energy
      b. Review Table 9-1
   B. Sympathetic or thoracolumnar division
      a. Counterbalance for parasympathetic effects
      b. Emergency system prepares for fight or flight
      c. Review Table 9-1
   C. Coordination of visceral activity
      a. Receptor end organs dispatch reports on environmental change
      b. Reports undergo preliminary grouping in thalamus
      c. Cerebral cortex is master switchboard; shunts most calls to hypothalamus which coordinates visceral responses
   D. Chemical mediators
      a. Acetylcholine released at all preganglionic fiber endings and at parasympathetic neuroeffector junctions; cholinesterase inactivates it; fibers called cholinergic
      b. Norepinephrine released as sympathetic neuroeffector junctions (except sweat glands); fibers called adrenergic; inactivated by COMT.

QUESTIONS FOR REVIEW

1. Visceral afferent fibers are primarily concerned with what activities?
2. Distinguish between somatic and visceral reflexes in terms of their efferent pathways.
3. What are the three major groups of autonomic ganglia?
4. The parasympathetic division is primarily concerned with what activities?
5. Why is the sympathetic division called our emergency system? Explain.
6. In general, what is the role of the hypothalamus in regard to sympathetic and parasympathetic activities?
7. Cover the right half of Table 9-1, then review the autonomic effects on the structures and activities listed in the left-hand column.

REFERENCES AND SUPPLEMENTAL READINGS

Best, C. H., and Taylor, N. B.: The Physiological Basis of Medical Practice, ed. 8. Baltimore, Williams & Wilkins, 1966.

DiCara, L. V.: Learning in the autonomic nervous system. Sci. Am., 222:30, Jan. 1970.

Guyton, A. C.: Textbook of Medical Physiology, ed. 4. Philadelphia, W. B. Saunders, 1971.

Truex, R. C., and Carpenter, M. B.: Human Neuroanatomy, ed. 6. Baltimore, Williams & Wilkins, 1969.

Table 9-1 Autonomic Effects of the Nervous System

Structure or Activity	Parasympathetic Effects	Sympathetic Effects
Pupil of the Eye	Constricted	Dilated
The Circulatory System		
Rate and force of heartbeat	Decreased	Increased
Blood vessels:		
in heart muscle	Constricted	Dilated
in skeletal muscle	*	Dilated
in abdominal viscera and the skin	*	Constricted
Blood pressure	Decreased	Increased
The Respiratory System		
Bronchioles	Constricted	Dilated
Rate of breathing	Decreased	Increased
The Digestive System		
Peristaltic movements of digestive tube	Increased	Decreased
Muscular sphincters of digestive tube	Relaxed	Contracted
Secretion of salivary glands	Thin, watery saliva	Thick, viscid saliva
Secretions of stomach, intestine, and pancreas	Increased	*
Conversion of liver glycogen to glucose	*	Increased
The Genitourinary System		
Urinary bladder		
Muscular walls	Contracted	Relaxed
Sphincters	Relaxed	Contracted
Muscle of the uterus	Relaxed; variable	Contracted under some conditions; varies with menstrual cycle and pregnancy
Blood vessels of external genitalia	Dilated	*
The Integument		
Secretion of sweat	*	Increased
Pilomotor muscles	*	Contracted (gooseflesh)
Medullae of adrenal glands	*	Secretion of epinephrine and norepinephrine

* = No direct effect.

The Sense Organs

"I have heard of thee by the hearing of the ear; but now my eye seeth thee."

Job 42:5

Man's internal and external environments constitute ever-changing worlds of beauty and of danger. However, were it not for millions of receptor end organs (aided by other components of the nervous system) he would be unable to appreciate the lovely things in life or to recognize danger in time to execute proper withdrawal responses. Needless to say, without his sense organs man would be completely helpless and unable to survive for any appreciable length of time.

Sensations may be classified as cutaneous, visceral, olfactory (smell), gustatory (taste), visual, auditory, and position sense.

SENSORY MECHANISMS

A sensory mechanism is a group of structures by means of which a change in the environment gives rise to a sensation. The three components of such a mechanism are: a sense organ or receptor; a pathway to the brain; and a sensory area in the cerebral cortex.

Sense organs, or **receptors,** are the peripheral endings of dendrites of afferent neurons. In these endings the property of excitability (p. 169) reaches its peak of development. The threshold, or minimal stimulus (least change necessary to affect the receptor), is very low. Each receptor, except those for pain, is specialized for the reception of a particular type of stimulus which is called its *adequate stimulus.* For example, light is the adequate stimulus for the receptors of vision, and sound waves for those of hearing.

The **afferent pathway to the brain** consists of those neurons that conduct nerve impulses from the periphery to appropriate sensory areas in the cerebral cortex. The first neuron in the pathway is the afferent neuron whose dendrites are specialized to serve as receptors; these fibers travel in various cranial and spinal nerves. From the axons of the latter neurons, impulses travel over one or more additional neurons before reaching the cerebral cortex.

Sensory areas in the cerebral cortex are well defined for most of our sensations (p. 202). Each sensory unit ends in such an area, and the end-result of the stimulus applied to a receptor is a sensation aroused in consciousness. However, activity does not stop here. Sensations are interpreted, analyzed, combined with sensations from other sensory units, and synthesized into perceptions. As sensations take on meaning, there may be voluntary or reflex movements, visceral reflexes, or storing of sensations in memory to be used in the future. In this manner various sensations contribute to the general fund of information about the outside world and to the knowledge of normal or pathologic activities within the body itself.

All parts of a particular sensory unit must be functioning in order to give rise to its characteristic sensations. If one sense is lost, other senses may be developed more highly. Such a condition exists in blindness, in which the patient learns to read and write through touch and muscle sense by the Braille method. Children who are deaf cannot learn to speak by hearing others speak; they learn through sight, by watching others speak, and by touch, as they feel the throat, the tongue, and the lips. Helen Keller, who was both blind and deaf, learned to speak through touch and muscle sense, by noting the position of tongue, cheeks, lips, and larynx when her teacher was speaking.

Characteristics of sensations

After-images. Sensations tend to persist in consciousness after the cessation of stimulation. This lingering sensation is called an *after-image.* For example, if we look at a bright light for a few seconds and then close our eyes we shall still see the light for a variable period of time.

Adaptation. Prolonged application of certain stimuli may result in a temporary loss of irritability of the receptor end organs. This phenomenon is called adaptation. For example, when we first enter a room we may be very aware of a stuffy odor, but within minutes the olfactory end organs in the nose become adapted, and we are no longer conscious of that odor. The speed with which adaptation occurs varies with different receptors.

Intensity. Sensations are more intense on some occasions than others. The intensity depends on the number of receptors stimulated and on the number of impulses that each afferent fiber transmits per second. With a stronger stimulus more receptors are stimulated, and the afferent fiber of each carries a greater number of impulses per second than when a weaker stimulus is applied.

CUTANEOUS SENSES

Touch, pressure, heat, cold, and pain comprise the cutaneous senses. The receptors for these senses are widely distributed in the skin and the connective tissues, but the number of each type varies. For example, pain receptors are the most numerous, while those for heat are least numerous. Because of such variations in the number of receptors, not all parts of the body are equally sensitive. An area that has few receptors is relatively insensitive, while sensitive areas have large numbers of receptors.

Touch, pressure, and temperature sense

Receptors for touch include free nerve endings around hair follicles and the corpuscles of Meissner in the papillary layer of the skin. **Receptors for pressure** are the corpuscles of Vater-

Pacini deep in the tissues beneath skin and mucous and serous membranes, and near tendons and joints. The adequate stimulus for these receptors is anything that moves a hair, or dents or deforms the tissue in which the receptors are located. **Receptors for cold** are the end bulbs of Krause which lie near the surface of the skin. The **receptors for heat** are the corpuscles of Ruffini which lie deep in the skin. These receptors are illustrated in Figure 10-1.

The **afferent pathways** begin with peripheral nerve fibers, spinal and cranial, that conduct impulses from the receptors to the spinal cord or to the brain stem. From here, second-order neurons conduct the impulses upward to the thalamus, whence they are relayed over third-order neurons to the cerebral cortex.

The **sensory area for touch, pressure, and temperature** is in the parietal lobe of the cerebral cortex, just posterior to the central fissure (see Fig. 8-5). Sensations of touch and pressure are very important in acquiring information about the outside world; children and adults are alike in their desire to handle objects when looking at them for the first time. The ability to recognize a point of stimulation (local sign) is very well developed, but after-images are short. Adaptation to touch and pressure stimuli is rapid. For ex-

ample, one becomes accustomed to clothing very quickly unless it is too tight for comfort.

Contrast is particularly important in temperature sense. To demonstrate this effect, place one hand in a basin of hot water and the other in a basin of cold water, for 30 seconds. Then place both hands in a basin of tepid water; it will feel cold to the one which has been in hot water and hot to the one which has been in cold water. In other words, the sensation of temperature aroused in each hand when placed in tepid water is greatly influenced by preceding events. Note also that after-images are long in temperature sense.

The skin of the face and hands is less sensitive to temperature changes than are those parts of the body that are covered with clothing. Mucous membranes, except in the mouth and rectum, are generally insensitive to temperature change. If hot food burns the mouth, the person usually swallows it quickly since he cannot feel it burn the esophagus; however, the mucous membrane may be damaged even though he is unaware of it.

Somatic pain

There are two general types of pain: somatic and visceral. Visceral pain is described on page 234. Somatic pain may be subdivided into superficial,

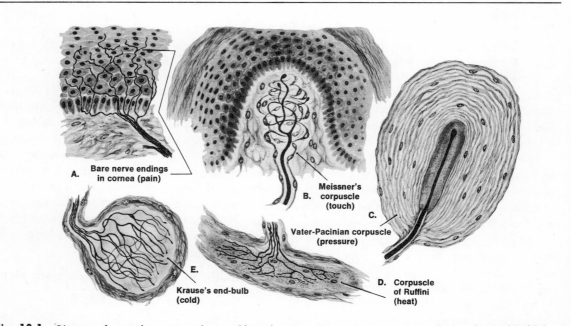

Fig. 10-1. Diagrams for touch, pressure, heat, cold, and pain receptors. (A.W. Ham: Histology, ed. 6. Philadelphia, J.B. Lippincott, 1969.

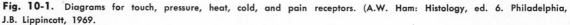

or cutaneous pain, and deep pain from muscles, tendons, and joints.

The **receptors for pain** are free nerve endings. They are scattered throughout the body, and it is estimated that there are several million of them. They respond to more than one variety of stimulus. An intense stimulus of any type, such as a hot object, can affect them, especially if it threatens harm to the body. Since pain is important in protecting us, it is the harmful nature of the stimulus that is noted rather than its specific quality.

The **afferent pathway for somatic pain** consists of peripheral nerve fibers that conduct impulses from receptors to the CNS. Central neurons then relay these impulses to the thalamus and to the cerebral cortex.

The **sensory area for the discrimination of somatic pain** is thought to be in the parietal lobe of the cerebral cortex, although when this region is stimulated in conscious patients during brain operations, they do not report a sensation of pain as one of the results. The fear and anxiety, usually associated with pain, probably are experienced in the prefrontal cortex.

Adaptation does not exist in the sense of pain. This is especially important to remember because pain is a warning signal of danger, and if we became used to it and ignored it, damage to the body would follow.

MUSCLE SENSE

Because conscious muscle sense impulses inform us of the degree to which the skeletal muscles are contracted and the tendons tensed, it is possible for us to adjust our muscular efforts to the work to be done. This sense also is called the *proprioceptive*, or *kinesthetic sense*, and it enables us to know the position of various parts of the body without the aid of vision. It is indispensable for the proper control of voluntary movements and for the correlation of various muscle groups, as in walking. When this sense is lacking, the movements become ataxic.

Muscle sense and touch combined enable us to judge the texture of cloth, the weight of objects, and the shape of objects even if we are blindfolded. The ability to judge these properties is called *stereognosis* (meaning *solid knowledge*). A lack of this ability is called *astereognosis*.

The **receptors for skeletal muscle** may be localized in muscular tissue, in tendons, or at the junctions of muscle and tendon. They are called proprioceptors.

The **afferent pathway for muscle sense** impulses consists of fibers in cranial and spinal nerves. Those impulses destined to reach consciousness travel in ascending tracts, or posterior columns, to the medulla where they then are relayed onward to the thalamus and cerebral cortex; those impulses which are destined for reflex adjustments travel to the cerebellum over spinocerebellar tracts.

The **sensory area for conscious muscle sense impressions** is in the general sensory area in the parietal lobe of the cerebral cortex, just posterior to the central fissure (see Fig. 8-5).

VISCERAL SENSATIONS

Stimuli and pathways

Adequate stimuli for visceral receptors fall into three groups: (1) dilatation or distention of a viscus, (2) spasms or strong contractions (such as colic), and (3) chemical irritants.

Impulses from visceral receptors travel over one of the following pathways: (1) afferent fibers which travel in company with parasympathetic nerves but are quite independent of them, (2) afferent fibers which travel with sympathetic nerves but are quite independent of them, and (3) afferent fibers which travel with somatic afferent nerves from the body walls and the diaphragm.

The dendrites that make up all three pathways belong to cell bodies in dorsal root ganglia of spinal nerves or comparable ganglia of cranial nerves. Visceral afferent fibers have no cell bodies or synapses in autonomic ganglia and must not be confused with visceral efferent fibers in whose company they travel for a part of their journey to the CNS.

Organic sensations

Visceral afferent impulses concerned with organic sensations are conducted by afferent fibers which travel with parasympathetic autonomic fibers. They give rise to the so-called organic sensations of hunger, thirst, nausea, and distention of the bladder and the bowel. The sensory areas in the brain have not yet been identified.

The **sensation of hunger** is projected to the stomach and is associated with powerful rhythmic contractions of the stomach musculature. These contractions are greatly increased by a low blood sugar level. Hunger is a disagreeable sensation that occurs periodically if food is not eaten. It may be accompanied by a feeling of weakness, trembling, and headache.

The **receptors** that are stimulated during hunger contractions have not been located. Perhaps the spindles in the stomach musculature serve this function.

Appetite is an agreeable desire for food and pleasure in thinking about it, seeing it, or smelling it. The sense of taste and smell contribute to one's appetite. You can have an appetite for your favorite dessert even at the end of a full meal.

The **sensation of thirst** is projected to the pharynx. It is a protective signal and warns of the need for the intake of fluid. It may be unquenchable after vomiting, diarrhea, or hemorrhage in which large amounts of fluid are lost. Thirst may be extreme in untreated diabetes mellitus (p. 485) and in diabetes insipidus (p. 477), because large amounts of urine are excreted and the patients are dehydrated.

Visceral pain

Visceral afferent impulses concerned with pain are conducted by afferent fibers which travel in the company of sympathetic (thoracolumbar) autonomic fibers. They reach the brain and give rise to somatic reflexes which involve skeletal muscle, visceral reflexes, or sensations of pain. An example of the first is the reflex muscular rigidity of the abdominal wall which accompanies appendicitis. An example of the second is the reflex secretion of mucus when the large intestine is distended. An example of the third is pain due to an irritating substance in the stomach.

The impulses that give rise to the sensation of pain require further consideration. The afferent neurons that conduct such impulses from receptors in the viscera to the spinal cord come in contact with central neurons in the spinal cord whose axons form the lateral spinothalamic tracts. Some of these impulses, such as those from serous membranes have a private pathway to the brain, somewhat like a private telephone line. Other impulses are not so fortunate; being on a party telephone line, they must share the central neurons with impulses from cutaneous areas. Thus the axons of the central neurons do double duty by con-

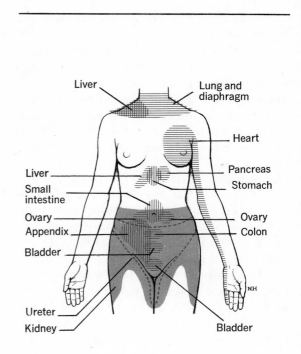

Fig. 10-2. Referred pain, anterior view. The cutaneous areas to which pain in the various organs is referred are indicated.

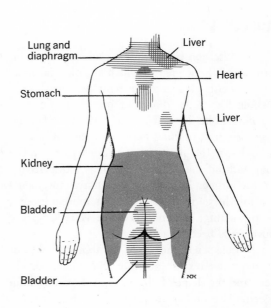

Fig. 10-3. Referred pain, posterior view.

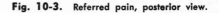

ducting impulses from visceral pain receptors and from cutaneous pain receptors to the same areas in the thalamus and the cerebral cortex. This arrangement is called convergence, and sometimes it can lead to confusion as you will see.

In the first case, in which there is a private wire to the cerebral cortex, the pain can be localized accurately as it is projected to the point of stimulation with ease. As an example of this, the patient with pleurisy can point to a certain spot where he experiences a sensation of pain on the chest, and the physician will hear a friction rub over this area, which means that the pleural membrane is inflamed in this particular spot.

In the second case, in which there is a party line, the sensation of pain is aroused in the brain as usual. However, it is projected to the cutaneous area from which impulses come to the same area in the brain. This happens since cutaneous pain is of more frequent occurrence than visceral pain, and the brain projects over the well-trod path. This phenomenon is called *referred pain*, which means that pain from a viscus is referred to a related cutaneous area.

The cutaneous areas to which visceral pains are referred are of great diagnostic importance. In angina pectoris (due to lack of oxygen in the heart muscle), the pain is referred to the left shoulder and down the left arm, instead of to the heart, where the difficulty really lies. In pneumonia the pain often is referred to the abdomen, and in some cases may be confused with appendicitis. Figures 10-2 and 10-3 show cutaneous areas to which pain from the various viscera is referred.

SMELL AND TASTE

The sense of smell

Receptors for the sense of smell are located in the mucous membrane lining (olfactory epithelium) of the upper part of the nasal cavity.

The olfactory cells are surrounded by supporting cells and glands. The latter secrete fluid which absorbs and dissolves gaseous particles of volatile, odorous substances. Only volatile substances in solution can stimulate the receptors of smell.

The **afferent pathway** is the olfactory, or first cranial nerve, fibers of which pass through the olfactory foramina of the ethmoid bone to reach the brain.

The **sensory area for smell** is thought to be located in the hippocampal gyrus on the medial aspect of the temporal lobe (see Fig. 8-5).

Adaptation is rapid in the sense of smell. Receptors are fatigued rapidly by persistent odors, but new odors may be detected at once. The sensations of smell which are aroused during ordinary breathing depend on diffusion from the moving air in the respiratory path into the still air of the upper portion of the nasal cavity. For a really good whiff, one must sniff to bring the air with its odorous particles into the upper nasal cavity to stimulate the receptors. Irritating substances, such as ammonium salts (smelling salts), stimulate the endings of the trigeminal or fifth cranial nerve and give rise to pain. Respiratory reflexes are initiated by such substances; therefore, they frequently are used to revive individuals who have fainted.

The sense of smell is less important in man than in lower animals, but it does play a part in initiating the flow of certain digestive juices at mealtime. Memory for odors is very keen, and we can recall odors that we have smelled only once.

Olfactory hallucinations (vivid sensations of smell when no stimulus is present) are a fairly common occurrence in persons who are mentally ill.

The sense of taste

Receptors for the sense of taste are the taste buds, which are located chiefly in papillae on the tongue. At the top of each bud there is an opening called the taste pore through which solutions enter in order to stimulate the receptors.

The **afferent pathway** is the glossopharyngeal, or ninth cranial nerve for the posterior third of the tongue, and a branch of the facial or seventh cranial nerve for the anterior two thirds of the tongue.

The **sensory area for taste** is thought to be located in the area that receives other sensory impulses from the mouth, at the base of the postcentral gyrus near the lateral fissure.

Substances must be in solution to be tasted, and one of the functions of saliva is to dissolve substances so that they may enter the taste pores and stimulate the receptors. Just as the smell of food serves as a stimulus for the secretion of certain digestive juices, so do taste sensations.

There are four primary tastes: sweet, sour,

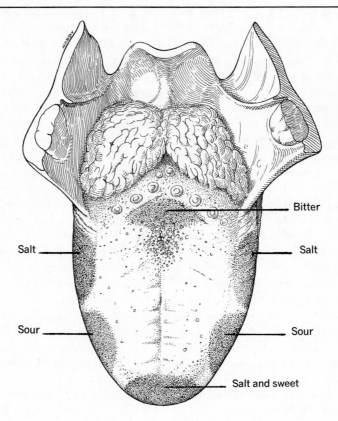

Bitter

Salt

Salt

Sour

Sour

Salt and sweet

Fig. 10-4. Regions of the tongue which are most sensitive to the various tastes. However, all four tastes can be perceived to some degree all over the tongue.

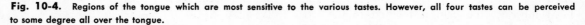

salt, and bitter. The tongue is not equally sensitive to all four tastes in all areas. The back of the tongue is more sensitive to bitter, the tip to sweet, the sides to sour, and both the tip and the sides to salt. The regions that are most sensitive to each of the primary tastes are shown in Figure 10-4.

The property which we call the taste of certain substances is really the odor; the sense of taste is greatly influenced by the sense of smell. If one has a cold, "foods taste differently," since they cannot be smelled.

VISION

The **receptors for vision** lie in the retina of the eyeball, the afferent pathway is the optic, or second cranial nerve, and the **sensory area for vision** is located in the occipital lobe of the cerebral cortex. In addition to these basic units, one must consider the associated structures which

make possible the total functioning of the sensory mechanism for vision.

Protection of the eye

The **orbit** is a cone-shaped cavity whose protective walls are formed by the union of seven cranial and facial bones: frontal, maxillary, zygomatic, lacrimal, sphenoid, ethmoid, and palatine (see Fig. 3-8). The eyeball occupies only the anterior one fifth of the orbital cavity, the posterior four fifths of the space being filled with muscles, nerves, blood vessels, the lacrimal gland, and a proportionately large amount of adipose tissue which serves as a protective cushion. Loss of this orbital fat causes the eyeball to present a sunken appearance (enophthalmus) as seen in persons who have suffered a severe weight loss during prolonged illness. The posterior three fourths of the eyeball is covered by fascia called the *capsule of Tenon;* the smooth inner surface of

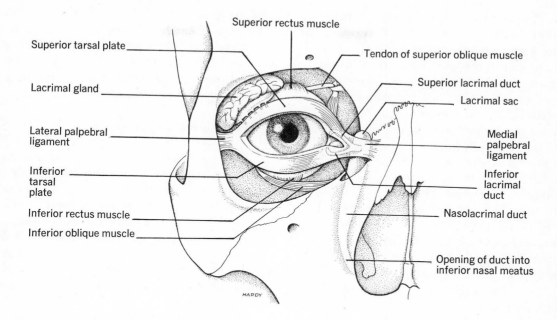

Fig. 10-5. The eye and its appendages, anterior view.

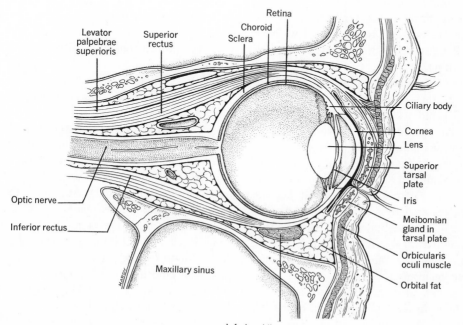

Fig. 10-6. The eye and its appendages, lateral view.

this capsule forms a socket in which the eyeball moves.

The **palpebrae,** or **eyelids,** are protective, movable curtains located in front of the eyeball. The space between them is called the *palpebral fissure.* The upper and the lower lids meet at the medial and the lateral angles of the eye; these meeting places are called the inner and the outer *canthi.* In each lid there is a *tarsus* or plate of dense connective tissue which gives shape to the lid. Modified sebaceous glands, called *meibomian glands,* are associated with the tarsal plate; a *chalazion* is a small tumor of the eyelid, formed by the distention of a meibomian gland with accumulated secretions. The *eyelashes* are hairs which are located at the margins of the eyelids. Sebaceous glands are associated with the follicles of the eyelashes; *hordeolum,* or sty, is an inflammation of one or more of these glands.

The **conjunctiva** is the mucous membrane which lines each eyelid and is reflected over a portion of the front part of the eyeball. It is a thin, transparent membrane, containing goblet cells which secrete mucus. When the conjunctiva becomes infected, the condition is called *conjunctivitis,* or pink-eye. Drops of silver nitrate are placed in the eyes of newborn babies to prevent the possibility of infection by the gonococcus during delivery through the birth canal of the mother. This treatment is required by law, since blindness follows such infection at birth (ophthalmia neonatorum).

The **lacrimal apparatus** consists of the lacrimal gland, its ducts and passages (Fig. 10-5). The lacrimal gland resembles an almond in size and shape. This gland secretes tears, which are carried to the conjunctival sac through about a dozen short ducts. Tears bathe the anterior surface of the eyeball and keep it moist at all times. They protect the cornea which would become cornified like the epidermis of the skin, if exposed to air. The tears also wash away any particles that enter the conjunctival sac. A bactericidal enzyme, lysozyme, in the tears offers protection against certain microorganisms.

Tears normally drain off through the lacrimal canals which begin as small openings, called the lacrimal *punctae,* at the inner canthus. Tears flow into the lacrimal sac and then into the *nasolacrimal duct* which passes downward through a bony canal to open into the lower portion of the nasal cavity. Inflammation and swelling of the nasal mucosa, as when one has a cold, causes obstruction of the nasolacrimal ducts and interferes with the normal drainage of tears; this accounts for the watering eyes which accompany a cold. During emotional crises and during irritation of the conjunctiva, tears are produced in such abundance that they overflow and run down the cheeks.

Extrinsic muscles of the eye

Each eye is equipped with six tiny muscles which arise from bones of the orbit and insert into the outer coat of the eyeball. Four straight, or *rectus, muscles* rotate the eyeball in the direction indicated by their names (superior, inferior, medial, and lateral); two *oblique muscles,* called superior and inferior, serve to rotate the eyeball on its axis. The extrinsic muscles of the eye are described fully on page 122 and illustrated in Figure 5-7 and 10-6.

The *levator palpebrae superioris* arises from the back of the orbit and passes forward to be inserted into the tarsal plate as shown in Figure 5-7. This muscle draws the upper lid upward and backward. The eyelids are closed by the *orbicularis oculi.*

Coats of the eyeball

The eyeball consists of three coats or layers of tissue which surround the transparent internal structures. There is an external fibrous layer, a middle vascular layer, and an inner layer of nervous tissue.

The **fibrous coat** of the eyeball consists of the sclera and the cornea. The **sclera,** also called the *white of the eye,* is composed of dense fibrous tissue. It is the outer, protective and supporting layer of the eyeball (Fig. 10-7). Anteriorly, the sclera is modified from a white, opaque membrane to the transparent cornea; the point at which this transition of sclera to cornea occurs is called the *limbus* (meaning *a border*). The **cornea** is composed of a special kind of dense connective tissue, and no blood vessels are present. The transparent cornea refracts light rays and allows them to enter the eyeball where they ultimately reach the retina.

The **vascular coat** of the eyeball is heavily pigmented and contains many blood vessels which help to nourish other tissues of the eye. This layer

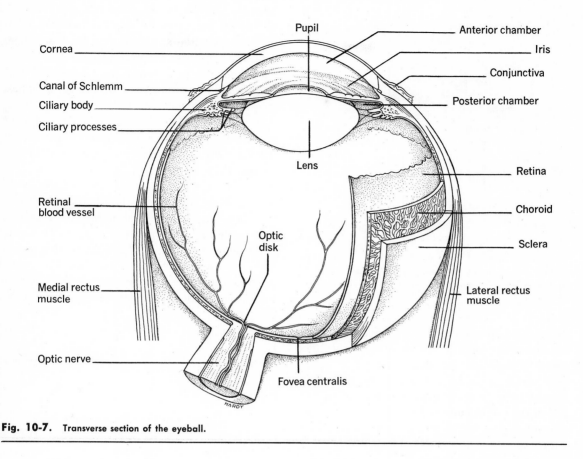

Fig. 10-7. Transverse section of the eyeball.

is subdivided into three belts or zones. (1) The **choroid** portion lies immediately under the sclera to which it is attached. Anteriorly, the choroid is thickened and modified to form (2) the **ciliary body,** the main mass of which is made up of the *ciliary muscle.* This muscle aids in the adjustment of the lens for vision of near objects. (3) The **iris** is a circular curtain attached at its outer circumference to the ciliary body and suspended between the cornea and the lens; its color varies in different individuals. The round hole in the center of the iris permits the entrance of light rays and is called the *pupil.*

The iris is made up of two sheets of smooth muscle, a circular one, called the *sphincter* muscle, and a radial one, called the *dilator* muscle. The function of the iris is to regulate the amount of light entering the eye. Contraction of the sphincter muscle causes the pupil to constrict or become smaller; this occurs when the light becomes brighter (light reflex) and when one looks

at a near object (accommodation reflex). In dim light, and when looking at a distant object, the pupil becomes larger, due to contraction of the dilator muscle.

Around the circumference of the cornea, where it meets the anterior surface of the iris, there is a sort of groove or recess in which the *canal of Schlemm* lies (Fig. 10-8). This is a drainage canal through which the aqueous humor leaves the eyeball and enters the blood.

The **nervous coat,** or **retina,** is the third and innermost layer of the eyeball. Posteriorly, the retina is continuous with the optic nerve; and anteriorly, it ends a short distance behind the ciliary body in a wavy border called the ora serrata. The retina is composed of two parts, the outer being pigmented and attached to the choroid layer, and the inner consisting of nervous tissue.

The nervous portion of the retina contains three orders of neurons. The first-order neurons

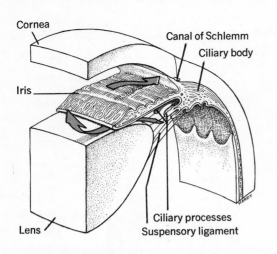

Fig. 10-8. Enlarged view of the anterior portion of the eyeball. Arrows indicate the flow of aqueous humor.

are photoreceptor cells called *rods* and *cones*; they lie next to the pigmented layer of the retina. The axons of first-order neurons converge on *bipolar cells* which are neurons of the second order. The bipolar cells, in turn, converge on *ganglion cells* which represent neurons of the third order. Axons of the ganglion cells form the *optic nerve* from each eye. Though it seems rather strange, light rays must pass through the layers of ganglion and bipolar cells in order to reach the rods and cones.

There are about 100 million rods and 7 million cones in each human retina. The rods are relatively slender in comparison with the cones (Fig. 10-9), but both contain light-sensitive chemicals which decompose on exposure to light. The products resulting from this decomposition excite the rod and cone cell membranes. As a result, signals flash over optic nerve fibers to the brain.

The rods are concerned with night vision, when we see the outlines of objects, but no color or detail. They are very sensitive to the movement of objects in the field of vision. The light-sensitive chemical in the rods is called *rhodopsin*, or *visual purple*. It is a combination of protein, *scotopsin*, and the carotenoid pigment *retinene*. When the retina is exposed to light, rhodopsin decomposes to scotopsin and retinene. The retinene, in turn, is converted to vitamin A. This process is reversed in darkness; that is, vitamin A is converted to

retinene which combines with scotopsin to reform rhodopsin. Thus one can see how a vitamin A deficiency may lead to *night blindness*. With reduced quantities of vitamin A in the blood there will be a similar reduction of vitamin A, retinene, and rhodopsin in the rods, thus decreasing their sensitivity. This condition is called night blindness because at night there is far too little light to permit adequate vision. A more serious impairment of vision may follow a separation of the retinal layer from the choroid; this condition is known as *detached retina*.

The cones are concerned with daylight vision, during which we see detail and color of objects. The light-sensitive chemicals in human cones have not yet been isolated. However, in certain lower animals the isolation has been accomplished, and the composition of the chemicals differs only slightly from that of rhodopsin. Apparently, the cone chemicals require brighter light for their breakdown.

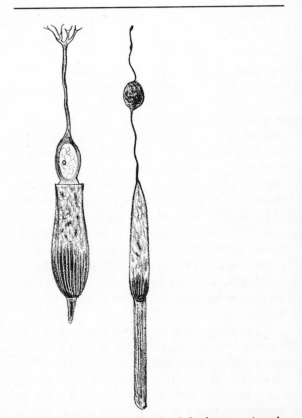

Fig. 10-9. Cone and rod cells of the human retina; the receptors for vision.

If white light (sunlight) is passed through a prism, it is divided into a band of colors which is called the *spectrum* (red, orange, yellow, green, blue, blue-green, and violet). The red rays are the longest, violet the shortest. Color vision is made possible by the presence of three light-sensitive pigments in three different kinds of cones. One type of cone responds maximally at the red wavelength, another at the green, and still another at the blue (Young-Helmholtz theory); other colors are perceived as a result of the blending of impulses from the three cone types. If all three types are stimulated equally, we have the sensation of seeing white.

After-images are long in vision. There are two types: (1) After-images which appear in the same color as the original stimulus are called *positive* after-images. A positive after-image can be brought to the attention by fixing the gaze on an electric light bulb for a few seconds and then closing the eyes; the after-image of the bulb appears in the same color as the stimulus. (2) *Negative* after-images are those which appear in a color complementary to the original stimulus. To produce a negative after-image, fix the gaze on a colored object for about 20 seconds, and then look at a white surface. The after-image appears in the complementary color.

Adaptation. In the eye, the term *adaptation* refers to the changes which occur in the retina when going from a lighted into a darkened room, or vice versa. You have all noted the difficulty in vision when you first step out into the sunlight when the ground is covered with snow, or when you first enter a dark movie theater after passing through a brightly lighted lobby. The sensitivity of the retina can be adjusted to correspond to the intensity of illumination, but it requires a little time. Adaptation is a very remarkable property of the retina; the range of adaptation, with the corresponding change in sensitivity of the retina, is far beyond that of any physical instrument. A dark-adapted retina is 10 billion times as sensitive to light as a light-adapted retina.

Near the center of the retina, as one looks into the eye, there is a small, yellowish spot called the *macula lutea*. The center of the macula lutea is depressed due to thinning of the retina; this region, known as the *fovea centralis* (see Fig. 10-7), is the area of most acute and detailed vision. The foveal region is composed entirely of cone cells.

A short distance to the nasal side of the fovea centralis a pale disk can be seen. This area, called the *optic disk*, is the point at which the optic nerve leaves the eyeball. The optic disk also is referred to as the *blind spot* since there are no rods or cones in the area.

The blood supply for the retina enters the eyeball as the central artery of the retina, a branch of the ophthalmic, which in turn is a branch of the internal carotid artery (p. 300). The central artery enters the eyeball in the center of the optic disk and divides into many branches.

Retinal Examination. In order to see the retina, reflected light is essential. The reason we do not see retinae when we look at people's eyes is that we are in our own light. We need a device that will enable us to throw more light into the eye and yet not cut it off as it is reflected to the source. The ophthalmoscope is such an instrument, being a mirror with a hole in it. The mirror enables the observer to throw more light into the eye, and the hole permits rays reflected from the patient's retina to enter the observer's eye. The appearance of the central artery and its branches contributes valuable information to the opthalmologist. Such pathologic conditions as high blood pressure and diabetes mellitus are accompanied by typical changes in the retinal vessels. A rise in the intracranial pressure, as with a brain tumor, may obstruct the flow of blood away from the eyeball through the central vein, causing a reddening and swelling of the optic disks; this condition is referred to as *papilledema*, or *choked disk*. If the intraocular pressure rises, the disk becomes cup-shaped as it is pushed backward; this condition is called *cupped disk*. The appearance of a normal retina is shown in Figure 10-10 and of a pathologic retina in Figure 10-11.

Transparent internal structures of the eye

Before reaching the retina, a light ray must pass through the cornea, the aqueous humor, the crystalline lens, and the vitreous body.

This involves the refraction or bending of light rays.

The **refraction of light rays** enables them to be focused on the retinae. When a ray of light passes from a medium of one density into a medium of a different density, such as from air to

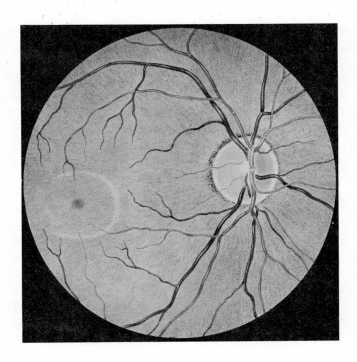

Fig. 10-10. Normal fundus, as seen in retinal examination with an ophthalmoscope.

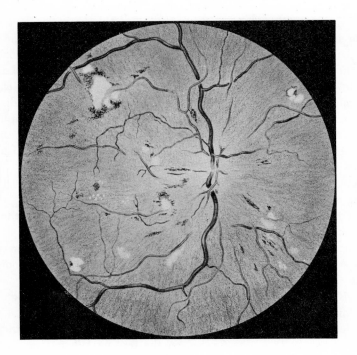

Fig. 10-11. Pathologic fundus, as seen in retinal examination. The macula is not evident, but one can see papilledema, petechial and flame-shaped hemorrhages, and interrupted arteriovenous crossings.

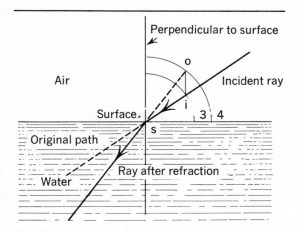

Fig. 10-12. Path of a ray of light passing from air into water. To construct the path, S is taken as the center, and two arcs with radii of four and three are drawn. A line is drawn from the point at which the ray crosses the inner arc, to the outer arc, parallel to the perpendicular, i to o. The line which connects o and S is projected into the water; this is the path of the ray after refraction. It is bent toward the perpendicular as it enters the denser medium.

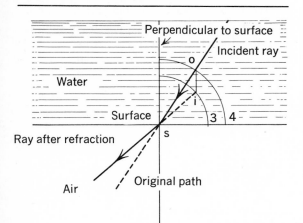

Fig. 10-13. Path of a ray of light passing from water into air. Arcs are drawn as in Fig. 10-12. A line is drawn from the point at which the ray crosses the outer arc to the inner, parallel to the perpendicular, o to i. The line which is projected through i and S indicates the path of the ray after refraction. It is bent away from the perpendicular as it enters the less dense medium.

glass, the ray is bent or refracted. If the ray passes into a denser medium it is bent toward the perpendicular, as shown in Figure 10-12. If it passes into a less dense medium it is bent away from the perpendicular as shown in Figure 10-13.

The crystalline lens is composed of proteinaceous but transparent fibers surrounded by a strong, elastic capsule. It is suspended from the

ciliary body by approximately 70 *suspensory ligaments* (Fig. 10-7); these ligaments attach radially around the capsule and pull the lens edges toward the ciliary body. Under normal resting conditions of the eye, the tension on the suspensory ligaments is such that the lens remains relatively flat. However, contraction of the ciliary muscle pulls the ciliary body forward, thus releasing some of the tension on the ligaments; as a result, the lens assumes a more spherical shape (like that of a balloon), due to the elasticity of its capsule. This gives the lens the increased refractive power needed for viewing objects that are close at hand. Conversely, for viewing objects at a distance, the ciliary muscle is relaxed, the suspensory ligaments are tense, and the capsule flattens the lens, thereby decreasing its refractive power. This process, whereby the strength of the lens is changed for viewing objects close at hand or far away, is called *accommodation* (Fig. 10-14).

A *cataract* is an opacity of the lens or its capsule; light rays cannot pass through the lens to reach the retina, and, consequently, the lens must be removed. After the lens is removed, corrective lenses must be used in order to see objects clearly. Opacity of the cornea likewise prevents light rays from reaching the retina, but this cannot be aided by use of lenses.

The space between the cornea and the lens is subdivided into two chambers. The *anterior chamber* is the space behind the cornea and in

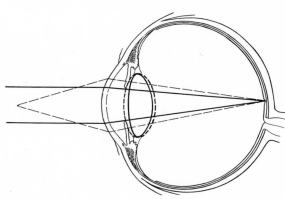

Fig. 10-14. Accommodation. The solid lines represent rays of light from a distant object, and the dotted lines represent rays from a near object. The lens is flatter for the former and more convex for the latter. In each case the rays of light are brought to a focus on the retina.

front of the iris. The *posterior chamber* is the small space behind the iris and in front of the lens and its suspensory ligaments (Fig. 10-8). Both of these chambers are filled with a watery fluid called the *aqueous humor*. This freely-flowing fluid is continually being formed and reabsorbed. It is believed that the formation of aqueous humor involves active secretion as well as filtration from the capillaries in tiny extensions of the ciliary body (ciliary processes). The fluid enters the posterior chamber, then flows through the pupil and into the anterior chamber, whence it drains into the canal of Schlemm for return to the bloodstream. The balance between the rate of formation and the rate of reabsorption of aqueous humor determines the total volume of fluid and the intraocular pressure; so long as these remain normal, the proper shape of the eyeball is maintained. The average intraocular pressure is about 20 mm. Hg, with a range from 15 to 25 mm. Hg. If there is interference with the drainage of aqueous humor, excess fluid accumulates within the eyeball and increases intraocular pressure, causing pain and interference with vision. This condition is called *glaucoma*.

The space posterior to the lens is filled with the **vitreous body** (Fig. 10-7), which has the consistency of jelly; it helps to support the retina and to maintain the shape of the eyeball. Loss of vitreous is a serious complication of eye surgery, since it may result in detached retina; unlike aqueous humor, vitreous cannot be reproduced by the person.

Sensory pathway for vision

Nerve impulses initiated by the rods and the cones are transmitted to the bipolar cells and thence to the ganglion cells, the axons of which form the **optic nerve.** On entering the cranial cavity, the optic nerves meet to form the **optic chiasma,** beyond which they are continued as the **optic tracts.** Visual impulses are ultimately transmitted to the occipital lobe of the cerebral cortex.

The optic nerve fibers from the medial halves of the retinae cross to the opposite side of the brain in the optic chiasma, while those from the lateral halves of the retinae remain uncrossed. As a result of this, the fibers from the right half of

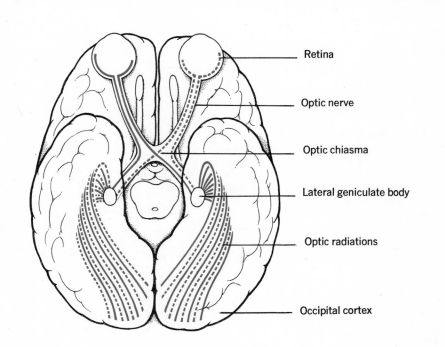

Retina

Optic nerve

Optic chiasma

Lateral geniculate body

Optic radiations

Occipital cortex

Fig. 10-15. Diagram of optic pathways. Note the crossing of fibers from the medial half of each retina.

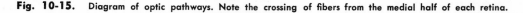

each eye carry impulses to the right occipital lobe, and fibers from the left half of each eye carry impulses to the left occipital lobe. The optic chiasma is shown by diagram in Figure 10-15.

Injury to one optic nerve produces total blindness in the corresponding eye. Lesions of the optic chiasma (e.g., pressure from an enlarged pituitary gland) most commonly involve the central area of fiber crossing and result in *tunnel vision* (*bitemporal hemianopsia*) as the two temporal visual fields are lost.

Binocular vision

Binocular vision refers to vision with two eyes. The advantages of binocular vision are (1) a larger visual field and (2) a perception of depth, or stereoscopic vision. There is a slight difference in the images on the two retinae; there is a right-eyed picture on the right retina and a left-eyed picture on the left retina. It is as if the same landscape were photographed twice, with the camera in two positions a slight distance apart. The two images blend in consciousness and give us an impression of depth or solidity. In order to experience stereoscopic vision, the images must fall on corresponding, or identical, points in the two retinae. Only under this condition is there blending in consciousness. Otherwise, one sees two views or dissimilar pictures, a condition which is called *diplopia*, or *double vision*.

If images are to be brought to identical points on the two retinae, various processes must be perfectly coordinated. For example, when one looks at a near object, there is a *convergence of visual axes*; this means that the eyeballs turn slightly toward one another. To test this movement, keep your eyes on the tip of a finger as you bring it slowly up to your nose. *Strabismus* is a deviation of one of the eyes from its proper direction, due to a muscle imbalance or a lack of coordination in the movements of the two eyes. Other processes to be coordinated include equal constriction or dilatation of the pupils and equal accommodation of the two lenses so that there will be uniform refraction and focusing of light rays on the retinae.

Failure of any of the aforementioned processes may result in defective vision. Since so many of us need glass lenses to correct defects in our eyes and to enable us to see objects near at hand when we grow older, we should under-

stand a few fundamental facts about glass lenses. The principal types are: (1) convex or positive lenses, which converge or bring together rays of light and add to the refractive power of the human eye; (2) concave or negative lenses, which diverge or spread apart rays of light and decrease the refractive power of the human eye; and (3) cylindrical lenses, which equalize refractive power in both the vertical and the horizontal planes, if we happen to have a spoon-shaped cornea instead of a spherical one, as is the case in *astigmatism*. A convex lens is shown in Figure 10-16 and a concave lens in Figure 10-17. Images on the retina are inverted, and it is by experience that we see them right side up.

If the refractive power of the eye is too great or the eyeball too long, the images fall in front of the retina; this condition is called *myopia*, or *nearsightedness*. The refractive power must be decreased by a concave or negative lens, as illustrated in Figure 10-18.

If the refractive power of the eye is too weak or the eyeball too short, the images fall back of the retina; this condition is called *hyperopia* (*farsightedness*). The refractive power must be in-

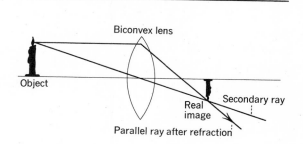

Fig. 10-16. Schematic view of formation of an image by a biconvex lens.

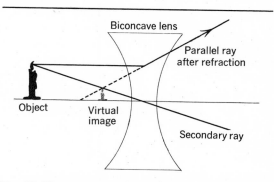

Fig. 10-17. Schematic view of formation of an image by a biconcave lens.

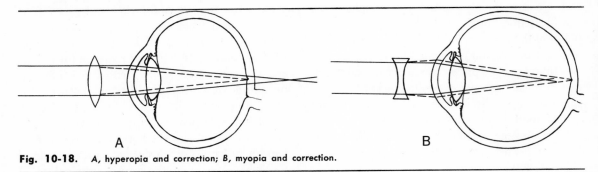

Fig. 10-18. *A, hyperopia and correction; B, myopia and correction.*

creased by a convex or positive lens (Fig. 10-18).

As one grows older the lens loses its elasticity or hardens, just as one's arteries harden with age. This condition is called *presbyopia*. The refractive power of the eye needs to be increased by convex lenses for near vision. Such individuals wear bifocals, which means that part of the lens is for distant vision and the remainder for near vision.

Reflexes of the Eye. Two pupillary reflexes are noted in an ordinary physical examination. One is the *light reflex*, which is tested by flashing a light into the eye and determining if the pupil becomes smaller. The second is the *accommodation reflex*, which is tested by having the patient shift the gaze from a distant to a near object, during which the pupil should become smaller. Figure 10-19 illustrates these reflexes. In syphilis of the CNS the light reflex may be lost, and the accommodation reflex retained; such a condition is called an *Argyll Robertson pupil*. The corneal and conjunctival reflexes are tested by touching these surfaces; the response is a blinking of the eye.

Effects of Drugs on the Size of the Pupil. *Homatropine*, which is used in eye examinations, makes the pupil larger by inhibiting the passage of impulses over the visceral efferent fibers of the oculomotor nerve. This means that the sphincter muscle of the iris and the ciliary muscle are paralyzed temporarily. *Epinephrine* and *cocaine* both lead to a dilatation of the pupil also, but through stimulation of the dilator or radial muscle of the iris. *Pilocarpine* leads to a constriction of the pupil by stimulation of the sphincter muscle of the iris. *Morphine* causes the pupil to become exceedingly small; it is called a *pinpoint* pupil. The effects of various drugs on the pupils are illustrated in Figure 10-20. Drugs that dilate the pu-

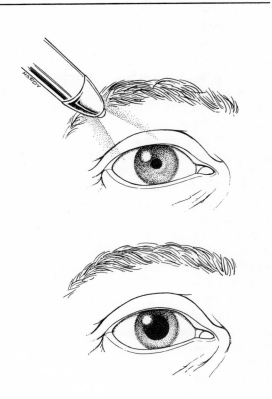

Fig. 10-19. Light and accommodation reflexes. *Above,* the pupil is constricted, as when exposed to light or in looking at a near object. *Below,* the pupil is dilated, as when light is withdrawn or in looking at a distant object.

pils are called *mydriatics,* and those that constrict the pupils are called *miotics.*

Optical Judgments. Judgments of size and distance depend mainly on the size of the retinal images, the amount of detail seen, the color, the

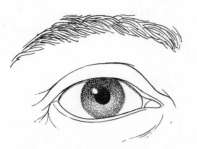

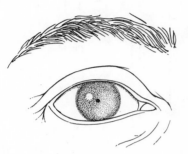

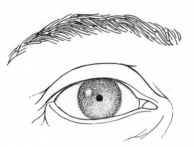

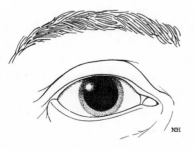

Normal

Morphine

Pilocarpine

Epinephrine, atropine, cocaine

Fig. 10-20. Reaction of pupil to drugs. In chronic morphinism, the pupil approaches a pinpoint; it is moderately constricted after pilocarpine; epinephrine (Adrenalin), atropine, and cocaine lead to a dilatation.

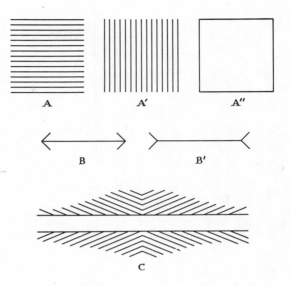

Fig. 10-21. Optical illusions. A, A' and A" are equal squares; A appears taller than A', and A" appears smaller than A or A'. Lines B and B' are of equal length; B' seems longer than B because of the obtuse angle. In C, the horizontal lines are parallel, but they do not appear so.

clearness of the outlines, and the relative positions of the two eyes.

Optical illusions are common experiences. An illusion is a false interpretation of a sensation. Examples of optical illusions are: (1) objects appear larger on a foggy night, since the outlines are hazy, and the objects appear to be farther away than they really are; (2) lines with obtuse angles at the ends appear longer than equally long lines with acute angles at the ends. Some optical illusions are illustrated in Figure 10-21.

HEARING AND POSITION SENSE

Hearing is the sense by which sounds are appreciated; position sense refers to the orientation of the head in space and the movement of the body through space, its balance and equilibrium.

Hearing is called the watchdog of the senses; it is the last to disappear when one falls asleep and the first to return when one awakens. Medical personnel should remember this fact and guard their remarks when a patient is going to sleep or when he is going under or coming out of anesthesia.

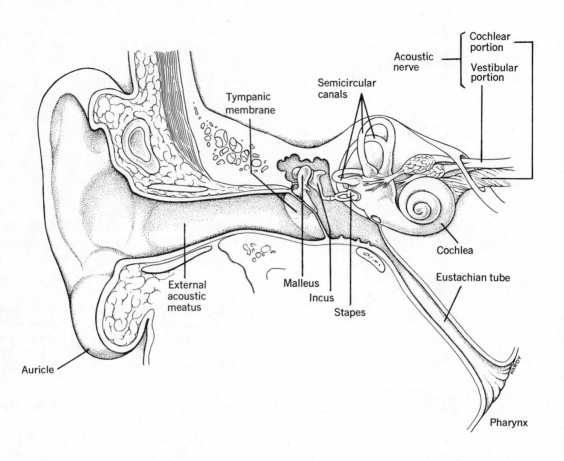

Fig. 10-22. Diagram of the ear, showing the external, middle, and internal subdivisions.

There are two sets of receptors in the ear: one set is concerned with hearing and the other with position sense. The eighth cranial or acoustic nerve serves as the sensory pathway to the temporal lobe of the cerebral cortex. Each ear consists of three distinct divisions: the external, middle, and inner ears as illustrated in Figure 10-22.

The external ear

The external portion of the ear, which receives sound waves, consists of the auricle and the external acoustic meatus. The **auricle** or **pinna**, is composed of cartilage, covered with skin. To some extent it serves to collect sound waves and direct them into the external meatus, but this function is of limited, if any, value in man. It is of greater importance in some of the lower animals.

The **external acoustic meatus** (auditory canal) extends inward, forward, and downward, from the auricle to the tympanic membrane (eardrum), a distance of approximately 1½ inches. This is an important fact for one to remember when using a stethoscope, since the ear pieces of that instrument are shaped to fit the normal curvatures of the cylindrical ear canal. The external acoustic meatus is formed partly by bone and partly by cartilage and is lined with skin. Ceruminous glands in the area secrete the brown ear wax, or *cerumen*, which normally serves as a protective agent. However, excessive amounts of this wax may block the ear canal and exert pressure on the tympanic membrane; this pressure is a common cause of earache and may even result in temporary deafness. In such cases, the removal of cerumen should be done by a doctor who gently syringes the canal with warm water.

The middle ear

The middle ear, or **tympanic cavity,** is a tiny, irregular cavity in the temporal bone. Its lateral wall is formed mainly by the *tympanic membrane* (drum membrane) that is located at the end of the external acoustic meatus. This thin fibrous membrane is covered with skin on its lateral aspect, but medially it is covered with mucous membrane which is continuous with that lining the entire middle ear. The tympanic membrane is so attached that it can vibrate freely with all audible sound waves that enter the meatus.

The middle ear is filled with air which reaches it from the upper part of the throat (nasopharynx) by way of the **auditory,** or **eustachian tube.** Posteriorly, the middle ear communicates with the *mastoid air cells* of the temporal bone. The mucous membrane which lines the middle ear is continuous with that of the pharynx and the mastoid air cells; thus it is possible for infection to travel along the mucous membrane from the nose or the throat to the middle ear (otitis media) and to the mastoid air cells (mastoiditis). When pus forms in the middle ear, the membrane should be lanced (myringotomy) to allow drainage and prevent the backing up of pus into the mastoid air cells. Early treatment is advisable for any middle-ear infection. If the membrane is not lanced it will rupture spontaneously, and when it heals the scar will be large and irregular instead of small and regular as occurs after lancing. The danger of permitting the infection to spread to the mastoid air cells lies in the nearness of these cells to the venous sinuses and the meninges of the brain.

The auditory tube serves to equalize the pressure in the middle ear with atmospheric pressure. It opens during swallowing and yawning. It tends to remain closed when the pressure is greater outside, as during rapid descent in an airplane. In some persons, after frequent infection, the tube is closed permanently by adhesions, and the hearing is greatly impaired. Whenever the pressure is unequal on the two sides of the tympanic membrane, it is not free to vibrate with sound waves.

Sometimes air is forced into the middle ear, by means of a bulb, to break down adhesions.

A chain of three tiny, movable bones extends across the cavity of the middle ear. Named according to their shape, these bones are the **malleus** (hammer), **incus** (anvil), and **stapes** (stirrup); their function is to transmit sound waves from the tympanic membrane to the inner ear. The malleus is attached by its handle to the tympanic membrane, and its head articulates with the incus which in turn joins the stapes. The footpiece of the stapes fits into the **oval,** or **vestibular window,** a small opening in the wall between the middle and the inner ears. Freely movable joints between these bones contribute to an ingenious little lever system which converts the vibrations of the tympanic membrane to intensified thrusts of the stapes against the fluid of the inner ear. However, the inner ear is protected from loud noise by two small muscles. The *tensor tympani* tightens the tympanic membrane, thus restricting its ability to vibrate; and the *stapedius* tends to pull the stapes away from the oval window, thus reducing the intensity of its thrusts into the inner ear.

The inner ear

The inner ear, or **labyrinth,** contains the receptors for hearing and for position sense. It consists of a bony labyrinth which contains a membranous labyrinth.

The **bony labyrinth** (Fig. 10-23) consists of a series of tiny canals and cavities which have been hollowed out of the petrous portion of the temporal bone. They contain a watery fluid which is called *perilymph.* There are three bony divisions: the *cochlea,* the *vestibule,* and the *semicircular canals.*

Within the bony labyrinth lies the **membranous labyrinth,** a series of fluid-filled sacs and tubes which have approximately the same shape as the surrounding bony walls. There are four membranous structures: the *cochlear duct* (in the cochlea), the *utricle* and the *saccule* (in the vestibule), and the *semicircular ducts* (in the semicircular canals). The fluid which fills these membranous structures is called *endolymph.* The general structure of the inner ear might roughly be compared to an automobile tire and its tube. The tough outer casing of the tire is comparable to

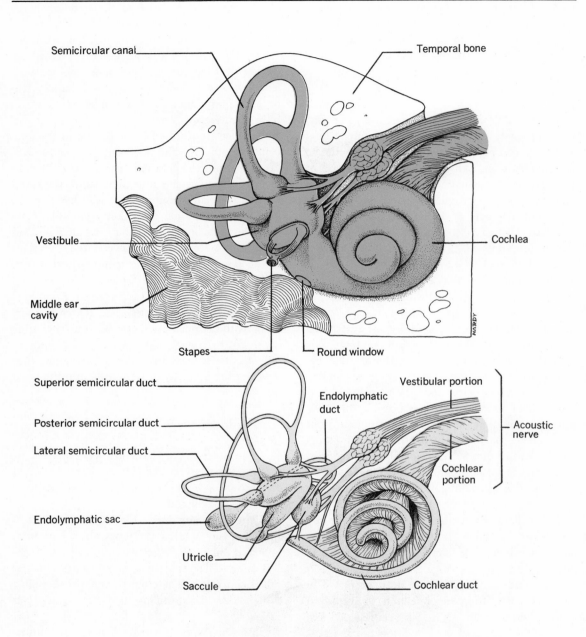

Fig. 10-23.   *Top, diagram of the bony labyrinth. Bottom, the membranous labyrinth as seen when removed from the bony labyrinth.*

the bony labyrinth and the rubber inner tube to the membranous labyrinth.

Hearing.   The **cochlea** is shaped like a snail shell, as shown in Figure 10-23. It is a bony tube which winds spirally around a central pillar of bone called the modiolus. The membranous **cochlear duct** resembles a lopsided triangle as it

lies within the bony cochlea (Fig. 10-24). It extends like a shelf across the bony canal to the sides of which it is attached. The portion of the bony canal above the cochlear duct is the *scala vestibuli,* and the portion below is the *scala tympani.*

The cochlear duct connects with the inner

wall of the bony canal by the osseous spiral lamina, and with the outer wall by the spiral ligament. The roof of the cochlear duct is thin and is called the *vestibular membrane*. The floor is composed of the *basilar membrane*. Resting on this basilar membrane is the **spiral organ of Corti**, the essential receptor end organ for hearing; like a long ribbon of cells, it extends from one end of the cochlear duct to the other. It is a very complicated structure, consisting of a supporting framework on which rest the important *hair cells*, so called because each one of them bears about 20 fine hairs on its free border. The *tectorial membrane* extends from the osseous spiral lamina and rests like a lid on the aforementioned hairs; vibration of the organ of Corti results in the hairs being bent against the fixed tectorial membrane. Nerve impulses initiated by the stimulated hair cells pass over dendrites whose cell bodies are located in the spiral ganglion in the modiolus (Fig. 10-24). Axons extending from this ganglion unite to form the *cochlear branch* of the eighth cranial or acoustic nerve.

In summary, sound waves which enter the external acoustic meatus cause the tympanic membrane to vibrate. These vibrations are transferred across the middle ear by the malleus, the incus, and the stapes. Since the stapes fits into the oval window like a piston it can move in and out with each vibration. In the transfer across the middle ear, sound waves are reduced in amplitude but increased in *force* by the lever system of bones. This makes possible the effective transmission of vibrations from air to fluid. However, as the stapes bulges into the fluid-filled vestibule of the inner ear, there must be a corresponding outward bulge, since fluids cannot be compressed; this protective function is assumed by the *round* or *cochlear window* which is a small membrane-covered opening in the bony wall located just below the oval window.

Each pistonlike thrust of the stapes into the vestibule sends a rippling wave through the perilymph of the inner ear. This sets into motion the basilar membrane and the organ of Corti. The resulting stimulation of hair cells arouses nerve impulses which are transmitted to the center for hearing in the temporal lobe of the cerebral cortex.

The Nature of Sound. Sound waves are vibrations of the air, or other elastic media, moving away from the source in all directions, like the ripples produced when one throws a stone into a pool of water. For example, when a guitar string vibrates it makes the air around it vibrate as well. The *frequency* or number of vibrations per second is often referred to as the cycles per second (cps.); this determines the *pitch* of a sound. For example, the note called middle C is produced by 256 cps. If the note is raised one octave the frequency is increased to 512 cps. In other words, the greater the frequency, the higher the pitch.

The ability to discriminate between high and low frequency sounds is related to the tension of the basilar membrane. For example, this membrane is relatively thin and tight at the end near the stapes (i.e., at the basal turn of the cochlea), but it is relatively thick and loose toward the apical end. Each thrust of the stapes into the perilymph initiates a wavelike ripple in the basilar membrane; this ripple travels from the tight end toward the loose end. High tones create their greatest crests at the basal end (where the membrane is tight), and low tones create their greatest crests toward the apex (where the membrane is loose). The position of the highest crest is very important since that determines which of the hair cells will be stimulated.

The human ear can detect sounds over a wide range of frequencies—from as low as 16 cps. to as high as about 20,000 cps., but there is a great deal of individual variation. Perception of high frequency sound waves is generally best in childhood; there is a gradual decrease as one gets older. An early sign of the aging process is a progressive loss of the ability to hear high tones.

Another important characteristic of sound is its *intensity*, or loudness (height of the waves). This usually is expressed in terms of decibels (dB), where zero dB is the intensity that one can barely hear (the threshold of hearing). A whisper is about 30 dB while normal conversation is 65 or 70 dB. Heavy automobile traffic rates about 80 dB, and the noise in a subway station reaches 105 dB. If one is subjected to sounds in the range of 120 to 140 dB, there will be discomfort or even pain. Jet aircraft produce such intense sound waves (160 dB) that ground personnel wear special ear shields to protect themselves from actual injury to the nerve cells of the inner ear. Figure 10-25 illustrates three sound waves; two pitches and

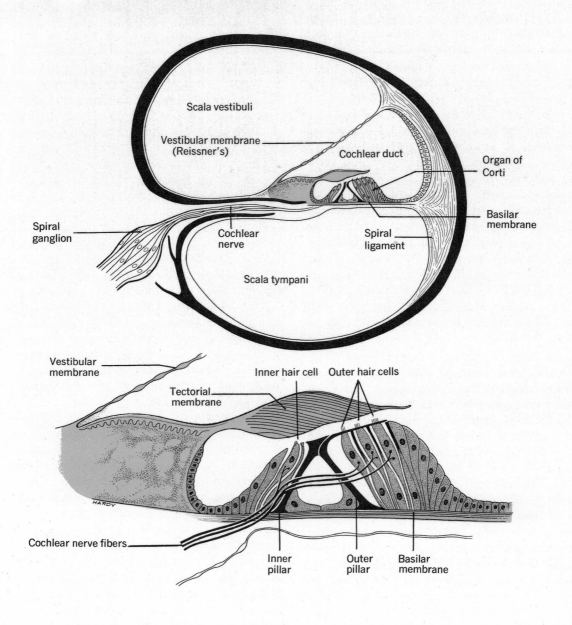

Fig. 10-24. *Top,* drawing of a portion of the cochlea. Note the relation of the cochlear duct to the scala vestibuli and tympani. *Bottom,* the spiral organ of Corti has been removed from the cochlear duct and greatly enlarged.

three degrees of loudness are represented.

An impairment of the sense of hearing is called *deafness*. There are three types of deafness: conduction, nerve, and central deafness.

Conduction deafness is caused by interference with the transmission of sound waves through the external or middle ear. For example, excessive amounts of cerumen (p. 248) may block the external acoustic meatus; there may be injury to the tympanic membrane; there may be ankylosis (stiffness of joints) of the bones of the middle ear; the stapes may be bound at the oval window by

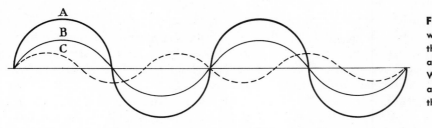

Fig. 10-25. Three sound waves. Waves A and B have the same pitch, but A has a greater intensity than B. Wave C has a different pitch and is least intense of the three.

new growths of bone; or the auditory tube may be closed by adhesions, thus preventing equalization of pressure on the two sides of the tympanic membrane. However, sound waves can be transmitted through the cranial bones to the fluid in the inner ear, and various hearing aids are so constructed as to take advantage of this fact.

Nerve deafness may follow degeneration of the sensory cells in the cochlear duct, tumors involving the cochlear branch of the acoustic nerve, and so on. Hearing aids are of no value in nerve deafness.

Central deafness may follow interference with the pathway of nerve impulses to the cerebral cortex, or it may be a manifestation of aphasia. This type of deafness is rare.

The instrument designed to measure one's ability to hear is called the *audiometer*. It is useful in testing hearing in relation to occupational fitness or disability in school children, in workmen who are subjected to excessive noise in their jobs, and in fitting hearing aids.

The sense of hearing is important in learning to speak and in modulating the voice; deaf persons cannot modulate their voices, since they cannot hear them. Children who are born deaf will be mute (unable to speak) unless speech is taught by some channel other than the ear. The sense of sight may be used, by watching the position of the lips and the tongue. It is quite unnecessary and cruel indeed for a child to be compelled to remain mute if his intelligence is sufficient to enable him to speak. Schools for the deaf and training in lip reading are important in preparing deaf persons for occupations in which their handicap does not prevent them from being able to support themselves.

Position Sense. The **vestibule** is a disklike cave in which float two membranous sacs, the utricle and the saccule, each of which is filled with endolymph. On the floor of the *utricle* is a patch of hair cells which constitute the vestibular receptors or maculae acustica. Entangled in the brushlike hairs are tiny particles of calcium carbonate that are called otoconia. Nerve impulses are initiated when the octoconia push or pull on the hairs and bend them in the direction of the pull of gravity. The *vestibular branch* of the acoustic nerve transmits impulses to the cerebral cortex, and we recognize the position of the head in space as it relates to the pull of gravity. The *saccule* has a similar patch of sensory epithelium, but its functions are not yet fully understood.

There are three **semicircular canals** in each ear; they are so arranged at right angles to one another that all three planes of space are represented (Fig. 10-23). Lying within these bony tunnels are the membranous **semicircular ducts** which loop through semicircles, beginning and ending at the utricle. At one end each duct has a dilatation, the *ampulla*, which contains a patch of hair cells or receptor end organs known as the *cristae acustica*. Movement of the endolymph bends the hairs and sets up impulses in the *vestibular branch* of the acoustic nerve. The cristae are stimulated by sudden movements, or by a change in rate or direction of movement. In other words, whenever movement begins or ends, accelerates or decelerates, or changes in direction, impulses are set up in dendrites of the vestibular nerve.

If a person is rotated in a chair, the endolymph in the semicircular canals is set into motion and stimulates the hair cells. A sense of movement either of self or surroundings (vertigo) occurs, together with a peculiar movement of the eyes called *nystagmus*. Nystagmus consists of a rapid movement of the eyes in one direction and a slow movement in the opposite direction; they appear to oscillate. Nausea may occur. We speak of these sensations which are set up by rotation as motion sickness. Many somatic reflexes accompany rotation; these involve the muscles of the neck, the trunk, and the extremities. Visceral reflexes occur, also. There is a fall in blood pressure and a change in heart rate, and the skin becomes pale, all of which involve visceral efferent nerves.

In summary, it may be stated that the semicircular canals are dynamic sense organs; they rouse sensations of starting and stopping movement, changing its speed (either acceleration or deceleration) and direction. They also originate many reflexes which involve skeletal, smooth, and cardiac muscles. The utricle is a static sense organ; it gives information regarding the orientation of the head in space, in relation to gravity and sets into action postural and righting reflexes. Injury to the semicircular canals or utricle leads to disturbances in maintaining one's orientation and balance. Vertigo is the most common symptom. Nystagmus is also a common sign of vestibular pathology.

Meniere's syndrome is a condition in which a person hears buzzing noises in the ears and also has severe attacks of vertigo. During an attack he cannot stand up and may even show rolling movements when lying down. It is caused by disturbances in the inner ear. Various surgical and medical procedures may relieve the distress. The hearing of subjective buzzing, high-pitched noises is called *tinnitus*.

SUMMARY

1. Introduction
A. Sensory mechanisms
   a. Sense organs, or receptors
   b. Afferent pathway to brain
   c. Sensory area in cerebral cortex
   d. All parts must function to give rise to sensation
B. Characteristics of sensations: after-image, adaptation, intensity

2. Cutaneous senses
A. Sense of touch, pressure, and temperature
   a. Receptors: free nerve endings; corpuscles and end bulbs
   b. Pathway: cranial and spinal nerves
   c. Cortex: parietal lobe of cerebrum
B. Somatic pain
   a. Receptors: free nerve endings
   b. Pathway: cranial and spinal nerves
   c. Cortex: parietal lobe of cerebrum (true discrimination of pain)

3. Muscle sense
A. Proprioceptive, or kinesthetic sense; awareness of position of body
B. Sensory mechanisms
   a. Proprioceptors in muscle and tendon
   b. Afferent pathway: cranial and spinal nerves
   c. Cortex: parietal lobe of cerebrum

4. Visceral sensations
A. Stimuli and pathways: distention, spasm, chemical irritants; fibers in cranial and spinal nerves
B. Organic sensations: hunger, thirst, nausea
C. Visceral pain
   a. Receptors in viscera: afferent fibers travel with sympathetic nerves
   b. Impulses reaching brain may result in somatic or visceral reflexes or pain
   c. Localized pain: neurons have direct line to brain
   d. Referred pain: impulses must share relay neurons with somatic areas

5. Sense of smell
A. Receptors: olfactory epithelium in nasal cavity
B. Pathway: olfactory, or first cranial nerve
C. Cortex: hippocampal gyrus; temporal lobe

6. Sense of taste
A. Receptors: taste buds on tongue
B. Pathway: glossopharyngeal (posterior third) and facial (anterior two thirds) nerves
C. Cortex: at base of postcentral gyrus near lateral fissure of cerebrum
D. Primary tastes: salt, sweet, sour, and bitter

7. Vision
A. Protection of eyeball: orbit; palpebrae; tarsus; meibomian glands; eyelashes; conjunctiva; lacrimal apparatus
B. Extrinsic muscles of eye
   a. Movement of eyeball: recti and oblique muscles
   b. Movement of eyelids: orbicularis oculi and levator palpebrae superioris
C. Coats of the eyeball
   a. Fibrous: sclera and cornea
   b. Vascular: choroid, ciliary body, and iris
   c. Nervous: the retina
      (1) Rods: receptors concerned with twilight vision
      (2) Cones: receptors concerned with daylight and color vision
      (3) Retinal examination with ophthalmoscope reveals: macula lutea, fovea centralis, optic disk, and blood vessels
D. Transparent internal structures of eye: aqueous humor, crystalline lens, and vitreous body
E. Sensory pathway for vision
   a. Rod and cone receptors sensitive to light; initiate

impulses which travel over optic nerves

   b. Optic chiasma: crossing point for fibers from medial halves of retinae

   c. Optic tracts: from chiasma to cerebrum

   d. Right occipital cortex receives impulses from right half of each eye; left cortex receives from left halves

F. Binocular vision: that with two eyes

   a. Provides a larger visual field and depth perception

   b. Images must be brought to identical points on the two retinae

   c. Processes necessary

     (1) Convergence of visual axes

     (2) Regulating size of pupil

     (3) Refraction of light rays

     (4) Accommodation

   d. Defective vision may be corrected by use of glass lenses

     (1) Myopia or near-sightedness: eyeball too long or refractive power of eye too great; images fall in front of retina

     (2) Hyperopia or far-sightedness: eyeball too short or refractive power of eye too weak; images fall back of retina

     (3) Presbyopia: loss of elasticity of lens

G. Reflexes of eye

   a. Light reflex

   b. Accommodation reflex

H. Effects of drugs on size of pupil

8. *Hearing and position sense*

A. External ear

   a. Auricle and external acoustic meatus

   b. Directs sound waves to tympanic membrane

B. Middle ear

   a. Lateral wall formed by tympanic membrane which vibrates with sound waves

   b. Mucosal lining continuous with that of nasopharynx via eustachian tube

   c. Communicates posteriorly with the mastoid air cells

   d. Malleus, incus, and stapes transmit sound waves from tympanum to oval window of inner ear; bony lever system

C. Inner ear or labyrinth

   a. Bony labyrinth consists of cochlea, vestibule, and semicircular canals

   b. Membranous labyrinth within bony labyrinth consists of cochlear duct, utricle, saccule, and semicircular ducts

   c. Fluid in bony canals: perilymph; in membranous structures: endolymph

   d. Spiral organ of Corti: within cochlear duct; receptor end organ for hearing

     (1) Hair cells stimulated by movements of perilymph as stapes move like piston in oval window

     (2) Nerve impulses transmitted over cochlear

branch of acoustic nerve to auditory area in temporal lobe of cerebral cortex

   e. Sound waves are vibrations of air; frequency determines pitch

     (1) Discrimination related to tension of basilar membrane

     (2) Intensity expressed in decibels

   f. Deafness: difficulty in external or middle ear (conduction type); in inner ear (cochlea) or acoustic nerve (nerve type); in cerebrum (central type)

   g. Maculae acustica: vestibular receptors for position of the head as it relates to the pull of gravity

   h. Cristae acustica: receptor end organs in ampullae of semicircular ducts; stimulated by movement of endolymph during changes in rate or direction of movement

   i. Menière's syndrome due to disturbance in inner ear

QUESTIONS FOR REVIEW

1. What are the cutaneous senses? Why are they important?

2. Distinguish between somatic and visceral pain. What is referred pain?

3. Why is conscious muscle sense very important?

4. Why does food taste differently when one has a cold?

5. Describe the structures that protect the eye.

6. What is the function of the ciliary muscle? of the lens?

7. List in proper sequence the structures through which a ray of light must pass to reach the rods and cones.

8. What is night blindness? Explain.

9. Describe the function of the cones.

10. What is the major physical defect in strabismus? astigmatism? hyperopia? myopia? cataract? glaucoma?

11. What are the processes that are essential for binocular vision?

12. List in proper sequence the structures through which vibrations must pass to reach the organ of Corti.

13. What is the function of the auditory tube?

14. Distinguish between the bony and membranous labyrinths.

15. What is the importance of the round window?

16. What determines the pitch of a sound? What is the role of the basilar membrane?

17. Distinguish between conduction deafness and nerve deafness.

18. What is the function of the utricle? the semicircular ducts?

REFERENCES
AND SUPPLEMENTAL READINGS

Best, C. H., and Taylor, N. B.: The Physiological Basis of Medical Practice, ed. 8. Baltimore, Williams & Wilkins, 1966.

Botelho, S. Y.: Tears and the lacrimal gland. Sci. Am., 211:78, Oct. 1964.

Guyton, A. C.: Textbook of Medical Physiology, ed. 4. Philadelphia, W. B. Saunders, 1971.

MacNichol, E. F. Jr.: Three-pigment color vision. Sci. Am., 211:48, Dec. 1964.

Pettigrew, J. D.: The neurophysiology of binocular vision. Sci. Am., 227:84, Aug. 1972.

Werblin, F. S.: The control of sensitivity in the retina. Sci. Am., 228:71, Jan. 1973.

Young, R. W.: Visual cells. Sci. Am., 223:80, Oct. 1970.

The Blood and Lymph

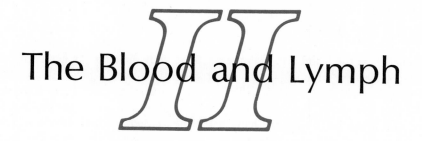

"The very life-blood of our enterprise."

King Henry IV • William Shakespeare

Each system of the human body plays an important part in maintaining homeostasis in the internal cellular environment, but the movement of blood through the circulatory subway system is of fundamental importance. Although blood as a whole normally does not leave the vessels in which it is contained (with the exception of certain female reproductive functions), it still provides an efficient pick up and delivery service for customer cells in every organ of the body. As was stressed in Chapter 2 (pp. 30 to 35), blood renders this service only while passing through the tiniest vessels of the circuit, the microscopic capillaries. Water and other diffusible substances are able to pass freely back and forth through the capillary walls, thus contributing to the formation of interstitial fluid which serves as the "middleman" between cap-

illary blood and the neighboring customer cells.

In summary, the maintenance of homeostasis in the internal cellular environment depends on both blood and interstitial fluid for the prompt delivery of supplies and the equally prompt removal of end products which result from the metabolic activities of all living cells.

Blood is classified as one of the connective tissues. However, its intercellular substance is a liquid, called plasma, and its cells, or corpuscles, are not in a fixed position as is the case in other tissues. Blood is circulating constantly through a closed circuit of tubes; it is pumped from the heart into arteries, from the arteries into capillaries, and from the capillaries it flows into veins for return to the heart.

FUNCTIONS OF THE BLOOD

The chief function of the blood is to keep the internal cellular environment of body tissues in that state of dynamic equilibrium which is called homeostasis. Various metabolic processes going on at all times within the body tend to change the chemical composition of the internal environment, such as the chemical changes that accompany contraction of muscles, conduction of nerve impulses, and secretion by glands. All of these forms of metabolism use food and oxygen and produce waste products. The circulating blood makes it possible for the various organs to take these changes in their stride and yet remain normal.

The blood transports oxygen, from the lungs to the tissues, and carbon dioxide, from the tissues to the lungs, to be eliminated. It carries food materials from the intestine to all parts of the body and returns waste to the kidneys to be excreted. The blood distributes the heat produced in active muscles and thus aids in the regulation of body temperature. It transports internal secretions from the glands in which they are produced to the tissues on which each exerts its effects. The buffers in the blood help to maintain acid-base balance. The blood also is involved in immunity to disease and in protecting the body against invading bacteria.

CHARACTERISTICS OF BLOOD

The **color** of whole blood (plasma and cells) is perhaps its most striking characteristic. With only one exception, arterial blood is a bright scarlet color due to its high oxygen content, and venous blood is a very dark red due to a lower oxygen content; the exception to this statement involves the pulmonary arteries and veins (in which the colors are reversed), since they carry blood to and from the lung capillaries for the elimination of carbon dioxide and the pickup of oxygen.

The **quantity** of circulating blood varies with the size of the person, the total volume being approximately 7 per cent of the body weight. For example, if you weigh 100 lbs. you have about 7 lbs. of blood, or 2.6 qts. (1 lb. of blood being equal to about 0.37 qt.).

The **viscosity** of blood refers to its resistance to flow. It is thick and sticky, and it normally flows with difficulty. It has been found that the viscosity of whole blood is about five times as great as that of water. Part of the viscosity is due to the cells, and part to the proteins of the plasma. When either of these factors is altered, the viscosity changes. For example, in certain diseases, such as malaria, red cells tend to stick together and to form clumps in the vessels. These clumps make it much more difficult for the blood to circulate; therefore, its speed of circulation is slow. This thick, highly viscous blood with its clumped cells is called *sludged blood.* Human beings are examined for the presence of sludge by looking at the capillaries of the conjunctiva with a special microscope.

Specific gravity refers to the weight of blood compared with that of water. The specific gravity of water is taken as 1.00 (1 cc. of water weighing 1 Gm. at 4° C.). The specific gravity of whole blood is between 1.055 and 1.065, with an average of 1.060, which means that 1 cc. of whole blood weighs 1.060 Gm.

The specific gravity of plasma is between 1.028 and 1.032. The cells are much heavier than the plasma. If clotting of blood is prevented, the cells tend to settle out, or sink to the bottom of a tube, as blood stands. They have no opportunity to settle out in the body, as the blood is kept moving in the vessels due to the pumping action of the heart.

The speed with which cells sink to the bottom of a tube is called the *sedimentation rate,* or *index* (Fig. 11-1). Expressed in millimeters per hour, the sedimentation rate normally ranges from 4 to 10 mm. per hour (Wintrobe); it depends on the condition of the blood and the general state of health of the individual. For example, the sedimentation rate is increased in acute infectious processes.

The **reaction,** or **pH,** of blood is slightly alkaline, the average figure being 7.4 under normal conditions. This pH is maintained within a remarkably narrow range (7.35 to 7.45), despite great fluctuations in the quantities of acid and base which are ingested in food, and equally great fluctuations in the amount of acid produced during metabolic activities in the tissues. The maintenance of optimal pH, or *acid-base balance,* is described on pages 464 to 468.

BLOOD STRUCTURE AND FORMATION

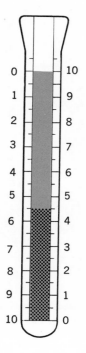

Fig. 11-1. Wintrobe tube. When blood is prevented from coagulating and allowed to stand undisturbed in this tube for one hour, the blood cells begin to settle out, and a clear layer of plasma is seen above them. The depth of the plasma is read on the scale at the left, in millimeters (1 cm. = 10 mm.); this is called the sedimentation index. Then the tube is centrifuged; this packs the blood cells. The height of the cell layer is read on the scale at the right; this is called the cell volume. As illustrated, the cell volume (hematocrit) is 45 per cent.

Whole blood consists of the liquid *plasma* in which are suspended the microscopic *formed elements*. Plasma constitutes a little more than half of the blood (53 to 58 per cent); it is made up of approximately 90 per cent water and 10 per cent dissolved solids. The formed elements are the erythrocytes, or red blood cells, the leukocytes, or white blood cells, and the thrombocytes, or blood platelets; these elements constitute a little less than half of the blood (42 to 47 per cent).

Hemopoietic tissue is a kind of connective tissue that is highly specialized for both the production of new blood cells and the removal of those which are old and worn out. To keep the total number of cells fairly constant, new ones must be added to the blood each day to replace the old ones which are removed. Hemopoietic tissue is found in red bone marrow (myeloid tissue) and in organs of the lymphatic system (p. 324). The primary cell, or forefather, of all types of formed elements is the *hemocytoblast*, or *stem cell*; for example, this cell is able to develop into the precursor of a red cell, a white cell, or a blood platelet according to the needs of the body (Fig. 11-2).

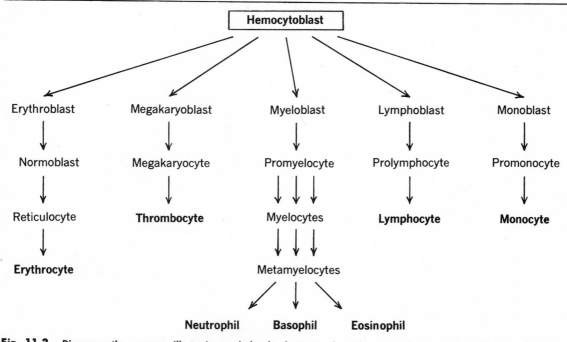

Fig. 11-2. Diagrammatic summary illustrating orderly development of erythrocytes, thrombocytes, and leukocytes from primary cell or forefather called hemocytoblast or stem cell.

The erythrocytes

The red cells of the blood are called erythrocytes, a term derived from the Greek words *erythros* (*red*) and *kytos* (*cell*). They are also referred to as red corpuscles (meaning *little bodies*), a term that is technically preferable to cells, since mature erythrocytes do not have nuclei. However, all of these terms (red blood cell, red blood corpuscle, and erythrocyte) are in common use, and students should be familiar with each of them.

Description. Red blood cells are shaped like biconcave disks (Fig. 11-3), which means that they are thinner in the center than around the edges. These tiny disks are very flexible, a feature that makes it possible for them to bend and twist when squeezing through the narrowest of capillaries and then to regain their original shape when they enter larger vessels.

The diameter of normal red cells varies from 6 to 9 microns, with an average of 7.5 (1 micron is 1/1,000 mm. or 1/25,000 inch). In normal blood, the red cells in a blood smear are all about the same size, as illustrated in Figure 11-4. If there is great variation in size, the condition is called *anisocytosis* (meaning *a condition of unequal cells*). If there is great variation in shape, it is called *poikilocytosis* (meaning *condition of varied cells*). The shape and the size of red cells

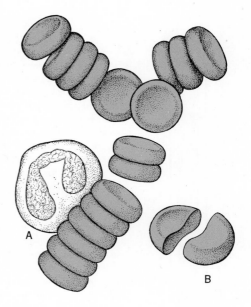

Fig. 11-3. Red blood cells grouped in rouleaux (like a stack of coins). One white blood cell (A) also is shown. (B) is a red cell split to show its biconcave shape.

is important in the diagnosis of various blood diseases.

Each red blood cell consists of an inner framework that contains hemoglobin, and an outer

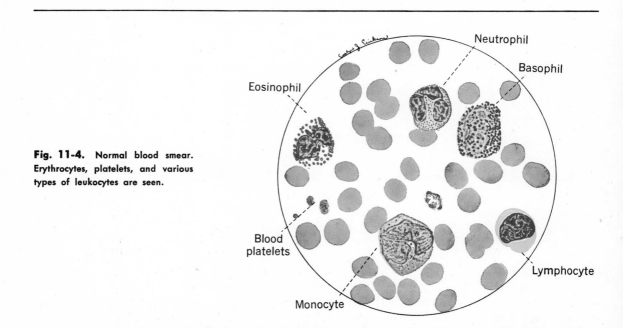

Fig. 11-4. Normal blood smear. Erythrocytes, platelets, and various types of leukocytes are seen.

limiting membrane. **Hemoglobin** is a conjugated protein, which means that it consists of a simple protein (*globin*) to which is attached an iron-containing pigment (*hematin*). By virtue of its iron, hemoglobin can combine with oxygen to form *oxyhemoglobin*, a compound that gives the bright scarlet color to oxygenated blood. We might note also that it is the presence of oxyhemoglobin in red blood cells passing through capillaries near the body surface that gives pinkness to the skin and varying degrees of redness to mucous membranes and to the lips.

Hemoglobin that has given up its oxygen is called *reduced hemoglobin*, and it is blue in color rather than red. Under normal conditions, not enough reduced hemoglobin is formed for the blue color to be visible, and blood in the systemic veins merely appears dark red in color. However, various diseases or abnormalities related to the cardiovascular and respiratory systems may result in the presence of such large quantities of reduced hemoglobin in the capillaries that the skin and mucous membranes have a distinctly blue color; this condition is called **cyanosis**. There must be at least 5 Gm. of reduced hemoglobin per 100 ml. of arterial blood to cause frank cyanosis.

The formation of oxyhemoglobin occurs while blood is passing through the pulmonary capillaries that surround the microscopic air sacs (al-veoli) in the lungs. When this blood eventually reaches systemic capillaries in all parts of the body, oxygen leaves the hemoglobin and diffuses out of the bloodstream and through the interstitial fluid to satisfy the needs of active cells. On the return trip to the lungs, hemoglobin does not travel empty-handed. In fact, one might even think of hemoglobin as being fickle, since as it gives up its oxygen, it immediately combines with a small amount of carbon dioxide, the gas that diffuses into the blood after being formed during oxidative processes in the tissue cells. In the lung capillaries, hemoglobin gives up carbon dioxide which then diffuses into the alveoli to be exhaled and eliminated from the body. At the same time, it picks up another supply of oxygen, and the entire process is repeated. The transport of respiratory gases is considered in greater detail on pages 368 to 371.

Hemoglobin also can unite with carbon monoxide gas, which is very unfortunate. In fact, if hemoglobin has a choice, it will select carbon monoxide as its partner in preference to oxygen, and the resulting union is a firm one. Since hemoglobin that is bound to carbon monoxide is unable to carry oxygen, it is really oxygen starvation that causes death in cases of carbon monoxide poisoning (e.g., following prolonged inhalation of automobile exhaust fumes).

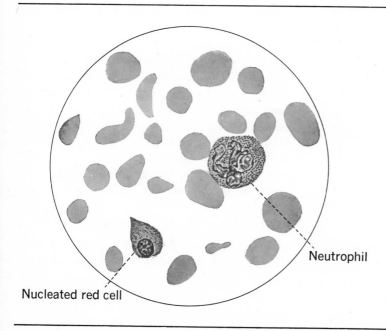

Fig. 11-5. Blood smear in pernicious anemia. Changes are evident in the erythrocytes and in the neutrophil.

Neutrophil

Nucleated red cell

Estimation of the *hemoglobin content* of the blood is an important part of every physical examination. In women the average is 14 Gm. per 100 ml. of blood (range 12 to 16 Gm.); in men the average is 16 Gm. per 100 ml. of blood (range 14 to 18 Gm.). The combined average is 15 Gm. per 100 ml. of blood.

The limiting membrane of a red blood cell is much like that of a nucleated cell, the important properties of which were described on page 22. Normally, the plasma and the red cells are isotonic; that is, they have the same osmotic pressure. If red cells are exposed to hypotonic solutions, water enters the cells more rapidly than it leaves, causing them to swell (see Fig. 2-15). If this swelling progresses to the point where the membrane ruptures, the cell will lose its hemoglobin; such a sequence of events is known as *hemolysis*, or *laking*. Red cells which have lost their hemoglobin can no longer transport oxygen and carbon dioxide. In the clinical situation, the *fragility test* is based on the behavior of red blood cells when placed in hypotonic sodium chloride solutions of varying strengths. Blood is drawn from the patient's vein (venipuncture), and one drop is placed in each of a series of tubes containing sodium chloride solutions ranging from 0.7 to 0.3 per cent. The cells swell to some extent in each solution, but the aim of the fragility test is to see at which concentration actual hemolysis begins and at which it is complete; these results then are compared with those of normal individuals. The fragility test is often helpful in the differential diagnosis of anemias, that is, in determining the exact type of blood deficiency that is causing the patient's trouble.

Hemolysis may be brought about not only by hypotonic salt solutions, but also by (1) the products of bacterial activity, (2) a transfusion of blood that is the wrong type (p. 270), (3) some snake venoms, and other agents.

Formation of Erythrocytes. The red blood cells have their origin in red bone marrow. In infancy and childhood every bone in the body contains active marrow to meet the intense needs that accompany normal growth processes. As the bones grow larger, there is more than enough red marrow to meet the demands of the body, and much of it is gradually replaced by fatty yellow marrow. In the adult, active red marrow is found in the thoracic bones, the vertebrae, the cranial bones, and in the proximal end of the femur and the humerus. Microscopic examination of this marrow is a valuable tool in the diagnosis of certain blood diseases.

Although millions of red blood cells are formed every second of every day of your life, their development proceeds in an orderly fashion through the following stages: from hemocytoblasts to erythroblasts, to normoblasts, to reticulocytes, to erythrocytes, or mature red blood cells (Fig. 11-2). Since reticulocytes are nearly mature red cells, it is normal to find a small number of them in the circulating blood (0.5 to 1.5 per cent of the total red cells). However, an increase in the reticulocyte count is one of the first indications that the production of red cells has been accelerated, usually to meet an increasing need on the part of body tissues. The presence of normoblasts in the circulating blood is distinctly abnormal and indicates such an urgent demand for erythrocytes that cells are being released from the red marrow before they are mature.

If red blood cells are to mature normally, one needs not only normal bone marrow, but also the presence of a special *maturation*, or *antianemic*, *factor*—vitamin B_{12}. This important vitamin sometimes is referred to as the *extrinsic factor* in blood formation, since it is ingested in the food which one eats. However, the absorption of vitamin B_{12} from the intestinal tract seems to depend on a special component of the gastric juice; this substance is called the *intrinsic factor*, since it is produced within the body by cells in the lining of the stomach. After its absorption from the intestine, vitamin B_{12} is carried in the bloodstream to the liver, where it is stored until needed in the bone marrow (see Fig. 11-6). Failure of the stomach lining cells to produce sufficient quantities of intrinsic factor may result in a disease called *pernicious anemia* (Fig. 11-5), one symptom of which is a marked reduction in the number of circulating red cells due to inadequate absorption of vitamin B_{12}.

Other essentials for the formation of normal red blood cells are found in a well-balanced diet. These include iron, copper, riboflavin, nicotinic acid, vitamin C, and plenty of protein, all of which serve as important raw materials in the red blood cell factories.

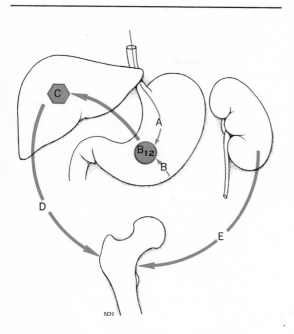

Fig. 11-6. Factors important in normal erythropoiesis: *A*, vitamin B_{12}, extrinsic factor ingested in food; *B*, intrinsic factor produced by gastric mucosal cells promotes absorption of B_{12}; *C*, B_{12} stored in liver until needed; *D*, B_{12} released to red bone marrow; *E*, erythropoietin produced by kidney stimulates erythropoiesis in bone marrow.

Functions. The chief function of red blood cells is to carry oxygen from the lungs to the tissues, and carbon dioxide back from the tissues to the lungs. The amount of oxygen that is dissolved in the plasma is small, and it would be impossible to supply the demands of the tissues if this were the only means of transport. The hemoglobin in the red cells, by means of its iron, makes it possible to carry an abundance of oxygen to tissues for ordinary requirements.

Life Span. After red blood cells leave the bone marrow, they circulate in the bloodstream for about 120 days, a figure based on studies of red cells which were tagged with radioactive iron. Each cell works very hard during its 4-month life span, loading and unloading oxygen about twice a minute and traveling an estimated 700 miles before wearing out. Worn-out red cells are destroyed by phagocytes (macrophages) in the spleen, the liver, and the red bone marrow (*reticuloendothelial system*). During the breakdown, iron is salvaged and returned to the bone marrow where it is used over again in making new hemoglobin.

The remainder of the hemoglobin molecule is converted into bile pigments which are excreted by the liver into the bile and eventually eliminated through the intestinal tract.

Number of Erythrocytes. Every second of every day, millions of red blood cells are destroyed, and other millions are formed to take their place. However, the body has an amazing capacity for maintaining a dynamic balance between these two processes, and the healthy adult has a relatively constant number of red cells circulating in his bloodstream at all times. The average red cell count falls within the range of 4½ to 5½ million per cu. mm. of blood; thus the total number for the entire body would be a staggering figure somewhere in the trillions. Women generally have a lower red cell count than do men. Determination of the *cell volume*, or *hematocrit* (see Fig. 11-1), also gives certain information about the number of red blood cells, since they constitute approximately 42 to 47 per cent of the total blood volume.

Normally, the rate of destruction of red blood cells equals the rate of production. However, if the number of red cells in the blood is reduced by hemorrhage or by an increased rate of destruction, the bone marrow can greatly increase its rate of production. The stimulus for such an increase is related to the reduced oxygen-carrying capacity of the blood and to the resultant tissue *hypoxia* (low oxygen content). It is believed that a lack of oxygen causes an increase in the formation of *erythropoietin*, a substance that directly stimulates red cell production; relatively large quantities of erythropoietin are formed in the kidney during hypoxia; lesser amounts are formed in the liver and possibly in other tissues.

Anemia is the term that refers to a blood deficiency in which the red cells are reduced in number, or are deficient in hemoglobin, or both. In any case, there is a decrease in the oxygen carrying capacity of the blood. With reduced amounts of oxygen being supplied to the tissues there is impairment of oxidative processes in the tissue cells, inadequate liberation of energy, and interference with normal cellular functions. In severe cases of anemia, a person may be bedridden and have only enough energy to maintain vital activities such as respiration and heartbeat. If the disease continues unchecked, death will be the end-result.

The leukocytes

The white blood cells are called leukocytes, a term derived from the Greek words *leukos* (*white*) and *kytos* (*cell*). They are less numerous than red blood cells but are easily recognized by their greater size (10 to 15 microns) and by the presence of nuclei. Although the red cells do their work in the blood, the chief functions of most white blood cells are performed only when they leave the blood and enter neighboring tissues; thus they use the bloodstream as a transport system, much as one might take a bus or subway to work.

The classification of white blood cells is based upon the presence or absence of granules in the cytoplasm. Thus we have two groups: granular leukocytes and nongranular leukocytes.

The Granular Leukocytes. In addition to granules in their cytoplasm, these white blood cells have distinctive nuclei that consist of a variable number of lobes. Because of such nuclear variations, the members of this group may be referred to as *polymorphonuclear* leukocytes, or *polymorphs*; however, in clinical practice this term most frequently is used in reference to one specific subgroup, the neutrophils. There are three subgroups of granular leukocytes, each being named according to the acid or basic staining characteristics of the cytoplasmic granules: neutrophils are those that do not stain well at the point of neutrality with either acid or basic dyes; eosinophils stain well with acid dye (eosin); and basophils stain well with basic dye.

1. *Neutrophils* have nuclei that are divided into three to five lobes; thus the term polymorphonuclear is particularly well suited to them. They are the most numerous of all the white blood cells, constituting from 60 to 70 per cent of the total number. Neutrophils serve as a mobile defense force against infection (as described under Functions); thus their numbers may be greatly increased on certain occasions. At such times, when there is a great need for neutrophils, immature forms may appear in the blood; these cells are called *band*, or *stab neutrophils*, because their nuclei are horseshoe-shaped.

2. *Eosinophils* have nuclei made up of two oval lobes, and their cytoplasmic granules stain with acid dye (eosin). They form only 1 to 3 per cent of the total number of white cells.

3. *Basophils* have nuclei that are bent into an S-shape, and their large cytoplasmic granules stain with basic dyes. They are the least numerous of the white cells, constituting approximately 0.3 to 0.5 per cent of the total number.

The Nongranular Leukocytes. These white cells have no granules in their cytoplasm. There are only two subgroups of nongranular leukocytes.

1. *Lymphocytes* have very large nuclei surrounded by a relatively thin layer of cytoplasm. They comprise from 20 to 35 per cent of the total number of white cells. Most lymphocytes are smaller than the other kinds of white blood cells.

2. *Monocytes* are the largest of all blood cells. They have large nuclei which may be oval or kidney-shaped, and there is an abundance of cytoplasm. Monocytes constitute only 3 to 8 per cent of the total number of white cells.

The various subgroups of leukocytes are illustrated in Figure 11-4.

Formation of Leukocytes. Leukocytes are formed in both types of hemopoietic tissue—red bone marrow and lymphatic tissue. The granular leukocytes originate in red bone marrow. The nongranular leukocytes are formed in lymphatic tissue which is found in such organs as the spleen, lymph nodes, and thymus. In a manner quite similar to that of the red blood cells, leukocytes develop in orderly stages as illustrated in Figure 11-2.

Functions. The broad, general function of the leukocytes is to protect the body against infection. For example, let us consider the case of a person who receives a cut through the skin of his finger. This is just like opening a door to admit the bacteria that are always present on the skin. Once inside, the bacteria begin to multiply and to damage tissue cells. A chemical substance is liberated by the injured cells, causing dilatation of the neighboring blood vessels and increasing the permeability of capillary walls. As a result, the skin *reddens* and feels *hot* as dilatation of the vessels brings more blood to the area. *Swelling* occurs as large quantities of fluid and protein escape from the highly permeable capillaries, and the pressure of these swollen tissues causes *pain*. This sequence of events constitutes the classic picture of *inflammation*.

From the very beginning of the inflammatory process, white blood cells are attracted to the area that is involved. Granulocytes and monocytes are particularly notable for their ability to leave the blood vessels and migrate to sites of inflammation or infection. Much like the amoeba, their protoplasm streams out into pseudopodia (false feet) enabling them to squeeze through the tiny openings in capillary walls and out into the neighboring tissues (diapedesis). This type of locomotion is slow indeed, but it enables the white cells to leave the bloodstream and enter into mortal combat with the bacterial invaders.

The major weapon of the leukocytes is **phagocytosis** (p. 25), a process whereby they engulf and actually digest bacteria, fragmented cells, and foreign particles. Granulocytes, the neutrophils in particular, seem to concentrate on the ingestion of bacteria; hence they are called microphagocytes. On the other hand, monocytes are especially good at the removal of larger particles such as broken-down tissue cells; thus they are called macrophagocytes. The ingested bacteria usually are killed and digested quickly, but occasionally the leukocytes themselves are destroyed by strong bacterial poisons. Therefore, following unusually severe battles, the combat area will be full of pus, which is a thick, semiliquid mass of living and dead leukocytes and bacteria, as well as battered tissue cell fragments. If the leukocytes defeat the bacterial invaders, they remove the debris from the battlefield, and healing begins. At this time, monocytes may become fibroblasts and aid in the repair of tissue. But if the bacteria overwhelm the leukocytes and continue to multiply, eventually they may get into the bloodstream and cause *septicemia* (*blood poisoning*).

The small lymphocytes also play an important role in defense of the body against disease. It is believed that these cells are capable of multiplying and forming certain precursor cells, which in turn give rise to *plasma cells* (*plasmacytes*); as previously described on page 39, plasma cells are the primary producers of antibodies (gamma globulins) in the presence of antigens (i.e., foreign substances such as bacteria). Antibodies, in turn, combine with the antigen and lead to its inactivation.

The function of eosinophils is not yet certain, but it has been suggested that they detoxify foreign protein that may gain entry to the body. Eosinophils increase in number in the circulating blood during allergic reactions, such as asthma, and when certain animal parasites invade the body (i.e., *Trichina*, the organism that may be ingested if one eats infected pork that is poorly cooked). Basophils, like eosinophils, also seem to be involved in allergic and stress reactions. Their granules are very similar to those in mast cells (p. 39), and it is believed that they, too, are a source of histamine and heparin.

Life Span. Because leukocytes do their work outside of the bloodstream, they remain in the circulation for a variable period of time; to some extent this is determined by the need for their services in neighboring tissues. Granular leukocytes remain in the blood for an average of about 10 hours before migrating into neighboring tissues. Lymphocytes circulate back and forth between the blood and the lymph streams for 90 to 100 days, depending on the need for them in other tissues. The length of time that monocytes remain in the blood is uncertain.

The ultimate fate of the leukocytes is variable. Some of these cells die in combat with bacteria; others may migrate into the intestine, where they are lost in the feces; and some probably are destroyed by cells of the reticuloendothelial system.

Number of Leukocytes. During the course of each day there may be marked variations in the number of circulating leukocytes. For this reason, the *absolute count*, or total number, of white blood cells may be anywhere from 5,000 to 10,000 per cu. mm. of blood in the healthy adult.

A *differential count* is an estimation of the number, or percentage, of each variety of leukocyte in the blood. It is made on a film of blood stained with Wright's stain. Each leukocyte seen is tabulated, until 100, or even 200, are counted, and then the percentage of each type is noted. For example, if a total of 200 have been counted and 140 of them are neutrophils, this means that 70 per cent of the leukocytes are neutrophils, and if 6 eosinophils are found, it means that 3 per cent of the total leukocytes are eosinophils. Variations in the percentage of certain types of leukocytes are characteristic of some diseases.

An increase in the number of leukocytes (above the upper limit of 10,000) is called *leukocytosis*; this occurs in most systemic and localized infectious processes, such as appendicitis or abscesses. It is a normal response to infection. On the other hand, a decrease in the number of leukocytes (below the lower limit of 5,000) is called *leukopenia*; this may occur in certain acute and chronic diseases, such as typhoid fever or tuberculosis. Leukopenia is also a constant finding in *radiation sickness*, the clinical result of excessive exposure to gamma rays. For example, victims of the atomic bomb explosions were exposed to intensive radiation and as a result suffered a marked depression of bone marrow function; the absolute white count in the more severe cases ranged from 1,500 to zero per cu. mm. of blood. There also was anemia, due to interference with the formation of red blood cells.

Leukemia is a disease which is characterized by a rapid and uncontrolled proliferation of leukocytes in the blood-forming organs; for this reason it is sometimes called *cancer of the blood*. The white cell count is greatly elevated, and large numbers of immature cells are found in the circulating blood. Chronic forms of this disease can be controlled for some time by the use of drugs that depress the activity of bone marrow and lymphatic tissue, but it usually terminates fatally within a few years. Acute leukemia is more rapidly fatal.

Infectious mononucleosis is an acute infectious disease that is probably caused by a virus. Commonly seen in young adults, it is characterized by fever, sore throat, and swollen cervical lymph nodes. As the white count rises, there is a decrease in the percentage of polymorphonuclear cells and an increase in the mononuclear series.

Agranulocytosis refers to a condition in which there is a deficiency of granulocytes (granular leukocytes). It is characterized by fever, prostration, ulcerative lesions in the mucous membranes of the mouth and the pharynx, and by a very low granulocyte count. It may occur in patients who have been taking thiouracil (for overactive thyroid gland), or sulfonamide drugs, or in individuals who become sensitized to various other drugs or chemical substances.

The blood platelets

Blood platelets, or *thrombocytes*, are the smallest (2 to 5 microns) of the formed elements (Fig. 11-4). These oval, granular bodies do not possess nuclei since they are merely fragments of large cells (megakaryocytes) found in red bone marrow.

Functions. The platelets perform several important functions in hemostasis, or the prevention of blood loss. When a blood vessel is dam-

aged, the lining losses its usual smoothness and nonwetability. As a result, platelets begin to stick to the walls and to each other, eventually forming a plug. If the vessel has only a small tear in its wall, the platelet plug by itself can stop the loss of blood. However, if there is a large tear, a blood clot also will be needed.

Platelets contain an activator substance, called *platelet factor*, that initiates blood clotting (p. 268). Once a blood clot has formed at the site of injury, platelets cause it to retract. During this process, the clot is changed from a soft mass to a firm one which helps to stop bleeding.

Finally, platelets contain *serotonin*, which is liberated as they disintegrate. Serotonin causes constriction of blood vessels and thereby further helps to control bleeding.

Life Span. Those platelets that escape disintegration at the site of injured tissues probably are destroyed in the reticuloendothelial system, particularly the spleen. Their survival time is thought to be about five to nine days.

Number. It is difficult to make an accurate platelet count, but the average range is probably somewhere between 150,000 and 350,000 per cu. mm. of blood. A marked reduction in the number of blood platelets is called *thrombocytopenia*, and a marked increase is called *thrombocytosis*.

The plasma

Plasma is the liquid portion of circulating blood, and its numerous constituents play an important part in the maintenance of homeostasis. In a healthy, fasting individual, the plasma is clear and straw-colored, but it is milky white following the ingestion of meals that contain fat. A few of the normal constituents of plasma will be described.

Water. The most abundant of all constituents is water, comprising from 91 to 92 per cent of the plasma. The functions of water are described on page 16.

Proteins. Plasma consists of 6 to 8 per cent proteins. Most of the plasma proteins are formed in the liver, but some are produced by plasma cells, the spleen, and the bone marrow, while others are released by disintegrating erythrocytes, leukocytes, and general tissue cells. Regardless of their location in the body, all the proteins serve to form a large, dynamic pool, or bank, which may be drawn on by any tissue that needs protein. Proteins are dynamic in the sense that a particular molecule may be synthesized in the liver, travel in the blood as part of the plasma pool, make up part of an enzyme in some cell, and eventually be incorporated into the hemoglobin of an erythrocyte. When this erythrocyte wears out and is broken down, the protein molecule may be available for use elsewhere in the body.

Plasma proteins have many functional and clinical uses, only a few of which will be mentioned at this time.

Albumin, the most abundant protein in plasma, is formed primarily by the liver. It is especially important because of its osmotic relationships, since it is the protein mostly responsible for the *colloid osmotic pressure* of the plasma. It helps to regulate the volume of plasma within the blood vessels by pulling in water from the interstitial fluid. When whole blood and plasma are not available, an albumin solution may be given intravenously to increase the blood volume following hemorrhage. Albumin is also of value in treating a patient who has lost large amounts of protein following extensive burns.

There are three major *globulin* fractions in the plasma: alpha, beta, and gamma. Alpha and beta globulins are formed primarily in the liver; they aid in the transport of other substances (by combining with them), and in the transport of protein itself from one part of the body to another. The gamma globulins (antibodies) are produced by plasma cells (p. 39). They resist infection, thus providing the body with immunity to certain diseases.

Fibrinogen also is made in the liver; it is in a liquid, or sol, form in circulating blood. When blood clots, fibrinogen changes from a liquid state to a solid state, or from a sol to a gel called *fibrin*. Fibrin films and fibrin foams are useful clinically to stop bleeding during general surgery, in making skin grafts, in surgical operations on blood vessels, and in neurosurgery (surgery of the CNS).

Prothrombin, which plays a part in the clotting of blood, is made in the liver. However, its production is governed by the availability of vitamin K; if this vitamin is lacking, the liver will be unable to produce sufficient amounts of prothrombin, and the blood clotting mechanism will be disturbed.

Nutrients. Carbohydrate is present in the blood chiefly in the form of glucose, a simple sugar, which is referred to as *blood sugar*. Glucose is carried to all tissues where it is used as a source of energy, or stored as reserve food until needed. Amino acids, the end products of protein digestion, likewise are carried to all tissues where they are used in the processes of building and repair. Lipids are present in plasma primarily as neutral fats, cholesterol, and phospholipids. They are carried to the tissues where they may be used to supply energy or stored as fat.

Inorganic Salts. Forming about 0.9 per cent of the plasma, inorganic salts occur primarily as chlorides, carbonates, and sulfates and phosphates of sodium, potassium, calcium, and magnesium. Iron, copper, iodine, and traces of many other elements also play a role in the preservation of homeostasis. The proper concentration and distribution of various salts is referred to as *electrolyte balance* (pp. 458 to 464).

Gases. Oxygen, carbon dioxide, and nitrogen are found in solution in the plasma, but only small amounts are transported in this manner.

Waste Products. Urea, uric acid, lactic acid, and creatinine are the most important of the metabolic waste products that are carried in the plasma to organs of elimination.

Special Substances. The endocrine glands, or glands of internal secretion, produce chemical substances called hormones; these substances are transported in the plasma to tissues in all parts of the body. Antibodies also are carried in the plasma.

Composition of Plasma in Disease. Laboratory tests which determine the quantities of various plasma constituents are valuable diagnostic tools. For example, in uncontrolled diabetes mellitus the blood sugar (glucose) is well above the average of 70 to 110 mg. per 100 ml. of blood, but if an overdose of insulin is given, the blood sugar may fall to dangerously low levels. In nephritis, one type of kidney disease, the nitrogenous waste products accumulate in excessive amounts in the blood. Functional disturbances in the endocrine glands also cause changes in the quantity of plasma constituents. For example, in overactivity of the parathyroid glands (hyperparathyroidism), the amount of calcium is greatly increased, but in cases of underactivity (hypoparathyroidism), the calcium level is low.

Coagulation of blood

The *coagulation*, or *clotting*, of blood is a protective mechanism that guards against excessive loss of the body's life-sustaining fluid. During the clotting process blood loses its liquid quality and forms a jellylike mass which helps to plug up leaks in the wall of an injured vessel. There are many theories regarding the specific chemical reactions which are involved in the clotting process, but such details are beyond the scope of this textbook. Consequently, the following discussion is limited to a brief consideration of the three general phases, or steps, in the clotting of blood as illustrated in Figure 11-7.

The **first phase** of coagulation involves the formation of *thromboplastin*, a substance that triggers the clotting mechanism. For example, let us suppose that your finger has been cut with a knife. As blood flows from severed vessels it passes over rough, or foreign, surfaces where the platelets tend to accumulate and to disintegrate with the release of a substance known as *platelet factor*. In the presence of calcium ions (Ca^{++}) and certain accessory substances, the platelet factor reacts with an antihemophilic factor to form thromboplastin. Neighboring tissue cells which were injured by the knife also will liberate thromboplastic substances.

The **second phase** of coagulation involves the conversion of prothrombin to *thrombin*. Prothrombin, one of the normal plasma proteins, is formed in the liver in the presence of vitamin K; a deficiency of this vitamin may result in a bleeding tendency due to inadequate prothrombin

formation. The thromboplastin formed in phase one serves to initiate the conversion of prothrombin, but the presence of calcium ions and certain accelerator factors is necessary for rapid completion of the reaction.

The **third phase** of coagulation involves the action of thrombin in converting soluble fibrinogen to insoluble *fibrin*. Fibrinogen is another protein which is produced in the liver and found circulating in the plasma. Fibrin first appears as a fine network of threads that gradually becomes denser, forming the structural substance of the clot. Erythrocytes are trapped in the fibrin mesh, but these red cells have nothing to do with actual coagulation, since clotting will occur in plasma after the blood cells have been removed. The clotting process draws to a close as blood platelets cause a retraction or shrinking of the clot, thus producing a firm but fairly elastic plug. As the clot shrinks, a clear yellow liquid is squeezed out; this liquid, called *serum*, is similar to plasma except for the absence of fibrinogen and certain other clotting elements that have been used in the coagulation process.

Fibrinolysis. Blood clots that have formed in the tissues and in small vessels are eventually dissolved by a process known as fibrinolysis. This process involves the activation of a precursor substance (*profibrinolysin*) to *fibrinolysin*. This active, proteolytic substance digests the fibrin threads in the clot. Fibrinolysis is an important process in clearing away clots that have formed in tissues after a blood vessel has ruptured. Also, it may prevent the clogging of small peripheral vessels where intravascular clots may occur.

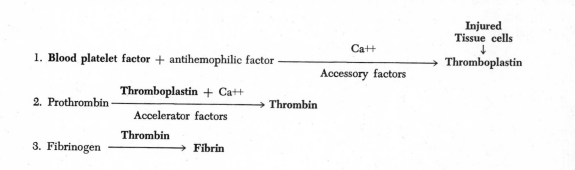

Fig. 11-7. Schematic representation of the three phases of blood clotting. Those substances which normally are not present in significant amounts in the circulating blood are shown in **boldface type.**

Coagulation Time. Blood taken from a normal person will clot within a predictable length of time, depending on the method that is used. For example, blood drawn into a capillary tube will clot in three to six minutes, but in a larger tube the range is from five to fifteen minutes (Lee-White method).

Bleeding Time. This test is performed to determine the time required for a small sharp incision to cease bleeding. The normal bleeding time is from one to three minutes. It is greatly prolonged in such conditions as thrombocytopenia, pernicious anemia, and acute leukemias.

Blood Preservation. Chemical agents that prevent the clotting of blood play an important role in hospital laboratories. The addition of citrate or oxalate solutions to blood samples will prevent clotting by combining with calcium ions, thus interfering with the formation of thrombin. However, when blood is to be used for transfusions and must be stored in blood banks until needed, the best preservative is an acid-citrate-dextrose solution (A.C.D.); this anticoagulant delays the breakdown of red cells and preserves blood for as long as three weeks.

Abnormal Coagulation of Blood. Intravascular coagulation, or *thrombosis*, refers to the abnormal clotting of blood within the vessels of a living person. A stationary blood clot is called a **thrombus**, and it may partially or completely block the vessel in which it is formed. Occasionally one of these clots may break loose from its place of formation and be swept away in the rapidly moving bloodstream until it plugs up a vessel elsewhere in the body; such a wandering clot is called an **embolus**.

One cause of thrombosis may be a roughening of the normally smooth endothelial lining of a blood vessel due to infection or to fatty degeneration (atherosclerosis) of the vessel wall. For example, *thrombophlebitis* is a term that means inflammation of a vein associated with the formation of a clot. Other conditions that favor thrombosis are stasis, or sluggish flow of blood through the vessels, and the presence of foreign material in the blood. The great danger associated with thrombosis and with embolism is that the blood flow to a vital organ may be shut off if clots plug up its vessels. For example, *coronary thrombosis* may be rapidly fatal if one of the vital arteries leading to the heart muscle is occluded. Another serious problem is *pulmonary embolism*, or an obstruction of one or both of the great arteries leading to the lungs. This may occur as a complication of prolonged bed rest when there is stasis of blood in the veins of the extremities. Such periods of sluggish blood flow favor the formation of clots which are only loosely attached to the vessel walls. When such a person gets out of bed, the contraction of his leg muscles may cause the clots to break loose and be swept onward to the pulmonary arteries. Fortunately, such complications are rare when early ambulation is possible.

Anticoagulant drugs are agents that retard or prolong the clotting process. They are of particular value in disease conditions that involve abnormal coagulation of the blood, such as thrombosis. However, great care must be exercised in the use of anticoagulants, since an overdose may result in fatal hemorrhage.

Commercial preparations of heparin are used when a rapid depression of blood coagulability is desired; injected intravenously, its effect is immediate, but short-lived. On the other hand, coumarin derivatives (e.g., Dicumarol) are anticoagulants that can be taken orally, but they do not take effect for about 48 hours; Dicumarol retards blood clotting by interfering with the formation by the liver of prothrombin and one of the accelerator factors, both of which are needed for coagulation. The dosage of Dicumarol is subject to individual variations, and for this reason frequent blood tests of prothrombin activity are advisable so that the dosage may be adjusted.

An estimation of prothrombin activity may be made in the laboratory by mixing a citrated sample of the patient's plasma with standardized amounts of thromboplastin and calcium ions; the time required for this plasma to clot is reported as the *prothrombin time*. If there is a normal amount of prothrombin and accelerator factor in the plasma, clotting will occur in 12 to 20 seconds. During treatment with Dicumarol, however, prothrombin activity may be only 15 to 25 per cent of normal, and the prothrombin time increased to 35 seconds or more.

Excessive bleeding

The escape of blood from the vessels in which it is contained is referred to as *hemorrhage*. The control of excessive bleeding may be accomplished by direct pressure at the site of injury, by hemostats (surgical clamps), by ligatures (ties), and by the application of tourniquets. Sometimes gauze applied over the bleeding surface will help to stop the flow by favoring the breakdown of platelets and hastening coagulation. Following the loss of a relatively small amount of blood (e.g., a blood donation of 500 ml.), the plasma volume is regained within a few hours as interstitial fluid is pulled in by osmosis from the tissue spaces, but the erythrocytes and the plasma proteins are restored more slowly. It takes about 50 days for a healthy blood donor to regenerate corpuscles and hemoglobin. The signs and symptoms of severe hemorrhage are described on page 345.

Hemophilia is the classic hereditary disease in which a sex-linked characteristic is transmitted by females, but in which bleeding difficulties occur exclusively in males. The underlying defect is still obscure, but it is thought to be a congenital absence of the antihemophilic factor (AHF) which plays such an important part in the formation of thromboplastin during the first phase of blood clotting. As a result, hemophilia is characterized by a prolonged coagulation time (up to two hours or more) and a tendency toward bleeding in the skin, the muscles, and the joints. The bleeding may occur spontaneously, or as a result of minor injuries.

A deficiency in one of the accessory factors, which also is needed in phase one of the clotting process, results in a disease that closely resembles hemophilia. However, it differs in that it can occur in either males or females.

Blood transfusions

A transfusion with whole blood is indicated primarily in those cases in which there has been an acute loss of both red cells and plasma, as in severe hemorrhage. Plasma is the next best substance for use in such emergencies, but it is no substitute for whole blood, since the oxygen-carrying red cells have been removed.

The person from whom blood is taken is called a *donor,* and the patient who receives blood is called a *recipient.* The safe transfusion of blood from donor to recipient depends on careful laboratory tests to ensure that the patient receives only *compatible* blood, or that which agrees with his own. The administration of unsafe, or *incompati-*

ble, blood may lead to serious or even fatal transfusion reactions.

Blood compatibility and the systems of classification are based on the presence or the absence of specific antigens (agglutinogens) in the red blood cells, and the presence or the absence of specific antibodies (agglutinins) in the plasma. Because of these factors, the red blood cells of one person may be *agglutinated,* or *clumped,* by the plasma of another person. Clarification of the relationships between compatible and incompatible blood groups won the Nobel prize for Karl Landsteiner and led to the establishment of the international ABO system of classification. According to this system, there are four human **blood groups,** or types: O, A, B, and AB.

There are two *antigens,* or *agglutinable substances,* called A and B, that may be present in red blood cells, making it possible for them to be agglutinated. Neither antigen is present in the red cells of group O individuals; A is present in group A, B in group B, and both A and B in group AB.

There are two *antibodies,* or *agglutinins,* called *alpha* and *beta,* that may be present in plasma, making it possible for plasma to agglutinate certain red cells. Both antibodies are present in the plasma of group O, beta in group A, alpha in group B, and neither alpha nor beta in group AB.

If the antigen A in the red cells of a donor comes in contact with the antibody or agglutinin alpha in the plasma of a recipient, or if antigen B in the red cells of a donor comes in contact with beta agglutinin in the plasma of a recipient, agglutination, or clumping, of cells will occur. This is followed by *hemolysis,* or rupture of red cell membranes, allowing hemoglobin to escape into the plasma. Such bloods are incompatible and lead to a transfusion reaction.

The hemoglobin that escapes into the plasma in the course of a transfusion reaction, is excreted in the urine. Masses of it may block the kidney tubules and make it impossible for the kidneys to function normally. Urticaria (hives), chills, fever, and hematuria (blood in the urine) are the indications of incompatible bloods. Thus the event to be avoided in transfusions is the destruction of the red blood cells of the donor by the plasma of the recipient. We need not worry about the possibility of destruction of the recipient's red cells by the donor's plasma, since the latter is immediately so greatly diluted by the plasma of the recipient that the agglutinins have little opportunity to destroy the recipient's red cells.

The Rh Factor. In addition to the ABO blood groups, there is yet another red cell factor which not only is very much concerned in transfusions, but also may play a role in pregnancy as well. This is the Rh factor, so named because it was first discovered in the blood of Rhesus monkeys. As with the A and the B antigens of the ABO system, the Rh factor is contained in the red blood cells. About 85 to 95 per cent of the population have Rh antigen in their red cells and are said to be *Rh positive*. The 5 to 15 per cent who lack this factor are said to be *Rh negative*. Human plasma normally does not contain corresponding Rh antibodies, but they may be formed in Rh-negative blood if Rh-positive blood is used for transfusion purposes; such an individual then would be sensitized to the Rh factor, and any subsequent injection of Rh-positive blood might result in a severe or even fatal transfusion reaction. Sensitization also may occur during pregnancy in Rh-negative women who are married to Rh-positive men, but we wish to emphasize the fact that the vast majority of such women have no difficulty with their pregnancies. However, it is possible for some Rh-negative women to become so sensitized during their first or second pregnancy that future pregnancies may result in injury to the baby. This sensitization may occur when an Rh-negative woman is carrying an Rh-positive baby in her uterus. Normally the bloodstreams of mother and child are completely separated by a thin partition of tissue through which only small nutrient and waste particles can pass. However, near the end of pregnancy, a small leak may develop in the membranous barrier, permitting Rh-positive red cells to pass from the child's bloodstream into the Rh-negative bloodstream of the mother; as a result, Rh antibodies are produced, and the mother is sensitized. These Rh antibodies do not harm the mother directly, since they will not attack her Rh-negative red cells, but in a future pregnancy they may harm an Rh-positive baby by diffusing into its bloodstream and destroying its red cells. Such a condition in the baby is called *erythroblastosis fetalis*. However, a new and dramatically effective vaccine (RhoGAM) is now available to prevent the sensitization of Rh-negative mothers and the subsequent erythroblastosis

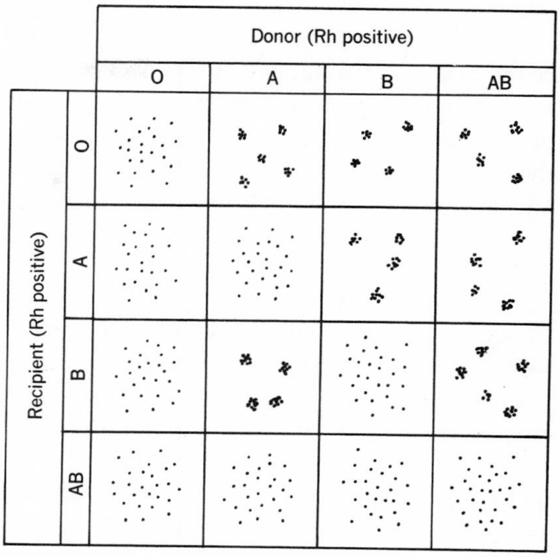

Fig. 11-8. Diagram of results obtained when cross-matching the main human blood groups, both donor and recipient being Rh positive. Even distribution of cells indicates safe or compatible blood for transfusion. Clusters indicate the clumped cells of incompatible blood which may cause transfusion reaction. Individuals in group O are called universal donors; their erythrocytes are not agglutinated by any plasma since they contain no antigen. Individuals in group AB are called universal recipients; since there is no antibody (agglutinin alpha or beta) in their plasma, erythrocytes from any group will not be agglutinated.

fetalis. Indeed, in the very near future, this serious medical problem may be completely eliminated.

Laboratory Tests. The safety of blood transfusions first requires determination of the ABO group and the Rh type. However, there are so many subgroups that a final test, the *cross-matching* of the patient's blood with that to be administered, also is required; this is done by mixing some of the donor's red cells with a sample of the recipient's plasma to see if clumping occurs. Figure 11-8 illustrates the results of cross-matching the four main blood groups of the ABO system. Blood to be used for transfusion purposes also is subjected to a serologic test for syphilis.

LYMPH

Definition and origin

Lymph is the fluid that is present in vessels of the lymphatic system (p. 321). As described previously (pp. 34 to 35), the fluid that filters out at the arterial end of a capillary has several courses available to it: (1) It may return to the blood-stream by osmosis; (2) it may move on into the tissue cells, becoming part of the intracellular fluid; (3) it may remain in the tissue spaces as part of the interstitial fluid reservoir; or (4) it may drain into the lymphatic capillaries; once it enters these vessels the fluid is called lymph.

Composition

Since lymph is derived from blood plasma we would expect it to have approximately the same composition as the interstitial fluid that filters through the vascular capillary wall. In other words, lymph has almost the same proportions of water and solutes as are found in plasma, but it has con-siderably less protein (with the exception of lymph that is drained from the liver, the latter having a protein content of about 6 per cent). Lymph that is drained from the intestinal area may contain large amounts of fat after meals; the suspended fat droplets give this lymph a milky white appearance. The cell content of lymph varies from 500 to 40,000 leukocytes per cu. mm. or more, depending on the number of lymph nodes it has traversed; these cells are predomi-nantly lymphocytes which have been formed in the lymphatic tissue.

Fate

Lymph ultimately enters the bloodstream after a journey through various lymphatic vessels. The circulation of lymph is described on pages 321 to 324.

SUMMARY

1. The blood

A. Functions
  a. Delivers vital substances to capillaries
  b. Picks up metabolic end products and delivers to organs for use or elimination
  c. Aids in regulation of body temperature
  d. Helps to maintain acid-base balance
  e. Protects against bacterial invasion and disease

B. Characteristics
  a. Color: bright red if well oxygenated; dark red if poorly oxygenated
  b. Quantity: approximately 7 per cent of body weight
  c. Viscosity: about five times thicker than water
  d. Specific gravity: between 1.055 and 1.065. Sedi-mentation rate ranges from 4 to 10 mm. per hour
  e. Reaction, or pH: slightly alkaline (7.35 to 7.45)

C. Structure and formation
  a. Hemopoietic tissue found in red bone marrow and lymphatic tissue
  b. Plasma constitutes 53 to 58 per cent of blood; formed elements 42 to 47 per cent

D. Erythrocytes: red blood cells
  a. Flexible, biconcave disks; framework contains hemoglobin
  b. Hemoglobin aids in transport of oxygen and carbon dioxide; average hemoglobin content: female 14 Gm.; male 16 Gm. per 100 ml. blood
  c. Red cell may rupture in hypotonic solution with loss of hemoglobin (hemolysis); basis of fragility test
  d. Formation: in red bone marrow; millions formed every second; requires vitamin B_{12} (extrinsic fac-tor); gastric juice component (intrinsic factor) essential for B_{12} absorption
  e. Destruction occurs in reticuloendothelial system: average life span 120 days
  f. Average range for number of red cells 4½ to 5½ million per cu. mm. of blood. Hypoxia causes increased formation of erythropoietin; stimulates bone marrow

E. Leukocytes: white blood cells
  a. Larger than red cells; all are nucleated
  b. Granular leukocytes (polymorphonuclear leuko-cytes): granules in cytoplasm; nuclei have lobes; subgroups: neutrophils, eosinophils, and baso-phils; formed in red bone marrow
  c. Nongranular leukocytes: no granules in cyto-plasm; subgroups: lymphocytes (formed in lym-phatic tissue), and monocytes (formed in bone marrow)
  d. General function: to protect body against in-fection; granulocytes and monocytes are active phagocytes; lymphocytes may be concerned with immunity reactions and give rise to plasma cells
  e. Life span may be 10 hours (granulocytes) to 100 days (lymphocytes); probably destroyed in reticuloendothelial system and in combat
  f. Absolute count falls within range of 5,000 to 10,000 per cu. mm. of blood

F. Blood platelets
  a. Merely cell fragments
  b. Function: help in control of bleeding
  c. Average count between 150,000 and 350,000

G. Blood plasma: liquid portion of circulating blood
  a. Composition: water (92 per cent); proteins (6 to 8 per cent); nutrients; inorganic salts; gases,

waste products, and special substances
- H. Coagulation of blood
  - a. Guards against excessive loss of blood
  - b. First phase: formation of thromboplastin
  - c. Second phase: conversion of prothrombin to thrombin
  - d. Third phase: conversion of soluble fibrinogen to insoluble fibrin
  - e. Serum is liquid squeezed out of clot as it shrinks
  - f. Fibrinolysis
  - g. Coagulation time: three to six minutes in capillary tube; five to fifteen minutes in larger tubes (Lee-White)
  - h. Bleeding time: one to three minutes
  - i. Blood preserved in blood bank by acid-citrate-dextrose solution (ACD)
  - j. Intravascular clotting: stationary clot—thrombus; wandering clot: embolus
    - (1) Anticoagulant drugs: Dicumarol and commercial preparation of heparin
    - (2) Prothrombin time averages 12 to 20 seconds
- I. Excessive bleeding
  - a. Severe loss may lead to death
  - b. Hemophilia; hereditary disease
- J. Blood transfusions
  - a. Treatment of severe hemorrhage; replaces red cells as well as fluid
  - b. Safe transfusion from donor to recipient depends on typing and cross-matching
  - c. Blood compatibility based on presence or absence of specific antigens and antibodies in red cells and plasma, respectively; ABO system and Rh factor of primary importance

2. Lymph
- A. Fluid present in lymphatic vessels; called interstitial fluid until it moves into the lymph capillaries
- B. Similar to interstitial fluid; water and solutes as in plasma, only small amount of protein. Lymph from intestinal area rich in fat after meals. Cell content varies; primarily lymphocytes
- C. Ultimately enters bloodstream

QUESTIONS FOR REVIEW

1. What is the chief function of the blood? Briefly, how does it carry out this function?
2. What are the major characteristics of blood?
3. Why are hemocytoblasts (stem cells) of such great importance?
4. Distinguish between oxyhemoglobin and reduced hemoglobin. What is cyanosis?
5. What is the significance of an increased number of reticulocytes in the circulating blood?
6. What factors are essential for the normal maturation of red blood cells?
7. Distinguish between granular and nongranular leukocytes in terms of subgroups, place of origin, and function.
8. What is the general sequence of events that constitute the picture of inflammation?
9. What would be the major disturbance in body processes associated with loss of
   a. plasma proteins
   b. red blood cells
   c. white blood cells
   d. blood platelets
10. How does plasma differ from whole blood? What are its major constituents?
11. Briefly, what are the three phases of blood clotting?
12. A friend tells you that his mother is taking Dicumarol because she has some problem with her blood. He wants to know what this medication is for, how it works, and if it is at all dangerous. What will you tell him?
13. If you have type AB, Rh negative blood, to whom could you donate blood? From whom could you receive blood?

REFERENCES AND SUPPLEMENTAL READINGS

Best, C. H., and Taylor, N. B.: The Physiological Basis of Medical Practice, ed. 8. Baltimore, Williams & Wilkins, 1966.

Child, J., Collins, D., and Collins, J.: Blood transfusions. AJN, 72:1602, Sept. 1972.

Clark, C. A.: The prevention of "Rhesus" babies. Sci. Am., 219:46, Nov. 1968.

Guyton, A. C.: Textbook of Medical Physiology, ed. 4. Philadelphia, W. B. Saunders, 1971.

Page, I. H.: Serotonin. Chicago, Year Book Medical Publishers, 1968.

Solomon, A. K.: The state of water in red cells. Sci. Am., 224:89, Feb. 1971.

The Heart

"He fashioned their hearts alike."
Psalms 33:15

Every minute of every hour of every day, the heart serves as a powerful muscular pump which beats out the steady rhythm of life. Contracting at a rate of 70 to 75 times each minute, the heart generates the all-important pressure necessary for the adequate circulation of blood through the vessels of the body. This amazingly tough muscle pumps over 3,000 gallons of blood every day as it helps to provide the capillaries with quantities of blood sufficient to meet the ever-changing cellular needs. While skeletal muscles can rest for relatively long periods of time, the heart muscle can never indulge in such luxury. For example, if you run up a flight of stairs, the leg muscles can rest as soon as the exercise is completed, but the sturdy heart has only a fraction of a second to rest before it is time for the next contraction.

STRUCTURAL FEATURES

Location and description

The heart is a hollow muscular organ that is slightly larger than your clenched fist. It is located in the *mediastinum*—the space between the lungs, in the thoracic cavity. About two thirds of the heart lies to the left of the midline, and its relation to other organs is illustrated in Figure 1-6. The heart is somewhat cone-shaped, with the base directed upward toward the right and the apex pointing downward to the left where it rests on the diaphragm. If you place your fingers in the space between the fifth and sixth ribs (about 2 inches below the nipple line of the left breast), you can feel the apex come in contact with the chest wall with each beat of the heart. An *apical pulse* is taken by counting the number of heartbeats while listening through a stethoscope placed over the apex of the heart.

Covering of the heart

The heart is covered by a slippery, loose-fitting sac called the **pericardium;** this protective cover consists of two parts: an outer fibrous portion and an inner serous portion.

The *fibrous pericardium* is a tough fibrous membrane; this layer is attached to the large vessels that enter or leave the heart, to the diaphragm, and to the sternum.

The *serous pericardium* consists of two distinct layers: (1) an outer or parietal layer that lines the fibrous pericardium and (2) an inner or visceral layer, the epicardium, which is closely adherent to the heart itself. Between the visceral and the parietal layers of the serous pericardium there is a potential space which contains a small amount of slippery pericardial fluid; this serous fluid serves as a lubricant to reduce friction resulting from movements of the heart as it beats.

An infection of the pericardium is called *pericarditis.* In some persons the inflammatory process may result in the formation of fibrous adhesions between the visceral and the parietal layers; such adhesions tend to constrict the heart and interfere with its activity. In severe cases of constrictive pericarditis, it may be necessary to perform a pericardectomy, or removal of the fibrous pericardium with its serous lining.

Walls of the heart

The walls of the heart are composed of three distinct layers. The outer layer is the *epicardium,* or visceral pericardium, as mentioned previously. The middle layer is called the *myocardium,* since it consists of interlacing bundles of cardiac muscle fibers. The arrangement of these muscle bundles is such that their contraction results in a wringing, or milking, type of movement that efficiently squeezes blood out of the heart with each beat. Inflammation of the heart muscle is called *myocarditis.*

Lining the inner surface of the myocardial wall is a thin layer of endothelial tissue which is called the *endocardium.* This delicate lining is continuous with that of the blood vessels. The endocardium also assists in forming the valves of the heart. Rheumatic fever is a disease that frequently attacks the heart, causing an inflammation of the lining (endocarditis), as well as myocarditis and pericarditis.

Heart chambers and valves

Study of the heart chambers and one-way valves is greatly simplified if one thinks of the heart as consisting of two separate pumps (right and left) which lie side by side and work in perfect harmony. The right side of the heart receives blood which is low in oxygen and pumps it to capillaries in the lungs. At the same time, the left side of the heart receives oxygenated blood which has passed through the lung capillaries, and pumps it into the aorta for distribution to all parts of the body.

The upper, or receiving, chambers of the heart are called the right and left atria. The lower, or dispatching, chambers are called the right and left ventricles. Grooves on the anterior and the posterior surfaces of the heart indicate the underlying dividing lines between chambers (Figs. 12-1 and 12-2). The main branches of the coronary arteries and veins lie in these grooves.

Right Atrium. The right atrium is larger than the left, but its walls are very thin. It receives blood from the upper portion of the body via the superior vena cava and from the lower portion of the body by way of the inferior vena cava. The venae cavae are called the *great veins.* The terminations of the great veins are shown in

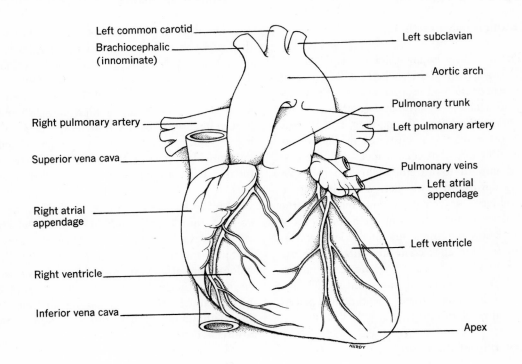

Fig. 12-1.   Anterior aspect of heart and great blood vessels.

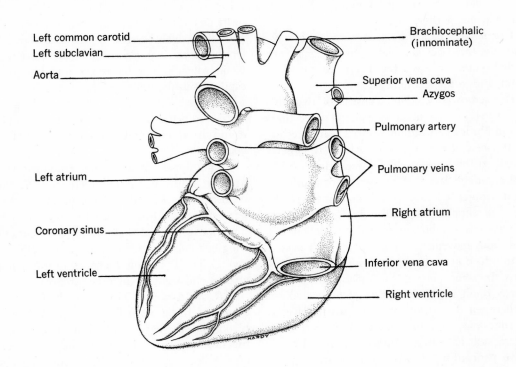

Fig. 12-2.   Posterior aspect of heart and great blood vessels.

Figures 12-2 and 12-4. The right atrium also receives blood from the heart muscle itself by way of the coronary sinus (Fig. 12-2).

The right atrium opens into the right ventricle; this opening is surrounded by a fibrous ring that is large enough to admit three or four fingertips. The three leaflets, or cusps, of the **tricuspid,** or **right atrioventricular valve,** are attached to the fibrous ring. These leaflets completely close off the atrium from the ventricle during contraction of the latter. The valve as it appears when looking down at it from the atrium is shown in Figure 12-3. The leaflets are composed of folds of endocardium, reinforced with a flat sheet of dense connective tissue.

Left Atrium.   The left atrium has thicker

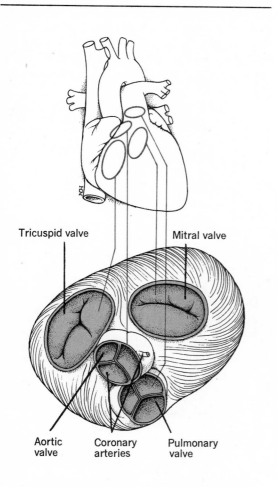

Fig. 12-3. Valves of the heart. In the lower drawing the atria have been removed and the heart tilted forward; thus the valves are seen from above.

walls than the right. It receives blood from the lungs by way of four pulmonary veins (Fig. 12-2). The opening between the left atrium and the left ventricle admits only two or three fingertips, and the two leaflets of the **bicuspid,** or **left atrioventricular valve,** are attached to the fibrous ring around this opening. Because of its resemblance to a bishop's miter, this valve also is called the **mitral valve.** The purpose of the atrioventricular valves of the heart is to keep the blood flowing in the proper direction. They permit it to flow from atria to ventricles, but prevent it from flowing in the reverse direction.

The **interatrial septum** is the partition that separates the right and left atria. During fetal life, there is an opening in this septum, called the *foramen ovale,* which permits a certain amount of blood to flow from right to left atrium, thus bypassing the lung circuit. After birth, when the lungs are functioning, this foramen normally closes; a thin, depressed area, called the fossa ovalis, indicates the site of closure. In some children, the foramen may remain open or patent; this congenital defect is called *patent foramen ovale.*

In addition to the main cavity of each atrium, there is a hollow, external projection that resembles a rooster's comb lying on its side. These protuberances are called the right and left auricular appendages. The internal surface of each appendage is uneven due to the presence of muscular ridges, but the main cavity of the adjacent atrium has a smooth inner surface.

Right Ventricle.   The right ventricle receives blood from the right atrium and, following closure of the atrioventricular valve, pumps that blood into the pulmonary artery; this artery arises from the base of the ventricle. Backflow of blood from the artery to the ventricle is prevented by the **pulmonary semilunar valve.** This valve consists of a set of three pocketlike flaps that face upward as they arise from the fibrous ring around the orifice of the artery. Blood flows freely out of the heart as the valve flaps are flattened against the arterial wall but is unable to return as the little pockets fill with blood and balloon out to block the opening (Fig. 12-3).

In contrast to the thin-walled atrium (which merely empties itself into the adjacent ventricular chamber), the wall of the right ventricle is thicker, since it must generate sufficient pressure to pump blood to the lungs. Also, the inner surface of the ventricle is very uneven, due to the presence of a latticework of muscular columns. Some of these columns project like papillae (nipples) into the ventricular cavity, and for this reason are called *papillary muscles*. From the summits of the papillae, you will see slender fibrous strands which extend upward to be attached to the edges of the leaflets of the atrioventricular valve. These strands are called *chordae tendineae* and are composed of dense fibrous connective tissue covered with endocardium. The function of the chordae tendineae is to keep the valve leaflets from turning inside out as they would do otherwise when the pressure is high in the ventricle. Papillary muscles and chordae tendineae are illustrated in Figure 12-4.

Left Ventricle. The left ventricle receives blood from the left atrium and, following closure of the atrioventricular valve, pumps it into the aorta. The aorta arises from the base of the left ventricle, and it is the largest artery in the body. Its orifice is guarded by the **aortic semilunar valve,**

which has the same pocketlike structural features as does the pulmonary valve.

While the right ventricle pumps blood only to the lungs, the left ventricle must generate sufficient pressure to pump blood to all other parts of the body; consequently, its walls are three times as thick and muscular as the right. The lower portion of the left ventricle forms the apex of the heart. Its inner surface is quite similar to that of the right ventricle except that the papillary muscles and chordae tendineae are much larger and stronger to withstand the greater pressure.

The **interventricular septum** is a thick partition that separates the right and left ventricles. Normally, there is no opening in this septum, but in certain cases of congenital heart disease (p. 320), there may be such a septal defect.

Valvular Disease. Two types of difficulty may arise in connection with heart valves. One type arises when there is narrowing of the fibrous ring to which the leaflets are attached, or when the leaflets themselves become thick and stiff and fail to open widely. This condition is called *stenosis,*

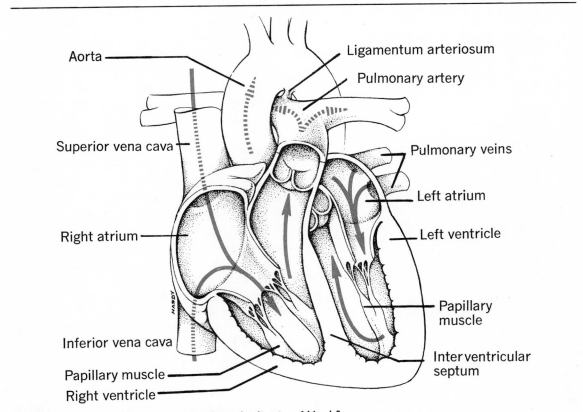

Fig. 12-4. Interior of the heart. Arrows indicate the direction of blood flow.

and when it exists there is difficulty in pumping the blood through the narrowed opening as rapidly as it should be going through. In the second type of difficulty, the edges of the valve leaflets are eroded, or the fibrous ring is stretched. This condition is called *insufficiency* or *regurgitation*; when it exists the valve cannot be closed completely, and blood leaks back through the valve when it is supposed to be flowing in the forward direction. These valvular difficulties are of frequent occurrence in bacterial endocarditis; some persons may have mitral stenosis, mitral regurgitation, aortic stenosis, aortic regurgitation, or tricuspid or pulmonary valve impairments. In some persons one valve may exhibit both types of difficulty; the leaflets may be thickened and stiff, and their edges may be eroded by the same disease process.

Modified cardiac muscle tissue

In certain areas of the myocardium, the muscle cells are modified and specialized to form the excitatory and conductive system of the heart; this system enables the atria and the ventricles to contract in orderly sequence.

As will be described later in this chapter, one very important property of cardiac muscle is its inherent ability to contract rhythmically in the absence of external stimuli. Those cells that have the most rapid inherent rhythm are found in the **S-A node** (sinoatrial node) of the conduction system; this knot of modified myocardium is located in the wall of the right atrium near the entrance of the superior vena cava. Since the heartbeat normally begins in this area, the S-A node is called the *pacemaker* of the heart.

A second knot of modified myocardium is the **A-V node** (atrioventricular node) which is located in the lower part of the interatrial septum near the opening of the coronary sinus. The A-V node is continuous with the **A-V bundle** (atrioventricular bundle) which serves as the connecting link between the atria and the ventricles; this modified myocardium also is known as the *bundle of His*, after its discoverer. The A-V bundle divides

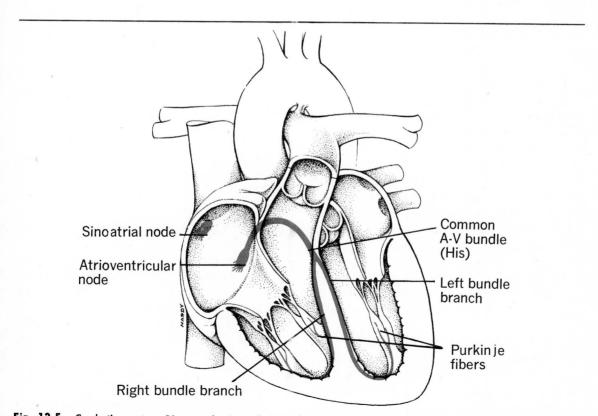

Sinoatrial node

Atrioventricular node

Common A-V bundle (His)

Left bundle branch

Purkinje fibers

Right bundle branch

Fig. 12-5. Conducting system. Diagram showing relations of the sinoatrial node, the atrioventricular node, the common atrioventricular bundle and its branches.

into right and left branches which extend downward into the interventricular septum where they become continuous with the Purkinje fibers; these fibers form a widespread network which penetrates both right and left ventricular muscle masses.

The importance of the A-V node and the A-V bundle becomes apparent when one considers the fact that the muscle mass of the two atria is completely separate and distinct from the muscle mass of the two ventricles; the rings of fibrous tissue which encircle the openings of the atrioventricular valves interrupt the continuity of muscle between the upper and the lower chambers. However, the A-V bundle serves as a bridge by means of which the rhythmic beat is transmitted from atria to ventricles. Figure 12-5 illustrates the conduction system of the heart. The relation of this system to the cardiac cycle is described on page 282.

Nerve supply

Although cardiac muscle possesses the property of rhythmicity, it also has an efferent and an afferent nerve supply (Fig. 12-6). These nerves play an important part in adjusting the rate of heartbeat to meet the ever-changing needs of the body.

Efferent Fibers. There is a double efferent nerve supply that transmits orders from cardiac centers in the medulla oblongata to the heart. These fibers are derived from both divisions of the autonomic system (see Chap. 9).

1. *Parasympathetic,* or *inhibitory fibers,* traveling in the vagus nerves, conduct impulses that slow and weaken the beat of the heart. Slowing of the heart rate provides longer periods of rest for the cardiac muscle cells. The vagus nerves to the heart are in tonic activity, which means that impulses are passing continuously to the heart from the center in the medulla. If the vagus nerves, which maintain a slow rate, are cut, the heart beats much faster than when the nerves are intact.

2. *Sympathetic,* or *accelerator fibers,* traveling in cervical and thoracic cardiac nerves, conduct impulses that speed and strengthen the heartbeat. Such acceleration is particularly important during times of physical or emotional stress. For example, note the change in your heartbeat the next time you run up a flight of stairs, or when

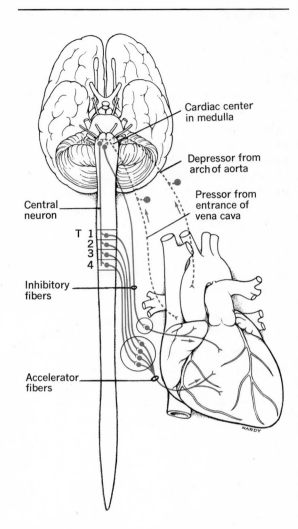

Fig. 12-6. Diagram showing the nerve supply of the heart. Afferent pathways are indicated by dashed lines, efferent pathways by solid lines.

the teacher unexpectedly asks you a question in class.

Afferent Fibers. The only sensation that may be aroused by impulses from the heart is that of pain. Whenever there is interference with the blood flow through the arteries of the heart, the resultant lack of oxygen in the myocardium serves to stimulate the receptors of certain visceral afferent fibers. These fibers then conduct impulses to the CNS, and the individual experiences a sensation of pain.

Other visceral afferent fibers conduct impulses that are initiated by pressure-sensitive receptors in the walls of the right atrium and adjacent portions of the venae cavae. These impulses may be involved in a reflex which causes acceleration of the heart during exercise (pp. 343 to 344).

Blood supply to the myocardium

One might think that since blood is flowing through the heart almost constantly, no other blood supply is needed, but the blood in the cardiac chambers comes in contact only with the endocardium and does not reach the myocardium. Consequently, like any other mass of muscular tissue, the thick myocardium needs its own blood supply.

The arteries that supply the myocardium are called the right and left *coronary arteries*. These two important blood vessels are the very first branches of the aorta, and their orifices may be seen in the aortic wall near the semilunar valve (see Fig. 12-3). The coronary arteries encircle the heart, and their many branches convey blood to capillaries in all parts of the myocardium. After delivering oxygen and nutrients to the cardiac muscle cells and picking up waste products, the

blood flows into the coronary veins. The principal coronary veins empty into the coronary sinus, which in turn conveys blood to the right atrium to be recirculated through the lungs. However, a few small coronary veins empty directly into both sides of the heart. The larger coronary vessels are illustrated in Figure 12-7.

The efficient contraction of heart muscle is dependent on a good blood supply, and any interference with the normal flow of blood through the coronary vessels may result in impaired cardiac function. For example, in some persons the smooth muscle in the walls of the coronary arteries may undergo spasmodic contractions; this in turn causes such a narrowing of the vessels that blood flow is temporarily shut off. In others, the walls of the coronary arteries undergo hardening (arteriosclerosis) or fatty degeneration (atherosclerosis); these conditions also interfere with the proper supply of oxygen and nutrients to the massive heart muscle. A cardinal symptom of oxygen-lack in the myocardium is *angina pectoris*, or paroxysmal pain that may be referred to the left shoulder and then radiate down the left arm; this pain may occur during physical exertion, or severe emotional stress. The mechanism of referred pain is described on page 235. Certain drugs, such as nitroglycerine and amyl nitrite, help to dispel anginal pain by causing a dilatation of the coronary arteries and thus improving blood flow to the myocardium.

When there is a complete and prolonged lack of blood supply to any area of heart muscle, there may be actual death of tissue. For example, in *coronary thrombosis* a blood clot plugs up an artery and prevents blood from reaching the tissues that lie beyond the point of obstruction. This area which has been cut off from its blood supply is known medically as an *infarct*, or an area of *myocardial infarction*. The pain associated with myocardial infarction is usually more severe than that of angina pectoris and is not relieved by nitroglycerin. This clinical picture commonly is referred to as a *heart attack*, or a *coronary*. Occlusion of a coronary artery may result in sudden death if a large area of myocardium is deprived of its blood supply.

If the victim survives the initial attack, the physician will attempt to assess the amount of damage, if any, to the myocardium. One test that may prove helpful is a laboratory determination

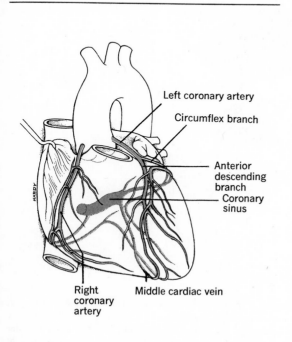

Left coronary artery

Circumflex branch

Anterior descending branch

Coronary sinus

Right coronary artery

Middle cardiac vein

Fig. 12-7. Diagram of the coronary arteries arising from the aorta and encircling the heart. The coronary sinus and some of the coronary veins also are shown.

of the amount of glutamic oxaloacetic transaminase in the serum (SGOT). Normally, this enzyme is present in the blood in the amount of 10 to 40 units per ml., but when cardiac muscle cells are injured they release additional quantities of enzyme into the bloodstream. Thus a mild heart attack may elevate the SGOT to 90 or 100 units, and a severe attack to as high as 500 units. Inasmuch as the rise is proportional to the severity of myocardial damage, it serves as a guide to the physician in prescribing treatment.

THE WORK OF THE HEART

The function of the heart is to pump into the arterial system that blood which returns to it from the venous system. The right side of the heart receives poorly oxygenated blood from the venae cavae and pumps it into the pulmonary artery for a trip to the lungs; the left side of the heart receives richly oxygenated blood from the pulmonary veins and pumps it into the aorta for a trip to all other tissues in the body. The pressure necessary for this vital task is generated by powerful contractions of the cardiac musculature. However, the rate and the force of these contractions, as well as the amount of blood ejected from the heart with each beat, will vary according to the changing needs of the body. The cardiac control mechanisms are described in Chapter 14.

The cardiac cycle

One cardiac cycle means one complete heartbeat, consisting of contraction (*systole*) and relaxation (*diastole*) of the muscular heart walls. The hearbeat is initiated by the pacemaker cells of the S-A node in the right atrium. These specialized cells generate an impulse that sweeps through the muscle mass of both atria, much like the ripple pattern produced when a pebble is dropped

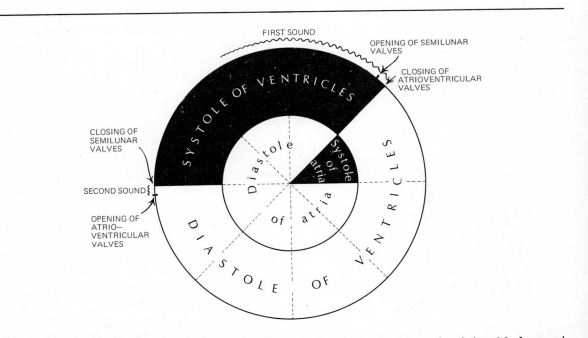

Fig. 12-8. Cardiac cycle. The heart rate represented in the diagram is 75 beats per minute; each cycle lasts 0.8 of a second; each sector represents 0.1 second. The inner circle represents the atria; atrial systole lasts 0.1 second, and atrial diastole, 0.7 second. The outer circle represents the ventricles; ventricular systole lasts 0.3 second, and diastole, 0.5. The atrioventricular valves close at the beginning of ventricular systole; the semilunar valves open shortly thereafter. The latter valves close at the beginning of ventricular diastole, and the atrioventricular valves open shortly thereafter. The first sound is due to the closing of the atrioventricular valves and systole of the ventricle; the second sound is due to the closing of the semilunar valves; it is of shorter duration than the first.

into a pool of still water; this impulse causes both atria to contract simultaneously.

When the atrial impulse reaches the A-V node and the related A-V bundle, conduction is slowed somewhat. However, once the impulse reaches the mass of ventricular muscle, it spreads rapidly from fiber to fiber throughout both ventricles, causing them to contract simultaneously. The slight delay in conduction over the node and the bundle branches provides sufficient time for the atria to complete their work before ventricular contraction begins.

The duration of one cardiac cycle is 0.8 second. As illustrated in Figure 12-8, atrial systole lasts 0.1 second, and diastole 0.7 second. In other words, the atria work 0.1 and rest 0.7 of each second. Ventricular systole lasts 0.3 second, and diastole 0.5, which means that the ventricles work 0.3 and rest 0.5 of each second. The entire heart rests 0.4 second, and the period during which both atria and ventricles are relaxed is called *diastasis*.

If the heart is to function efficiently, the events of the cardiac cycle must follow each other in orderly sequence, and the conduction system of the heart must be in good working order. Impairment of myocardial function is the causative factor in *heart failure,* a broad term which means that the heart is unable to maintain an adequate output of blood in relation to the needs of the body.

Cardiac output

Under average conditions, each ventricle pumps out about 70 ml. of blood with each contraction; this is called the *stroke volume*. If the stroke volume is multiplied by the number of heartbeats per minute, one arrives at the *minute volume,* or *cardiac output*. For example, if the stroke volume is 70 ml. and the heart is beating at a rate of 72 times per minute, the cardiac output is 5,040 ml. In other words, each ventricle is pumping about 5 L., or qts., of blood every minute. During strenuous exercise the heart pumps more rapidly, and the cardiac output may be increased to 15 or 20 L. per minute, thus helping to meet the needs of the hard-working muscles. Cardiac output varies during many physiologic activities and in disease.

There are various methods whereby cardiac output may be estimated, but the most accurate figures are obtained during *cardiac catheterization* (p. 288), when the oxygen content of blood samples taken directly from the right side of the heart can be compared with the oxygen content of blood which has passed through the lungs.

Variations in heart rate

There are many factors which influence the heart rate, but **exercise** probably is the best known. Busy muscles require increased delivery of oxygen and swift removal of waste products if they are to function at peak efficiency; consequently, the heart beats rapidly to pump more blood through the muscle circuits.

The **size** of an individual tends to influence heart rate; generally speaking, the larger the size, the slower the heartbeat. This is true also of lower animals, as shown by the following examples: elephants, about 20; rabbits, 150; and mice, 700 beats per minute.

Age is another factor in determining heart rate. In a fetus the rate is between 120 and 160; in an infant, 110 and 130; in a child, 72 and 92; and in adults, between 65 and 80 beats per minute.

Sex has an influence on heart rate in that women have rates between 70 and 80, while the range for men is between 65 and 70 beats per minute.

Certain **hormones** have a definite effect on heart rate. Epinephrine, a secretion of the adrenal glands, causes an increase in both the rate and the strength of heartbeat. An increase in the amount of thyroid hormone (thyroxine) in the blood also speeds up the heart rate.

When the body **temperature** rises, as during fever, the heart rate is faster; when the temperature is low, as during hypothermia, the heart beats very slowly.

Blood pressure changes have a marked influence on heart rate. This factor is described in detail in Chapter 14.

Properties of cardiac muscle

Rhythmicity. As previously described on page 102, cardiac muscle has the ability to generate spontaneous and rhythmic impulses that are independent of any stimulus from nerves. In an embryo, the heart begins to beat long before any nerves have grown out to it. This property of *rhythmic contractility* is so highly developed in cardiac muscle that if a heart is kept under proper conditions, it will continue to beat after removal from the body.

In order to keep a heart beating in the laboratory, it must be bathed by a balanced solution of sodium, potassium, and calcium chlorides. If the solution is not balanced properly, the heart will soon stop beating; likewise, if the electrolytes (p. 458) in one's plasma are not in proper balance, the activity of the heart is altered and may even cease.

Refractory Periods. The *absolute* refractory period is the time during which an excitable tissue will not respond to any stimulus, no matter how strong the latter is. The absolute refractory period is very long for heart muscle, since it lasts throughout systole. Because of the long absolute refractory period, no summation or tetanic contractions are possible in heart muscle. This is an excellent protection, since the heart would be worthless as a pump if it went into a tetanic, or sustained contraction.

When the absolute refractory period is over, the *relative* refractory period begins, and the heart gradually recovers its excitability. The relative refractory period coincides with the early diastolic period of the ventricle. A strong stimulus given at this time can cause a contraction before the next rhythmic contraction is due. The early beat is smaller in amplitude than a regular beat and is called a *premature systole.* Thus it is possible to make a beat appear ahead of time, by applying a strong stimulus early in diastole.

A premature systole is followed by a longer diastole, to compensate for its lack of rest after the previous beat. The long diastole, is called a *compensatory pause.* Such a pause, following a premature beat is illustrated in Figure 12-9.

An occasional premature beat is of frequent occurrence in the human heart. The cause is not known, but it is some stimulus within the heart muscle itself. The premature beat is small and passes unnoticed, but the person may be conscious of the compensatory pause and say, "My heart has just skipped a beat." The heart does not skip a beat, but the premature contraction is not recognized, and only the compensatory pause comes to the attention.

The All-or-Nothing Response. This refers to the fact that if a heart beats, it gives the strongest contraction that is possible under the conditions existing at that moment. There is no such thing as gradation in heartbeats, as is possible in skeletal muscle contractions. In skeletal muscle it is possible to have different numbers of motor units active, but in the heart the muscle mass behaves as a single cell when stimulated. Thus we say that it obeys the *all-or-nothing response.*

Starling's Law of the Heart. When cardiac muscle fibers are stretched, or elongated (within physiologic limits), their contraction is more forceful. Thus if a large quantity of blood flows into the heart, the fibers will be stretched, the force of contraction will be increased, and a larger stroke volume results. If only a small amount of blood enters the heart, the contraction will be less forceful, and a smaller stroke volume results. This direct relationship between the volume of blood and the force of contraction is called Starling's law of the heart.

Electrical Activity. Every beat of the heart is accompanied by an electrical change, or *action potential* (p. 102). Electrolytes in the body fluids and tissues act as conductors to the skin

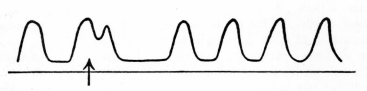

Fig. 12-9. In this tracing, a heart was stimulated at the time indicated by the arrow. A premature systole followed; the next diastole was of longer duration than usual. This long rest period is called the compensatory pause.

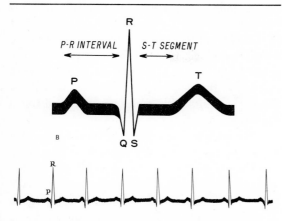

P-R INTERVAL   S-T SEGMENT

R

P   T

B

Q S

R

P

Fig. 12-10. *Top,* diagram of the electrical pattern of one cardiac cycle. The P wave represents activity related to atrial contraction; the QRS complex is related to ventricular contraction, and the T wave represents recovery phase. *Bottom,* normal electrocardiogram showing the electrical pattern of several cardiac cycles.

surface; metal leads applied to the skin then conduct the current to an instrument (electrocardiograph) that makes a graphic tracing of the activity. This tracing, called an *electrocardiogram* (ECG), is a valuable diagnostic tool in various types of heart disease.

As illustrated in Figure 12-10, the electrical activity during each cardiac cycle is characterized by five wave deflections that are arbitrarily designated as P, Q, R, S, and T. Any pathologic process that disturbs the electrical activity of the heart will produce characteristic changes in one or more of the waves.

1. The *P wave* represents the electrical activity associated with depolarization of the atria. Normal P waves indicate that the impulse originated in the S-A node and then spread through the atria; if the P waves are missing, or wandering, it suggests that the impulse originated outside the S-A node.

2. The *P-R interval* is that period of time from the start of the P wave to the beginning of the QRS complex; it indicates the time required for the original impulse to reach the ventricles. If this interval is prolonged, it suggests that there is a conduction delay in the A-V node.

3. The *QRS waves* represent the activity associated with depolarization of the ventricles; it reflects the time required for the impulse to spread through the A-V bundle and its branches. If the duration of this impulse is increased, it

shows that the ventricles have been stimulated in a delayed, abnormal manner, such as might occur in a bundle branch block.

4. The *S-T segment* represents the period between completion of ventricular depolarization and the start of repolarization. If there is some injury to the muscle, as in acute myocardial infarction, the S-T segment may be elevated or depressed.

5. The *T wave* represents the recovery phase following contraction. If there is tissue injury or ischemia, repolarization is abnormal, and the T wave may be inverted.

The electrocardiogram provides information regarding the electrical activity of the heart, but it does not reflect the actual physical state or pumping function of this vital organ. One should keep in mind the fact that the electrocardiogram may be perfectly normal in the presence of severe heart disease unless the pathologic process disturbs the heart's electrical activity.

In some persons, the atria do not beat normally but show a pathologic type of contraction called *fibrillation*; such contractions are extremely rapid, fluttery, and ineffective. The electrocardiogram in atrial fibrillation shows no P waves at all, and the R waves appear at a rapid and irregular rate. While atrial fibrillation is compatible with life, ventricular fibrillation generally is a terminal event, since the chaotic contractions do not pump blood into the great arteries; this is one of the major hazards of cardiac surgery, particularly when the operation is conducted under hypothermia, since severe cooling tends to increase the incidence of fibrillation.

Digitalis is a drug that is used frequently to improve myocardial function in atrial fibrillation and other forms of heart disease. It not only increases the force of each beat, but also reduces the rate by slowing impulse conduction between the atria and the ventricles. However, excessive amounts of digitalis may completely stop A-V conduction; this condition, called *heart block,* also may be caused by disease or injury to the conduction system of the heart. Ventricles which are thus isolated, or functionally separated from the atria, will beat at their own slow rate instead of following the atrial beat. Thus, the electrocardiogram in heart block is characterized by R waves which appear about 20 times per minute, and the P waves about 72 times per minute.

Pressure changes within the heart

The amount of blood and the pressure which it exerts in the heart chambers varies during each phase of the cardiac cycle. Knowledge of such pressure differences is essential to a full understanding of valve action. The valves of the heart are inert structures that are opened and closed only because of differences in pressure exerted by the blood on their two surfaces.

Let us begin with the moment at the end of ventricular systole, when the atria are relatively full of blood and the atrioventricular (A-V) valves are still closed, and trace the events step by step. As the blood continues to enter the atria from the veins, the pressure in the atria soon exceeds that in the relaxed ventricles, and the A-V valves open, due to the greater pressure on the atrial surface of the leaflets.

When the A-V valves open, blood flows rapidly into the ventricles, and the valve leaflets float in midstream. The two atria contract simultaneously and pump the blood remaining in them into the ventricles; this is followed by ventricular systole.

As soon as the ventricles begin to contract, the A-V valves close, due to the increasing pressure on the ventricular surface of their leaflets. The leaflets are kept from being pushed up into the atria, like an umbrella being prevented from turning inside out, by the contraction of the papillary muscles; these muscles hold the chordae tendineae taut.

Ventricular systole continues with a very sharp rise in intraventricular pressure, until it becomes great enough to force the semilunar valves open. The extent to which the pressure rises to accomplish this varies in the two ventricles. The pressure in the right ventricle rises to only about one sixth that of the left ventricle, and the pressure in the pulmonary artery is only about one sixth that in the aorta, but the valves open simultaneously as soon as the pressure in the corresponding artery is exceeded.

As soon as the semilunar valves are open, the ventricles begin to empty, the right into the pulmonary artery and the left into the aorta. Ventricular systole continues for about 0.3 second, during which time most of the blood is pumped out. With the onset of ventricular diastole, the intraventricular pressure falls rapidly.

As soon as the pressure in the right ventricle falls below that in the pulmonary artery and that in the left ventricle falls below the pressure in the aorta, the blood tends to come back into the ventricles as well as to go forward. The backward flow closes the semilunar valves. The elasticity of the large arteries then comes into play and forces the blood onward.

Meanwhile, the atria have been filling with blood coming in from the veins. As the ventricular diastole continues, the pressure within the ventricles falls below that of the atria, the A-V valves open, and the cardiac cycle begins again. The above events are repeated over and over with each beat, as long as you live.

The opening and the closing of the valves are indicated in Figure 12-8. One should study valvular action in connection with pressure changes in order to understand it clearly. The sole function of the valves is to keep blood flowing in an onward direction. When the A-V valves close at the beginning of ventricular systole, they prevent the passage of blood from the ventricles, back into the atria. When the semilunar valves are forced open early in ventricular systole, they allow blood to flow from the left ventricle into the aorta and from the right ventricle into the pulmonary artery. When ventricular systole is completed, the semilunar valves are closed because the elasticity of the large arteries tends to force blood in both directions, and that which flows back toward the heart closes the semilunar valves. Blood enters both atria from the related veins throughout ventricular systole and diastasis. When ventricular diastole occurs, the ventricular pressure falls, the weight of blood in the atria forces the A-V valves to open, and blood enters the ventricles.

Heart sounds

If you listen (auscultate) with a stethoscope over the heart you will hear two sounds with each heartbeat. The first is longer and louder than the second and sounds like *lubb*; it is due to the closure of the A-V valves and contraction of the mass of ventricular muscle. The second sound is shorter and softer and is due to the closure of the semilunar valves; it sounds like *dup*. The time in the cycle in which each sound in heard is indicated in Figure 12-8.

Since valve closure is the major contributor to

heart sounds, one would expect to hear abnormal sounds if the valves failed to function properly. For example, pathologic *heart murmurs* are abnormal sounds which frequently accompany rheumatic heart disease and congenital anomalies of the heart (p. 320). If valves fail to close tightly (insufficiency) and blood leaks back (regurgitation), or if a valve orifice is narrowed and constricted (stenosis), the resulting turbulence in blood flow will be heard through the stethoscope as murmurs.

In *mitral insufficiency* the first heart sound is altered as blood leaks back into the left atrium during contraction of the ventricle; this regurgitation of blood results in a systolic murmur. However, in *mitral stenosis* a murmur is heard just prior to ventricular contraction, when the atrium is pumping blood through the constricted valve, and is called a presystolic murmur. In *aortic insufficiency*, blood leaks back into the left ventricle from the aorta during ventricular diastole and is responsible for a diastolic murmur. In *aortic stenosis* a systolic murmur is heard as blood is forced through the constricted semilunar valve. The tricuspid and the pulmonary valves may be similarly involved, but the valves of the left side of the heart are the ones most commonly affected by pathologic changes in adult life.

In the diagnosis of pathologic heart conditions, the physician makes use of electrocardiograms, x-ray pictures of the chest, fluoroscopic examinations, heart sounds, and other findings of the physical examination. Figure 12-11 is an x-ray picture of a normal heart, and Figure 12-12 illustrates left ventricular enlargement, or hypertrophy. Hypertrophy of the left ventricle occurs in such conditions as aortic regurgitation and high blood pressure (hypertension).

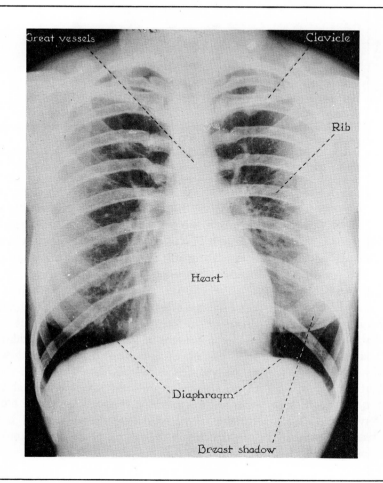

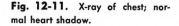

Fig. 12-11. X-ray of chest; normal heart shadow.

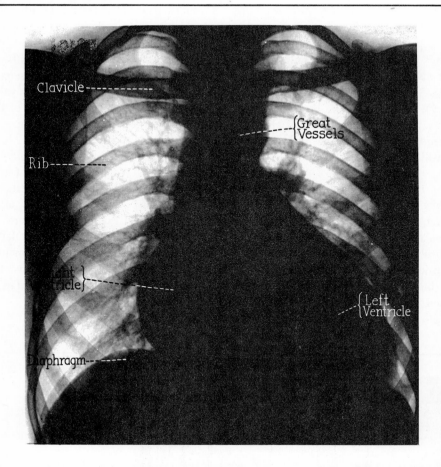

Fig. 12-12. X-ray of chest; abnormal heart shadow. The left ventricle is markedly enlarged; the right ventricle is moderately enlarged. The great vessels are moderately dilated.

Cardiac catheterization

The insertion of a long thin tube, or catheter, into the heart is referred to as cardiac catheterization. The catheter is passed into the venous system through one of the superficial veins (basilic, or cephalic) of the left forearm near the bend of the elbow; veins in the right forearm may be used, but those on the left permit entry into the heart from a better angle. While looking through a fluoroscope, the doctor gradually advances the catheter until its tip enters the right atrium, passes through the tricuspid valve into the right ventricle, and then into the pulmonary artery (Fig. 12-13). Blood samples for chemical analysis are taken from each chamber and from the pulmonary artery. In addition, by measuring pressures in the different chambers, the presence of defective valves can be detected. This is a painless procedure, and general anesthesia usually is not required for mature patients.

The importance of cardiac catheterization in the diagnosis of various types of congenital heart disorders has been well established, and it is a routine procedure in many hospitals. In addition to its value in clinical medicine, cardiac catheterization has become equally important in the study of cardiovascular physiology in experimental laboratories. For example, chemical analysis of blood samples taken from the pulmonary artery is an essential step in the process of calculating cardiac output.

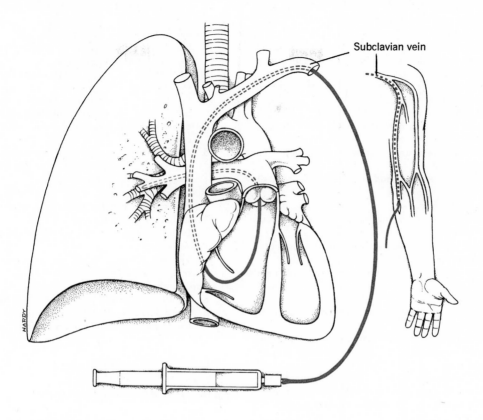

Subclavian vein

Fig. 12-13. Route of cardiac catheter. (See text for description.)

SUMMARY

1. Structural features

A. Hollow, muscular organ in mediastinum; base directed upward to right, apex downward to left

B. Covering
  a. Pericardial sac, loose-fitting
  b. Outer fibrous portion lined with serous, parietal pericardium
  c. Serous, visceral pericardium closely covers heart
  d. Pericardial space between parietal and visceral layers contains slippery fluid

C. Walls of heart
  a. Outer layer: epicardium (visceral pericardium)
  b. Muscular layer: myocardium
  c. Lining: endocardium

D. Heart chambers and valves
  a. Right atrium receives from venae cavae and coronary sinus; empties into right ventricle
  b. Right atrioventricular (tricuspid) valve prevents backflow from right ventricle to right atrium
  c. Left atrium receives from pulmonary veins; empties into left ventricle
  d. Left atrioventricular (bicuspid, mitral) valve prevents backflow from left ventricle to left atrium
  e. Interatrial septum is partition between atria
  f. Right and left auricular appendages lie on top of atria
  g. Right ventricle pumps blood into pulmonary artery; backflow from artery prevented by pulmonary semilunar valve
  h. Left ventricle pumps blood into aorta; backflow from aorta prevented by aortic semilunar valve; left ventricular walls thicker than right; must generate more pressure to pump into larger circuit
  i. Interventricular septum is partition between ventricles

E. Modified cardiac muscle tissue
  a. Forms conduction system of heart
  b. S-A node called pacemaker; heartbeat normally begins here

c. A-V node and bundle serve as connecting link between atria and ventricles; transmit beat across gap in muscular continuity

F. Nerve supply
   a. Efferent fibers: parasympathetic in vagus nerve, slow the heart; sympathetic in cardiac nerves speed the heart
   b. Afferent fibers stimulated by lack of oxygen in myocardium; pain

G. Blood supply to myocardium
   a. Right and left coronary arteries branch to capillaries in heart wall
   b. Coronary veins drain capillaries; most empty into coronary sinus
   c. Interference with blood supply may cause heart attack or coronary

2. *Work of the heart*

A. Sole function to pump into arterial system that blood which returns to it from venous system
   a. Right side of heart receives poorly oxygenated blood, pumps to lungs
   b. Left side of heart receives highly oxygenated blood from lungs, pumps to other parts of body

B. Cardiac cycle
   a. Systole and diastole (one complete heartbeat); requires 0.8 second
   b. Impulse begins at S-A node, sweeps over atria; atrial systole lasts 0.1 second; diastole 0.7 second
   c. A-V node and bundle conduct impulse to ventricles as atria contract; ventricular contraction follows, lasts 0.3 second; ventricular diastole 0.5 second

C. Cardiac output
   a. About 70 m. of blood with each beat; called stroke volume
   b. At 72 beats per minute, minute volume, or cardiac output, is about 5 L. May be increased to 15 or 20 L. during exercise

D. Factors causing variations in heart rate
   a. Size
   b. Age
   c. Sex
   d. Hormones
   e. Body temperature
   f. Blood pressure

E. Properties of cardiac muscle
   a. Rhythmicity: ability to generate spontaneous and rhythmic beat
   b. Refractory periods: absolute, will not respond to stimulus; relative, gradually recovers excitability; strong stimulus at this time may cause premature systole followed by compensatory pause
   c. All-or-nothing response gives strongest contraction possible under existing conditions; beats as one unit
   d. Starling's law: direct relationship between volume of blood and force of contraction

e. Electrical change: action potentials accompany contractions; can be recorded; electrocardiogram

F. Pressure changes within heart
   a. Blood moves from area of greater pressure to area of lower pressure
   b. Contraction of atria forces blood into empty ventricles; A-V valves open
   c. Contraction of ventricles raises pressure so that blood flows into arteries; semilunar valves open and A-V valves close to prevent backflow into atria
   d. After ventricles empty, semilunar valves close as pressure higher in arteries; elasticity of arteries forces blood onward to capillaries

G. Heart sounds
   a. First sound lubb, due to contraction of ventricles and closure of A-V valves
   b. Second sound dup, due to closure of semilunar valves
   c. Valvular disease leads to abnormal sounds called murmurs

H. Cardiac catheterization
   a. Insertion of catheter into chambers of heart
   b. Blood samples and blood pressure measurement facilitates diagnosis of heart defects

QUESTIONS FOR REVIEW

1. Distinguish between the visceral and the parietal pericardium. What is the function of the pericardium?
2. List, in proper sequence, the structures through which a drop of blood will flow as it travels from the right atrium to the aorta.
3. What are the major differences between the right and the left sides of the heart?
4. What is the function of the foramen ovale?
5. Why are the chordae tendineae important for proper A-V valve action?
6. Describe the action of the semilunar valves.
7. What is the purpose of the S-A node? of the A-V node and bundle?
8. Stimulation of the vagus nerve has what effect on the heart rate?
9. What vessels convey blood to the myocardium? What will happen if one of them is obstructed by a blood clot?
10. Describe one cardiac cycle.
11. If the heart is beating 75 times per minute, what is the cardiac output?
12. Why will a heart continue to beat after its extrinsic nerves are cut?
13. What is the absolute refractory period?

14. Why do we say that the heart obeys the all-or-nothing response?
15. What is Starling's law of the heart?
16. What information is given by electrocardiograms?

17. Briefly, describe the pressure differences that are essential to valve action in the heart.
18. What causes the first heart sound? the second heart sound?

REFERENCES AND SUPPLEMENTAL READINGS

Adolph, E. F.: The heart's pacemaker. Sci. Am., 216:32, March 1967.

Best, C. H., and Taylor, N. B.: The Physiological Basis of Medical Practice, ed. 8. Baltimore, Williams & Wilkins, 1966.

Guyton, A. C.: Textbook of Medical Physiology, ed. 4. Philadelphia, W. B. Saunders, 1971.

Lehman, Sr. J.: Auscultation of heart sounds. AJN, 72: 1242, July 1972.

Sher, A. M.: The electrocardiogram. Sci. Am., 205:132, Nov. 1961.

The Major Blood Vessels and Lymphatics

"Everywhere he feels his heart because the vessels run to all his limbs"

The Beginning of the
Secret Book of the Physician (Egyptian, C. 1550 B.C.)
Translated from the German Version By Cyril P. Bryan
(1931)

The circulatory system also is known as the cardiovascular system, since it consists of the heart and the blood vessels. As described in Chapter 12, the muscular heart serves as a pump which ejects into the arterial portion of the system that blood which returns to it from the venous system. The blood vessels (arteries, capillaries, and veins) sometimes are referred to as the plumbing of the circulatory system, since they are tubular structures that transport the life-sustaining blood to and from all parts of the body.

The lymphatic circulatory system is structurally and functionally related to the blood circulatory system. Lymph vessels, like blood vessels, are widely distributed throughout the body, but they are concerned with the transport of lymph rather than blood.

THE MAJOR BLOOD VESSELS

The arteries

Arteries are elastic, muscular tubes through which blood flows from the heart to the capillaries. The great arteries which arise from the ventricles give off branches that divide and subdivide, on and on, becoming smaller and smaller as they extend to various parts of the body. The smallest arteries, called *arterioles*, end in the microscopic capillary networks which permeate every organ of the body.

With each ventricular contraction, blood is pumped under pressure into the pulmonary artery and the aorta, causing them to become stretched, or distended, with blood. Between beats of the heart these elastic tubes recoil, forcing the blood onward in a pulsating wave that sweeps through the entire arterial system; consequently, when an artery is cut, the blood flows in spurts which reflect the ventricular beat of the heart. Obviously, such vessels need strong walls to withstand the pressure of the blood that surges within them.

Coats. The thickness of arterial walls varies with the caliber of the vessel. As large arteries branch into smaller ones, the walls become pro-

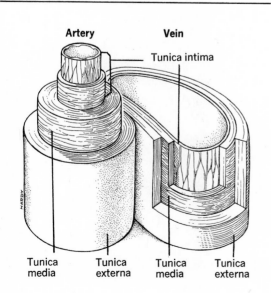

Fig. 13-1. Medium-sized artery and vein showing relative size and thickness of the three coats, or tunics.

gressively thinner, but in any case there are three coats, or layers, of tissue in the wall (Fig. 13-1).

1. The inner layer, or *tunica intima*, consists of endothelium which rests on an elastic membrane.

2. The middle layer, or *tunica media*, constitutes the bulk of the arterial wall. It is composed primarily of elastic connective tissue in large arteries, such as the aorta, but in smaller arteries, smooth muscle predominates; contraction or relaxation of this muscle makes possible the constriction or the dilatation of the vessel.

3. The outer layer, or *tunica externa* (*adventitia*), consists largely of fibrous connective tissue which gives strength to the wall. Although this layer has some give to it as blood surges through the vessel, it resists rupture even when the blood pressure is elevated. The thickness of arterial walls prevents them from collapsing when cut; consequently, arterial bleeding may be difficult to control by simple pressure if a large vessel is cut.

As you might expect, there can be abnormal changes in the walls of the arteries. Sometimes there is such a change in structure that the wall becomes weak and undergoes a marked dilatation in some portion. The dilated sac which results is called an *aneurysm*. The most frequent site of aneurysm formation is in the ascending portion of the aorta, but it may occur elsewhere. Aneurysms pulsate with each beat of the heart, and one great danger is the possibility that the weakened wall will rupture, causing a fatal hemorrhage. In aortic aneurysm there also is the danger of suffocation if the mass exerts sufficient pressure on the neighboring trachea.

Arteriosclerosis, hardening of the arteries, is a more common affliction of the vessels, and one that most of us will experience as we grow older. Degenerative changes in the arterial walls result in a decrease in elastic tissue and, in some cases, excessive deposits of calcium. These changes also may be accompanied by *atherosclerosis*, a fatty degeneration and lipid infiltration of the vessel walls. Many disturbing changes in function may follow such structural alterations.

Blood Supply. Nourishment of tissue cells in the aforementioned layers depends on tiny blood vessel networks which permeate the walls of all but the smallest of arteries; these nutrient blood vessels are called the *vasa vasorum* (meaning *vessels of the vessels*).

Nerve Supply. Arteries are well supplied with nerve fibers. *Afferent nerves* are found which carry impulses to the CNS and lead to reflex changes in heart rate and size of blood vessels; sometimes this is called the *intrinsic control* of the cardiovascular system. Thus when there is a change in pressure in the blood vessels in one part of the body, messages are sent to the CNS which immediately set into operation reflexes to compensate for the original change. These reflexes help to maintain normal conditions, or a steady state (homeostasis), throughout the vascular system.

Efferent nerves, or *vasomotor nerves*, of the autonomic division of the nervous system supply the smooth muscle in the walls of small arteries and arterioles. Contraction of this muscle results in *vasoconstriction*, a decrease in the caliber of the vessel as its lumen becomes smaller; as a result there is a reduction in the amount of blood flowing through the vessel. On the other hand, relaxation of the muscle results in *vasodilatation*, an increase in the caliber of the vessel as its lumen becomes larger, permitting an increased flow of blood through the vessel. These opposing responses play a vital role in shunting blood from inactive, or resting, tissues to areas where there is a greater need for service. The vasomotor control mechanism is described in Chapter 14.

Anastomoses. In many regions of the body there are communications, or anastomoses, between adjacent arteries, so that if the blood supply from one source is cut off, another source is immediately available. Knowledge of this fact enables one to understand why gangrene (death of tissue) does not follow the ligation (tying off) of arteries during surgical operations. In certain parts of the body such anastomoses are not present; the arteries in these areas are called *end arteries*, and if they are obstructed, gangrene will occur.

The veins

Veins conduct blood from the capillaries to the heart. In addition, they are able to constrict or to expand, thus serving to store variable quantities of blood depending on the needs of the body.

In general, veins tend to follow a course that is more or less parallel with the artery which supplies the area to be drained. However, there are many more veins than arteries, and anastomoses between veins are common. It has been estimated that about 59 per cent of the total blood volume is contained in the systemic veins. The smallest veins, called *venules*, receive blood from the capillaries and convey it to slightly larger veins, which in turn empty into veins that increase progressively in size from the tissues to the heart.

Coats. Like the arteries, the veins have three coats, or layers, of tissue in their walls (tunica intima, tunica media, and tunica externa), but the tunica media consists only of a thin layer of smooth muscle and a few elastic fibers. Consequently, veins have thinner walls than do the arteries, a fact that is correlated with the relatively lower pressure of the blood that flows within them. However, since the venous system is highly distensible, the active constriction of veins helps to preserve the venous blood pressure and to shift blood from the periphery toward the heart. When veins are cut, their thin walls tend to collapse, and blood escapes in a steady stream. A section of a medium-sized vein is shown in Figure 13-1.

Another structural feature of many veins is the presence of *valves* which help to prevent the backflow of blood. Each valve consists of a pocketlike fold of endothelium that is strengthened by

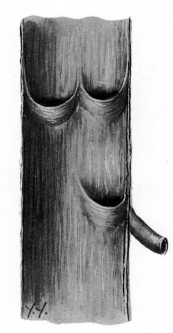

Fig. 13-2. Portion of femoral vein opened to show valves.

connective tissue; there are usually two such semilunar folds placed opposite one another on the vein wall (Fig. 13-2). These valves are so attached that blood can flow freely toward the heart, but any backflow causes the pockets to balloon out as they fill with blood, thus temporarily blocking the venous channel and preventing blood from returning to the capillaries. Valves are most abundant in the veins of the extremities, particularly the lower extremities, since these vessels must conduct blood upward against the powerful forces of gravity. Some veins, such as the venae cavae, have no valves at all; others may have only one or two.

In persons who stand a great deal of the time, such as teachers and dentists, the valves of the leg veins seem to wear out and, eventually, are unable to resist the pressure of the long column of blood which they are called upon to support. In these instances, the superficial veins become tortuous and dilated, and we speak of them as *varicose veins*. Varicose veins may appear during pregnancy, due to the pressure in the pelvis which offers some interference with the return of blood from the lower extremities.

Blood and Nerve Supply. The blood and nerve supply to the walls of veins is similar to that of the arteries. Networks of tiny blood vessels, the *vasa vasorum*, permeate the venous walls and nourish the cells therein. Nerve fibers are also distributed to the veins as to the arteries, but they are somewhat less abundant.

The capillaries

Capillaries are microscopic vessels through which blood flows from the arterioles to the venules (Fig. 13-3). Their walls are only one endothelial cell in thickness, or, in other words, they are tubular continuations of the innermost lining cells of the arteries and the veins. Tiny openings, or intercellular spaces, facilitate exchanges between the circulating blood in the capillary and the interstitial fluid which bathes the neighboring tissue cells. The maintenance of homeostasis in the liquid cellular environment is described in Chapter 2, pages 28 to 35.

Another anatomic feature of importance is the diffuse branching of arterioles. One arteriole may branch into a number of capillaries, thus contributing to a fantastic, spreading network that

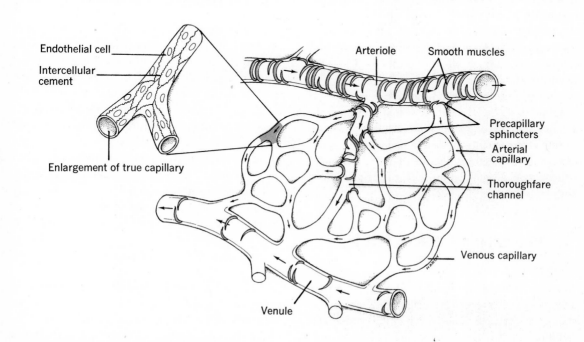

Endothelial cell
Intercellular cement
Enlargement of true capillary
Arteriole
Smooth muscles
Precapillary sphincters
Arterial capillary
Thoroughfare channel
Venous capillary
Venule

Fig. 13-3. Diagram of a capillary bed. Precapillary sphincters are relaxed, thus permitting flow of blood through true capillary network. A greatly magnified portion of capillary wall is shown in the inset.

permeates every organ of the body. However, not every capillary in the body is open at the same time. The entrances into each capillary network are guarded by tiny rings of smooth muscle which are called *precapillary sphincters*; by contracting or relaxing, these muscular gatekeepers can regulate the flow of blood through their individual capillaries according to tissue needs in the immediate area. This is indeed a fortunate arrangement for us, since there is insufficient blood to completely fill every vessel in the body at the same time. The distribution of blood according to body needs is discussed in Chapter 14.

In summary, we wish to emphasize this cardinal principle: *The capillaries are the focal point of the entire cardiovascular system!* The heart is merely a pump which serves to keep blood moving through the arterial expressways that lead to the all-important capillaries, and the veins merely collect blood from the capillaries and return it to the heart. The whole system is geared to the task of supplying the capillaries with sufficient quantities of blood to meet the constantly changing needs of the cells.

Sinusoids are microscopic, capillarylike blood vessels found in such organs as the liver, the spleen, and the bone marrow. They differ from true capillaries in that their walls are formed by irregularly placed cells, some of which are phagocytic; the walls of the sinusoids also are more permeable; thus large particles (such as proteins) readily pass through them to enter or leave the bloodstream.

Arteriovenous anastomoses are found in the distal parts of the extremities. When they are open a large amount of blood flows through the skin directly from small arteries to small veins, without going through capillaries. This arrangement is important in connection with the regulation of body temperature (p. 427).

THE PULMONARY CIRCULATION

The blood from all parts of the body and from the walls of the heart itself enters the right atrium, passes through the right A-V valve into the right ventricle, then through the pulmonary semilunar valve into the pulmonary trunk, into the right and left pulmonary arteries and their branches, and into the pulmonary capillaries where

it picks up a fresh load of oxygen and gives off some carbon dioxide. The oxygenated blood then enters the pulmonary veins and flows to the left atrium. This short path through the lungs is called the pulmonary circulation and is illustrated in Figure 13-4.

The **pulmonary trunk** arises from the right ventricle, passes upward, backward and to the left, then divides into the right and the left pulmonary arteries.

The right pulmonary artery passes transversely toward the base of the right lung; it divides into three branches at the root (hilus) of the lung, and one branch goes to each lobe of the right lung. The left pulmonary artery is shorter than the right; it passes to the root of the left lung, divides into two branches, and sends one branch to each of the two lobes of the left lung.

From the arteries the blood enters the *pulmonary capillaries*, where it takes up oxygen and gives up carbon dioxide and then is collected into the four *pulmonary veins*. Two pulmonary veins arise from the root of each lung. They pass to the posterior surface of the left atrium, into which they open.

THE SYSTEMIC CIRCULATION

From the left atrium blood passes through the left A-V valve into the left ventricle; then it is pumped through the aortic semilunar valve into the aorta. It is distributed to all of the systemic arteries of the body, then to the systemic capillaries, where oxygen and nutritive materials are delivered to the tissues and carbon dioxide and waste materials are picked up. From the capillaries the blood passes to the veins and eventually returns to the right atrium to begin its circuit all over again. This longer path through all parts of the body (except the air sacs of the lungs) is called the *systemic circulation*; it is illustrated in Figure 13-4.

Arteries of the systemic circuit

The arteries which arise from the aorta may supply one specific organ, or they may give off branches to various structures as they extend to distant regions. Only the major systemic arteries and the most important branches will be mentioned in this text. However, one should keep in

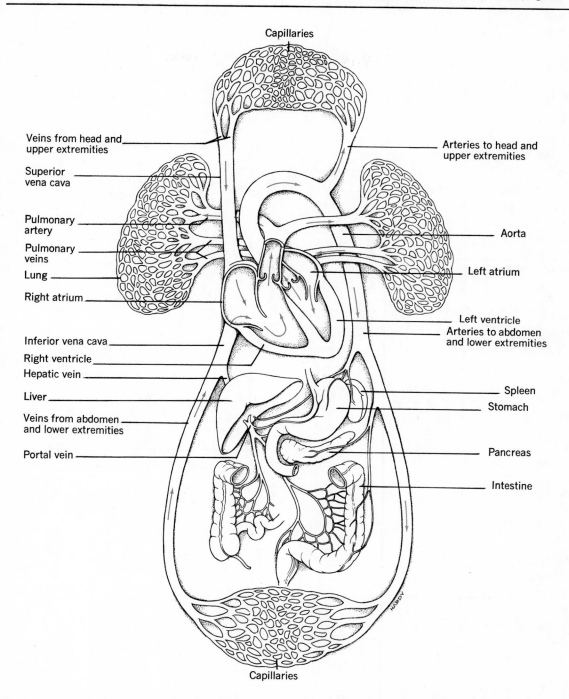

Capillaries

Veins from head and upper extremities

Superior vena cava

Pulmonary artery

Pulmonary veins

Lung

Right atrium

Inferior vena cava

Right ventricle

Hepatic vein

Liver

Veins from abdomen and lower extremities

Portal vein

Arteries to head and upper extremities

Aorta

Left atrium

Left ventricle
Arteries to abdomen and lower extremities

Spleen

Stomach

Pancreas

Intestine

Capillaries

Fig. 13-4. Diagram of the pulmonary and systemic circulations. The pulmonary circulation includes the pulmonary arteries, capillaries, and veins. The systemic circulation includes all the other arteries, capillaries, and veins of the body.

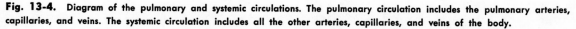

mind the fact that these are only the main highways, and each has many side roads which also lead to capillary networks. Most of the arteries are named according to the regions through which they pass, or for the structures that they supply with blood. In any case, you will find that frequent

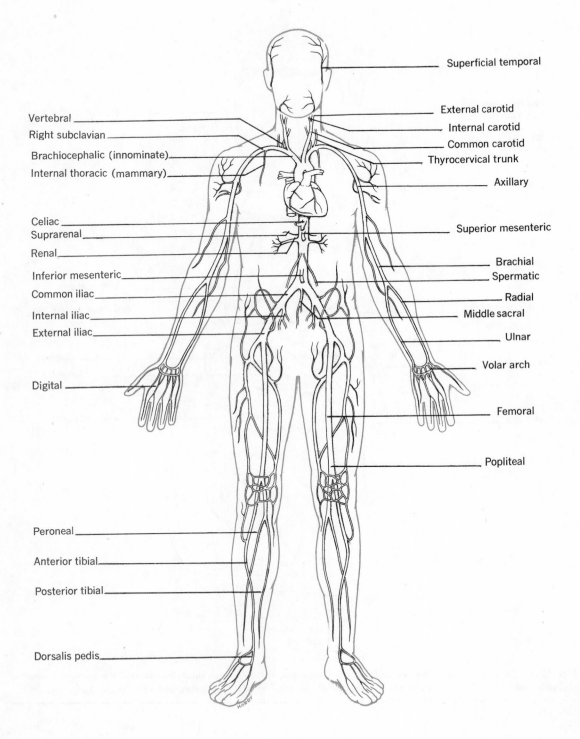

Vertebral

Right subclavian

Brachiocephalic (innominate)

Internal thoracic (mammary)

Celiac

Suprarenal

Renal

Inferior mesenteric

Common iliac

Internal iliac

External iliac

Digital

Peroneal

Anterior tibial

Posterior tibial

Dorsalis pedis

Superficial temporal

External carotid

Internal carotid

Common carotid

Thyrocervical trunk

Axillary

Superior mesenteric

Brachial

Spermatic

Radial

Middle sacral

Ulnar

Volar arch

Femoral

Popliteal

Fig. 13-5. Schematic drawing of the arterial system.

reference to the diagrams will be most helpful in locating the various arterial trunks.

The **aorta** arises from the base of the left ventricle, arches over to the left side of the vertebral column, and passes downward along the vertebral column to the level of the fourth lumbar vertebra. At this level, it terminates by dividing into the common iliac arteries. The aorta is shown in Figure 13-5.

For convenience, the aorta may be divided into the ascending aorta; the aortic arch, which extends to the left side of the fourth thoracic vertebra; the thoracic aorta, which extends from the arch to the diaphragm; and the abdominal aorta, which extends from the diaphragm to the fourth lumbar vertebra.

The Ascending Aorta. The only branches of the ascending aorta are the right and the left *coronary arteries* (Fig. 13-6). These arteries arise just above the origin of the aorta, and their orifices can be seen immediately above the leaflets of the aortic semilunar valve. The coronary arteries and their branches conduct blood to capillaries in the myocardium.

The Aortic Arch. The arch of the aorta curves to the left and slightly backward (Fig. 13-6) before turning downward to the level of the fourth thoracic vertebra, where it becomes continuous with the thoracic aorta. A short cord, the ligamentum arteriosum (see Fig. 12-4), extends from the upper border of the pulmonary artery to the undersurface of the aortic arch. This is the remains of the *ductus arteriosus*, a fetal vessel that permitted blood to flow from the pulmonary artery into the aorta, thus bypassing the pulmonary circuit. The arteries which arise from the arch of the aorta are the brachiocephalic, the left common carotid, and the left subclavian arteries.

The **brachiocephalic (innominate) artery** is the largest of the aforementioned branches, but it has a relatively short course. Ascending obliquely upward, this unpaired trunk divides into the right common carotid and the right subclavian arteries.

The **common carotid arteries** conduct blood to capillaries in the head and the neck. The left common carotid arises from the aortic arch, but its counterpart on the right side is a branch of the brachiocephalic artery (Fig. 13-6). These arteries course upward in the neck along either side of the trachea; by pressing gently in this area you can feel them pulsating under your fingertips. At the upper border of the thyroid cartilage (Adam's apple), each common carotid bifurcates (divides) into an external and an internal carotid artery. Near this point of bifurcation there is a slight dilatation in the internal carotid artery called the *carotid sinus*; specialized nerve endings in this area respond to changes in the arterial blood pressure. The functions of the carotid sinus are described on pages 342 to 343.

1. The *external carotid artery* gives off many branches as it ascends on its own side to the level of the neck of the mandible; the names of these branches are generally indicative of the areas which they supply (e.g., thyroid, pharyngeal, lingual). The terminal branches of the external carotid artery are the *superficial temporal* and the *internal maxillary* arteries. The superficial temporal artery is one of the arteries in which you can feel the pulse easily. Place your fingertips immediately in front of the ear at the level of the eye and note

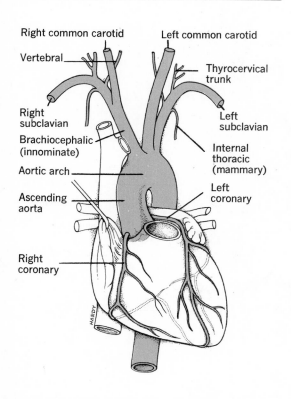

Fig. 13-6. Branches of the ascending aorta and the arch of the aorta.

Right common carotid

Vertebral

Right subclavian

Brachiocephalic (innominate)

Aortic arch

Ascending aorta

Right coronary

Left common carotid

Thyrocervical trunk

Left subclavian

Internal thoracic (mammary)

Left coronary

the pulsation of this vessel. The internal maxillary artery is the larger of the two terminal branches; it supplies the deep structures of the face and sends a large branch (the middle meningeal artery) to the dura mater of the brain.

Some of the structures supplied by the external carotid artery and its many branches are the thyroid gland and the muscles of the neck, the pharynx, the tongue, the teeth and the salivary glands, the muscles of facial expression and mastication, part of the nose and the ear, and most of the dura mater. Figure 13-7 shows a few of the branches of the external carotid artery.

2. The *internal carotid artery* ascends perpendicularly upward to enter the cranial cavity by way of a canal in the temporal bone. Its main branches are: (a) the *ophthalmic* artery, supplying the eyeball and other structures in the orbital cavity and sending branches to the nose and the forehead; (b) the *posterior communicating* artery, forming part of the circle of Willis (Fig. 13-8) as it courses backward to anastomose with the posterior cerebral artery; (c) the *anterior cerebral* artery, supplying part of the frontal and the parietal lobes of the cerebrum; and (d) the *middle cerebral* artery, supplying the basal ganglia, the internal capsule, and most of the lateral surface of the brain.

The *circle of Willis* is an arterial anastomosis at the base of the brain. The two internal carotid arteries conduct blood to the anterior portion of the circle, and the two vertebral arteries (branches of the subclavians) conduct blood to the basilar artery which leads to the rear of the circle. The anterior and the posterior communicating arteries are connecting links. Careful study of Figure 13-9

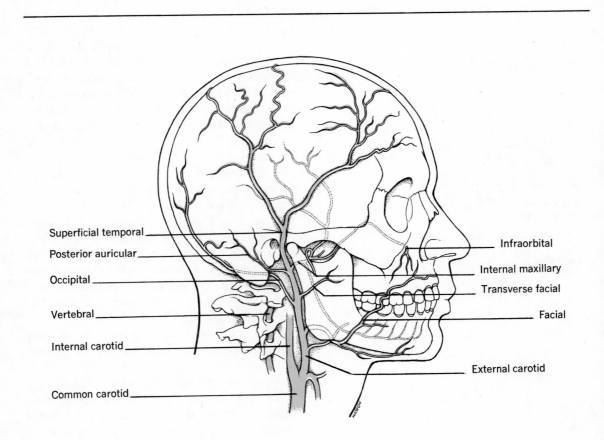

Fig. 13-7. Branches of the right external carotid artery. The internal carotid artery ascends to the base of the brain. The right vertebral artery also is shown as it ascends through the transverse foramina of the cervical vertebrae.

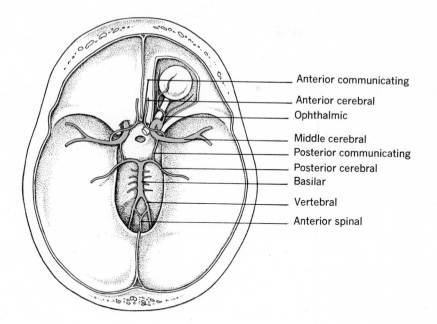

Fig. 13-8. Blood supply to the brain (circle of Willis) and to the eyeball shown in relation to the base of the skull. The brain has been removed; the right orbital cavity has been opened.

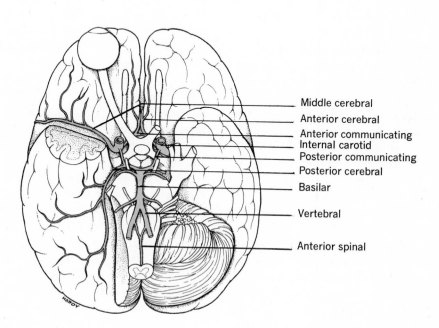

Fig. 13-9. The circle of Willis as seen at the base of a brain which has been removed from the skull.

will indicate that the circle is formed by the anterior communicating artery, the proximal portions of the anterior cerebrals, the internal carotids, the posterior communicating arteries, and the proximal portions of the posterior cerebral arteries. The blood supply of the brain is derived mainly from the three pairs of cerebral arteries which may be considered as branches of the circle of Willis.

Cerebral vascular accident (CVA). Of all the affections of the brain, cerebral thrombosis and cerebral hemorrhage are the most frequent. Such vascular accidents usually are the result of pathologic changes in the walls of cerebral blood vessels. For example, hemorrhage within the brain substance usually is related to an underlying arteriosclerosis; a sudden strain with a rise in blood pressure may rupture a weakened vessel that has no give to it. Cerebral vascular accident, also called *stroke*, or *apoplexy*, most frequently occurs in the region of the internal capsule, which is supplied by branches of the middle cerebral artery. Because there are only a few, small anastomoses between these branches, there is little chance of a collateral circulation supplying adequate amounts of oxygen and glucose to the affected area. As a result, that portion of the brain which is deprived of blood undergoes necrosis or death; paralysis of one side of the body follows such necrosis, if the person survives.

The **subclavian arteries** are shown in Figures 13-5 and 13-6. The left subclavian artery arises directly from the aortic arch, but the right subclavian is a branch of the brachiocephalic artery as described previously. One can feel the pulsations in the subclavian artery by pressing inward on the tissues just above the medial third of the clavicle. At the outer border of the first rib, the subclavian artery becomes the axillary artery. Major branches of each subclavian artery and the structures which they supply are as follows.

1. The *vertebral* artery ascends in the neck through the transverse foramina of the cervical vertebrae (see Fig. 13-7), enters the skull through the foramen magnum, and unites with its counterpart from the opposite side to form the basilar artery at the base of the brain (see Fig. 13-8). Branches of the vertebral artery supply the deep muscles in the neck, parts of the spinal cord, and the medulla oblongata; the basilar artery supplies the cerebellum and the pons before branching to

form the *posterior cerebral* arteries which supply the occipital lobes of the cerebrum.

2. The *thyrocervical trunk* (see Fig. 13-6) is short and thick, dividing almost immediately into branches which supply the thyroid gland, the trachea, the larynx, and other structures of the neck.

3. The *internal thoracic (mammary)* artery (see Fig. 13-6) arises from the undersurface of the subclavian and descends into the thorax. Its branches supply the thymus and the pericardium in the mediastinum, the pectoral muscles, the mammary glands, and the skin of the anterior thorax. The *superior epigastric* artery is a terminal branch of the internal thoracic; it supplies the muscles and the skin of the anterolateral abdominal wall.

The **axillary artery** is the continuation of the subclavian into the axillary region (see Fig. 13-5). It begins at the outer border of the first rib and courses downward to the lower border of the teres major muscle, where it becomes the brachial artery of the arm. While passing through the axillary region, it gives off branches which supply tissues in the upper part of the thorax, the region of the shoulder, and the axilla.

The **brachial artery** is the continuation of the axillary artery into the upper extremity (Fig. 13-10). It sends branches to the humerus, the muscles, and the skin of the arm. At the bend of the elbow, the brachial artery lies in front of the humerus, midway between its two epicondyles, where you can feel its pulsations under your fingertips. About 1 cm. below the bend of the elbow, the brachial artery ends by dividing into its two terminal branches, the radial and the ulnar arteries. These vessels supply the forearm, the hand, and the fingers.

1. The *radial artery* (Fig. 13-10) continues the course of the brachial artery down the lateral aspect of the forearm, but it is smaller in caliber than the ulnar artery. Pulsations in the radial artery are easily felt on the thumb side of the wrist where the artery passes in front of the radius. In the palm of the hand, the radial artery turns medially across the metacarpal bones and terminates by anastomosing with the deep volar branch of the ulnar artery to form the *deep volar (palmar) arch.*

2. The *ulnar artery,* larger of the two terminal brachial branches, courses down the medial

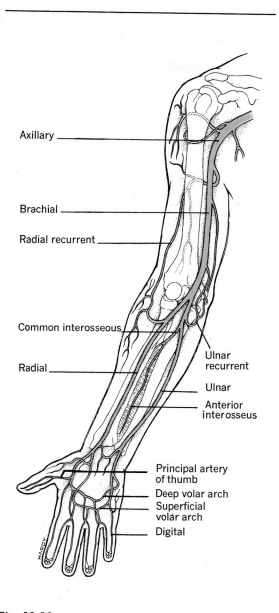

Axillary

Brachial

Radial recurrent

Common interosseous

Radial

Ulnar recurrent

Ulnar

Anterior interosseus

Principal artery of thumb

Deep volar arch

Superficial volar arch

Digital

Fig. 13-10. Major arteries of the right upper extremity, anterior view.

aspect of the forearm and the wrist. Crossing the palmar surface of the hand it forms the main component of the *superficial volar arch* and terminates by anastomosing with the superficial volar branch of the radial artery.

The **thoracic aorta** extends downward from the aortic arch, through the posterior mediastinal cavity, to the aortic hiatus (opening) in the diaphragm at the level of the twelfth thoracic vertebra. This great arterial trunk gives off branches which supply the viscera and the walls of the thoracic cavity (see Fig. 13-5).

A. *Visceral branches* of the thoracic aorta:

1. The **pericardial arteries** are distributed to the posterior pericardium.

2. The **bronchial arteries** are the nutrient arteries of the lungs and are distributed mainly to the walls of the bronchi.

3. The **esophageal arteries,** four or five in number, form a chain of anastomoses along the esophagus as it passes through the thorax.

B. *Parietal branches* of the thoracic aorta:

1. The **intercostal arteries** (nine or ten pairs) supply the intercostal muscles, the pleurae, the muscles, and the skin of the thoracic wall. Posterior branches supply the vertebral column, the spinal cord, and the muscles and the skin of the back.

2. The **superior phrenic arteries** are distributed to the upper surface of the diaphragm.

The **abdominal aorta** is the continuation of the thoracic aorta through the abdominal cavity. It extends from the diaphragm to the level of the fourth lumbar vertebra where it ends by dividing into the right and the left common iliac arteries. Both visceral and parietal branches arise from this arterial trunk.

A. *Visceral branches* of the abdominal aorta: of these branches the celiac, the superior, and the inferior mesenteric arteries are unpaired, while the suprarenal, the renal, and the spermatic or ovarian arteries are paired.

1. The **celiac artery (celiac axis)** is a relatively short, thick trunk which arises from the anterior surface of the abdominal aorta a short distance below the diaphragm. It ends by dividing into three large branches: the left gastric, the hepatic, and the splenic arteries (Fig. 13-11).

a. The *left gastric artery* supplies the esophagus and part of the stomach. It anastomoses with esophageal branches from the thoracic aorta.

b. The *hepatic artery* is the nutrient artery of the liver. In addition, it gives rise to the right gastric and the gastroduodenal arteries (which supply part of the stomach, the duodenum, and the pancreas), and the cystic artery which supplies the gallbladder.

c. The *splenic artery,* largest branch of the celiac, conducts blood to the spleen. In addition, its branches supply part of the stomach and the pancreas.

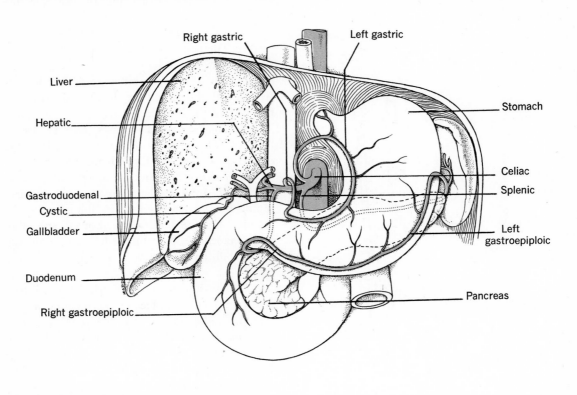

Fig. 13-11.   The celiac artery and its branches.

2. The **superior mesenteric artery** arises from the front of the aorta just below the celiac axis (Fig. 13-12); after extensive branching it ends near the junction of the large and small intestines by anastomosing with one of its own branches, the iliocolic artery. The superior mesenteric artery supplies the entire length of the small intestine (except the duodenum) and part of the large intestine (i.e., the cecum, the appendix, the ascending colon, and the right half of the transverse colon).

3. The **middle suprarenal arteries** are paired vessels which arise, one from either side of the aorta, opposite the superior mesenteric artery (see Fig. 13-5). They convey blood to the suprarenal (adrenal) glands which are located just above the kidneys.

4. The **renal arteries** are two large vessels which arise from either side of the aorta immediately below the suprarenals (see Fig. 13-5). They conduct blood to the kidneys.

5. The **spermatic (testicular) arteries** (in the male) and the **ovarian arteries** (in the female) are paired vessels which arise from the aorta a little below the renal arteries. The spermatic arteries descend through the inguinal canals to supply the testes. The ovarian arteries descend into the pelvis, where they supply the ovaries and anastomose with the uterine arteries.

6. The **inferior mesenteric artery** is an unpaired vessel which arises from the front of the aorta just above its division into the common iliacs (Fig. 13-12); its terminal branch is the superior hemorrhoidal (rectal) artery which descends along the rectum. The inferior mesenteric artery supplies the remaining portions of the large intestine (i.e., the left half of the transverse colon, the descending colon, the sigmoid colon, and most of the rectum).

B. *Parietal branches* of the abdominal aorta:

1. The **inferior phrenic arteries** are two small vessels which arise from the first part of the abdominal aorta just above the celiac axis. They supply the undersurface of the diaphragm and send branches to the suprarenal glands.

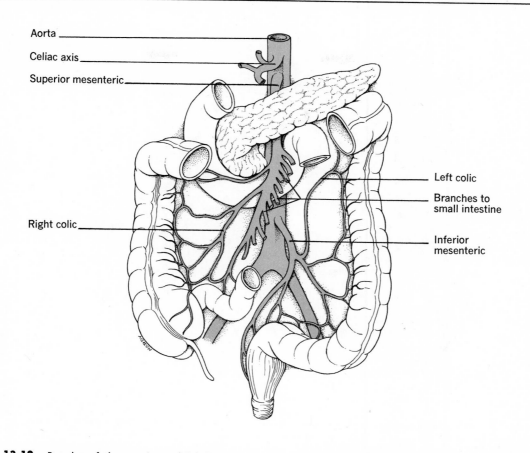

Aorta

Celiac axis

Superior mesenteric

Right colic

Left colic

Branches to small intestine

Inferior mesenteric

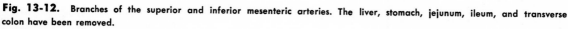

Fig. 13-12. Branches of the superior and inferior mesenteric arteries. The liver, stomach, jejunum, ileum, and transverse colon have been removed.

2. The **lumbar arteries,** usually arranged in four pairs, arise from the back of the aorta (Fig. 13-5). They supply the muscles of the anterior and the posterior abdominal walls, the spinal cord, the lumbar vertebrae, and the muscles and the skin of the back.

3. The **middle sacral artery** is an unpaired vessel which arises from the back of the terminal portion of the aorta (see Fig. 13-5). It descends in the midline to supply the sacrum and the coccyx.

Terminal Branches of the Aorta. At the level of the fourth lumbar vertebra, the aorta terminates by dividing into the right and the left *common iliac arteries.* These vessels pass laterally downward to the level of the sacroiliac joint on their respective sides; at this point each terminates by dividing into the internal and the external iliac

arteries (see Fig. 13-5). The common iliacs give off small branches to the psoas muscles, the peritoneum, and the ureters.

The **internal iliac arteries** (*hypogastric arteries*) pass into the pelvis where each divides into anterior and posterior divisions (Fig. 13-13). These divisions give off branches to the urinary bladder, the rectum, the internal and the external organs of reproduction, the buttocks, and the medial side of the thigh.

The **external iliac arteries** are shown in Figure 13-5. Each courses obliquely downward to a point beneath the inguinal ligament; from this point each continues as the femoral artery of its own side and enters the thigh. The external iliac artery gives off branches to the psoas major muscle and the muscles of the abdominal wall. The *inferior epigastric artery,* a relatively large branch of the

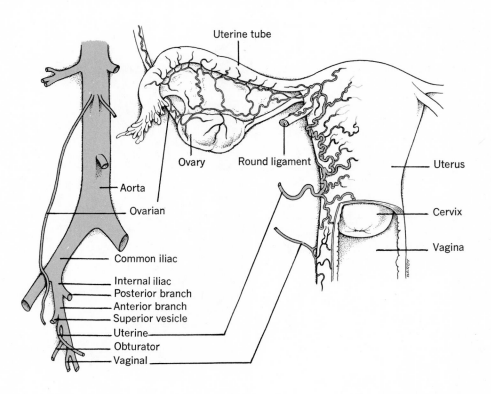

Fig. 13-13. Arterial blood supply to ovary, uterus, and vagina. Cut-away section shows cervix and interior of vagina.

external iliac, ascends through the abdominal wall to anastomose with the superior epigastric artery above.

The **femoral artery** is the continuation of the external iliac artery below the inguinal ligament (Fig. 13-14). As it passes in front of the hip joint, you can feel its pulsations under your fingertips by pressing firmly in the groin. This vessel courses downward in the thigh to the popliteal space where it continues as the popliteal artery. Branches of the femoral artery supply the bone, the muscles, and the skin of the thigh, and also contribute to anastomoses around the hip and the knee joints.

The **popliteal artery** is the continuation of the femoral artery as shown in Figure 13-14. It courses through the popliteal space behind the knee, giving off muscular, articular, and cutaneous branches. This artery ends by dividing into the anterior and the posterior tibial arteries of the leg.

1. The **anterior tibial artery** is one of the ter-minal branches of the popliteal artery (Fig. 13-14). It courses downward in the leg, between the tibia and the fibula, to the front of the ankle joint, where it becomes the dorsalis pedis artery. It supplies muscular and cutaneous branches to the front of the leg.

The *dorsalis pedis artery* is the continuation of the anterior tibial beyond the ankle joint, and pulsations in this vessel may be felt as it passes forward along the top of the foot (Fig. 13-14). Physicians frequently palpate this artery to deter-mine if there is adequate circulation to structures in the leg and the foot. The dorsalis pedis not only supplies superficial tissues in the foot, but one of its branches, the deep plantar artery, de-scends into the sole of the foot where it unites with a branch of the posterior tibial artery to complete the plantar arch.

2. The **posterior tibial artery** gives off mus-cular and cutaneous branches as it descends in the back of the leg to the ankle (Fig. 13-14); it

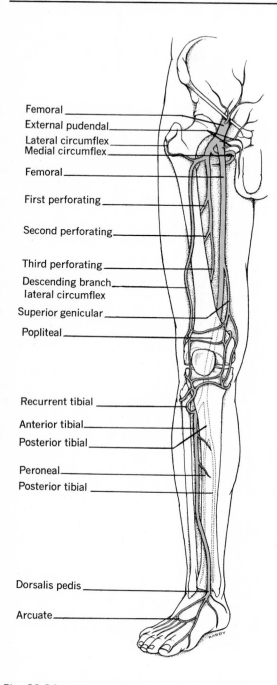

Femoral

External pudendal

Lateral circumflex

Medial circumflex

Femoral

First perforating

Second perforating

Third perforating

Descending branch
lateral circumflex

Superior genicular

Popliteal

Recurrent tibial

Anterior tibial

Posterior tibial

Peroneal

Posterior tibial

Dorsalis pedis

Arcuate

Fig. 13-14. Major arteries of right lower extremity.

digital branches arise from this arch. The *peroneal artery* is a large branch of the posterior tibial which supplies structures in the lateral aspect of the leg.

Veins of the systemic circuit

The systemic veins drain the vast capillary networks which are supplied by the systemic arteries. However, the veins are generally larger and more numerous than the arteries. There are three types of systemic venous channels in the body; superficial veins, deep veins, and cranial venous sinuses. The *superficial veins* lie just beneath the skin and return blood from it as well as from the superficial fascia. The *deep veins* accompany the principal arteries and, in most cases, have the same name as the artery with which they travel; they also have many connections with the superficial veins. The *cranial venous sinuses*, found only within the skull, consist of channels formed by separations of the two layers of dura mater (p. 208). The general scheme of the venous circulation is illustrated in Figure 13-15.

Blood ultimately is returned to the right atrium of the heart by way of three veins: (1) the coronary sinus, which drains the heart muscle; (2) the superior vena cava, which drains the upper portion of the body; and (3) the inferior vena cava, which drains the lower portion of the body. For convenience, the tributaries of the venae cavae will be grouped according to the regions which they drain (e.g., veins of the head and the neck, veins of the upper extremities).

The Coronary Sinus and Its Tributaries. The coronary sinus occupies the right half of the groove lying between the left atrium and the left ventricle (see Fig. 12-2). It opens into the right atrium on the posterior surface, below the opening of the inferior vena cava. It receives blood from the great, small, and middle cardiac veins, the posterior vein, and the oblique vein. Since only a few small veins open directly into the right side of the heart, it is evident that the coronary sinus drains most of the heart muscle.

Veins of the Head and the Neck. The cranial venous sinuses, the internal jugular veins, and the external jugular veins are the major systemic venous channels which drain capillary networks in the head and neck.

terminates by dividing into the medial and the lateral plantar arteries. The lateral plantar artery unites with the deep plantar branch of the dorsalis pedis to form the plantar arch; metatarsal and

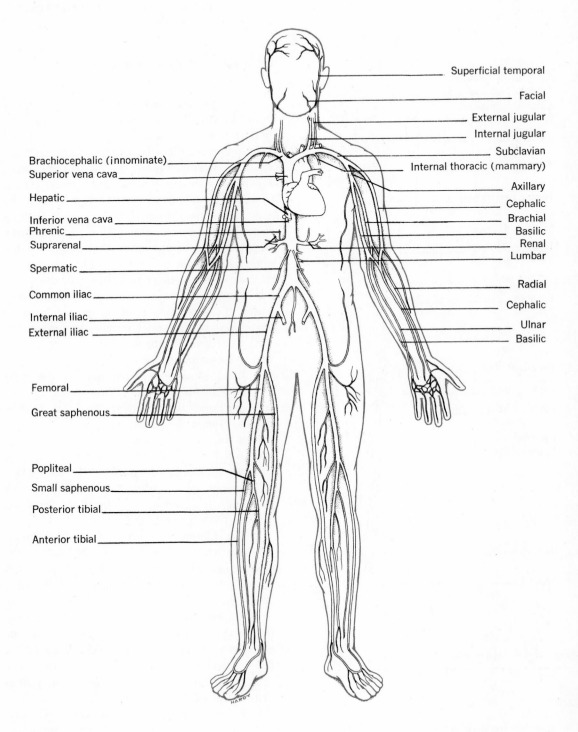

Fig. 13-15.   Schematic drawing of the venous system.

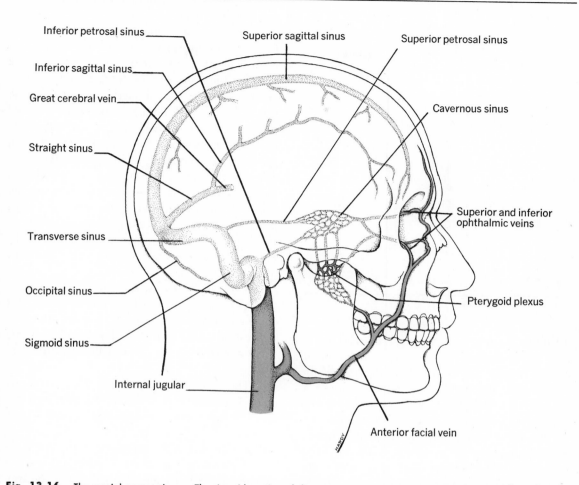

Inferior petrosal sinus

Inferior sagittal sinus

Great cerebral vein

Straight sinus

Transverse sinus

Occipital sinus

Sigmoid sinus

Internal jugular

Superior sagittal sinus

Superior petrosal sinus

Cavernous sinus

Superior and inferior ophthalmic veins

Pterygoid plexus

Anterior facial vein

Fig. 13-16. The cranial venous sinuses. The sigmoid portion of the transverse sinus continues on as the internal jugular vein.

The **cranial venous sinuses** are a series of channels that lie between the two layers of dura mater; they drain the structures in the orbital and the cranial cavities and communicate with veins outside of the cranium by numerous emissary veins that pass through the cranial wall. The venous sinuses receive blood from the ophthalmic, the cerebral, the cerebellar, and the meningeal veins. None of these vascular channels possesses valves, and their walls are extremely thin, due to the absence of muscular tissue.

There are many cranial venous sinuses, some of which are median and unpaired, and others which consist of lateral pairs. The major unpaired venous sinuses are: (1) superior sagittal, (2) inferior sagittal, (3) straight, and (4) occipital. The

major paired venous sinuses are: (1) transverse (lateral), (2) superior petrosal, (3) inferior petrosal, and (4) cavernous. Some of these sinuses are illustrated in Figure 13-16 as they convey blood to one of the internal jugular veins.

The **internal jugular veins** are continuations of the *sigmoid* portions of the two transverse sinuses. They begin at the base of the skull and descend in the neck, just lateral to the common carotid arteries. At the level of the sternoclavicular joint, each unites with the subclavian vein of its own side to form the brachiocephalic (innominate) veins (see Fig. 13-18). The internal jugular veins collect blood from the brain, the eyes, the superficial parts of the face, and the neck.

The facial area that is most susceptible to

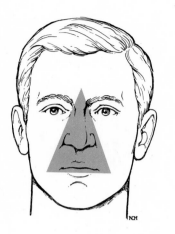

Fig. 13-17. The dangerous triangle.

infection is the so-called *dangerous triangle*, an area having its base at the opening of the mouth and its apex at a point just above the level of the eyebrows in the midline of the forehead (Fig. 13-17). This triangular area is well supplied with blood and lymphatics, but has very little connective tissue and practically no fascial planes; hence two important mechanisms for localizing infections are lacking here. Skeletal muscles in the area are attached to the skin, and as a result of the constant motion associated with changes in facial expression, purulent material is rapidly disseminated throughout the triangle.

Drainage of blood from the triangle is primarily by way of the facial vein which normally empties into the internal jugular vein. However, the facial vein has a tributary (the angular vein) which anastomoses with the ophthalmic veins at the inner angle of the eye. When purulent material is present, this venous drainage is extremely important for two reasons: (1) these veins have no valves; thus there is no directional control of blood flow, and (2) the superior ophthalmic vein is a main tributary of the cavernous venous sinus (Fig. 13-16). The swelling associated with inflammation promotes drainage via the intracranial route, and the normally slow rate of blood flow through the cavernous sinus favors *septic thrombosis*. This is an extremely serious complication with a mortality rate of 80 per cent. Thus one

should *never pluck nasal hairs or squeeze pimples* found in the dangerous triangle.

The **external jugular veins** are formed near the angle of the mandible, one on either side. Each collects blood from the exterior of the cranium, the deep parts of the face, and the neck before ending in the subclavian vein of the same side. Figure 13-18 illustrates the right external jugular vein. When the head is turned to the side, or when one wears a tight collar, this vein will stand out, along the edge of the sternocleidomastoid muscle. Sometimes you will see pulsations, just above the clavicle, in the external jugular vein; these pulsations are called the *central venous pulse* and are reflected changes in pressure from the right atrium, since there are no valves at the entrance of the superior vena cava into the right atrium.

Veins of the Upper Extremity. These veins are subdivided into superficial and deep sets, but there are many anastomoses between the two sets. Valves are present in all of these veins, but are more numerous in the deep set than in the superficial.

The **superficial veins** of the upper extremity show considerable individual variation, but the chief ones are the cephalic and the basilic (Fig. 13-19).

1. The *cephalic vein* begins in the dorsal venous network on the thumb side of the hand and ascends along the lateral border of the forearm. In front of the elbow joint (antecubital fossa) it gives rise to the *median cubital vein*, which crosses over to join the basilic vein. The cephalic vein then ascends along the lateral aspect of the arm, passes between the deltoid and the pectoralis major muscles, and ends in the axillary vein just below the clavicle. It collects blood from small veins all along its course.

2. The *basilic vein* begins in the dorsal venous network on the little finger side of the hand and ascends along the medial side of the forearm and the arm. It pierces the fascia of the arm and joins the deep brachial vein to form the axillary vein of that extremity. It receives blood from small veins all along its course.

The **deep veins** of the upper extremity follow the course of the arteries and usually are enclosed in the same sheath. They also are generally arranged in pairs, called the *venae comitantes* (meaning *companion veins*), one lying on either

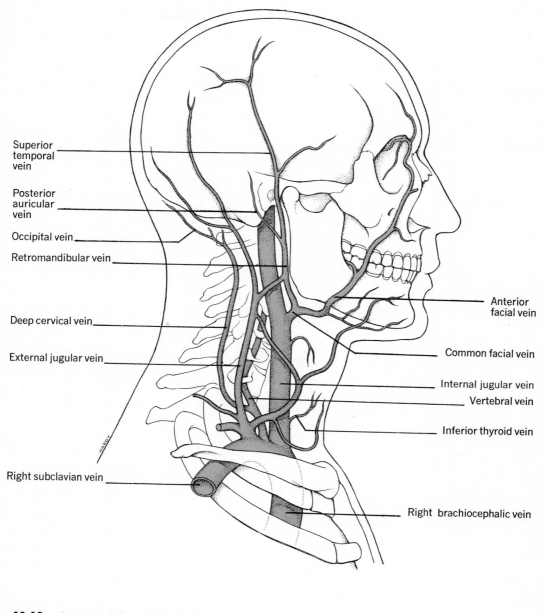

Superior temporal vein

Posterior auricular vein

Occipital vein

Retromandibular vein

Deep cervical vein

External jugular vein

Right subclavian vein

Anterior facial vein

Common facial vein

Internal jugular vein

Vertebral vein

Inferior thyroid vein

Right brachiocephalic vein

Fig. 13-18.   The veins of the head and neck.

side of the artery with which they travel. They receive blood from the deep structures of the upper extremity.

1. The *radial veins* are the upward continuation of veins of the deep volar arch.

2. The *ulnar veins* are formed by the union of veins from both the deep and the superficial

volar arches. They join the radial veins at the elbow.

3. The *brachial veins* are the companion veins of the brachial artery and are formed at the elbow joint by the union of the radial and the ulnar veins. At the level of the lower border of the pectoralis major muscle, the brachial and the

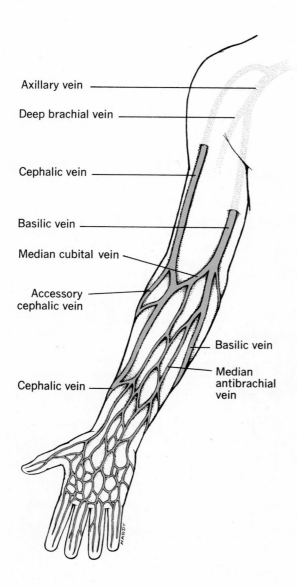

Axillary vein

Deep brachial vein

Cephalic vein

Basilic vein

Median cubital vein

Accessory cephalic vein

Basilic vein

Median antibrachial vein

Cephalic vein

Fig. 13-19. Superficial veins of the anterior aspect of the right upper extremity.

basilic veins unite to form the axillary vein.

4. The *axillary vein,* formed by the union of the brachial venae comitantes and the basilic vein, becomes the subclavian vein at the outer border of the first rib.

5. The *subclavian vein* (Fig. 13-20) is the end of the venous system of the upper extremity. It extends to the sternal end of the clavicle, where

it unites with the internal jugular to form the brachiocephalic vein of that side.

Veins of the Thorax. The thoracic vessels which pour their contents into the superior vena cava are the brachiocephalic veins and the azygos system of veins.

The **brachiocephalic (innominate) veins** are formed by the union of the internal jugular and subclavian veins on either side (Fig. 13-20). These large venous trunks descend to the level of the first right costal cartilage, where they unite to form the **superior vena cava.** In addition to the internal jugular and the subclavian veins, the major tributaries of each brachiocephalic vein are the vertebral, the internal thoracic, and the inferior thyroid veins. The left brachiocephalic also receives the left highest intercostal veins from the upper three or four intercostal spaces.

The **azygos system of veins** (Fig. 13-21) not only collects blood from various structures in the thorax, but also serves as the connecting link between the superior and the inferior vena caval systems. In cases of obstruction of the inferior vena cava, the azygos veins are able to return blood from the lower portion of the body by virtue of their collateral communications with the common iliac, the renal, and the lumbar veins.

1. The *azygos vein* lies in front of the vertebral column, slightly to the right of the midline. It usually begins as the continuation of the right ascending lumbar vein, which in turn has connections with the inferior vena cava, the right common iliac, and the lumbar veins. The principal tributaries of the azygos vein are the right intercostal veins, the hemiazygos veins, several esophageal, mediastinal, and pericardial veins, and the right bronchial vein. At the level of the fourth thoracic vertebra, the azygos empties into the superior vena cava.

2. The *hemiazygos* and the *accessory hemiazygos veins* also form a channel in front of the vertebral column, but they are slightly to the left of the midline. The hemiazygos begins as the continuation of the left ascending lumbar vein and communicates with the left renal vein instead of with the inferior vena cava; it receives blood from the lower four or five intercostal veins on the left side and from some of the esophageal and mediastinal veins before crossing the midline to the right side, where it joins the azygos. The accessory hemiazygos receives blood from the three or four

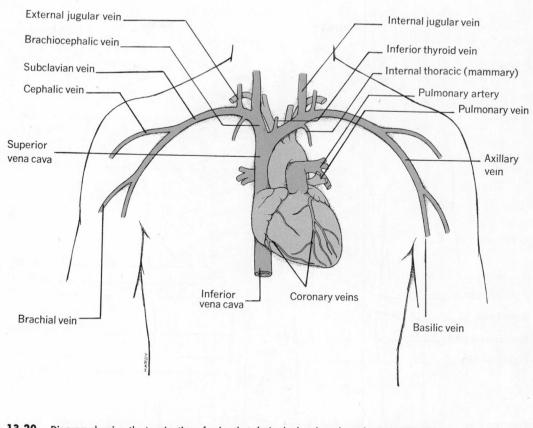

External jugular vein

Brachiocephalic vein

Subclavian vein

Cephalic vein

Superior vena cava

Brachial vein

Internal jugular vein

Inferior thyroid vein

Internal thoracic (mammary)

Pulmonary artery

Pulmonary vein

Axillary vein

Inferior vena cava

Coronary veins

Basilic vein

Fig. 13-20. Diagram showing the termination of veins that drain the head, neck, and upper extremities.

intercostal veins above the highest tributary of the hemiazygos, and from the left bronchial vein before crossing the midline to join the azygos vein at the level of the eighth thoracic vertebra.

Veins of the Lower Extremity. These veins are subdivided into superficial and deep sets, as are those of the upper extremity. Valves are present in both sets, but they are more numerous in the deep veins.

The **superficial veins** of each lower extremity are the great and the small saphenous veins and their many tributaries. These vessels are of considerable clinical importance because of the frequency with which they become varicosed.

1. The *great saphenous vein* is the longest vein in the body. It begins at the medial end of the dorsal venous arch of the foot, passes in front of the medial malleolus, and then courses upward along the medial aspect of the leg and the thigh

(Fig. 13-22). It receives tributaries from superficial tissues all along its course and has many connections with the deep veins as well. The great saphenous empties into the femoral vein in the region of the groin (see Fig. 13-23).

2. The *small saphenous vein* begins at the lateral end of the dorsal venous arch of the foot, passes behind the lateral malleolus and courses upward under the skin in the back of the leg. It receives blood from the foot and from the posterior superficial portion of the leg and then empties into the popliteal vein in back of the knee (see Fig. 13-22).

The **deep veins** of the lower extremity accompany the arteries and receive blood from the structures along their course. As in the upper extremity, many of these veins are arranged in pairs, called the venae comitantes.

1. The *posterior tibial veins* are formed be-

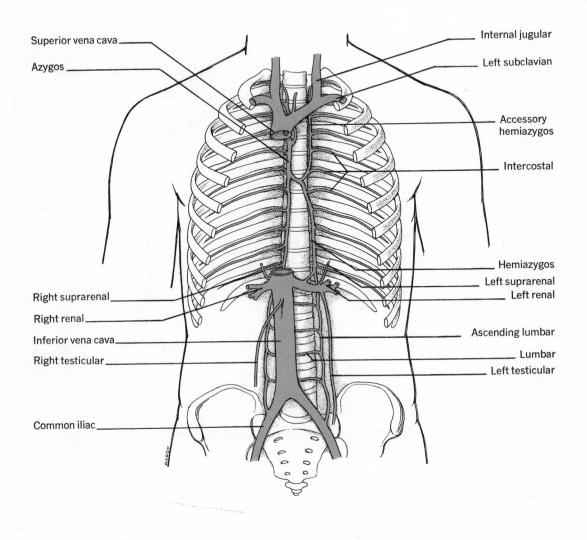

Superior vena cava

Azygos

Internal jugular

Left subclavian

Accessory hemiazygos

Intercostal

Hemiazygos

Left suprarenal

Left renal

Right suprarenal

Right renal

Inferior vena cava

Right testicular

Ascending lumbar

Lumbar

Left testicular

Common iliac

Fig. 13-21. Diagram of the azygos system of veins. The inferior vena cava and its tributaries are also shown. The heart has been removed.

hind the medial malleolus by the union of the medial and the lateral plantar veins. Coursing upward in the leg, they receive the peroneal veins and other tributaries before uniting with the anterior tibial veins just below the knee (Fig. 13-23).

2. The *anterior tibial veins* are the upward continuation of the venae comitantes of the dorsalis pedis artery. They unite with the posterior tibial veins to form the popliteal vein.

3. The *popliteal vein,* formed by the union of the anterior and the posterior tibial veins at the lower border of the popliteal space, receives blood from muscular and articular tributaries as well as from the small saphenous vein. Just above the knee, the popliteal vein becomes the femoral vein.

4. The *femoral vein* is the upward continuation of the popliteal vein, and it receives many tributaries which drain the deep structures of the thigh (Fig. 13-23). After receiving the great saphenous vein in the region of the groin, the femoral vein passes under the inguinal ligament and continues as the external iliac vein.

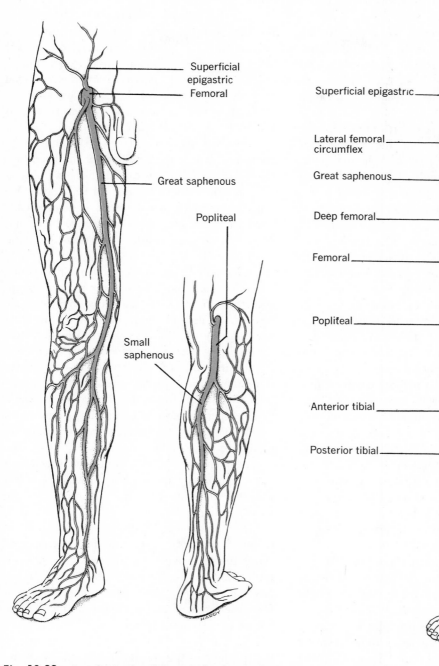

Fig. 13-22. Superficial veins of the right lower extremity, medial and posterior aspects.

Fig. 13-23. Deep veins of right lower extremity.

Veins of the Pelvis. The pelvic veins accompany the pelvic arteries and drain the general areas supplied by those arteries.

The **external iliac veins** are direct continua-

tions of the right and the left femoral veins. Each begins at the inguinal ligament, curves upward to the level of the sacroiliac joint, and there unites with the internal iliac vein to form the common iliac vein of its own side (see Fig. 13-15). They receive blood from the inferior epigastric and the deep circumflex iliac veins.

The **internal iliac veins** (hypogastric veins) receive blood from the pelvic wall, the pelvic viscera, the external genitalia, the buttocks, and the medial aspect of the thigh. Each unites with the external iliac vein of its own side, as described above.

The **common iliac veins** are formed by the union of the internal iliac and the external iliac veins (Fig. 13-15). Each receives iliolumbar veins, and the left receives, in addition, the middle sacral vein. At the level of the fifth lumbar vertebra, the common iliacs unite to form the inferior vena cava.

Veins of the Abdomen.   The **inferior vena cava** (Figs. 13-15 and 13-21) is the largest vein in the body. It is formed by the union of the two common iliac veins and courses upward along the front of the vertebral column, on the right side of the aorta. After passing through a groove on the posterior surface of the liver, this vein perforates the diaphragm and ascends to enter the lower back part of the right atrium. The inferior vena cava receives blood from the common iliac veins and from the following vessels as well.

1. The *lumbar veins,* four in number on each side, arise in the muscles of the posterior abdominal wall (see Fig. 13-21). They receive tributaries from the skin and the muscles of the back, the vertebral column, and the spinal cord. The lumbar veins are interconnected on either side by an ascending lumbar vein which forms the origin of the corresponding azygos, or hemiazygos vein, as previously described.

2. The *spermatic (testicular) veins* (in the male) receive blood from the testes and neighboring structures. They ascend through the inguinal canal and upward into the abdominal cavity. The right spermatic vein empties directly into the inferior vena cava, but its counterpart on the left empties into the left renal vein (Fig. 13-15). The *ovarian veins* (in the female) receive blood from the ovaries and neighboring structures. They ascend into the abdominal cavity and end in the same way as the spermatic veins.

3. The *renal veins* (Fig. 13-21) empty into the inferior vena cava after receiving blood from the kidneys and the ureters. The left renal vein receives, in addition, the left spermatic, or ovarian, the left suprarenal, and the left inferior phrenic veins.

4. The *suprarenal veins* receive blood from the suprarenal (adrenal) glands. The right vein empties into the inferior vena cava, and the left empties into the left renal vein (Fig. 13-21).

5. The *inferior phrenic veins* receive blood from the undersurface of the diaphragm and convey it to the inferior vena cava. The left vein sends a tributary to the left renal vein.

6. The *hepatic veins* (Figs. 13-15 and 13-24) are formed within the substance of the liver; they convey blood from the capillarylike sinusoids of this vascular organ to the inferior vena cava. One might think of the hepatic veins as doing double duty, since the liver sinusoids receive blood not only from the hepatic artery, but from the portal system as well.

The **portal system of veins** includes those veins that drain blood from the spleen, the stomach, the intestines, the pancreas, and the gallbladder. In general, these veins have the same names as the arteries which supply the aforementioned organs, but instead of emptying their blood directly into the inferior vena cava, they deliver it to the portal vein for a detour through the liver. The portal vein itself is formed by the union of the superior mesenteric and the splenic veins.

1. The *superior mesenteric vein* begins in the right iliac fossa and courses upward toward the liver. In general, it drains those structures which were supplied by the superior mesenteric artery (i.e., the small intestine and the first half of the large intestine).

2. The *splenic vein* drains the spleen, and, as it courses from left to right toward the liver, it receives the following tributaries: (a) *gastric veins* from the stomach; (b) *pancreatic veins* from the pancreas; and (c) the *inferior mesenteric vein* from the last half of the large intestine (except the lower portion of the rectum).

3. The *portal vein,* formed by the union of the splenic and the superior mesenteric veins, receives the following additional tributaries before it enters the substance of the liver: (a) the left *gastric (coronary) vein* of the stomach, which receives blood from the esophagus as well; (b) the *paraumbilical veins,* which anastomose with veins of the anterior abdominal wall; and (c) the *cystic vein* from the gallbladder. Figure 13-24 illustrates the portal vein and its tributaries.

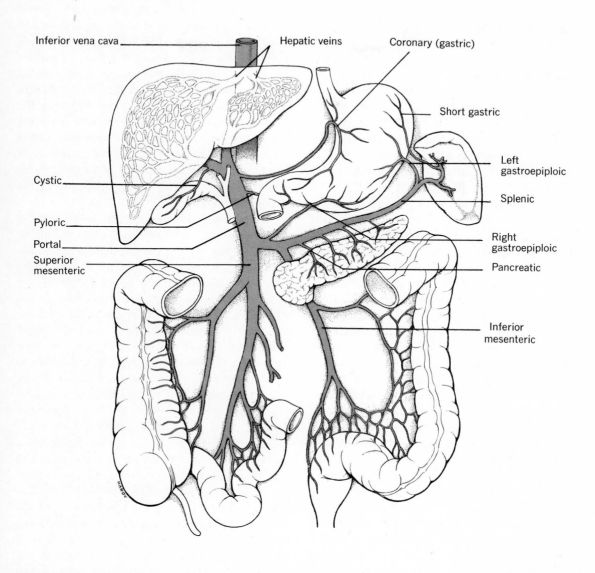

Fig. 13-24. The portal system. Various organs have been moved somewhat out of position to expose the portal vein and its tributaries.

On reaching the liver, the portal vein divides into right and left branches which accompany the corresponding branches of the hepatic artery as they ramify throughout the liver substance; these vessels deliver blood to the microscopic sinusoids that lie between the plates of liver cells (see Fig. 16-16). Within the sinusoids, the arterial blood from the hepatic artery mixes with the venous blood from the portal vein. This mixed blood flows through the sinusoids and eventually reaches the hepatic veins which empty into the inferior vena cava.

In summary, we re-emphasize the fact that the liver sinusoids receive blood from two sources: the hepatic artery and the portal vein. The blood in the hepatic artery is well oxygenated, thus meeting the need of the hard-working liver cells for a constant supply of oxygen. The portal vein, on the other hand, contains blood that is poorly oxygenated, since it has already passed through capillary networks in the digestive organs and the spleen. However, portal blood is rich with substances that have been absorbed from the gastrointestinal tract; this unique arrangement makes it possible for the liver to inspect all such materials before they are sent on to other tissues in the body. For example, some substances may be temporarily stored in the liver until needed; others are modified for easier oxidation or disposal in tissues elsewhere in the body, and harmful substances are detoxified. In other words, the liver stands guard by monitoring all absorbed materials before they are allowed to enter the general circulation for dispersal to all parts of the body.

Portal hypertension. Any interference with the flow of blood through the liver may result in portal hypertension, an abnormal increase in the pressure of the blood within the portal system of veins. Obstruction of the portal system occurs fairly frequently. It may be due to cirrhosis of the liver (an inflammatory process that results in structural changes in liver tissue), various tumors of the liver, or it may be due to back pressure in the inferior vena cava and the hepatic veins in certain types of heart disease. As blood backs up in the portal circuit, its pressure rises, and a collateral circulation is established; this primarily involves communications between the inferior mesenteric and the hemorrhoidal (rectal) veins that empty into the internal iliac veins, and communications between the gastric veins and the esophageal veins that empty into the azygos system. The esophageal veins frequently become dilated and tortuous (*esophageal varices*); rupture of these veins may lead to fatal hemorrhage. Back pressure in the paraumbilical veins may cause the abdominal veins around the umbilicus to dilate, forming a radiating pattern that is called the *caput medusae*.

Surgical intervention in the treatment of portal hypertension may involve a *portacaval shunt*. In this procedure an anastomosis is established between the portal vein and the inferior vena cava so that some of the portal blood can bypass the obstructed liver.

FETAL CIRCULATION

In the fetus, the lungs, the gastrointestinal tract, and the kidneys do not carry out the functions that they perform after birth. Instead, the fetus derives all of its oxygen and nutritive material from the mother's circulation and depends on her for the elimination of carbon dioxide and other waste products. This exchange of materials between the maternal and the fetal circulation takes place within the *placenta*, a highly vascular structure that is sometimes called the *afterbirth*, since it is cast out of the uterus following the birth of the baby. However, there is no direct connection between the circulatory system of the fetus and that of the mother. Fetal capillaries in the placenta are surrounded by pools of maternal blood, and all exchanges take place across the capillary membrane.

The vessels that connect the fetus with the placenta are the **umbilical arteries** and the **umbilical vein** (Fig. 13-25); these vessels lie in the umbilical cord. Blood leaves the fetus by way of the two umbilical arteries, which branch off the internal iliac arteries. After circulating through the capillaries in the placenta, where it picks up nutrients and oxygen and eliminates waste, blood returns by way of the umbilical vein; this vein enters the abdomen of the fetus at the umbilicus and ascends to the liver where it divides into two branches. One branch of the umbilical vein joins the portal vein and enters the liver; the other branch, called the *ductus venosus* continues onward to the inferior vena cava, thus permitting a certain amount of blood to bypass the liver. The umbilical vein and the ductus venosus are obliterated within a few days after the baby is born.

Circulation through the lower portion of the body is like that in the postnatal infant. As the blood returns from the lower portion of the body of the fetus, it is mingled with the blood that is returning from the placenta laden with oxygen and nutritive materials. The mixing takes place in the inferior vena cava, and the blood then enters the right atrium.

Circulation through the upper portion of the body also is like that in the postnatal infant. After circulating through the upper portion of the body, the blood is returned to the right atrium by the superior vena cava. In the right atrium, blood from the inferior vena cava mixes to some extent with that from the superior vena cava. However, a valve in the inferior vena cava guides a steady stream of blood from the right atrium through the *foramen ovale* (p. 277) and into the left atrium, thus bypassing the lungs. That blood

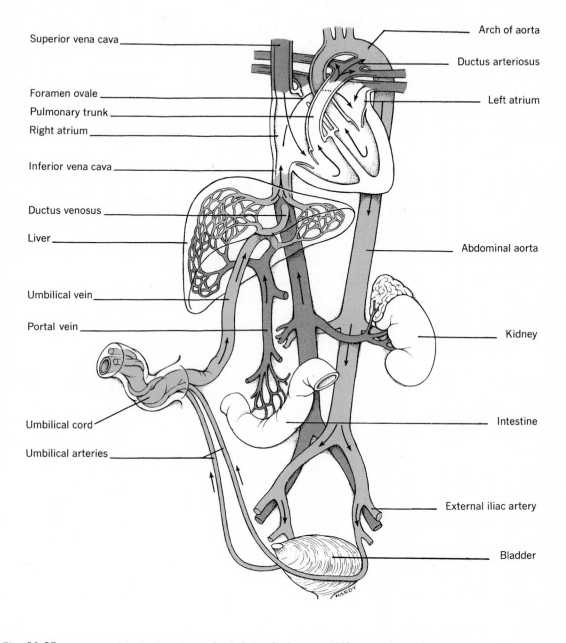

Superior vena cava

Foramen ovale

Pulmonary trunk

Right atrium

Inferior vena cava

Ductus venosus

Liver

Umbilical vein

Portal vein

Umbilical cord

Umbilical arteries

Arch of aorta

Ductus arteriosus

Left atrium

Abdominal aorta

Kidney

Intestine

External iliac artery

Bladder

Fig. 13-25. Diagram of the fetal circulation shortly before birth; course of blood is indicated by arrows.

which descends into the right ventricle is pumped into the pulmonary artery, but since the fetal lungs are collapsed, only a small amount of blood passes through the pulmonary circuit; most of the blood in the pulmonary artery flows through the *ductus arteriosus* (p. 299) which opens into the arch of the aorta. Normally, the foramen ovale and the ductus arteriosus close soon after the baby is born and breathing is established.

The blood in the aorta, a mixture of oxygenated and poorly oxygenated blood, is carried to all parts of the body through its systemic branches.

However, on reaching the bifurcation of the common iliac arteries, part of the blood flows into the internal iliac arteries, into the umbilical arteries, and back to the placenta for another exchange of nutrients and wastes. Figure 13-25 is a diagram of the fetal circulation; note that the umbilical vein is the only fetal vessel that carries fully oxygenated blood.

Congenital heart disease

Improper or inadequate development of the fetal heart and related blood vessels may result in various congenital defects. For example, there may be atrial or ventricular septal defects, stenosis of the great arteries or of their valves, failure of the ductus arteriosus to close after birth. However, through constantly improving diagnostic and surgical technics many of these defects now can be corrected. In addition to the electrocardiogram and cardiac catheterization, another diagnostic aid in congenital heart disease is *angiocardiography*; this involves injecting a radiopaque fluid into a vein or into a chamber of the heart and then taking a rapid series of x-ray pictures of the heart and neighboring vessels.

The **tetralogy of Fallot** was one of the hopeless congenital anomalies of the heart prior to the advent of cardiac surgery, and it usually terminated the patient's life by age 12. Today this condition can be corrected, or greatly improved, by surgical intervention. There are four characteristic features of this anomaly: (1) pulmonary stenosis, (2) a high ventricular septal defect, (3) an overriding aorta, and (4) right ventricular hypertrophy. The **pulmonary stenosis** obstructs blood flow to the lungs and results in *hypertrophy of the right ventricle* due to the overload. If the obstruction is severe enough, the pressure of backed-up blood in the right ventricle will become greater than the pressure in the left ventricle, causing unoxygenated blood to flow through the **ventricular septal defect** into the left ventricle and into the *overriding aorta*. Consequently, the blood that is delivered to the systemic capillaries is poorly oxygenated and contains so much reduced hemoglobin that marked cyanosis is present. Because of their cyanotic appearance, these little patients frequently are referred to as *blue babies*.

When such youngsters try to run or to play, their oxygen-starved muscles tire very quickly; this

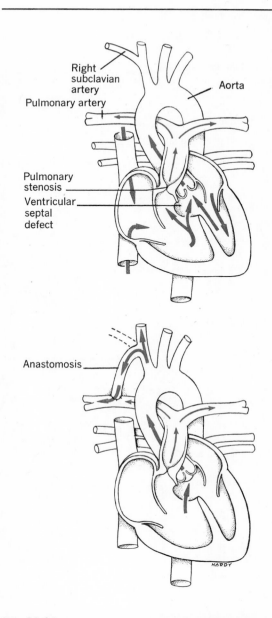

Fig. 13-26. Diagram of the circulation before (*top*) and after (*bottom*) the Blalock-Taussig operation as described on pages 320 to 321.

causes them to assume a characteristic squatting position on their haunches as they rest and try to catch their breath.

Tremendous relief of both cyanosis and disability may be obtained by the Blalock-Taussig operation (Fig. 13-26). This surgical procedure consists of joining the end of a systemic artery (usually the right or left subclavian) with the side of the right or the left pulmonary artery. As a result, some of the mixed arterial-venous blood in the aorta flows through the shunt into the pulmonary artery, greatly increasing the volume of blood that reaches the lungs for oxygenation. Although this operation is merely palliative and does nothing to correct the original anatomic defects, it is of great benefit to those patients for whom open-heart surgery is not available.

THE LYMPHATIC SYSTEM

The lymphatic system is anatomically and physiologically interrelated with the cardiovascular system. It not only helps in the removal of interstitial fluid from the intercellular spaces, but also plays an important role in protecting the body from bacterial invasion. The structures of the lymphatic system include: (1) lymphatic vessels

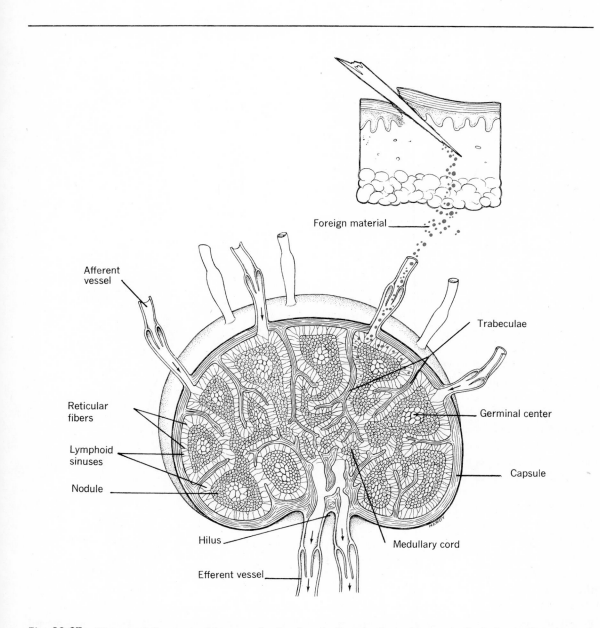

Fig. 13-27. Diagram of the structural features of a lymph node. Bacteria that gain entry to the body are filtered out of the lymph as it flows through the node.

which are concerned with the transportation of lymph, and (2) related lymphatic organs that filter lymph, produce lymphocytes, and carry out other specific functions.

The lymphatic vessels

Those vessels that are concerned with the transport of lymph include lymphatic capillaries, collecting vessels, and terminal ducts.

The **lymphatic capillaries** are microscopic, blind tubes which anastomose freely and form vast networks in the intercellular spaces. These networks constitute the beginning of the lymph drainage system, and are widely distributed throughout the body; however, they are absent from the CNS, the meninges, the eyeball and orbital fat, the internal ear, cartilage, and the epidermis. The walls of lymphatic capillaries consist of a single layer of thin, flat endothelial cells, and they are much more permeable than are the walls of blood capillaries. For this reason, large particles (escaped plasma protein, foreign substances such as soot and dirt) are able to pass through the lymphatic capillary wall and be carried away in the lymph. Those lymphatic capillaries which are located within the intestinal walls are called *lacteals* because of the milky appearance of their lymph following the absorption of fats. The various lymphatic capillary networks are drained by an elaborate system of collecting vessels.

Lymphatic collecting vessels arise in the capillary networks which they drain. They course through the loose subcutaneous tissues, through the connective tissue between muscles and other organs, and often form networks around the blood vessels which they accompany. Lymphatic vessels are extremely delicate, and their walls are so transparent that one can see the lymph within. Another feature of these vessels is the presence of numerous valves which prevent the backflow of lymph; these valves occur at such close intervals that the vessel has a knotted, or beaded, appearance.

Lymph nodes (p. 324) are situated along the course of the lymphatic vessels and serve to filter the lymph before it is returned to the bloodstream. A number of *afferent lymphatic vessels* penetrate the capsule of each lymph node and break up into lymphoid sinuses (Fig. 13-27); the sinuses lead into one or two *efferent lymphatic*

vessels which carry lymph away from the node. A variable number of nodes may be traversed in a similar manner before the efferent vessels unite to form collecting trunks. Each collecting trunk drains a definite area of the body and eventually empties into one of two terminal vessels; these vessels are the thoracic duct and the right lymphatic duct.

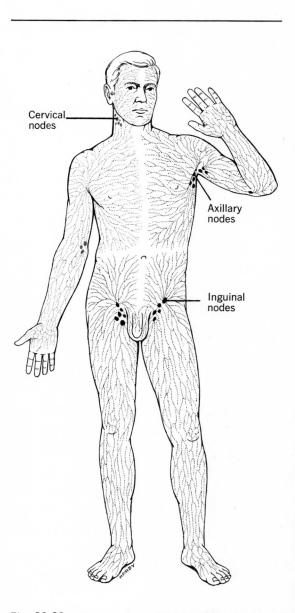

Fig. 13-28. Diagram of superficial lymphatic vessels and cervical, axillary, and inguinal nodes. Unshaded area drained by thoracic duct, shaded area drained by right lymphatic duct.

The **thoracic duct** is the larger of the two terminal vessels. As illustrated by the unshaded areas in Figure 13-28, it receives lymph from all parts of the body below the diaphragm and from the left half of the body above the diaphragm. The thoracic duct begins at the level of the second lumbar vertebra. At its origin there is usually a dilated portion known as the *cisterna chyli*; collecting trunks from the lower extremities, the abdominal and the pelvic viscera, and the body wall empty into this structure. Coursing upward along the spinal column, the thoracic duct receives intercostal trunks as it passes through the thorax; at the root of the neck, it is joined by collecting trunks from the left upper extremity and the left side of the head, the neck, and the upper thorax.

The thoracic duct ends by opening into the angle of junction of the left subclavian and the left internal jugular veins; a pair of valves at the terminal end prevents the passage of venous blood into the duct. The thoracic duct is shown in Figure 13-29.

The **right lymphatic duct** (Fig. 13-29) is a short duct that may be formed by the union of trunks from the right upper extremity and the right side of the head, the neck, and the upper thorax; it empties into the right brachiocephalic vein near its origin. However, this terminal duct frequently is missing; when it is, the various collecting trunks empty directly into the right subclavian vein near its junction with the right internal jugular vein. The shaded area in Figure

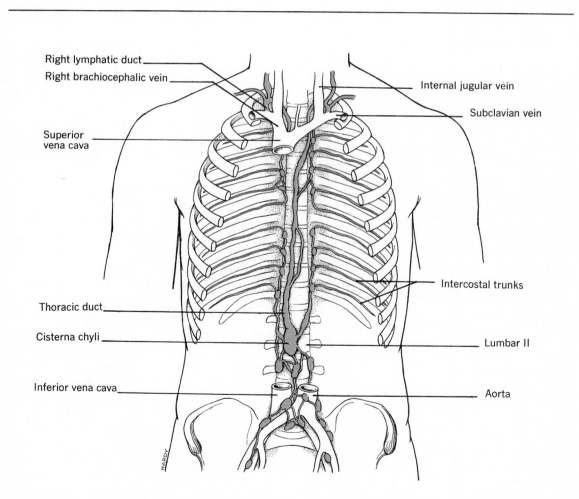

Fig. 13-29. Diagram illustrating course of the thoracic duct and the right lymphatic duct. Deep lymphatic vessels and nodes also are shown.

13-28 illustrates the portions of the body drained by the right lymphatic duct.

In summary, the function of the lymphatic vessels is to return to the bloodstream the water, escaped plasma protein, newly formed protein, and products of cellular metabolism that cannot be reabsorbed by the blood capillaries; bacteria and other harmful substances generally are filtered out of the lymph stream as it flows through the lymph nodes.

Interference with the drainage of lymph from any part of the body may result in a marked swelling of the tissues; this is called *lymphedema*. Such a condition may occur in the upper extremity following surgical removal of a breast and the axillary lymph nodes in the treatment of cancer; deep x-ray therapy following surgery may damage the remaining lymphatic vessels and further disrupt the flow of lymph.

The flow of lymph

Although there is no pumping mechanism comparable to the heart, lymph is kept moving along by the contraction of smooth muscle in the walls of the lymphatic vessels. This flow is aided by the contractions of skeletal muscles, the peristaltic movements of the smooth muscle in visceral walls, and the pulsating arteries; all of these activities serve to squeeze, to compress, or to massage the neighboring lymphatic vessels, thus forcing the lymph onward toward the veins. Backflow of lymph is prevented by the numerous valves that are present in the vessels.

Breathing movements also play a part in the movement of lymph. With each inspiration the pressure within the thoracic cavity is decreased, and the resulting suction causes lymph to flow upward from the abdominal portion of the thoracic duct.

Lymphatic tissue

Lymphatic tissue is the second type of hemopoietic tissue (p. 259). It consists of a meshwork of reticular fibers in which are found primitive *stem cells* that are capable of mitotic division and of differentiation. Some stem cells differentiate to form *phagocytic cells* of the reticuloendothelial system, others become free *lymphoblasts* that can go on to form *lymphocytes*. As previously described on page 265, lymphocytes can give rise to

precursor cells that in turn become *plasma cells,* the important producers of antibodies.

Thus the very nature of the cells to be found in lymphatic tissue emphasizes its primary function of defending the body against bacteria and other disease-inducing agents that may break through the skin or mucous membrane barriers. The distribution of lymphatic tissue throughout the body is quite logical if one keeps in mind the protective role that it plays.

Lymphatic nodules, designed to filter interstitial fluid, are found under wet epithelial surfaces that are exposed to contamination, such as the digestive tract and the respiratory and the genitourinary passages. The simplest forms are merely irregular accumulations of lymphoid cells under the mucous membranes; larger accumulations of these cells produce definite round nodules, which may occur singly (solitary nodules) or in groups (aggregated nodules). *Aggregated nodules,* or *Peyer's patches,* are particularly abundant in the distal portion (ileum) of the small intestine; they may become secondarily infected in cases of typhoid fever and tuberculosis, occasionally contributing to perforation of the intestinal wall. Still larger masses of lymphatic tissue form a protective *tonsillar ring* in the pharynx; the pharyngeal tonsil, the lingual, and the palatine tonsils are described on pages 353 to 354.

Lymph nodes, designed to filter lymph, are small oval-, or bean-shaped, bodies varying in size from that of a pinhead to that of an almond. Although they may occur singly, most nodes are clustered in groups along the course of lymphatic vessels that drain specific areas of the body; these groups are referred to as regional lymph nodes. Figures 13-28 and 13-29 illustrate some of the superficial and the deep lymphatic vessels and their regional nodes.

Each node consists of a mass of lymphatic tissue that is enclosed in a fibrous capsule; little partitions (trabeculae) extend inward from the capsule and divide the node into compartments. Lymph flows through the node by way of the *lymph sinuses,* a system of microscopic channels that permeate the lymphatic tissue; great numbers of newly formed lymphocytes enter the lymph stream as it passes germinal centers near the periphery of the node. A section of a lymph node is shown in Figure 13-27.

In addition to the production of lymphocytes,

lymph nodes also function as filtration plants (Fig. 13-27). Each lymph sinus is crisscrossed by reticular fibers to which cling numerous phagocytic cells. As lymph flows through these living filters, bacteria and other foreign particles are first trapped by the network of fibers and then destroyed by the phagocytes. This function of the lymph nodes is of particular importance when one has a localized infection; by filtering bacteria from the lymph stream, the nodes prevent the infection from spreading to other parts of the body. However, as the bacteria and the phagocytes engage in mortal combat, the lymph nodes themselves may become inflamed (*lymphadenitis*). One should be aware of the fact that the presence of an infection in almost any part of the body is accompanied by enlargement and tenderness of the lymph nodes which drain that region. Another complication that may follow a severe infection is an inflammation of the lymphatic vessels (*lymphangitis*). When an infection is spreading along the superficial lymphatic vessels its course may be revealed by red streaks in the skin. This is a danger signal, since it indicates that unless the infection is stopped, the invading organisms eventually will reach the bloodstream, and *septicemia* (*blood poisoning*) may follow.

In most instances this vast lymphatic network provides a valuable line of defense against infection and disease, but occasionally it works to the body's disadvantage. In cases of *carcinoma* (*cancer*), malignant cells frequently break away from the primary, localized tumor and enter the lymph stream; on reaching the regional lymph nodes the cancer cells set up secondary growths, or *metastases*. It is for this reason that in treating carcinoma, the surgeon removes not only the original cancerous organ, but as many of the surrounding lymph nodes as possible. However, if the disease is not treated in its early stages, it will continue to spread throughout the body until the victim dies.

The **spleen**, designed to filter blood, is the largest unit of lymphatic tissue in the body. It is located in the left hypochondriac region, just below the diaphragm and posterolateral to the stomach. The shape of the spleen is somewhat variable, since it is accommodated to the surfaces of the adjacent viscera, but it generally is an elongated and ovoid body. It is soft, friable, highly vascular, and dark purple.

The spleen is covered by a capsule composed of fibrous and elastic connective tissue, together with a few smooth muscle fibers. At the hilus, a fissure near the medial border, little strands of the tissue of the capsule extend into the substance of the organ; these strands (trabeculae) form a supporting framework for vessels and nerves. To a lesser extent, trabeculae extend inward from the capsule itself, as shown in Figure 13-30. The spaces within the framework are filled with a highly vascular lymphatic tissue called the *splenic pulp*. Two kinds of pulp are readily visible: red and white.

The *white pulp* consists of ordinary lymphatic tissue, some being distributed as tiny islands (nodules) in the soft red pulp that occupies the remaining space. These lymphatic nodules are the major sites for lymphocyte production in the spleen. White pulp is also distributed along the sheaths of arteries as they leave the trabeculae and enter the red pulp. Plasma cells are produced in the white pulp when antigens (e.g., bacteria) gain entrance to the bloodstream. The *red pulp* consists of a meshwork of reticular fibers containing vast numbers of red blood cells and reticuloendothelial cells (both fixed and free macrophagocytes); many lymphocytes may be seen in this pulp. Monocytes and granular leukocytes also are present, but they are less numerous. It now remains to be seen how blood enters and leaves this living filtering device.

The blood supply of the spleen is derived from the splenic artery, a vessel noted for its large size in proportion to the size of the organ. This large artery divides into six or more branches that enter the hilus and diverge in various directions throughout the substance of the spleen. Each branch gives off smaller and smaller branches that ultimately deliver blood to highly permeable capillaries in the red pulp; here the blood trickles slowly through the reticular meshwork and the macrophagocytes can perform their functions (the removal of old red cells, bacteria, and so forth). The filtered blood then flows through venous sinusoids that empty into veins of the red pulp. The splenic vein finally conveys blood from the spleen to the portal vein, as previously described.

Although the spleen is not essential to life, it does perform many important functions. In the fetus it is a major site for the production of red blood cells, but after birth it is limited to the

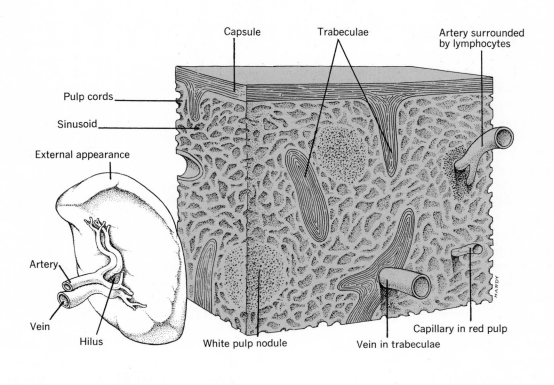

Fig. 13-30. Section of spleen, showing general arrangement of splenic tissue.

formation of lymphocytes, plasma cells, and some monocytes. The phagocytic reticuloendothelial cells remove most of the old, worn-out red blood cells from the circulation. The iron is extracted from the hemoglobin and liberated into the bloodstream for reuse by the body. The splenic cells also form a pigment, *bilirubin*, from the hemoglobin of the old red cells; bilirubin enters the bloodstream and is carried to the liver for elimination in the bile. Inasmuch as the spleen is able to produce plasma cells, it is a source of antibodies that offer protection against various diseases. The spleen also has a definite capacity to store blood and then to liberate that blood at times when it would be helpful to the body; however, this function is not as important in man as it is in certain lower animals.

Surgical removal of the spleen (splenectomy) ordinarily is tolerated quite well, as other organs in the body can take over splenic functions. For example, other lymphatic tissue will produce

lymphocytes in sufficient numbers to meet the body's needs, and phagocytic cells in other parts of the reticuloendothelial system will destroy old red blood cells, bacteria, and foreign substances. Blood vessels in the skin, the liver, the lungs, and the portal system are more than adequate to provide blood storage space when the body is at rest.

The **thymus** is a pinkish-gray mass of lymphatic tissue that is situated partly in the root of the neck and partly in the upper thorax beneath the sternum. It is relatively large in children, reaches its maximum size at the age of puberty, and then undergoes involution, slowly becoming smaller as one ages.

The thymus is enclosed in a connective tissue capsule from which trabeculae extend into its substance. It is divided into two pyramidal-shaped lobes, which, in turn, are subdivided into numerous lobules. Each lobule displays two zones of tissue, an outer cortex and an inner medulla;

the cortical portion consists of a meshwork filled with many small lymphocytes, and the medullary portion contains cell nests, or *thymic corpuscles.* The blood supply of the thymus is derived from branches of the thyroid and the internal thoracic arteries; its nerve supply is derived from both divisions of the autonomic system. Lymphatic vessels from the thymus drain into the anterior mediastinal lymph nodes.

Functions. The thymus plays a key role in the development of *immunological competence*— the ability of certain cells to recognize and to attack foreign substances that may gain entry to the body. Dr. Robert A. Good and his associates at the University of Minnesota found that the body has a dual system of immunity, one dependent upon the thymus, the other governed by lympho-epithelial tissue located elsewhere in the body (possibly clusters of cells in the intestinal tract). Consequently, fetal lymphoid stem cells can be stimulated to differentiate into two distinctly different systems of immune lymphocytes.

The first system includes those cells that are completely dependent on the thymus gland for their full differentiation. These thymus-dependent lymphocytes, or T cells, migrate by way of the bloodstream to the lymph nodes, spleen, and other lymphoid tissue throughout the body, thus helping to populate them with immunologically competent cells. There is considerable evidence that the thymus also produces a hormone that activates the proliferation of cells within lymphoid tissue. However, this so-called thymic hormone has not yet been isolated. T cells circulate in the blood and lymph. Whenever they come in contact with foreign substances, such as certain bacteria, fungi, and foreign protein (e.g., a kidney transplant), the T cells produce antibodies that destroy the invader. This is known as cell-mediated immunity.

The second system of immune lymphocytes is independent of thymic influence. Here we find the immobile plasma cells that synthesize **anti**-bodies and secrete them into the blood and lymph. These antibodies can inactivate most invading bacteria and some viruses. This is known as humoral immunity.

SUMMARY

1. Blood vessels
A. Arteries
   a. Elastic muscular tubes; convey blood to capillaries
   b. Recoil forces blood onward in pulsating waves; reflect ventricular beat
   c. Coats: tunica intima, tunica media, tunica externa
   d. Aneurysm: dilated sac resulting from weakened arterial wall
   e. Arteriosclerosis: decrease in elastic tissue in wall; atherosclerosis: fatty degeneration and lipid infiltration of vessel wall
   f. Blood supply: vasa vasorum
   g. Nerve supply: vasomotor (constrictors and dilators); afferent fibers related to reflex cardiac and vasomotor control
   h. Anastomoses: communications between adjacent vessels
B. Veins
   a. Conduct blood from capillaries to heart; more numerous than arteries
   b. Thinner walled than arteries; have same three tunics; valves prevent backflow of blood
   c. Blood and nerve supply as in arteries, but less abundant
   d. Varicose veins: dilated and tortuous superficial vessels

C. Capillaries
   a. Microscopic vessels between arterioles and venules; walls single cell thick
   b. Exchanges between blood and interstitial fluid take place through capillary wall; focal point of cardiovascular system
   c. Precapillary sphincters regulate flow of blood through capillaries according to need; not enough blood to fill all vessels at same time
   d. Sinusoids: capillarylike vessels in liver, spleen, bone marrow
   e. Arteriovenous anastomoses bypass the capillaries

2. Pulmonary circulation
A. Concerned with oxygenation of blood; relatively short circuit
B. Pulmonary trunk from right ventricle sends right and left branches to capillaries in each lung; blood oxygenated here, and CO_2 eliminated
C. Pulmonary veins (two from each lung) convey blood from capillaries to left atrium

3. Systemic circuit: arterial
A. Oxygenated blood pumped into aorta and thence to all parts of body by way of systemic branches which lead to capillaries
B. Aorta: largest artery in body; arises from left ventricle, arches over to left side of vertebral column, descends to level of fourth lumbar vertebra, bifurcates into common iliac arteries

C. Ascending aorta: coronary arteries arise just above semilunar valve; extend to capillaries in myocardium
D. Aortic arch: in adult, connected to pulmonary artery by ligamentum arteriosum; in fetus, connected to pulmonary artery by ductus arteriosus
  a. Branches
    (1) Brachiocephalic; branches to form right common carotid and right subclavian
    (2) Left common carotid
    (3) Left subclavian
  b. Common carotids supply capillaries in head and neck; branch to form external and internal carotids
  c. Circle of Willis: arterial anastomosis at base of brain; branches (cerebral arteries) supply capillaries in brain
  d. Subclavians supply head, thorax, and upper extremities; major branches include vertebrals, thyrocervical trunk, and internal thoracic arteries; continue into extremity as axillary, brachial, radial, and ulnar arteries
E. Thoracic aorta: from aortic arch to diaphragm
  a. Visceral branches: pericardial, bronchial, and esophageal arteries
  b. Parietal branches: intercostal and superior phrenic arteries
F. Abdominal aorta: from diaphragm to fourth lumbar vertebra
  a. Visceral branches: celiac (left gastric, hepatic, and splenic), superior mesenteric, middle suprarenal, renal, and spermatic (or ovarian), and inferior mesenteric arteries
  b. Parietal branches: inferior phrenic, lumbar, and middle sacral arteries
G. Terminal branches of aorta
  a. Common iliacs bifurcate to form internal and external iliac arteries; supply pelvic viscera and lower extremities
  b. External iliac continues in thigh as femoral, to popliteal, to anterior and posterior tibial arteries; dorsalis pedis is continuation of anterior tibial across top of foot

4. Systemic circuit: venous
A. Blood returned from capillaries to heart via superficial and deep veins and cranial venous sinuses
B. Coronary sinus and tributaries return blood from myocardium to right atrium
C. Veins of head and neck
  a. Cranial venous sinuses drain capillaries of brain, empty into internal jugular veins
  b. Internal jugular veins drain blood from brain, eyes, superficial face, and neck
  c. External jugular veins drain deep face and neck
D. Veins of upper extremity
  a. Superficial: cephalic and basilic
  b. Deep: radial, ulnar, brachial, axillary, and subclavian
E. Veins of thorax
  a. Brachiocephalic veins formed by union of internal jugular and subclavian on either side; brachiocephalics unite to form superior vena cava after receiving vertebral, internal thoracic, and inferior thyroid veins
  b. Azygos system of veins collects blood from intercostal, mediastinal, pericardial, and bronchial veins; empty into superior vena cava
F. Veins of lower extremity
  a. Superficial: great and small saphenous
  b. Deep: anterior and posterior tibials, popliteal and femoral veins
G. Veins of pelvis
  a. External iliacs join internal iliacs to form common iliac veins
  b. Common iliac veins unite to form inferior vena cava
H. Veins of abdomen
  a. Inferior vena cava: largest vein in body; courses upward on right side of aorta, perforates diaphragm, and ascends to enter right atrium
  b. Tributaries of inferior vena cava: lumbar, spermatic, or ovarian, renal, suprarenal, inferior phrenic, and hepatic veins
  c. Portal system of veins drains gastrointestinal tract, pancreas, and spleen; provides detour through liver which monitors all absorbed materials; liver drained by hepatic veins which empty into inferior vena cava
  d. Portal hypertension results from interference with blood flow through liver

5. Fetal circulation
A. Fetus derives oxygen and nutrients from mother's blood; eliminates waste at same time as blood passes through placental capillaries
  a. No direct connection between mother and fetus; each has separate circulatory system
  b. Maternal blood pooled around capillaries in placenta
  c. Umbilical vein conveys blood from placenta to fetus; umbilical arteries convey blood from fetus to placenta
B. Circulation through upper and lower portions of fetal body like that in postnatal individual
C. Foramen ovale and ductus arteriosus provide bypasses of lung circuit; only small amount of blood passes through lung capillaries since lungs not inflated; fetal structures close after birth

6. Lymphatic system
A. Interrelated with cardiovascular system; aids in removal of interstitial fluid and in protection against infection and disease
B. Lymphatic vessels
  a. Lymphatic capillaries: highly permeable, blind tubes in tissue spaces; receive water, escaped protein, foreign particles
  b. Lymphatic collecting vessels drain capillary networks; valves prevent backflow of lymph
  c. Terminal vessels: thoracic duct drains collecting vessels below diaphragm and on left side of body above diaphragm; right lymphatic duct drains right side of body above diaphragm; both return lymph to bloodstream
C. Flow of lymph depends on massaging action of con-

tracting muscle tissue and on breathing movements

D. Lymphatic tissue

    a. Lymphatic nodules found under wet epithelial surfaces exposed to contamination

    b. Lymph nodes produce lymphocytes; filter foreign particles from lymph

    c. Spleen produces lymphocytes, filters foreign particles from blood, destroys old red blood cells and saves iron, and forms antibodies

    d. Thymus plays key role in development of immunological competence of body; thymus-dependent lymphocytes involved in cell-mediated immunity

QUESTIONS FOR REVIEW

1. How does blood flow in the arteries differ from that in the veins? Explain.
2. What is an aneurysm? Why may it be dangerous?
3. What are varicose veins? What may cause them?
4. What is the function of precapillary sphincters?
5. Distinguish between the pulmonary circulation and the systemic circulation.
6. Imagine yourself in a microscopic submarine ready to travel from the left ventricle of the heart to various parts of the body. Through what blood vessels would you pass to reach capillaries (a) in the frontal lobe of the brain, (b) in the right hand, (c) in the spleen, (d) in the kidney, (e) in the pelvic viscera, and (f) in the top of your foot?
7. Still in your submarine, through what veins would you pass on the return trip to the heart from capillaries in the above locations?
8. What is the significance of the double blood supply to the liver?

9. Why is it dangerous to squeeze a pimple in the nose?
10. What fetal structures are concerned with bypassing the lungs?
11. What is a blue baby?
12. How is the lymphatic system interrelated with the cardiovascular system?
13. Distinguish between the thoracic duct and the right lymphatic duct.
14. What lymphatic structures are designed to filter interstitial fluid? What lymphatic structures are designed to filter lymph?
15. What are the major functions of the spleen?
16. What is the role of the thymus in cell-mediated immunity?

REFERENCES AND SUPPLEMENTAL READINGS

Best, C. H., and Taylor, N. B.: The Physiological Basis of Medical Practice, ed. 8. Baltimore, Williams & Wilkins, 1966.

Eckberg, T. J.: The dangerous triangle. JAMA, 209:561, July 28, 1969.

Good, R. A.: Progress toward a cellular engineering. JAMA, 214:1289, Nov. 16, 1970.

Goss, C. M. (ed.): Gray's Anatomy of the Human Body, ed. 28. Philadelphia, Lea & Febiger, 1966.

Ham, A. W.: Histology, ed. 6. Philadelphia, J. B. Lippincott, 1969.

Mayerson, H. S.: The lymphatic system. Sci. Am., 208:80, June 1963.

Wood, J. E.: The venous system. Sci. Am., 218:86, Jan. 1968.

The Circulating Blood

"And now I see with eye serene
The very pulse of the machine."
She Was A Phantom of Delight. William Wordsworth

The general function of the cardiovascular system is to transport life-sustaining blood from one part of the body to another. In the preceding chapters we have shown that this system consists of a muscular pump (the heart) which is coupled into a closed circuit of tubes (the arteries, capillaries, and veins). Although each of these structural units plays an important role in the circulation of blood, the capillaries are the focal point of the entire system. In fact, all of the physiologic mechanisms which control heart rate and the internal caliber of blood vessels are geared to the task of supplying the capillaries with adequate amounts of blood to meet the ever-changing needs of the cells.

As the circulating blood flows through the capillary networks, each organ of the body is supplied with the materials it needs for growth, repair, maintenance, and its own particular functions. For example, the brain is absolutely dependent on oxygen and glucose; if these are not supplied continuously by the circulating blood, consciousness is lost. Muscles need oxygen, glucose, and amino acids, as well as the proper ratio of sodium, calcium, and potassium salts in order to contract normally. The glands need sufficient supplies of raw materials from which to manufacture their specific secretions. While these deliveries are being made, the blood picks up the end products of cellular metabolism and carries them to other parts of the body for use or for elimination. All such functions depend on the integration of cardiovascular activities which serve to maintain optimum blood pressure and blood flow through the capillary networks.

BLOOD PRESSURE AND BLOOD FLOW

Blood pressure refers to the force exerted by the blood against the walls of the vessels in which it is contained. Since a liquid will flow only from an area of higher pressure to an area of lower pressure, it necessarily follows that the blood pressure will be higher in some parts of the system than in others. The establishment of this pressure gradient depends on the high head of pressure generated during contraction of the ventricles of the heart.

Since there are three types of blood vessels in the vascular system, there are three blood pressures to be considered; these are arterial, capillary, and venous pressures. Although arterial pressure is the pressure most frequently determined in clinical situations, one must not overlook the fact that blood is also under pressure in the capillaries and the veins. However, in view of the previously mentioned decreasing pressure gradient that governs the flow of liquids, it naturally follows that capillary pressure is lower than arterial pressure, and venous pressure is lower than capillary pressure. In other words, the farther a liquid flows from its pressure source the lower its pressure becomes.

Blood pressure is expressed in terms of millimeters of mercury *above* the atmospheric pressure, which is 760 mm. Hg. For example, if one's blood pressure is said to be 100 mm. Hg, it really means 860 mm. Hg. Obviously, if the pressure within the blood vessels were not above atmospheric pressure, there would be no bleeding from a wound; instead, air would enter the severed vessels.

Arterial pressure

Arterial blood pressure is the force exerted by the blood against the walls of the arteries. It is highest in the aorta since this great vessel is closest to the powerful left ventricular pump. The aorta also offers little resistance to the flow of blood. As blood flows away from the heart, there is a progressive decrease in the pressure in the arterial tree. This decrease is related to the progressive increase in resistance resulting from the changing architecture of the vascular bed. As illustrated in Figure 14-1, a very significant drop in pressure occurs at the level of the arterioles where the peripheral resistance (p. 335) is greater than at any other part of the system. Thus blood is delivered to the delicate capillaries under greatly reduced pressure (Fig. 14-2).

Blood pressure is much higher in the long systemic arteries than it is in the short pulmonary circuit, but in either locale the pressure varies during the cardiac cycle. With each ventricular contraction a great spurt of blood is pumped into the large arteries, causing their elastic walls to

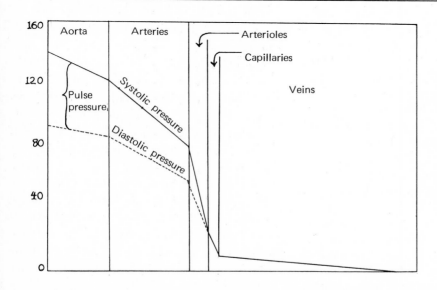

Fig. 14-1. Diagram showing the gradient of pressure in the various parts of the vascular system, between the aorta and the great veins. The gradient is steepest in the arterioles. The pulse pressure disappears as the capillaries are reached.

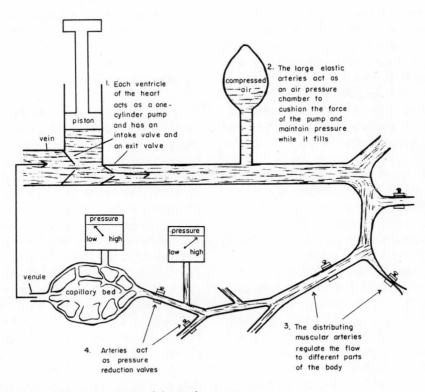

Fig. 14-2. Diagram illustrating the mechanics of the circulatory system.

become stretched and distended with blood. In spite of the intermittent delivery of blood into the arterial tree, the flow is even and steady through the all-important capillaries. This is possible because much of the force of each ventricular contraction is stored as potential energy in the stretched arterial walls. Then, during relaxation of the ventricles, the stored energy is released as the elastic walls recoil and keep the blood flowing onward to the capillaries (backflow is prevented by closure of the semilunar valves). Thus we are dealing with two distinct pressure phases in the arteries: (1) *systolic pressure*, generated during contraction, or systole, of the ventricles, and (2) *diastolic pressure*, which is that maintained by the recoil of stretched elastic tissue in arterial walls during ventricular relaxation, or diastole.

Systemic Arteries. In healthy young adults at rest, the average systolic pressure range in the *systemic arteries* is 115 to 120 mm. Hg; the diastolic range is slightly lower, or 75 to 80 mm. Hg. However, a systolic pressure of 90 to 140, and a diastolic pressure of 60 to 90 are considered to be within the normal range. When recording a blood pressure, the systolic pressure precedes the diastolic, in this manner: 115/75 or 130/90, whatever the case may be. The difference between systolic and diastolic blood pressures is called the *pulse pressure*; it averages about 40 mm. Hg.

The pulse pressure is affected by two major factors: (1) the amount of blood that is ejected during systole (stroke volume), and (2) how much the arteries can stretch when blood is ejected from the heart (distensibility, or compliance of the arterial tree). For example, an increase in heart rate can result in a decrease in the stroke volume, due to the shortened filling time for the heart, in which case the pulse pressure decreases accordingly. On the other hand, when the peripheral resistance is decreased, allowing a rapid flow of blood from the arteries to the veins, the increased venous return to the heart will result in an increased stroke volume and an increase in pulse pressure.

The distensibility of the arterial tree may be decreased by an elevated systemic blood pressure, or by pathologic changes that make the arterial walls more rigid than normal. In either case, there is resistance to the ejection of blood, and the arterial pressure rises to a high level during systole, then falls greatly during diastole. Thus the pulse pressure is increased.

When the arterial blood pressure is significantly above average for the individual involved, he is said to have *hypertension*, or high blood pressure. There are no hard and fast figures, but most physicians will agree that a *sustained systolic pressure of 140 mm. Hg* (or more), or a *sustained diastolic pressure of 90 mm. Hg* (or more) in a person at rest, constitutes a condition which is called hypertension.

In many cases, the cause of hypertension is unknown, and the person is said to have *primary*, or *essential hypertension*; overweight, heredity, and nervous tension are thought to be possible predisposing factors. *Secondary hypertension* is so named because the elevated pressure stems from a known cause; for example, it may accompany kidney disease and endocrine disorders. The accelerated, or fulminating, form of high blood pressure is called *malignant hypertension*; this is extremely serious, and many pathologic changes occur, often involving the brain and the kidneys. In severe cases, the pressure may rise to levels of 250/150 to 270/170 mm. Hg. The greatest dangers faced by hypertensive persons are cerebral hemorrhage and heart failure.

Occasionally a systolic pressure is consistently below the normal range, and such a person is said to have *hypotension*. The hypotensive person may complain of fatigue, lack of physical endurance, and episodes of fainting, due to lack of adequate blood flow to the brain.

Pulmonary Arteries. The average systolic pressure in the pulmonary arteries is 22 mm. Hg, and the diastolic is 8 mm. Hg. It is essential that the pulmonary pressure be maintained at this relatively low level to insure a correspondingly low pressure value in the pulmonary capillaries, as described below.

Capillary pressure

Capillary blood pressure is the force exerted by the blood against the walls of the capillaries.

It is about 32 mm. Hg at the arterial end of the systemic capillaries, and 12 mm. Hg at the venous end. The importance of the capillary blood pressure in the formation of interstitial fluid is discussed on page 31. Since this pressure is the driving force in the filtration of liquid through the capillary walls, you may wonder why the air sacs, or alveoli, of the lungs are not filled with fluid which has filtered out of the pulmonary capillaries. Normally, there is no fluid in the alveoli, due to the fact that the pulmonary vessels are a very low pressure system. The pressure in the pulmonary capillaries is about 7 mm. Hg; thus there is only a slight hydrostatic force tending to push fluid out of the capillary. Opposing this is the colloid osmotic pressure of the plasma, about 25 mm. Hg, which is a large force tending to pull fluid into the capillaries. However, should there be a marked increase in the capillary hydrostatic pressure, large quantities of fluid may accumulate in the interstitial spaces and in the alveoli; this condition, known as *pulmonary edema*, is commonly caused by left heart failure and the consequent damming of blood in the pulmonary vessels.

Venous pressure

Venous blood pressure is the force exerted by the blood against the walls of the veins. It varies with the location of the vein and with the position of the body (i.e., standing, sitting, or lying). However, for all practical purposes, the venous pressure gradually decreases as blood travels through the peripheral veins toward the heart. The *central venous pressure* (*CVP*) is that within the venae cavae and the right atrium of the heart. It is most accurately determined by inserting a catheter, through one of the antecubital veins, then attaching a manometer which measures the pressure in centimeters of water. The average pressure in the venae cavae is 6 to 12 cm. of H_2O, and in the right atrium, 0 to 4 cm. of H_2O. However, the absolute value of a CVP is much less informative than the upward or downward trend that it takes. Thus monitoring of the CVP is helpful in the management of acute circulatory failure since it reflects the dynamic interrelationship between cardiac activity, vascular tone, and effective blood volume.

An increase in venous blood pressure is a cardinal feature of the clinical syndrome known as *heart failure.* Since the failing heart is unable to generate an effective pumping pressure, blood tends to back up in the veins. This results in an abnormally high filtration pressure in the capillary beds, and fluid begins to accumulate in the interstitial spaces (edema). On the arterial side, a reduction in blood flow through the kidneys may lead to the retention of sodium and water, which further contributes to edema formation. However, if the heart is able to maintain an adequate output of blood to meet the basic needs of the body, it is said to be *compensated.* If the heart is unable to maintain an adequate circulation, it is said to be *decompensated.* In cardiac decompensation, the symptoms of pulmonary congestion include pulmonary edema, dyspnea (labored breathing), and cough as the air sacs of the lungs fill with interstitial fluid; as a result of systemic venous engorgement, there is cyanosis, generalized edema of dependent parts, and enlargement of the liver. These signs and symptoms are indicative of a condition that is referred to as *congestive heart failure.* Digitalis is a valuable drug that acts directly on the heart muscle and improves myocardial efficiency by increasing the force of each contraction; the ventricles empty more efficiently, and general circulation shows a marked improvement.

Gravity and Venous Return. When one assumes the upright position, the force of gravity tends to interfere with the return of blood from lower portions of the body. However, there are several factors that help to overcome this force and to promote venous return to the heart. One, the *muscular pump,* involves the contraction of various skeletal muscles which squeeze the veins and force blood away from the areas that are being compressed. As you will recall, veins have valves that permit blood to flow only in the direction of the heart; thus everytime you take a step, or even tense your muscles, a certain amount of blood is propelled toward the heart.

Another aid to venous return involves the changes in pressure within the thoracic cavity during inspiration and expiration. During normal breathing, the intrathoracic pressure is always negative or subatmospheric; however, it becomes even more negative when the thorax enlarges during the act of inspiration (p. 364). This expands the thin-walled intrathoracic veins, producing a suction effect that causes blood to flow into the thorax and then into the heart. During expiration, the venous inflow is reduced as the intrathoracic pressure is increased.

A third factor that promotes venous return is venoconstriction. This active contraction of smooth muscle fibers in the vein walls moves blood toward the heart and helps to prevent pooling in the highly distensible venous system. Veins are supplied with nerve fibers from the sympathetic division of the ANS.

A great strain is placed on the circulatory system of those persons who are required to stand quietly in one position for prolonged periods of time. A good example of this is to be found in the operating room; during long surgical procedures, doctors and nurses may stand in one place for several hours without relief. At such times, blood tends to accumulate (stagnate) in the extremities, causing them to feel uncomfortably full and swollen; the person may unconsciously begin to shuffle his feet and shift his weight in an attempt to get his muscular pump working. In extreme cases, so much blood may be pooled in the lower portions of the body that venous return becomes completely inadequate; the heart does not receive sufficient quantities of blood to supply the brain, and the person *faints.*

The essential cause of fainting is a lack of oxygen in the brain, due to a decreased blood supply. Under such conditions the person loses consciousness and falls. This is nature's remedy, since consciousness is soon regained in the horizontal position as blood flow to the brain improves, and venous return is no longer a problem; well-meaning bystanders who prop the patient in a sitting position tend to defeat the entire purpose of the faint.

Fainting usually occurs only when other compensatory mechanism are inadequate. For example, as arterial blood pressure falls in the aorta and the carotid sinuses, reflex cardio-acceleration and peripheral vasoconstriction (p. 343) may restore the pressure and prevent fainting. However, when a person begins to feel faint, he should lie down if this is possible; if not, he should sit down, bend over, and get his head between his knees. Attempting to walk out of a room when feeling faint may result in a preventable fall.

The maintenance of arterial blood pressure

Several factors are involved in the maintenance of arterial blood pressure at its usual high level. These are the rate and the force of the heartbeat, the elasticity of large arteries, the peripheral resistance, the quantity of blood in the vascular system in relation to the capacity, and the viscosity of the blood.

Rate and Force of Heartbeat. The faster the rate of the beat of the heart and the greater its force of contraction, the higher the arterial pressure will be. The heart is the pump, and if it changes, the pressure in the system is bound to change. A decrease in the rate or force of the beat of the heart leads to a lowering of the arterial blood pressure.

Elasticity of the Large Arteries. The elastic recoil of distended arterial walls is responsible for the maintenance of arterial blood pressure between beats of the heart (diastolic pressure). If the large arteries lose their elasticity, the pressure is unusually high during systole and falls to a low level during diastole. With rigid

arteries, the flow into the capillaries is intermittent, coming in spurts with each ventricular systole and ceasing during diastole.

Figure 14-3 illustrates the steady flow in an elastic system in contrast with the intermittent flow in rigid tubes.

Peripheral Resistance. The resistance offered by arterioles to the flow of blood is called peripheral resistance. By offering resistance to flow, arterioles maintain a high pressure behind them in the larger arteries, and a lower pressure in front of them in the capillaries and the veins. It has been stated that they "act as nozzles that gently spray blood into the capillary beds."

The amount of resistance to flow varies with the size of the lumens of the arterioles. If the lumens are wide, there is little resistance to flow. In this case, the pressure on the arterial side becomes lower and that on the venous side higher than usual. If the lumens are narrow, the arterioles offer more resistance to the flow of blood, and in

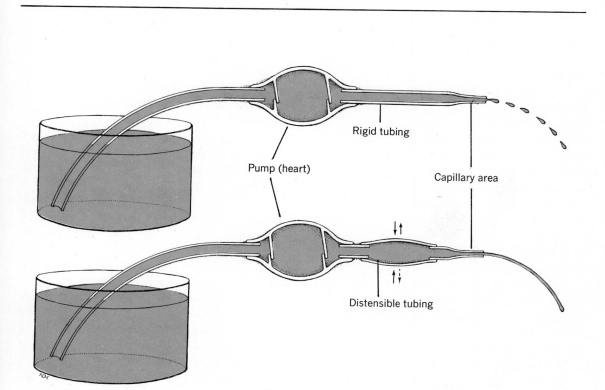

Fig. 14-3. Diagram showing the effects of elastic vessels. If the vessels were rigid, the blood flow would be intermittent, but since they are elastic, the flow is continuous. The recoil keeps blood flowing between beats of the heart.

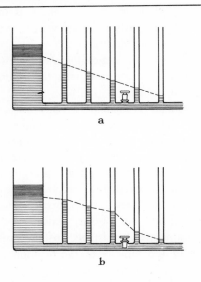

Fig. 14-4. Diagram of pressure conditions in the vascular system. *A,* a steady fall in pressure is noted with an increase in distance from the reservoir. *B,* the effect of increasing the peripheral resistance (vasoconstriction) is shown by the screw which represents the arteriolar area. The pressure rises on the arterial side and falls on the venous side.

this case, the pressure on the arterial side becomes higher and that on the venous side lower than before.

The arteriolar region of the vascular system may be thought of as a reducing valve, or screw, in the outflow tube, as indicated in Figure 14-4. The size of the lumens of the arterioles, which really means the peripheral resistance, is controlled reflexly by vasomotor nerves that supply the smooth muscle in the walls of the arterioles.

By the time the blood enters the venules from the capillaries, it exerts very little pressure on the walls of the veins. This means that the force imparted to it by the heart and by the elasticity of the walls of the large arteries is almost spent.

Quantity of Blood. The volume of blood in the vascular system in relation to the capacity of the system is exceedingly important. If a considerable amount of blood is lost suddenly by hemorrhage, the blood pressure falls. The walls of the arteries are not put on stretch, and the supply to the vital organs, such as the heart and the brain, is not sufficient to maintain homeostasis.

Whenever the quantity of blood is altered, protective mechanisms are brought into action as a result of the attempt to maintain the arterial blood pressure at the usual level. The heart rate is faster, and the arterioles are markedly constricted after blood is lost.

Viscosity of Blood. The presence of red blood cells and plasma proteins gives blood its viscosity, or *thickness.* If the number of red cells is increased, in proportion to the plasma, the viscosity is increased, and the blood pressure is higher. This is noted in polycythemia, in which there is a very high red cell count. If the number of red cells is decreased, as in anemia, the viscosity is low, and the arterial blood pressure is below the average.

If the concentration of plasma proteins is low, the blood pressure is low. If albumin solutions are injected, viscosity increases, and blood pressure rises.

Factors influencing arterial blood pressure

There are several physiologic factors, aside from those described above, which influence the arterial blood pressure; these include age, weight, sex, emotions, and exercise.

Age. In newborn babies the systolic pressure ranges between 20 and 60 mm. Hg. By the end of the first month, the pressure has risen to between 70 and 80 mm. Hg. It continues to rise slowly during childhood and reaches the adult level about the age of puberty. Blood pressure may remain at the same level until a person reaches 50 or 60 years of age, or may rise at any time if changes occur in the walls of large arteries (e.g., arteriosclerosis).

Walls of arteries deteriorate earlier than walls of veins. The arterial walls are under constant tension and are not really well supplied with blood. The capillaries which supply arterial walls collapse when the walls are stretched (which is practically all of the time), and the lymphatic drainage is not efficient. It is often said that a man is "as old as his arteries." This means that if arterial walls are damaged by the age of 30, a man may be 70 so far as his blood pressure is concerned.

Weight. Excess weight is important, because often it is associated with the onset of high blood pressure, especially after the age of 40 years. However, at any age, the person who is overweight imposes an additional burden on his cardiovascular system.

Sex. After puberty, men have higher average blood pressures than women. However, after menopause, the blood pressure in women tends to be slightly higher than in men.

Emotion. Fear, anger, excitement have a marked influence on the blood pressure, as a result of increased activity of the sympathetic nervous system and the increased secretion of epinephrine. The rise in blood pressure is due to an increase in the rate and force of the heartbeat, and to peripheral vasoconstriction. If a person is having his blood pressure taken for the first time, the procedure should be explained to him; otherwise, the pressure may be elevated, due to fear or anxiety. Severe emotional crises also have been known to cause cerebral vascular accidents (p. 206) as the sudden, sharp increase in blood pressure ruptures an artery in the brain.

Exercise. Of all the physiologic conditions, strenuous exercise has the most powerful effect upon the arterial blood pressure. The systolic pressure may rise to a height of 180 or 200 mm. Hg during vigorous muscular effort. One of the major contributing factors is the increase in stroke volume that results from skeletal muscle contraction and from heightened respiratory activity, both of which result in increased venous return. Except in the well-trained athlete, this is invariably associated with a marked increase in heart rate (150 to 180 beats per minute); thus there is a very real element of danger in running for a bus, or in carrying out any vigorous bout of exercise if the individual is not in good physical condition.

Method of measuring arterial blood pressure

An important part of any physical examination is a determination of the arterial blood pressure. The instrument used for this purpose is called a *sphygmomanometer*; it consists of a cloth-covered inflatable rubber cuff which communicates with a mercury manometer and a pressure bulb. To measure blood pressure, the cuff is wrapped around the arm above the elbow and fastened snugly in place. By means of the bulb, air is pumped into the cuff, thus putting pressure on the arm. Sufficient pressure is applied to compress

the brachial artery. This shuts off the blood flow to the portion of the upper extremity distal to the cuff.

The rubber cuff has a second outlet; this leads to a glass tube which contains mercury. The mercury rises in the tube as air is pumped into the cuff. Along the side of the tube in which the mercury rises is a scale ruled in millimeters. The distance to which the mercury rises indicates the amount of pressure necessary to stop the flow of blood in the forearm.

When the brachial artery has been compressed completely, the air is slowly released from the cuff by means of a valve on the bulb. As the air is released, a point is reached at which the pressure of the blood in the artery is just equal to that of the air in the cuff. Below this point, the blood pressure exceeds the air pressure in the cuff, and blood begins to flow into the forearm again.

One listens with a stethoscope for sounds over the course of the brachial artery. The bell is placed just below the bend in the elbow. As air is gradually released from the cuff, the pressure falls to a point where a pulse wave and a spurt of blood

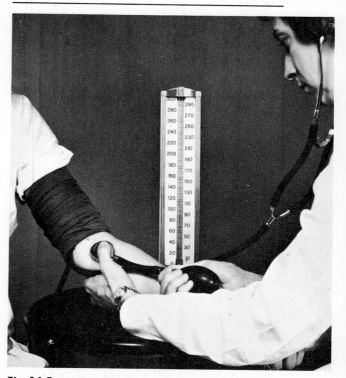

Fig. 14-5. Method of taking blood pressure by auscultation.

can break through the partially compressed artery at the height of ventricular systole. The temporary turbulence in the blood as it passes through this narrow opening results in a distinct tapping sound which can be heard through the stethoscope with each pulse wave. The height of the column of mercury when the sound is first heard corresponds to the *systolic blood pressure*. As additional air is released from the cuff, the tapping sounds first become progressively louder, then suddenly diminish in intensity and develop a rather muffled quality just before disappearing altogether; the reading on the mercury scale at the time of the sudden muffling of sound corresponds to the *diastolic pressure*. Complete disappearance of the sounds indicates that the cuff pressure has fallen below the diastolic pressure, and there is no longer any turbulence in the blood as it flows freely through the decompressed artery. This method of taking the blood pressure is illustrated in Figure 14-5.

Velocity of blood flow

The distance that a particle of blood travels in a given time is called the *velocity*, or *speed*, of the blood flow. It is expressed in millimeters per second. Velocity varies in different parts of the vascular system. It is about 300 mm. per second in the aorta, slows down to 0.5 mm. in the capillaries, and speeds up to 150 mm. per second in the great veins.

Blood flows most rapidly where the cross-section area of the vascular system is least; in the systemic circuit this is the aorta. Every time an artery branches, the cross-section area of its two branches combined is greater than that of the original vessel. There are countless divisions before the capillaries are reached, and, as a result, the combined cross-section area of the systemic capillaries is about 600 times as great as the cross-section area of the aorta. Therefore, the blood flows only 1/600 as fast as in the aorta (1/600 ×

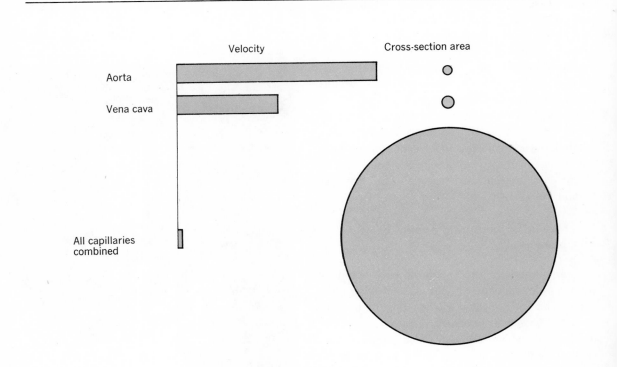

Fig. 14-6. Comparison of velocities and cross-section areas in different parts of the vascular system. In the aorta, where the cross-section area is small, the velocity is most rapid. In the venae cavae, the cross-section area is about twice as great as it is in the aorta, and the velocity is about half as rapid. In the capillaries, the combined cross-section area is about 600 times as great as it is in the aorta, and the velocity is correspondingly slowed.

300 = ½ mm.). The combined cross-section area of the superior and the inferior venae cavae is about twice that of the aorta, and the speed of blood flow is half as fast. The relationship between cross-section area and velocity in different parts of the vascular system is illustrated in Figure 14-6.

Circulation time depends on velocity of blood flow. It is the time required for a particle of blood to start in the right atrium, go through the pulmonary circuit to the left heart, then through a systemic circuit, and back to the right atrium. It takes about 23 seconds for the average round trip (e.g., to the foot and back). This means that it takes about 28 heartbeats to drive a given unit of blood around the two circuits.

Measurement of the circulation time in part of a vascular circuit may be a helpful diagnostic tool. This is accomplished by the injection of a substance into a peripheral vein and notation of the time it takes for that substance to arrive at another point in the circuit. For example, the doctor may inject a bitter tasting substance (decholin) into the patient's median cubital vein and note the number of seconds that pass before the patient experiences a bitter taste in his mouth; average circulation time in the *arm to tongue* circuit is between 10 and 16 seconds. In a similar manner, the *arm to lung* circulation time may be determined by the injection of an aromatic substance such as ether; within four to eight seconds after injection, the patient will smell ether as it diffuses from the bloodstream into the air sacs of his lungs and is exhaled through the nose. You will note that the arm to tongue circulation time is slightly greater than that required for the arm to lung circuit; this is to be expected, since the blood must travel completely through the pulmonary circuit and return to the left heart before it can be pumped to capillaries in the tongue. However, in either case, the circulation time is increased considerably whenever there is any circulatory obstruction; for example, in cases of congestive heart failure, the circulation time may be increased up to three times the normal.

The arterial pulse

The arterial pulse is a wave of distention or expansion and recoil that sweeps over the arterial system with each systole of the left ventricle. The pulse wave grows weaker as it passes over the arterial system and disappears completely in the capillaries. The pulse wave travels rapidly, at a rate of 6 to 9 м. per second over the large arterial branches. It reaches the periphery before the blood that was pumped from the left ventricle at the same time. In other words, we feel the wave of distention in the wall of the radial artery because blood is ejected before that blood gets out of the aorta.

Since the arterial pulse generally is indicative of heart action, its rate and characteristics are noted several times a day on hospitalized patients. Most frequently, the radial artery is used for this purpose, but the pulse may be felt in any artery that lies near the surface of the body and over a bone or other firm tissue. In counting the pulse, place the fingertips over the artery, press lightly and note the waves of distention; the average pulse *rate* is between 70 and 80 waves, or *beats*, per minute. One also should note the pulse *rhythm* (regular or irregular intervals between beats), *volume* (large or small pulse), and *tension* (compressibility of artery—hard or soft pulse).

If the heart rate is very rapid, one is said to have *tachycardia*; on the other hand, a very slow rate is called *bradycardia*. Tracings representing

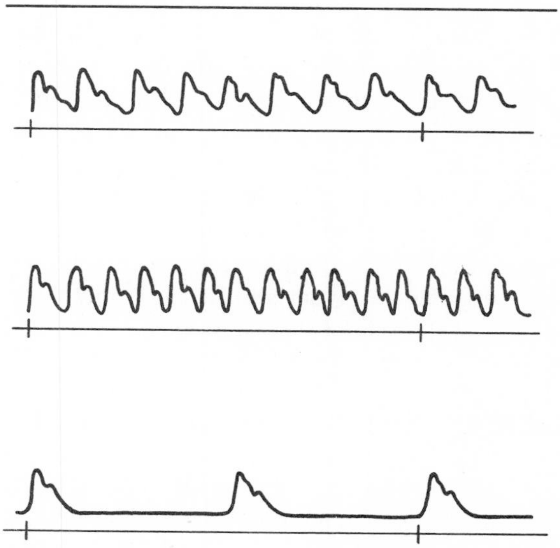

Fig. 14-7. Radial pulse tracings. *Top,* pulse in normal individual; the notch on the descending limb indicates the time of closing of the semilunar valves. *Center,* pulse in hyperthyroidism; the rate is rapid (tachycardia). *Bottom,* pulse in heart block; the rate is very slow.

normal, rapid, and slow pulses are illustrated in Figure 14-7. If the pulse is very weak and rapid with one wave running into the next, it is described as a *thready*, or *running pulse*. In *atrial fibrillation*, the pulse is totally irregular; this means that in addition to irregular time intervals between each two pulses, strong and weak beats are intermingled. Since the ventricles of the heart are contracting at an abnormally rapid rate, they do not always have time to fill with blood; as a result, the pulse rate is slower than the heart rate because some of the systoles do not eject quantities of blood sufficient to distend the arteries and start a pulse wave. In such cases, it is helpful to take an *apical-radial pulse*. To carry out this procedure, one person listens through a stethoscope placed over the cardiac area of the chest and counts the heart rate; at the same time a second person counts the radial pulse at the wrist. If the apical beat is 160 per minute and the radial pulse is only 90 per minute, the patient is said to have a *pulse deficit* of 70. In the normal individual, the rate of the heart counted at the apex is the same as the rate counted in a peripheral artery, such as the radial.

The central venous pulse

The arterial pulse dies out in the capillaries, and there are no pulsations in the peripheral veins. However, in the large veins near the heart, a venous pulse is noted. This is called the *central venous pulse*, and it is due to pressure changes which are reflected from the right atrium. By using suitable apparatus, the central venous pulse may be recorded, and it gives the same type of information as the electrocardiogram in some cardiac disorders. You can see the pulse in the external jugular vein of some individuals.

REGULATION OF THE CIRCULATION

The maintenance of optimal blood pressure and blood distribution throughout the body requires that continual adjustments be made in the rate and the force of the heartbeat and in the internal caliber of the blood vessels. Many of these adjustments can be made by three basic *intrinsic control mechanisms*: (1) the ability of the heart to respond automatically to an increased

input of blood (Starling's law, p. 284); (2) the ability of terminal arterioles and precapillary sphincters to adjust blood flow in response to the needs of neighboring tissues (p. 344); and (3) the ability of the circulation, working with the kidneys, to regulate extracellular fluid volume and blood volume, which in turn influence cardiac output and arterial blood pressure.

In addition to the intrinsic mechanisms, nature provides nervous and humoral regulations that increase the effectiveness of control. The initiation of neural adjustments depends primarily on nerve impulses which are dispatched from cardiac and vasomotor reflex centers within the CNS. These centers are stimulated by various afferent nerve impulses and by chemical substances in the circulating bloodstream.

Cardiac reflex centers

There are two cardiac reflex centers in the medulla oblongata of the brain stem. When stimulated, the *cardio-inhibitory* center dispatches impulses over parasympathetic fibers in the vagus nerves and causes the heart to beat more slowly. Located nearby is the *cardio-accelerator* center which dispatches impulses over sympathetic fibers and causes the heart to increase its rate. A very cooperative relationship exists between these centers, as stimulation of one is associated with depression of the other.

The activities of these centers are greatly dependent on afferent impulses which stream in from all parts of the body, including the heart itself. Thus maintenance of the normal resting heart rate, and changes in the rate under various physiologic conditions are in large measure either reflex in nature, or due to impulses from the cerebral cortex.

Vasomotor reflex centers

Since there is great variation in the amount of blood required in different parts of the vascular system during activity and during rest, it follows that there must be ways of regulating blood pressure and blood flow to meet the changing needs of the body. This is accomplished primarily by the vasomotor center in the medulla oblongata. The lateral portions of this center are *vasoconstrictor areas*, while the medial portion is a *vasodilator area.* The activity of these medullary centers can be modified by still higher centers in the hypothalamus and the cerebral cortex. Thus afferent impulses from all parts of the body can influence arterial blood pressure and flow.

As previously described on page 335, peripheral resistance is determined by the caliber of the blood vessels and is of fundamental importance in the maintenance of arterial blood pressure. For example, if all of the blood vessels in the body were to dilate simultaneously to their maximum capacity, there would be no peripheral resistance and the blood pressure and the cardiac output would fall to zero. Thus it is imperative that most of the blood vessels be at least partially constricted much of the time. This is accomplished through the tonic activity of the sympathetic nervous system. The parasympathetic system is primarily concerned with control of the heart rate through fibers traveling in the vagus nerve. However, there are a few parasympathetic vasodilator fibers that will be described later.

Sympathetic Vasoconstrictor Fibers. Essentially all blood vessels, except the capillaries, are innervated by sympathetic vasoconstrictor fibers. The activity of these fibers in the regulation of peripheral resistance is controlled through the medullary vasomotor center. The vasoconstrictor areas are tonically active, which means that they are constantly dispatching one or two nerve impulses every second; this serves to maintain *vasomotor tone,* a state of partial contraction of smooth muscle in arteriolar walls. When an increase in peripheral resistance is required, the

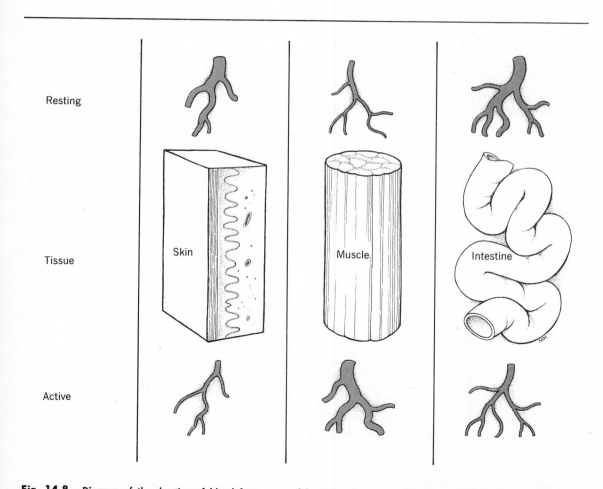

Fig. 14-8. Diagram of the shunting of blood from areas of lesser need to areas of greater need. During periods of rest, the blood supply to the skin and intestine is abundant; the blood supply to skeletal muscle is limited. During muscular activity, less blood flows to the skin and intestine, but the blood supply to the skeletal muscles is abundant.

number of impulses transmitted along these fibers becomes greater, thus increasing the degree of vasoconstriction and raising the arterial blood pressure. On the other hand, when less vasoconstriction is needed, the vasodilator area sends inhibitory impulses to the vasoconstrictor areas, thereby decreasing their rate of firing and allowing the blood vessels to relax. The resulting vasodilatation and decrease in peripheral resistance lowers arterial blood pressure.

Although the sympathetic vasoconstrictor nerves are widely distributed, some tissues have more than others. For example, these fibers are particularly abundant in the skin and in the abdominal viscera, but there is poor distribution in skeletal and cardiac muscle. This helps to explain how blood can be shunted away from the abdominal viscera to the skeletal muscles during a bout of physical exercise (Fig. 14-8).

Constriction of the veins has little effect on the overall peripheral resistance. However, it does have the important effect of decreasing their capacity which means that there will be an increase in the venous return to the heart. As a result, the heart will pump with greater effectiveness (Starling's law), and the cardiac output will be increased. This, too, will raise arterial blood pressure.

Sympathetic Vasodilator Fibers. Peripheral nerves supplying the blood vessels in skeletal muscles carry sympathetic vasodilator fibers as well as constrictor fibers. However, the chemical transmitter substance released at vasodilator fiber endings is *acetylcholine.* This substance acts on the smooth muscle of the blood vessels to cause vasodilatation, in contrast to the vasoconstrictor effect of norepinephrine, which is released at vasoconstrictor fiber endings.

The functional significance of the sympathetic vasodilator system of fibers is uncertain. However, since they originate in the cerebral cortex, it has been suggested that they are activated in response to cerebral activity at the onset of exercise. The resulting vasodilatation in the skeletal muscles would allow an anticipatory increase in blood flow before the muscles actually required increased supplies of oxygen and glucose.

Parasympathetic Vasodilator Fibers. The parasympathetic vasodilator fibers probably do not influence peripheral resistance because of the relatively small numbers of blood vessels that are involved. Distribution of these fibers is limited to cranial and sacral areas such as the brain, salivary glands, throat, and the external genitalia.

There are no known parasympathetic vasoconstrictor fibers.

Cardiovascular pressoreceptor reflexes

Various afferent nerve impulses may stimulate the cardiac and the vasomotor centers, but the most important ones are initiated by specialized receptor end organs in the wall of the aortic arch and in the carotid sinuses. These end organs are called *pressoreceptors,* or *baroreceptors,* since they are highly sensitive to changes in the pressure of the blood against the vessel walls; their afferent fibers travel in the vagus and the glossopharyngeal nerves, respectively (Fig. 14-9).

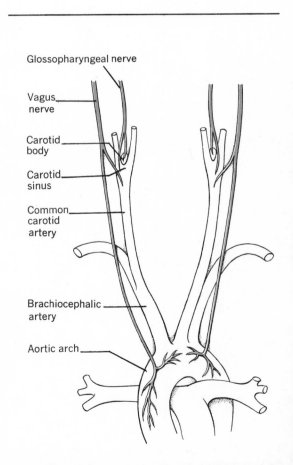

Fig. 14-9. Diagram of the innervation of the arch of the aorta, the carotid sinus, and the carotid body.

These receptor end organs are discharging nerve impulses continuously, but the *rate of discharge* varies directly with the pressure of the blood within the vessel. If the arterial blood pressure rises above normal levels, the pressoreceptors greatly increase their rate of discharge and the medullary centers are bombarded with nerve impulses; this leads to stimulation of the cardio-inhibitory center and inhibition of the accelerator center, thus causing a reflex slowing of the heart rate. At the same time, there is inhibition of the vasoconstrictor center which results in a reflex vasodilatation of peripheral blood vessels. Both of these reflexes serve to lower the arterial blood pressure. On the other hand, if the arterial pressure begins to fall below normal levels, the pressoreceptors are less strongly stimulated and discharge fewer impulses; consequently, the cardio-inhibitory center is less stimulated and the accelerator center is less inhibited, the end-result being a reflex increase in the heart rate. At the same time, the vasoconstrictor center is strongly stimulated and reflex vasoconstriction occurs in the peripheral blood vessels. Both of these reflexes tend to raise the arterial blood pressure.

The pressoreceptors are of fundamental importance in regulating the activity of the heart and the blood vessels and in keeping the blood pressure within normal limits. The effects of changing arterial pressures and the variation in

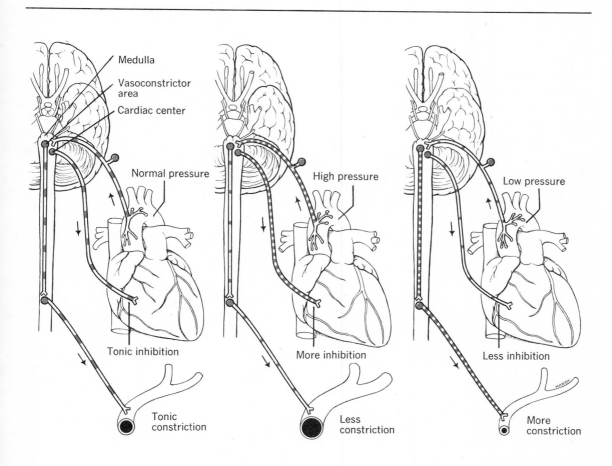

Fig. 14-10. Diagram of effects of changing pressure in arch of aorta. Afferent and efferent impulses travel over fibers in vagus nerve; arrows indicate direction of transmission. Nerve impulses are shown as shaded segments, and the spacing of segments indicates frequency of impulses. With high pressure, the heart rate slows and there is less vasoconstriction; with low pressure, the heart rate increases and there is more vasoconstriction.

number of nerve impulses are shown in Figure 14-10.

Atrial pressoreceptors may be involved in the reflex increase in heart rate that occurs when there is an increase in venous return to the heart. It has been postulated that an increase in right atrial pressure stimulates pressoreceptors in the atrial wall. The pressoreceptors then initiate nerve impulses that stimulate the cardio-accelerator center in the medulla. This is known as the *Bainbridge reflex.* However, physiologists question the actual importance of this reflex since an increase in right atrial pressure has a direct, stretching effect on the atrial wall which, in itself, will cause an increase in heart rate.

Chemical control of circulation

The chemical composition of the circulating blood plays an important part in regulating cardiovascular activities. The concentration of carbon dioxide, oxygen, and certain hormones produced by the endocrine glands is of particular interest in this respect. These chemical agents may exert their effect in one of three ways: (1) stimulation of chemoreceptors in the aortic and the carotid bodies, (2) direct action on the vasomotor centers, or (3) direct action on the arteriolar walls.

Stimulation of Chemoreceptors. Chemoreceptors are highly specialized receptor end organs which are sensitive to variations in the chemical composition of the blood. Clusters of these little nerve endings are found in the arch of the aorta and in the carotid bodies, which are close to the carotid sinuses (Fig. 14-8). These peripheral chemoreceptors are more sensitive to a decrease in blood oxygen than they are to excesses of carbon dioxide and hydrogen ions. Thus whenever the oxygen concentration in the arterial blood becomes too low (hypoxia), the chemoreceptors are excited. Nerve impulses are dispatched to the vasoconstrictor areas in the medullary center, resulting in the constriction of arterioles and venous reservoirs and raising the arterial blood pressure. This reflex is most powerful when there has been a marked decrease in the arterial blood pressure, and it serves to increase the amount of oxygen carried to the tissues. However, under ordinary circumstances, the chemoreceptors are far more important in the control of breathing (p. 373) than in the regulation of blood pressure.

Direct Action on the Vasomotor Centers. A high concentration of carbon dioxide in the blood flowing through capillaries in the vasomotor center results in a marked increase in arterial blood pressure. The exact mechanism is uncertain, but it is believed that carbon dioxide has a direct stimulatory effect on the neurons in the vasoconstrictor areas.

The *CNS ischemic response* is a powerful emergency control measure that is called into action whenever blood flow to the brain is decreased to dangerously low levels. As a result of inadequate blood flow through the medulla, there is an intense stimulation of the vasoconstrictor areas, resulting in a rapid and marked elevation of the arterial blood pressure. This control system serves to restore blood flow and prevent damage to the vital neurons in the brain. It is purely an emergency system and not a routine mechanism for regulating arterial pressure.

Direct Action on Arteriolar Walls. The smooth muscle of terminal arterioles and the precapillary sphincters has a high degree of intrinsic tone that is independent of nerve impulses. However, this state of contraction may be decreased or increased depending on whether more blood or less blood is needed in the capillary networks. In this way each tissue can regulate its own blood flow according to the metabolic needs of the moment. For example, when skeletal muscles are active, they are using oxygen and producing carbon dioxide. Both factors, oxygen depletion and carbon dioxide accumulation, cause relaxation of precapillary sphincters, thus permitting more blood to flow through the active tissues. When the work period ends and the metabolic needs are reduced, the sphincters contract accordingly.

Another example of direct action on arteriolar walls is found in the cerebral circulation. A most striking feature of these vessels is their sensitivity to carbon dioxide. An increase in the level of carbon dioxide in arterial blood provokes marked cerebral vasodilatation and an increase in cerebral blood flow. On the other hand, a decrease in the level of carbon dioxide, such as produced by rapid and deep breathing (hyperventilation), causes vasoconstriction and a decrease in cerebral blood flow. This explains why one may feel dizzy or even lose consciousness after extreme hyperventilation; so much carbon dioxide is blown off that blood flow to the brain is impaired as a result of the cerebral vasoconstriction.

Finally, certain *hormones* produced by the endocrine glands also play a role in the regulation of cardiovascular activities. For example, epinephrine increases the rate and the force of the heartbeat and causes vasoconstriction of arterioles in the skin and the viscera. The endocrine glands and their secretions are described in Chapter 20.

Control by higher centers

Nerve pathways from many parts of the cerebral cortex and the hypothalamus converge on the medullary cardiovascular centers. Consequently, impulses traveling to the brain over *any* afferent nerve fibers ultimately may have some effect on cardiac and vasomotor activity. For example, a frightening sight or sound can cause the heart to beat more rapidly and the skin to turn pale (cardio-acceleration and vasoconstriction). On the other hand, feelings of embarrassment or shame can make one blush (vasodilatation). Persons who are angry, fearful, anxious, too warm, too cold, suffering intense pain, may experience marked changes in cardiovascular activities.

The purpose of all the cardiac and vascular reflexes is to keep the heart rate and the arterial blood pressure as near the normal level as possible and thereby maintain homeostasis. Sufficient examples have been presented to impress on the reader the beauty of the self-regulation in the cardiovascular system. There can be no changes in the heart rate or in blood pressure in any part of the system which do not set into action reflexes to restore conditions to normal so long as one remains "in circulatory health."

Hemorrhage

Hemorrhage refers to an escape of blood from the vessels in which it is normally confined. An *external hemorrhage* is easily seen as the blood flows from the body, but an *internal hemorrhage* is concealed as blood escapes into the surrounding tissues or into a body cavity (e.g., the stomach, the peritoneal cavity). In either case, the body's compensatory mechanisms are quickly activated. If the injury is relatively slight, bleeding may be controlled by local vasoconstriction and the clotting mechanism (p. 268), but in extensive lesions the

hemorrhage may be so great that it overwhelms the clotting process, and large quantities of blood may be lost before the flow can be stopped.

The **signs and symptoms of hemorrhage** are directly related to the amount of blood that escapes from the vascular system. A relatively small hemorrhage (up to 500 ml.) usually is tolerated quite well by most persons. On the other hand, a loss of 30 to 40 per cent of the total blood volume is a serious threat to life and should be treated immediately by transfusions (p. 270). There are three key links in the chain of events that trigger the body's protective negative feedback mechanisms in response to a severe hemorrhage: (1) a great *reduction in blood volume* leads to (2) *decreased venous return*, and (3) *reduced cardiac output*. As the cardiac output falls, so does the arterial blood pressure; this results in reflex acceleration of the heart and vasoconstriction in the skin and the abdominal viscera. The vasoconstrictor mechanism attempts to restore arterial blood pressure by reducing the capacity of the vascular system; it also helps in the redistribution of circulating blood, so that larger amounts go to the CNS, the heart, and the lungs, and correspondingly less to those organs which are not immediately essential to life (i.e., the skin and the abdominal viscera). A temporary restoration of blood volume may be promoted by the discharge of red blood cells from storage depots such as the liver and the spleen; at the same time, decreased filtration pressure in the capillaries leads to a *relative* increase in the colloid osmotic pressure, and great quantities of interstitial fluid are pulled into the bloodstream. The latter results in dehydration of the tissues and accounts for the symptom of thirst that usually follows severe hemorrhage.

If the blood loss continues, the signs and symptoms increase in severity as the compensatory mechanisms are overwhelmed. In addition to pale, cold skin and falling blood pressure, the pulse is rapid and thready, generalized sweating may be observed, and the patient is very restless as nature endeavors to make use of the muscular pump mechanism to increase venous return. Breathing movements become rapid and labored (air hunger), there is loss of consciousness, and the patient will die of circulatory shock if the bleeding is not controlled and if treatment has not been started promptly. One should be familiar with the signs and symptoms of hemorrhage, since early recognition of the condition may save a life.

Circulatory shock

Shock is a form of circulatory failure in which the cardiac output is so greatly reduced that tissues are no longer adequately perfused with blood.

The resulting oxygen deficit interferes with energy conversion at the cellular level and will ultimately cause death if treatment is ineffective.

Causes of Shock. Basically, shock results from an *inadequate cardiac output;* therefore, it may be caused by any factor that can reduce the cardiac output. These factors can be classified in two major categories: (1) those that tend to de-

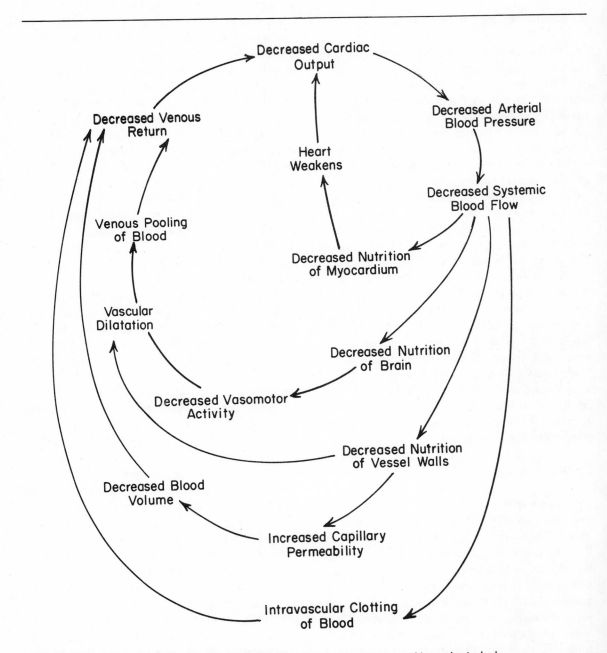

Fig. 14-11. Positive feedback mechanisms that may develop in the progressive stage of hemorrhagic shock.

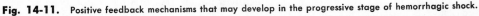

crease the venous return, for example, hemorrhage and (2) those that decrease the ability of the heart to pump blood, for example, myocardial infarction.

Hemorrhagic shock. In the light of our preceding discussion of the body's response to a severe hemorrhage, let us look at the three stages of hemorrhagic shock.

In the *nonprogressive*, or *compensated stage*, the body's protective, negative feedback mechanisms are maximally activated in a matter of seconds or minutes. The increased heart rate, the peripheral vasoconstriction of arterioles and venous reservoirs all help to improve the cardiac output despite the loss of blood. Over a slightly longer period of time, other compensatory mechanisms help to restore blood volume. These include the osmotic movement of fluid from the interstitial tissue spaces into the bloodstream, and the activation of the renin-angiotensin mechanism, due to reduced blood flow to the kidneys (p. 440).

Progressive hemorrhagic shock will occur if the body's compensatory mechanisms are unable to cope with the problem, or if the bleeding continues. In this stage the patient is faced with actual deterioration of his cardiovascular system, due to decreased nutrition of the tissues. As illustrated in Figure 14-11, various types of *positive feedback mechanisms* develop, causing a vicious cycle of progressively decreasing cardiac output. That is, shock breeds more shock. The *intensity* of these mechanisms is of major importance in determining whether or not the shock will respond to treatment, or continue its progression.

Irreversible hemorrhagic shock is the terminal stage. Progression to this point means that it is too late for transfusions or for any other form of treatment, and death is inevitable. The most detrimental factor in all types of shock probably is the lack of oxygen in the tissues, and particularly in the myocardium. It is believed that the primary factor in irreversible shock is deterioration of the heart muscle itself, due to inadequate supplies of oxygen. Thus it has been said that "oxygen deficit not only stops the machine, but wrecks the machinery as well."

SUMMARY

1. **Circulating blood delivers vital substances and removes wastes in all parts of body; integration of all cardiovascular activities of fundamental importance**
2. **Blood pressure and blood flow**
A. Blood pressure is force exerted by blood against vessel walls
   a. Liquids flow only from areas of greater to lesser pressure
   b. Heart establishes initial high head of pressure
   c. Blood pressure measured in millimeters of mercury above atmospheric pressure
B. Arterial pressure
   a. Highest in aorta
   b. Progressive decrease; great drop at arteriolar level
   c. Systolic generated during ventricular contraction; systemic: 115 to 120 mm. Hg; pulmonary: 22 mm. Hg
   d. Diastolic due to recoil of stretched arteries; systemic: 75 to 80 mm. Hg; pulmonary: 8 mm. Hg
   e. Pulse pressure: difference between systolic and diastolic; average 40 mm. Hg; affected by stroke volume and distensibility of vessels
   f. Hypertension: sustained pressure of 140/90 mm. Hg or more
C. Capillary pressure: driving force in formation of interstitial fluid
   a. Systemic: 32 mm. Hg at arterial end; 12 at venous end
   b. Pulmonary: average 7 mm. Hg; colloid osmotic pressure 25 mm. Hg keeps alveoli dry
D. Venous pressure: varies from 12 mm. Hg down to zero at heart
   a. CVP in venae cavae 6 to 12 cm. H_2O
   b. CVP in right atrium 0 to 4 cm. H_2O
   c. Increased in heart failure
E. Gravity and venous return
   a. Muscular pump compresses veins; blood forced onward
   b. Negative intrathoracic pressure; suction effect
   c. Active venoconstriction; valves prevent backflow
   d. Fainting due to lack of oxygen in brain
F. Maintenance of arterial blood pressure
   a. Depends on
      (1) Rate and force of heartbeat
      (2) Elasticity of large arteries
      (3) Peripheral resistance
      (4) Quantity of blood
      (5) Viscosity of blood
   b. Influenced by age, weight, sex, emotions, exercise
G. Velocity of blood flow
   a. Most rapid where cross-sectional area of system is the least: arteries
   b. Slowest where cross-sectional area greatest: capillaries

c. Picks up speed again after entering veins, but slower than in arteries

d. Arm to tongue circulation time 10 to 16 seconds; arm to lung 4 to 8

e. Increased in cases of circulatory obstruction

H. Arterial pulse

   a. Wave of distention and recoil which sweeps over arterial system

   b. Reflects ventricular beat

   c. Disappears completely in capillaries

   d. Radial artery frequently used for checking pulse rate, rhythm, volume, and tension

I. Central venous pulse due to pressure changes reflected from right atrium

3. Regulation of the circulation

A. Intrinsic control mechanisms

   a. Heart response to increased input

   b. Local control of blood flow

   c. Work with kidneys to regulate blood volume

B. Cardiac reflex centers in medulla

   a. Inhibitory and excitatory

   b. Stimulation of one associated with depression of other

C. Vasomotor reflex centers in medulla: vasoconstrictor and vasodilator areas

   a. Activities modified by hypothalamus and cerebral cortex

   b. Regulation of peripheral resistance and distribution of blood

   c. Sympathetic vasoconstrictor fibers to all vessels except capillaries

     (1) Tonic activity maintains vasomotor tone

     (2) Increased discharge to increase vasoconstriction

     (3) Constriction of veins increases venous return

     (4) Most abundant in skin and abdominal viscera

   d. Sympathetic vasodilator fibers to vessels in skeletal muscle

     (1) Liberate acetylcholine at endings

     (2) Anticipatory increase in blood flow

   e. Parasympathetic vasodilator fibers

     (1) No influence on peripheral resistance; few in number

     (2) Supply vessels in cranial and sacral areas

D. Cardiovascular pressoreceptor reflexes

   a. Nerve endings sensitive to pressure changes; aortic arch and carotid sinuses

   b. Rise in arterial pressure leads to reflex slowing of heart and peripheral vasodilatation; lowers pressure

   c. Fall in arterial pressure leads to increased heart rate and peripheral vasoconstriction; raises pressure

   d. Atrial pressoreceptors and Bainbridge reflex; debatable

E. Chemical control of circulation

   a. Stimulation of chemoreceptors in aortic arch and carotid bodies; respond to decrease in blood oxygen; vasoconstriction results

   b. Direct action on vasomotor center

     (1) Carbon dioxide stimulates vasoconstrictor areas

     (2) CNS ischemic response; emergency system to raise pressure

   c. Direct action on arteriolar walls and precapillary sphincters; local control of flow dependent on metabolic activity

   d. Hormone epinephrine speeds heart; peripheral vasoconstriction

F. Control by higher centers

   a. Hypothalamus and cerebral cortex

   b. Any afferent nerve impulse may ultimately affect vasomotor center

   c. Sights, sounds, temperature, emotions can alter cardiovascular activity

G. Hemorrhage

   a. External: flows from body; internal: into a body cavity, concealed

   b. Key links in chain that trigger response to hemorrhage

     (1) Reduction in circulating blood volume

     (2) Decreased venous return

     (3) Reduced cardiac output

   c. Signs and symptoms of hemorrhage: falling blood pressure; pale, cold skin; thirst; rapid, thready pulse; generalized sweating; restlessness; air hunger; loss of consciousness; and, finally, death

H. Circulatory shock

   a. Caused by inadequate cardiac output; oxygen deficit in tissues

   b. Nonprogressive or compensated stage; negative feedback mechanisms

   c. Progressive stage; positive feedback mechanisms

   d. Irreversible stage; deterioration of heart muscle

QUESTIONS FOR REVIEW

1. What is blood pressure? Why is it important?

2. Distinguish between systolic and diastolic pressure.

3. What is the pulse pressure? How is it affected by an increase in stroke volume?

4. Why is hypertension potentially dangerous?

5. Under normal conditions, the alveoli or air sacs of the lungs are free of fluid. Why?

6. Why does a person with decompensated heart failure have a cough and difficulty in breathing?

7. What factors help to overcome gravity and promote venous return to the heart? Why do individuals faint?

8. What are the five major factors that are involved in maintaining arterial blood pressure?

9. What effect do emotions have on the blood pressure? why?

10. Why does the blood pressure rise during vigorous exercise?

11. In what vessels does blood flow most rapidly? most slowly? Why?

12. What is the arterial pulse? Why is it noted several times a day on hospitalized patients?
13. If you were asked to take an apical-radial pulse how would you do it? What would the results tell you?
14. What and where are the cardiac reflex centers? What is the relationship between them?
15. How does stimulation of the medullary vasodilator center result in peripheral vasodilatation?
16. What is vasomotor tone? Why is it important?
17. How does constriction of the veins help to raise arterial blood pressure?
18. How does the action of acetylcholine differ from that of norepinephrine at the vasomotor nerve endings?
19. What is the function of the arterial pressoreceptors?
20. How do oxygen and carbon dioxide in the blood exert their effects in regulating cardiovascular activities?
21. What is the CNS ischemic response?
22. Why does a patient who has suffered a severe hemorrhage have a rapid pulse? pale skin? low blood pressure? a sensation of thirst?
23. What is the basic cause of shock? What are the major factors that can do this?

REFERENCES AND SUPPLEMENTAL READINGS

Beeson, P. B., and McDermott, W.: Cecil-Loeb, Textbook of Medicine, ed. 13. Philadelphia, W. B. Saunders, 1971.

Best, C. H., and Taylor, N. B.: The Physiological Basis of Medical Practice, ed. 8. Baltimore, Williams & Wilkins, 1966.

Betson, C., and Ude, L.: Central venous pressure. AJN, 69:1466, July 1969.

Chapman, C. B., and Mitchell, J. H.: The physiology of exercise. Sci. Am., 212:88, May 1965.

Guyton, A. C.: Textbook of Medical Physiology, ed. 4. Philadelphia, W. B. Saunders, 1971.

Schumer, W., and Sperling, R.: Shock, its effect on the cell. JAMA, 205:75, July 22, 1968.

The Respiratory System

"The breath of life"
Paradise Lost • John Milton

Respiration *is a broad term that is used in reference to the exchange of gases between a living organism and its environment. In aerobic organisms, respiration involves the taking up of oxygen and the giving off of carbon dioxide. A single-celled creature, such as the amoeba, has no great respiratory problem, since the exchange of oxygen and carbon dioxide takes place by simple diffusion through its cell membrane. However, in multicellular organisms as complex as man, provision must be made for the trillions of cells that are far removed from the external environment. In solving this problem, nature has devised an ingenious respiratory plan that involves the combined services of two elaborate systems (respiratory and cardiovascular); this plan is conveniently divided into two phases which are referred to as external and internal respiration.*

External respiration *involves the exchange of gases between the circulating blood and the air. For this exchange, one needs a large, moist surface where air and blood can come in close contact; such an area for diffusion is provided by the lungs. In addition, there must be a passageway through which fresh air can be moved into and stale air expelled from the lungs. The final requirement for external respiration is an air pump (like a bellows) such as the thoracic cage, with muscles to operate it and nerves to control it, as air is constantly renewed in the lungs.*

Internal respiration *involves the exchange of gases between the circulating blood and the various tissue cells as they use oxygen and produce waste carbon dioxide. One may think of the tissue cells as tiny chemical factories in which oxygen makes possible the reactions by which energy is released from the foods that you eat. A discussion of the many aspects of internal respiration is not within the scope of this chapter, which is primarily concerned with the organs and the mechanics of external respiration, as well as with the transport of respiratory gases in the blood.*

ORGANS OF THE RESPIRATORY SYSTEM

The organs which are concerned with external respiration are the nose, pharynx, larynx, trachea, bronchi, and the lungs, in addition to the thoracic cage and the respiratory muscles and nerves.

The nose

The nose consists of an external and an internal portion. The *external portion* of the nose protrudes from the face and is highly variable in shape. The upper part of this triangular structure is held in a fixed position by the supporting nasal bones which form the bridge of the nose, but the lower portion is movable because of its pliable framework of fibrous tissue, cartilage, and skin. Numerous sebaceous glands are found in the skin that covers the nose; obstruction of these glands may be followed by an inflammatory process and the formation of pimples. Squeezing pimples in the region of the nose is a dangerous procedure as previously described on page 310 (the dangerous triangle).

The *internal portion* of the nose lies within the skull, between the base of the cranium and the roof of the mouth, in front of the nasopharynx. The skull bones that enter into the formation of the nose include the frontal, the sphenoid, the ethmoid, the nasal, the maxillary, the lacrimal, the vomer, the palatine, and the inferior conchae; these bones are described in Chapter 3.

The Nasal Septum. The narrow partition that divides the nose into right and left nasal cavities is called the nasal septum. As shown in Figure 3-17, the anterior portion of this septum is composed primarily of cartilage and skin, while the middle and the posterior portions consist of bone (chiefly the vomer and the perpendicular plate of the ethmoid). In some individuals the nasal septum is markedly deflected to one side so that the affected nasal cavity is almost completely obstructed; this condition, called *deviated nasal septum*, can be corrected by surgery.

The Nasal Cavities. These cavities open to the outside by means of the anterior *nares,* or *nostrils;* posteriorly, they open into the nasopharynx by means of the posterior nares, or *choanae.* The *vestibule* of each cavity is the dilated portion just inside the nostril. Anteriorly, the

vestibule is lined with skin and presents a ring of coarse hairs which serve to trap dust particles. Posteriorly, the lining of the vestibule changes from skin to a highly vascular ciliated mucous membrane, called the nasal mucosa, which lines the rest of the nasal cavity. The total surface area of the nasal mucosa is greatly increased by the fact that it covers three plates of bone which project from the lateral wall of each nasal cavity. Arranged one above the other like sagging shelves, these bony projections are the *superior,* the *middle,* and the *inferior conchae,* or *turbinates* (Fig. 15-1).

The surface of the nasal mucosa is covered with mucus which is secreted continuously by goblet cells in the epithelial layer. As air passes over this wet vascular membrane, it is warmed or cooled (according to the outside temperature), moisture is added, and dust is collected by the mucus; the ciliated cells sweep particles of dust backward into the throat where they may be swallowed and then eliminated through the gastrointestinal tract. The **olfactory region** of the nasal mucosa is confined to the superior conchae and adjacent parts of the septum. The special sense of smell is described on page 235.

In summary, the nose serves not only as a passageway for air going to and from the lungs, but also as an *air-conditioner* and as the sense organ for smell. The importance of breathing through the nose is obvious as it serves to moisten and filter and to warm or cool the air that is on its way to the lungs.

The Paranasal Sinuses. Air spaces in certain bones of the skull are called paranasal sinuses. As described in Chapter 3, and illustrated in Figures 3-14 and 3-15, these spaces take the names of the bones in which they lie; thus they are called the maxillary, the frontal, the ethmoidal, and the sphenoidal sinuses. Each sinus cavity communicates with the nasal cavity on its own side, and each is lined with ciliated mucous membrane that is continuous with the nasal mucosa. The openings of the paranasal sinuses are shown in Figure 15-2.

The Nasolacrimal Ducts. As described on page 238, the nasolacrimal ducts open into the nasal cavity just below the inferior concha on each side (Figs. 10-5 and 15-2); the mucous membrane lining is continuous with that of the nasal mucosa. Tears, which are constantly being secreted by the

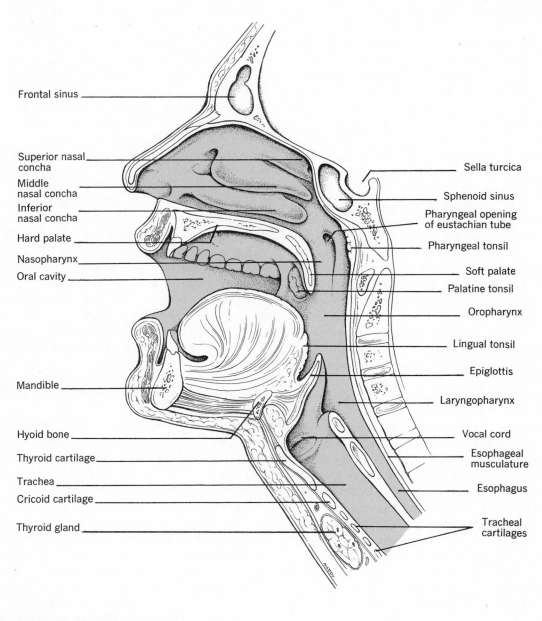

Fig. 15-1. Sagittal section of head.

lacrimal glands to cleanse the surface of the eye, drain into the nose and serve as an added source of moisture to humidify the air.

Rhinitis. A common cold or rhinitis is an inflammation of the mucous membrane of the nose. It is characterized by an acute congestion of the mucous membrane and increased secretion. It is difficult to breathe through the nose because of the swelling of the mucous membrane and accumulated secretions.

The pharynx

The pharynx or throat is a tubular passageway that is attached to the base of the skull and extends downward to the esophagus. Its walls are

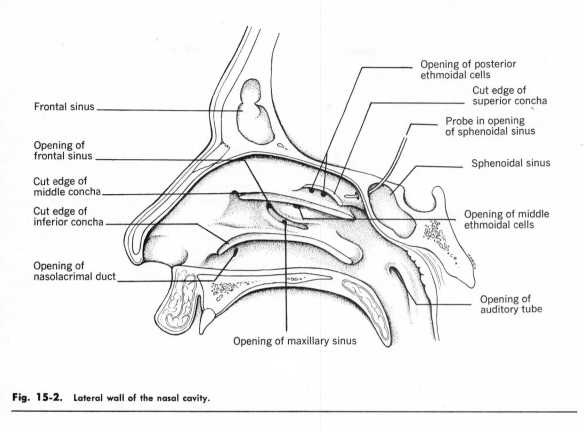

Frontal sinus

Opening of
frontal sinus

Cut edge of
middle concha

Cut edge of
inferior concha

Opening of
nasolacrimal duct

Opening of posterior
ethmoidal cells

Cut edge of
superior concha

Probe in opening
of sphenoidal sinus

Sphenoidal sinus

Opening of middle
ethmoidal cells

Opening of
auditory tube

Opening of maxillary sinus

Fig. 15-2. Lateral wall of the nasal cavity.

composed of skeletal muscle, and the lining consists of mucous membrane. The pharynx is subdivided from above downward into three parts: *nasal*, *oral*, and *laryngeal*.

The Nasopharynx. The superior portion of the pharyngeal cavity is called the nasopharynx; it lies behind the nasal cavities and above the level of the soft palate. This portion of the pharynx is purely respiratory in function, and its ciliated mucosal lining is continuous with that of the nasal cavities. During the act of swallowing, food is prevented from entering the nasopharynx by the action of various muscles which elevate the soft palate and block off the nasal area.

There are four openings in the walls of the nasopharynx. Anteriorly, it communicates with the nasal cavities by way of the two *posterior nares*. In the lateral walls of the nasopharynx are the openings of the two *auditory*, or *eustachian tubes* (Fig. 15-1), which communicate with the middle ear. The mucous membrane lining the auditory tube is continuous with that of the nasopharynx;

thus it is possible for an infection of the throat to spread along this tube to the middle ear (otitis media). Such an extension of infection is relatively common in children, since their auditory tubes are shorter, wider, and straighter than those of the adult.

The *pharyngeal tonsil* is a mass of lymphatic tissue that is located on the posterior wall of the nasopharynx (Fig. 15-1). During childhood this mass is frequently enlarged, at which time it is called *adenoids* (meaning *glandlike*). This enlargement of the pharyngeal tonsil may obstruct the upper air passage to such an extent that the child must breathe through his mouth. Mouth-breathing is undesirable because the air is not properly warmed, moistened, and filtered before reaching the lungs.

The Oropharynx. The middle portion of the pharyngeal cavity is called the oropharynx; lying behind the mouth, it extends from the soft palate to the level of the hyoid bone (Fig. 15-1). The only opening in the wall of the oropharynx is the *fauces*, or archway, into the mouth. The fauces is bounded by the soft palate above, the dorsum of the tongue below, and the *glossopalatine arches* (*anterior pillars*) laterally. Each glossopalatine arch extends downward from the palate to the base of the tongue on its own side; just behind these arches lie the *pharyngopalatine arches* (*posterior pillars*) which extend downward from the palate to the sides of the pharynx.

The *palatine tonsils* are prominent oval masses of lymphatic tissue situated between the glossopalatine and the pharyngopalatine arches on either side of the oropharynx. These tonsils are relatively larger in the child than in the adult, but their shape and size show considerable individual variation. Inflammation of the palatine tonsils is called *tonsillitis*, and the removal of tonsils is called a tonsillectomy. Acute inflammation of the tonsil and peritonsillar tissue with pus formation is referred to as *peritonsillar abscess*.

Underlying the mucous membrane at the root of the tongue are collections of lymphatic tissue which compose the *lingual tonsil*. In addition to the formation of lymphocytes, the various masses of tonsillar tissue form a protective circular band of lymphatic tissue (Waldeyer's ring) which guards the opening into the respiratory and digestive tubes. This tonsillar ring consists of the two palatine tonsils laterally, the lingual tonsil below, and the pharyngeal tonsil above.

The Laryngopharynx. The inferior portion of the pharyngeal cavity is called the laryngopharynx. As it extends downward from the level of the hyoid bone, the laryngopharynx opens into the larynx anteriorly and continues as the esophagus posteriorly (Fig. 15-1). In summary, the nasopharynx is purely respiratory in function, but the oropharynx and the laryngopharynx are common to both the digestive and the respiratory systems, since they serve as passageways for both food and air.

The larynx

The larynx, or voice box, serves as a passageway for air between the pharynx and the trachea. It lies in the midline of the neck, below the hyoid bone and in front of the laryngopharynx. The unique structure of the larynx enables it to function somewhat like a valve on guard duty at the entrance to the windpipe, controlling air flow and preventing anything but air from entering the lower passages. Exhalation of air through the larynx is controlled by voluntary muscles and so enables the larynx to become the organ of voice. During the act of swallowing, the larynx is pulled upward and forward in such a way that the air passageway is closed off, and food is shunted into the esophagus. Should food or other foreign matter enter the larynx, a cough reflex is set up in an attempt to expel it; a common expression for such a mishap is that one has "swallowed down the wrong throat."

The larynx is a triangular box composed of nine cartilages that are joined together by ligaments and controlled by skeletal muscles. The larynx is lined with ciliated mucous membrane (except the vocal folds), and the cilia move particles upward to the pharynx.

Cartilages. There are three single cartilages in the larynx (cricoid, thyroid, and epiglottis), and three paired cartilages (arytenoid, corniculate, and cuneiform). The most prominent of these are illustrated in Figures 15-1 and 15-3.

The *thyroid* cartilage, shaped like a shield, is the largest in the larynx. It is composed of two broad plates, a right and a left, which meet and fuse in the midline in front like the bow of a ship. The plates form a more acute angle with each other in the male than in the female, and the ventral edges form the subcutaneous laryngeal prominence known as the *Adam's apple*.

The *epiglottis* is thin and leaf-shaped and extends above the thyroid cartilage in front of the entrance to the larynx. The aryepiglottic folds of mucous membrane are attached to its sides. As the larynx moves upward and forward during the act of swallowing, the free edge of the epiglottis is moved downward, thus helping to close the opening into the larynx.

The *cricoid* cartilage is shaped like a signet ring with the band in front and the signet at the back of the larynx. It lies at the lower end of the larynx and is connected with the first cartilaginous ring of the trachea.

The *arytenoid* cartilages, which are above the cricoid cartilage, furnish attachment for the vocal ligaments. The position and the tension of these ligaments are altered by changes in position of the arytenoid cartilages. The apex of each of the arytenoids is covered by one of the *corniculate* cartilages.

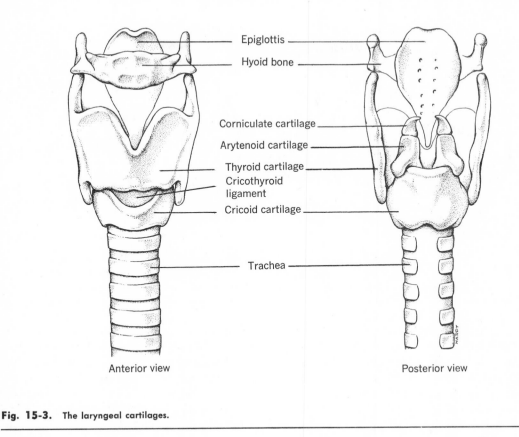

Epiglottis

Hyoid bone

Corniculate cartilage

Arytenoid cartilage

Thyroid cartilage

Cricothyroid ligament

Cricoid cartilage

Trachea

Anterior view

Posterior view

Fig. 15-3. The laryngeal cartilages.

The *cuneiform* cartilages are small, rodlike bodies found in the aryepiglottic folds, just anterior to the *corniculate* cartilages.

Ligaments.   There are extrinsic and intrinsic ligaments in the larynx. The extrinsic ligaments connect the thyroid cartilage and the epiglottis with the hyoid bone and the cricoid with the trachea. The intrinsic ligaments connect the cartilages of the larynx with each other. For example, the thyroid and cricoid cartilages are connected anteriorly by the *cricothyroid ligament*. This ligament is of clinical importance, since it may be pierced for emergency tracheotomy with little fear of bleeding. However, because of the proximity of the vocal cords, it is not used for prolonged intubation (tracheostomy, p. 357) since scar formation may interfere with the mobility of the cords.

Cavity of the Larynx.   The cavity of the larynx extends from the entrance to the lower border of the cricoid cartilage where it is continuous with that of the trachea. Two pairs of

mucous membrane folds project like shelves into the laryngeal cavity and narrow it to an anteroposterior slit in two places. The upper paired folds are called the *false vocal cords*, indicating that they do not function in the production of sound. The lower paired folds are called the *true vocal cords;* each of these mucosal folds encloses a strong band of connective tissue (the vocal ligament), and each plays an important part in the production of sound (Fig. 15-4). Near the free edge of each vocal cord the mucous membrane is designed for rapid vibration and repeated contact; in contrast with the remainder of the air passageways, which has ciliated mucosa, the outer layer of the vocal mucosa consists of stratified squamous nonkeratinizing epithelium and contains no mucous glands or blood vessels.

The vocal cords form the *glottis*, and the opening between them is called the *rima glottidis* (meaning *slit of the glottis*); this is the narrowest part of the laryngeal cavity, and any obstruction

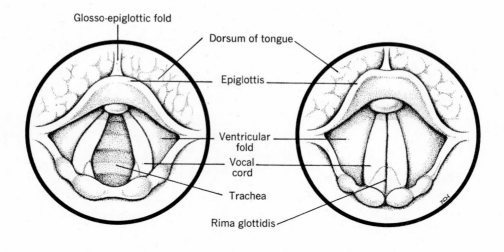

Fig. 15-4.   Interior of the larynx, as seen with laryngoscope: *left,* rima glottidis wide open; *right,* rima closed.

(foreign body, edema) can easily lead to death by suffocation if not relieved promptly. The tension of the vocal cords and the size of the slit between them are controlled by muscles which act on the cords directly, and by other muscles that affect them indirectly by moving the arytenoid cartilages to which they are anchored.

The vagus nerves contain afferent fibers from the mucous membrane of the larynx, the latter being so sensitive that any irritation initiates a cough reflex. The vagus nerves also supply efferent fibers to the skeletal muscles of the larynx; the *recurrent laryngeal nerve* is a vagal branch of particular importance in this respect, since it supplies all but one of the laryngeal muscles.

A *laryngospasm* is a sudden, violent, involuntary contraction of the muscles of the larynx. It is seen most commonly in young children as croup and is characterized by a sudden crowing inspiration followed by a stoppage of respiration for several seconds, with cyanosis, then ending with long, loud, whistling inspirations.

Laryngoscopy refers to a visual examination of the cavity of the larynx with a tubular instrument called the laryngoscope. This procedure is helpful in the diagnosis of various laryngeal diseases.

Speech.   As expired air passes through the larynx, it can be made to set the vocal cords into vibration, and this, in turn, sets the whole column of air above the vocal cords into vibration. The pharynx, the mouth, the nasal cavities, and the paranasal sinuses are resonating cavities. The muscles of the neck, the face, the tongue, and the lips are used to modify the sounds into words. The speech center in the cerebral cortex integrates the activity of all the muscles concerned so that intelligible sounds or words may be spoken. The loudness of the voice depends on the volume and the force of the expired air and on the amplitude of vibrations of the vocal cords. The pitch of the voice is determined by the number of vibrations per second and the length and the tension of the vocal cords. Since women and children usually have shorter vocal cords than do men, their voices tend to be higher pitched. At puberty, the larynx in males increases in size, the cavity becomes larger, and the vocal cords become longer and thicker. As a result, there is a marked change in the quality and the pitch of the voice.

Laryngectomy, or surgical removal of the larynx, is necessary in cases of carcinoma of the larynx. Such patients are unable to speak in the usual way and may learn esophageal speech. This type of speech is accomplished by swallowing air into the esophagus and expelling it with an eructation or belch; the sound that is produced is modified by the palate, the tongue, and the lips.

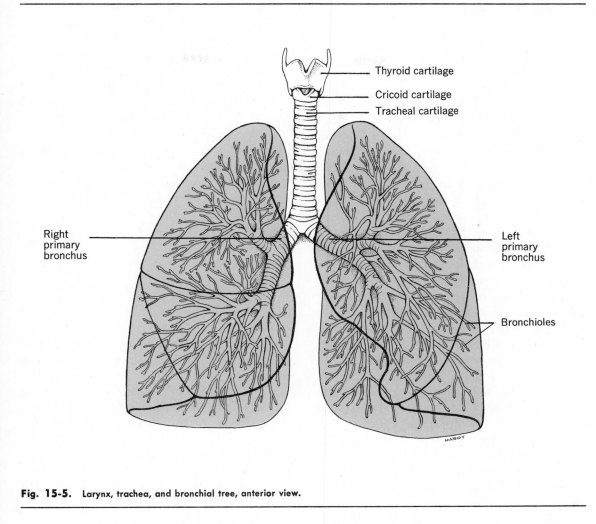

Thyroid cartilage
Cricoid cartilage
Tracheal cartilage

Right
primary
bronchus

Left
primary
bronchus

Bronchioles

Fig. 15-5. Larynx, trachea, and bronchial tree, anterior view.

The trachea

The trachea, or windpipe, is a flexible, tubular structure that extends from the larynx downward through the midline of the neck and into the thorax. It is about 11 cm. in length and lies in front of the esophagus. Despite the flexibility of the trachea, its walls are very strong, due to the presence of 16 to 20 U-shaped tracheal cartilages; these cartilages are placed horizontally, one above the other, at very close intervals (Fig. 15-5). The open end of the U is at the back, and the gap is bridged by connective tissue and smooth muscle. This arrangement permits expansion of the esophagus during swallowing of a *bolus* (meaning *a morsel*) of food, since it can encroach on the posterior wall of the trachea. The cartilage rings

keep the trachea open at all times for the passage of air to and from the lungs. The trachea is lined with ciliated mucous membrane containing goblet cells. The dust particles stick to the mucus, and the cilia move them upward to the pharynx.

Tracheostomy is a surgical procedure whereby an opening is made in the trachea to provide a free passageway for air. This may be a life-saving operation in cases of respiratory obstruction. For example, certain inflammatory conditions related to the larynx may produce a severe edema of the glottis; this condition can easily lead to death by suffocation if an alternate airway is not established below the obstruction. After this operation, the patient requires special care, for the tube which has been inserted into the trachea must be kept free of mucus. Patients with tracheostomy tubes are particularly susceptible to pulmonary infections, since the air is not conditioned before it reaches the lungs.

The bronchial tree

The trachea ends by dividing into right and left **primary bronchi** which extend to the lungs. Each bronchus enters the lung of its own side through a slit, called the *hilus*. The right primary bronchus is more nearly vertical, shorter and wider than the left. Consequently, when foreign bodies enter the respiratory passages, they are likely to be found in the right primary bronchus. As soon as each primary bronchus enters its respective lung it branches to form smaller or *secondary bronchi*, one for each lobe of the lung (i.e., three on the right and two on the left). The secondary bronchi continue to branch, forming still smaller tubes or *bronchioles* as shown in Figure 15-5.

Structurally, the primary bronchi are similar to the trachea in that their walls are strengthened by the same type of cartilaginous rings and their lining is ciliated mucous membrane. However, progressive structural changes accompany the continued subdivision of the bronchi. As they become smaller, less and less cartilage is present in their walls, and more and more smooth muscle appears. The smooth muscle completely encircles the lumen; it is arranged in right and left spirals. If you take two narrow ribbons and wind them around a pencil, winding one in a clockwise direction and the other in a counterclockwise direction, you will duplicate the spiral arrangements of smooth muscle in the small bronchi. The contraction of this smooth muscle narrows the lumen of the tube and throws the mucous membrane into longitudinal folds. The mucous membrane is ciliated and contains goblet cells; the mucus and the cilia help to keep the bronchi free of foreign particles.

Parasympathetic fibers which travel in the vagus nerves, cause contraction of the smooth muscle in the walls of the bronchioles. Fibers from the sympathetic division cause relaxation of the smooth muscle and, therefore, dilate the lumens of the bronchioles.

Bronchitis is an infection of the mucous membrane of the bronchi, in which fluid collects in the passages and is expelled by coughing.

Bronchiectasis is the condition in which there is pathologic dilatation of a bronchus. It is characterized by spasms of coughing and the expectoration of mucopurulent material.

Bronchoscopy is the visual examination of the interior of the primary bronchi.

Asthma is characterized by attacks of difficult breathing, coughing, wheezing, raising mucoid sputum, and a sense of constriction of the chest. There is spasm of the bronchial muscles and edema of the mucous membrane. It is usually due to sensitivity to inhaled or ingested substances. The attacks are relieved by the use of drugs, such as epinephrine, which relax the constricted muscle and cause dilatation of the bronchioles.

The lungs

The lungs are the essential organs of respiration, since it is here that the exchange of gases between the blood and air takes place. They lie within the thoracic cavity, one on either side of the heart and the other contents of the mediastinum.

The *thoracic cavity* is the space above the diaphragm, within the walls of the thorax. It is closed, below, by the diaphragm; above, by the scalene muscles and the fascia of the neck; and, on the sides, the front, and the back, by the ribs, the intercostal muscles, the vertebrae, the sternum, and the ligaments. The size of the thoracic cavity changes constantly during the act of breathing.

The *mediastinum* is the space between the lungs. The contents of this cavity include the heart and pericardium; the thoracic aorta; the pulmonary artery and veins; the venae cavae and azygos vein; the thymus, lymph nodes, and vessels; the trachea, the esophagus, and the thoracic duct; and the vagus, cardiac, and phrenic nerves.

Description.   The lungs are cone-shaped organs which fill their own halves of the thoracic cavity and lie against the rib cage both anteriorly and posteriorly. The right lung is shorter and broader than the left. The concave *base* of each lung rests on the diaphragm, and the *apex* extends into the root of the neck for a distance of about 2.5 cm. above the first rib. The medial surface of each lung is also concave and shows the imprint of the heart. Above and behind the cardiac imprint is the hilus, or slit, through which pass those structures that make up the *root* of the lung; the root

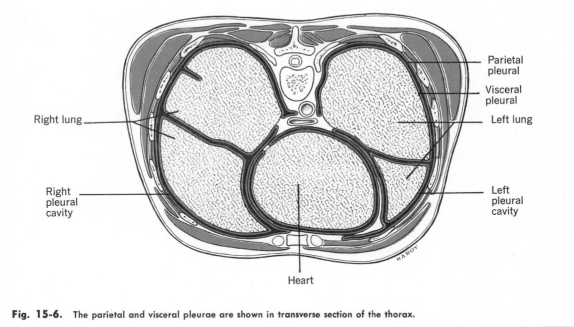

Right lung

Right pleural cavity

Parietal pleural

Visceral pleural

Left lung

Left pleural cavity

Heart

Fig. 15-6. The parietal and visceral pleurae are shown in transverse section of the thorax.

structures include the primary bronchus, the pulmonary and the bronchial blood vessels, the lymphatics, and the nerves. The lungs are freely movable except at the hilus where they are anchored by the root and the pulmonary ligaments.

The Pleura. Each lung is invested by a transparent, serous membrane called the *pleura*. This membrane is arranged in the form of a closed sac and consists of two layers—a visceral layer and a parietal layer. The *visceral pleura* is closely applied to the surface of the lung and dips into the fissures between its lobes. The *parietal pleura* lines the inner surface of the chest wall, covers the upper surface of the diaphragm, and is reflected over the structures of the mediastinum (Fig. 15-6). The two layers of pleura are continuous with one another around the root of each lung.

Although the space between the visceral and the parietal pleurae is called the *pleural cavity*, it must be emphasized that this is purely a *potential* space. Normally, the pleural layers are held in close contact with one another by the cohesive effect of a thin film of serous fluid; this fluid also serves to lubricate the surfaces so that they can glide smoothly one over the other. To visualize the effect of the pleural fluid, place a thin film of

water between two glass slides; note how the wet surfaces cling together, yet permit a slippery side-to-side movement.

Pleurisy is an inflammation of the pleura; it may manifest itself as *dry* pleurisy, or as pleurisy with *effusion*. In dry pleurisy, the pain that accompanies every breath is due to friction between the two layers of pleura which are unable to glide smoothly over one another. In pleurisy with effusion there is a marked increase in the amount of fluid in the pleural cavity; if the fluid is purulent, the person is said to have *empyema*.

Structure. The lungs are light, porous, and spongy in texture, and they will float when placed in water. Within the spongy tissue are the secondary bronchi and the bronchioles which pipe air to and from the respiratory units of the lung. An important feature of lung tissue is its great elasticity; this property is of fundamental importance during the act of breathing.

Deep fissures divide each lung into *lobes*. The right lung has three lobes (superior, middle, and inferior), while the left lung has only two lobes (superior and inferior). The individual lobes are further subdivided into small *lobules*.

Lobules are irregular in shape and size, but

each is supplied with a bronchiole. As a bronchiole enters a lobule it branches repeatedly to form *terminal bronchioles*. Each terminal bronchiole, in turn, branches into two or more *respiratory bronchioles*, which give rise to *alveolar ducts* and sacs. Alveolar sacs are clusters of cup-shaped, thin-walled *alveoli* (Fig. 15-7). Each alveolus is surrounded by a network of capillaries and supported by elastic and reticular fibers. It has been estimated that there are about 300 million alveoli in the lungs, and the surface that they present for the diffusion of gases is about 70 square meters.

Special cells in the alveolar walls secrete *pulmonary surfactant*, a substance that coats the alveolar lining and serves to lower surface tension between the moist surfaces. As you know, it is surface tension that causes any two wet surfaces to adhere to one another; thus surfactant makes it possible for the alveoli to expand with relative ease each time you inhale. *Atelectasis*, or incomplete expansion of the lungs, is responsible for the death of many newborn infants, particularly those born prematurely. The collapse of the lungs and the filling of the air spaces with fluid is called the *respiratory distress syndrome*, or *hyaline membrane disease*. It is believed that this disease is due to inadequate quantities of surfactant in the lungs.

The Respiratory Membrane. Air in the alveoli is separated from blood in the pulmonary capillaries by a number of structures that are referred to collectively as the respiratory membrane. This membrane consists of a layer of surfactant, an epithelial cell in the alveolar wall, a tiny interstitial space, the capillary basement membrane, and finally, an endothelial cell in the capillary wall. Although this seems to be a large number of layers, the overall thickness of the respiratory membrane is less than 1 micron (1/1000 mm.), and the exchange of gases normally occurs very rapidly. However, if for any reason the thickness of the membrane is increased (e.g., pulmonary edema), there will be a corresponding decrease in the rate of gas exchange.

Blood Supply. There are two separate arterial blood supplies for the lungs. The *pulmonary arteries* and their branches give rise to pulmonary capillaries in the walls of alveoli, through which exchanges of respiratory gases take place. The *bronchial arteries* and their branches supply the walls of the bronchi, the walls of blood vessels,

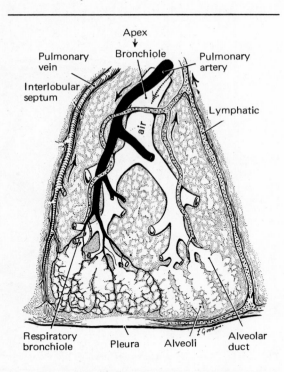

Fig. 15-7. Schematic drawing of a lobule of the lung. Only one set of blood vessels and lymphatics is shown.

the lymph nodes, and the pleurae with blood. They are the nutrient arteries for the lung tissue and play no part in oxygenation of the blood.

Lymphatic vessels are present in the visceral pleura and in the dense connective tissue around bronchioles, bronchi, arteries, and veins.

Disorders of the Lungs. *Pulmonary emphysema* is a serious lung disorder, characterized by the obstruction of bronchioles, the trapping of air in passages distal to the obstruction, disruption of those passages, and a loss of pulmonary elasticity. The end-result is a state of chronic hyperinflation, as the victim is unable to empty his lungs on forced expiration; in severe cases the chest is enlarged in all dimensions (*barrel chest*). Pulmonary emphysema is frequently associated with chronic bronchitis, asthma, and other conditions that may cause inflammatory damage to the bronchioles; it is a chronic, slowly progressive disease, causing irreversible structural damage to the lung.

Pneumonia is the condition in which there is inflammation of lung tissue or of the walls of the bronchi. The former is called *lobar pneumonia* and the latter *bronchopneumonia*. There is also a virus pneumonia which is similar to influenza. When fluid collects in the alveoli, causing the lung to become solid and firm, the condition is called *consolidation*.

Tuberculosis is an infectious disease characterized by the formation of tubercles. Pulmonary tuberculosis is treated by special drugs and by giving the infected lung as much rest as possible. In mild cases this may be accomplished by keeping the patient in bed. In advanced cases, the lung may be temporarily or permanently collapsed by various surgical procedures.

THE MECHANICS OF BREATHING

Breathing is the act of taking fresh air into, and expelling stale air from, the lungs. As noted previously, the external phase of respiration requires not only a large moist surface where blood and air can come in close contact, but also a bellows arrangement such as the lungs and the thoracic cage, with muscles to operate it and nerves to control it. Consequently, it is essential that one understand the mechanical and physical principles involved in the act of breathing.

Air pressure and flow

The basic principle underlying the movement of any gas is that it travels from an area of higher pressure to an area of lower pressure, or from a point of greater concentration to a point of lower concentration. This principle applies not only to the flow of air into and out of the lungs, but also to the diffusion of oxygen and carbon dioxide through alveolar and capillary membranes. The respiratory muscles and the elastic lungs make possible the necessary changes in the pressure gradient so that air first flows into and then is expelled from the air passages. However, before studying the phases of a respiratory cycle (inspiration and expiration) one should be familiar with the following pressures: atmospheric, intrapulmonic, and intrapleural.

Atmospheric Pressure. The pressure exerted against all parts of the body by the surrounding air is called atmospheric pressure. It averages 760 mm. Hg at sea level. Any pressure that falls below atmospheric pressure is called a *subatmos-*

pheric, or *negative pressure*, and represents a partial vacuum.

Intrapulmonic Pressure. The pressure of air within the bronchial tree and the alveoli is called intrapulmonic pressure. During each respiratory cycle, this pressure fluctuates below and above atmospheric pressure as air moves into and out of the lungs.

Intrapleural, or Intrathoracic Pressure. The pressure that exists between the two layers of pleura (i.e., in the pleural cavity) is called the intrapleural pressure. During normal breathing the intrapleural pressure is always subatmospheric, or *negative*. This is due to the fact that the lungs are always tending to collapse, due to their elasticity, and this elastic recoil is continuously exerting a pull on the thoracic walls. The intrapleural pressure undergoes changes during each respiratory cycle (e.g., 751 to 756 mm. Hg), but it is still negative. It becomes positive only when one is coughing, straining at stool, and so on.

Inspiration

Inspiration is the act of taking air into the lungs. It involves a series of four events.

1. The diaphragm and the external intercostal muscles contract when impulses from the central nervous system (CNS) come to them by way of the phrenic and the intercostal nerves, respectively.

2. As the diaphragm contracts, its dome moves downward and enlarges the thoracic cavity from top to bottom. The contraction of the external intercostal muscles raises the ribs and, at the same time, rotates them slightly, thereby pushing the sternum forward. This enlarges the thoracic cavity from side to side and from front to back.

3. The enlargement of the thoracic cavity in its three dimensions is accompanied by a decrease in the intrapleural and the intrapulmonic pressures, which may be explained as follows. As the thoracic cavity enlarges, the parietal pleura tends to pull away from the visceral pleura, thus lowering the intrapleural pressure (i.e., it becomes more negative). However, as noted previously, the cohesion between the two pleural layers, (due to the thin film of pleural fluid), is very great, and separation normally does not occur. Instead, the visceral pleura, and the elastic lung to which it is

attached, expand with the enlarging thoracic cavity. As the lung increases in size, the intrapulmonic pressure immediately falls below atmospheric pressure. In other words, if one increases the volume of a gas container, there will be a decrease in the pressure of the gas within that container (Boyle's law).

4. Since the lungs communicate freely with the outside air by way of the trachea and the pharynx, air will flow into the lungs until the intrapulmonic pressure is equal to that of the atmosphere.

This series of events is necessary to complete the inspiratory phase of the respiratory cycle.

Expiration

Expiration is the act of expelling air from the lungs. It also involves a series of four events.

1. The diaphragm and the external intercostal muscles relax, due to a cessation of impulses from the CNS.

2. As these muscles return to their resting position, and the elastic lungs recoil, the thoracic cavity is decreased in its three diameters. This action is entirely passive during normal quiet expiration.

3. As the thoracic cavity and the lungs decrease in size, there is a corresponding increase in the intrapleural pressure (i.e., it becomes less negative) and in the intrapulmonic pressure which quickly rises above atmospheric pressure. In other words, if one decreases the volume of a gas container, there will be an increase in the pressure of the gas within that container.

4. Since the lungs communicate freely with the outside, air flows out until the intrapulmonic pressure again is equal to atmospheric pressure.

This series of events is necessary to complete the expiratory phase of the respiratory cycle (Fig. 15-8).

Summary of pressure changes

Intrapulmonic pressure is below atmospheric pressure during inspiration, equal to atmospheric pressure at the end of inspiration, above atmospheric pressure during expiration, and again equal to atmospheric pressure at the end of expiration.

This series of changes in intrapulmonic pressure is repeated with each respiratory cycle. Whenever the size of the thoracic cavity remains constant for a few seconds, or in a position of rest, the intrapulmonic pressure is equal to atmospheric pressure.

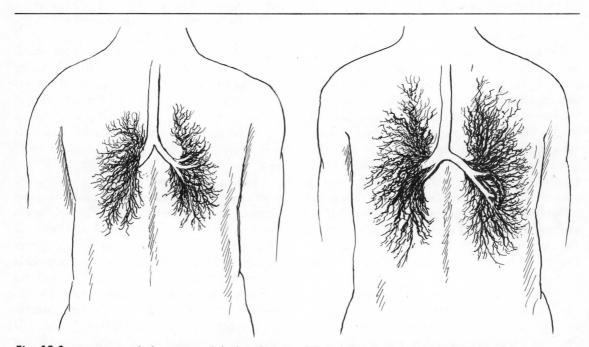

Fig. 15-8. Drawings made from X-rays of the bronchial tree: *left,* in full expiration; *right,* in full inspiration.

Although the *intrapleural pressure* is always negative, or below atmospheric, during normal breathing, it undergoes changes with each respiratory cycle, becoming lower, or more negative, during inspiration and higher, or less negative, during expiration. The negative intrapleural pressure not only plays a part in respiration, but it also has a decided effect on venous return of blood to the heart (p. 334). This pressure expands the great veins and has a sort of sucking effect that assists the inflow of blood into the thoracic veins. Since the negative pressure is greater during inspiration, more blood is flowing into the thoracic veins at the same time that air is flowing into the lungs. In some individuals, this increased return of venous blood to the heart results in a noticeable increase in the pulse rate during inspiration.

Modified breathing movements

Normally, both the diaphragm and the external intercostal muscles are used in inspiration. However, in some individuals the diaphragm predominates (abdominal type of breathing), and in others the external intercostals are more active (costal type of breathing).

In the terminal stages of many diseases, breathing may become periodic. The most common type of periodic breathing is called *Cheyne-Stokes respiration*. In this type there is a period of *apnea* (temporary cessation of breathing) lasting several seconds; then breathing begins, at first, very short and shallow; then each breath becomes a bit deeper until a maximum is reached; and then it tapers off until apnea occurs again. Each time the period of apnea occurs, one may think that the person has expired. This type of breathing indicates a respiratory center that is badly damaged by hypoxia or toxins.

In *asthma* there is prolongation of the expiratory phase of respiration, and the ratio of inspiration to expiration is altered, since there is difficulty in getting air out of the lungs. Wheezing in asthmatic persons is due to air being forced out through constricted bronchioles.

Hiccough is due to spasmodic contraction of the diaphragm, which causes a sudden inspiration. A sudden closure of the glottis shortens inspiration and gives origin to the characteristic sound. Often it is due to irritation of the stomach.

Pneumothorax refers to the presence of air in the pleural cavity. It may occur spontaneously as a result of any lung disease that leads to rupture of an alveolus or a bronchiole, and it always follows perforating wounds of the chest (accidental or surgical). In any case, collapse of the lung occurs, and breathing movements are of no avail on the affected side.

Artificial respiration

Artificial respiration is administered routinely by anesthetists during the course of lung and heart operations and often during long abdominal operations. It is administered during resuscitation after drowning, electric shock, asphyxiation, and in poliomyelitis.

In the operating room an anesthetist compresses the bag of the anesthetic machine by hand, to administer artificial respiration by way of a catheter in the trachea, or a mask over the nose and the mouth, attached to the gas machine. Artificial respiration is easy to manage by this method, and there should be a portable gas machine in all emergency stations for the administration of oxygen by artificial respiration.

If no gas machine is at hand, or the rescue squad has not arrived, artificial respiration is administered by the mouth-to-mouth method (Fig. 15-9). This method has been found to be superior to the older manual methods in all age groups. It provides adequate ventilation, requires no equipment, and can be continued for long periods of time.

Changes in pulmonary volume

Spirometry. The volume of air that moves in and out of the lungs with inspiration and expiration is measured with an instrument called a spirometer (Fig. 15-10). The record produced by a person breathing in and out of a spirometer is called a *spirogram*. From this record the following information can be obtained:

1. **Tidal volume.** This is the volume of air moved in and out of the lungs with each respiratory movement. The average tidal volume is 500 ml.

2. **Inspiratory reserve.** This is the volume of air, in excess of the tidal, that can be inhaled by the deepest possible inspiration. The average inspiratory reserve is 3,000 ml.

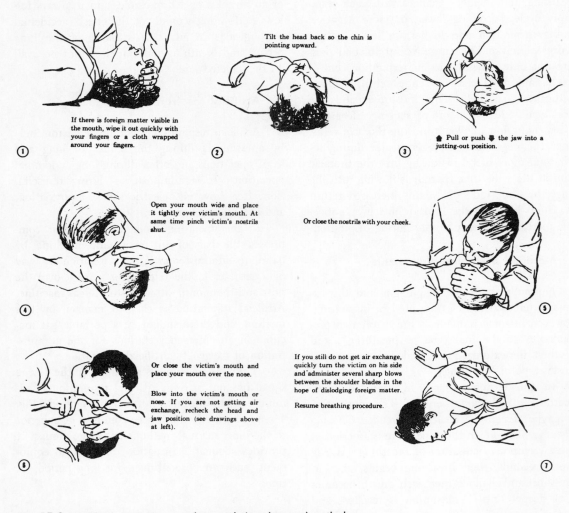

Fig. 15-9. Artificial respiration, mouth-to-mouth (mouth-to-nose) method.

3. **Expiratory reserve.** This is the volume of air, in excess of the tidal, which can be exhaled by the deepest possible exspiration. It is about 1,200 ml.

4. **Vital capacity.** This may be defined in two ways. (A) It is the volume of air that can be exhaled by the deepest possible expiration after the deepest possible inspiration. (B) It is the sum of the tidal, the inspiratory reserve and the expiratory reserve air. The average vital capacity is 4,700 ml.

Vital capacity is reduced in various cases of heart and lung disease. It is also reduced in any condition in which there is muscular weakness,

since the patient is not capable of the increased muscular effort required in testing vital capacity. Reduced vital capacity is shown in Figure 15-10.

5. **Minute respiratory volume.** This is determined by multiplying the tidal volume by the number of respirations per minute. If a person has a tidal volume of 500 ml. and breathes 12 times a minute, the minute respiratory volume would be 6,000 ml. or 6 L.

6. **Maximal breathing capacity.** This is determined by having a person breathe as rapidly and as deeply as possible for a number of seconds, usually 15. Multiplying by four, the volume of air moved in and out of the lungs in one minute can

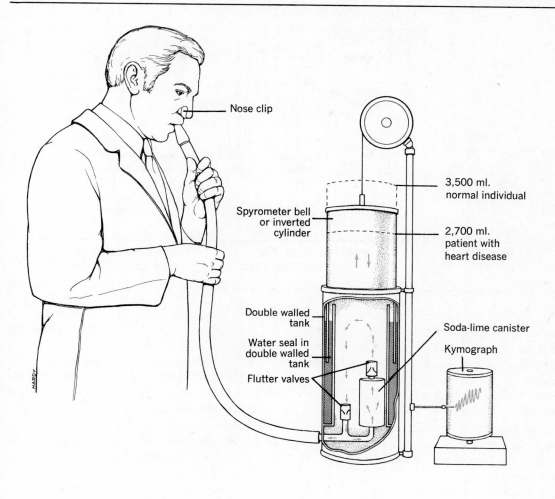

Nose clip

Spyrometer bell
or inverted
cylinder

3,500 ml.
normal individual

2,700 ml.
patient with
heart disease

Double walled
tank

Water seal in
double walled
tank

Flutter valves

Soda-lime canister

Kymograph

HARDY

Fig. 15-10. Measuring vital capacity with a spirometer. In a normal person the vital capacity might be 3,500 ml., but an individual with heart disease might have a vital capacity of only 2,700 ml., or less, as the congested pulmonary capillaries encroach upon the space normally occupied by alveolar air.

be calculated. This may be as high as 100 liters per minute during a period of strenuous exercise.

Other pulmonary volumes. Volumes which cannot be determined directly from the spirogram are those involving air left in the lungs following expiration.

Residual air is the amount of air remaining in the lungs after the deepest possible expiration, and it cannot be removed by voluntary effort. The average volume of residual air is 1,200 ml., and it is eliminated only by collapse of the lungs. The *functional residual capacity* of the lungs is calculated by adding the expiratory reserve volume and the residual volume.

The sum of all the volumes is called *total lung capacity* and normally is about 5,900 ml. in healthy young men. The various pulmonary volumes are illustrated diagrammatically in Figure 15-11.

Minimal air is that remaining in the lungs after they have collapsed. If a baby has taken even one breath, the lungs retain some air that is trapped in the alveoli as the elastic lungs recoil. Minimal air may be important in medicolegal practice to determine if a baby lived even briefly or was born dead.

Dead space air is the volume of air that fills the passages such as the larynx, trachea, and the

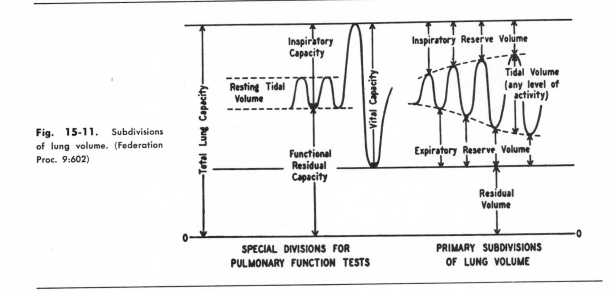

Fig. 15-11. Subdivisions of lung volume. (Federation Proc. 9:602)

SPECIAL DIVISIONS FOR PULMONARY FUNCTION TESTS

PRIMARY SUBDIVISIONS OF LUNG VOLUME

bronchial tree. Its name is derived from the fact that it plays no part in the gaseous exchange of respiration. It is approximately 150 ml. for young men and 110 ml. for young women. The amount increases slightly with age and varies with certain physiologic conditions. It will be noticed that the dead space referred to includes only the passageways and not the alveoli. For this reason it is more specifically called the *anatomic dead space*. When some of the alveoli are not functioning, they too must be considered as dead space. Their function may be impaired if they are ventilated but have no blood supply, or if the ratio between ventilation and blood flow is so low that they are only partially functional. When dead space in the alveoli is added to anatomic dead space, the total is called *physiologic dead space*. In normal people the anatomic and physiologic dead space are essentially the same because all of the alveoli are considered to be functioning in the normal lung. Any impairment of the function of the alveoli as a result of respiratory disease will cause an increase in the physiologic dead space.

The work of breathing

When one is in good health, little thought is given to the work that is involved in breathing. However, there are certain factors that offer resistance to ventilation of the lungs, and work is required to overcome them. These factors are compliance and airway resistance.

Compliance. This is a term used to describe the elastic properties of the lungs and the chest wall. It is dependent on elastic tissue and on the surface tension of the fluid lining the alveoli. Restrictive disorders of the chest wall or of the lung itself can cause a decrease in compliance. As a result, there is a reduction in total lung volume, vital capacity, and inspiratory capacity. These patients have increased work of breathing because stiff structures are more difficult to move.

Airway Resistance. The airways serve as conduits, the walls of which offer resistance to the flow of gases. However, during normal quiet breathing, only a slight amount of energy is needed to move air along the respiratory passageways. On the other hand, during a forced expiration, there is a great increase in airway resistance, due to the fact that the intrapleural pressure becomes temporarily positive and leads to compression of those airways in the thorax. Thus the work load is increased accordingly. This factor is of considerable importance in patients suffering from chronic obstructive pulmonary disease, since their resistance to airflow is high even at rest.

In summary, to move the chest wall and the lungs and to move air through the airways, work is required. To do this work, the respiratory muscles must consume oxygen. In normal individuals, the work of breathing is relatively small, except when they are exercising vigorously. However, in patients with obstructive lung disease the work of breathing may be increased to five or ten times its normal value. Under such conditions the amount of oxygen consumed just for breathing activities may be a large fraction of the total oxygen consumption. Patients with reduced compliance also have increased work of breathing, but they tend to take rapid shallow breaths, which reduces their oxygen consumption. However, gas exchange may be impaired if breathing becomes too shallow.

THE RESPIRATORY GASES

The air we breathe has a remarkably uniform composition insofar as oxygen (O_2), carbon dioxide (CO_2), and nitrogen (N_2) are concerned. Of these gases, only oxygen and carbon dioxide are of physiologic significance under ordinary circumstances, nitrogen serving merely to dilute the oxygen.

Composition of respired and alveolar air

The amounts of respiratory gases vary in inspired, expired and alveolar air, as indicated in Table 15-1. Nitrogen and water vapor are not included in this table, but they would bring each total to 100 per cent.

Table 15-1 Concentrations of Respiratory Gases

	Oxygen	Carbon Dioxide
Inspired air	20.93%	0.04%
Alveolar air	14.00%	5.50%
Expired air	15.70%	4.40%

You will note that expired air contains less oxygen and more carbon dioxide than inspired air, and that alveolar air contains less oxygen and more carbon dioxide than expired air. Expired air is really a mixture of alveolar and inspired air from the dead space. The composition of alveolar air is fairly constant, and only one fifth of the air in the lungs is renewed by each inspiration. In this way, sudden and marked changes in the composition of alveolar air are prevented.

In addition to the differences in content of oxygen and carbon dioxide, there are physical differences between inspired air and expired and alveolar air. Expired and alveolar airs are saturated with water vapor and warmed to body temperature. The amount of water vapor and the temperature of inspired air vary markedly in different localities and in different seasons. One loses heat by warming the inspired air whenever its temperature is below body temperature; one also loses heat by saturating inspired air whenever it is not saturated as it enters the respiratory passages.

The exchange of gases

The diffusion of gases takes place because of pressure gradients. For example, the amount of oxygen in either air or blood is referred to as oxygen tension, or partial pressure, and is written as Po_2. The term partial pressure comes from *Dalton's law* (the law of partial pressures), which explains that the pressure exerted by a gas in a mixture of gases is equal to the pressure which the same quantity of that gas would exert alone. The total pressure of a mixture of gases must then be equal to the sum of the pressure of the gases which make up the mixture. The atmosphere exerts a pressure of 760 mm. Hg at sea level, and oxygen makes up about 20.93 per cent of the gases in the atmosphere. Therefore, the partial pressure exerted by oxygen is 20.93 per cent of 760, or 159.1 mm. Hg under standard conditions. Likewise, the partial pressure of carbon dioxide is 0.04 per cent of 760, or 0.3 mm. Hg.

Under ordinary conditions, inspired air contains very little water vapor, but as this air passes through the nose and pharynx, it becomes fully saturated with water. At normal body temperature (37° C.), the partial pressure of water vapor is 47 mm. Hg. Since the total gas pressure must equal atmospheric pressure, this means that only 713 mm. Hg (760 to 47) is available for the sum of the partial pressures of oxygen, carbon dioxide, and nitrogen in the tracheobronchial tree and alveoli. For example, the Po_2 of alveolar air would be 14.00 per cent of 713, or 100 mm. Hg. The partial pressures of gases in inspired air and alveolar air are given in Table 15-2.

Table 15-2   Partial Pressures of Gases Entering the Lungs (in mm. Hg)

	Dry Inspired Air	Alveolar Air
P_{O_2}	159.1	100.0
P_{CO_2}	0.3	40.0
P_{H_2O}	0.0	47.0
P_{N_2}	600.6*	573.0

* Includes small amounts of rare gases.

External Respiration. The partial pressures of the respiratory gases in the bloodstream are such that the venous blood has a lower P_{O_2} than the alveolar air, and the P_{CO_2} is higher in venous blood than in alveolar air. The pressure gradients then promote the transfer, or inward diffusion, of oxygen from alveolar air to the blood and the outward diffusion of carbon dioxide from the venous blood to the alveolar air. The result of this exchange is to convert venous blood to arterial blood. These exchanges occur very rapidly, so that arterial blood leaving the lungs has almost the same partial pressure of gases as those found in the alveolar air. The partial pressure of the gases in venous and arterial blood is given in Table 15-3.

Table 15-3   Partial Pressures of Blood Gases (in mm. Hg)

	Venous Blood	Arterial Blood
P_{O_2}	40	100
P_{CO_2}	46	40
P_{N_2}	573	573

There are several factors which determine how much oxygen diffuses into the blood of the lung capillaries. An obvious one is the oxygen pressure gradient between alveolar air and venous blood. Anything that decreases alveolar P_{O_2}, such as high altitude, will decrease the oxygen pressure gradient, and thus decrease the amount of oxygen entering the venous blood in the lung capillaries. Another way in which oxygen diffusion may be

decreased is by decreasing the surfaces available for diffusion, as would occur in emphysema, where structural changes in the lungs reduce the total respiratory surface area. A third factor is any condition that reduces the minute volume. A very slow respiratory rate with shallow inspirations may markedly reduce the minute volume and therefore decrease the amount of oxygen which is available for diffusion.

Internal Respiration. The exchange of gases between the blood in systemic capillaries and the interstitial fluid also depends on pressure gradients. Metabolically active cells are constantly using oxygen and producing carbon dioxide. Thus the P_{O_2} of the surrounding interstitial fluid will be lower than the P_{O_2} of the arterial blood, while the P_{CO_2} will be higher. Consequently, oxygen will diffuse out of the blood to supply the cells, and carbon dioxide will be carried away as it diffuses into the blood. Arterial blood thus becomes venous blood and returns to the heart for another trip to the lungs.

Transport of gases in the blood

On entering the bloodstream, both oxygen and carbon dioxide immediately go into simple physical solution in the plasma. However, since a fluid can hold only a small amount of gas in solution, most of the oxygen and the carbon dioxide quickly enter into chemical combinations with other blood constituents. A few of the more prominent chemical reactions are shown schematically in Figures 15-12 and 15-13. For a detailed discussion of this subject one should consult a textbook in biochemistry.

Oxygen Transport. As previously noted, the amount of oxygen that is in solution in the blood is extremely small. Actually, it is such a small amount that one would need about 120 liters of blood to circulate through the tissues of the body each minute to supply adequate oxygen. Approximately 0.5 ml. of oxygen is in solution in every 100 ml. of blood, and yet we know that there is about 20 ml. of oxygen in every 100 ml. of blood. This additional amount of oxygen is carried in chemical combination with hemoglobin stored in the red blood cells. Each gram of hemoglobin can unite with approximately 1.3 ml. of oxygen, and with 15 grams of hemoglobin in every 100 ml. of blood, we can see that 19.5 ml. of oxygen is

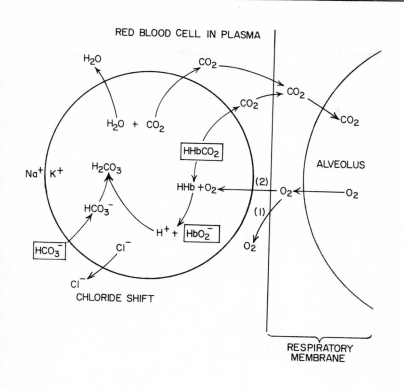

RED BLOOD CELL IN PLASMA

CHLORIDE SHIFT

RESPIRATORY MEMBRANE

ALVEOLUS

Fig. 15-12. Exchange of gases between a red blood cell (RBC) in a pulmonary capillary and an alveolus. As O_2 diffuses across the respiratory membrane and enters the bloodstream, a small amount (1) is carried in solution in the plasma, but most of it (2) enters the RBC and combines with reduced hemoglobin (HHb) to form oxyhemoglobin (HbO_2-) for transport to the tissues. Simultaneously, CO_2 diffuses out of the bloodstream and across the respiratory membrane into the alveolus. As the P_{CO_2} of the plasma falls, additional CO_2 is released from its transport forms, carbamino-hemoglobin ($HHbCO_2$) and bicarbonate ions (HCO_3-). Note that as bicarbonate ions move into the RBC, chloride ions move out to maintain electrical neutrality; this is known as the chloride shift. C.A. stands for carbonic anhydrase which speeds the dehydration of H_2CO_3. Boxes indicate the major forms in which CO_2 and O_2 are transported to and from the lungs.

Fig. 15-13. The exchange of gases between a red blood cell in a systemic capillary and a tissue cell. CO_2 diffuses through the interstitial fluid (ISF) into the bloodstream where a small amount is carried in solution in the plasma (1), but most of it enters the RBC where it follows the chemical pathway (2) leading to the formation of bicarbonate ions (HCO_3-). The majority of these ions leave the RBC and travel in the plasma as $NaHCO_3$, but some also remain in the RBC as $KHCO_3$. Negative chloride ions move into the RBC to main-electrical neutrality. The remaining CO_2 (3) combines with reduced hemoglobin (HHb) and is carried to the lungs as carbamino-hemoglobin ($HHbCO_2$). Simultaneously, O_2 diffuses out of the bloodstream and through the ISF to the tissue cell. Boxes indicate the major forms in which CO_2 and O_2 are transported to and from the lungs. C.A. stands for carbonic anhydrase which speeds the hydration of CO_2.

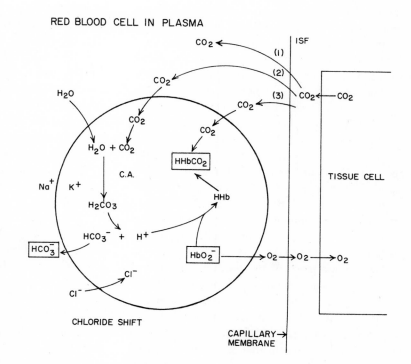

RED BLOOD CELL IN PLASMA

ISF

TISSUE CELL

CHLORIDE SHIFT

CAPILLARY MEMBRANE

carried in each 100 ml. of blood as oxyhemoglobin, while 0.5 ml. of oxygen is carried in solution. The oxygen capacity of blood is usually expressed as volumes per cent, and so it is said that the oxygen capacity of the blood is 20 volumes per cent.

The combination of oxygen with hemoglobin is influenced by the partial pressure of oxygen. When the partial pressure of oxygen is increased, so is its combination with hemoglobin. Conversely, when the partial pressure of oxygen is lowered, oxygen is released from hemoglobin. Hemoglobin that has lost its oxygen is called *reduced hemoglobin*. The relationship between oxygen and hemoglobin is best described by the *oxygen dissociation curve*, in which the percentage of hemoglobin saturation is plotted against the partial pressure of oxygen (Fig. 15-14). The curve shows us that the higher the P_{O_2}, the greater the degree of saturation of hemoglobin with oxygen, and that when the P_{O_2} is low, as it is in venous blood, there is less oxygen bound to hemoglobin. Hemoglobin releases oxygen in such an environment. Two other factors influence the release of oxygen from hemoglobin. One of these is the presence of carbon dioxide. As shown in Figure 15-14, a high P_{CO_2} causes the oxygen dissociation curve to shift to the right. This tells us that less than normal amounts of oxygen are combined with hemoglobin in an environment with a high partial pressure of carbon dioxide. This shift to the right is caused mainly by the increased acidity that results from the presence of carbon dioxide. The effects of high partial pressures of carbon dioxide are similar to the effects of a lowered pH. A second factor which influences the dissociation of oxygen from hemoglobin is temperature. As temperature rises, less oxygen is combined with hemoglobin. Conversely, when temperature is lowered, the oxygen dissociation curve shifts to the left. This tells us that at lower temperatures more than normal amounts of oxygen are combined with hemoglobin.

The importance of the unloading of oxygen from hemoglobin at low partial pressures of oxygen is obvious. Oxygen will be given up to supply tissue cells. An additional factor which promotes the release of oxygen, such as an elevated carbon dioxide level, also means that oxygen will be given up as it passes through the capillaries that supply tissue cells. Increases in temperature and in acid

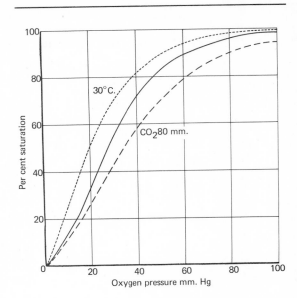

Fig. 15-14. Graph showing oxygen dissociation curves. Note that when the carbon dioxide is increased, the normal curve (solid line) is shifted to the right (dashed line) and, therefore, for any partial pressure of oxygen the hemoglobin binds less oxygen. Also note that a decreased temperature causes the curve to shift to the left (dotted line) so for any partial pressure of oxygen the hemoglobin binds more oxygen.

production during contraction of skeletal muscle in exercise will both act to increase the unloading of oxygen which is necessary if the muscle is to continue to contract.

Hypoxia is a term that refers to any state wherein an inadequate amount of oxygen is available to the cells; it is generally classified on the basis of one of four causes.

1. *Anemic hypoxia* is due to a reduction in the oxygen-carrying capacity of the blood. For example, in anemia and hemorrhage there is a reduced amount of hemoglobin available for oxygen transport.

2. *Stagnant hypoxia* is due to a slow rate of circulation of the blood, as in heart disease. Oxygen is not delivered to the cells with sufficient rapidity to meet their needs.

3. *Hypoxic hypoxia* refers to decreased oxygen saturation of the blood, due to insufficient oxygen in the alveolar air. This may occur in high altitude flying and in cases of respiratory obstruction.

4. *Histotoxic hypoxia* occurs when poisons, such as cyanide, enter the body. Such poisons

destroy the enzyme systems present in cells, and the oxygen that is present cannot be used in the absence of these systems.

Carbon Dioxide Transport. Since carbon dioxide reacts with water to form carbonic acid, the transport and elimination of this gas are very important factors in acid-base balance (p. 464), as well as in respiratory physiology.

As carbon dioxide diffuses into the plasma of systemic capillary blood, most of it enters the red blood cells. However, a small amount remains in the plasma. Of this carbon dioxide in the plasma, some is carried in simple solution, and some is hydrated to form carbonic acid:

$$CO_2 + H_2O \rightleftharpoons H_2CO_3 \rightleftharpoons H^+ + HCO_3^-$$

Since there is no catalyst in the plasma, the above reaction is very slow, and the hydrogen ions are easily handled by the plasma buffer systems (p. 465).

The major buffering of carbon dioxide occurs within the red blood cells. Here we find *carbonic anhydrase*, a special enzyme that speeds the hydration of carbon dioxide so that large quantities of carbonic acid are formed. Fortunately for us, hemoglobin is available to serve as a buffer for the hydrogen ions. The great numbers of bicarbonate ions resulting from this reaction readily diffuse outward through the red cell membrane (which is highly permeable to negative ions); this leaves the red cells with a net positive electric charge. To re-establish an equilibrium, negative chloride ions (Cl^-) in the plasma move to the interior of the red cells; this exchange of ions is called the chloride shift. The process is reversed when blood reaches the lungs. Carbon dioxide diffuses out of the plasma and out of the red blood cells as the CO_2 reaction described above goes to the left.

Carbon dioxide also is carried in combination with hemoglobin, the resulting compound being called *carbamino-hemoglobin* ($HHbCO_2$). Thus hemoglobin does double duty; it not only serves to carry oxygen to the tissues, but also helps in the removal of carbon dioxide.

The total amount of carbon dioxide in 100 ml. of venous blood is about 48 ml.; of this, about 2.7 ml. is carried in solution in the plasma, 42.3 ml. is carried as bicarbonate, and the remaining 3 ml. is carried as carbamino compound.

Nitrogen. Although there are no known metabolic reactions involving molecular nitrogen (N_2) in the human body, it is found in solution in the blood and in other tissues. Under ordinary circumstances this nitrogen is of no physiologic significance, but in the caisson worker, the deep-sea diver, or even the adventurous weekend scuba diver, dissolved nitrogen becomes a problem of immediate concern. These persons must breathe air that is under greatly increased pressure, and, as the pressure increases, more nitrogen goes into solution in the plasma as well as in the interstitial fluid. Excessive amounts of dissolved nitrogen may produce giddiness and other effects similar to those that follow the consumption of alcohol; this condition, called *nitrogen narcosis*, may imperil the life of the scuba diver who swims alone and has no contact with the surface.

Decompression sickness, also known as *caisson disease* or *the bends*, may follow too rapid ascent or return to normal pressure. In such cases, the excess nitrogen is released in the form of bubbles in the blood, and the severity of symptoms depends on where the bubbles become trapped. These bubbles may lodge in capillaries all over the body, but the most common symptom of decompression sickness is pain in the muscles and the joints. More serious symptoms include dyspnea, chest pain, and muscular paralysis. Treatment of decompression sickness depends on recompression; the victim must be placed in a special chamber where the air pressure can be elevated until nitrogen once again is forced into solution. If decompression is carried out slowly, the nitrogen gradually is eliminated through the lungs and causes no further trouble.

Decompression sickness is not limited to tunnel workers and divers. In this age of high-flying jet aircraft, pressurized cabins protect pilots from the same hazard of changing pressures. Without this protection, the nitrogen that normally is present in the blood would be released as bubbles during the rapid ascent from atmospheric pressure on the ground to the much lower pressure that prevails at high altitudes.

Monitoring of Blood Gases. Various electronic and other instruments have been devised for continuous monitoring of the Po_2, hemoglobin saturation, Pco_2, and pH of the blood. Such information is valuable in the management of patients with certain cardiopulmonary diseases. For example, when a patient is in a mechanical respirator (iron lung), it is essential that adequate ventilation be maintained to assure proper oxygenation of the blood; thus it is helpful to know the arterial oxygen saturation. Also, the blood gases and pH may signal a change in a patient's condition long before it is reflected in the routine blood pressure, pulse, and temperature.

THE CONTROL OF BREATHING

Whether one is sitting quietly in a chair or playing a vigorous game of tennis, the breathing movements are controlled so as to meet the immediate oxygen requirements of the body. However, the respiratory muscles contract at the proper time and with the proper force only if they are so stimulated by nerve impulses from the CNS. Unlike the heart muscle, which can contract automatically, the skeletal muscles of respiration are completely dependent on their nerve supply. Therefore, in view of the constantly changing needs of the body, it follows that there must be a central control unit that is capable of coordinating and correlating the activities of the respiratory muscles; this vital unit, or respiratory center, and the many factors that influence it will be described briefly.

The respiratory center and efferent pathways

The respiratory center, consisting of inspiratory and expiratory neurons, is located in the medulla oblongata and in the pons. Although nature has endowed the medullary center with an intrinsic rhythm, its activity is influenced by facilitatory and inhibitory centers in the pons. Connections with the spinal cord and with higher centers in the hypothalamus and cerebral cortex also are of importance in the control of breathing.

The medullary center is under constant bombardment by nerve impulses as it receives information from receptors in various parts of the body. These receptors report, either directly or indirectly, on respiratory conditions in their area. In response to this stimulation, the respiratory center dispatches the appropriate orders for contraction of the respiratory muscles.

During normal, quiet breathing (called *eupnea*), the only muscles that contract to bring about inspiration are the diaphragm and the external intercostal muscles. Nerve impulses dispatched from the respiratory center reach the diaphragm by way of the phrenic nerves, and the intercostal muscles by way of the intercostal nerves. Quiet expiration normally is a passive act as the inspiratory muscles relax, due to the cessation of impulses from the respiratory center.

In order to breathe more deeply, as during mild exercise, there must be more forceful contractions of the above muscles. More and more motor units are brought into action, and there are more impulses per unit of time conducted over each axon. There is a wide range of gradation possible in the depth of breathing, since any number of motor units may be brought into activity. In order to breathe more rapidly, there are shorter intervals of rest between bursts of impulses traveling over the phrenic and the intercostal nerves.

Following strenuous exercise, or in certain diseases that are accompanied by difficult breathing (called *dyspnea*), the respiratory center will call many more muscles into action as both *forced inspiration* and *forced expiration* are needed to meet the body's needs. Any muscle capable of lifting the ribs may be used during forced inspiration; some of these are the sternocleidomastoids, the scalenes, the levatores costarum, and the pectoralis major and minor muscles.

As noted previously, expiration is normally passive, but in forced expiration it becomes active, and muscles that depress the ribs are called into action. These include the internal intercostals, the quadratus lumborum, and the muscles of the abdominal wall. In addition to pulling on the ribs, the abdominal wall muscles compress the abdominal viscera and force the diaphragm upward, thus decreasing the volume of the thorax and helping to expel air more rapidly.

The *respiratory rate* in an average adult is between 12 and 18 breaths per minute. In the newborn, the rate is about 40; in the infant, 30; and in a five-year-old child, about 25 per minute. However, there is considerable individual variation at all ages.

Voluntary control of breathing

Nerve impulses originating in the motor areas of the cerebral cortex and traveling to the respiratory center enable one to consciously alter the rate and the depth of breathing. For example, during the act of speaking or singing, breath control is of considerable importance, and if a person wishes to swim under water he may hold his breath voluntarily for a short period of time. However, voluntary control is limited, and the respiratory center will ignore messages from the cortex when it is necessary to meet the body's basic needs. To prove this point to yourself, try to talk or to sing in a normal manner after a bout of strenuous exercise, or try to hold your breath for a prolonged period of time.

Chemical factors in the control of breathing

As metabolic activity in the body increases, usually there is a proportionate increase in the rate and depth of breathing; this sequence of events is quite logical in view of the fact that busy cells require more oxygen (O_2) and form more carbon dioxide (CO_2) and hydrogen ions ($H+$). On the other hand, as metabolic activity declines, the cells need less O_2 and form less CO_2 and $H+$, thus breathing activity is decreased. The respiratory center is able to issue the appropriate orders for changes in the rate and depth of breathing because it is alerted to any significant changes in the composition of the cerebrospinal fluid and the circulating blood. The alerting mechanism involves special sensory nerve endings, called *chemoreceptors*. These receptors, when stimulated by chemical changes in their natural liquid environment, initiate afferent nerve impulses that influence the activity of the respiratory center. The chemoreceptors are of two types: central and peripheral.

Central chemoreceptors, located on or near the ventral surface of the medulla, are strongly influenced by the $H+$ and CO_2 content of the cerebrospinal fluid; to a lesser degree, they also may be influenced by the CO_2 in arterial blood. CO_2 diffuses readily through the walls of brain capillaries and into the cerebrospinal fluid that bathes the central nervous system. As previously described, carbon dioxide reacts with water to form carbonic acid, and some of the acid molecules dissociate to yield hydrogen ions. As a result, nerve impulses are dispatched to the nearby respiratory center, and one begins to breathe more

deeply and more rapidly. As the excess carbon dioxide is blown off, and the amount of carbonic acid is reduced proportionately ($H_2CO_3 \rightarrow CO_2 + H_2O$), breathing returns to normal. On the other hand, if the blood level of carbon dioxide falls below normal levels there will be a temporary cessation of breathing (apnea). However, decreased carbon dioxide tension does not persist for any length of time, since carbon dioxide is produced constantly by metabolic processes in all cells, and it soon accumulates to the required level to stimulate the respiratory center again. Thus one can see that breathing plays a vital role in the maintenance of acid-base balance. As carbon dioxide is blown off with each expiration there is a corresponding reduction in the amount of acid in the blood.

This chemical control mechanism is responsible for the fundamental rhythm of the respiratory center, and it keeps us breathing both day and night without our giving the matter any conscious thought. It is this mechanism that comes to the rescue of the distraught mother whose child, in a fit of temper, is holding his breath until he becomes "blue in the face." If only the mother can remain calm for a few seconds, the child will breathe whether he wishes to do so or not, as the chemical control overcomes the voluntary effort, and the air in his rebellious little lungs will be freshened.

Various drugs may affect the respiratory center, and one of the most important of these is *morphine*. Morphine has a depressing effect on the respiratory center, and respiratory failure may follow an overdose of morphine. When an injection of morphine is given to any patient it should be charted at once; otherwise it may be repeated by another person, and a serious depression of respiration might ensue.

If morphine is given to a woman in labor just before delivery, the respiratory center of the newborn may be so depressed that it cannot function, and respiration cannot be established. The question of sedation and anesthesia is a vital one in obstetrics.

Peripheral chemoreceptors, located in the aortic and carotid bodies, play a role in the regulation of circulation (p. 344), but they are of even greater importance in regulating the oxygen content of the blood; under certain extreme conditions they also help to regulate the CO_2 and $H+$

content of the blood. Whenever the amount of O_2 dissolved in the plasma falls, there is a corresponding decrease in the O_2 content of the interstitial fluid bathing the chemoreceptor cells; this lack of O_2 serves to stimulate those cells, and nerve impulses are dispatched to the respiratory center. When the respiratory center is stimulated by such incoming impulses, it orders an increase in the depth and rate of breathing. This reflex is important, not only in normal daily living, but also when the oxygen supply to the body is decreased in heart disease, and in lung disease, such as emphysema.

Physical factors in the control of breathing

Although the chemical factors that regulate breathing are of fundamental importance, there are various physical factors that also are of significance. Some of these factors and the respiratory reflexes that they initiate will be described briefly.

Inflation of the Lungs. Lung inflation serves to stimulate receptor end organs which are sensitive to stretch or distention. These special nerve endings, located in the alveolar ducts, initiate a stream of nerve impulses which travel to

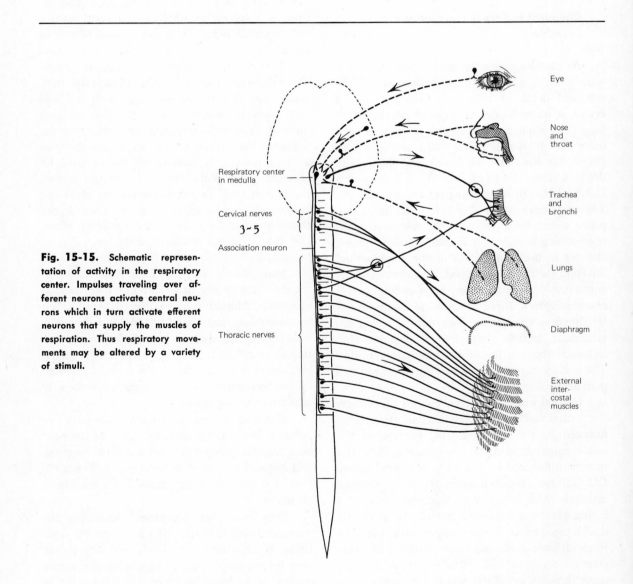

Fig. 15-15. Schematic representation of activity in the respiratory center. Impulses traveling over afferent neurons activate central neurons which in turn activate efferent neurons that supply the muscles of respiration. Thus respiratory movements may be altered by a variety of stimuli.

Eye

Nose and throat

Trachea and bronchi

Lungs

Diaphragm

External intercostal muscles

Respiratory center in medulla

Cervical nerves
3~5

Association neuron

Thoracic nerves

the CNS by way of the vagus nerves. The inspiratory center is inhibited by these afferent impulses, and, as a result, impulses are no longer sent to the diaphragm and the external intercostal muscles. With relaxation of these muscles, passive expiration follows. This reflex, called the *Hering-Breuer* reflex, is of importance in the rhythmic control of normal, quiet breathing. If the vagus nerves are cut, inspiration is greatly prolonged and of increased depth; however, the respiratory center eventually ceases to discharge efferent impulses, and expiration will occur. This indicates that there are other factors which continue to operate in the absence of the vagus nerves.

Changes in Blood Pressure. Pressoreceptors in the aortic and the carotid sinuses are affected by sudden, marked changes in blood pressure. When arterial blood pressure rises suddenly, impulses are sent in from the pressoreceptors by way of the vagus and the glossopharyngeal nerves to the CNS. Such impulses depress the activity of the respiratory center, and respiration becomes slower and more shallow, temporarily.

When the arterial blood pressure falls suddenly, as after a severe hemorrhage, the number of impulses reaching the CNS from the pressoreceptors decreases, and the respiratory center increases its activity, thereby increasing the rate and probably the depth of respiration. However, it must be kept in mind that the pressoreceptors are not nearly as important in the control of respiration as they are in the control of circulation (pp. 342 to 343).

Irritation of the Respiratory Passages. Any irritating substance can stimulate sensitive receptors in the mucous membrane lining. An irritating substance in the nose, such as pepper, induces a *sneeze* which is a violent expiration of air through the nose. The presence of a crumb in the larynx or the trachea induces a spell of coughing. A cough is a sudden, violent expulsion of air after a deep inspiration and temporary closure of the glottis to build up pressure in the lower respiratory passages. Reflexes of this type are protective in nature and tend to remove the irritating substances from the body.

Temperature. The temperature of the blood flowing through the respiratory center also may influence breathing. With an increase in body temperature, as during fever or severe muscular exercise, there is an increase in the respiratory rate. On the other hand, a subnormal temperature, as during hypothermia, results in a decreased respiratory rate.

Incidental Reflexes. Reflex activity may be initiated by receptors of various kinds in other parts of the body. You may see something that makes you gasp or hold your breath momentarily. You may hear a shriek that either speeds up your respiratory rate or stops all respiratory movement for a few seconds. When you touch something cold and slimy, you may gasp. These are some of the ways in which respiratory reflexes can be aroused by a variety of stimuli which affect many types of receptors.

Figure 15-15 illustrates the respiratory center and some of the afferent pathways by which impulses reach the CNS and alter activity in the respiratory system. The efferent pathways for all respiratory reflexes are the phrenic and the intercostal nerves, which are the final common paths for such reflexes.

SUMMARY

1. **Respiration refers to exchange of gases between organism and environment**
A. External respiration: exchange between blood and air
B. Internal respiration: exchange between blood and tissue cells
2. **Organs of respiratory system**
A. Nose
  a. Serves to warm, moisten, and filter air
  b. Paranasal sinuses and nasolacrimal duct open into nose
B. Pharynx: tubular passageway from nose
  a. Nasopharynx: superior portion; closed during swallowing; pharyngeal tonsil on posterior wall
  b. Oropharynx: middle portion; lies behind mouth with which it communicates by fauces; palatine tonsils between pillars; lingual tonsil at root of tongue
  c. Laryngopharynx: inferior portion; opens into larynx anteriorly; continues as esophagus posteriorly
C. Larynx: stands guard over trachea and serves as organ of voice
  a. Closed during act of swallowing
  b. cartilages: connected by ligaments
  c. Vocal cords from glottis; tension of cords con-

trolled by muscles during speech
D. Trachea and bronchi
   a. Trachea: a flexible tubular structure reinforced by cartilage rings; lined with ciliated mucosa
   b. Bronchial tree
      (1) Trachea divides into primary bronchi which then divide to form secondary bronchi
      (2) Division continues to bronchioles; less cartilage, more smooth muscle
E. Lungs: essential organs of respiration
   a. In thoracic cavity
   b. Cone-shaped; base rests on diaphragm; apex extends above first rib
   c. Right lung shorter and broader; three lobes; left lung, two lobes
   d. Visceral pleura applied to lung; parietal pleura lines thorax, covers diaphragm, reflected over mediastinum. Pleural cavity: potential space between visceral and parietal layers
   e. Light, porous and spongy; elastic; lobes divided into lobules
   f. Each lobule supplied with bronchiole; branches to terminal bronchiole, to respiratory bronchioles, to alveolar ducts, and to alveoli
   g. Surfactant, lowers surface tension in alveoli; prevents atelectasis
   h. Respiratory membrane separates alveolar air from capillary blood; less than 1 micron thick
   i. Blood supply: pulmonary and bronchial arteries

3. **Mechanics of breathing**
A. Air pressure and flow
   a. Gases travel from area of higher to lower pressure
   b. Atmospheric pressure: exerted by outside air
   c. Intrapulmonic pressure: that within bronchial tree and alveoli
   d. Intrapleural pressure: that within pleural cavity; negative, or subatmospheric pressure
B. Inspiration: act of taking air into lungs
   a. Contraction of inspiratory muscles
   b. Enlargement of thoracic cage
   c. Decrease in intrapleural and intrapulmonic pressures
   d. Air flows in until equal to atmospheric pressure
C. Expiration: act of expelling air from lungs
   a. Relaxation of inspiratory muscles; recoil of stretched tissues
   b. Decrease in size of thoracic cage
   c. Increase in intrapleural and intrapulmonic pressures
   d. Air flows out until equal to atmospheric pressure
D. Modified breathing movements
   a. Abdominal
   b. Costal
   c. Cheyne-Stokes
   d. Hiccough

4. **Changes in pulmonary volume**
A. Spirometry record shows:
   a. Tidal volume
   b. Inspiratory reserve
   c. Expiratory reserve
   d. Vital capacity
   e. Minute respiratory volume
   f. Maximal breathing capacity
B. Other pulmonary volumes
   a. Residual air
   b. Minimal air
   c. Dead space air
C. The work of breathing
   a. Normally is minimal
   b. Decreased compliance and/or increased airway resistance means more work

5. **Respiratory gases**
A. Composition of respired and alveolar air
   a. Expired has less O_2 and more CO_2 than inspired air
   b. Alveolar air fairly constant; only one fifth of air in lungs renewed by each inspiration
   c. Expired air and alveolar air warm and moist
B. Exchange of gases
   a. Depends on pressure gradients
   b. External respiration: oxygen from alveoli to blood in pulmonary capillaries; carbon dioxide from pulmonary capillaries to alveoli
   c. Internal respiration: oxygen from systemic capillaries to interstitial fluid; carbon dioxide from interstitial fluid to systemic capillaries
C. Transport of gases in blood
   a. Oxygen: small amount in solution in plasma; most in combination with hemoglobin; release influenced by low P_{O_2}, high P_{CO_2} and temperature
   b. Hypoxia: inadequate amount of oxygen available to cells
   c. Carbon dioxide transport: small amount carried in solution in plasma; small amount carried as carbamino-hemoglobin; large amount carried as bicarbonate in plasma
   d. Nitrogen of significance only in persons breathing air under pressure

6. **Control of breathing**
A. Respiratory center and efferent pathways
   a. Located in medulla and pons; coordinates and correlates breathing activities
   b. Dispatches orders to respiratory muscles stimulating contraction
   c. Forced inspiration involves any muscle capable of lifting ribs
   d. Forced expiration an active process; muscles that depress ribs and compress abdominal viscera are stimulated
B. Voluntary control
   a. Nerve impulses from cerebral cortex stimulate center; enable one to control rate and depth of breathing
   b. Limited by the basic needs of the body, which come first
C. Chemical factors in control
   a. Stimulation of central chemoreceptors by carbon dioxide and hydrogen ions in cerebrospinal fluid of fundamental importance; resultant increase in rate and depth of breathing enables one to

blow off CO_2, reduce acid in blood. Limits voluntary control
  b. Peripheral chemoreceptor reflexes; impulses from aortic and carotid bodies when O_2 level falls increase breathing rate
D. Physical factors in control of breathing
  a. Inflation of lungs; Hering-Breuer reflex
  b. Changes in blood pressure
  c. Irritation of respiratory passages
  d. Temperature of blood
  e. Incidental reflexes initiated by receptors elsewhere in body

QUESTIONS FOR REVIEW

1. Distinguish between internal and external respiration.
2. What is the significance of enlarged pharyngeal tonsil?
3. What is the primary function of the larynx?
4. A foreign body obstructing the glottis may cause death. Why?
5. Why are patients with tracheostomy tubes particularly susceptible to pulmonary infection?
6. Why are aspirated foreign bodies more frequently found in the right bronchus than the left?
7. What is the meaning and significance of pneumothorax?
8. How does pulmonary surfactant prevent atelectasis?
9. Briefly, describe the structure and function of the respiratory membrane.
10. Distinguish between the pulmonary and the bronchial arteries in terms of their function.
11. How does intrapulmonic pressure differ from intrapleural pressure?
12. Describe the sequence of events that occur when you inhale and exhale.
13. Why is the mouth-to-mouth method of resuscitation superior to older methods?
14. What information would you need to calculate your minute respiratory volume?
15. What is the significance of a decrease in lung compliance?
16. How does expired air differ from inspired air?
17. What is meant by the term partial pressure of a gas?
18. What factors determine how much oxygen will diffuse into the lung capillaries?
19. How is oxygen transported in the blood?
20. What is the significance of the chloride shift in CO_2 transport?
21. During strenuous exercise, what muscles are involved in forced inspiration and expiration?
22. Why is it impossible to hold your breath indefinitely?
23. Distinguish between central and peripheral chemoreceptors in terms of location and function.
24. What physical factors can influence breathing activities?

REFERENCES AND SUPPLEMENTAL READINGS

Avery, M. E., Wang, N. S., and Taeusch, H. W.: The lung of the newborn infant. Sci. Am., 228:75, April 1973.

Best, C. H., and Taylor, N. B.: The Physiological Basis of Medical Practice, ed. 8. Baltimore, Williams & Wilkins, 1966.

Clements, J. A.: Surface tension in the lungs. Sci. Am., 207:121, Dec. 1962.

Comroe, J. H., Jr.: Physiology of Respiration. Chicago, Year Book Medical Publishers, 1965.

———: The lung. Sci. Am., 214:57, Feb. 1966.

Goss, C. M. (ed.): Gray's Anatomy of the Human Body, ed. 28. Philadelphia, Lea & Febiger, 1966.

Guyton, A. C.: Textbook of Medical Physiology, ed. 4. Philadelphia, W. B. Saunders, 1971.

Ham, A. W.: Histology, ed. 6. Philadelphia, J. B. Lippincott, 1969.

Smith, C. A.: The first breath. Sci. Am., 209:27, Oct. 1963.

The Digestion of Food

16

"You can't ignore the importance of a good digestion.
The joy of life . . . depends on a sound stomach . . ."

Under Western Skies • Joseph Conrad

Foods are substances taken into the body for one of three purposes: to yield energy, to build and repair tissues, and to regulate various metabolic processes. However, the mere act of swallowing a mouthful of food does not mean that it is "in the body"; not until the food material has been absorbed from the lumen of the digestive tube can it be said to have entered the body.

Some food substances can be absorbed with little or no change in the form in which they were ingested, but others must undergo extensive changes before they can enter the blood- or lymph stream. It is the function of the digestive system to prepare foods for absorption so that they may be transported to cells in all parts of the body. Complex foods, such as meat, potatoes, and gravy, must be broken down mechanically and chemically to small particles that can pass through membranes. Thus the digestive system is comparable in some ways to a mill as it modifies or refines food substances.

Let us consider briefly how it is possible for

the protein and the starch in a grain of wheat to become part of the body. First the wheat must be ground and purified, the flour separated by chemical and physical processes and then retained while the waste is rejected. After the flour is made into bread, the bread is eaten and mechanically and chemically broken into small pieces during various digestive processes. The tiny end products of protein digestion are amino acids; those of starch digestion are simple sugars, such as glucose. These useful products of digestion are absorbed into the bloodstream and carried to all parts of the body where they are used by the cells. Thus the original wheat has been subjected to two milling processes, first in making it into flour, and second in preparing the flour for absorption into the circulating blood.

It is the purpose of this chapter first to consider the various food groups and then the machinery of the digestive system mill and the way in which it prepares food for absorption. Chapter 17 will describe the absorption of food and the way in which it is utilized by the cells.

FOOD FOR THE MILL

The following discussion of foods is quite limited; for a more thorough presentation, one should consult a textbook of biochemistry.

Foods may be divided into two general classes: inorganic and organic. The former comprises inorganic salts and water, and the latter comprises carbohydrates, lipids, proteins, and accessory foodstuffs, or vitamins.

Inorganic

Inorganic Salts. The salts in foods are primarily combinations of the following cations: sodium, potassium, calcium, iron, copper, magnesium, and manganese, and the following anions: chloride, bicarbonate, phosphate, sulfate, and iodide. Proper amounts of inorganic salts are necessary for various body functions, such as maintenance of osmotic relationships, excitability of muscle and nerve.

Water. The most abundant of all constituents is water, and life can continue for only a few days without it. Body fluids and electrolytes are discussed in detail in Chapter 19.

Organic

The importance of the organic compounds was previously described in Chapter 2 (pp. 17 to 18). At this time we shall present a brief outline of their groups, or classes, with examples of each.

Carbohydrates. These compounds contain carbon, hydrogen, and oxygen. There are three groups of carbohydrates.

Monosaccharides, or simple sugars, such as glucose (dextrose), fructose, and galactose, are soluble in water. Glucose is the principal form in which carbohydrate is present in the blood.

Disaccharides, or double sugars, are also water-soluble. This group includes sucrose (cane sugar), lactose (milk sugar), and maltose (malt sugar).

Polysaccharides are multiple sugars, such as plant starch, cellulose, dextrin, and glycogen (animal starch). They are insoluble in water and are a common form of storage for carbohydrate.

Lipids. Lipids are also organic compounds, composed of carbon, hydrogen, and oxygen, but they are generally insoluble in water. They consist of a group of substances that are classified together because of their solubility in fat solvents such as benzene or ether. For convenience, lipids may be divided into three main groups: simple, compound, and steroid.

Simple lipids include the true fats, which consist of fatty acids and glycerol. They also are known as *triglycerides* because each molecule of glycerol is combined with three molecules of fatty acid.

Compound lipids include the phospholipids and glycolipids; they are lipids combined with other substances.

Steroids include the sterols, bile acids, vitamin D, and various hormones. Cholesterol is a well-known representative of this group.

Proteins. Proteins are organic compounds composed of carbon, hydrogen, oxygen, and nitrogen; some contain sulfur and phosphorus. They are large molecules consisting of enormous numbers of amino acids, and they are in a constant state of change.

The proteins in the plasma, the liver, and the cells of the tissues form a large protein pool, in which the proteins of all parts of the body are in dynamic equilibrium. An amino acid that is a component part of a plasma protein today may be in a liver cell tomorrow, in a muscle fiber the next day, in an enzyme on the fourth day, and so on.

The proteins that enable one to maintain a good state of nutrition at any age are called *complete* proteins. Most animal proteins are complete proteins. This means that they contain all of the *essential* amino acids, that is, those that the body cannot synthesize. Most plant proteins lack one or more of the essential amino acids.

There are two main classes of protein: simple and conjugated. *Simple proteins* are those that consist only of amino acids; albumin and globulin are examples of simple proteins. *Conjugated proteins* are those that consist of other substances in addition to amino acids. Hemoglobin is an example of a conjugated protein.

VITAMINS

Vitamins, which are organic compounds present in many natural foodstuffs, are required for growth and other processes. They produce no energy but play a role in transformations of energy and in the regulation of metabolism.

Most of the vitamins now known take part in enzymatic or oxidative reactions. Some act as catalysts, which means that they speed up reactions and enable them to take place at body temperature; in the absence of catalysts, the same reactions would require a much longer time and a much higher temperature. Some vitamins are precursors (forerunners) of enzymes. Some work as assistants to enzymes. Vitamins aim at various targets; some act on epithelial tissues, some on bone, and some aid the absorption of foods from the lumen of the intestine.

We receive most of the vitamins needed by the body from the food we eat. A few vitamins are synthesized by bacterial action in the large intestine and are absorbed into the blood.

If the supply of vitamins is insufficient, difficulties arise. There are several causes of vitamin deficiency, three of which will be mentioned: (1) a low intake, as a result of poor eating habits, loss of appetite, or the use of special diets in the treatment of obesity, hypertension, and allergy; (2) failure of absorption, as in prolonged vomiting or diarrhea, the absence of bile salts from the intestine (which affects the absorption of the fat-soluble vitamins), or the use of mineral oil as a laxative or in low-calorie salad dressings (mineral oil dissolves the fat-soluble vitamins and carries them out with it in the feces); (3) liver disease, in which there may be a failure to convert precursors to active vitamins or a failure to store the vitamins.

Fat-soluble vitamins

The best known fat-soluble vitamins are A, D, E, and K. They are soluble only in fats and are absorbed with them.

Vitamin A. The ultimate source of vitamin A is the carotenoid pigments of dark green leafy vegetables, yellow vegetables, and yellow fruits. These pigments are converted to vitamin A, probably during absorption, and stored in the liver. Thus fresh liver and fish oils (e.g., cod-liver oil) are excellent sources of this vitamin.

Except for its role in the formation of rhodopsin, or visual purple (p. 240), the basic function of vitamin A in the metabolism of the body is unknown. However, it is essential for the normal growth of most body cells, and particularly those making up the different types of epithelium.

Night blindness is one of the early symptoms of vitamin A deficiency. Later, the epithelial structures of the body tend to become keratinized; thus the skin is rough and dry, and the cornea may become opaque. In children, there is interference with normal growth and development.

Vitamin D. There are several forms of vitamin D. One is the natural vitamin D which is formed by ultraviolet light irradiation of 7-dehydrocholesterol in the skin. Another is a synthetic compound, calciferol, which is formed by irradiation of ergosterol, a sterol occurring in animal and plant tissue.

Vitamin D promotes the active transport of calcium through the intestinal wall; thus it increases calcium absorption. This vitamin is also involved in the normal calcification of bones, as well as in the development and maturation of teeth. However, the precise chemical manner in which vitamin D does its work is unknown.

In children, a lack of vitamin D results in *rickets.* The bones are soft and fragile; thus skeletal malformations are common; there is delayed closure of the fontanels and retarded eruption of teeth. In adults, the condition is called *osteomalacia*; it is characterized by softening of the bones, leading to deformities of the thorax, vertebral column, pelvis, and legs. There may be frequent fractures.

Vitamin E. There are several compounds (tocopherols) which are known to have vitamin E activity. They are found most abundantly in unmilled cereals, vegetable oils, eggs, and in the green leaves of many vegetables; they are stored in all tissues.

Compounds with vitamin E activity are known to be antioxidants, and it is believed that they play a protective role in preventing the oxidation of unsaturated fats. At present, their specific chemical activity in metabolic processes is unknown. However, like all vitamins, they are essential for normal growth. In certain lower animals, vitamin E is essential for normal reproduction.

It is questionable whether any significant vitamin E deficiency occurs in human beings. However, it has been suggested that children with cystic fibrosis may show a deficiency, due to the excessive loss of fats in the feces (steatorrhea).

Vitamin K. This vitamin is found in liver, spinach, and cabbage. However, the synthesis of vitamin K by bacteria in the large intestine is of fundamental importance, and a dietary source usually is not necessary.

It is believed that vitamin K acts as a co-enzyme involved in the synthesis of prothrombin (and possibly certain other blood-clotting factors) in the liver. Anticoagulant drugs, such as coumadin, interfere with this activity. Storage of vitamin K is relatively slight.

The outstanding sign of vitamin K deficiency is a tendency to hemorrhage. This may occur in the newborn infant until bacteria become established in the intestine. It also occurs in liver disease and in drug therapy that disrupts the normal bacterial flora in the intestine.

Water-soluble vitamins

The water-soluble vitamins are those of the B complex and vitamin C. These vitamins are soluble only in water and are absorbed with it.

The *vitamin B complex* includes twelve chemical substances that are grouped together because of their close association in the tissues; however, only four of them (thiamine, niacin, riboflavin, and B_{12}) will be described. Some of the B vitamins are synthesized in the large intestine, but most of them occur naturally in liver, kidney, lean meat, eggs, yeast, whole-grain cereals, and milk products.

Thiamine. This vitamin functions as a co-carboxylase in the metabolic systems of the body. Working with a protein decarboxylase, thiamine is essential for the decarboxylation of pyruvic acid and certain other alpha-keto acids so that they can enter the citric acid cycle (pp. 418 to 419).

As you might expect, a prolonged thiamine deficiency may have serious consequences. For example, there may be greatly impaired function of the central nervous system, since it depends almost entirely on the metabolism of carbohydrates for its energy. In addition, there is weakening of cardiac muscle that may lead to heart failure. Gastrointestinal symptoms include anorexia, indigestion, constipation, and weight loss. The clinical condition resulting from prolonged thiamine deprivation is called *beriberi*.

Niacin (Nicotinic Acid). Niacin functions in the forms of two coenzymes, NAD and NADP, which are important hydrogen acceptors in the electron transport chain described on page 419. Thus niacin is concerned with the utilization of foodstuffs for energy, and a deficiency of this vitamin can depress function throughout the body.

With a moderate deficiency, there is weakness of all types of muscle tissue and poor glandular secretion. However, severe niacin deprivation results in actual death of tissues and the clinical condition called *pellagra*. Lesions in the CNS, the skin, and the gastrointestinal tract have led to the description of pellagra as the disease of "D's": dementia, dermatitis, diarrhea, and death.

Riboflavin. This vitamin functions in combination with phosphoric acid to form several coenzymes, two of which are FMN and FAD. These coenzymes serve as hydrogen carriers, working with NAD and NADP in the electron transport chain described on page 419. Thus, like niacin, riboflavin is concerned with the utilization of foodstuffs for energy.

Mild riboflavin deficiency is rather common, and its most characteristic lesion is cheilosis (inflammation and cracking at the corners of the mouth); there also may be glossitis and digestive disturbances. Severe deficiencies apparently do not occur in human beings, but in experimental animals the effects are much the same as in niacin deficiency.

Vitamin B_{12} (Cyanocobalamin). Vitamin B_{12} is believed to act as a coenzyme in several metabolic reactions. For example, it facilitates the conversion of amino acids into other substances, and it aids in the reduction of ribonucleotides to deoxyribonucleotides (a major step in the formation of genes, p. 19). These activities help to explain the ability of vitamin B_{12} to promote the maturation of red blood cells and to promote growth.

The best known result of vitamin B_{12} deficiency is pernicious anemia (p. 262). These patients also may have degenerative changes within the CNS, due to demyelination of nerve fibers of the spinal cord.

Vitamin C (Ascorbic Acid). This vitamin is found in citrus fruits and in many vegetables, such as broccoli, potatoes, cabbage. The precise chemical function of vitamin C is unknown, but it is a strong reducing compound that probably can be reversibly oxidized and reduced in the body. Physiologically, the major function of this important vitamin centers on the maintenance of normal intercellular substances throughout the

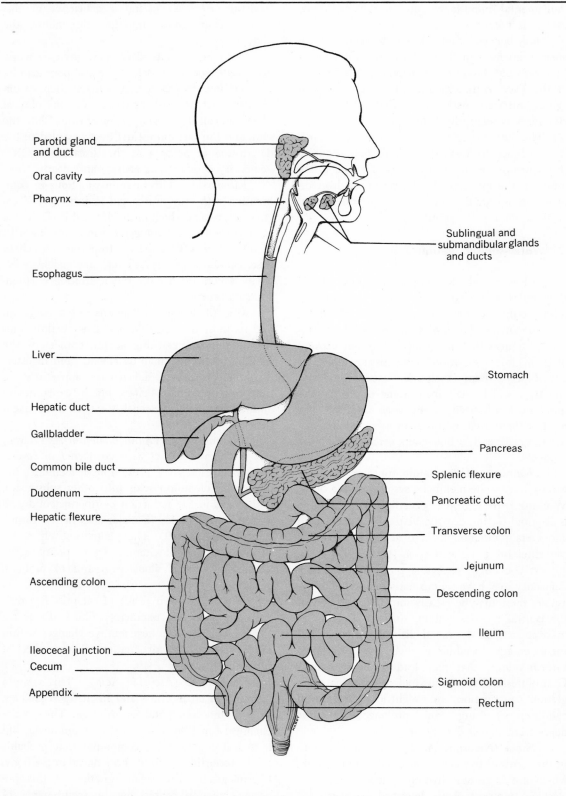

Parotid gland and duct

Oral cavity

Pharynx

Sublingual and submandibular glands and ducts

Esophagus

Liver

Stomach

Hepatic duct

Gallbladder

Pancreas

Common bile duct

Splenic flexure

Duodenum

Pancreatic duct

Hepatic flexure

Transverse colon

Ascending colon

Jejunum

Descending colon

Ileum

Ileocecal junction

Cecum

Appendix

Sigmoid colon

Rectum

Fig. 16-1.   Diagram of the digestive system, showing the digestive tube, or alimentary canal, and accessory organs.

body; this includes the formation of collagen, intercellular cement material, bone matrix, and tooth dentin. Vitamin C also plays a role in the absorption of iron.

When there is a deficiency of vitamin C, intercellular material is not deposited normally, and the resulting clinical condition is called *scurvy*. Some of the effects include failure of wounds to heal, cessation of bone growth, and petechial hemorrhages due to fragility of blood vessel walls. In severe scurvy, muscle cells may fragment, the gums are sore and spongy, teeth may loosen, and there may be internal bleeding.

THE DIGESTIVE TUBE

The digestive tube, or *alimentary canal*, is a long muscular tube lined with mucous membrane, extending from the lips to the anus. The total length is about 29 or 30 feet, of which 1½ feet are above the diaphragm. Each part of the tube is called by a different name—for example, mouth, pharynx, esophagus, stomach, small intestine, and large intestine. Each part has its own special anatomic characteristics and performs its own particular functions, yet each makes an essential contribution to the process of making food available to the cells of the body. The functions of the digestive tube are enhanced by its accessory organs: the tongue, teeth, salivary glands, liver, and pancreas (Fig. 16-1).

Structural layers

There are four main layers in the wall of the digestive tube: (1) mucous membrane lining, (2) submucous layer, (3) muscular layer, and (4) an external layer. The general plan is shown in Figure 16-2. Variations in the structural layers will

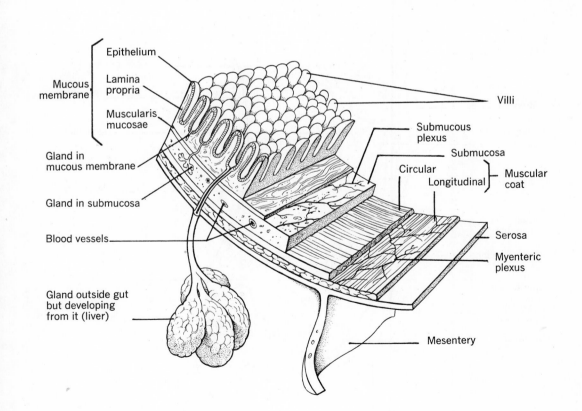

Fig. 16-2. Diagram of the four main layers of the wall of the digestive tube: mucosa, submucosa, muscular, and serosa (below the diaphragm).

be described as they occur throughout the tube.

The *mucous membrane*, or *mucosa*, is the innermost layer of the digestive tube. Its epithelium varies in structure according to the work being done in the part of the tube that it lines. For example, it is designed for protection (stratified squamous) in the mouth, the pharynx, the esophagus, and the anal region, but from the stomach to the anus, it is designed for processes of secretion and absorption (columnar). In addition to two thin layers of smooth muscle tissue (the muscularis mucosae), the mucosa contains many glands, blood vessels, and lymphatics.

The *submucous layer* serves to connect the mucous membrane with the main muscular layer. It is composed of loose connective tissue and contains networks of blood vessels. Nerve fibers and ganglion cells are interwoven to form a submucous plexus (Meissner's plexus) in which are found components of both divisions of the autonomic system.

The *muscular layer* (*muscularis externa*) is the third main layer of the wall of the digestive tube. It is composed of two very substantial sheets of smooth muscle, an inner circular and an outer longitudinal sheet. Its function is to mix the contents of the lumen with the digestive juices and to move them through the alimentary canal.

A plexus of nerve fibers and ganglion cells called the *myenteric*, or *Auerbach's plexus*, is found between the two sheets of smooth muscle. This plexus resembles the submucous plexus in that both divisions of the autonomic system are represented in it.

The *external layer* of the tube is *fibrous* in nature above the diaphragm and *serous* below the diaphragm. The serous membrane below the diaphragm is known as the visceral peritoneum.

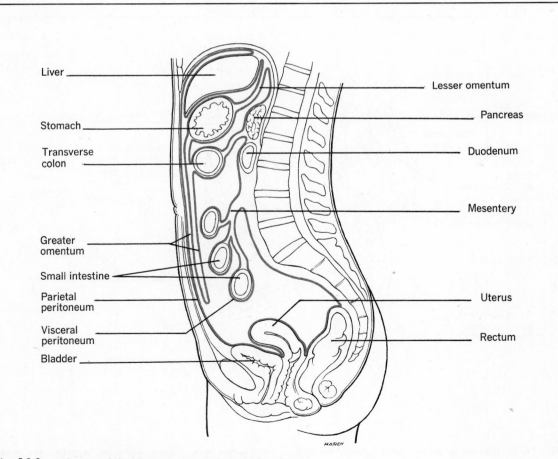

Fig. 16-3. Reflections of the peritoneum as seen in sagittal section.

The peritoneum

Although the peritoneum is not limited to the digestive tube, it seems appropriate to discuss it at the present time. The peritoneum is the largest serous membrane in the body, and it consists of a double-walled sac. The outer layer of this sac, called the *parietal peritoneum*, lines the abdominal cavity and then is reflected (turned or bent back) over the viscera, forming the *visceral peritoneum*, or inner layer of the sac. The space between these layers is called the *peritoneal cavity*. However, this is merely a potential space, and it contains nothing but a small amount of fluid which permits the viscera to glide freely on each other or against the wall of the abdominal cavity with a minimum of friction.

The *mesentery* is a fan-shaped extension of the peritoneum which encircles the intestine and serves to attach it to the posterior abdominal wall. Other folds of peritoneum that serve to connect organs within the abdominal cavity are called ligaments. The *greater omentum* is a special ligament, arranged like an apron of fat, that connects the stomach and the transverse colon. A second special ligament is the *lesser omentum*, which connects the stomach and the first part of the duodenum with the liver. The complicated reflections of the peritoneum, as seen in sagittal section, are illustrated in Figure 16-3.

Inflammation of the peritoneum is called *peritonitis*. It is accompanied by rigidity of the muscles of the abdominal wall, tenderness over the abdomen, paralysis of the intestine, vomiting, and pain.

The mouth

The mouth, or *buccal cavity*, is the first portion of the digestive tube. The mouth cavity proper is bounded laterally and in front by the teeth, above by the hard and the soft palates, and below by the tongue (Fig. 16-4).

The *lips* are composed of connective tissue and striated muscle (orbicularis oris). The outer surface is covered with skin, and the inner surface is lined with mucous membrane. The free margins are covered with modified skin, the derma of which contains so many capillaries that it gives a red color to the lips.

The *cheeks* also consist primarily of striated muscle (buccinator). They are covered with skin and lined with mucous membrane.

The *roof of the mouth* is formed by the bony hard palate anteriorly, and by the soft palate posteriorly. The hard palate is formed by the palatine processes of the maxillae and the horizontal plates of the palatine bones. The movable soft palate is attached to the posterior margin of the hard palate. It is drawn upward during the act of swallowing; this closes the nasopharynx and prevents food from entering the nasal cavities. The lower, or posterior, margin is free and forms an arch from one side of the pharynx to the other, with the uvula projecting from the middle of the arch, as shown in Figure 16-4. The soft palate is composed of interlacing striated muscle fibers, covered with mucous membrane. The passage from the mouth to the pharynx is called the *fauces*.

The Tongue. The tongue is an accessory organ and is composed of interlacing bundles of striated muscle fibers. The mucous membrane on the undersurface is thin; it forms a fold, called the *frenulum linguae*, that extends from near the tip of the tongue to the floor of the mouth. Tonguetie (ankyloglossia) refers to the condition in which the frenulum linguae is abnormally short and prevents free movement of the tongue. This may interfere with sucking during infancy and with speech later.

The mucous membrane on the dorsum of the tongue is thick, and that over the anterior two thirds, or body of the tongue, is studded with *papillae*, as shown in Figure 16-5. The papillae are of three types: filiform, fungiform, and vallate.

The *filiform papillae* are tall narrow structures arranged in parallel rows across the tongue. In certain animals, such as the cat, these papillae are so horny that the tongue feels like sandpaper. In man, the filiform papillae are not very highly developed, but they do permit one to lick substances, such as ice cream, in a satisfactory manner. These papillae also contain nerve endings for tactile sense (touch).

The *fungiform papillae* are scattered among the filiform papillae, but they are not nearly as numerous. Most contain taste buds (p. 235).

The *vallate papillae* are arranged in a V-shaped line between the body and the root of the tongue (i.e., the posterior third). All of them contain taste buds.

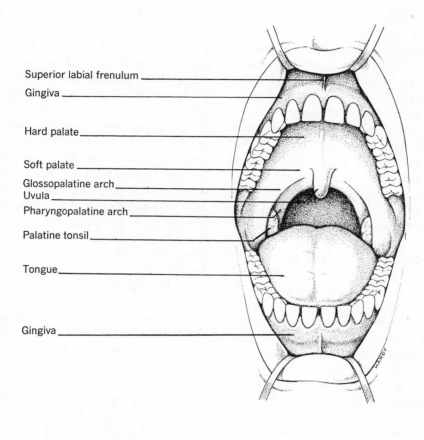

Superior labial frenulum
Gingiva
Hard palate
Soft palate
Glossopalatine arch
Uvula
Pharyngopalatine arch
Palatine tonsil
Tongue
Gingiva

Fig. 16-4. Structures in the mouth, or buccal cavity.

Underlying the mucous membrane at the root of the tongue are collections of lymphatic tissue, which compose the lingual tonsil.

The tongue is important in mastication (chewing), deglutition (swallowing), and speaking. Inflammation of the tongue is referred to as *glossitis*.

The Teeth. The teeth also are accessory organs. Each person has two sets of teeth during his lifetime. The first set is called the *milk*, or *deciduous*, teeth. These begin to erupt (called *dentition*) at about six months of age, and one appears about each month thereafter until all 20 have erupted. The deciduous teeth are lost between 6 and 13 years of age.

The second set of teeth is called the *permanent set*. These appear between 6 and 17 years (except the wisdom teeth). There are 32 in the permanent set. They are named from their shape or function. Beginning at the midline and progressing laterally, in each jaw, the teeth are called: central incisor, lateral incisor, canine, two premolars, or bicuspids, and three molars.

The *incisors* are shaped like chisels and serve to cut food.

The *canines* are shaped like dogs' teeth; they tear and shred food.

The *premolars* crush and tear food.

The *molars* crush and grind food.

The last molar tooth in each jaw is called a *wisdom tooth*; the four wisdom teeth may erupt between 17 and 25 years of age, or may never erupt at all. The teeth are shown in Figure 16-4.

The two chief parts of a tooth are the crown and the root.

The *crown*, or portion beyond the gum, is

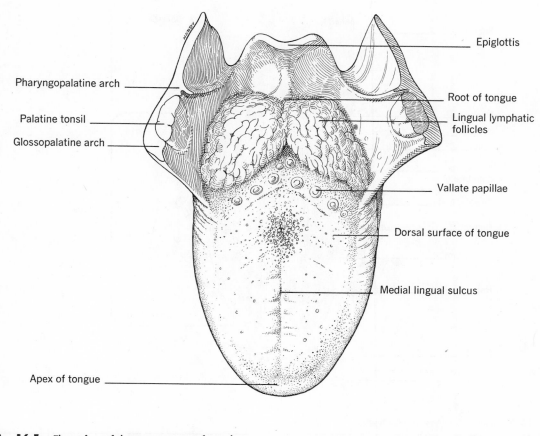

Epiglottis

Pharyngopalatine arch

Palatine tonsil

Glossopalatine arch

Root of tongue

Lingual lymphatic follicles

Vallate papillae

Dorsal surface of tongue

Medial lingual sulcus

Apex of tongue

Fig. 16-5. The surface of the tongue as seen from above.

covered with enamel, which is the hardest substance in the body. The bulk of the tooth is composed of dentin, which is calcified connective tissue. The central cavity in the shell of dentin is called the *pulp cavity* and is filled with soft, pulpy connective tissue. Blood vessels and nerves (branches of the trigeminal) enter the tooth by way of the root canal. The parts of a tooth are shown in Figure 16-6.

The *roots* of the teeth fit into sockets in the alveolar processes of the mandible and the maxillae. The roots are covered with cementum and are bound to the bone of the sockets by a strong connective tissue membrane, the *periodontal* membrane. The *gingiva,* or gum, is the part of the mucous membrane attached to the periosteum.

The teeth are laid down in the embryo, so the diet of the mother should contain an abun-

dance of calcium and vitamin D during pregnancy to insure proper development of teeth.

The Salivary Glands. There are three pairs of salivary glands which pour their secretions into the mouth: parotid, submandibular, and sublingual. These accessory organs manufacture saliva.

The *parotid glands* lie below and in front of the ears, between the mastoid processes of the temporal bones and the rami of the mandible. The location is shown in Figures 16-1 and 16-7. The duct (Stensen's) passes through the buccinator muscle and pours its secretion into the vestibule of the mouth opposite the upper second molar tooth, on each side. Inflammation of the parotid glands is called *parotitis* or mumps.

The *submandibular* (*submaxillary*) *glands* lie in contact with the inner surface of the mandible and the ducts (Wharton's) open into the floor

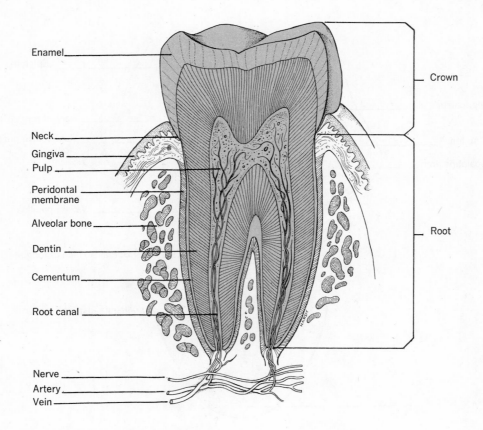

Enamel

Neck

Gingiva

Pulp

Peridontal
membrane

Alveolar bone

Dentin

Cementum

Root canal

Nerve

Artery

Vein

Crown

Root

Fig. 16-6.   Longitudinal section of a tooth.

of the mouth beside the frenulum linguae, behind
the lower incisors (Fig. 16-7).

The *sublingual glands* lie beneath the mucous
membrane of the floor of the mouth and open by
means of several ducts in the floor of the mouth.

The secretions from the salivary glands may
be thin and watery or thick and viscid, according
to the type of stimulus which arouses them.

The pharynx

The various subdivisions of the pharynx are
described in connection with the respiratory system (pp. 352 to 354). At this time we shall mention only the muscular layers that are concerned
with the act of swallowing.

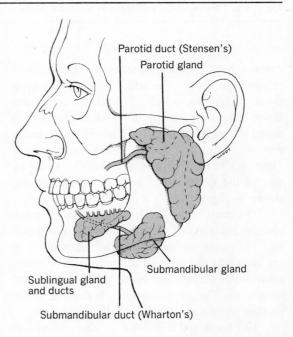

Parotid duct (Stensen's)
Parotid gland

Submandibular gland

Sublingual gland
and ducts

Submandibular duct (Wharton's)

Fig. 16-7.   The salivary glands and their ducts.

Beneath the mucous membrane and fibrous tissue of the pharynx, there is a thick layer of skeletal muscle. This muscular layer forms the superior, middle, and inferior *constrictor muscles* of the pharynx. When a bolus of food enters the pharynx, it is grasped by the constrictor muscles and moved onward into the esophagus. Another series of muscles raises the pharynx during swallowing and draws the sides laterally to increase its width.

The esophagus

The esophagus is a muscular, collapsible tube, about 10 inches long, which lies behind the trachea. It begins at the end of the laryngopharynx and passes through the mediastinum and the diaphragm to end in the stomach.

The muscular walls of the upper third of the esophagus contain striated muscle which is gradually replaced by smooth muscle as the tube descends toward the stomach. However, the aforementioned striated muscle is not under conscious control; it is supplied by autonomic fibers which travel in the vagus nerve and supply the smooth muscle as well. At the lower end of the esophagus, just above its junction with the stomach, the circular layer of the muscular wall is slightly hypertrophied. Although this is not a true anatomic sphincter, it functions as such by remaining tonically constricted until the third stage of swallowing (p. 401); thus the term *gastroesophageal constrictor* is more meaningful than cardiac sphincter.

The stomach

Just below the diaphragm, the alimentary canal expands to form the stomach, a pouchlike structure that serves as a reservoir for food during the early stages of digestion. It lies in the upper part of the abdominal cavity, in the epigastric and left hypochondriac regions (see Fig. 1-7).

Divisions. The esophagus opens into the upper end of the stomach through the *cardiac orifice*, an opening so named because of its proximity to the heart. The part of the stomach that lies above and to the left of the cardiac orifice is called the *fundus*, while the central portion is known as the *body*. The pyloric portion of the stomach is that beyond the body; it communicates with the small intestine by way of the *pylorus*, an opening that is surrounded by a muscular ring called the *pyloric sphincter*. Hypertrophic *pyloric*

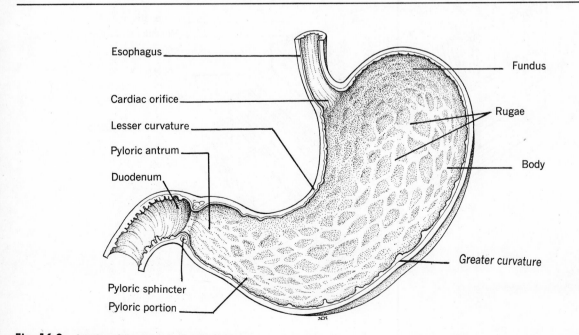

Fig. 16-8. Interior of the stomach, showing rugae.

stenosis occurs most frequently in young male infants; in this condition the pyloric sphincter is in a state of persistent spasm, and hypertrophy of the sphincter results. The condition must be treated surgically. The medial border of the stomach is called the *lesser curvature*, and the lateral border the *greater curvature*. Figure 16-8 illustrates the divisions of the stomach.

Structural Variations. When the stomach is empty, the mucous membrane lining is thrown into folds called *rugae*. These folds gradually smooth out and disappear as the stomach becomes distended with food.

The epithelium of the mucous membrane consists of a single layer of columnar cells; this layer not only covers the surface of the membrane, but also extends down into numerous little wells called the gastric pits. Opening into the bottom of each pit are two or three simple tubular glands; these gastric glands produce the *gastric juice*, a secretion that is delivered into the pits, which, in turn, conduct it to the surface of the mucosa.

Gastric juice contains hydrochloric acid, digestive enzymes, and mucus. The hydrochloric acid is produced by *parietal cells*, located near the opening of a gland into its gastric pit. The enzymes are produced by *zymogenic*, or *chief cells*, located in the base or body of a gland. Protective mucus is produced by cells in the neck of each gland and by cells in the surface epithelium. The function of gastric juice is described on page 403. A special component of gastric juice, the intrinsic factor, is described on page 262 in relation to the absorption of vitamin B_{12} and the maturation of red blood cells.

The main muscular layer of the stomach varies from the general structural plan in that it is composed of three layers of smooth muscle: an inner oblique, a middle circular, and an outer longitudinal layer. The ringlike pyloric sphincter is formed by a thickening of the circular muscle fibers.

The outer serous layer of the stomach is composed of visceral peritoneum. The various layers

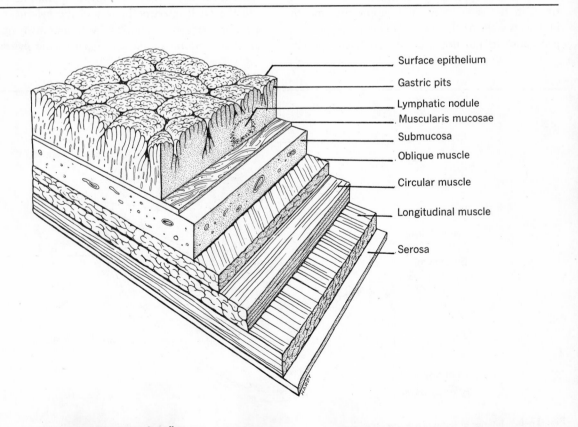

Surface epithelium
Gastric pits
Lymphatic nodule
Muscularis mucosae
Submucosa
Oblique muscle
Circular muscle
Longitudinal muscle
Serosa

Fig. 16-9. *Layers of the stomach wall.*

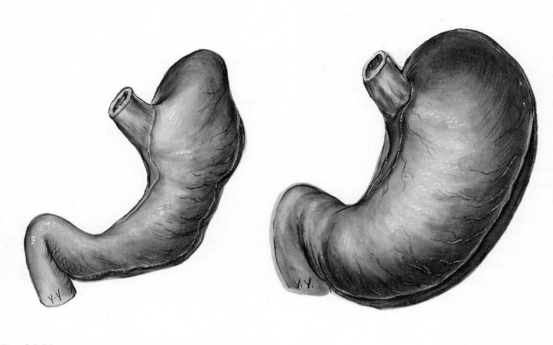

Fig. 16-10. *Left,* stomach before, and *right,* after a meal.

of the wall of the stomach are illustrated in Figure 16-9.

The extensibility of the smooth muscle of the stomach is well known to all of us. We can eat a fairly large quantity of food at one meal, and the stomach stretches to accommodate it without an increase in pressure. The difference in size before and after a meal is illustrated in Figure 16-10.

The small intestine

The greatest amount of digestion and absorption takes place in the small intestine, a convoluted (coiled) tube extending from the stomach to the large intestine. It is approximately 23 feet in length and 1 inch in diameter. The fan-shaped mesentery attaches the coiled intestinal loops to the posterior abdominal wall.

Divisions. The small intestine consists of three divisions: duodenum, jejunum, and ileum.

The *duodenum,* about 10 inches in length, is the shortest, widest and most firmly attached portion of the small intestine. It is arranged in the shape of the letter "C," and the head of the pancreas lies in the concavity which it forms.

The *jejunum* forms the next two fifths of the small intestine and extends from the duodenum to the ileum.

The *ileum* forms the remainder of the small intestine; it terminates by joining the large intestine at right angles. A muscular sphincter (ileocecal valve) guards this exit gate.

Structural Variations. The mucous membrane of the small intestine differs markedly from that in the remainder of the alimentary canal. To increase the epithelial surface, and thus provide a large area for absorption of nutrients, the mucosa is arranged in circular folds and is studded with villi.

The *circular folds* (*plicae circulares*) are folds of mucosa that project into the lumen of the intestine, forming ridges that somewhat resemble an old-fashioned washboard (Fig. 16-11).

The *villi* are tiny, fingerlike projections of mucous membrane that are barely visible to the naked eye. Much like a forest of trees on a series of hills and dales, the villi are crowded so close together over the mucosal surface that they give it the appearance of velvet. The villi are highly specialized for the absorption of nutrients, and each contains a blood capillary and a lymph vessel (lacteal). Figure 16-12 shows villi enlarged many times.

Most of the epithelial cells lining the small intestine are tall, columnar cells that have numerous microvilli on their free border; the primary function of these cells is absorption, and the microvilli greatly increase the surface area for such activity. Distributed among these cells are true goblet cells (p. 37), which secrete mucus.

There is an abundance of lymphatic tissue just below the epithelium of the mucosa. This may appear as solitary nodules or in groups of nodules called *Peyer's patches*. The *intestinal glands (crypts of Lieberkühn)* also are found in this area. They open into the depressions between villi, and their function is to secrete enzymes which aid in the digestion of food.

The *duodenal glands (Brunner's glands)* are located in the submucosal layer; they are largest and most numerous near the pylorus. Their function is to secrete protective mucus.

Intestinal obstruction may be due to: (1) hernial-sac incarceration or retention; (2) foreign bodies, especially at the ileocecal junction; (3) stricture from an ulcer, peritonitis, or tumor; (4) volvulus, a knotting or twisting of the intestine;

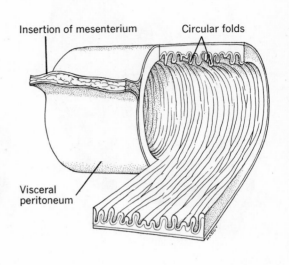

Fig. 16-11.   Circular folds of the jejunum.

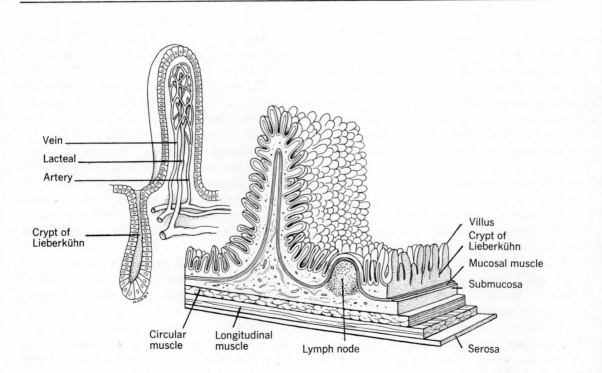

Fig. 16-12.   Drawing of the mucous membrane of the small intestine, showing the numerous villi on one circular fold. At the *left* is an enlarged drawing of a single villus.

(5) intussusception, or telescoping, of one part of the intestine into another portion of it; and (6) bands that arise from adhesions. Any intestinal obstruction is exceedingly serious, since it upsets the water and salt balance of the body and cuts off the blood supply to a part of the intestinal wall.

Paralytic ileus is a condition in which the bowel is paralyzed. It may occur after any abdominal or pelvic operation and has the same effect as other types of obstruction.

Ileostomy refers to the surgical formation of a fistula, or an artificial anus, through the abdominal wall into the ileum. This may be necessary for those who suffer from cancer of the colon or the rectum.

The large intestine

The large intestine is about 4½ to 5 feet in length and 2½ inches in diameter as it extends from the end of the ileum to the anus. It is larger in caliber than the small intestine and has a sacculated appearance.

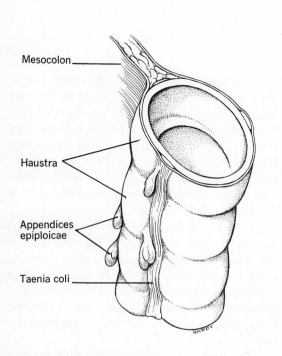

Mesocolon

Haustra

Appendices epiploicae

Taenia coli

Fig. 16-13. A portion of the descending colon, showing haustrae, taenia coli, and appendices epiploicae.

Divisions. The large intestine may be divided into the cecum, the colon, the rectum, and the anal canal (see Fig. 16-1).

The *cecum* is a blind pouch about 2 to 3 inches long which hangs down at the junction of the ileum and the colon. The ileocecal valve lies at the upper border of the cecum and prevents the return of feces from the cecum into the small intestine. The vermiform appendix arises from the cecum about 1 inch below the ileocecal valve. It is a twisted and coiled tube whose walls contain all the layers of the intestine. Inflammation of the appendix, called *appendicitis*, is quite common. Since the appendix is a blind tube, wastes accumulate which cannot be moved by peristaltic contractions. Muscular rigidity, pain, and vomiting are common. The chief danger is that of rupture, followed by peritonitis. If the appendix becomes inflamed, it should be removed (appendectomy) before it ruptures.

The function of the appendix has not yet been determined. However, studies carried out on the appendix of the rabbit suggest that this organ, like the thymus (p. 326), may play a role in the development of immunological competence.

The *colon* is subdivided into ascending, transverse, descending, and sigmoid portions.

The *ascending colon* extends from the cecum to the undersurface of the liver, where it turns left to form the right or hepatic flexure.

The *transverse colon* crosses the upper part of the abdominal cavity from right to left and turns downward to form the left, or splenic, flexure.

The *descending colon* extends from the splenic flexure to the brim of the pelvis, where it turns toward the midline to become the sigmoid colon (Fig. 16-13).

The *sigmoid (pelvic) colon* extends from the descending colon at the level of the pelvic brim to the rectum. As shown in Figure 16-1, the sigmoid colon is S-shaped as the name implies. *Colostomy* refers to the surgical formation of an artificial anus or opening from the colon to the exterior.

The *rectum* extends from the sigmoid colon to the anal canal and is about 5 or 6 inches in length. It descends along the hollow of the sacrum and the coccyx to the tip of the coccyx. Cancer may occur in any part of the colon, but it is more commonly found in the rectum.

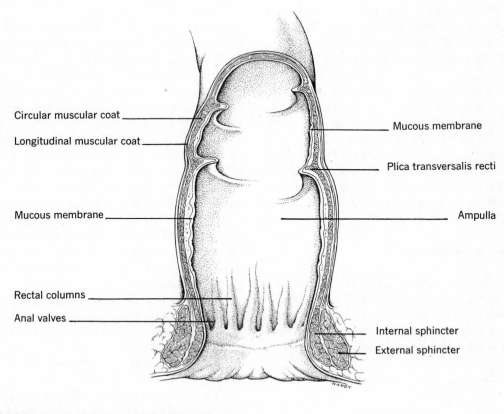

Circular muscular coat

Longitudinal muscular coat

Mucous membrane

Rectal columns

Anal valves

Mucous membrane

Plica transversalis recti

Ampulla

Internal sphincter

External sphincter

Fig. 16-14. Interior of the rectum and anal canal.

The *anal canal* is the terminal portion of the large intestine. It extends from the rectum to the anus (opening to the exterior) and is about 1 to 1½ inches in length. The anal canal passes between the two medial borders of the levator ani muscles, which compress it to a mere slit. Powerful sphincter muscles guard the anal opening; these muscles are described below.

Structural Variations. The mucous membrane of the large intestine differs from that of the small intestine. There are no villi, and the epithelium contains large numbers of goblet cells. The glands (crypts of Lieberkühn) secrete mucus but no digestive enzymes. The mucosa of the upper end of the anal canal differs from that of other parts of the large intestine in that it is arranged in vertical folds, called *rectal columns* (Fig. 16-14). Beneath these columns are hemorrhoidal (rectal) veins which may become tortuous and dilated; this condition is called *hemorrhoids.*

The main muscular layer differs from that in other parts of the alimentary canal in that the longitudinal muscle fibers are concentrated into three flat bands, called the *taeniae coli* (see Fig. 16-13). These bands are not as long as the large intestine; consequently, the latter is gathered (shirred) into sacculations, or pouchlike structures, known as *haustra*, which are characteristic of this portion of the digestive tube.

In the anal canal, the circular muscle fibers are greatly thickened, forming the *internal anal sphincter*; this ring of smooth muscle helps to guard the exit of the alimentary canal, but it is not under control of the will. The *external anal sphincter* is a cylinder of skeletal muscle that also helps to close the anus, but it is under voluntary control.

The visceral peritoneum forms little pouches containing fat. These pouches hang from the large intestine and are called *appendices epiploicae.*

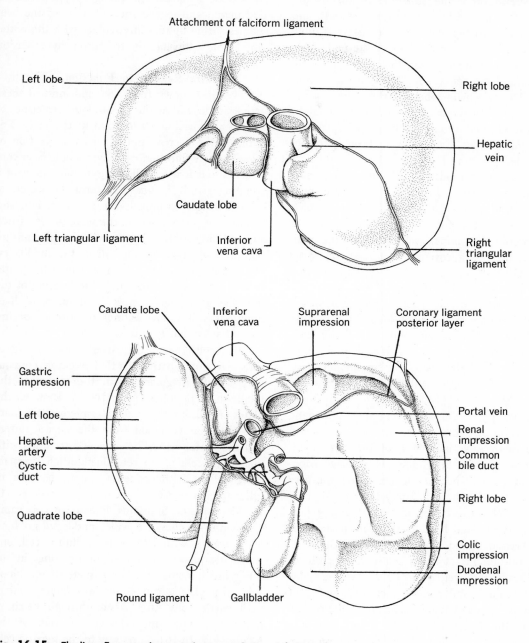

Attachment of falciform ligament

Left lobe

Right lobe

Hepatic vein

Caudate lobe

Left triangular ligament

Inferior vena cava

Right triangular ligament

Caudate lobe

Inferior vena cava

Suprarenal impression

Coronary ligament posterior layer

Gastric impression

Left lobe

Portal vein

Renal impression

Hepatic artery

Cystic duct

Common bile duct

Right lobe

Quadrate lobe

Colic impression

Duodenal impression

Round ligament

Gallbladder

Fig. 16-15. The liver. *Top,* superior, posterior aspect. *Bottom,* inferior surface.

THE ACCESSORY ORGANS

The tongue, the teeth, and the salivary glands are accessory organs related to the mouth; they are described on pages 385 to 388. The liver, with its biliary apparatus, and the pancreas remain to be discussed.

The liver and biliary apparatus

The liver is the largest organ of the body, weighing about 3 pounds in an adult. It is located in the upper part of the abdominal cavity, under the dome of the diaphragm, in the epigastric and right hypochondriac regions (see Fig. 1-7).

Description. The superior, or diaphragmatic, surface of the liver is relatively smooth and convex (Fig. 16-15). The inferior surface is more concave and shows impressions of the underlying organs, such as the stomach, the kidney, the duodenum, and the colon. The liver consists of four lobes. The two main lobes are the right and the left, separated by the falciform ligament. The right lobe is subdivided into right lobe proper, quadrate, and caudate lobes. Each lobe, in turn, is further subdivided into tiny lobules.

The liver is almost completely covered with visceral peritoneum. In certain areas this peritoneum is thrown into folds (ligaments) which help to attach the liver to the diaphragm and to the anterior abdominal wall. Beneath the peritoneum there is a covering of fibroelastic tissue called *Glisson's capsule*, which accompanies the blood vessels into the interior of the liver. The porta is a fissure which extends transversely across the left portion of the right lobe. At the porta there is a tree trunk of connective tissue, continuous with the capsule, which extends up into the substance of the liver. The tree branches extensively and furnishes internal support for the liver.

At the porta, the portal vein and the hepatic artery enter the tree of connective tissue and branch with each branching of the tree. Lymphatic vessels and bile ducts, which have tributaries in every branch of the tree, leave the liver at the porta. The hepatic artery supplies the liver with arterial blood; hence, it supplies a large part of the oxygen used by the liver. The portal vein brings venous blood that has traversed the walls of the intestinal tract and is rich in absorbed food. Blood leaves the liver by way of the hepatic veins, which are not found in the branches of the tree of connective tissue. These veins begin in the central veins of the liver, which unite to form the sublobular and then the hepatic veins, which, in turn, empty into the inferior vena cava as it passes through a groove on the posterior surface of the liver.

A unit of liver substance is called a *liver lobule*. The lobules are separated by a small amount of interlobular connective tissue in which are found terminal branches of the hepatic artery, the portal vein, and the beginnings of bile ducts and lymphatic vessels.

Cirrhosis is a condition in which parts of the liver undergo structural changes as a result of destruction and regeneration of some of the liver lobules. Often there is interference with the portal circulation which leads to portal hypertension (p. 318).

Structure Related to Function. The liver has over 100 known functions, and most of these activities depend upon the unique structure of the liver, as well as its location in the body. For example, the liver is famous for its production of bile. Thus we find that each liver lobule is so constructed that it meets all of the requirements of a secreting gland (i.e., cells to form the secretion, a blood supply to provide the raw materials, and a system of ducts whereby the secretion is carried away). The versatile hepatic cells also play an important role in monitoring all of the food-laden blood that is brought to them by way of the portal system of veins (p. 316); thus the location of the liver has great significance. A summary of liver functions will follow a brief description of important structural features.

The substance of each liver lobule consists of thin plates of hepatic cells that resemble cords as they radiate irregularly from the center of the lobule to its periphery (Fig. 16-16). Between the plates of cells are the *sinusoids*, microscopic channels that receive blood from the hepatic artery and the portal vein. The sinusoids are lined with phagocytic cells of the reticuloendothelial system; these cells (Kupffer's cells) engulf bacteria or other foreign particles that may gain entrance to the bloodstream. Sinusoids differ from true capillaries in that they are wider and tend to have leaky walls, inasmuch as their lining cells are irregularly placed. Although the openings in the lining are too small for the passage of red blood cells, the plasma can come into direct contact with hepatic cells. After passing through the sinusoids, blood flows into the central vein of the lobule (Fig. 16-16) on its way to the hepatic vein and thence to the inferior vena cava. One should keep in mind the fact that the millions of sinusoidal spaces in the liver constitute a vast spongework, which can hold large quantities of blood at any given time; thus the liver must be considered an important reservoir, which can influence the circulating blood volume as well as the cardiac output.

The Bile Drainage Apparatus. This drainage system begins as microscopic, cleftlike spaces

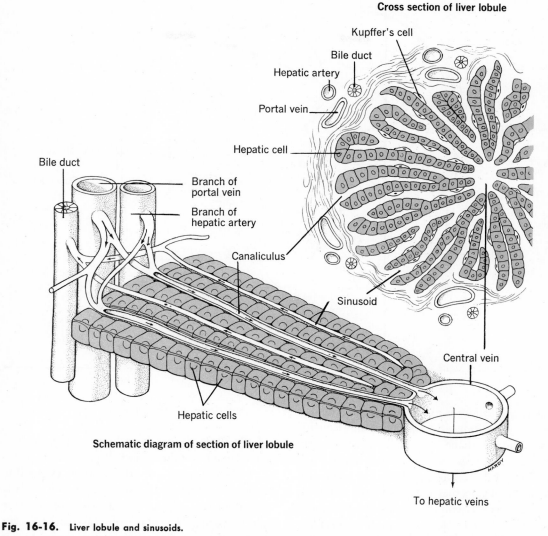

Cross section of liver lobule

Kupffer's cell

Bile duct

Hepatic artery

Portal vein

Hepatic cell

Bile duct

Branch of
portal vein

Branch of
hepatic artery

Canaliculus

Sinusoid

Central vein

Hepatic cells

Schematic diagram of section of liver lobule

To hepatic veins

Fig. 16-16.   Liver lobule and sinusoids.

(canaliculi) between adjacent rows of hepatic cells within the plates (Fig. 16-16). As bile is produced by the hepatic cells it enters the canaliculi and drains toward the periphery of the lobule; here the canaliculi drain into bile ductules, which drain, in turn, into larger ducts. Ultimately, all bile is collected into one large duct from each lobe. Two main trunks, one from the right lobe and one from the left lobe, unite to form the *common hepatic duct*. The hepatic duct descends to the right for a distance of about 1½ inches, at which point it is joined at an acute angle by the *cystic duct* from the gallbladder to form the *common*

bile duct. The common duct descends behind the head of the pancreas, where it usually joins the pancreatic duct, forming a single dilated tube known as the *ampulla of Vater;* the ampulla opens into the duodenum at the *duodenal papilla* (Fig. 16-17). Muscle tissue associated with the ampulla forms a very weak sphincter (of *Oddi*), but the bile duct (just before it fuses with the pancreatic duct) is guarded by a strong sphincter muscle (of *Boyden*). When this sphincter is closed, bile is prevented from entering the intestine and, as a result, passes by way of the cystic duct to the gallbladder, in which it is stored until needed.

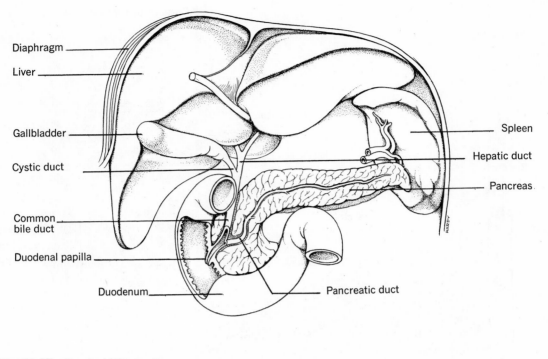

Fig. 16-17. Liver and biliary system.

The *gallbladder* is a musculomembranous sac with an average capacity of 40 to 50 ml. Roughly the size and the shape of a small pear, this organ is located in a fossa on the undersurface of the liver (Fig. 16-17). The gallbladder is lined with mucous membrane that is thrown into tiny folds when the organ is empty. Connective tissue and smooth muscle fibers form the framework of the wall. Except where it is attached to the liver, the external coat of the gallbladder is serous membrane (visceral peritoneum).

The function of the gallbladder is to store and concentrate bile. When food enters the duodenum, a hormone (cholecystokinin) is released by cells in the duodenal mucosa. Carried in the bloodstream to the gallbladder, this hormone stimulates the smooth muscular wall to contract and dispatch bile to the duodenum when it is needed. The importance of bile is discussed under liver function.

Cholecystitis is inflammation of the gallbladder, and it often is accompanied by obstruction of the cystic duct. The presence of gallstones is referred to as *cholelithiasis*.

Functions of the liver

Of all the organs in the human body, the liver is the largest and the busiest. It performs an amazing variety of functions, many of them essential to life itself. Although the liver possesses a remarkable ability to regenerate and thus replace tissue lost by partial resection (surgical removal), we must stress the fact that severe injury to, or disease of, the liver may be rapidly fatal. A few of the broad areas of liver function are described briefly.

The Production of Bile. Bile, which is considered to be both a secretion and an excretion, is formed continuously by the hepatic cells; the average daily volume of bile is about 600 to 800 ml. It is a greenish-yellow fluid that contains water, bile salts, bilirubin, cholesterol, and various inorganic salts. Of these constituents, only the *bile salts* (formed from cholesterol) have important functions to perform in the digestive tube. Upon entering the small intestine, the bile salts have two important actions. First, they serve to *emulsify fats* by so lowering the surface tension that large fat globules are broken into tiny particles; this action is valuable, because the fat-splitting enzymes can attack many small units more efficiently than one large unit. Secondly, bile salts help in the *absorption of fatty acids* from the intestinal tract (p. 412).

Bilirubin, or bile pigment, is excreted by the liver; this yellowish substance is one of the major end-products resulting from the decomposition of hemoglobin. As previously described (p. 326), when worn-out red blood cells are removed from the circulating blood their hemoglobin is broken down by cells of the reticuloendothelial system; bilirubin, bound to a plasma protein, travels in the bloodstream to the liver, where it is conjugated (made water-soluble) and excreted into the bile.

Between meals, bile is stored in the gallbladder, in which it becomes very concentrated because of the absorption of water, sodium, chloride, and other small electrolytes. The mechanism of emptying of the gallbladder is described on page 398.

Metabolic Functions. The liver plays a vital role in the metabolism of the three major food groups. A brief summary of these functions is presented below. Further details can be found in Chapter 17.

Carbohydrate metabolism. The liver helps to maintain a normal concentration of glucose in the blood. For example, after meals when the glucose level tends to rise, the liver converts glucose to glycogen and stores it until needed; when the blood glucose level falls, glycogen is changed back to glucose and released to the blood. Other simple sugars, galactose and fructose, coming to the liver, are converted to glucose, which then may be stored or used immediately as required by the body. If the blood glucose level continues to fall, the liver has the ability to form glucose from noncarbohydrate sources, such as amino acids and the glycerol portion of fats; this activity is referred to as *gluconeogenesis.*

Lipid (fat) metabolism. Although the metabolism of fats probably can take place in most cells of the body, certain activities occur much more rapidly in the liver. For example, fatty acids are an important source of energy, but they must first be broken down to small molecules that can enter the citric acid cycle and be oxidized (p. 421); it is believed that about 60 per cent of all the preliminary breakdown of fatty acids in the body

occurs in the liver. Other specific functions are the formation of large quantities of cholesterol and phospholipids, the formation of lipoproteins, and the synthesis of fat from glucose and amino acids.

Protein metabolism. The liver plays its most vital role in the metabolism of protein. For example, except for the gamma globulins, essentially all of the plasma proteins (p. 266) are formed by the hepatic cells. The liver forms important chemical substances, such as phosphocreatine, from amino acids. It synthesizes certain amino acids, referred to as nonessential amino acids because it is not essential that they be in the diet, so long as the liver can provide them. The liver deaminizes amino acids (p. 424) so that they can be used for energy or can be converted to glucose or to fat. It removes ammonia from the blood by converting it to urea for elimination by the kidney.

Miscellaneous metabolic functions. The liver is capable of forming vitamin A from precursor substances in certain vegetable foods. It also stores large quantities of vitamin A, vitamin D, and vitamin B_{12}. Iron is stored in the liver in the form of ferritin. Various hormones (e.g., gonadal and adrenocortical) are inactivated in the liver and made more soluble, so that they can be excreted in the bile and in the urine. The liver also is concerned with the detoxification of harmful compounds, such as alcohol and certain drugs, so that these substances can be eliminated in the bile and urine.

Other functions of the liver include the formation of heparin, an anticoagulant, and the phagocytic action of the Kupffer cells, as previously described on page 396.

Tests of liver function

The liver performs such a variety of functions that no one test can give information about the liver as a whole. Only a few of the available tests are mentioned at this time.

Serum Enzymes. Patients with liver disease may have considerable elevation of serum transaminases and dehydrogenases. This is due to the fact that these intracellular enzymes escape from injured liver cells and enter the bloodstream.

Bilirubin (Van Den Bergh Test). An abnormally high level of bilirubin in the blood may be indicative of liver disease, of obstruction of the biliary tract, or of increased rate of red blood cell destruction. However, if for any reason the rate of bilirubin excretion fails to equal that of production, the level of bilirubin in the blood increases. As the blood level of bilirubin rises (above 2 mg. per 100 ml.) there is a proportionate accumulation of bilirubin in the tissues, staining them and giving rise to the characteristic yellow pigmentation known as *jaundice* (icterus). Although any of the body tissues may become stained with bilirubin, jaundice makes its earliest appearance in the sclera of the eye and in the skin of the face and upper trunk.

Bromsulphalein (BSP) Test. Injected intravenously, BSP dye is eliminated almost exclusively by the liver and is a very sensitive indicator of hepatic disease. Normally, less than 5 per cent of the dye is retained after 45 minutes.

Gallbladder Visualization. The person ingests a special dye that is excreted by the liver into the bile; the dye-laden bile enters the gallbladder, where it is concentrated. Because the dye is opaque to x-rays, it outlines the gallbladder and makes it visible. The person then is given a fatty meal, and additional x-ray films are taken to see whether or not the gallbladder has emptied properly.

The pancreas

The pancreas is both an exocrine and an endocrine gland; it is located in the epigastric and left hypochondriac regions of the abdomen. As illustrated in Figure 1-7, the head of this hammershaped organ rests in the concavity of the duodenum, the body extends toward the spleen, and the tail is in contact with the spleen. Fresh pancreas appears white, with a tinge of pinkish color, and the surface has a lobulated appearance.

Closer examination of the pancreas shows that it is composed of lobules in which the *exocrine* secretory units, called *acini*, resemble grapes hanging on stems. Each acinus consists of a single row of epithelial cells arranged around the lumen of a duct; these acinar cells are concerned with the production of *pancreatic juice*, the exocrine secretion of this important gland. As will be described later, the pancreatic juice contains various enzymes that are important in the digestion of foods. Figure 16-17 illustrates the duct system

whereby pancreatic juice reaches the duodenum; the main *pancreatic duct* (of Wirsung) usually unites with the common bile duct, while an accessory duct (of Santorini) may open directly into the duodenum.

The *endocrine* secretory units consist of irregular clumps of cells referred to as the *islets of Langerhans*; these islets, scattered throughout the pancreas, contain extensive capillary networks that serve to carry off the endocrine secretions (hormones). Two types of cells—beta and alpha—may be distinguished in the islets. The beta cells produce *insulin*, a hormone that is essential for normal carbohydrate metabolism; hypofunction of these cells causes *diabetes mellitus*. The alpha cells produce *glucagon*, which raises the blood sugar level. Additional information on the endocrine function of the pancreas can be found on page 484.

DIGESTIVE PROCESSES

Attention already has been called to the fact that complex foods must undergo specific changes before they can be absorbed from the digestive tract; these foods are carbohydrates, fats, and proteins. Others, such as water, vitamins, and minerals, are ready for immediate absorption. Two types of activity are involved in preparing complex foods for absorption: mechanical and chemical.

Mechanical Activity. This involves purely physical processes. These include the chewing of food, moving it along the digestive tube, mixing it with the various digestive juices, spreading it over the absorptive surfaces, and, finally, the elimination of the residue.

Chemical Activity. This activity involves the process of *hydrolysis*, the splitting of complex compounds into simple fragments by the addition of water. This process depends much on the presence of *digestive enzymes*, the organic catalysts that greatly increase the speed of hydrolytic reactions. Each enzyme acts on a certain type of food and requires a certain pH if it is to operate with greatest efficiency.

It is customary to name enzymes after the food acted upon by adding the suffix *-ase*. Thus *amylase* acts on starches, *lipase* acts on fats, and *protease* acts on proteins. However, some of the older, original names, such as pepsin and trypsin, still are in use. The names of the enzymes, the foods on which they act, and the products formed as a result of their various activities are presented in Table 16-1.

Digestion in the mouth

Saliva is the first secretion encountered by the food eaten. It consists of a mixture of the secretions of the three pairs of salivary glands, as well as mucus and cells from the surface epithelium of the mouth. Saliva contains watery mucin, inorganic salts, and a digestive enzyme called *salivary amylase* (ptyalin). However, the mechanical functions of saliva probably are of greater importance than is its somewhat limited chemical contribution. Saliva protects the lining of the mouth against drying, facilitates speech, and lubricates food so that it can be chewed and swallowed; chemically, saliva begins the digestion of starches.

The secretion of saliva begins when one thinks about food, sees it, smells it, or hears it talked about; this preliminary flow of saliva is called the *psychic* secretion. When food is taken into the mouth, it dissolves to some extent in the saliva and stimulates the taste buds; as a result of this stimulation, saliva continues to flow while one actually is eating. The average daily secretion of saliva is 1200 ml.; its pH is between 6.3 and 7.0, a favorable range for the action of salivary amylase.

Mechanical activity in the mouth depends on saliva, the tongue, the teeth, and the muscles of mastication. As one chews, or masticates food, it is ground into small particles and thoroughly mixed with saliva, which serves as an important lubricant.

Chemical activity in the mouth depends on the presence of *salivary amylase*; this enzyme attacks complex starch molecules and breaks them down to dextrins and maltose. Dextrin molecules are intermediate in size between starch and maltose; because food does not remain in the mouth for a long period of time, and because salivary action is inhibited by the acid gastric juice, the conversion of all starches to dextrins and of all dextrins to maltose may not occur until they reach the small intestine.

Role of the pharynx and esophagus

Food is carried from the mouth to the stomach by the act of swallowing, or deglutition. For convenience of description, this is divided into three stages.

The *first stage* of swallowing consists of collecting the bolus, or mass of food, on the dorsum of the tongue. Then a sudden elevation of the tongue throws the mass into the pharynx. The first stage of swallowing is under voluntary control.

The *second stage* of swallowing is the passage of the bolus through the pharynx. The air passages are shut off, as previously described (p. 354) and respiration is inhibited. The bolus is grasped by the constrictors of the pharynx and passed along into the esophagus. Figure 16-18 indicates the changes that occur in the position of the soft palate and the larynx during swallowing. The second stage is involuntary, although muscles of the pharynx are striated.

The *third stage* is the passage through the esophagus. In man, the muscle in the upper portion of the esophagus is striated, but not under the control of the will. The muscle in the lower portion of the esophagus is smooth. It exhibits waves of contraction called *peristalsis* (meaning *to contract around*). A typical peristaltic wave is preceded by relaxation. The function of peristaltic waves is to move food along the digestive tube in a caudal direction (Fig. 16-19). The gastroesophageal constrictor relaxes as a peristaltic wave in the esophagus approaches. Each wave takes about 5 or 6 seconds to traverse the esophagus.

The only secretion produced by glands in the pharynx and the esophagus is mucus, which facilitates swallowing of a bolus of food by making it slippery. No enzymes are produced in this portion of the digestive tube.

Digestion in the stomach

Gastric juice is the highly acid secretion produced by glands in the mucosa of the stomach (p. 390). The secretion of this digestive fluid occurs in three phases.

The *cephalic phase* is stimulated by the sight, smell, taste, or even by the thought of food; this phase derives its name from the fact that the related receptors and afferent nerves are located in the head. Efferent impulses travel over the

Breathing   Swallowing

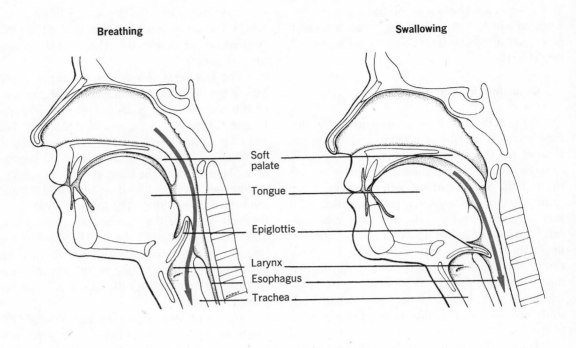

Soft palate

Tongue

Epiglottis

Larynx

Esophagus

Trachea

Fig. 16-18. Diagram showing elevation of soft palate and larynx during swallowing.

vagus nerves and promote the reflex secretion of gastric juice. The cephalic phase is important in that it assures the presence of gastric juice as soon as food enters the stomach; thus one can see the value of preparing appetizing food that is attractively arranged.

The *gastric phase* occurs because of the presence of food in the stomach; this food is thought to stimulate secretion in two ways. First, owing to its actual bulk, there is a mechanical stimulus involved in stretching the wall of the stomach. Second, a chemical stimulus is provided by the presence of partially digested protein and certain food extractives. These mechanical and chemical factors both serve to stimulate the production of a hormone, *gastrin*, by cells in the pyloric portion of the gastric mucosa. Gastrin enters the bloodstream and is carried to glands in the fundus and body of the stomach, where it stimulates the production of gastric juice.

The *intestinal phase* of gastric secretion involves the presence of chyme in the duodenum. (After food has become mixed with gastric juice

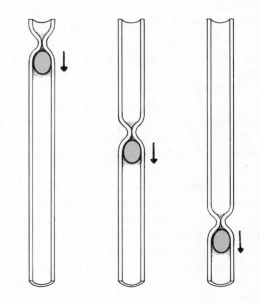

Fig. 16-19. Diagram of the mechanism of peristalsis. Area of contraction is preceded by an area of relaxation.

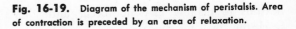

it is referred to as *chyme.*) Small amounts of gastric juice continue to be secreted as long as chyme is present. The mechanism is not well understood, but it is thought that a hormone, produced by the intestinal mucosa, is carried to the stomach by way of the blood. The total amount of secretion for all phases averages about 2,000 ml. each day.

Mechanical activity in the stomach depends upon rhythmic contractions of the smooth muscular wall, which serve to mix the food with gastric juice. The waves of contraction begin as shallow ripples near the cardiac end of the stomach and become deeper as they move toward the pylorus.

As digestion proceeds, small amounts of liquid chyme are emptied from the stomach into the duodenum. The rate of emptying of the stomach is influenced by such factors as the fluidity of the chyme, the amount of chyme already present in the duodenum, the acidity of the chyme, and the type of food in the chyme. For example, the presence of fat in the chyme stimulates cells in the duodenal mucosa to secrete a hormone (*enterogastrone*) that is carried to the stomach by way of the bloodstream; enterogastrone not only inhibits the secretion of gastric juice, but also inhibits the muscular contractions of the stomach wall. On the other hand, carbohydrates in the chyme have only a mild delaying effect; protein has an intermediate effect.

Chemical activity in the stomach depends primarily on the secretion of *pepsin* (gastric protease); this enzyme is highly active in an acid medium (optimal pH = 2.0); thus the secretion of hydrochloric acid is equally important for the digestion of protein in the stomach. On the other hand, gastric acidity inhibits the activity of the incoming salivary amylase; therefore, inasmuch as food is layered as it enters the stomach, salivary digestion continues only in the center of the stomach contents, or until such time as the food is mixed thoroughly with the highly acid gastric juice.

Pepsin is secreted in an inactive form, called *pepsinogen*, which is converted to the active enzyme by the hydrochloric acid (HCl) in the stomach; if pepsin were secreted in its active form it would digest the protein in the very cells that produced it. Once activated in the lumen of the stomach, pepsin attacks amino acid linkages in the center of protein molecules, thus liberating protein fragments rather than amino acids.

The surface epithelial cells tend to be protected from the corrosive effects of gastric juice by the mucous secretion that coats the lining of the stomach; however, it is believed that these cells do have a high rate of mortality, and that they are replaced by cells from deeper parts of the gastric pits (p. 390). A defect in the mucosal defense may result in *gastric ulcer* formation.

Gastric lipase, secreted in very small quantities, is of questionable value in the adult. Inasmuch as it principally acts on emulsified fats, it has practically no effect on the large fat globules that make up the bulk of the fat in one's diet.

Feeding Other Than by Mouth. A person who cannot take food by mouth may receive fluids by intravenous injection, by gavage, or by gastrostomy. *Gavage* means a feeding that is given by way of a tube passed through the nose or the mouth and the esophagus into the stomach. If the esophagus is obstructed, a surgeon may perform a *gastrostomy;* this involves the creation of an opening through the abdominal wall into the stomach.

Gastric Analysis. This procedure refers to the study of stomach contents after a test meal or the injection of histamine, which stimulates the parietal cells to secrete hydrochloric acid; if no acid is present, the patient is said to have *achlorhydria;* this is a finding in pernicious anemia. Less than the normal amount of acid is called *hyposecretion* and occurs in cancer of the stomach. If there is an unusually high content of acid, the patient is said to have *hyperacidity,* or hypersecretion; this may be associated with ulcer formation.

Gastric lavage refers to repeatedly washing out the stomach, as after one has taken poison or an overdose of sleeping pills.

Nausea. Nausea, meaning *seasickness,* is a term that has been in use for 3,000 years. It is a feeling of discomfort, accompanied by salivation, aversion to food, weakness, trembling, and perspiration. Usually it precedes vomiting. Nausea is common in the early months of pregnancy, in gastrointestinal disturbances from various causes, and disease of the inner ear (especially of the semicircular canals).

Vomiting. Morning sickness of pregnancy, brain tumors, and upsets or infections of the digestive system are usually accompanied by vomiting. In the act of vomiting, there is compression of the stomach against the liver and the diaphragm by the muscles of the abdominal wall. The gastro-esophageal constrictor relaxes, and the contents of the stomach are ejected. Emetics are substances which are given to produce vomiting when the stomach needs to be emptied, in cases in which the individual has taken poison or some other harmful substance.

Digestion in the small intestine

The greatest amount of digestion occurs in the small intestine for the following reasons. First, food entering the intestine already has undergone an extensive preliminary breakdown owing to mechanical and chemical activities in the mouth and stomach. Second, the small intestine receives secretions that provide all of the enzymes necessary for the complete digestion of each major food group. Third, food remains in the small intestine for a relatively long period of time, thus giving the enzymes ample opportunity to complete their tasks.

Three secretions are emptied into the small intestine: pancreatic juice, intestinal juice, and bile. The rates of production and flow of these secretions are influenced by nerve impulses and by hormones produced in the mucosa when acid chyme enters the duodenum.

Pancreatic Juice. The exocrine secretion of the pancreas (p. 400), is produced very slowly between meals. However, the entry of acid chyme into the duodenum results in a marked increase in the rate of production owing to stimulation of the pancreas by two hormones: secretin and pancreozymin; these hormones are released by cells in the duodenal mucosa and carried to the pancreas by way of the bloodstream.

Secretin causes the pancreas to produce a thin, watery juice that is rich in sodium bicarbonate; sodium bicarbonate helps to neutralize hydrochloric acid in the chyme. Because the duodenal mucosa is not as well equipped as the stomach to withstand the corrosive effect of gastric juice, neutralization of the hydrochloric acid is an important protective mechanism. With development of a *duodenal ulcer,* there is usually an accompanying hypersecretion of hydrochloric acid, but this is only one of many possible factors that could be involved. Another important effect of this secretion is the provision of a favorable duodenal pH

(6.0 to 7.0) so that the pancreatic enzymes may operate efficiently.

Pancreozymin is a hormone that stimulates the pancreas to produce a thick, viscid secretion that is rich in amylase, lipase, and three proteases; the digestive effects of these enzymes are described under chemical changes.

Nervous mechanisms in the regulation of pancreatic secretion are far less important than hormonal regulation. However, during the cephalic phase of gastric secretion, vagus nerve impulses are transmitted to the pancreas, as well as to the stomach. These impulses stimulate the production of enzyme-rich pancreatic juice, which is stored temporarily in the acini.

The total secretion of pancreatic juice averages about 1,200 ml. per day.

Intestinal Juice. Intestinal juice, or *succus entericus,* is produced by the glands of the small intestine (p. 392). The flow of this juice is initiated by distention of the intestine, by the presence of food, and by parasympathetic (vagus) stimulation. On the other hand, sympathetic nerve stimulation inhibits intestinal secretion, particularly of the mucus-producing duodenal glands. This tends to leave the mucosa unprotected and may be a factor in the development of duodenal ulcer.

Enzymes associated with intestinal juice include enterokinase, proteases ("erepsin"), sucrase, maltase, and lactase; in addition there are very small quantities of amylase and lipase. The digestive effects of these enzymes are described under chemical changes.

Bile. Formation of bile and the mechanism of emptying of the gallbladder were previously described on pages 398 to 399. Two duodenal hormones are involved: *secretin,* which stimulates the production of bile in the liver, and *cholecystokinin,* which provokes evacuation of the gallbladder. Although bile contains no digestive enzymes, the bile salts play an important role in the digestive process, as they emulsify fats. Table 16-2 summarizes the gastro-intestinal hormones.

Mechanical activity in the small intestine involves segmental and pendular mixing contractions, as well as the propulsive movements of peristalsis.

Segmentation consists of a squeezing and mixing of the intestinal contents by means of ringlike contractions of the circular muscle; at times the intestine resembles a chain of little link sausages owing to the local constrictions. Segmental contractions occur at the rate of eight or nine per minute in the duodenum, and at progressively slower rates in the jejunum and ileum. Segmentation produces no onward movement of the intestinal contents, but it does serve to mix them with the various digestive juices and to bring them into contact with the mucosa, so that absorption can take place. Segmental contractions in no way depend upon nerve impulses, but their intensity can be increased by parasympathetic nerve impulses or decreased by sympathetic impulses.

Pendular movements consist of small constrictive waves that sweep forward, then backward, over a few inches of the intestine. The function of these movements is uncertain, but a certain amount of mixing of intestinal contents must occur.

Peristalsis is responsible for the onward movement of the intestinal contents. These wavelike contractions may begin in any part of the small intestine and move along for variable distances. Those that travel for relatively short distances are called peristaltic *waves*; those that travel longer distances are referred to as peristaltic *rushes*. If one applies a stethoscope to the abdomen, these propulsive waves can be heard as long continued rumbles.

When one begins to eat, and food enters the stomach, peristaltic activity of the small intestine is greatly increased; the existence of this *gastroenteric reflex* is one reason for not giving food by mouth immediately after surgical operations on the small intestine, as the increased activity might result in a tearing of tissues at the suture lines. The rate of movement of chyme through the intestine varies with its composition. However, the "head" of an average meal usually reaches the ileocecal valve in three to four hours, and the last of the meal leaves the ileum about eight or nine hours after ingestion.

Although smooth muscle of the intestinal tract is able to initiate spontaneous contractions, the integration of peristaltic activity depends upon autonomic nerve fibers. Parasympathetic fibers conduct impulses that stimulate peristalsis, while those of the sympathetic division inhibit peristalsis.

Chemical activity in the small intestine is of fundamental importance in the breakdown of all three major food groups. Although some digestion occurs in the mouth and stomach, it should be emphasized that the pancreatic and intestinal enzymes (together with the emulsifying action of bile) are capable of initiating and carrying through the entire digestive process.

Carbohydrate enters the duodenum partly in the form of dextrins and maltose (produced by the action of salivary amylase), starches and glycogen (that escaped digestion in the mouth), or as sucrose and lactose (cane sugar and milk sugar ingested with the food). Pancreatic *amylase* attacks starch, dextrins, and glycogen, converting them to maltose (a disaccharide). Intestinal enzymes then complete the digestion of carbohydrate to simple sugars (monosaccharides): *maltase* splits maltose to two molecules of glucose; *sucrase* splits sucrose to one molecule of glucose and one of fructose; *lactase* splits lactose to one molecule of glucose and one of galactose.

Fats are attacked by pancreatic *lipase* and split into glycerol and fatty acids. Although lipase can and does attack large fat globules, its action is much more effective when fats first are emulsified (split into tiny particles) by the action of bile salts.

Protein and protein fragments resulting from the action of pepsin are attacked by a series of enzymes in the small intestine, where they are broken down to amino acids. Generally speaking, the pancreatic proteases initiate the digestion of large protein molecules and fragments, splitting them into still smaller fragments. Although a few amino acids are split off by the pancreatic enzymes, the intestinal proteases do the bulk of the work in completing the digestion of protein fragments to their constituent amino acids.

Because the pancreatic proteases will attack native protein, they are secreted in inactive forms to prevent digestion of the pancreas itself; these forms (trypsinogen and chymotrypsinogen) become activated only after they enter the small intestine. *Enterokinase*, an enzyme in the intestinal juice, triggers the activation process by converting trypsinogen to active *trypsin*. Trypsin, in turn, activates chymotrypsinogen to active *chymotrypsin*.

Intestinal proteases (aminopeptidase and dipeptidase) do not attack native protein and can be produced in their active form. They complete the digestive processes initiated by the pancreatic enzymes; thus amino acids are made available for absorption.

Digestion in the large intestine

Activity in the large intestine is limited to mechanical and bacterial activity. The only secretion produced in this portion of the digestive tube is mucus. In addition to its role as a protector of the intestinal wall, mucus serves to hold fecal matter together and to facilitate its movement toward the rectum. There are no enzymes secreted by the glands of the large intestine.

Mechanical activity involves rather sluggish mixing and propulsive movements. The mixing movements are similar to the segmentation contractions of the small intestine; these contractions promote the reabsorption of water as the intestinal contents are churned about and brought into contact with the mucosa.

Propulsive movements, or mass movements, propel the fecal contents toward the anus; these movements occur only a few times each day, usually after meals (*gastrocolic reflex*). Mass movements are most abundant after breakfast, and many persons take advantage of this reflex as they establish the habit of emptying the bowel each morning. The parasympathetic autonomic nerves are the nerves of emptying, as they stimulate contraction of the smooth muscle in the wall of the colon and cause relaxation of the sphincters of the anal canal. The sympathetic nerves are the nerves of retention, as they promote relaxation of the muscular wall of the colon and contraction of the sphincters of the anal canal.

Bacterial Action. The large intestine is filled with bacteria, which act on the undigested residues. They produce fermentation of carbohydrate and putrefaction of proteins. Some of the split products are eliminated with the feces, and others are absorbed. Those that are absorbed are phenol, indol and skatol; these are conjugated with other substances in the liver, and then excreted by way of the urine.

Bacteria that inhabit the large intestine synthesize vitamin K, which is then absorbed, provided that bile is present in the intestine. Some components of the vitamin B complex are produced by intestinal bacteria, and these likewise

are absorbed and used by the body. Patients who are taking sulfonamide or antibiotic drugs may develop a vitamin deficiency if a sufficient number of these helpful bacteria are destroyed.

Defecation (Emptying of the Rectum). The sigmoid colon is filled with feces by the mass movements of the large intestine. The rectum remains empty until just before defecation. As the feces enter and stretch the rectum the desire to defecate is aroused. If unheeded, the desire disappears (due to adaptation of stretch receptors), and the feces remain in the rectum where they become hard and dry; defecation then becomes difficult and painful. Infrequent, or difficult evacuation of the feces is referred to as *constipation*.

Contractions of the smooth muscle in the walls of the rectum are accompanied by relaxation of the internal and external sphincters and contraction of the voluntary muscles of the abdominal walls and pelvic floor. The levator ani muscles assist in expelling the last of the fecal contents.

Composition of Feces. The feces contain water and mucus as well as great numbers of bacteria. Also present are epithelial cells that have been sloughed off the intestinal mucosa, various food residues, such as undigestible cellulose, and various food substances which have escaped digestion and absorption in the small intestine. Elements such as calcium, potassium, iron, nitrogen, and magnesium also are lost in the feces. The bile pigments (bilirubin and biliverdin) give the normal brown color to feces. If for any reason bile fails to enter the intestinal tract, the feces will be clay colored or chalky in appearance. If old blood from the upper part of the gastrointestinal tract is present, the stools have the color and the consistency of tar (ulcer or cancer of stomach). Fresh blood appears as red streaks in the stools and often is due to hemorrhoids or to cancer of the rectum. Thus, stool examinations are of considerable diagnostic importance.

Diarrhea. In this condition there are frequent bowel movements, and the fecal material is liquid, since it is moved so rapidly through the digestive tube that absorption cannot occur. Usually it is due to the intake of spoiled food or infection in the digestive system. Laxatives and cathartics may lead to diarrhea.

Vomiting and diarrhea can become very serious conditions if they persist for a few days. One loses large volumes of liquid and salts in both conditions. Normally, about 8,000 ml. of fluid is poured into the digestive tube from the blood in a day (in the form of digestive juices). As stated previously, this is taken back into the blood by absorption, so that it is not lost to the body. If either vomiting or diarrhea occurs, the patient becomes dehydrated due to the loss of part of the digestive juices. If vomiting persists for a couple of days, alkalosis may occur, since a large amount of HCl is lost in the vomitus. If diarrhea persists, with a loss of pancreatic, as well as intestinal juices, acidosis may occur due to the loss of $NaHCO_3$. These conditions are especially serious in infants.

SUMMARY

1. **Digestive system prepares food for absorption**
2. **Foods**
A. Inorganic: inorganic salts, water
B. Organic
   a. Carbohydrate: mono-, di-, and polysaccharides; energy
   b. Lipids: simple, compound, and steroid; stored energy and special compounds
   c. Protein: simple and conjugated; building blocks
3. **Vitamins**
A. Required for growth and other processes; take part in various reactions
B. Most obtained in diet, some synthesized in intestine
C. Fat-soluble vitamins
   a. A: formation of rhodopsin; growth of epithelium
   b. D: absorption of calcium; development of bones and teeth
   c. E: antioxidant
   d. K: formation of prothrombin in liver
D. Water-soluble vitamins
   a. Thiamine: decarboxylation of pyruvic acid; CH_2O metabolism
   b. Niacin: NAD, NADP, hydrogen acceptors; energy metabolism
   c. Riboflavin: FMN, FAD; hydrogen carriers; energy metabolism
   d. B_{12}: maturation of RBC's
   e. C: maintenance of intercellular substance
4. **The digestive tube**
A. Structural layers
   a. Mucous membrane lining
   b. Submucous layer
   c. Muscular layer
   d. External: fibrous above diaphragm and serous below
B. Peritoneum
   a. Double-walled sac: outer layer lines cavity (parietal peritoneum); inner layer covers organs (visceral peritoneum); peritoneal cavity is space between
   b. Mesentery: attaches intestine to posterior abdominal wall
   c. Greater omentum connects stomach and transverse colon
   d. Lesser omentum connects stomach with liver
C. Mouth or buccal cavity
   a. Lips and cheeks covered with skin, lined with mucosa; muscular walls
   b. Roof of mouth: bony hard palate anteriorly; soft palate posteriorly
   c. Tongue aids chewing, swallowing, speaking. Papillae on dorsum contain taste buds
   d. Teeth concerned with chewing food; first set called deciduous, 20 in number; second set permanent, 32 in number; named for shape or function; chief parts are crown and root
   e. Three pairs of salivary glands secrete saliva
      (1) Parotid
      (2) Submandibular
      (3) Sublingual
D. Pharynx
   a. Constrictor muscles aid swallowing
E. Esophagus
   a. Muscular tube; behind trachea
   b. Conveys food to stomach
F. Stomach
   a. Food reservoir; in epigastric and left hypochondriac regions
   b. Divisions: cardiac orifice; fundus and body; pylorus surrounded by pyloric sphincter; greater and lesser curvatures
   c. Structural variations: mucosal folds form rugae; special mucosal cells secrete gastric juice; three layers of smooth muscle highly extensible
G. Small intestine
   a. Coiled tube: stomach to large intestine
   b. Divisions: duodenum, jejunum, and ileum; ileocecal valve guards opening into large intestine
   c. Structural variations: circular folds and villi; each villus contains a blood and a lymph capillary; duodenal glands secrete mucus; intestinal glands secrete digestive enzymes
H. Large intestine
   a. Extends from ileum to anus
   b. Divisions: cecum and appendix; colon (ascending, transverse, descending, sigmoid); rectum; anal canal, guarded by sphincters
   c. Structural variations: no villi; glands secrete mucus but no enzymes; longitudinal muscle fibers in bands (taeniae coli), form pouches, or haustra; circular muscle in anal canal forms internal anal sphincter (involuntary); skeletal muscle forms ex-

ternal anal sphincter (voluntary)

5. Accessory organs: tongue, teeth, salivary glands related to mouth
A. Liver and bile ducts
   a. Located under dome of diaphragm; epigastric and right hypochondriac regions
   b. Description: superior surface smooth and convex; inferior concave; four lobes (right, left, quadrate, caudate), subdivided into lobules; attached to diaphragm and body wall by folds of peritoneum; Glisson's capsule, accompanies vessels into interior of liver; connective tissue extends from porta throughout liver, provides internal support; hepatic artery supplies oxygenated blood; portal vein supplies blood rich in end products of digestion; hepatic veins drain liver and empty into inferior vena cava
   c. Microscopic appearance: branching plates of hepatic cells radiate from center of lobule; sinusoids between cords receive blood from hepatic artery and portal vein, are lined with phagocytic cells; much blood can be held in sinusoids (reservoir effect)
   d. Biliary apparatus: bile canaliculi originate between liver plates; drain into larger ducts, finally into hepatic duct which joins cystic duct from gallbladder to form common bile duct that empties into duodenum at duodenal papilla. Gallbladder a musculomembranous sac on undersurface of liver; receives bile for storage and concentration
B. Functions of liver
   a. Production of bile
   b. Metabolic functions
     (1) Carbohydrate: regulation of blood sugar level
     (2) Lipid: synthesis; breakdown fatty acids for oxidation
     (3) Protein: deaminizes amino acids, forms new ones; forms plasma proteins
     (4) Miscellaneous: forms vitamin A; stores vitamins A, D, and B_{12}; stores iron; steroid hormone inactivation; phagocytosis; forms heparin; detoxifies
C. Pancreas
   a. Head rests in concavity of duodenum, body and tail extend toward spleen
   b. Composed of lobules and secretory units which produce pancreatic juice; secretion emptied into duodenum via pancreatic ducts
   c. Groups of irregular cells (islets of Langerhans) secrete hormones insulin and glucagon

6. Digestive processes
A. In the mouth
   a. Mechanical: mastication and insalivation
   b. Chemical: salivary amylase converts starch to dextrin and maltose
B. Deglutition or swallowing
   a. Formation of bolus
   b. Passage of bolus through pharynx; constrictor muscles

   c. Passage of bolus through esophagus; peristaltic waves
C. In the stomach
   a. Secretion of gastric juice in three phases: cephalic, gastric, intestinal
   b. Mechanical: mixing; forms chyme
   c. Chemical: see Table 16-1 on page 409
D. In the small intestine
   a. Pancreatic juice: flow stimulated by presence of chyme in duodenum; hormones secretin and pancreozymin; vagal impulses
   b. Intestinal juice: flow stimulated by distention, presence of food, vagal impulses
   c. Bile: secretin stimulates production; cholecystokinin evacuates gallbladder
   d. Mechanical: segmentation, pendular, peristalsis
   e. Chemical: see Table 16-1 on page 409
E. In the large intestine
   a. Mechanical: segmentation; mass movements
   b. Bacterial action: synthesis of vitamin K and some B complex; putrefaction, fermentation
   c. No digestive enzymes
   d. Defecation: emptying of rectum and elimination of waste
     (1) Feces contain water, mucus, bacteria, various food residues, elements and bile pigments

QUESTIONS FOR REVIEW

1. What is meant by the term "protein pool"?
2. What are the physiological functions of the fat-soluble vitamins A, D, and K?
3. Prolonged thiamin deficiency would have what effects?
4. What do niacin and riboflavin have in common?
5. Scurvy has what effects? What vitamin will prevent scurvy?
6. Identify the following structures: peritoneum, mesentery, gingivae, submandibular glands, gastric mucosa, duodenum, Glisson's capsule, and duodenal papilla.
7. Distinguish between incisors and molars.
8. What is parotitis? What is pyloric stenosis?
9. What is the function of the ileocecal valve?
10. What structures contribute to the great surface area of the small intestine?
11. What is the chief danger of appendicitis?
12. Through what structures would a drop of bile pass as it traveled from a canaliculus to the duodenum?
13. Of what importance are the bile salts?
14. Briefly describe the sequence of events that occur during the act of swallowing.
15. Distinguish between the exocrine and endocrine secretions of the pancreas.
16. Briefly, what is the role of the liver in carbohydrate metabolism? in protein metabolism?
17. What is the enzyme in saliva? What does it do?
18. What stimulates the flow of gastric juice?

19. Why is pepsin secreted in the form of pepsinogen?
20. Why does the greatest amount of digestion occur in the small intestine?
21. Describe the mechanism of emptying of the gall-bladder.
22. What are the functions of the mechanical activities in the small intestine?

23. What enzymes are found in pancreatic juice? What foodstuffs do they attack? What are the end products?
24. Of what importance are the mass peristaltic movements in the large intestine?
25. Of what value are the bacterial inhabitants of the large intestine?

REFERENCES AND SUPPLEMENTAL READINGS

Beeson, P. B., and McDermott, W.: Cecil-Loeb, Textbook of Medicine, ed. 13. Philadelphia, W. B. Saunders, 1971.

Best, C. H., and Taylor, N. B.: The Physiological Basis of Medical Practice, ed. 8. Baltimore, Williams & Wilkins, 1966.

Davenport, H. W.: Why the stomach does not digest itself. Sci. Am., 226:87, Jan. 1972.

Dudrick, S. J., and Rhoads, J. E.: Total intravenous feeding. Sci. Am., 226:73, May 1972.

Goss, C. M. (ed.): Gray's Anatomy of the Human Body, ed. 28. Philadelphia, Lea & Febiger, 1966.

Guyton, A. C.: Textbook of Medical Physiology, ed. 4. Philadelphia, W. B. Saunders, 1971.

Ham, A. W.: Histology, ed. 6. Philadelphia, J. B. Lippincott, 1969.

Table 16-1 Summary of Digestive Activities

Glands	Secretions	Acts On	Products Formed
Salivary	Salivary amylase (ptyalin)	Starch and glycogen	Dextrins
		Dextrins	Maltose
Gastric	Hydrochloric acid	Activates pepsinogen	Pepsin
	Gastric protease (pepsin)	Protein	Protein fragments (partially digested protein)
	Gastric lipase (?)	Emulsified fats	Fatty acids and glycerol
Pancreas (acinar cells)	Pancreatic amylase (amylopsin)	Starch, glycogen and dextrins	Maltose
	Pancreatic lipase (steapsin)	Emulsified fats	Fatty acids and glycerol
	Pancreatic proteases (trypsin and chymotrypsin)	Protein (intact or fragmented)	Protein fragments and amino acids
Intestinal (crypts of Lieberkuhn)	Enterokinase	Activates trypsinogen	Trypsin
	Proteases ("erepsin")	Protein fragments	Amino acids
	Maltase	Maltose	Glucose
	Sucrase	Sucrose	Glucose and fructose
	Lactase	Lactose	Glucose and galactose
Liver	Bile salts (no enzymes)	Large fat globules	Emulsified fats (tiny droplets)

Table 16-2 Gastrointestinal Hormones

Hormone	Where Produced	Agents Stimulating Production	Action
Gastrin	Gastric mucosa of pyloric region	Distention and protein derivatives	Stimulates production of gastric juice
Enterogastrone	Duodenal mucosa	Fatty acids, fats, and certain sugars	Inhibits secretion of gastric juice and motility of stomach
Secretin	Duodenal mucosa	Acidity of duodenal contents	Stimulates production of bile and of bicarbonate-rich pancreatic juice
Pancreozymin	Duodenal mucosa	Chyme	Stimulates production of enzyme-rich pancreatic juice
Cholecystokinin	Duodenal mucosa	Fats, fatty acids, dilute HCl	Stimulates emptying of gallbladder

The Absorption and Utilization of Food

17

We have shown, in Chapter 16, how complex foods are mechanically and chemically broken down to absorbable forms. Through the action of powerful digestive enzymes, carbohydrate, fat, and protein have been hydrolyzed to simple sugars, fatty acids, glycerol, and amino acids. The stage now is set for these nutrients to enter the body, in the truest sense of the word, and to follow various metabolic pathways that are open to them.

ABSORPTION

Absorption involves the transfer of the products of digestion through the epithelial lining of the digestive tube into the blood or the lymph streams. Two of the processes involved in this activity are *osmosis* and *diffusion*. However, physical processes alone cannot account for all of the intricacies of absorption, and *active transport* must be given credit for much of the action. These processes are described on pages 22 to 25.

From the mouth and stomach

There is no absorption of food from the mouth. However, certain drugs, such as nitroglycerin, can pass through the oral epithelium into the bloodstream.

Absorption from the stomach is very limited. Alcohol, a small amount of water, and a few drugs are thought to be absorbed from this part of the digestive tube.

From the small intestine

The greatest amount of absorption takes place from the small intestine for several important reasons: (1) food is in an absorbable form inasmuch as all of the enzymes required for complete digestion are readily available; (2) the circular folds, the villi, and the microvilli provide a great surface area for absorption (in addition, villi are in constant motion owing to the activity of smooth muscle fibers in the mucosa; this helps to stir the intestinal contents and promote maximal contact between the absorptive surface and the nutrients); (3) blood and lymph vessels are abundant; (4) food remains in the small intestine for a relatively long period of time.

Absorption of Organic Materials. The *simple sugars*—glucose, galactose, and fructose—are absorbed by an active transport carrier mechanism. They enter the bloodstream in the intestinal capillaries as shown in Figure 17-1.

Fatty acids and *glycerol*, products of fat digestion, enter the epithelial cells of the intestinal mucosa separately. Glycerol enters as such, but bile salts help to ferry the fatty acids across the cell membranes; energy is not required for this process. During their passage through the epithelial

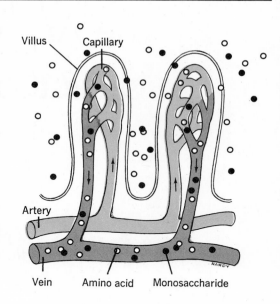

Fig. 17-1. Diagram of absorption of end products of carbohydrate and protein digestion; monosaccharides and amino acids enter the capillaries of the villi.

cells, fatty acids and glycerol are resynthesized to microscopic droplets of neutral fats (Fig. 17-2). Most of these neutral fats then enter the lymphatic capillaries (lacteals) in the villi and are carried off in the lymph which ultimately flows into the bloodstream. Small quantities of short-chain fatty acids may be absorbed directly into the bloodstream rather than being converted to chylomicrons (p. 421), the latter being too large to enter a blood capillary.

Amino acids are absorbed by an active transport system and enter the blood capillaries, as shown in Figure 17-1. For all practical nutritional purposes, protein is absorbed only in the form of amino acids. However, extremely small quantities of *whole protein* may be absorbed at times, probably by the process of pinocytosis (p. 25); inasmuch as protein can act as an antigen, this may have some relation to food allergies.

Absorption of Inorganic Materials. *Sodium* and some *chloride* are absorbed by active transport; however, the transport of sodium (Na^+) creates sufficient electrical potential to pull chloride (Cl^-) with it. *Potassium* is absorbed passively. *Calcium* absorption is related to body needs; parathyroid hormone and vitamin D are thought

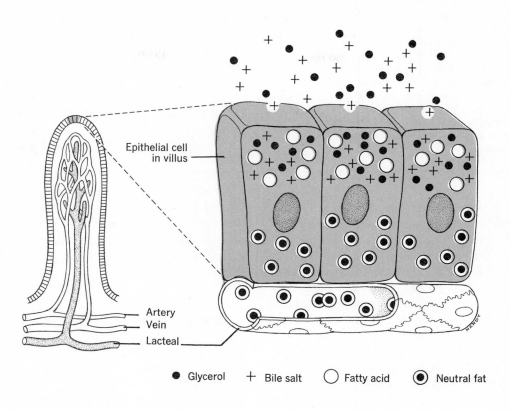

● Glycerol   + Bile salt   ○ Fatty acid   ◉ Neutral fat

Fig. 17-2. Diagram of absorption of fat. Glycerol enters the epithelial cells as such; fatty acids combine with bile salts which help to ferry them into the epithelial cells. Within the cells, the fatty acids are freed, and they unite with the glycerol to form neutral fat which enters the lacteals. Some short chain fatty acids may be absorbed directly into the bloodstream.

to play an important role in regulating the absorption of this important ion (p. 461). *Iron* also requires a special mechanism for transport, which depends, in part, on body needs (p. 463).

Water is absorbed principally by osmosis. As the inorganic salts and nutrients are absorbed, the intestinal fluid becomes hypotonic, which means that water then will be absorbed by the osmotic process (p. 32).

From the large intestine

Great quantities of water are absorbed in the large intestine, as witnessed by the fact that fecal material entering the cecum is in a liquid state, but by the time it reaches the rectum it is normally a soft, semisolid mass. There is no significant absorption of food in this part of the alimentary canal.

METABOLISM

Metabolism refers to all the changes in foodstuffs from the time they are absorbed from the small intestine until they are excreted as waste products from the body. Metabolism consists of two phases: *anabolism* (meaning *a building up*) and *catabolism* (meaning *to throw down*). In other words, during the phase of anabolism, substances are built up, while during the phase of catabolism, substances are broken down by oxidative processes with the release of energy, as heat and work, or the storage of energy, in the form of high energy bond phosphate.

The processes of metabolism are carried out by enzymes within the cells. The substrate for a single enzyme in an enzyme system is produced by the action of the enzyme preceding it, and the product of this single enzyme becomes the sub-

Fig. 17-3. Schematic representation of the action of an enzyme system. Each enzyme in the system is dependent on the enzymes preceding it for its substrate. A lack of any enzyme in the system could prevent the action of all enzymes following it and cause an accumulation of its normal substrate. For example, if enzyme "c" is absent, enzyme "d" has no substrate and normal end products could not be produced. In addition, substrate C would accumulate.

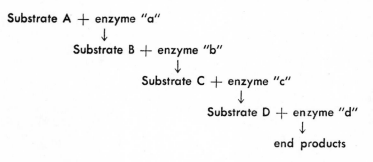

Substrate A + enzyme "a"
↓
Substrate B + enzyme "b"
↓
Substrate C + enzyme "c"
↓
Substrate D + enzyme "d"
↓
end products

strate for the enzyme following it (Fig. 17-3). Thus the action of the entire system depends upon the proper functioning of each component enzyme. This system can be compared to the production line of a bicycle manufacturing plant, in which the final product is the end result of the work done by the individual members on the line. Worker A receives a bicycle frame which has been made by the workers ahead of him. It is his responsibility to attach a brace on the front end so that worker B next to him in the line can attach a front wheel. If worker A is absent, worker B cannot attach the wheel and the production of the bicycle stops at this point. If all of the workers responsible for making the frame continue to work, bicycle frames which cannot go through the rest of the assembly line begin to pile up in the plant.

So it is with enzyme systems. If even one enzyme in the system is missing or malfunctioning, the effects may be very serious; such persons are said to have *inborn errors in metabolism.* For example, phenylketonuria is a condition in which the amino acid phenylalanine cannot be properly metabolized in the body, due to the absence of a single enzyme (phenylalanine hydroxylase). The toxic substances produced in the enzyme system prior to the action of phenylalanine hydroxylase cannot continue through the system. If these products continue to accumulate, they will interfere with brain development and mental retardation will result. Today pediatricians test young infants for the presence of phenylketonuria at about six weeks of age so that treatment can be instituted before mental retardation develops.

Without their normal complement of enzymes, cells cannot reproduce, they cannot convert food substances to energy or to new protoplasm.

Many of the enzymatic reactions, in turn, depend on the presence of certain vitamins (p. 379) and on certain minerals (p. 464); many are regulated by hormones which are produced by the endocrine glands.

Energy metabolism

Energy is involved in every body activity. The beating of the heart, the digestion and the absorption of food, the building and repair of tissue cannot occur without energy. In fact, the very organization of each living cell depends on complicated chemical reactions that require a constant supply of energy. Without energy, there is disorganization of cells, illness, and death. Thus the energy needs of the body must take precedence over all others.

To supply the constant demand for energy, nature has devised an ingenious process, called *biologic oxidation,* that enables one to procure energy from food without burning up body tissue at the same time. Because of the catalytic action of intracellular enzymes, food can be burned at low temperatures which are compatible with the life of the cell. Biologic oxidation is a stepwise process that has been compared with the orderly progression of a boat through a series of locks in a river, rather than plunging over the falls.

Biologic oxidations are similar to other types of oxidation in that there is a loss of electrons, either by combination with oxygen or, more commonly, by the removal of hydrogens. The transfer of these hydrogens, or their electrons, through intermediary acceptors and ultimately to oxygen, yields the energy released by cellular oxidations. The biochemical agents involved in these reactions are enzymes, coenzymes, and hydrogen acceptors. The enzymes that make possible the removal of hydrogen from a substrate are *dehydrogenases,* while those that act on oxygen and cause it to ultimately take part in the electron-transport system (p. 419) are called *oxidases.* Certain coenzymes such as nicotinamide dinucleotide (NAD) can be alternately reduced and oxidized and act as hydrogen acceptors or as hydrogen donors. NAD accepts hydrogens from a metabolite, becoming NADH$_2$, transfers the hydrogens to other hydrogen acceptors, and returns to its original oxidized state, NAD.

As illustrated in Figure 17-4, the end products of digestion may ultimately be oxidized by way of the citric acid cycle and the electron transport system (p. 419). In addition to CO$_2$ and H$_2$O, a tremendous amount of energy is liberated. Much of this energy is stored temporarily in the high-energy phosphate bonds of

ATP (p. 418), and the remainder helps to maintain normal body temperature.

ATP frequently is compared to a cocked and loaded gun in that it is always ready to fire and release its energy when needed by the cells. In Chapter 4 we discussed the utilization of the energy derived from ATP during muscle contraction, but this is only a small part of the total picture. ATP is found in every living cell, where it provides the energy required not only for mechanical work, but also for chemical work, such as the building and repair of protoplasm.

Carbohydrates, fats, and proteins all contain stored energy that can be transferred to ATP following biologic oxidation in the citric acid cycle and the electron transport system. This potential energy value of food is referred to as the caloric, or heat value. A large calorie, abbreviated Cal., is the amount of heat required to raise the temperature of 1 Kg. (2.2 lbs.) of water 1° C.

Carbohydrates are very easily burned, and they serve as a quick source of energy. Each gram of carbohydrate releases 4 Cal. of heat when it is oxidized. Fats have an even higher caloric value as they yield 9 Cal. per Gm. when oxidized. Under normal conditions, only a relatively small amount of protein is used to supply energy, but each gram yields 4 Cal. on oxidation. When food intake is severely restricted, as in starvation, the oxidation of body protein is of major importance. Since energy needs take precedence over all others, the very protoplasm of the cells is oxidized to meet those needs. It is obvious that death will be the ultimate result if food intake is not resumed.

Basal metabolism refers to the energy output required to keep a person alive; that is, to maintain his body temperature and to keep his respiration and circulation going. It is related to one's surface area, being about 1,000 Cal. per square meter of body surface per day. This is about 38 to 40 Cal. per square meter per hour for men, and 36 to 38 for women. Variations of plus or minus 10 per cent are permitted in the normal range. This means that on the basis of 40 Cal. for men, the limits are 36(40−4) and 44(40+4) Cal. per square meter per hour. The basal metabolic rate, or BMR, commonly is reported as a percentage of normal. For example, if a man's BMR is recorded as +30, it means that it is 30 per cent higher than the normal for a person of his size, age, and sex.

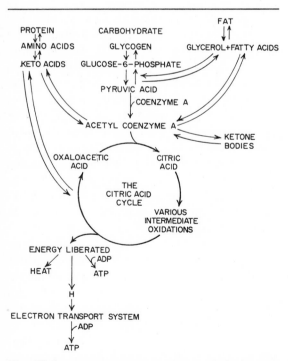

Fig. 17-4. Diagrammatic representation of the ways in which all major foods may enter the citric acid cycle (Krebs cycle) for biologic oxidation and related energy liberation. The double arrows also indicate the pathways whereby one food group can be converted to another: protein to carbohydrate or fat, carbohydrate to fat and some amino acids. See text for details.

In the laboratory, the basal metabolic rate can be determined by an indirect method; this means that only the oxygen consumption is measured. The subject, with his nose clamped, breathes into a special chamber by means of a rubber tube leading to his mouth. After a given period of time, the amount of oxygen used from the chamber is measured. Each liter of oxygen consumed is equivalent to a certain number of calories, and from this the BMR can be calculated. During the test, the subject must be at complete rest and have had nothing to eat since dinner the previous evening.

Several factors influence the basal metabolic rate. For example, it varies with age, being higher in children, since they are growing rapidly. It is higher in men than in women, the former generally being larger and having more muscle tissue. It varies also with the state of nutrition, being low in the presence of malnutrition. Abnormal activity of the thyroid gland may cause marked changes in the BMR, since the thyroid hormone plays an important role in determining the rate at which chemical reactions proceed in the cells. If this gland is overactive, the basal metabolic rate is elevated beyond the normal limits; if underactive, the basal rate is depressed below the normal limits. However, the BMR measures only the overall oxygen consumption; it may be increased in unrecognized infections, congestive heart failure, and various other conditions not related to the thyroid. For this reason, determination of the amount of protein-bound iodine (PBI) in the plasma is of greater diagnostic value than is the BMR with regard to thyroid function (p. 478). Recently, since the availability of radioactive iodine, the function of the thyroid gland can be evaluated more directly by measuring its ability to take up iodine from the blood. This radioactive iodine uptake test is commonly used in medical facilities equipped to handle radioactive materials.

The *total caloric requirement* is subject to great individual variation, since it depends on the energy expended when one is eating, working, or playing, in addition to the basic needs previously described. For the average sedentary student, a maximum of 2,400 Cal. is sufficient, but a man at very hard muscular work may require 6,000 to 8,000 Cal. to meet his energy needs. In any case, caloric intake should be balanced by energy output. Difficulty arises when one who follows a sedentary occupation takes in the amount of food needed by a heavy laborer.

Obesity, or overweight, is frequently due to an excessive intake of food, beyond the needs of the body. The excess is stored as adipose tissue. For weight control, the caloric intake should be reduced, yet the diet can be balanced and contain essential salts and vitamins. Peculiar diets are usually not satisfactory, and reducing pills may be harmful.

Carbohydrate metabolism

Carbohydrate metabolism includes all the reactions undergone in the body by carbohydrates which have been ingested as part of the diet and by those which are formed in the body from other sources, such as pyruvic acid, lactic acid, amino acids, and glycerol. However, the main purpose underlying all carbohydrate metabolism is the ultimate provision of *energy* through oxidative reactions. Although a small amount of carbohydrate is utilized in building tissues, some of the compounds formed are vitally important. For example, ribose and deoxyribose essential for the construction of RNA and DNA can be synthesized from glucose.

Utilization of Glucose. In discussing the important facts of carbohydrate metabolism, we shall list the various pathways that are open to glucose following its absorption from the small intestine into the bloodstream (Fig. 17-5).

1. Some glucose may enter the tissues immediately and be oxidized to carbon dioxide and water, with the release of energy needed to maintain body temperature and to enable one to do muscular work.

2. Glucose that is not needed for the moment will be converted to glycogen (glycogenesis) for temporary storage in the liver and in muscle cells. ATP provides the necessary energy for the synthesis of glycogen. Slightly less than one pound of carbohydrate can be stored in the body.

3. When the glycogen storage bins are filled, another portion of glucose is changed to fat and stored as adipose tissue. There are fat depots in the subcutaneous tissues, around the kidneys, in the mesentery and the greater omentum, as well as in various other regions of the body.

4. If there is a large intake of carbohydrate within a short period of time, such as when one

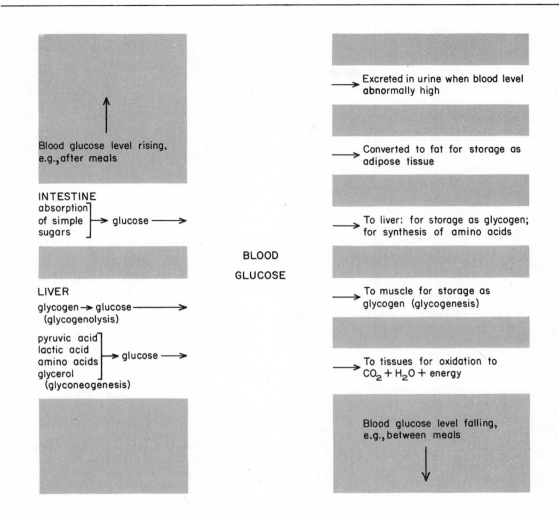

Fig. 17-5. Diagrammatic summary of carbohydrate metabolism. Note that as the blood glucose level begins to fall below normal levels, the liver converts glycogen and other substances to glucose and releases it into the blood. As the glucose level rises above normal levels, glucose is stored as glycogen and fat; in extreme cases, it is eliminated by the kidneys.

eats a generous amount of candy at one sitting, some of the glucose may be eliminated by the kidneys. When the factories that normally handle glucose are flooded by a sudden large intake, the excess spills over into the urine (alimentary glycosuria). A similar spillage may occur in cases of uncontrolled diabetes mellitus. Because of a deficiency of insulin, diabetic patients are unable to metabolize carbohydrate efficiently, their blood sugar level is high, and much glucose is lost in the urine.

5. When the blood sugar level begins to fall, as it usually does between meals, the liver responds in two ways. First, it converts its stores of glycogen into glucose for release into the bloodstream (glycogenolysis). Second, liver cells form glucose molecules from noncarbohydrate sources (glyconeogenesis). For example, some of the lactic acid and the pyruvic acid produced during muscle contraction is carried in the bloodstream to the liver and there converted to glucose. Amino acids and glycerol also can be utilized by the liver in forming blood sugar when necessary to meet the energy needs of the body.

The average range for blood glucose is between 80 and 110 mg. per 100 ml. of blood. Normally, the blood glucose level varies continually, but within rather narrow limits, due to the homeostatic mechanisms described above.

Catabolism of Carbohydrates. There are two catabolic processes involved in the liberation of energy from carbohydrates. The first, or **anaerobic phase**, results in the breakdown of glycogen with the formation of pyruvic and lactic acids; oxygen is not required in this process, and only a few ATP molecules are formed directly.

The anaerobic phase of carbohydrate metabolism includes both the utilization and the production of ATP, as well as the transfer of hydrogens and their energy to the hydrogen acceptor, NAD. Energy becomes available for heat or for ATP production as the hydrogens, or their electrons, are transported through the electron transport system. Let us follow a molecule of glucose through the anaerobic phase while referring to Figure 17-6. If glucose, a six-carbon compound, is to be used directly as it enters the cell, it must first be activated by being combined with phosphate. Phosphate and the energy necessary for the reaction are supplied by ATP. Thus the system must be primed, and priming requires energy.

As the active glucose, glucose-6-phosphate, continues on its path, it is converted to fructose-6-phosphate. Then a second phosphate from an ATP is added to form fructose-1,6-diphosphate. The six-carbon compound, fructose-1,6-diphosphate, splits in half, and both of the resulting three-carbon compounds continue through the process. Each of these compounds receives a phosphate and loses two hydrogens to the hydrogen acceptor, NAD. Immediately, this added phosphate is transferred to ADP to produce ATP. The tremendous energy required to attach the phosphate to ADP is liberated from the metabolite and trapped by the high-energy bond between ADP and the third phosphate.

ADP + phosphate + energy → ADP ~ P (ATP)

The two three-carbon compounds with their remaining phosphate undergo further chemical changes, finally yielding their phosphate and enough energy for the production of more ATP, and thus become pyruvic acid.

If molecular oxygen is not available to the cell which is anaerobically metabolizing glucose, the pyruvic acids are reduced to lactic acid by accepting the hydrogens from the $NADH_2$ produced earlier. If molecular oxygen becomes available, the process is reversed: lactic acid is oxidized to pyruvic acid, which undergoes a series of

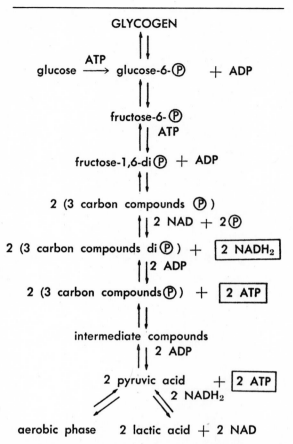

Fig. 17-6. Schematic diagram of *anaerobic metabolism* of glucose. The symbol ⓟ is used to represent phosphate. Some of the energy of glucose is used to produce 4 molecules of ATP. However, two molecules of ATP are used in converting glucose to glucose-6- ⓟ, and fructose-6- ⓟ to fructose-1,6-di ⓟ. The *net* yield is 2 ATP. Some of the energy of glucose contained in hydrogen is transferred to the hydrogen acceptor NAD. If oxygen is available, this energy is released later when the hydrogens go through the electron transport system. If no oxygen is available, $NADH_2$ is used to reduce pyruvic acid to lactic acid. See text for details.

changes such that large amounts of additional energy are made available for heat or for the formation of more ATP. This series of reactions is termed the **aerobic phase** and depends ultimately on the presence of oxygen. Figure 17-7 illustrates some of the details of the process.

Pyruvic acid, by the loss of 2H and CO_2 and the addition of coenzyme A, is oxidized to acetyl CoA. Acetyl CoA now becomes the central figure in the ensuing reactions. It loses coenzyme A and condenses with oxaloacetic acid to become citric acid—and the **citric acid cycle**

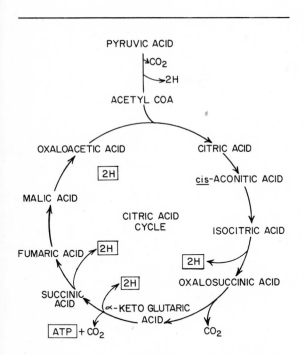

Fig. 17-7. Schematic diagram of the *aerobic metabolism of pyruvic acid.* As pyruvic acid is oxidized to acetyl CoA, CO_2 is given off and 2H are transferred to a hydrogen acceptor. Acetyl CoA then condenses with oxaloacetic acid to form citric acid and begin the citric acid cycle. During one complete turn of the cycle, two molecules of CO_2 are given off, one molecule of ATP is formed directly and eight atoms of H are carried into the mitochondrial inner membrane for passage through the electron transport system. See text for details.

is underway. Citric acid is oxidized by a series of steps, changing from one compound to another. At various steps in this cycle, hydrogens with their energy are released from the metabolite and transferred to the hydrogen acceptors. It is believed that NAD, or its phosphorylated form, NADP, and flavin adenine dinucleotide (FAD) are the principal hydrogen acceptors for this system. You will note in Figure 17-7 that a total of eight hydrogens are released from intermediary metabolites and attached to hydrogen acceptors with each turn of the cycle. Also one-ATP is formed when α-ketoglutaric acid is converted to succinic acid.

The liberation of the energy contained in the hydrogens now attached to NAD and FAD is accomplished through an ingenious system which transfers the hydrogens, or their electrons, from one acceptor to another. This system has been called the **electron transport system,** the cytochrome system, and the respiratory chain. A proposed scheme for this system is shown in Figure 17-8. In this scheme, NAD acts as the first hydrogen acceptor, receiving hydrogens from various metabolites during anaerobic and aerobic metabolism. The $NADH_2$ thus produced enters the inner membrane of the mitochondria where all the necessary enzymes and acceptors involved in

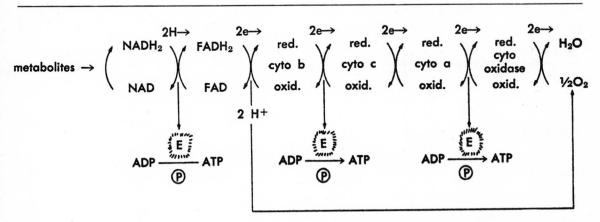

Fig. 17-8. Diagrammatic representation of the proposed scheme for the electron transport system. The $NADH_2$ formed during the oxidation of metabolites transfers its hydrogens to FAD which then becomes $FADH_2$. The energy E released at this point, coupled with phosphate ⓟ, is used to convert ADP to ATP and is thus trapped. The hydrogens on $FADH_2$ lose their electrons (e) to cytochrome b and become hydrogen ions, $2H^+$. These electrons are now transferred from one cytochrome to another, alternately reducing and oxidizing the cytochromes as the electrons are gained or lost. Finally, cytochrome oxidase transfers these electrons to oxygen which then reacts with the $2H^+$ released from $FADH_2$ to produce water. At two points in this transfer of electrons by the cytochromes, energy is released and trapped in ATP. Thus, an $NADH_2$ entering the system would provide enough energy for the formation of 3 ATP. See text for details. (After Lehninger).

the release of energy from these hydrogens are located. $NADH_2$ transfers the hydrogens to FAD. Thus $NADH_2$ is oxidized to NAD, FAD is reduced to $FADH_2$, and enough energy is released to convert one molecule of ADP to ATP, thus trapping this energy. Since phosphate is also necessary to convert ADP to ATP, this reaction is often referred to as *oxidative phosphorylation.*

$FADH_2$ releases the hydrogens, which then pass their electrons to an electron acceptor, the iron-containing compound *cytochrome b,* and become $2H+$. These $2H+$ ultimately react with oxygen and the two electrons to become water. In the meantime, the two electrons which were transferred to cytochrome b are now transferred to cytochrome c, thence to cytochrome a, and finally to cytochrome oxidase. The cytochrome oxidase transfers the two electrons to oxygen, which then unites with the $2H+$ and forms water. Enough energy is released when the electrons are transferred from cytochrome b to cytochrome c, and again from cytochrome a to cytochrome oxidase, to produce two more molecules of ATP.

Thus, by a bucket brigade operation, two hydrogens from the metabolite release enough energy to produce three molecules of ATP, and ultimately appear in water. It is important to note that, although oxidation is accomplished by dehydrogenation reactions, oxygen is the final hydrogen acceptor, and humans cannot survive in an oxygen free atmosphere for more than a few minutes.

At one point in the citric acid cycle, the hydrogens released from the metabolite are accepted not by NAD, but by FAD. The $FADH_2$ thus formed enters the electron transport system, but the first step, $NADH_2 + FAD \rightarrow NAD + FADH_2$, is no longer necessary, and $FADH_2$ enters the system at step 2. You will note in Figure 17-8 that enough energy was released when hydrogens were transferred from $NADH_2$ to FAD to generate one molecule of ATP. If $FADH_2$ enters the system directly, this ATP is *not* formed, and the total output of ATP from $FADH_2$ is only two.

Let us now summarize the total number of ATP molecules that are produced during the various steps in aerobic metabolism. In the citric acid cycle, three molecules of $NADH_2$ are produced. As the hydrogens in each $NADH_2$ are transported through the electron transport system, enough energy is liberated to produce three molecules of ATP, a total of nine for the three compounds. One $FADH_2$ is also produced in the citric acid cycle, and enough energy is liberated from this compound in the electron transport system to produce two more molecules of ATP. One ATP was produced directly as α-ketoglutaric acid was oxidized to succinic acid. The total number of ATP molecules ultimately produced by one turn of the citric acid cycle is twelve; nine from the $3NADH_2$, two from the $FADH_2$, and one directly. Therefore, each acetyl CoA, regardless of the source (glucose, fatty acids, or amino

			Directly	ATP PRODUCTION Electron Transport System	TOTAL
anaerobic	glucose ↓	4 ATP-2 ATP 2 $NADH_2$	2	6	
aerobic	2 pyruvic acid ↓	2 $NADH_2$		6	8
	2 acetyl CoA				6
	citric acid cycle ↓	6 $NADH_2$ 2 $FADH_2$ 2 ATP	2	18 4	
	$CO_2 + H_2O$				24 38

Fig. 17-9. Summary of ATP production from the complete oxidation of one molecule of glucose. Some ATP is generated directly during both anaerobic and aerobic metabolism. However, the major production of ATP depends upon the release of energy as hydrogens, or their electrons, are transported through the electron transport system.

acids), completely oxidized in the citric acid cycle will yield enough energy for the production of twelve molecules of ATP.

However, if these same foodstuffs enter the system as pyruvic acid, three additional ATP molecules will be produced. When pyruvic acid is converted to acetyl CoA, two hydrogens are accepted by NAD to produce $NADH_2$, which then releases energy in the electron transport system. Thus the complete oxidation of pyruvic acid yields 15 molecules of ATP.

A summary of the total ATP production from the complete oxidation of a molecule of glucose is given in Figure 17-9. This energy represents about 40 per cent of the total energy available from glucose. The remaining 60 per cent is liberated as heat which is distributed to all parts of the body by way of the bloodstream. The regulation of this heat in maintaining body temperature is described on pages 427 to 429.

Lipid metabolism

Absorbed lipids are transported in the blood partly as chylomicrons, and partly as lipoprotein complexes. In addition, there may be free fatty acids loosely combined with albumin. Chylomicrons are composed primarily of triglycerides (simple lipids) combined with a small amount of cholesterol, phospholipid, and protein; the lipoprotein complexes contain less triglyceride and more cholesterol, phospholipid, and protein. The sources of these lipids are: absorption from the small intestine, mobilization of depot fat, and synthesis in the liver.

Utilization of Lipids. Lipids are vital constituents of all living cells, and, when oxidized, they also are an important source of energy. The storage of lipids in the form of adipose tissue provides insulation and mechanical protection as well as energy reserves. The liver serves as a central clearing house in lipid metabolism, as it is the principal site for the synthesis of lipids and also for their degradation (i.e., making them less complex by splitting off one or more groups).

The *storage of fats* (simple lipids) involves the formation of adipose tissue in various parts of the body. Because one can store only a small amount of carbohydrate (less than one pound), these fat deposits constitute the bulk of stored, energy-producing food in the body. Depot fat is

in a continuous state of change, which means that it is constantly being deposited and remobilized; however, deposits may exceed withdrawals when there is an excessive intake of food.

Under normal conditions, the liver is able to oxidize any fat that is conveyed to it; thus there is no accumulation. However, should large quantities of fat be mobilized from the depots and transferred to the liver, there may be an accumulation resulting in the condition known as *fatty liver*; this may occur in such conditions as untreated diabetes mellitus, starvation, and the consumption of a high-fat diet.

The *oxidation of fatty acids* involves a complex series of reactions that result in the formation of a number of molecules of acetyl CoA, the two-carbon compound previously described. For example, one typical fatty acid (palmitic) consists of a chain of 15 carbon atoms attached to the carbon of a carboxyl group (—COOH), giving a total of 16 carbons:

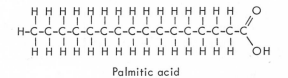

Palmitic acid

The first step in the breakdown of the above molecule involves its activation; this is accomplished by a coupling with coenzyme A (CoA) to form fatty acyl CoA:

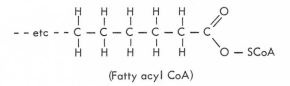

(Fatty acyl CoA)

Next, through a series of reactions involving the removal of hydrogen atoms and the addition of water, the compound splits between the alpha and beta carbons (the alpha carbon is that attached to the —COOH group, the beta is that next to alpha). The long portion of the chain then combines with a new molecule of CoA, while the short portion remains combined with the original CoA, thus forming a molecule of acetyl CoA:

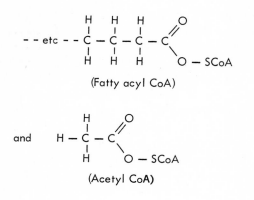

(Fatty acyl CoA)

and

(Acetyl CoA)

The beta-oxidation of fatty acids takes place mainly in the liver cells, but it can occur in other tissues as well. In any event, the process is repeated until the entire molecule of fatty acid has been broken down, and eight molecules of acetyl CoA have been formed. Inasmuch as each triglyceride molecule usually contains three long-chain fatty acids, one can see why fats have a greater caloric yield than do carbohydrates (i.e., one molecule of glucose provides only two molecules of acetyl CoA).

Although the liver cells generate far more acetyl CoA than is required for their own metabolic needs, this valuable compound normally does not go to waste. Instead, when there is an excess, the liver condenses two molecules of acetyl CoA to form one molecule of *acetoacetic acid*. This keto acid diffuses into the bloodstream and is carried to active tissues throughout the body. Upon entering a tissue cell, the process is reversed and two molecules of acetyl CoA are formed; these molecules then enter the citric acid cycle and are oxidized with the liberation of energy. Normally, the amount of acetoacetic acid in the blood rarely exceeds 1 mg. per cent, but in certain metabolic disorders it may accumulate and lead to ketosis.

Ketosis. The so-called ketone bodies include *acetoacetic acid* along with two substances that are derived from it, *beta-hydroxybutyric acid* and *acetone*. When these substances accumulate in the blood (ketonemia) and are found in the urine (ketonuria), a state of *ketosis* exists.

Ketosis may be severe when carbohydrate metabolism is depressed and the body is deriving most of its energy from the metabolism of fats. For example, in untreated diabetes mellitus, a lack of insulin prevents the transport of glucose into the cells; thus the body must turn to fat as an alternative source of energy. As a result, large quantities of acetyl CoA are generated, and ketone bodies pour out of the liver for delivery to the cells. Unfortunately the tissue cells are limited in their ability to metabolize ketone bodies; thus these substances begin to accumulate and ketosis develops. In a similar manner, ketosis can occur in starvation, or in consumption of a high-fat diet.

Acetoacetic acid and beta-hydroxybutyric acid are relatively strong organic acids; thus an excessive accumulation may cause severe *metabolic acidosis* (p. 468) for two reasons. First, their mere presence in the blood tends to deplete the basic, or alkaline, half of the buffer systems (p. 464). Second, and even more important, large quantities of these acids are accompanied by sodium during their excretion by means of the kidney; this not only decreases the amount of base available to buffer acids, but also leads to dehydration, because water is lost whenever sodium is lost. Acetone is eliminated in two ways. Because it is a volatile substance, small quantities are blown off in expired air from the lungs, and persons with severe ketosis have the odor of acetone on their breath. Acetone also is eliminated in the urine. By a simple urine test for acetone, the diabetic may be forewarned that a compensated acidosis exists. Various laboratory procedures are then utilized in analyzing the total concentration of ketone bodies, thus enabling a physician to determine the degree of ketosis that exists. Treatment must be started promptly, because uncontrolled ketosis ultimately leads to coma and death.

Biosynthesis of Lipids. Although the major pathway for acetyl CoA is its oxidation by means of the citric acid cycle, this two-carbon compound actually serves as a bridge between catabolism and anabolism. In other words, acetyl CoA serves not only as a source of energy, but also as the major precursor in the synthesis of fatty acids, acetylcholine, cholesterol, and other steroids. Free glycerol serves as the precursor for the triglycerides and some of the phospholipids. A summary of lipid metabolism is presented in Figure 17-10.

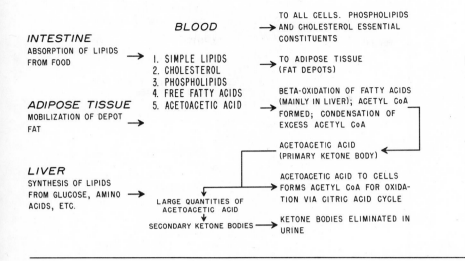

Fig. 17-10. Diagram of lipid metabolism. Sources of blood lipids are indicated on the left. The means of utilizing these lipids and related substances are indicated on the right.

INTESTINE
ABSORPTION OF LIPIDS FROM FOOD →

ADIPOSE TISSUE
MOBILIZATION OF DEPOT FAT →

LIVER
SYNTHESIS OF LIPIDS FROM GLUCOSE, AMINO ACIDS, ETC. →

BLOOD
1. SIMPLE LIPIDS
2. CHOLESTEROL
3. PHOSPHOLIPIDS
4. FREE FATTY ACIDS
5. ACETOACETIC ACID

→ TO ALL CELLS. PHOSPHOLIPIDS AND CHOLESTEROL ESSENTIAL CONSTITUENTS

→ TO ADIPOSE TISSUE (FAT DEPOTS)

BETA-OXIDATION OF FATTY ACIDS (MAINLY IN LIVER); ACETYL CoA FORMED; CONDENSATION OF EXCESS ACETYL CoA

ACETOACETIC ACID (PRIMARY KETONE BODY)

LARGE QUANTITIES OF ACETOACETIC ACID

SECONDARY KETONE BODIES →

ACETOACETIC ACID TO CELLS FORMS ACETYL CoA FOR OXIDATION VIA CITRIC ACID CYCLE

KETONE BODIES ELIMINATED IN URINE

Protein metabolism

The primary function of protein is to supply the amino acids necessary for building new protoplasm, blood proteins, enzymes, and many of the hormones that regulate the chemical processes of the body. Proteins also are burned for energy, but this is not their chief function.

There are three sources of amino acids in the blood: absorption from the small intestine, synthesis by liver and tissue cells, and catabolism of tissue protein. The amino acids in the blood constitute a pool that can be drawn on for all purposes of protein metabolism. In other words, the story of protein metabolism is essentially the story of what happens to amino acids as they enter and leave this dynamic plasma pool. This process is shown in Figure 17-11. (Fig. 17-11).

Amino Acid Synthesis. All amino acids are needed by the body, but only certain ones can be formed within the body. Those which cannot be synthesized are called *essential* amino acids, and they must be obtained from ingested food. On the other hand, *nonessential* amino acids can be synthesized in the liver if they are missing from the diet. For example, the liver cells can transfer the amino group ($-NH_2$) from an amino acid to a keto acid and so form a different amino acid (transamination). These amazing cells also can form an amino group from some other nitrogen-containing compound (e.g., ammonia) and add it to a keto acid, forming an amino acid (amination). In this way new substances are formed.

The amino acid nitrogen in the body is shifting continually from one amino acid to another, being added to the pool, then withdrawn from the pool and incorporated into tissue, liver, or plasma protein, then split and again returned to the pool. Urea is the chief nitrogenous waste product. Lesser ones are creatinine and ammonium salts.

Protein Synthesis in Body Tissue. As blood flows through the vast capillary networks of the body, each tissue selects the specific amino acids needed to build the kinds of protein required for growth, maintenance, and proper function of that tissue. A child, who is growing rapidly and building tissues, requires great quantities of amino acids as new protoplasm is being formed throughout his body. As one gets older, amino acids still are required to replace the protoplasm that breaks down under the wear and tear of daily living.

Enzymes and hormones are other examples of protein synthesis by the body tissues. While every cell in the body can produce the enzymes that are necessary to catalyze its own internal chemical reactions, only special cells in the gastrointestinal tract and the pancreas can synthesize the extracellular enzymes needed to digest food. The production of certain hormones also requires the services of highly specialized cells. A good example of this is the synthesis of insulin by islet cells in the pancreas; no other cells in the body can produce this vital substance.

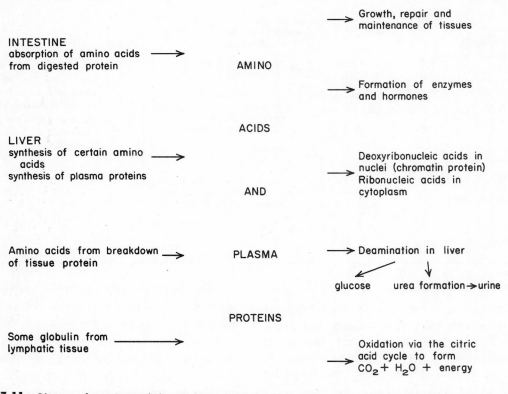

Fig. 17-11. Diagram of protein metabolism. Sources of amino acids and plasma proteins are indicated on the left, and the means of utilizing amino acids are indicated on the right.

The most fascinating aspect of the whole subject of protein synthesis by the tissue cells concerns the mechanism whereby cells know what to build from the amino acids in the blood—that is, why muscle cells build sarcoplasm rather than hemoglobin. In addition, for every type of protein there is a blueprint that indicates the exact sequence of amino acids; that is, one arrangement of amino acids results in a molecule of albumin, but another sequence is required to produce globulin. Deoxyribonucleic acid and ribonucleic acid are of fundamental importance in all protein synthesis, as previously described on page 28.

Catabolic pathways are followed by protein molecules when they are needed as a source of energy, or if there is ingestion of a high protein diet. In either case, amino acids lose their amino group by transamination, or by deamination, in the liver. *Deamination* is the enzymatic removal of the amino group with the formation of ammonia and a keto acid. The ammonia is converted to urea and eliminated in the urine. The keto acids are converted by any of a variety of processes to ketone bodies, especially acetoacetic acid, to pyruvic acid, to acetyl CoA, and to various intermediates in the citric acid cycle (e.g., succinic acid, fumaric acid, and α-ketoglutaric acid). When energy needs must be met, these compounds yield their energy through the citric acid cycle. However, if energy is not required and if structural protein supplies are adequate, these substances are converted to glycogen or to fat for storage; thus it is possible to get fat on proteins.

If one goes without food temporarily (a fasting state), the oxidation of body fat provides about 85 per cent of the required energy, protein supplying only 15 per cent. On the other hand, in starvation, when the fats have been used up, *all* energy comes from protein catabolism; emaciation and death will follow if food intake is not resumed.

Nitrogen Balance.   This refers to the state

of protein balance in the body. If the nitrogen intake in food is equal to the nitrogen output in waste products, one is said to be in nitrogen balance; this is the healthy state of an average adult.

A *positive* nitrogen balance exists when the intake of nitrogen is greater than the output. This occurs in growing children who are building new tissues, in the pregnant woman whose baby is growing in the uterus, and in the postoperative patient who is repairing tissues. In such cases, less deamination is going on, since the amino acids are needed by the body for growth and repair of protoplasm.

In *negative* nitrogen balance the intake is less than the output. This is a serious situation, since nitrogen must be coming from the breakdown of the person's own tissues, and life cannot continue for long in such a state. Negative nitrogen balance occurs in such cases as starvation and in wasting diseases.

THE BODY TEMPERATURE

The maintenance of a fairly constant body temperature in man is another example of homeostasis in which many activities are correlated to accomplish one purpose. As long as one is alive, heat is produced. The amount produced in health varies greatly in accordance with the activity of skeletal muscles, the type of food eaten, the temperature of the surroundings, and the condition of certain endocrine glands. If an adult man rested for 24 hours and took no food, he would produce about 1,700 Cal. of heat. If he were moderately active, he would produce about 3,000 Cal., and if he did heavy work, he would produce about 7,000 Cal. Not only are there such wide variations in health, but the variations may be equally wide in infectious diseases and glandular disturbances. Whatever the amount of heat produced in the body, the cooperation of the circulatory system is needed to distribute the heat throughout the body.

Since heat is produced continuously during life, it must be lost continuously. If there were no heat loss, the temperature of the body under basal conditions would rise by 1.8° F. per hour; with moderate activity it would rise by 3.6° F. per hour. It is evident that not only must there be continued loss of heat, but the amount lost must vary also in accordance with changes in heat production. Most of the heat loss takes place through the skin and is regulated by the amount of blood flowing through the cutaneous vessels and also by the activity of the sweat glands.

In order to maintain a balance between heat production and heat loss, fine adjustments are needed. A device somewhat like a thermostat is required, but the regulating mechanism in the body differs from a thermostat in that heat production cannot be turned off completely, no matter how hot the external environment becomes. Heat production in the body cannot drop below the *basal* level.

The mechanisms for temperature regulation are so efficient in health that only when one does exceedingly heavy work or when the environmental conditions interfere with heat loss does the body temperature depart from the normal range. Since one adjusts so well to both external and internal threats, a departure from normal body temperature has come to be one of the cardinal signs of disease.

Heat production

Heat production is the result of the oxidation of foodstuffs in all cells and may be thought of as a by-product in the production of useful energy for the activities of the body. Thus heat production can vary widely in response to the body's need for the heat itself. The production of heat cannot be decreased below the basal need for energy even when the environmental temperature is so high that life may be in danger. Neither can heat production be increased sufficiently to protect a person against extreme cold, and one can freeze to death.

Activity of Muscles. Since much of the body is composed of skeletal muscle, most body heat is produced by metabolic activity of these muscles. When the skeletal muscles are relaxed, as during rest, heat production is lowest but still provides about 25 per cent of the total heat produced. With strenuous exercise, heat production by skeletal muscles increases greatly. If the body temperature falls and one does not voluntarily take some action to correct this condition, a reflex mechanism will initiate shivering to increase heat production.

Activity of Glands. Glandular secretory activity also produces heat. Since the liver is the largest and most active gland in the body, it accounts for about 20 per cent of the heat production in the body under basal conditions and is a very important source of heat when skeletal muscles are relaxed, as during sleep.

As previously mentioned (p. 416), thyroid hormones stimulate metabolism in all cells. If the amounts of these hormones are increased above normal, there will be increased heat production.

Sympathetic Stimultion. Both epinephrine and norepinephrine have a direct effect on metabolic rate and thus play a role in heat production. Heat production may be increased by as much as 20 to 30 per cent with maximum sympathetic stimulation (p. 225).

Food Intake. The increased heat production that accompanies intake of food is the result of several factors. There is increased activity of the smooth muscle tissue in the gastrointestinal tract and marked activity of the glands that secrete the digestive enzymes. In addition, the foods exert what is known as the *specific dynamic action* of food. The increase in heat varies with the type of food eaten, being greatest with protein foods. This increased heat production usually lasts for several hours after the food has been consumed. One of the requirements for basal metabolism tests is that no food be eaten during the preceding 14 to 18 hours.

Body Temperature. As body temperature rises, heat production also rises, due to an increase in the metabolic rate, the average increase being about 7 per cent for each degree Fahrenheit rise. A rise in temperature disturbs metabolic patterns, and the resulting disorganization of brain function is evident by the delirium that accompanies high fevers.

Heat loss

The major channel of heat loss is the skin, but some loss also occurs through the lungs and the excreta. Heat is lost from the body by the physical processes of radiation, conduction, convection, and vaporization.

Radiation. Radiation is the transfer of heat through the air from the surface of one object to the surface of another without physical contact. One loses heat by radiation only if the surroundings are cooler than the body. If the surroundings are warmer, heat will be transferred to the body;

this may be a distinct disadvantage under certain circumstances and may interfere with the physiologic regulation of temperature. Under ordinary conditions, in a temperate climate, the greatest loss of body heat occurs through radiation from the skin. The amount and type of clothing that one wears, as well as the amount of heat brought to the skin by the circulating blood, will greatly influence the radiation process.

Conduction. Conduction refers to the transfer of heat from one object to another by direct physical contact. Man loses heat by conduction to the air that is in contact with the skin (if it is cooler than the skin), to clothing, to air in the respiratory passages, to cold food and beverages in the alimentary canal, and to cold furniture. Since man protects himself by clothing, the loss of heat by conduction through the skin may not be great. Air is a poor conductor of heat, and the protective value of clothing is proportional to the thickness of the layer of still air trapped in the meshes or between the layers of material.

Convection. The movement of air is known as convection, and heat may be lost from the body by convection air currents. As heat is conducted to the air surrounding the body, it warms that air; the heated air rises and is replaced by denser, cool air. Electric fans, cool winds, and the like cause the warm air to be replaced more rapidly. Clothing, on the other hand, reduces heat loss by convection.

Vaporization. Vaporization refers to loss of heat when water evaporates or changes from the liquid to the vapor state. Each gram of water that evaporates (at body temperature) requires 0.58 Cal. of heat.

Some of our body heat is lost by saturating the expired air with water vapor from the lining of the respiratory tract, but the amount lost in this way depends on the temperature and humidity of the inspired air. The amount of water lost by vaporization from the lungs is about 300 ml. daily.

The greater part of heat loss by vaporization is through the skin. A continuous loss of heat through the insensible loss of water occurs, since one loses from 300 to 800 ml. of water daily by this means. The loss of heat by evaporation of perspiration (sensible loss) varies with the temperature and the relative humidity of the external air. The maximum rate of vaporization of sweat is about 1.6 L. per hour, such as occurs during work in a hot dry atmosphere.

The flexibility of the skin temperature in man is the most important device for adjusting heat loss. The water of the plasma takes up heat in the warmer parts of the body, such as the liver and the skeletal muscles, and gives it up in cooler regions, such as the skin. The amount of heat distributed to a cutaneous area is determined by the rate of blood flow through the area. If the vessels are constricted, the blood flow is curtailed, and there is less heat to be eliminated by radiation, conduction, convection, and vaporization. If there is vasodilation, the skin is warmer, and more heat will be eliminated. The amount of blood flowing through the skin can be varied markedly; it can be decreased to almost zero or increased to as much as 12 per cent of the total blood flow. In the fingers the blood flow can be changed from about 1 ml. to 90 ml. per 100 Gm. of tissue per minute. The blood flow through the hands and the feet is particularly significant, since these are very important areas for dissipation of heat. Not only can the blood flow through arterioles, capillaries, and venules be varied, but there are also arteriovenous anastomoses in the extremities through which blood can be shunted from arteries to veins without going through capillaries. This makes possible a great increase in the amount of blood flowing through the area concerned.

Temperature regulation

The hypothalamus is the integrating center for the regulation of body temperature. A heat-sensitive area in the anterior portion of the hypothalamus protects the body from overheating. The sweat glands are stimulated to produce more sweat, the vasodilator nerves to the skin area are stimulated, and the sympathetic centers in the posterior hypothalamus are inhibited. Thus vasoconstrictive tone is reduced, resulting in increased blood flow to the skin. Damage to the anterior portion of the hypothalamus gives rise to fever, with permanent inability to regulate body temperature on exposure to hot environments.

A cold-sensitive area in the posterior hypothalamus protects the body against overcooling, by vasoconstriction, by inhibition of the sweat glands, and by stimulation of the skeletal muscles. Thus tone is increased and finally shivering en-

sues. It has been estimated that heat production can increase by as much as 50 per cent even before shivering starts. With damage to the posterior hypothalamus, there is no shivering and no vasoconstriction on exposure to cold.

The main heat-regulating centers are apparently controlled in two ways, by the temperature of the blood reaching these areas and reflexly by the receptors for cold and warmth located in the skin. These centers operate like a thermostat, responding to temperature changes by stimulating heat-producing, or heat-conserving, mechanisms when they are cold, or heat-loss mechanisms when they are warm. This thermostat operates so efficiently that in a state of normal health, the internal, or *core temperature*, rarely varies more than a few tenths of a degree. Changes in the thermostat due to abnormal conditions will cause the core temperature to vary by several degrees.

The major mechanisms for heat gain and heat loss are summarized in Figure 17-12. In a cold environment, the heat of metabolism is lost by radiation (R), convection (C), and evaporation of insensible perspiration ($E_{insens.}$). To maintain body temperature, heat production must equal heat loss. This can be accomplished by an increase in metabolism or a decrease in losses. The first line of defense against a change in body temperature is *vasoconstriction*: the cutaneous area acts like a thick outer shell, preventing heat loss at the skin surface. If heat loss continues to be greater than heat production, the second line of defense mechanisms comes into play. These include several mechanisms for the increased production of heat, increased muscle tone, shivering, and increased muscle activity, as well as mechanisms for the conservation of heat; one adds clothing and seeks shelter. If heat of metabolism is now balanced by heat loss, $M - R - C - E_{insens.} = 0$, the core temperature will not change.

In a hot environment heat is lost only by evaporation. Because of the high external temperature, both radiation and convection are adding heat to the body. Therefore, heat loss must increase if core temperature is to be maintained. In this circumstance, the first line of defense against a change of core temperature is *vasodilatation*: the cutaneous area acts like a thin covering through which the large amounts of heat being brought to the surface by the blood can be readily transmitted. If metabolic heat brought

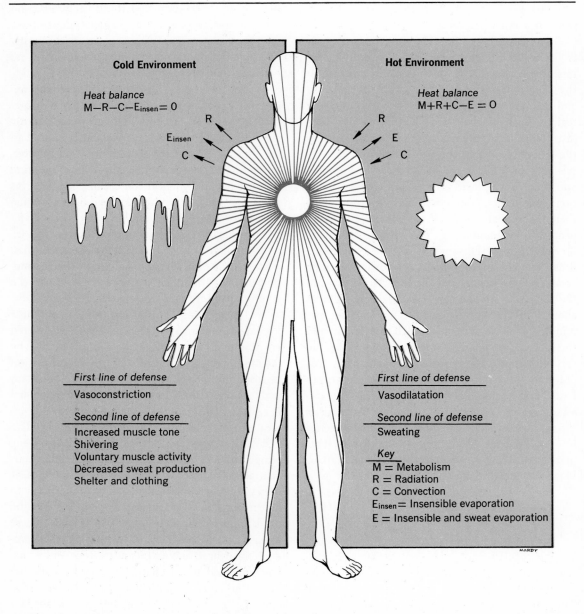

Fig. 17-12. Schematic representation of the temperature regulating mechanisms that operate to maintain heat balance when the body is subjected to cold or to hot environments. See text for details.

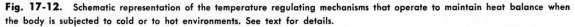

to the surface is sufficient, environmental temperature will not exceed surface temperature and heat may now be lost by radiation and convection. If environmental temperature still exceeds surface temperature, heat will continue to be gained. Now sweating, the second line of defense against a rise in body temperature, comes into play. In a dry atmosphere, large amounts of sweat will evaporate,

resulting in the loss of large amounts of heat. If heat lost by evaporation now balances that gained by metabolism, radiation and convection—$M + R + C - E = 0$—the core temperature will not change.

In a humid environment, sweat cannot evaporate sufficiently, and it pours off the body as a liquid. However, the heat loss is minimal, and the core temperature will begin to rise. A normal body temperature can be maintained even in extremely hot environments, 240° F. to 260° F., if the air is dry, if large amounts of replacement water are consumed, and if the body is stripped of clothing; whereas in a humid environment, with a temperature of only 120° F., the core temperature may begin to rise in a very short time.

Physiologic variations in body temperature

The range of temperature in healthy young adults varies between 96.5° to 99.3°, with an average of approximately 98.6° F.

Temperature regulation is poorly developed in newborn babies and is especially lacking in premature babies. In infants, a fit of screaming can cause a marked rise in body temperature, and warm and cold baths can alter the temperature more than in adults.

In elderly persons, the temperature is usually subnormal. The body is less active, the circulation is feeble, and there is less power of compensation for changes in external temperature. These persons are intolerant of extremes of external temperature.

There is a daily, or diurnal, variation in body temperature, which is established at about one year of age. The temperature is lowest in the early morning, after several hours of rest in bed, when one is inactive and when one is not digesting food. The temperature is highest in the late afternoon and the early evening, after a day of activity and after the evening meal. If the temperature is taken every hour over a 24-hour period, as much as 3° F. variation may be found, but the usual variation between the early morning and the late afternoon is about 1° F.

There is some variation in basal temperature with the menstrual cycle. Ovulation is thought to occur at the low point of the monthly temperature curve; the basal temperature then rises, reaching a peak within 24 hours after ovulation.

Abnormal variations in body temperature

High External Temperatures. **Heat exhaustion.** A person subjected to high external tem-peratures may perspire so much that heat exhaustion occurs. This can happen because temperature regulation has priority over the maintenance of water and salt balance, and sweating continues even though it produces severe dehydration and marked sodium chloride loss. Heat exhaustion is characterized by severe muscle cramps, pallor, low blood pressure, vertigo, weakness, nausea, vomiting, and fainting. The symptoms are due to sodium chloride depletion and can be prevented by taking enough sodium chloride and water to make up for the excessive loss by perspiration.

Heat stroke. This is seen most frequently in the person who has a preexisting acute, or chronic, disease. When first exposed to excessive heat, these persons perspire, and the loss of water leads to a decrease in plasma volume, with a decrease in cardiac output. This means a decreased blood flow through the skin. As the cutaneous blood supply becomes inadequate, sweat secretion diminishes and then ceases completely as circulatory failure supervenes.

When a major pathway of heat loss, namely, evaporation of sweat, is closed, the body temperature rises with explosive suddenness, and the general condition of the person becomes very critical. The skin is dry and flushed. As the body temperature rises to dangerously high levels, the patient becomes comatose. The absence of sweat is the most striking characteristic of heat stroke. Death invariably occurs if the body temperature rises above 114° F., due to coagulation of the protein of protoplasm and irreversible inactivation of enzymes.

Low External Temperatures. If the body is subjected to excessive cold, or if there is too great a loss of heat through insufficient clothing or by immersion in cold water, the body temperature falls markedly. Along with this, there is a feeling of exhaustion and an uncontrollable desire to sleep. Respirations slow, blood pressure falls, dissociation of oxyhemoglobin decreases, and coma follows. When the body temperature falls below 94° F., temperature regulation is impaired, and it is lost completely when the body temperature falls below 82° F. One case has been reported in which there was recovery after the body temperature had fallen to 64° F. The woman, while intoxicated, went to sleep in the street in very cold weather. Her body was stiff when found, but there was a faint sound, which was interpreted as a heart sound, occurring about 12 times a minute. With excellent care she recovered, but both legs and several fingers had to be amputated, since these tissues had been irreparably damaged, probably due to the formation of ice crystals within the cells.

During surgery on the heart, the great vessels, or the intracranial vessels, the brain can be protected by artificially lowering the body temperature (*hypothermia*). As body temperature falls, metabolism decreases and the cells' demand for oxygen decreases. Thus circulation may be interrupted for a much longer time without damage.

Fever. Fever is an elevation of body temperature beyond the normal range. The most frequent cause of fever is a bacterial or a viral infection. Less commonly, it may be a drug or a brain lesion disturbing heat-regulating mechanisms, a foreign protein, or one liberated during destructive processes such as might occur in a

severe heart attack. Dehydration fever is seen occasionally, especially in children with low blood plasma volumes.

The agents responsible for infectious fever are proteins, protein breakdown products, or other large molecules, such as polysaccharides, collectively called *pyrogens*. It is believed that pyrogens act by resetting the hypothalamic thermostat upward, so that all of the heat-producing and heat-conserving mechanisms immediately become active. There is vasoconstriction, decreased sweat gland activity, increased muscle tone, and finally shivering. As additional heat is produced and conserved, the body temperature begins to rise. When the temperature reaches the setting ordered, the mechanisms for heat production and heat loss are again equalized, and temperature is regulated at that level. The temperature will remain there until a new factor breaks the disease process and the thermostat is turned down. Now the mecha-

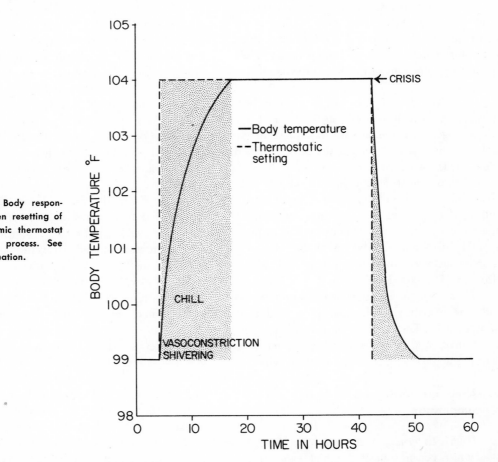

Fig. 17-13. Body responses to a sudden resetting of the hypothalamic thermostat by a disease process. See text for explanation.

nisms for heat loss are activated; vasodilatation and sweating continue until the temperature reaches the new lower setting.

Figure 17-13 shows the pyrogen effect during a disease process. As the pyrogen is released, the hypothalamic thermostat is abruptly set upward from 99° F. to 104° F., activating all of the heat-producing and heat-conserving mechanisms. The patient complains of feeling cold, and his skin is pale and dry. Even after the temperature begins to rise, additional covering and hot water bottles cannot make the patient feel warm. His teeth begin to chatter, and he may shiver so violently that the bed shakes. This response, called a *chill*, will continue until the body temperature has reached the new thermostat setting of 104° F. At that point, the chill stops, the patient no longer complains of feeling cold, and his skin is again warm and moist. Thirty-eight hours after the onset of the chill, some factor breaks the

disease process, and the thermostat is abruptly reset down to 99° F. Now the mechanisms for heat loss are activated. The patient feels hot, his skin is flushed, and he sweats profusely until the temperature reaches the new setting. This sudden drop is called the *crisis*.

More commonly, body temperature that was elevated by pyrogenic resetting of the thermostat falls gradually, often with periods of increased temperature followed by a drop in temperature. This is called *lysis* and is probably due to a jiggling of the thermostatic setting (Fig. 17-14).

In fever due to an infection, there is increased breakdown of body protein, probably due to the increased metabolism in all cells; tissue wasting may result when the fever is prolonged. It can be prevented if enough proteins and carbohydrates are provided to meet the greatly increased demand. Thus the old adage, "feed a fever," is correct.

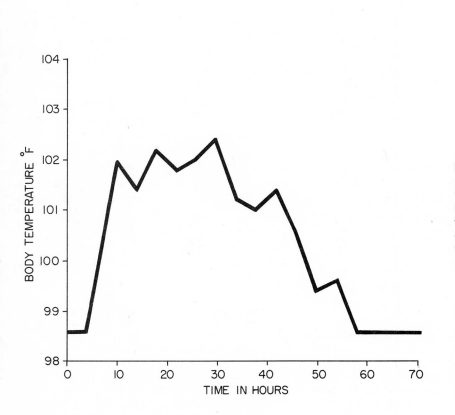

Fig. 17-14. Typical temperature chart of a patient whose fever is falling by lysis. Note the fluctuations in temperature due to a "jiggling" of the hypothalamic thermostat. See text for details.

The significance of fever is not known. It is possible that bacteria do not multiply well at higher temperatures, or perhaps the body manufactures antibodies more readily at higher temperatures. It is known that phagocytosis is increased during fever. All of these responses would tend to aid in control of the infectious agent.

Such beneficial effects of fever as may exist are overshadowed by its harmful effects. Many drugs have the capacity to lower body temperature when it is elevated above normal. These *antipyretics* (such as aspirin, quinine, and salicylates) are used extensively in combating fever. These drugs are thought to increase the osmotic pressure of the blood, resulting in fluids being drawn into the bloodstream and increasing the volume of the radiator system for carrying heat to the surface.

Certain other drugs cause a fall in temperature, even when it is within the normal range, as well as in fever. Morphine and general anesthetics, for example, cause decreased cerebral activity, and the response to changes in environmental temperatures is deficient. Alcohol causes vasodilatation, with the result that large amounts of heat are brought to the surface even in cold environments and the body temperature may fall. The skin feels warm, however, and this leads to the false assumption that alcohol will warm you up in a cold environment.

Taking the temperature

The close observation of body temperature in ill persons, especially small children and others whose temperature regulating mechanisms are unstable, is important, since the optimal body temperature in man extends over a limited range. The cellular enzymes are very sensitive to temperature change and may be inactivated by high body temperatures.

The temperature may be taken in the mouth, the rectum, or the axilla, the patient's age and condition usually determining the method. Small children, individuals who are delirious or unconscious, and individuals who have mouth lesions should have their temperature taken by rectum. When neither the oral nor the rectal method is practical, the axilla may be used.

The mouth temperature is the most variable, being affected by eating hot or cold foods, smoking, mouth breathing, and talking. To get an accurate mouth temperature, one should place the thermometer under the tongue, close the lips but not the teeth, and refrain from talking for the 3 minutes that the thermometer should remain in place.

The rectal temperature is about 0.5° to 0.7° F. higher than that of the mouth and is the best guide to the internal temperature of the body. The thermometer should be inserted to a depth of about 2 inches and held in place for 3 minutes.

The axillary temperature is about 1° F. lower than the mouth temperature. It may be influenced by the presence of sweat; thus the axilla should be dried gently. The thermometer should be gripped in the axilla for about 10 minutes.

Variations in body temperature due to infectious disease are often indicative of the progress of the disease, and the temperature record should be accurate. When possible, the temperature should be taken the same way each time, or a note should be made if an alternate method is used.

SUMMARY

1. ***Absorption***
A. Stomach: slight
B. Small intestine: greatest; large surface area; adequate blood supply; long time contact; foods in diffusible form
   a. Into blood: active transport, sugars, amino acids, some ions, for example, Na+; passive, some fatty acids, some ions, for example, Cl−
   b. Into lacteals: fatty acids and glycerol; enter lymph stream
C. Large intestine: much water

2. ***Metabolism: all changes in foods from time of absorption until waste matter excreted***
A. Types: anabolism and catabolism
B. Enzymes required: enzymes systems
C. Energy metabolism: energy needs take precedence over all others
   a. Biological oxidation: dehydrogenation
   b. Potential energy of food: caloric, or heat value
   c. Basal metabolism: energy output to stay alive; special conditions
   d. Caloric requirement: varies with activity

D. Carbohydrate: main purpose to provide energy
  a. Glucose pathways: tissues for immediate oxidation; stored as glycogen or adipose tissue; if in excess, excreted in urine
  b. Catabolism
    (1) Anaerobic phase: breakdown to pyruvic acid and lactic acid; H released; ATP formed
    (2) Aerobic phase: pyruvic acid converted to acetyl CoA, which enters citric acid cycle; H released; ATP formed
    (3) Hydrogens release energy in electron transport system; ATP formed

E. Lipid: vital constituent of all cells; supply energy
  a. Lipid pathways: tissues for synthetic processes; stored as adipose tissue; oxidized for energy
  b. Beta-oxidation yields acetyl CoA; excess converted to ketone bodies; ketosis, acidosis

F. Protein: build and repair tissue
  a. Amino acid synthesis: nonessential synthesized by transamination or amination; essential cannot be synthesized
  b. Protein synthesis: all tissues
    (1) Builds new protoplasm; repairs old protoplasm; builds enzymes; builds some hormones, for example, insulin
    (2) Under control of DNA and RNA
  c. Catabolism: gets rid of excess; provides energy
    (1) Lose amino group: deamination; keto acid, ammonia
    (2) Keto acid: new amino acid; glucose; ketone bodies; oxidized by citric acid cycle; converted to fat
    (3) Ammonia: converted to urea; excreted by kidney
  d. Nitrogen balance: intake equals output
    (1) Positive: growth; recovery from tissue destruction, for example, burns
    (2) Negative: starvation

3. **Body temperature**

A. Heat production: result of metabolism of foodstuffs
  a. Cannot be decreased below basal level; cannot be increased to prevent freezing
  b. Activity of muscles: large amount of heat production; lowest during sleep, highest during exercise
  c. Activity of glands: large amount by liver; some by all glandular activity
  d. Sympathetic stimulation: direct effect on metabolic rate
  e. Food intake: increased smooth muscle activity; gland activity; specific dynamic action of foods
  f. Body temperature: increased metabolic rate; 7 per cent for each degree.

B. Heat loss
  a. Radiation: transfer through air; warmer to cooler
  b. Conduction: transfer by direct contact; warmer to cooler; air poor conductor
  c. Convection: movement of air
  d. Vaporization: evaporation of water; varies with temperature and humidity of air

C. Temperature regulation: balance between heat production, or gain, and heat loss
  a. Center of integration: hypothalamic thermostat
  b. Cold environment: vasoconstriction; increased muscle tone, shivering; increased muscle activity; shelter and clothing
  c. Hot environment: vasodilatation; sweating
  d. Physiologic variations in temperature: age; diurnal variations

D. Abnormal variations in temperature
  a. Heat exhaustion: extreme sweating; dehydration, salt loss
  b. Heat stroke: sweating ceases; explosive rise in temperature
  c. Fever: abnormal elevation
    (1) Thermostat reset by pyrogens
    (2) Chill: feels cold, dry skin, shivers, body temperature rises
    (3) Crisis: sudden drop in temperature
    (4) Lysis: gradual drop in temperature
    (5) Antipyretics: reduce fever; reduce body temperature

E. Temperature taken through the mouth, the rectum, or the axilla

QUESTIONS FOR REVIEW

1. In what form are carbohydrates absorbed from the intestinal tract?
2. Define metabolism.
3. What is the most abundant and economical source of energy?
4. The doctor reports that your BMR is +8. What does this mean to you?
5. Why is the citric acid cycle called a cycle?
6. How many ATP can be produced from an amino acid which is deaminized to pyruvic acid?
7. Can you get fat from a high protein diet? Why?
8. Why does a person feel cold during a chill even though the body temperature is elevated?

REFERENCES AND
SUPPLEMENTAL READINGS

Best, C. H., and Taylor, N. B.: The Physiological Basis of Medical Practice, ed. 8. Baltimore, Williams & Wilkins, 1968.

Guyton, A. C.: Textbook of Medical Physiology, ed. 4. Philadelphia, W. B. Saunders, 1971.

Langley, L. L.: Homeostasis. New York, Reinhold Publishing Corp., 1965.

Lehninger, A. L.: Bioenergetics. New York, W. A. Benjamin, Inc., 1965.

Orten, J. M., and Neuhaus, O. W.: Biochemistry, ed. 8. Saint Louis, C. V. Mosby Co., 1970.

Rafelson, M. E., Benkley, S. B., and Hayashi, J. A.: Basic Biochemistry, ed. 3. New York, The Macmillan Co., 1971.

Sodeman, W. A., and Sodeman, W. A. Jr.: Pathologic Physiology. Philadelphia, W. B. Saunders, 1967.

Tepperman, J.: Metabolic and Endocrine Physiology, ed. 2. Chicago, Year Book Medical Publishers, 1968.

The Work of the Kidneys

". . . what matters most for our internal chemical equilibrium is, not what we eat and drink, but what our kidneys retain. They are truly the master chemists of our sea within."

Sea Within • William B. Snively, Jr., M.D.

The various metabolic activities of all living cells not only provide for the growth, the repair, and the maintenance of tissues, but also result in the production of waste products which must be removed from the body. One of these waste products (carbon dioxide) is volatile, like smoke from a furnace, and is eliminated in the air that is exhaled from the lungs. Certain other waste products, such as food residues, are comparable to ashes in the furnace and are eliminated by way of the large intestine. Water is eliminated through all channels of excretion (lungs, skin, large intestine, and kidneys). The remaining waste products (some inorganic salts and most of the nitrogenous waste products) require the special services of the kidneys for their elimination.

Waste products are formed continuously and must be eliminated continuously. In previous chapters we have considered the work of the skin, the lungs, and the large intestine as they contributed to the removal of waste products. It now remains to discuss the work of the kidneys as they remove nitrogenous waste products from the blood and concentrate them in a fluid called urine. The elimination of urine, in turn, requires two ducts (the ureters) to convey urine from the kidneys to the urinary bladder, where it is stored temporarily, and another duct (the urethra) to carry urine from the bladder to the exterior of the body.

The production of urine is not the only function performed by the kidneys. The amount of water that they can eliminate is not limited to that needed to keep waste products in solution, but can vary widely beyond this amount. This means that in health, one's kidneys can produce either very dilute or very concentrated urine, and this flexibility or adjustability in elimination of water makes it possible for the kidneys to play a very important role in maintaining water balance. They serve as a spillway for water not needed by the body.

The amount of inorganic salts, particularly sodium chloride, excreted by the kidneys can vary widely also, and this indicates the essential function of the kidneys in the maintenance of salt balance. The kidneys also have the ability to manufacture ammonia and to substitute it for sodium and in this way to help maintain the acid-base balance of the body.

In summary, the kidneys contribute a generous share to homeostasis or maintenance of the constancy of the internal environment by removing and excreting waste and by conserving useful substances and returning them to the plasma. Without normally functioning kidneys, it would be impossible to preserve the normal composition of the plasma and the interstitial fluid.

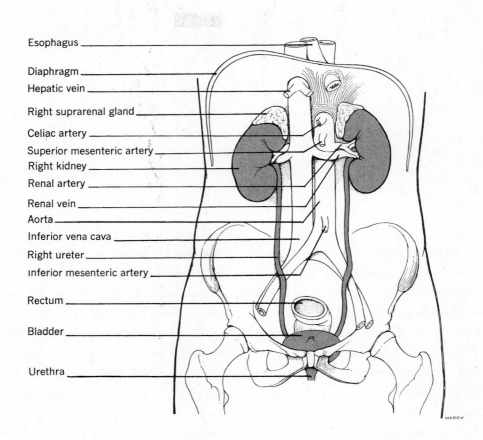

Esophagus

Diaphragm

Hepatic vein

Right suprarenal gland

Celiac artery

Superior mesenteric artery

Right kidney

Renal artery

Renal vein

Aorta

Inferior vena cava

Right ureter

Inferior mesenteric artery

Rectum

Bladder

Urethra

Fig. 18-1. Posterior abdominal wall showing position of blood vessels, kidneys, and ureters. Relationships between ureters, urinary bladder, and pelvic bones also are shown.

THE KIDNEYS

Location

The kidneys are located in the posterior part of the abdominal cavity, one on either side of the vertebral column, behind the peritoneum. They rest on the psoas major and the quadratus lumborum muscles and on part of the diaphragm. The location is shown in Figure 18-1, and the surface projections in Figure 1-8.

Fixation

The kidneys are not fixed in a rigid position against the abdominal wall, since they move with the diaphragm during inspiration. However, they are embedded in a mass of *adipose tissue* around which is found a supporting layer of fibrous tissue called the *renal fascia*. The large renal arteries and veins also give a measure of support to the kidney.

Gross appearance

The kidney presents a bean-shaped appearance; its lateral border is convex, and the medial is concave. In the central portion of the concave border there is a deep longitudinal fissure, called the *hilus*, where vessels and nerves enter and leave and from which the ureter descends toward the urinary bladder. The kidney is covered with a fibrous capsule, or tunic, that provides a firm, smooth covering.

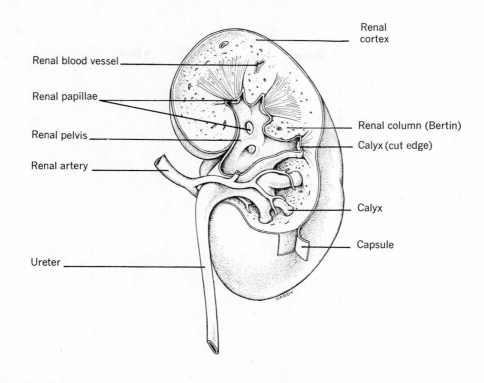

Renal cortex

Renal blood vessel

Renal papillae

Renal pelvis

Renal artery

Renal column (Bertin)

Calyx (cut edge)

Calyx

Capsule

Ureter

Fig. 18-2. Diagram of internal structure of kidney, showing relations of renal pelvis and calyces to pyramids in medullary region.

Longitudinal section

If a longitudinal or coronal section is made through the kidney, as illustrated in Figure 18-2, three general regions may be observed: the renal cortex, the renal medulla, and the renal pelvis.

The **renal cortex,** or outer portion of the kidney, is granular and reddish-brown in appearance. It arches over the pyramids of the medulla and dips in between adjacent pyramids; these inward extensions of cortical substance are called the *renal columns* (of Bertin).

The **renal medulla** is darker in color and consists of striated, cone-shaped masses called the *renal pyramids.* The pyramids vary in number, but the average is twelve. The base of each of these pyramids is directed toward the cortex, and the free end or apex of each projects centrally toward the renal pelvis where it forms a *papilla.* The summit of each papilla resembles a sieve, because it is studded with a variable number of openings; urine flows through these openings into

an extension (calyx) of the renal pelvis.

The **renal pelvis** is a funnel-shaped sac that forms the upper expanded end of the ureter. It receives urine from all parts of the kidney by way of the *calyces,* which are cup-shaped extensions of the sac. Each kidney has a variable number of calyces, since a single calyx may surround more than one papilla. An x-ray study of the renal pelves and calyces is shown in Figure 18-3.

Kidney stones (renal calculi) may form in the pelves of the kidneys; they frequently cause pain and hematuria (blood in the urine).

Microscopic appearance

The microscopic unit of structure is the **nephron,** and there are over 1 million of them in each human kidney. Each nephron consists of a renal corpuscle and a renal tubule.

A **renal corpuscle** consists of a glomerulus and a surrounding capsule, called the capsule of Bowman, as shown in Figure 18-4.

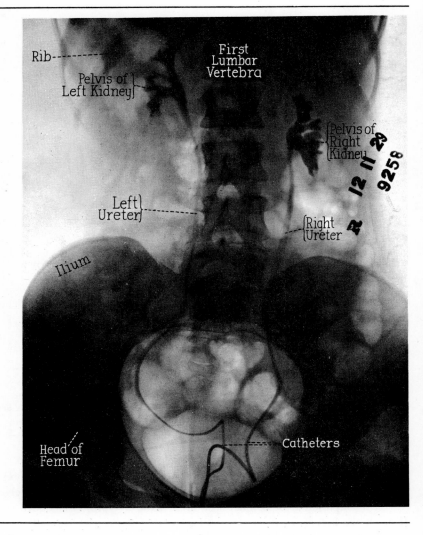

Fig. 18-3. Retrograde pyelogram. Sodium iodide was injected through catheters, which are shown in position. The pelves of the kidneys and the ureters are shown. The right kidney is a little lower than the left, which is normal. (X-ray taken through back.)

The *glomerulus* is a tuft of capillaries. The vessel that conveys blood to the glomerulus is called the *afferent arteriole*, and it divides to form the capillary loops. These loops then unite to form the *efferent arteriole* that drains blood from the glomerulus. Note that this vessel is not called a vein, even though it drains blood from capillaries. The efferent arteriole has the structure of a typical arteriole, and it conveys blood from the glomerulus to the region of the renal tubule, where it arborizes to form the peritubular capillaries (see Fig. 18-4).

The afferent arteriole is not a typical arteriole, and its diameter is greater than that of the efferent vessel. The tunica intima and tunica externa are poorly developed; thus the diameter of the afferent arteriole is large because its tunica media consists of a very substantial layer of smooth muscle. A further modification involves a cluster of unusual cells in the tunica media just before this arteriole gives rise to the glomerulus; these cells are known as *juxtaglomerular cells*, because they are near to or adjoining the glomerulus. The juxtaglomerular cells are involved in the production of a chemical substance called *renin*. The importance of renin will be discussed on page 440.

The *capsule of Bowman* is the beginning of a renal tubule, and it is invaginated by the glomerulus. The visceral layer of the capsule is intimately attached to the glomerulus, but there are large pores or openings in it. The outer or parietal layer is intact. The space between the two layers contains fluid that has filtered out of

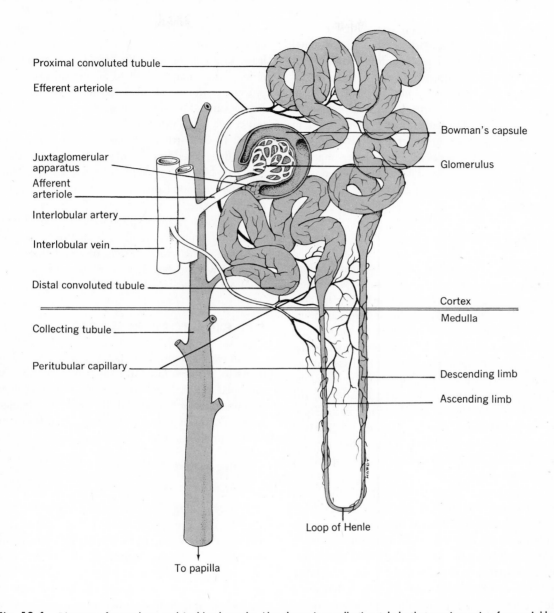

Proximal convoluted tubule

Efferent arteriole

Juxtaglomerular apparatus

Afferent arteriole

Interlobular artery

Interlobular vein

Distal convoluted tubule

Collecting tubule

Peritubular capillary

Bowman's capsule

Glomerulus

Cortex

Medulla

Descending limb

Ascending limb

Loop of Henle

To papilla

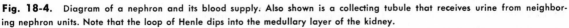

Fig. 18-4. Diagram of a nephron and its blood supply. Also shown is a collecting tubule that receives urine from neighboring nephron units. Note that the loop of Henle dips into the medullary layer of the kidney.

the glomerulus; this fluid is referred to as the *glomerular filtrate.*

A **renal tubule** is called by different names in its tortuous course. It begins in the capsule of Bowman and then becomes the *proximal convoluted tubule,* the *loop of Henle,* and, finally, the *distal convoluted tubule,* which is the end of the nephron unit. The distal convoluted tubule

empties its contents into a branch of a straight collecting tubule, or duct. The collecting tubule courses through a renal pyramid and joins other collecting tubules to form one of the papillary ducts that open into a calyx at the summit of a papilla. Many nephron units drain into each collecting tubule. Figure 18-4 illustrates the various structures described above.

Location of Microscopic Structures. All of the renal corpuscles are confined to the cortex. The proximal and the distal convoluted tubules are also located in the cortex, close to the renal corpuscles. However, the descending limb of each loop of Henle dips downward for a variable distance into the medulla, bends on itself to form the loop, and then returns to the cortex as the ascending limb. The collecting tubules begin in the cortex and unite at short intervals with one another to form a series of larger tubes that descend into the medulla. These tubes are responsible for the striated appearance of the renal pyramids.

Blood and nerve supply

Blood is conveyed to the hilus of each kidney through a *renal artery*. On entering the kidney, the renal artery branches to form the *interlobar arteries;* these vessels pass between adjacent pyramids and give off small branches as they course toward the cortex. As the interlobar arteries approach the cortex, they arch over the bases of the medullary pyramids where they become the *arcuate arteries*. Branches of the arcuate arteries, the *interlobular arteries*, extend into the cortex. Some of the interlobular arteries give off the afferent arterioles of the glomeruli (Fig. 18-4), while others send terminal branches to the capsule of the kidney.

After passing through the glomeruli, blood flows through the efferent arterioles into capillary beds that surround the renal tubules; these microscopic networks are called the *peritubular capillaries*. On leaving the peritubular capillaries, blood flows into stellate veins and then into the interlobular veins; from this point, the veins that drain the kidney follow the same general course as the arteries and lie beside them.

Reduced Blood Supply. Any condition that markedly reduces the blood supply to the kidneys will cause the juxtaglomerular cells (p. 438) to release *renin*. On entering the bloodstream, renin acts on a substance called angiotensinogen, converting it to angiotensin I. The latter, in turn, is converted to *angiotensin II*, which helps to raise the blood pressure in two ways. First, it helps to increase peripheral resistance by causing a generalized vasoconstriction of arterioles in other parts of the body. Second, angiotensin stimulates the adrenal cortex (p. 482) to increase its secretion of *aldosterone*. Aldosterone acts on the renal tubule cells and promotes the reabsorption of sodium; because sodium is accompanied by water, this leads to an increase in the circulating blood volume. These changes lead to an increase in blood pressure and in blood flow through the vital organs, including the kidney. Such a response is protective when the reduction in renal blood flow is due to hemorrhage. However, when the renal blood flow is reduced by obstruction of the renal artery or by certain kidney diseases, an abnormal elevation of blood pressure may ensue (renal hypertension).

The kidneys are supplied with visceral afferent fibers that transmit impulses from the kidneys to the CNS. The blood vessels of the kidneys are supplied with vasoconstrictor fibers from the thoracolumbar division of the autonomic system.

THE EXCRETORY PASSAGES

In each kidney, urine flows from the papillary ducts to the pyramids into the calyces leading to the renal pelvis. From the pelvis, urine flows into the ureter which leads to the urinary bladder. Urine ultimately passes to the exterior of the body by way of the urethra.

The ureters

Location. The ureters are tubes, about 10 or 12 inches long, that convey urine from the kidneys to the urinary bladder. Lying behind the parietal peritoneum, each ureter descends along the posterior abdominal wall to the pelvic brim of its own side. After crossing the brim of the pelvis, the ureters pass along the lateral wall of the pelvis and follow a downward and medial course along the pelvic floor to reach the urinary bladder (Fig. 18-1). They enter the bladder by passing obliquely through its muscular wall. There is no valve at the ureteral openings, but as the pressure of urine within the bladder begins to rise, the flexible ends of the ureters are compressed against their firm muscular supports in the bladder wall; this prevents backflow of urine.

Description. Each ureter is lined with mucous membrane that is continuous with the mucosal lining of the renal pelvis above and the urinary bladder below. The muscular portion of the ureteral wall consists of a longitudinal and a circular layer of smooth muscle; contractions of these muscular layers produce the peristaltic waves that move urine toward the urinary bladder. The adventitia, or outer layer of the ureteral wall, consists of fibrous connective tissue and some elastic fibers.

Visceral afferent and efferent nerve fibers from both divisions of the autonomic system also are supplied to the ureters. Kidney stones that escape from the renal pelvis into the ureter may cause excruciating pain (renal colic) as they move toward the urinary bladder.

The urinary bladder

Location. The urinary bladder is a thick-walled, muscular sac that lies behind the pubis, below the parietal peritoneum. It serves as a reservoir for urine and is emptied periodically. Its location is shown in Figure 18-1.

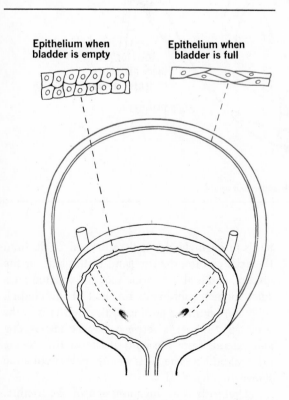

Epithelium when bladder is empty

Epithelium when bladder is full

Fig. 18-5. Transitional epithelium in empty and distended bladder. Note flattening of epithelial cells when bladder wall is stretched.

Description. The mucous membrane lining of the bladder is thrown into folds, or *rugae*, when the bladder is empty; these folds disappear as the bladder becomes distended with urine. The epithelium of the mucosa is of the transitional type; that is, its appearance varies with the state of emptiness or fullness of the bladder as shown in Figure 18-5. The *trigone* is a small triangular area in the floor of the bladder. Its anterior angle is formed by the internal opening of the urethra, and its posterolateral angles by the openings of the ureters. There are no rugae in the trigone, and its mucosal surface is always smooth.

Cystitis is an infection of the mucosa of the bladder. Pain and frequency (desire to empty the bladder every few minutes) are two symptoms of this condition.

The muscular body of the bladder is known as the *detrusor muscle*; it is composed of an interwoven network of smooth muscle fibers. At the internal urethral opening, muscle fibers arch obliquely over the bladder outlet, or bladder neck. Although these fibers do not constitute a true anatomic sphincter, they help to maintain urinary continence. Figure 18-6 is a diagrammatic representation of innervation of the bladder and the urethra.

A *cystoscopic* examination is one in which the interior of the bladder is examined visually. In certain conditions it is necessary to wash out the bladder, and such a procedure is called a bladder *irrigation*. Figure 18-7 shows the appearance of the urinary bladder when distended with different amounts of fluid.

The urethra

The urethra is the tube that conveys urine from the urinary bladder to the exterior. In men it has both an excretory and a reproductive function, but in women it is concerned only with excretion.

The Female Urethra. The urethra in the female is only 1 to 1½ inches in length. As it curves obliquely downward and forward from the bladder, the posterior wall of the urethra is united firmly with the anterior wall of the vagina. The external orifice of the urethra is located directly in front of the vaginal opening as shown in Figure 21-11. The wall of the urethra is composed of mucous membrane and a muscular layer. An external sphincter of striated muscle surrounds

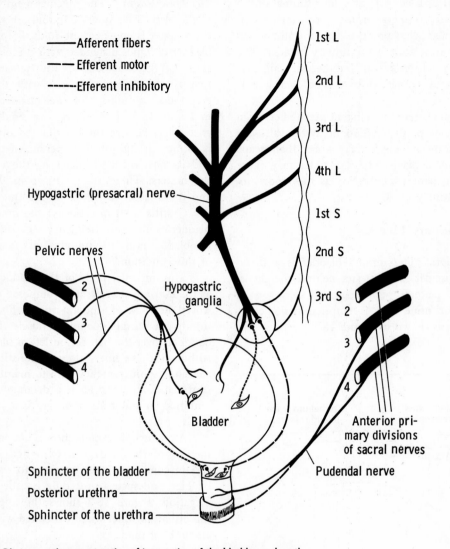

——Afferent fibers
— —Efferent motor
-----Efferent inhibitory

1st L

2nd L

3rd L

4th L

1st S

2nd S

3rd S

Hypogastric (presacral) nerve

Pelvic nerves

2

3

4

Hypogastric ganglia

2

3

4

Bladder

Anterior primary divisions of sacral nerves

Sphincter of the bladder

Posterior urethra

Sphincter of the urethra

Pudendal nerve

Fig. 18-6. Diagrammatic representation of innervation of the bladder and urethra.

the orifice and enables one to voluntarily control the escape of urine.

The Male Urethra. The urethra in the male is made up of three portions: the prostatic, the membranous, and the cavernous portions.

The *prostatic* portion is about 1 inch long and extends from the urinary bladder to the pelvic floor. During its course, it is completely surrounded by the prostate gland. When this gland enlarges, as it commonly does in elderly men, it compresses the prostatic urethra and interferes with the flow of urine (p. 499).

The *membranous* portion of the urethra ex-

tends through the pelvic floor, or body wall, from the apex of the prostate gland to the bulb of the urethra, and is about ½ inch long. A striated muscular sphincter surrounds this part of the urethra.

The *cavernous* portion of the urethra is in the penis (p. 499). It is between 5 and 6 inches long and extends from the bulb of the urethra to the external orifice. The parts of the male urethra are shown in Figure 21-1.

Urethritis is an inflammation of the urethra. It may be followed by a *stricture*, or narrowing of the urethra, due to adhesions. When this condition exists, frequency and dribbling of urine may

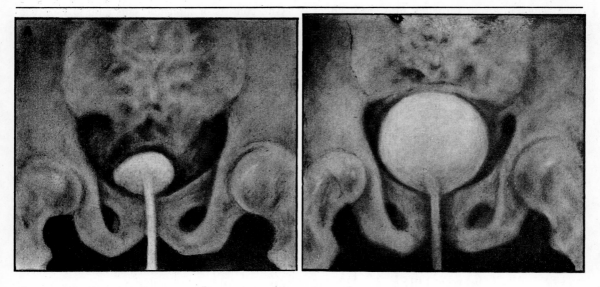

Fig. 18-7. Cystograms of male bladder, showing about 20 ml. of sodium iodide solution (A) and 500 ml. (B) within the bladder.

be noted, due to incomplete emptying of the bladder.

Micturition

This refers to the emptying of the urinary bladder. When about 250 ml. of urine has accumulated in the bladder, stretch receptors in the wall are stimulated by distention, and impulses are sent over visceral afferent nerve fibers to the CNS which arouse a desire to empty the bladder. Then there is a removal of voluntary control, which is essential to the release of a chain of reflexes. The act of micturition is actually started by a relaxation of the voluntary muscles of the perineum.

The first reflex appears as a contraction of the detrusor muscle that becomes more and more powerful. This contraction of the wall of the bladder is accompanied by a relaxation of the neck of the bladder.

As urine enters the urethra other receptors are stimulated which initiate additional reflexes. Further contraction of the wall of the urinary bladder occurs, accompanied by a relaxation of the external sphincter of the urethra.

Although the emptying of the urinary bladder is reflex in nature, it is initiated by an effort of the will and a voluntary removal of restraint. It can be stopped at various stages. The external sphincter of the urethra can be closed voluntarily, but it is opened as a part of a chain of coordinated reflexes.

If the urinary bladder is not emptied when the desire to do so first is noted, it will be distended further. Micturition can be inhibited voluntarily until about 600 ml. of urine has accumulated. By this time a sensation of pain may be aroused, and micturition becomes urgent.

Pathologic distention of the bladder may occur in both the conscious and the unconscious patient. In extreme cases, the bladder may distend to the level of the umbilicus, but this is rare when medical attention is available. Retention of urine may follow an obstruction or an infection of the urethra, an enlargement of the prostate gland, or damage to the nerve supply of the bladder.

Catheterization of the bladder refers to emptying it by insertion of a tube or catheter through the urethra into the bladder. It is done to relieve distention or for continuous drainage in incontinent patients. It is done also when a catheterized specimen is requested for analysis. In women, the urine eliminated spontaneously may be contaminated by vaginal discharge; therefore a specimen may be obtained by catheter to rule out such a possibility.

Sterile equipment and technic must always be

used when inserting catheters into the bladder. The introduction of bacteria may result in cystitis or even a pyelitis. *Pyelitis* is an infection of the renal pelvis that may follow infection of the bladder, since it is easy for infection to spread along the mucous membrane of the ureters from that of the bladder.

CHARACTERISTICS OF URINE

Urine is a watery solution of nitrogenous waste and inorganic salts that is removed from the plasma and eliminated by the kidneys. The color usually is amber, but varies with the amount produced. The reaction is acid on a mixed diet and alkaline on a vegetable diet. The specific gravity usually is between 1.016 and 1.020, but may vary from 1.002 to 1.040 in normal kidneys, according to whether the urine is very dilute or very concentrated. The quantity of urine is about 1.5 L. daily. An increase in daily output that is persistent is called *polyuria*, and a temporary increase is called *diuresis*. A decrease below the average amount is called *oliguria*.

Table 18-1 Composition of Urine in a Normal Man

	Element	Amount in Grams
Inorganic constituents	NaCl	15.000
	K	3.3
	SO$_4$	2.5
	PO$_4$	2.5
	NH$_3$	0.7
	Mg	0.1
	Ca	0.3
	Fe	0.005
	Other	0.2
Organic constituents	Urea	30.000
	Creatinine	1.0
	Uric acid	0.7
	Hippuric acid	0.7
	Indican	0.01
	Acetone bodies	0.04
	Undetermined	2.945

The composition varies with the waste products to be removed and with the need for maintenance of homeostasis. About 95 per cent of urine is water. The average amount of solids excreted per day is about 60 Gm.; of this, about 25

Gm. consists of inorganic salts and 35 Gm. consists of organic substances. The composition of urine of an adult man on a mixed diet is shown in Table 18-1. The normal pH range is 4.8 to 7.5; the average is about 6.

FORMATION OF URINE

The work of the kidneys centers on maintaining the normal composition, volume and pH of body fluids. In accomplishing this task, each kidney removes from the blood variable amounts of water, organic, and inorganic constituents, as well as toxic substances, and eliminates them in the urine.

Urine formation begins in the renal corpuscles by the process of glomerular filtration. It is completed in the renal tubules where processes of osmosis, diffusion, tubular reabsorption, tubular secretion, and tubular excretion all help in the conversion of glomerular filtrate to urine. Some of these processes involve an expenditure of energy by the tubular cells as certain substances are moved by active transport.

Glomerular filtration

As previously described, each glomerulus is a tuft of capillaries to which blood is brought by an afferent arteriole and from which blood is drained by an efferent arteriole of smaller diameter. This anatomic arrangement favors the maintenance of a relatively high blood pressure throughout the capillary loops in the glomerulus. The pressure of the fluid in Bowman's capsule is low, and the walls of the capillaries and the inner layer of the capsule are thin and permeable. Consequently, the glomerulus is an ideal filtering apparatus as water and true solutes are pushed out of the bloodstream into the capsular beginning of the renal tubule. Since large particles, such as red blood cells and most plasma proteins, are unable to pass through a normal membrane, they remain in the bloodstream as it flows out of the glomerulus.

More than 500 ml. of blood enter each kidney every minute, but the amount of filtrate produced depends on such factors as the glomerular blood pressure and the permeability of capillary and capsular walls, among others. An average amount of glomerular filtrate is 120 ml. per minute or approximately 170 L. in 24 hours. The fact that only 1 to 1.5 L. of urine is eliminated each day points to a very large reabsorption of fluid during the tubular phase of urine formation.

Tubular reabsorption

As the glomerular filtrate flows through the renal tubule, many constituents are reabsorbed into the bloodstream. This is particularly true of substances that are needed by the body. For example, water, glucose, amino acids, sodium, chloride, bicarbonate, and other tiny particles are mechanically filtered through the glomerular membrane along with various waste products. However, by highly complicated mechanisms of active transport, in addition to simple osmosis and diffusion, the useful substances are removed from the tubular fluid and returned to the blood as it flows through the peritubular capillaries.

Active Transport. This process requires the expenditure of energy and the presence of highly specific enzymes, as certain substances are pumped across the tubule cells and into the surrounding interstitial fluid. At the same time, these transported substances are prevented from diffusing back to the tubular lumen whence they came. Thus they enter the peritubular capillary and are carried away in the bloodstream.

Some constituents of the glomerular filtrate, such as *potassium*, are reabsorbed almost completely, regardless of their concentration. Others, such as *glucose*, have a well-defined limit, or renal threshold, which means that excessive amounts are eliminated in the urine. For example, if one eats a large amount of candy within a short period of time, the blood is loaded with glucose, and large amounts are filtered into the renal tubules; the resulting overload on the active transport mechanism is responsible for the appearance of glucose in the urine. Other important threshold substances include amino acids, acetoacetic acid, vitamins, and uric acid.

Large quantities of filtered *sodium* (Na^+) are delivered constantly to the tubules, chiefly as chloride (Cl^-) and bicarbonate (HCO_3^-). However, Na^+ is reabsorbed efficiently, and less than 1 per cent is excreted in the urine. The active reabsorption of Na^+ represents the most massive operation carried out by the kidney, aside from the reabsorption of water which moves more or less

passively. About 80 per cent of the filtered sodium is reabsorbed in association with chloride, and about 20 per cent with bicarbonate.

Passive Transport. The reabsorption of water is accomplished by the simple process of *osmosis* (p. 32). As sodium and other solutes are transported from the lumen of the tubule, their concentration in the surrounding interstitial fluid is increased as their concentration in the tubular fluid is diminished. Thus there is a net diffusion of water from the tubule into the peritubular interstitial fluid. Diffusion from the interstitial fluid into the peritubular capillary blood is facilitated by the relatively high colloid osmotic pressure of that blood; two factors are operative in maintaining this effective pulling pressure. First, unlike water, the plasma proteins are not readily filtered through the glomerular membrane; thus they remain in the blood as it leaves the glomerulus and flows into the peritubular capillary. Second, the hydrostatic or pushing pressure of the blood is reduced as it flows toward the venous end of the peritubular capillary bed. Because of these factors, all but about 1 or 1.5 L. of the water that filters through the glomerular capillary membrane ultimately is reabsorbed and returned to the bloodstream. The role of antidiuretic hormone (ADH) will be described presently.

Another form of passive transport involves the establishment of *electrical gradients*. For example, when positive ions (e.g., Na^+) are actively transported out of the tubule, the surrounding interstitial fluid becomes electropositive; thus negative ions (e.g., Cl^-) will follow the positive ions in the reabsorptive process.

Urea is the chief nitrogenous constituent of the blood, constituting over one half of the NPN (nonprotein nitrogen). This substance also is the chief organic constituent of the urine. Although urea is a waste product, some of it can and does diffuse from the tubular fluid back into the bloodstream. As you might expect, this backward diffusion is most pronounced when urine flow is decreased; thus, in kidney disease the blood urea nitrogen (BUN) may greatly exceed the normal levels of 10 to 15 mg. per cent.

Tubular secretion

The kidney plays an important role in regulating the acid-base balance of the body (p. 464).

Through the ability of its tubular cells to secrete varying amounts of hydrogen ions and ammonia, the kidney can increase or decrease the acidity of the urine. These secretory activities occur primarily in the distal portion of the tubules and are part of the final phase of urine formation.

Hydrogen ions (H^+) are secreted into the lumen of the tubule in exchange for sodium (Na^+) ions. Na^+, accompanied by bicarbonate (HCO_3^-), returns to the bloodstream where it serves as part of the basic buffer system described on page 464.

As the acidity of the urine increases, *ammonia* (NH_3) is formed by the tubule cells by deamination of amino acids. The NH_3 diffuses into the tubule lumen and unites with H^+ to form the nondiffusible ammonium radical (NH_4^+); the NH_4^+ then combines with chloride (Cl^-) to be excreted as ammonium chloride. The excretion of NH_4Cl, rather than $NaCl$, is an additional way of conserving sodium for the body and thus safeguarding its total available base.

Another secretory function of the renal tubule is the regulation of the amount of *potassium* (K^+) in the body. Since almost all of the filtered K^+ is reabsorbed in the proximal tubule, that which appears in the urine primarily results from the active secretion of K^+ by the distal tubule cells in exchange for sodium. This not only serves to eliminate excess K^+, but is an additional way of conserving sodium.

Tubular excretion

By processes of active transport, the tubule cells can transfer substances from the blood into the tubular lumens. In contrast to the secretory processes, the substances to be removed include those that are of no use to the body (e.g., creatinine) as well as those that are completely foreign to the body (e.g., penicillin and certain dyes). In the latter case, tubular excretion supplements glomerular filtration in that it facilitates the elimination of compounds that could not otherwise be disposed of by metabolic processes.

Tubular excretion of foreign substances is clinically important in that it provides the basis for certain laboratory tests of kidney function. For example, the dye *phenolsulphonphthalein* (commonly called *P.S.P.*) is injected into the body, and urine specimens are collected at timed intervals.

Normal kidneys eliminate 60 to 80 per cent of the injected dye within a specified time, but impaired kidneys eliminate proportionately less according to the amount of tubular damage.

Hormonal effects

Chapter 20 is devoted to the endocrine glands and their chemical messengers, or hormones. However, it seems advisable at this time to introduce two particular hormones that greatly influence the formation of urine.

The Antidiuretic Hormone (ADH). This hormone is a neurosecretion produced by nerve cell bodies in the hypothalamus and passed along axons to the posterior lobe of the pituitary gland for storage. The hypothalamus is concerned not only with the production of ADH, but also with its release from the pituitary into the bloodstream.

As the name implies, *antidiuretic* means that this hormone is *against diuresis*, or against the excretion of urine; its chief effect is on the kidney tubules in which it accelerates the rate of water reabsorption, possibly by widening the pores in cell membranes. ADH acts on cells in the loop of Henle, in the distal convoluted tubule, and in the collecting ducts.

The role of the hypothalamus in regulating water balance was previously described on page 196, but a brief summary may prove helpful at this time. The rate at which ADH is released from the posterior lobe of the pituitary gland depends on the concentration of solutes (e.g., salt) in the circulating blood as it flows through the hypothalamus. If the concentration of solute is high, the hypothalamus orders the pituitary to release ADH; as a result, the reabsorption of water is accelerated until the concentration of solute is reduced to normal levels. On the other hand, if the solute concentration in the blood is too low, the secretion of ADH is inhibited; as a result, more water is lost in the urine until the solute concentration is increased to normal levels.

Aldosterone. Aldosterone is sometimes called a salt and water hormone, and is secreted by cells in the cortex of the adrenal glands (p. 482). This hormone plays an important part in the active transport of both sodium and potassium. When the concentration of aldosterone is high, sodium is retained in excessive amounts in the body, and potassium is excreted in greater amounts than usual. On the other hand, when there is a deficiency of aldosterone, as in Addison's disease (p. 483), potassium is retained while excessive sodium is lost from the body and with it an equivalent amount of water; as a consequence, body fluid is progressively reduced to lethal levels.

Urinalysis

Routine analysis of the urine involves a description of the color and turbidity, determination of the specific gravity, pH, and the presence or absence of abnormal constituents. A microscopic examination of the sediment may also be ordered.

Specific Gravity. Normal kidneys are able to vary the specific gravity of urine from 1.002 to 1.040. If the kidneys are diseased, the variation in specific gravity of urine is limited. In fact, the specific gravity may be fixed and remain about 1.010, regardless of variations in intake of water. Tests of flexibility of specific gravity are easy to perform, and the information derived is valuable, since lack of ability to concentrate the urine is one of the earliest signs of kidney damage.

Abnormal Constituents. These constituents include blood, pus, albumin, sugar, and ketone bodies.

Blood. The presence of blood in the urine is called *hematuria*. It may be due to contamination from vaginal discharge, cystitis, tumors, and stones in the kidney, the ureter, or the bladder. Whenever blood is found in the urine, additional examinations are imperative to determine the origin of the blood.

Pus. The presence of pus in the urine is called *pyuria*. It may be due to infection of the bladder or of the renal pelvis.

Albumin. The presence of albumin in the urine, called albuminuria or proteinuria, may be a symptom of serious kidney disease. However, before such a diagnosis can be made, other causes of albuminuria must be ruled out; for example, the urine may have been contaminated with vaginal discharge, or there may be an existing bladder infection. Orthostatic albuminuria also must be ruled out. This is a condition in which albumin is present in specimens of urine collected during the day, when the person is in the upright position; the specimens collected after resting in bed are free of albumin. This is purely a circulatory phenomenon; in the standing position there is

enough pressure on the renal veins to cause a disturbance in filtration. There is no disease of the kidneys present. If all other causes for albuminuria, such as those listed above, have been ruled out, it is safe to conclude that there is pathology of the kidneys in the form of glomerulonephritis (primarily involvement of glomeruli) or nephrosis (primarily involvement of tubules).

Sugar. The presence of sugar in the urine is called *glycosuria*; it may be *alimentary glycosuria*, due to large intake of sugar. It may be due to emotional upsets, such as during examinations (epinephrine causes excessive glycogenolysis, with marked elevation of blood sugar to above renal threshold concentration). If glycosuria is found in a pregnant or lactating woman, lactosuria must be ruled out. When all other causes have been ruled out, diabetes mellitus may be suspected, and additional tests (glucose tolerance) are imperative.

Ketone Bodies. The presence of acetone, acetoacetic acid, or beta-hydroxybutyric acid (the so-called ketone bodies) in the urine is called *ketonuria*. However, a small number of ketone bodies may be present in the urine of healthy persons, particularly in the fasting state. As previously described (p. 422), ketone bodies tend to accumulate when carbohydrate metabolism is depressed and the body is deriving most of its energy from the metabolism of fats (e.g., untreated diabetes mellitus and starvation).

Microscopic Examination. When a drop of abnormal urine (which has been centrifuged) is examined, erythrocytes, tiny stones (renal calculi), and casts may be noted. The casts are tiny masses of material that have hardened in the tubular lumens, and they assume the shape of the lumen. The presence of casts indicates kidney injury.

Diuresis

Diuresis refers to an increased excretion of urine. It may be caused by a variety of factors, a few of which will be described.

Water diuresis follows the intake of a large amount of water within a short period of time. This lowers the solute concentration in the blood and inhibits the secretion of antidiuretic hormone (ADH), as mentioned previously. Most of the water will be eliminated by normal kidneys within two hours after it has been taken in.

Osmotic diuresis is due to the presence of large amounts of waste products in the glomerular filtrate.

Urea is a natural diuretic. When one eats a large amount of protein, the diuretic effect of urea is noted.

Glucose in an untreated diabetic is a waste product and is responsible for the large amount of urine (polyuria) eliminated by such persons.

Waste products in the lumens of the tubules hold enough water (by osmosis) to keep themselves in solution and cause diuresis by opposing the reabsorption of water.

Mercurial diuretics include those drugs that in some way promote the excretion of sodium which, in turn, takes water with it.

Digitalis, used in treating certain types of heart disease, has a diuretic effect, since it improves the blood flow through the kidney.

Caffeine causes a dilatation of the renal blood vessels, and, since more blood flows through the kidney, more urine is formed.

Ethyl alcohol inhibits the release of ADH, thus more urine is formed.

Dialysis

Dialysis is the process of separating crystalloids from colloids in solution by virtue of the difference in their rates of diffusion through a semipermeable membrane: crystalloids pass through readily, colloids very slowly or not at all. The two types of dialysis are hemodialysis and peritoneal dialysis.

Hemodialysis. The equipment used for hemodialysis frequently is referred to as the artificial kidney. Briefly, it consists of a long cellophane tube arranged in coils and submerged in a temperature-controlled dialyzing solution; the chemical composition of this solution usually approximates normal extracellular fluid, but it may be altered in various ways as necessary to meet the specific needs of the patient.

Dialysis begins when blood from one of the patient's arteries is pumped through the tubing. Because of their large size, red blood cells and plasma proteins are unable to pass through the membrane; thus they remain in the bloodstream. Crystalloid substances that need to be removed from the blood (e.g., potassium and urea in the case of kidney failure, or barbiturates in the case of drug overdosage) diffuse across the cellophane membrane from the area of higher concentration in the blood to the area of lower concentration in the dialyzing solution. The solution is changed at intervals to prevent the accumulation of these substances and to maintain the diffusion gradient. After passing through the length of tubing, the blood is returned to the patient's circulation by way of a vein.

Hemodialysis may mean the difference between life and death for a person who has taken an overdose of certain drugs (e.g., barbiturates), for a patient whose kidneys have temporarily stopped producing urine (e.g., in acute nephritis), and for various other conditions related to renal excretion.

Peritoneal Dialysis. In peritoneal dialysis, the dialyzing fluid is introduced into the peritoneal cavity. The peritoneum serves as the semipermeable membrane across which dialysis occurs. However, results usually are not as good as are those of hemodialysis.

SUMMARY

1. *Kidneys*
A. Location: posterior abdomen; retroperitoneal
B. Fixation: adipose tissue and renal fascia
C. Gross appearance: bean-shaped; hilus in concave medial border
D. Longitudinal section
   a. Renal cortex: outer portion of kidney
   b. Renal medulla: composed of pyramids
   c. Renal pelvis: funnel-shaped sac with extensions called calyces
E. Nephron: the unit of structure

   a. Renal corpuscle: glomerulus and Bowman's capsule; renal tubule: proximal portion, loop of Henle, distal portion
   b. Drain into collecting tubules which join to form papillary ducts that open into calyces of pelvis
   c. Most of nephron unit in cortex; loop of Henle and collecting tubules in medulla
F. Blood supply: renal artery branches to interlobar arteries to arcuate to interlobular to afferent arteriole of glomerulus; efferent arteriole of glomerulus to peritubular capillaries to stellate vein to inter-

lobular veins to arcuate veins to interlobar veins to renal veins
  a. Reduced blood supply: renin-angiotensin mechanism

2. Excretory passages
A. Ureters
  a. Retroperitoneal tubes; convey urine to bladder
  b. Muscular layers produce peristaltic waves to move urine; mucosal lining continuous with that of kidney and bladder
B. Urinary bladder
  a. Thick-walled muscular sac; behind pubis; reservoir for urine
  b. Mucosa in folds (rugae) in empty bladder
  c. Muscular coat: the detrusor muscle
C. Urethra: conveys urine from bladder to exterior
  a. Female: excretory function only
  b. Male: prostatic, membranous and cavernous portions; both excretory and reproductive functions
D. Micturition: act of emptying the urinary bladder; series of reflexes following release of voluntary control

3. Characteristics of urine
A. Watery, amber-colored solution of nitrogenous waste and inorganic salts
  a. Specific gravity averages 1.016 to 1.020; acid reaction on mixed diet
  b. Quantity about 1.5 L. daily
B. Composition varies with products to be removed to maintain homeostasis

4. Formation of urine
A. Glomerular filtration
  a. High blood pressure in glomerulus; water and true solutes filter into Bowman's capsule; red cells and most protein remain in blood
  b. Average 170 L. daily
B. Tubular reabsorption: osmosis, diffusion, and active transport as substances are returned to the bloodstream
  a. Some substances have renal threshold; excess is eliminated
  b. Active transport requires expenditure of energy and presence of specific enzymes
  c. Passive transport: osmosis; electrical gradients
C. Tubular secretion: active transport and diffusion
  a. Kidney important in acid-base balance by increasing or decreasing acidity of urine
  b. H^+ secreted into lumen of tubule in exchange for Na^+; latter combines with HCO_3^- and returns to blood; conserves base
  c. Ammonia formed by tubule cells also conserves base; excreted as ammonium salt, thus saving Na^+
  d. Excess potassium actively secreted by tubule cells in exchange for Na^+
D. Tubular excretion: active transport of substances of no use to body as well as those foreign to body (e.g., penicillin and P.S.P.)
E. Hormonal effects
  a. ADH accelerates reabsorption of water by tubules; under control of hypothalamus
  b. Aldosterone important in active transport of Na^+ and K^+
F. Urinalysis: examine for color, turbidity, pH, presence of abnormal constituents; microscopic examination of sediment
G. Diuresis (increased excretion of urine): water diuresis; osmotic diuresis; drugs
H. Hemodialysis: removal of crystalloid waste products from blood as it flows through cellophane tubing submerged in special dialyzing solution

QUESTIONS FOR REVIEW

1. Describe the location and fixation of the kidneys.
2. Through what structures would a drop of water pass after it filters out of a glomerulus and travels to the urinary bladder?
3. What is the kidney's response to reduced blood flow following a hemorrhage?
4. What is the function of the detrusor muscle?
5. In what ways does the female urethra differ from the male?
6. What is polyuria? oliguria?
7. What stimulus initiates emptying of the bladder?
8. Why is the glomerulus an ideal filtering apparatus?
9. What is meant by the term "renal threshold"?
10. What two factors are operative in maintaining an effective osmotic pressure in the peritubular capillaries?
11. Briefly, describe tubular secretion.
12. What is the physiologic response to an increased secretion of ADH?
13. What is the importance of tubular excretion?
14. What is the significance of an extremely low specific gravity of urine?
15. Why do large amounts of glucose in the glomerular filtrate act as a diuretic?
16. What is the purpose of dialysis in the treatment of renal failure?

REFERENCES AND
SUPPLEMENTAL READINGS

Best, C. H., and Taylor, N. B.: The Physiological Basis of Medical Practice, ed. 8. Baltimore, Williams & Wilkins, 1966.

Beeson, P. B., and McDermott, W.: Cecil-Loeb, Textbook of Medicine, ed. 13. Philadelphia, W. B. Saunders, 1971.

Goss, C. H. (ed.): Gray's Anatomy of the Human Body, ed. 28. Philadelphia, Lea & Febiger, 1966.

Guyton, A. C.: Textbook of Medical Physiology, ed. 4. Philadelphia, W. B. Saunders, 1971.

Ham, A. W.: Histology, ed. 6. Philadelphia, J. B. Lippincott, 1969.

Pitts, R. F.: Physiology of the Kidney and Body Fluids. Chicago: Year Book Publishers, 1963.

Fluids and Electrolytes. Acid-Base Balance

19

Virginia G. Braley, Ph.D.

"By a small sample we may judge of the whole piece"

Don Quixote
Miguel De Cervantes

A body cell can be likened to a factory that produces materials either for its own efficient operation or for distribution to other factories for their use. To perform its functions properly, the factory requires a constant supply of raw materials, an efficient internal system for utilizing these materials, and a constant outlet for its products and for waste materials resulting from the manufacturing process. Cells function in a liquid environment provided by the body fluids. These fluids, consisting of dissolved and suspended materials, are essential for the efficient operation of the machinery. Being mobile, they also serve as the purveyor of supplies as well as the distributor of the products and the wastes resulting from cellular activity.

BODY WATER

Water is the most abundant constituent of the body, accounting for some 50 to 60 per cent of adult body weight and as much as 75 to 77 per cent of body weight in infants under one month of age. By approximately age 17, the adult percentage is attained.

Water is uniquely suited to its body functions: 1) It is an excellent solvent for a great variety of substances, including nutrients and waste products; 2) it provides for ionization of electrolytes; 3) its high specific heat and heat of vaporization make it especially suitable as a temperature regulator; and 4) it acts as a reagent in many chemical reactions occurring in the body.

Intake

Normally, the volume of body water remains remarkably stable, and, over a period of time, intake equals output. Water intake includes not only the water consumed in beverages, but that contained in solid foods and that formed during metabolism of organic foodstuffs. The water taken in by beverages and food, *exogenous water*, may vary tremendously from one individual to another, or from day to day in a single individual. However, an average adult living in a moderate climate and receiving a mixed diet consumes about 2,000 to 2,500 ml. daily. Approximately 1,000 ml. is obtained from beverages and 1,500 ml. from solid and semisolid foods. Even such dry food as uncooked rice contains a considerable amount of water.

Water formed during metabolism of organic foodstuffs is *endogenous water*. Since metabolism varies with body temperature, the amount of exercise being performed, and other factors (p. 416), the amount of endogenous water available will also vary on a day to day basis. In a healthy adult performing a moderate amount of exercise, an average of 300 to 350 ml. of endogenous water will be available daily.

Output

Body water is eliminated through the four channels of excretion: the lungs, the gastrointestinal tract, the skin, and the kidney.

Lungs. Expired air is nearly saturated with water. The amount lost through the lungs varies with the humidity and the temperature of inspired air, as well as the rate and depth of respiration. As a rule, about 300 to 350 ml. of water is thus lost in a single day.

Gastrointestinal Tract. The amount of water lost through this channel is usually quite small, averaging approximately 150 ml. daily. However, with vomiting or diarrhea, this loss may be great.

Skin. Water lost via the skin can be considered under two categories, *insensible perspiration* and *sensible perspiration*. The insensible loss occurs independently of sweat secretion. The skin is not impervious to water, and there is a constant diffusion of moisture from its deeper layers to the dry surface where the water evaporates. Insensible loss depends largely on environmental humidity. Sensible perspiration refers to loss of water by production of sweat. Sweating is an emergency mechanism for regulating body temperature when the heat produced by metabolic processes is excessive. The amount of sweat will, therefore, vary with exercise and with body temperature. In a moist atmosphere, sweat may be more visible than in a dry atmosphere, but the amount of water lost is the same. Despite its role as a safety factor, sweating can become a hazard when body water supplies are low, because the body continues to lose sweat to maintain its temperature. In the normal adult performing moderate exercise in a comfortable environment, approximately 500 ml. of water is lost in both sensible and insensible perspiration.

Kidneys. Water loss via the kidneys varies with the supply of body water. The kidneys serve as a spillway for water not needed by the body. If the amount of excess water is great, such as might occur when a large quantity is consumed in a short period of time, the kidneys will excrete a very dilute urine whose specific gravity might approach 1.002. If the amount of excess water is small, the kidneys will concentrate the urine, and the specific gravity may approach 1.040. Under abnormal conditions, such as vomiting or diarrhea, less water is available, and the kidneys respond promptly to curtail water loss via urine. However, the minimal water loss via the kidneys varies with the quantity of waste that they must excrete. Water excretion must be sufficient to keep these wastes in solution. This minimal loss is called the *obligatory loss*.

Balance between intake and output

In a state of normal health, the balance between intake and output is remarkably stable. If the kidneys are functioning normally, one suffers no harmful effects from consuming either much or little water, provided there is enough for regulation of body temperature and excretion of wastes. A typical intake and output record for an adult eating a balanced diet and exercising moderately in a comfortable environment is shown below.

Intake		Output	
Exogenous water		Lungs	350 ml.
Beverages	1,000 ml.	Gastrointestinal	
Solid food	1,500 ml.	tract	150 ml.
Endogenous		Skin	500 ml.
water	300 ml.	Kidney	1,800 ml.
Total	2,800 ml.	Total	2,800 ml.

In the presence of certain cardiac diseases, severe weight reduction diets, renal abnormalities, or hormonal imbalance, water balance may be disturbed, and too much water may be retained or lost. Under these circumstances, the physician may wish to evaluate the water balance and will ask the patient to keep an accurate intake and output record. Items recorded under "intake" should include the quantity of all beverages and the amount and kinds of food consumed over a 24-hour period. Items recorded under "output" should include the amount of urine excreted in 24 hours and an estimate of the amount of sweat, if it has been excessive. Deviations from normal, such as elevation in body temperature, vomiting or diarrhea, and bleeding or discharge from a wound should be noted, since these can affect water balance.

DISTRIBUTION OF BODY FLUIDS

Body water is divided into two compartments, that within the cells being *intracellular fluid* (ICF) and that outside the cells being *extracellular fluid* (ECF). These compartments are separated by cell membranes and differ in both volume and composition. Figure 19-1 shows the distribution of body fluids.

Extracellular fluid

The extracellular fluid is anatomically separable into two parts by blood vessel walls, the intravascular part being plasma and the extravascular, *interstitial fluid* (ISF). Normally, the plasma volume tends to remain stable while the ISF volume may vary. Since capillary membranes are not selectively permeable to small particles, ions and small molecules can exchange rapidly between the plasma and the ISF. Thus the electrolyte content of these two parts is essentially the same, whereas the protein content differs because most protein molecules are too large to cross the capillary barrier. The protein molecules create the colloid osmotic pressure (COP) of the two parts. Dynamic equilibrium between the plasma and the ISF is largely determined by pressures acting on the fluids, rather than on electrolyte concentration differences (p. 458 to 461). The hydrostatic pressures of the blood and ISF tend to oppose each other, since they are pushing in opposite directions across the capillary wall. Similarly, the colloid osmotic pressures of the

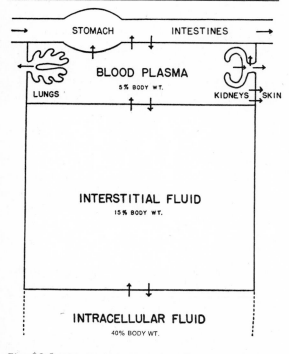

Fig. 19-1. Diagram showing the fluid compartments of the body. The intracellular compartment is not complete as indicated by the dotted lines; it is about two-and-one-half times as large as the other two compartments combined.

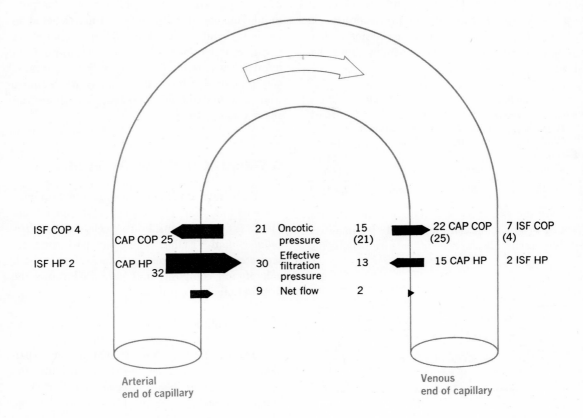

ISF COP 4				
CAP COP 25	21	Oncotic pressure	15 (21)	22 CAP COP (25)	7 ISF COP (4)
ISF HP 2				
CAP HP 32	30	Effective filtration pressure	13	15 CAP HP	2 ISF HP
	9	Net flow	2		

Arterial
end of capillary

Venous
end of capillary

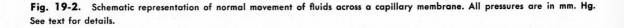

Fig. 19-2. Schematic representation of normal movement of fluids across a capillary membrane. All pressures are in mm. Hg. See text for details.

blood and ISF tend to oppose each other because they are "pulling" in opposite directions. The difference in hydrostatic pressures is the *effective filtration pressure*, while the difference in colloid osmotic pressures is the *oncotic pressure*. A

schematic representation of normal exchange of fluids between the plasma and the ISF is given in Figure 19-2.

A change in either effective filtration pressure or oncotic pressure can upset the dynamic

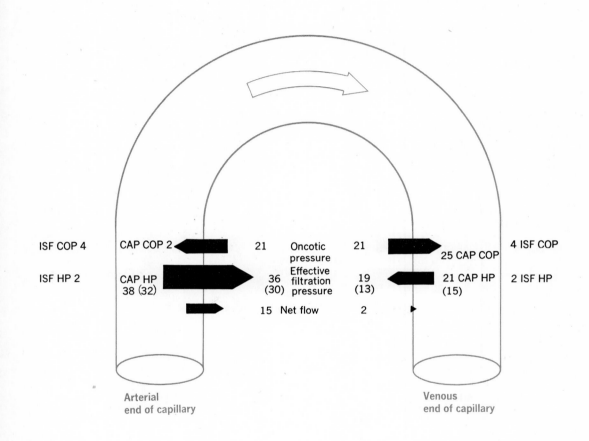

ISF COP 4   CAP COP 2   21   Oncotic pressure   21   25 CAP COP   4 ISF COP

ISF HP 2   CAP HP 38 (32)   36 (30)   Effective filtration pressure   19 (13)   21 CAP HP (15)   2 ISF HP

15   Net flow   2

Arterial end of capillary   Venous end of capillary

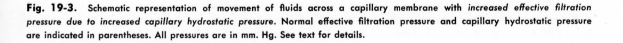

Fig. 19-3. Schematic representation of movement of fluids across a capillary membrane with *increased effective filtration pressure due to increased capillary hydrostatic pressure.* Normal effective filtration pressure and capillary hydrostatic pressure are indicated in parentheses. All pressures are in mm. Hg. See text for details.

equilibrium between the blood and the ISF. If effective filtration pressure far exceeds oncotic pressure, fluid coming into the interstitial spaces cannot be returned to the blood and tends to accumulate—a condition known as *edema.*

Edema. Several conditions that tend to alter effective filtration pressure or oncotic pressure can account for fluid accumulation. A general, or local, increase in capillary hydrostatic pressure will increase effective filtration pressure, while a

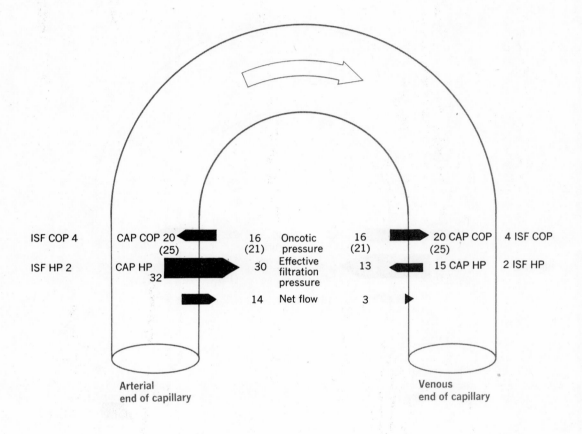

Fig. 19-4. Schematic representation of movement of fluids across a capillary membrane with *decreased oncotic pressure due to the loss of plasma proteins.* Normal oncotic pressure and capillary colloid osmotic pressure are indicated in parentheses. All pressures are in mm. Hg. See text for details.

decrease in plasma proteins or an increase in protein content of the ISF will result in a decrease in oncotic pressure. Both increased effective filtration pressure and decreased oncotic pressure will cause increased outward flow of fluid at the arteriolar end of the capillary and decreased inward flow at the venous end. A general, or local, increase in capillary pressure might arise in certain cardiac conditions or in venous obstruction, and would result in increased effective filtration pressure

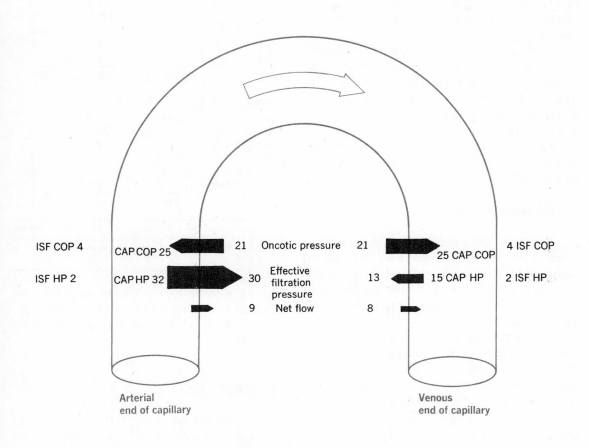

Fig. 19-5. Schematic representation of movement of fluids across a capillary membrane with *decreased oncotic pressure due to a shift of plasma proteins* into the interstitial fluid. Normal oncotic pressure and colloid osmotic pressure are indicated in parentheses. All pressures are in mm. Hg. See text for details.

(Fig. 19-3). Several conditions can cause a decrease in oncotic pressure. In some kidney diseases, for example, large quantities of blood proteins are lost in the urine with a resultant decrease in oncotic pressure (Fig. 19-4). Conditions that

cause a protein shift from the blood into the ISF would also cause a decrease in oncotic pressure by increasing the ISF · COP and reducing the cap COP. In some allergic conditions, such as hives and asthma, the chemical substances released from affected cells increase capillary membrane permeability. Thus more blood proteins escape into the interstitial spaces. An obstruction of lymph flow would prevent the small amount of protein that normally escapes into the ISF from being returned to the blood through the lymph channels, and as a result the blood proteins would be gradually decreased and the ISF proteins increased (Fig. 19-5).

If all factors remain equal, there is a tendency for edema formation to be self-limiting. As fluid accumulates in the interstitial spaces, the ISF · HP rises, finally reaching a high enough pressure to offset the increased net flow into these spaces, and the edema fluid remains stable.

Electrolytes also play an important role in edema formation. The role of sodium is discussed on page 459.

Intracellular fluid

The intracellular fluid differs in several respects from the extracellular fluid, one major difference being the composition of the two compartments, notably the electrolyte content. It is extremely difficult to measure the electrolytes of the two fluid compartments, and numerous techniques have been used to measure or to estimate the concentration of these substances, some of which are shown in Table 19-1. You will note important differences in concentrations. For example, sodium ions (Na^+) and chloride ions (Cl^-) are predominantly extracellular ions, while potassium ions (K^+), phosphate ions ($HPO_4^=$), magnesium ions (Mg^{++}) and, to a lesser extent bicarbonate ions (HCO_3^-) are predominantly intracellular ions. This disparity in ion concentration across the cell membrane is essential to normal cell function and is possible only because cell membranes are selectively permeable and have mechanisms for active transport (pp. 24 to 25). Thus the shift of fluids in and out of cells depends to a large extent on the concentration gradients or the electrical gradients that can be established and maintained. Although there are differences in the concentration of individual ions in the two compartments, the total number of positive ions (cations) in the compartment is equal to the total number of negative ions (anions) in that same compartment so that electrical neutrality is maintained. Ion shifts from

Table 19-1 Electrolyte structure of extracellular and intracellular fluids

	ECF		ICF
	Plasma mEq./L.	ISF mEq./L.	mEq./Kg.
Cations			
Na	142	145	11
K	4	4	162
Ca	5	5	2
Mg	2	2	28
	153	156	203
Anions			
Cl	101	114	3
HCO$_3$	27	31	10
HPO$_4$	2	2	102
SO$_4$	1	1	20
Protein	16	1	65
Organic acids	6	7	3
	153	156	203

These values are within the normal or estimated ranges given in the literature. Although they will change with actual conditions, the electrical neutrality of each compartment is maintained.

one compartment to another can upset this electrical neutrality; when these occur, electrical potentials can develop across the cell membranes (p. 102).

ELECTROLYTES

The electrolytes in the body fluids make up only a small fraction of the body weight, yet are essential for normal function. They have a variety of functions to perform: 1) They maintain electroneutrality of the body fluids; 2) they help to maintain osmotic equilibrium between ECF and ICF; 3) they help regulate neuromuscular activity; and 4) they maintain proper conditions for chemical reactions within the body. Electrolytes also have special functions, some of which will be described.

Sodium

The total amount of sodium in the body cannot be measured accurately in the living subject. However, studies have indicated that about one-fourth of the sodium is nonexchangeable—sequestered in bone and possibly in other tissues and not available for exchange between compartments. The exchangeable sodium is found primarily in ECF with the remainder being found in ICF and bone.

Plasma sodium averages about 142 mEq. per L. It accounts for about 90 per cent of all the cations in ECF and usually does not vary more than 5 mEq. per L. above or below the mean value. Variations greater than this can affect many physiologic activities, so mechanisms for regulating sodium concentration are of prime importance in maintaining balance.

Maintaining Sodium Balance. Sodium intake by different individuals and by the same individual varies tremendously from day to day. Since few healthy individuals make any attempt to control their intake of salt or of sodium-containing foods, the burden of maintaining sodium balance rests with the body's ability to get rid of excess sodium or to conserve sodium, if body stores are low. Usually, the sodium output in sweat and feces is small, and the kidneys assume the major role in sodium balance (p. 445).

Isotonicity of body fluids depends mainly on the concentration of sodium and its most abundant anion, chloride. Concentration of sodium and chloride in the fluids is maintained primarily by loss, or retention, of water. Loss of salt is accompanied by loss of water, retention of salt by retention of water. It is commonly said, "Water goes where the salt is." Thus retention of large amounts of salt may result in *edema*. In this situation, withdrawal of sodium from the diet results in decreased sodium concentration and aids in reducing cardiac or nephritic edema. When salt intake is reduced, the resulting hypotonicity of the plasma stimulates mechanisms that increase water loss via the kidneys (p. 196). Edematous fluid may also be removed from the body by diuretic drugs, most of which decrease sodium reabsorption by the kidney tubules. Thus much of the sodium filtered from the blood at the glomerulus is lost in the urine carrying water with it.

Sodium is essential in the establishment of *electrical potentials* across cell membranes (p. 102). Diffusion, or transport, of ions across membranes creates ionic differences that result in an electrical potential. The active transport of sodium by the sodium pump is described on pages 24 to 25.

Sodium plays a major role in the *maintenance of acid-base balance*. This role is so important for the maintenance of homeostasis that it is discussed on page 464.

Sodium depletion. Sodium depletion by the strictest definition is a decrease of total body sodium. A decrease of sodium in the plasma—*hyponatremia*—may or may not reflect sodium depletion, but it is usually responsible for the changes in body activity that accompany a sodium decrease.

The most important cause of sodium depletion is excessive loss of sodium rather than decreased intake. Severe depletion usually results from massive losses from the alimentary canal. Although most secretions into the alimentary canal contain sodium in concentrations similar to that of ECF, normally little sodium is lost to the body by this route because almost all is reabsorbed in the intestinal tract. However, in diarrhea with the loss of large amounts of watery stool, sodium will be carried out of the intestinal tract before it can be reabsorbed and a severe sodium depletion can result.

Sweating is a less common cause of sodium depletion, but can be a significant factor if profuse and if the fluids lost are replaced by water alone. A decrease in the hormone aldosterone can cause urinary losses (p. 482); a more common cause is osmotic diuresis. The kidneys must excrete enough water to keep the solutes being excreted in solution. When the quantity of solutes is large, the *obligatory loss* of water and salt may be excessive. For example, large amounts of sugar are often lost in the urine of diabetics, and the increased solute load increases the obligatory loss of water and sodium by the kidney, thus bringing about sodium depletion.

Decreased cardiac output, decreased systemic blood pressure, and loss of skin elasticity, all reflect changes in tonicity of body fluids resulting from decreased sodium concentration in the ECF. The decrease in blood pH that is often seen in severe hyponatremia reflects sodium's role in maintenance of acid-base balance, while muscle weakness and decreased reflex activity reflect sodium's role in maintenance of normal neuromuscular activity.

Sodium excess. Edema is the most common manifestation of *sodium excess*. However, if water retention is prevented, edema is prevented and the excess sodium remains in the plasma resulting in *hypernatremia*. In the healthy individual with normal kidneys, dietary intake of sodium plays only a minor role in hypernatremia. Hypernatremia occurs most frequently when renal mechanisms for sodium excretion are disturbed, and is manifested chiefly in central nervous system responses: increased reflex activity, increased body temperature, mental impairment, irritability, and coma.

Potassium

Body potassium cannot be measured accurately in the living subject. Since potassium is essentially an intracellular ion, this problem is particularly difficult, and different techniques yield different results. Plasma potassium, however, can be measured accurately and averages about 4.5 to 5.0 mEq. per L. Its regulation is vitally important since alterations can result in profound neuromuscular changes.

Maintaining Potassium Balance. Daily potassium intake does not vary as widely as that of sodium, and in the normal diet is approximately 50 to 150 mEq. daily. Potassium is excreted by the same routes as sodium: sweat, feces, and urine. Normally little potassium is lost in sweat though more is lost in feces than is true of sodium. Fecal potassium probably represents potassium that has entered the intestinal tract in exchange for sodium, rather than potassium that was not absorbed in the tract. The final maintenance of potassium balance rests with the kidney.

Potassium performs the same functions intracellularly that sodium performs extracellularly. With sodium it plays a major role in maintaining electrical potentials, thereby affecting neuromuscular activity (p. 102), and it is involved in acid-base balance (p. 464). Potassium's physiologic action can probably best be inferred by examining the effects of increases and decreases in body levels.

Potassium depletion. Potassium depletion may or may not be accompanied by changes in plasma potassium concentration. True depletion develops only with a net loss of potassium, while a decrease in plasma potassium, *hypokalemia*, may occur with a shift of potassium from the ECF to the ICF. Decreased intake can cause a mild deficit, since the mechanisms for potassium conservation are not as efficient as those for so-

dium. Severe depletion results from abnormal losses rather than decreased intake.

Increased potassium loss in urine can arise with normal excretion when excessive potassium is released from cells or the kidneys are excreting it inappropriately. When cells are being destroyed, as in tissue wasting, potassium released from the injured cells causes a transient hyperkalemia. Influenced in part by aldosterone (p. 482), the kidneys respond by excreting the excessive plasma potassium, and the end result is a decrease in total body potassium with a normal or somewhat elevated plasma concentration. Normally, the kidney stimulated by certain diuretics (p. 448) will exchange large amounts of potassium for the increased amounts of sodium reaching the tubules, regardless of the plasma potassium level.

It is difficult to assess the body's response to potassium depletion per se because it is usually associated with other electrolyte disturbances. Cardiac arrhythmias and weakness of skeletal muscle are commonly observed in mild hypokalemia, reflecting potassium's role in neuromuscular function, and severe depletion can cause widespread damage to cell function and structure.

Potassium excess. Potassium excess is manifested in *hyperkalemia*. This toxic state develops when the kidneys are unable to excrete potassium adequately or when potassium is very rapidly released from the cells. Healthy individuals rarely ingest enough potassium to produce toxic levels since transient hyperkalemia is quickly corrected by excretory mechanisms. In the hospital, potassium is frequently added to intravenous solutions to correct a potassium deficit. Involved personnel must take care to see that it is introduced into the plasma very slowly and in small quantities so that the potassium in the plasma does not reach toxic levels before it can enter the cells. Lethal doses have accidentally been administered to patients by rapid intravenous injection—in fact, cases are recorded in which death occurred within five minutes of the rapid injection of just 25 mEq.

If the level of aldosterone falls (p. 482) or the kidneys are severely damaged, as in acute renal failure, they may be unable to excrete adequate amounts of potassium. In either case, a serious hyperkalemia may result.

Body response to hyperkalemia is related to potassium's effect on cardiac and skeletal muscle.

A plasma potassium concentration in the range of 7 to 13 mEq. per L. is usually lethal. First there is cardiac arrhythmia, and finally there is cardiac arrest. Although the toxic effects on cardiac muscle are of overriding concern, many patients have some degree of skeletal muscle paralysis as well, and respiratory paralysis is occasionally the immediate cause of death.

Calcium and phosphate

All but 1 per cent of body calcium—estimated to be 1900 Gm.—is deposited in bone tissue as crystalline salts composed primarily of calcium and phosphate, and the remainder is in the plasma, ISF, and soft tissues. The major fraction of calcium that accounts for its physiologic effects is the ionizable calcium in plasma—roughly half the total of 4.5 to 5.2 mEq. per L. normally present there. The remainder is bound to protein and other substances in nonionizable form.

The major portion of phosphate ions is found within the cells, and these ions regulate phosphate function. They are, in fact, the major anions in the ICF and probably represent the single most important mineral constituent in cellular activity. The small amount of phosphate ions in the plasma are important in acid-base balance.

Maintaining Calcium and Phosphate Balance. Calcium and phosphate are absorbed from the intestine. Calcium is poorly absorbed because it occurs in a relatively insoluble form in most foods, while phosphate is readily absorbed unless there is too much dietary calcium with consequent formation of insoluble calcium phosphate salts. Hence phosphate absorption depends largely on calcium absorption.

Many foods, including meat and some vegetables, contain calcium in relatively large amounts but are not good dietary sources, because the calcium is in insoluble, nondiffusible form. Milk and milk products containing an abundance of readily available calcium are the major sources of dietary calcium. However, the rate at which even this readily available calcium is absorbed can vary. For example, the presence of sugar in the intestinal tract favors calcium absorption, since the acids produced by sugar fermentation increase solubility of the calcium salts, whereas the presence of fats inhibits calcium absorption by forming insoluble calcium compounds. The presence of large amounts of phosphates also decreases calcium absorption by forming insoluble calcium salts. Apparently by fostering active transport of calcium through the intestinal mucosa, vitamin D enhances calcium absorption.

However, even if all necessary conditions for calcium absorption are present, some may not be absorbed. The *calcium ion concentration in the ECF* greatly determines the rate of calcium absorption. A slight decrease in this concentration will greatly increase the absorption rate while a slight increase will decrease absorption. One regulating factor is parathyroid hormone secreted by the parathyroid glands, which directly stimulates the intestinal mucosa to increase calcium absorption and thereby increase the concentration of calcium ions in ECF—an excellent example of a negative feedback system in which the inciting factor is itself regulated by the system.

Calcium output is through the intestinal tract and the kidneys, the major part being excreted in feces. Fecal calcium is that which was not absorbed and which was secreted into the intestinal tract. The little calcium lost through the kidneys is partly under the control of parathyroid hormone which, again, exerts negative feedback to control calcium concentration in ECF. Should calcium concentration decrease even slightly, increased secretion of parathyroid hormone not only stimulates the intestinal mucosa to increase calcium absorption, but also stimulates the kidney tubules to increase calcium ion reabsorption. Phosphate is also excreted through the intestinal tract and the kidneys.

Thus *calcium balance* depends on a number of factors. During periods of growth, in pregnancy, and when a calcium depletion is being corrected, factors that enhance absorption and inhibit excretion come into play, and intake exceeds output; a state of *positive* calcium balance then exists. There are also times when output exceeds intake, and a *negative* calcium balance exists. Lactation greatly increases the demand for calcium, and intake may not keep pace with loss. Decreased intake as in starvation, and decreased absorption due to a vitamin D deficiency also cause a negative balance. Calcium imbalances may not be evident immediately because of the shift of calcium ions from one compartment to another.

The concentration of calcium ions in the ECF regulates calcium function, and as we have seen, it is partly controlled by parathyroid hormone acting on the intestinal mucosa and the kidney tubules. However, this hormone also stimulates the calcium shift from bone tissue to ECF. Thus, if intake is inadequate to maintain balance, calcium ions may enter the blood from bone tissue, and calcium concentration will be maintained at a minimal level for a period of time. The body responds to an increase in calcium ions so rapidly that the absence of parathyroid hormone alone could not account for it. It is believed that another hormone opposing parathyroid hormone is responsible; this hormone, which has been isolated from the thyroid gland, is called *thyrocalcitonin* (p. 478).

Functions of Calcium. Calcium is indispensable for normal body function. Calcium and phosphate are required for bone formation (p. 53); calcium is an essential cofactor in blood clotting (p. 268); it is required for production of milk by the female mammary glands, and for the maintenance of neuromuscular excitability.

Calcium ions decrease nerve membrane permeability to sodium and thereby help to maintain normal resting potential. It is generally believed that positively charged calcium ions are bound to protein and line the pores of the nerve membrane. The calcium ions repel sodium and other positive ions inhibiting their passage through the membrane. If some calcium ions are dislodged by neurotransmitters, sodium ions begin to move inward, causing further dislodgement until virtually no calcium ions remain to block the movement of sodium. As the sodium ions flow into the membrane they carry more and more positive charges to the inside; finally the inside becomes positive to the outside. At this point sodium flow stops, calcium again binds to protein in the pores, and the membrane once again becomes impermeable to sodium.

Calcium also binds to protein in muscle fiber, but its contribution to normal excitatory response is apparently quite different from that described above. When a skeletal muscle is in a relaxed state, calcium is bound to the sarcoplasmic reticulum. It is believed that when a signal for contraction passes through the innermost parts of the fiber, calcium ions are released from their binding sites and in some way activate the ATPase in myosin which in turn causes ATP to release the needed energy. Instantly these calcium ions become bound to the sarcoplasmic reticulum again, and the fiber returns to the relaxed state. Although it is known that calcium enters skeletal muscle rapidly upon membrane excitation, it is thought to be superfluous under normal circumstances because there is ample calcium stored in the sarcoplasmic reticulum to initiate the contractile process. However, in cardiac muscle, extracellular calcium apparently is indispensable, perhaps because the sarcoplasmic reticulum of cardiac muscle fibers is different from that of skeletal muscle fibers and calcium may not be released as efficiently from internal binding sites.

Abnormalities in Calcium Balance. A *deficit* in body calcium, due to decreased intake or decreased absorption from the intestine, can cause abnormal bone mineralization. In response to a slight reduction in calcium concentration, the parathyroid gland secretes its hormone, thereby increasing calcium absorption from the intestine and making more calcium ions available from bone. In infantile rickets, the parathyroids are often enlarged because they are chronically overstimulated. *Hypocalcemia*—reduced plasma concentration of calcium—may be so severe as to cause extreme immediate effects. The nervous system becomes progressively more excitable as the membrane becomes increasingly permeable to sodium, and at a certain critical level of calcium, usually 7 mg. per 100 ml., the nerve fibers become so excitable that they begin to discharge spontaneously. Impulses pass to skeletal muscles and cause severe tetanic spasms called *tetany* (p. 481). Severe hypocalcemia rarely gives rise to other acute responses because the tetany may be rapidly fatal. An increased secretion of parathyroid hormone, most commonly due to a parathyroid tumor, can cause *hypercalcemia*. In this situation, there is nervous system depression resulting in reduced reflex activity, and depression of muscle contractility resulting in skeletal muscle weakness, constipation, and loss of appetite. Because some calcium is excreted in the urine, a mild hypercalcemia can induce kidney stones, as the calcium combines with phosphate or other anions and precipitates.

Functions of Phosphate. As previously mentioned, *phosphate* is probably the single most important mineral constituent required by cells.

It is essential in production of 1) high-energy phosphate compounds such as ATP, ADP (p. 418), 2) nucleic acids, DNA and RNA (p. 19), 3) hexose phosphates in intermediary carbohydrate metabolism (p. 418), 4) phospholipids, 5) phosphate enzymes and coenzymes such as NAD and cytochromes (p. 419), 6) calcium phosphate salts for bone formation (p. 56), and 7) phosphate salts for the phosphate system in acid-base balance (p. 464).

Hyperphosphatemia probably evokes no discernible body response, whereas in *hypophosphatemia*, there may be bone changes such as rickets or osteomalacia, and the acid-base balance may be disturbed. Phosphate deficiency is unlikely if calcium intake is sufficient.

Magnesium

Magnesium is an essential element that is found primarily in muscle and bone. It apparently has an effect on tissue irritability and is a cofactor in various enzyme reactions. Although magnesium occurs abundantly in all green vegetables, its absorption is influenced by the calcium concentration. There is evidence that calcium and magnesium are absorbed via a common pathway. If the pathway is being used by increased amounts of one of these elements, the other cannot use the pathway, and its absorption is decreased. The effect of parathyroid hormone and calcitonin on magnesium absorption and excretion is the same as that on calcium (p. 462).

A magnesium *deficit* usually arises from inadequate intake associated with chronic malnutrition, or surgical removal of large segments of the small intestine. In *hypomagnesemia*, hyperirritability and convulsions are observed; *hypermagnesemia* can develop when large quantities of magnesium are administered parenterally during medical treatment.

Iron

Iron is an essential constituent of hemoglobin (p. 261), of myoglobin (muscle hemoglobin that stores oxygen in muscle), and of several enzymes and cytochromes involved in biologic oxidations (p. 414). Only small amounts of iron are present in the human body, 3 to 5 Gm., the major portion being in hemoglobin.

Iron is considered to be a one-way substance in that the iron that is absorbed is retained and used over and over. Only minute quantities are excreted in the intestinal tract, while almost none is found in the urine.

Intestinal absorption of iron is influenced by several factors, including the acidity of intestinal tract contents, the presence of reducing agents, such as vitamin C (ascorbic acid), the chemical state of the iron in food, and, to a great extent, on the body's demand for iron. It is believed that iron is absorbed into the intestinal mucosa as ferrous iron (Fe^{++}). The Fe^{++} in food is released, due to the acidity of the gastric contents, and the ferric iron (Fe^{+++}) is reduced to Fe^{++} by the presence of ascorbic acid or of other reducing agents. As the Fe^{++} enters the mucosal cells of the jejunum and duodenum, it is oxidized to Fe^{+++}, combined with protein and phosphate, and stored as *ferritin*. When iron is needed, the Fe^{+++} in ferritin is reduced to Fe^{++} and enters the blood. Immediately, the Fe^{++} is oxidized to Fe^{+++} and attached to a protein, *transferrin*. Most iron is transported in the blood in this form. Upon reaching the peripheral circulation, Fe^{+++} is released to ISF, where it is reduced to Fe^{++} and enters the cells. The iron from transferrin and the endogenous iron from the breakdown of hemoglobin and tissue ferritin are utilized first by red bone marrow in manufacturing hemoglobin. Iron is then taken up by other cells for storage or for the manufacture of iron-containing complexes. Thus the level of plasma iron is primarily controlled by the balance between the iron available from these sources and that being taken up by bone marrow and other tissues. Excretion plays a minor role in stabilizing the plasma level of iron.

It has been estimated that the average American diet supplies about 15 mg. of iron daily, adequate for the adult male but inadequate for those individuals with increased demands. *Iron deficiency anemia* is not uncommon in young children, women of childbearing age, and individuals who have abnormal losses of iron due to chronic or acute bleeding episodes.

Trace elements

Several minerals are present in exceedingly small amounts—which makes their role in man difficult to assess. On the basis of current studies, it is known that copper, manganese, zinc, cobalt, and molybdenum are essential activators of enzyme systems, or are part of enzyme structure.

Halogen salts, chlorides, iodides, fluorides, and bromides also occur in the human body. Chlorides have already been described as the major anions in ECF. The function of iodides in the thyroid gland has been extensively studied, but little is known about their role elsewhere in the body. The physiologically active substances manufactured in the thyroid gland are iodine compounds (p. 478). Fluorides, in small amounts, apparently are effective in preventing dental caries, but large amounts may be toxic and have deleterious effects on teeth and bones. Fluoridation of the water supply remains controversial and is bitterly opposed in many communities. Bromides are absorbed, distributed, and eliminated in the same manner as chlorides and, in large amounts, may displace chlorides. An excess of bromide is not uncommon because it is contained in sedative drugs that may be freely purchased. Mental and neurologic disturbances may occur when large doses are taken.

Arsenic, aluminum, silver, and other elements occur in trace amounts in body tissue and are thought to be incidental because no physiologic action has been discovered for them.

ACID-BASE BALANCE

Acid-base balance refers to maintenance of the normal hydrogen ion concentration in body fluids. Since even slight changes in this concentration can have profound, often devastating effects, the maintenance of this delicate balance is vital.

Hydrogen ion concentration $[H+]$ of a solution is expressed as pH. The expression pOH indicates the hydroxyl ion concentration $[OH-]$, but it is seldom used, since the sum of pH and pOH always equals 14, and pOH can be inferred from pH. For example, a solution of pH 5 would have a pOH of 9 (pH 5 + pOH 9 = 14). Since pH is an expression of a number with a negative power, the smaller the pH, the greater the actual number of hydrogens. At a pH of 7, the hydrogen ion concentration is 1×10^{-7} or 0.0000001; at a pH of 6, the hydrogen ion concentration is

1×10^{-6} or 0.000001. Thus at pH 6 there are ten times as many $H+$ as at pH 7, and the solution is acid (Table 2-2).

The normal pH of blood is 7.35 to 7.45. At this pH, $[OH-]$ slightly exceeds the $[H+]$, and the reaction is alkaline. This is true of body fluids with two exceptions: gastric juice has a pH of 0.9 to 1.2, due to the presence of hydrochloric acid, and urine usually has a pH of about 6, due to the acids being excreted. The mechanisms by which the critical balance is maintained between acids and bases include chemical buffers, the respiratory system, and the kidneys.

Chemical buffers

Chemical buffers, usually present as buffer pairs in both ECF and ICF, maintain a fairly constant hydrogen ion concentration in the presence of acids or alkalis. The buffer pairs consist of a weakly dissociated acid and a salt of that acid, for example, carbonic acid (H_2CO_3) and a bicarbonate salt ($NaHCO_3$, or $KHCO_3$), or of two salts, one weakly acid and one weakly basic, for example, monosodium phosphate (NaH_2PO_4) and disodium phosphate (Na_2HPO_4).

Table 19-2   Chemical Buffer Systems

System	Buffer Pairs	
	ECF	ICF
Carbonate	$\dfrac{H_2CO_3}{NaHCO_3}$	$\dfrac{H_2CO_3}{KHCO_3}$
Phosphate	$\dfrac{NaH_2PO_4}{Na_2HPO_4}$	$\dfrac{KH_2PO_4}{K_2HPO_4}$
Protein	$\dfrac{H\ protein}{Na\ proteinate}$	$\dfrac{H\ protein}{K\ proteinate}$
		$\dfrac{HHb}{KHb}$
		$\dfrac{HHbO_2}{KHbO_2}$

Examples of compounds comprising the buffer pairs of the carbonate, phosphate, and protein systems. These systems operate extracellularly and intracellularly. Since sodium is the chief cation in ECF, the buffers are essentially sodium salts. The ICF buffers are essentially potassium salts. The hemoglobin (HHb) and oxyhemoglobin ($HHbO_2$) systems are examples of specific protein systems operating intracellularly.

Amino acids and proteins formed from them contain both an acid carboxyl group ($-COOH$)

and a basic amino group ($-NH_2$); hence, they can act as acids, or as bases, as the need arises. To simplify the picture, let us represent the proteins as buffer pairs, the weak acid with the $-COOH$ functioning as H^+ protein, and the salt of that weak acid with the $-NH_2$ functioning as Na^+ or K^+ proteinate. Thus the three major buffer systems are the carbonate system, the phosphate system, and the protein system (see Table 19-2).

The Carbonate System. The carbonate system is the least efficient one at body fluid pH, yet it is a most important system in blood plasma. Normally it occurs in the ratio of approximately twenty parts of $NaHCO_3$ to one part of H_2CO_3, and as long as this ratio is maintained, the carbonate system effectively prevents major changes in blood pH. How might changes occur?

Assume a carbonate buffer solution in a test tube under the same conditions as prevail in plasma, that is, pH 7.35 to 7.45 and a ratio of 20 $NaHCO_3$:1 H_2CO_3. If a strong acid, such as hydrochloric acid (HCl), is introduced into the system, the HCl immediately reacts with the basic half of the buffer pair, $NaHCO_3$

$$HCl + NaHCO_3 \rightarrow H_2CO_3 + NaCl$$

Hydrochloric acid is a strong acid and ionizes readily in solution, furnishing many H^+. These H^+ combine with bicarbonate ions, HCO_3^- to produce H_2CO_3 which ionizes weakly. Hence, many of the H^+ available from the HCl are now in the weakly ionizing H_2CO_3, unavailable as ions to lower the pH of the solution.

If a strong base such as sodium hydroxide (NaOH) is introduced into the same carbonate buffer solution, the following reaction will occur.

$$NaOH + H_2CO_3 \rightarrow NaHCO_3 + H_2O$$

Sodium hydroxide ionizes almost completely in solution, thereby furnishing many OH^- to raise the pH of a nonbuffered solution. However, with the weak acid, H_2CO_3, present in the buffer solution, these OH^- combine with the H^+ from H_2CO_3 to produce essentially nonionized water, and the number of OH^- available to alter the pH is greatly reduced.

All of these buffer systems operate in an identical manner to reduce the concentration of H^+ or OH^- introduced, so that drastic changes in pH cannot occur (Table 19-3).

Table 19-3 Buffer Systems Reactions

Carbonate System	$H_2CO_3 + NaOH \rightarrow NaHCO_3 + H_2O$
	$NaHCO_3 + HCl \rightarrow H_2CO_3 + NaCl$
Phosphate System	$NaH_2PO_4 + NaOH \rightarrow Na_2HPO_4 + H_2O$
	$Na_2HPO_4 + HCl \rightarrow NaH_2PO_4 + NaCl$
Protein System	$HHb + NaOH \rightarrow NaHb + H_2O$
	$KHb + HCl \rightarrow HHb + KCl$

Examples of reactions that occur when a strong acid, HCl, or a strong base, NaOH, is introduced into a solution buffered by the carbonate, phosphate, or protein hemoglobin system. In every instance, the $[OH^-]$ from the highly ionized (strong) base is effectively reduced by the formation of nonionized water, and the $[H^+]$ from the highly ionized (strong) acid is reduced by the formation of a weakly ionizing acid or a weakly ionizing acid salt.

Function of the Buffer Systems. Now that we have described the chemistry of the buffer systems, how do they function in the body? When strong acids or bases are introduced into body fluids, weak, less highly ionized compounds are formed, and the pH is altered only slightly. Although H_2CO_3 and $NaHCO_3$ make a buffer pair and are weakly ionizing compounds, if introduced into the blood in large quantities, they must also be buffered, or the pH of the fluid will change beyond safe limits. Carbon dioxide produced during metabolism is constantly being carried into the blood and into the red blood cells, where it reacts with water to produce H_2CO_3 ($CO_2 + H_2O \rightarrow H_2CO_3$). Although weakly ionizing, this large quantity of H_2CO_3, if not buffered, will provide enough H^+ to alter the pH. The oxyhemoglobin system in the red blood cell acts to buffer the H_2CO_3

$$KHbO_2 + H_2CO_3 \rightarrow KHCO_3 + HHbO_2$$

In ISF and ICF, the phosphate and protein systems are responsible for buffering H_2CO_3.

It is obvious that if these compounds in the buffer pairs react continuously and are not replaced, they would eventually be depleted, and extreme changes in pH could occur. The carbonate system is intimately related to the two other mechanisms of homeostasis—respirations and kidney excretions.

Respiratory homeostatic mechanisms

With each expiration, carbon dioxide and water are expelled from the lungs (p. 367). As the acids from metabolic processes enter the blood, a large portion of them are buffered by the carbonate system with the production of H_2CO_3. This H_2CO_3 is transported in various forms by the blood to the lungs, where it is changed to H_2O and CO_2. As the CO_2 is exhaled, or blown off, H_2CO_3 concentration is reduced. By this elaborate system, metabolic acids entering the blood are buffered by the carbonate system so that the excess $H+$ are carried in the weak acid H_2CO_3. As H_2CO_3 concentration is increased, the ratio of the buffer pairs is altered. The increased levels of H_2CO_3 stimulate respirations so that more CO_2 is blown off, and H_2CO_3 is expelled. If bases have been added to the blood, the H_2CO_3 level may be low; thus respirations are reduced, less CO_2 is blown off, and H_2CO_3 is conserved. In this way, respiratory mechanisms regulate the acid half of the carbonate pair, H_2CO_3.

As acids enter the blood and are buffered by $NaHCO_3$, not only does the H_2CO_3 level rise, but the $NaHCO_3$ level falls and the ratio is further altered. The loss of H_2CO_3 by respiratory mechanisms helps to restore the ratio, but if the $NaHCO_3$ is not increased, the respiratory mechanisms will eventually be unable to restore the ratio. Thus we can see that unless there is a mechanism for restoring $NaHCO_3$ levels, acid-base balance cannot be regulated constantly. The kidney provides for the regulation of the $NaHCO_3$ level.

Renal mechanisms of acid-base balance

The kidney excretes large quantities of acids without losing too much bicarbonate. Acids entering the blood are buffered by $NaHCO_3$ and converted to sodium salts, and HCO_3^- is lost to the system. If there were no provision for retrieving the bicarbonate, the buffer system would stop functioning.

Then how is sodium bicarbonate conserved and retained? It is believed that the process involves an exchange of $H+$ for $Na+$ so that $H+$ are excreted and $Na+$ are returned to the blood as $NaHCO_3$. A simplified version of this proposed

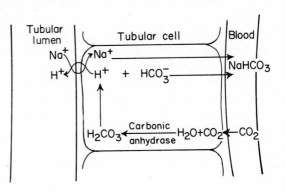

Fig. 19-6. Schematic representation of proposed mechanism for HCO_3^- retrieval by the kidney tubular cells. CO_2 diffuses into the cell from the blood and, in the presence of carbonic anhydrase, combines with water to form H_2CO_3. The H_2CO_3 ionizes to $H+$ and HCO_3^-. The $H+$ are secreted into the tubular fluid in exchange for $Na+$ which then enters the ECF with HCO_3^-. See text for details.

process can be summarized as follows: CO_2 from metabolic processes enters the renal tubular cell and in the presence of carbonic anhydrase reacts with water to produce H_2CO_3 which ionizes to $H+$ and HCO_3^-. The $H+$ are secreted into the tubular lumen in exchange for $Na+$ and the $Na+$ and HCO_3^- are then returned to the ECF. Figure 19-6 is a schematic representation of this simplified summary.

The $H+$ which are secreted into the tubular lumen may react with any of the basic buffers present in the tubular fluid and then be excreted in the urine. In addition to these mechanisms, renal tubular cells can generate ammonia from certain amino acids so that acids which would normally be excreted as sodium salts (e.g., sodium chloride) are instead excreted as ammonium salts so that $Na+$ may be returned to the ECF. Figure 19-7 summarizes the fate of the $H+$ secreted into the tubular lumen in exchange for $Na+$.

Alterations in acid-base balance

There are times when the quantity of acids or bases entering the bloodstream exceeds the body's capacity to eliminate them, and wide deviations in blood pH occur. We have already seen that the normal ratio of $NaHCO_3:H_2CO_3$ must be maintained for effective buffering action.

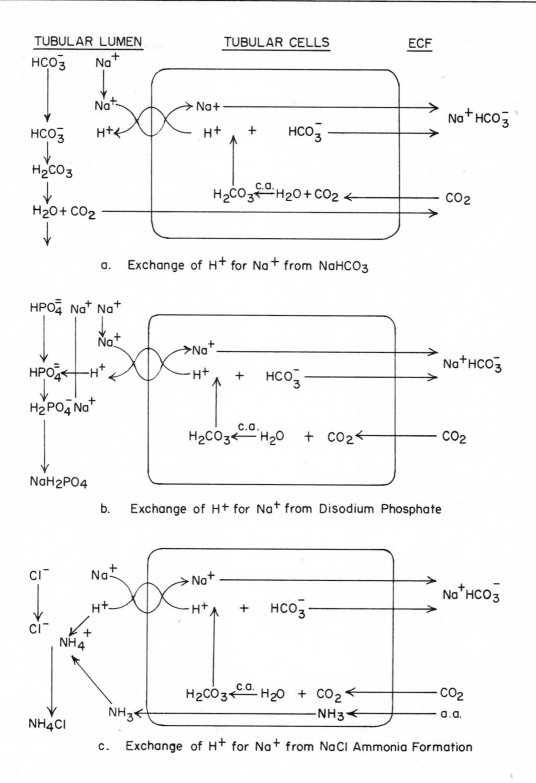

a. Exchange of H^+ for Na^+ from $NaHCO_3$

b. Exchange of H^+ for Na^+ from Disodium Phosphate

c. Exchange of H^+ for Na^+ from NaCl Ammonia Formation

Fig. 19-7. Schematic representation of the fate of H^+ secreted into tubular fluid in exchange for Na^+. (a) buffered by $NaHCO_3$; (b) buffered by Na_2HPO_4; (c) buffered by NH_3 formed by tubular cells. See text for details.

A quantitative change in either $NaHCO_3$ or H_2CO_3 can cause an imbalance. When the ratio is changed in favor of the acid buffers, a state of *acidosis* exists, while a change in favor of the basic buffers results in *alkalosis*.

Acidosis. A deficit of $NaHCO_3$ due to ingestion, retention, or increased production of acids, or to abnormal losses of $NaHCO_3$ causes *metabolic acidosis*. It is commonly associated with diabetes mellitus and starvation. Infantile diarrhea can cause an abnormal loss of base. In some renal disorders, the kidney partially loses its capacity to excrete acids. Except in the case of renal disorders, an increase in excretion of acid in the urine and the formation of ammonia will compensate for the decrease, and reestablish the ratio of H_2CO_3:$NaHCO_3$. The ratio is more quickly reestablished by the blowing off of CO_2. If these compensatory mechanisms can maintain the blood pH within normal limits, a *compensated acidosis* exists; if they cannot, *uncompensated acidosis* results, and the blood pH falls. If these homeostatic mechanisms continue to fail, the patient becomes progressively drowsy, stuporous, comatose, and finally dies.

An increase in H_2CO_3 can also change the ratio toward acidosis. This type of acidosis is referred to as *respiratory acidosis*, and is due to a reduction in CO_2 expelled by the lungs. Emphysema and drug-induced respiratory depression may be causes. In this situation, the kidney provides the only compensatory mechanisms, and there is increased ammonia formation, increased excretion of acids, and increased renal reabsorption of bicarbonate. The clinical picture is similar to that of metabolic acidosis, which may be overshadowed by the pronounced hypoxia.

Akalosis. The ratio of H_2CO_3:$NaHCO_3$ can be altered in the opposite direction with a decrease in H_2CO_3 or an increase in $NaHCO_3$, resulting in alkalosis. If the $NaHCO_3$ level is *actually* increased by ingestion of an excessive amount, or if the $NaHCO_3$ is *relatively* increased by the loss of acids, such as HCl from the stomach, a state of *metabolic alkalosis* exists. The compensatory mechanisms are increased renal excretion of alkali and decreased ammonia formation, as well as depressed respirations to conserve CO_2. If these mechanisms are sufficient to prevent an increase in blood pH, the alkalosis is said to be *compensated*; if they are not, the blood pH rises, and the alkalosis is *uncompensated*. The patient has increased neuromuscular excitability; tetany can develop.

Abnormally high losses of CO_2 can also cause a shift toward alkalosis. Any condition that produces hyperventilation would cause this increased loss, thus, the term *respiratory alkalosis*. Lack of oxygen at high altitudes, as may occur in mountain climbers, can contribute to an increase in respirations with a loss of CO_2. Anxiety and hysteria are contributing factors also, and are occasionally observed in airplane passengers. Compensatory mechanisms for alkalosis lie entirely with the kidney which decreases ammonia formation and increases bicarbonate excretion.

SUMMARY

1. Body water

A. Intake: exogenous, endogenous

B. Output: lungs, intestinal tract, skin, kidneys

C. Balance: intake equals output

2. Distribution of body fluids

A. Extracellular fluid: 20 per cent of body weight

  a. Vascular compartment: high protein

  b. Interstitial compartment: low protein

  c. Dynamic equilibrium: determined by pressure differences; accumulation of ISF—edema

B. Intracellular fluid: 40 per cent of body weight

  a. Cell membranes selectively permeable

  b. Establish electrolyte concentration differences

3. Electrolytes

A. Small amounts: variety of essential functions

B. Sodium: major cation in ECF

  a. Intake: variable; rarely controlled

  b. Output: kidney has major role

  c. Major functions: tonicity of fluids; electrical potentials; acid-base balance

  d. Hyponatremia: decreased intake, excessive loss; decreased cardiac output, acid-base imbalance, decreased CNS activity

  e. Hypernatremia: retention by kidney; mental impairment, coma

C. Potassium: major cation in ICF

  a. Intake: less variable than sodium

  b. Output: feces, major control by kidney

  c. Major functions: same as sodium, but intracellularly

  d. Hypokalemia: increased loss, decreased intake, shifts between compartments; cardiac arrhythmias, weakness of skeletal muscle

    e. Hyperkalemia: kidney dysfunction, rapid shift from cells; cardiac arrhythmias, paralysis

D. Calcium and phosphate: major portion in bone; ions responsible for activity

    a. Intake: absorption influenced by variety of factors, (e.g., vitamin D)

    b. Output: kidneys, major portion in feces

    c. Major functions of calcium: bone formation, blood clotting, neuromuscular activity

    d. Major functions of phosphate: bone, phosphate compounds, (e.g., ATP), acid-base balance

    e. Hypocalcemia: decreased intake; tetany

    f. Hypercalcemia: parathyroid tumor; decreased muscle activity

    g. Hypophosphatemia and hyperphosphatemia: rare, few discernible responses

E. Magnesium: in muscle and bone

    a. Intake: reciprocal with calcium

    b. Output: kidneys and feces

    c. Major functions: tissue irritability, cofactor for enzymes

    d. Hypomagnesemia: decreased intake (e.g., alcoholism); tetany

    e. Hypermagnesemia: parenteral infusion; anticonvulsant

F. Iron: small amounts, reused repeatedly

    a. Intake: regulated by body's need

    b. Output: feces; minor role in balance

    c. Major functions: hemoglobin, some enzymes, cytochromes

    d. Decreased levels: inadequate intake; iron-deficiency anemia

G. Trace elements: minute quantities; variety of functions

4. Acid-base balance: H+ concentration 7.35 to 7.45 in most body fluids

A. Chemical buffers: prevent major change in pH; occur in pairs

    a. Carbonate system: most important in blood; $NaHCO_3:H_2CO_3$

    b. Phosphate system: mainly renal tubules; $Na_2HPO_4:NaH_2PO_4$

    c. Proteins: basic —NH_2, acidic —COOH; major buffer of CO_2 in blood; KHb:HHb

B. Respiratory mechanisms: these conserve or blow off CO_2

C. Renal mechanisms: exchange of H+ for Na+, conservation of $NaHCO_3$; production of NH_3

D. Alterations in acid-base balance: acidosis, alkalosis

    a. Metabolic: increased production, or increased loss, of acids or bases

    b. Respiratory: increased or decreased loss of CO_2

    c. Compensated: mechanisms maintain pH in normal limits

    d. Uncompensated: pH exceeds normal limits

QUESTIONS FOR REVIEW

1. What is the major difference between the two parts of the extracellular fluid compartment? Why is this important?

2. In radical surgery for breast cancer, the lymph nodes on the affected side are usually removed. Why would this cause edema of the arm?

3. How can ions such as sodium and potassium flow uphill against a concentration difference?

4. Why might a diabetic patient who is excreting large amounts of glucose in the urine develop a sodium deficit?

5. Why would there be few body responses to decreased calcium absorption for quite a period of time?

6. Why is the carbonate system the most important chemical buffer system in the blood?

7. By what mechanisms does the kidney help to regulate acid-base balance?

8. What is uncompensated metabolic acidosis? How might this occur?

REFERENCES AND SUPPLEMENTAL READINGS

Best, C. H., and Taylor, N. B.: The Physiological Basis of Medical Practice, ed. 8. Baltimore, Williams & Wilkins, 1968.

Black, D. A. K.: Essentials of Fluid Balance, ed. 4. Oxford, Blackwell Scientific Publications, 1967.

Brooks, S. M.: The Sea Inside Us: Water in the Life Processes. New York, Meredith Press, 1968.

Brooks, S. M.: Basic Facts of Body Water and Ions, ed. 2, New York, Springer Publishing Co., 1968.

Chidsey, C. A.: Calcium metabolism in the normal and failing heart. Hospital Practice, 7:65, August 1972.

Guyton, A. C.: Textbook of Medical Physiology, ed. 4. Philadelphia, W. B. Saunders, 1971.

Muntwyler, E.: Water and Electrolyte Metabolism and Acid-Base Balance. Saint Louis, C. V. Mosby Co., 1968.

Pitts, R. P.: Physiology of the Kidney and Body Fluids, ed. 2. Chicago, Year Book Medical Publishers, 1968.

The Endocrine System

"... for I am fearfully and wonderfully made."

Psalms 139:14

The maintenance of homeostasis is highly dependent upon the two great communication systems of the body—the nervous system and the endocrine system. The most rapid communication service is provided by the nervous system, which is roughly comparable to the telephone system of a great city; nerve impulses, traveling swiftly to and from the central nervous system switchboards over peripheral nerve fiber wires, enable one to make rapid adjustments in both the external and the internal environments. The endocrine system, in contrast to the nervous system, provides a slower form of communication, because its messages are carried in the bloodstream in the form of highly specialized chemical substances (hormones). However, the functions of these two communication systems are closely coordinated; each one depends on the other for its optimum performance. For example, certain hormones, such as epinephrine, have a profound effect on the central nervous system. On the other hand, discovery of the relationships between the central nervous system and the pituitary gland led to the development of a whole new subdivision of endocrinology, called neuroendocrinology. In this chapter, we shall consider the many and varied activities of the endocrine communication system.

ENDOCRINE GLANDS

The endocrine system consists of hormone-producing cells, which are called endocrine glands. These glands are relatively simple structures, consisting of clumps, or cords, of secretory cells that lie close to capillary networks and are supported by delicate reticular fibers. Because endocrine glands have no ducts, their secretions are discharged into the capillaries rather than onto the surface, or into a cavity, of the body; thus these glands also are referred to as *ductless glands*, or the *glands of internal secretion*.

As described on page 37, and illustrated in Figure 2-19, some of the endocrine glands are derived from epithelium that grows inward from the covering, or lining surfaces, of the body; certain other endocrine glands are derived from the nervous system and from other embryonic tissues. However, regardless of their derivation, endocrine glands have one important feature in common—their secretory products are carried primarily in the bloodstream.

The glands of the endocrine system do not form an anatomic system in the sense that they are in continuity, as are the organs that comprise other systems. However, even though the individual glands are located in widely separated re-

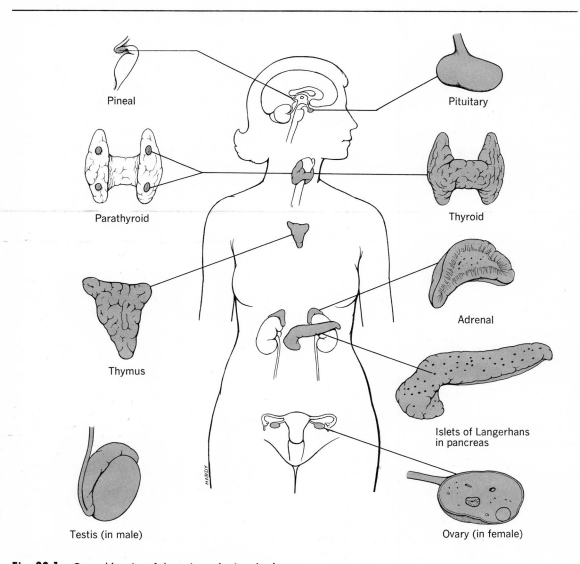

Pineal

Pituitary

Parathyroid

Thyroid

Thymus

Adrenal

Islets of Langerhans in pancreas

Testis (in male)

Ovary (in female)

Fig. 20-1.   General location of the major endocrine glands.

gions of the body (Fig. 20-1), they do form a system from a functional point of view (i.e., the functioning of one affects the functioning of others).

Hormones

The secretions produced by endocrine glands are called *hormones,* a term derived from words meaning *to arouse* or *to spur on.* Hormones are specific organic substances that are transported to various parts of the body, where they regulate the rates of specific processes. In carrying out their regulatory functions, some hormones are essential to life itself and others are not. For example, certain adrenal cortical hormones regulate the vital fluid and electrolyte balance of the body, but hormones produced by the ovaries and testes are considerably less important to mere survival of the individual.

When studying the function of hormones, one should not lose sight of the fact that the action of one hormone can affect the activities of others. For example, one hormone may reinforce another, two or more hormones may act in sequence, and one hormone may sensitize tissues to the action of another.

The production of each hormone is regulated in a particular way; for example, by the level of the blood sugar (insulin), by the level of sodium chloride in the blood (antidiuretic hormone), by the level of blood calcium (parathormone), by the nervous system (epinephrine) or by the blood level of a hormone from another gland. An example of the last is a tropic or regulatory hormone, such as the thyrotropic hormone of the anterior lobe of the pituitary, which regulates the production of thyroxine by the thyroid gland. The glands whose activities are regulated by tropic hormones are called *target organs.* The blood level of a hormone from a target gland reciprocally controls the production of the corresponding tropic hormone.

In order to have mental and physical health and to preserve homeostasis, a delicate and intricate balance must be maintained in the endocrine system. Too little secretion by any one gland not only leaves its own particular function undone, but also affects the functions of other hormones. Too much secretion by any one gland, such as occurs in adenomata or tumors of glands, likewise disturbs its own function and those of other hormones. This applies to all the components of the endocrine system, which comprises the pituitary, the pineal, the thyroid, the parathyroids, the adrenals, the gonads, the islets of Langerhans of the pancreas, and the mucosa of the stomach and the duodenum, which produces several hormones.

Cyclic AMP

Cyclic AMP (3',5'-adenosine monophosphate) is a tiny molecule that plays a key role in regulating the speed of chemical processes at the cellular level. The discovery of this substance by Dr. E. W. Sutherland added a new dimension to endocrinology and gave birth to the concept of a *second messenger* system—hormones being the first messengers.

It is believed that the target cells for most hormones have highly specific receptor sites in their outer limiting membranes. This specificity of receptor sites determines which hormones will affect which target cells. For example, receptor sites in cells in the thyroid gland recognize thyroid stimulating hormone and bind it to the sites; other hormones that may be in the bloodstream are not so recognized, thus they have no effect on the thyroid cells.

The binding of a hormone to its receptor site may in some way increase the activity of an enzyme (adenyl cyclase) within the cell membrane. As a result, cytoplasmic ATP on the inner side of the membrane is converted to cyclic AMP which then orders the cell to respond in a characteristic way. In the case of a thyroid gland cell, the result would be secretion of thyroid hormone. In a liver cell, cyclic AMP might order the conversion of glycogen into glucose, but in the kidney it might increase the permeability of a renal tubular cell to water. Several hormones (e.g., insulin) may lower the cellular level of cyclic AMP, but the mechanism whereby they do this is uncertain.

In summary, hormones serve as the first messengers as they travel from their cells of origin to the cells in their target tissues. Here the hormones alter the intracellular level of cyclic AMP which serves as the second messenger as it does the work of the hormones within the target cells. Cyclic AMP mediates many of the actions of a great variety of hormones and regulates the release of most, or possibly all, of the others. Cyclic AMP is inactivated by the enzyme phosphodiesterase.

Methods of study

Various methods of study have been used to collect information regarding the endocrine glands and, with the passage of time, these methods have become progressively more discriminating and precise. In addition, new techniques have evolved to meet the needs of investigators in the field of neuroendocrinology. A brief summary of methods and techniques is as follows:

(1) Clinical observation of persons with endocrine disorders correlated with findings at autopsy (both gross and microscopic); (2) surgical removal of a gland, or depression of its secretion by x-ray or radioactivity, to determine the effects of a lack of a particular hormone; (3) the administration of a hormone (natural or synthetic) to normal human beings and animals, as well as to those who have a deficiency of that hormone; (4) chemical analysis of body fluids and tissues to determine the presence or absence of hormones— the use of radioactive isotopes has been particularly valuable in study of the thyroid gland; (5) the placement of either destructive or stimulating electrodes in certain parts of the brain (e.g., the hypothalamus) has been of particular value in studying the interrelations of the nervous system and the endocrine system.

Pathology

A detailed presentation of the effects of hypofunction and hyperfunction of the various endocrine glands is not within the scope of this textbook. However, a few results of dysfunction will be mentioned briefly following the description of each gland.

THE PITUITARY GLAND

General description

The pituitary gland, or *hypophysis cerebri*, lies at the base of the brain, to which it is attached by a thin stalk (Fig. 8-2). The ovoid body of this gland is enclosed by dura mater and rests in the sella turcica of the sphenoid bone; this arrangement provides maximum protection for what is probably the most important single endocrine gland. Pituitary hormones dominate the activities of the thyroid gland, the adrenal cortices, and the gonads. In addition, they exert important independent effects that will be described shortly.

The pituitary gland has two primary lobes or divisions: anterior and posterior. The anterior lobe is called the *adenohypophysis* because of its glandular nature. Structurally, the anterior lobe consists of irregular cords, or nests, of large, polyhedral cells supported by delicate reticular fibers; sinusoids (capillarylike blood vessels) lie between the nests of cells (Fig. 20-2). There are relatively few nerve endings in this part of the pituitary gland.

The posterior lobe of the pituitary is an outgrowth of the hypothalamus; thus it is called the *neurohypophysis*, because it is derived from the nervous system. Thousands of axons descend into this lobe from cell bodies located in the hypothalamus. Capillary networks are found near the endings of the nerve fibers.

Role of the hypothalamus

Almost all secretory activity of the pituitary gland is controlled by that small but complicated area of the brain known as the hypothalamus (p. 196). In response to various messages coming from the internal and external environments, the hypothalamus sends signals down the pituitary stalk and orders appropriate changes in the secretory activity of the anterior lobe, and the discharge of stored hormones by the posterior lobe of this gland.

Control of the Anterior Lobe. A special arrangement of nerve fibers and tiny blood vessels provides the pathway for hypothalamic control of secretory activity in the anterior lobe; these vessels and nerve fibers are referred to as the *pituitary portal system*, or the hypothalamico-hypophyseal portal system (Fig. 20-2). Branches of the internal carotid arteries supply blood to a rich capillary plexus in the proximal portion of the pituitary stalk; blood from these capillaries then flows into a number of tiny venules, called portal vessels, that course down the stalk to empty into

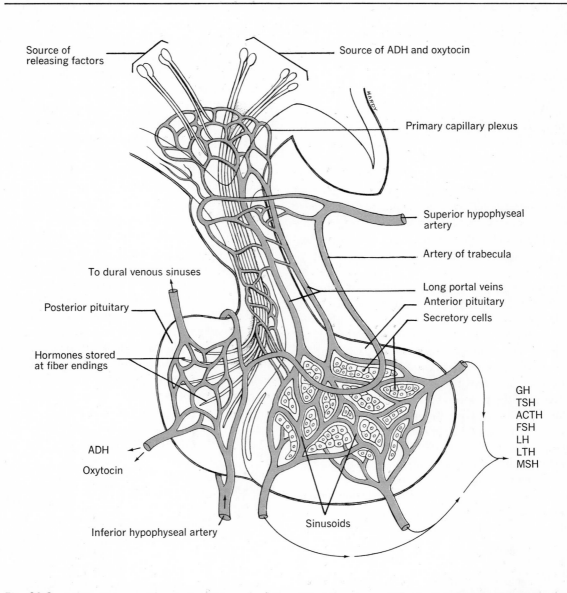

Fig. 20-2. Highly diagrammatic and schematic representation of hypophyseal nerve fiber tracts and portal system. Releasing factors produced by cell bodies in hypothalamus trickle down axons to proximal part of stalk where they enter the primary capillary plexus and are transported via portal vessels to sinusoids in adenohypophysis for control of secretions. ADH and oxytocin, produced by other cell bodies in hypothalamus, trickle down axons for storage in neurohypophysis until needed.

the sinusoids of the anterior lobe.

Special nerve cell bodies at the base of the hypothalamus produce hormones, called *neurosecretory substances*, or *releasing factors*. These substances trickle down the axons for storage at nerve fiber endings near the capillary plexus in the stalk. When appropriate messages reach the

hypothalamus these special neurons are stimulated, causing their fiber endings to secrete one or more of the stored releasing factors. These factors then enter the pituitary portal system, described above, and are transported to the anterior lobe, where they serve to promote the secretion of its hormones.

Control of the Posterior Lobe. In contrast to the anterior lobe, which receives only a few nerve fibers, the posterior pituitary is distinguished by its abundant nerve supply. As previously described on page 196, thousands of axons descend into the posterior lobe from highly specialized nerve cell bodies in the hypothalamus (Fig. 20-2). These hypothalamic nerve cell bodies produce two neurosecretions (antidiuretic hormone and oxytocin) that trickle down the axons and are stored at the fiber endings in the posterior lobe. When the hypothalamus detects a need for these substances, nerve impulses are dispatched over the appropriate fibers and the stored hormones are released for entry into the neighboring capillaries. Thus the posterior lobe is under direct nervous control.

Hormones of the anterior pituitary (Adenohypophysis)

At least six physiologically important hormones are produced by cells in the anterior lobe of the pituitary gland. These include growth hormone, thyrotropin, adrenocorticotropin, prolactin, and two gonadotropins: follicle stimulating and luteinizing. In addition to these major hormones, a seventh distinct endocrine substance, melanocyte-stimulating hormone (MSH), is found in the pituitary; however, its physiologic importance has not yet been established in man.

With the exception of MSH, the anterior pituitary hormones are concerned with normal growth and metabolism, and, in addition, they exert an important measure of control over certain other endocrine glands. All seven of these hormones are protein or polypeptide in nature.

Growth Hormone (GH). Growth hormone, also called *somatotropin,* is concerned with the growth of bones, muscles, and other organs of the body. It steps up the active transport of amino acids into cells, increases the rate of protein synthesis, and promotes cell division. By acting on the cartilage cells in the epiphyses of long bones, this hormone plays an important part in controlling the growth of the skeleton. Growth hormone also promotes the utilization of fats as an energy source (ketogenic effect), while decreasing the rate of utilization of carbohydrates (diabetogenic effect).

Regulation of growth hormone secretion involves somatotropin-releasing factor (SRF) formed in the hypothalamus and transported to the anterior lobe by way of the pituitary portal system. It is believed that depletion of cellular proteins is a major stimulus for the secretion of SRF.

Thyrotropin, or Thyroid-stimulating Hormone (TSH). Thyrotropin promotes the growth and the secretory activity of the thyroid gland. Regulation of thyrotropin production is related reciprocally to the blood level of thyroid hormone (p. 479) and to the formation of a thyrotropin-releasing factor (TRF) in the hypothalamus.

Adrenocorticotropin (ACTH). This important hormone controls the growth and the secretory activity of the adrenal cortex. Control of ACTH production is related reciprocally to the blood level of cortical hormones (p. 483) and to the formation of corticotropin-releasing factor (CRF) in the hypothalamus in response to various biological stresses (e.g., pain, hypoglycemia, hypoxia, bacterial toxins, extremes of heat or cold).

Prolactin, or Lactogenic Hormone. This hormone also has been called luteotropin (LTH) since it may play some role in maintaining the corpus luteum of the ovary. However, it is best known for its ability to stimulate the secretion of milk by prepared mammary glands (i.e., made responsive by action of ovarian hormones) at the end of pregnancy.

Gonadotropins. Gonadotropins are hormones that control the growth, development, and functioning of the ovaries and testes. These hormones are described in Chapter 21, and only a brief summary of their activities is presented here.

1. *Follicle-stimulating hormone (FSH):* stimulates development of ovarian follicles and the secretion of estrogen; stimulates development of seminiferous tubules and maturation of spermatozoa in male.

2. *Luteinizing hormone (LH):* works with FSH in final stages of follicular growth; provokes ovulation; stimulates formation of corpus luteum and secretion of progesterone. Male hormone called *interstitial cell-stimulating hormone (ICSH);* stimulates secretion of testosterone.

Melanocyte-stimulating Hormone (MSH). This hormone may have a certain effect on the melanin granules in pigmented skin. Although its physiologic importance has not yet been established, there is some evidence that it plays a role in controlling skin pigmentation in man. The chemical structure of MSH is similar to that of ACTH, and both of these hormones can cause increased pigmentation of the skin under certain circumstances.

Hormones of the posterior pituitary (Neurohypophysis)

The two hormones found in the posterior lobe of the pituitary actually are produced by special nerve cell bodies in the hypothalamus. These substances (antidiuretic hormone and oxytocin) trickle down the axons of their respective cell bodies and are stored at the fiber endings in the posterior pituitary until needed.

Antidiuretic Hormone (ADH). This hormone also is referred to as *vasopressin,* because, when present in large quantities, it causes contraction of smooth muscle in the walls of blood vessels and can raise the blood pressure. However, in routine daily living, the antidiuretic effect is more pronounced, because even small quantities of ADH in the bloodstream promote the reabsorption of water from the renal tubules and thus cause decreased excretion of water by the kidneys.

The role of the hypothalamus in regulating the secretion of ADH is described on page 196, and the specific action of ADH is described in more detail on page 446.

Oxytocin. Oxytocin has two important effects. First, it causes contraction of the muscle of the uterus, particularly toward the end of gestation. Second, oxytocin plays an important role when a mother nurses her baby; it causes milk to be expressed from the glandular tissue by stimulating the contraction of surrounding myoepithelial cells. When the hypothalamus receives afferent nerve impulses, initiated by dilatation of the cervix or by suckling stimuli on the nipple of the breast, it orders the release of oxytocin from the posterior lobe of the pituitary. Under certain conditions, commercial preparations of this hormone (e.g., Pitocin) are used clinically to induce uterine contractions during and after childbirth.

Pathology

Hypofunction of the Anterior Lobe. If a child is deficient in anterior pituitary hormones, the most obvious symptom is *dwarfism* or a failure

Fig. 20-3. Effects of pituitary disease, showing a giant and a dwarf; the woman is of normal height.

to grow (Fig. 20-3). In addition, varying degrees of hypothyroidism, adrenal cortical insufficiency, and hypogonadism may occur.

Hypofunction in the adult may result in a condition known as *Simmonds' disease.* It is characterized by premature senility, weakness, emaciation, mental lethargy, and wrinkled, dry skin. In addition, there is loss of pubic and axillary hair and, sometimes, loss of the teeth.

Hyperfunction of the Anterior Lobe. The excessive quantity of growth hormone dominates this picture. If the condition develops before puberty, *gigantism* results; there is overgrowth of the entire body, and persons with this condition frequently reach a height of 7 or 8 feet (Fig. 20-3).

If hyperfunction occurs after the epiphyses of the long bones have closed, only the face, the hands, and the feet are affected, and these become larger. The condition is called *acromegaly* (Fig. 20-4). The lower jaw becomes prominent, and the lower teeth separate; as the face enlarges, the features become coarser. The skin is greasy and moist, and pigmentation may be increased.

Hypofunction of the Posterior Lobe. This condition is characterized by the output of a large volume of dilute, sugar-free urine caused by the lack of antidiuretic hormone. The average daily output usually is 4 or 5 L., but in some cases

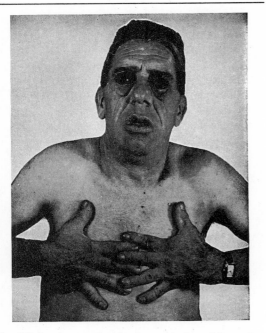

Fig. 20-4. Acromegaly, showing protrusion of lower jaw, heavy lips, and "spade" hands.

it may be as great as 15 or 20 L., depending primarily on how much water the person drinks.

The absence of sugar in the urine led to the name of *diabetes insipidus* for this condition. The name stems from the ancient practice of physicians who diagnosed diabetes by tasting the urine; in diabetes mellitus ("sugar diabetes") the urine tasted as sweet as honey, but in diabetes insipidus the lack of sugar gave it an insipid taste.

Persons suffering from diabetes insipidus may have a great thirst (polydipsia) that is essentially a compensatory symptom for the loss of a large volume of urine (polyuria).

Hyperfunction of the posterior lobe is not recognized clinically.

THE PINEAL GLAND

General description

The pineal gland, or *pineal body,* is a small cone-shaped structure that extends posteriorly from the third ventricle in the diencephalon, lying just above the roof of the midbrain. Its functions have been under intensive study in re-

cent years, and investigators hope that the mystery surrounding this organ soon will be solved. Inasmuch as study of the human pineal gland presents obvious technical difficulties, much of the laboratory work has been done on rats. However, a mounting body of evidence indicates that the mammalian pineal gland is an intricate and highly sensitive biological clock, which participates in regulating the gonads, or sex glands.

The pineal hormone

In some way not yet completely understood, the amount of light in the environment influences the production of *melatonin,* a hormone that apparently is produced only by pineal cells. This hormone inhibits the activity of the gonads. It is believed that nerve impulses from the retina are relayed over sympathetic nerve fibers to the pineal gland, where they influence the secretion of melatonin. Thus, like the hypothalamus, the pineal serves as a neuroendocrine transducer (i.e., a structure that converts a nervous input into a hormonal output). Light stimulates sexual maturation in the rat, but apparently inhibits such maturation in the human being. For example, a study of girls who either were born blind or lost their sight in the first year of life showed that blindness was associated with an early onset of puberty.

Further evidence that the pineal mechanism affects human gonads is found in cases of true pineal tumors; youngsters with such tumors have a delayed onset of puberty. On the other hand, tumors that destroy the pineal gland cause precocious puberty.

THE THYROID GLAND

General description

The thyroid gland is located immediately below the larynx in the anterior middle portion of the neck (Fig. 20-5). Lying on either side of, and anterior to, the trachea, this gland consists of two lobes united by a strip called the *isthmus.* It is covered with a capsule of connective tissue; partitions, or septa, extend inward from the capsule and divide the gland into irregular lobules.

The structural units of the thyroid gland are tiny sacs known as follicles. Each follicle is formed

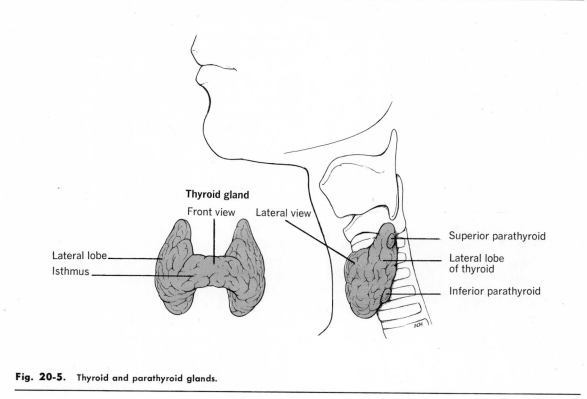

Fig. 20-5.   Thyroid and parathyroid glands.

by a single layer of epithelial cells surrounding a cavity that contains a secretory product known as *colloid*. Colloid is a viscous homogeneous fluid consisting mainly of a glycoprotein-iodine complex called *thyroglobulin*. Between the follicles is a network of reticular fibers with an extensive capillary bed, lymphatics, and nerves.

The thyroid hormones

The thyroid gland produces three hormones: *thyroxine, triiodothyronine,* and *thyrocalcitonin,* the last secretion being a relatively recent discovery. Thyroxine represents over 95 per cent of the circulating thyroid hormone; triiodothyronine represents less than 5 per cent. Despite its relatively low concentration, triiodothyronine passes out of the bloodstream more readily than does thyroxine; it has a more rapid action, and may account for as much as one fourth to one third of the total metabolic effect of the two secretions.

Thyroxine and triiodothyronine are stored in the follicular colloid as part of the thyroglobulin molecules. When these hormones are needed by the body, an enzyme in the colloid splits the thyroglobulin molecules and frees the hormones for entry into the bloodstream. Upon entering the bloodstream, thyroxine and triiodothyronine combine with specific plasma proteins and are transported to all parts of the body as protein-bound iodinated compounds. Thus laboratory tests (p. 480) that determine the amount of protein-bound iodine (PBI) in the plasma reflect the level of circulating thyroid hormone.

Functions.   Thyroxine and *triiodothyronine* control the metabolic rate of the body—that is, the rate of oxidation in the cells and the resulting heat production. They are essential for normal growth and development of the body, including mental development and the attainment of sexual maturity. They have widespread effects on all aspects of metabolism, such as promoting protein synthesis and protein breakdown, maintaining normal blood levels of cholesterol and fatty acids, increasing the rate of glucose absorption and utilization, and promoting gluconeogenesis. They also assist in maintaining fluid and electrolyte balance.

Thyrocalcitonin (or *calcitonin*) aids in maintaining the proper level of calcium in the blood. It does so by counteracting the effects of parathyroid hormone (p. 480) and inhibiting the resorption of calcium from the bones; this serves to decrease the blood level of calcium.

Control of Secretion. *Thyroid-stimulating hormone (TSH)* from the anterior pituitary controls the growth and secretory activity of the thyroid gland. It increases the size and number of follicular cells and increases their ability to take up iodide. TSH also promotes the breakdown of thyroglobulin in the follicles with the resultant release of thyroxine and triiodothyronine.

The secretion of TSH by the anterior pituitary is regulated by the concentration of thyroxine and triiodothyronine in the circulating blood. When the concentration of these hormones is decreased, the pituitary cells are stimulated, both directly and indirectly, by way of the hypothalamus to secrete more TSH. On the other hand, an increase in the amount of these hormones in the blood serves to inhibit the production of TSH.

The stimulus for release of thyrocalcitonin apparently is an elevation of the level of calcium in the blood.

Pathology

Iodine Deficiency. If the iodine intake is insufficient, a simple or colloid goiter occurs. The gland is enlarged due to an increase in the number and the size of epithelial cells. Usually the gland, because of its increased size, makes enough hormones so there are no symptoms of hypofunction, and the only evidence of pathology is the enlargement of the gland. In some persons the gland may be so large that it exerts pressure on the trachea and leads to choking sensations.

The best treatment for simple goiter is prevention. Persons living in areas where iodine is lacking in the soil and water find the use of iodized table salt a most convenient solution to the problem.

Hypothyroidism. Persons who fail to produce sufficient quantities of thyroxine and triiodothyronine are said to be *hypothyroid.* When the thyroid gland is underactive in an infant, a condition called *cretinism* develops. The first signs of cretinism appear at about six months of age, when it is noted that both mental and physical development are slow. Thereafter the infant does

Fig. 20-6. Cretinism. At the time this photograph was taken, the patient was 42 years of age, weighed 88 pounds, and was 45 inches tall; her I.Q. was 12. The patient died four months after the photograph was taken; no thyroid tissue could be found at autopsy.

not sit up, walk, and talk at the usual age. The mental development is so retarded that the child may become an idiot. The physical growth is so retarded that the child remains a dwarf.

The teeth are poorly developed, and ossification of bones is delayed. The tongue is large and protrudes from the mouth. The skin is thickened and dry. There is poor tone in the skeletal muscles. The secondary sexual characteristics do not develop. The body temperature is low, the blood pressure low, the heart rate slow, and the basal metabolic rate is about 25 per cent below the normal range. A typical cretin is shown in Figure 20-6.

When pronounced hypofunction of the thyroid gland appears in an adult, the condition is called *myxedema.* A myxedematous patient undergoes a complete change of personality and mental ability, becoming slow, lethargic, and stupid. The hair falls out, and the skin becomes thick and dry. Obesity is common, and the basal metabolic rate is low.

Hyperthyroidism. Persons who produce excessive quantities of thyroxine and triiodothyronine are said to be *hyperthyroid*. In such cases the thyroid gland usually is enlarged to some extent. Most commonly hyperthyroidism is part of a syndrome known as *Graves' disease*. It is characterized not only by the signs and symptoms of hyperthyroidism, but also by protrusion of the eyeballs (exophthalmos). It is believed that one cause of this disease is an excessive production of thyroid-stimulating hormone (TSH) by the anterior pituitary.

Some of the signs and symptoms of hyperthyroidism are as follows: high basal metabolic rate, rapid respiration, sweating, increased appetite, weight loss, tachycardia and increased cardiac output, fine tremors of skeletal muscles, high level of protein-bound iodine in the blood, restlessness, nervousness, and emotional instability. The eye changes include the aforementioned protruding eyeballs and a widening of the space between the lids (Fig. 20-7).

Treatment of hyperthyroidism centers on reducing the amount of circulating thyroid hormone. This may be accomplished by the use of drugs that depress the activity of the thyroid gland (e.g., propylthiouracil) or by removing a portion of the gland (subtotal thyroidectomy).

Laboratory Tests. Determination of the PBI (*protein-bound iodine*) is a diagnostic test of considerable clinical importance in regard to thyroid function. The normal PBI range is between 4 and 8 mcg. per 100 ml. of plasma. Decreased levels (as low as 0.8 mcg.) are found in hypothyroidism, and increased levels (as high as 24 mcg.) are found in hyperthyroidism. The BMR (*basal metabolic rate*) also is decreased in hypothyroidism and increased in hyperthyroidism, but it should be noted that the BMR measures only the overall oxygen consumption and not thyroid function specifically; for example, the BMR is increased in various infections, in pregnancy and in other conditions not related to thyroid disorders.

The *radioiodine uptake* is another test of thyroid function. The patient swallows a measured dose of radioactive iodine in distilled water; 24 hours later a radiant energy counter is placed near the thyroid gland to determine the amount of radiation from that area. In hypothyroidism the uptake usually ranges from zero to 15 per cent; in normal glands, from 15 to 50 per cent; in hyperthyroidism, 50 to 100 per cent.

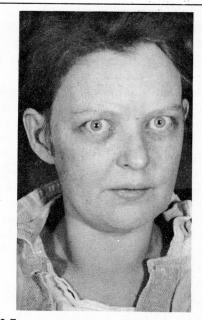

Fig. 20-7. Graves' disease; exophthalmic goiter. Note increased lid space and diffuse enlargement of the thyroid gland. The pulse rate was 134, and the BMR was plus 69 per cent.

THE PARATHYROID GLANDS

General description

The parathyroid glands are located on the posterior surface of the thyroid gland (see Fig. 20-5). They lie beneath the capsule and actually may be embedded in the substance of the thyroid gland. Usually there are four parathyroid glands.

Parathyroid hormone

Parathyroid hormone (parathormone) is a principal regulator of the calcium concentration in the body fluids. Thus the parathyroid glands are essential to life, since their secretion helps to preserve the normal excitability of muscle and nerve tissue. For example, excitability is increased when the blood calcium is low, and depressed when the calcium level is high.

Parathyroid hormone has direct effects in three areas that participate most prominently in maintaining the blood calcium level: bones, intestine, and kidneys. The hormone apparently does its job by influencing the rate at which calcium is transported across membranes. For example, when blood calcium is low, parathyroid hormone stimulates the breakdown, or resorption, of bone, thus mobilizing calcium so that it can diffuse into the bloodstream. At the same time, the hormone increases the absorption of calcium from the intestinal tract. In the kidneys, parathyroid hormone promotes the reabsorption of calcium from glomerular filtrate in the renal tubules, and it inhibits the reabsorption of phosphate ions which are then excreted in the urine. This is helpful in understanding the reciprocal relationship that exists between calcium and phosphate (i.e., if the blood level of one begins to fall, the other rises, because the product of the two is a constant).

Control of Secretions. The secretion of parathormone is triggered by a decrease in the concentration of blood calcium ions below the normal level of 10 mg. per cent. Contributing factors include low calcium diets, lack of vitamin D, pregnancy, lactation, and kidney disease in which there is retention of phosphates.

As previously described (p. 478), thyrocalcitonin serves to counteract the effects of parathyroid hormone and to lower the blood calcium level. Thus these two hormones each have important roles to play in maintaining calcium homeostasis.

Pathology

Hypoparathyroidism. The symptoms of this disorder are related to the low blood calcium level that results from a lack of parathormone. The most striking change is increased excitability of the nervous system with latent, or frank, *tetany.* One early manifestation of tetany is carpopedal spasm, a sharp flexion of the wrist and ankle joints as illustrated in Figure 20-8; this is accompanied by muscle twitchings and cramps. Generalized convulsions may occur, leading to respiratory paralysis and death by asphyxiation unless the blood picture is corrected by the intravenous injection of calcium salts. Fortunately, this condition is not common, but it may develop following accidental removal of the parathyroid glands during thyroidectomy.

Hyperparathyroidism. This relatively rare disease usually is caused by a tumor in one or more of the parathyroid glands. With the continual production of an excess of parathyroid hormone, the blood level of calcium begins to rise. In severe cases, the blood calcium level may rise to 15 mg. per cent or higher, as parathyroid hormone promotes withdrawals from the body's calcium bank in bone tissue. As a result, x-ray studies may show demineralized, cystic areas in the bones (osteitis fibrosa cystica), and multiple, spontaneous fractures are frequent.

Another complication of hyperparathyroidism is the formation of kidney stones: this occurs because with a high calcium level in the blood, more calcium is excreted in the urine. Other symptoms include decreased excitability of nerve tissue and weakness of skeletal muscles.

ADRENAL GLANDS (SUPRARENAL)

General description

The adrenal glands are located above the kidneys. Figures 1-8 and 18-1 show the location of these glands. Each consists of an outer portion called the cortex and an inner portion called the medulla. The cortex makes up the bulk of the

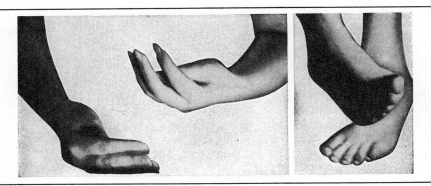

Fig. 20-8. The hands and feet of a patient with tetany, showing the typical position of the hands and the feet assumed in carpopedal spasm.

gland; it is bright yellow in the outer area and reddish brown in the inner area. The medulla is thin and gray.

The *adrenal cortex* consists of three distinct layers or zones. The outer zone (zona glomerulosa) secretes hormones known as the *mineralocorticoids*. The middle zone (zona fasciculata) and the inner zone (zona reticularis) are concerned with the production of the *glucocorticoids* and the *sex hormones*.

The *adrenal medulla*, like parts of the hypothalamus and the pineal gland, is a neuroendocrine transducer. Preganglionic fibers of the sympathetic division of the autonomic nervous system (ANS) provide the nervous input that stimulates liberation of the *catecholamines* (epinephrine and norepinephrine) by the medullary cells.

Hormones of the adrenal cortex

The specialized cells of the adrenal cortex produce three types of steroid hormones, some of which are essential to life.

1. The **mineralocorticoids.** These hormones are of fundamental importance in the maintenance of fluid and electrolyte balance. They also are called the salt and water hormones, since they regulate the absorption and the excretion of sodium, chloride, potassium, and water.

Aldosterone is the most important mineralocorticoid secreted by the adrenal cortex. It acts on the cells of the renal tubules of the kidneys where it promotes the reabsorption of sodium and the elimination of potassium.

Control of secretions. The production of aldosterone is controlled primarily by the intake of sodium, but it is influenced also by potassium and by the volume of fluid within the body. For example, there is an increase in the secretion of aldosterone when the body is deprived of sodium, or has an excess of potassium; there may also be an increase during blood loss (p. 440) and dehydration. Adrenocorticotropic hormone (ACTH) exerts only a slight amount of control over the secretion of aldosterone.

2. The **glucocorticoids,** or *sugar hormones*, constitute the second type of adrenal cortical hormones. *Cortisol* (hydrocortisone) accounts for about 95 per cent of the glucocorticoid activity. Corticosterone and cortisone are less abundant members of this group.

The glucocorticoids not only have important effects on the metabolism of all three groups of foodstuffs, but they are also concerned with one's capacity to withstand stresses of many types. For example, the secretion of glucocorticoids is increased in such stressful situations as heavy muscular exercise, exposure to cold, hypoglycemia, hypoxia, trauma, infection, allergy.

The hormones of this group are concerned with preservation of the carbohydrate reserves of the body. They have a catabolic effect on protein metabolism, as they promote the mobilization of amino acids from the cells. The amino acids then are available for conversion to glucose (gluconeogenesis). In addition to promoting gluconeogenesis, the glucocorticoids also promote the deposition of glycogen in the liver, the mobilization of depot fat and the oxidation of that fat in the liver. Thus these hormones serve as functional antagonists to insulin because they conserve carbohydrate.

Although aldosterone is the main salt-retaining hormone of the adrenal cortex, the glucocorticoids also have an effect on fluid and electrolyte balance. This is particularly true when they are present in excessive quantities.

Of major clinical importance is the fact that glucocorticoids have an *anti-inflammatory effect*. Not only do they tend to suppress inflammatory reactions, but they also tend to limit the destructive effects of such reactions. Although injections of glucocorticoids are highly beneficial in controlling such conditions as rheumatoid arthritis, excessive quantities may work to one's disadvantage in other situations in which the inflammatory response (p. 264) is part of the normal mechanism for combating infection and promoting healing. The glucocorticoids also have an *anti-allergic effect* as they block the related inflammatory responses.

The exact mechanism of the anti-inflammatory effect is uncertain. However, it is probably related to the ability of the glucocorticoids to stabilize cell membranes and to prevent the disruption of lysosomal membranes (lysosomes are cellular organelles that contain powerful digestive enzymes, as described on page 21).

Other effects that follow the injection of glucocorticoids include a reduction in the size of lymphatic organs and a decrease in the number of circulating lymphocytes and eosinophils.

Control of secretions. The production of glucocorticoids is primarily under the control of adrenocorticotropin (ACTH). For example, a decrease in the blood level of glucocorticoids stimulates the anterior pituitary to release ACTH, which, in turn, stimulates the adrenal cortex to increase its secretory activity. As the level of glucocorticoids rises, it serves to shut off the secretion of ACTH. The blood level of glucocorticoids is believed to stimulate the pituitary cells directly as well as indirectly by means of the hypothalamus, which secretes a corticotropin-releasing factor (CRF) into the pituitary portal system.

In addition to being chemically stimulated by the blood flowing through it, the hypothalamus also produces CRF in response to nerve signals from various parts of the body. For example, stress situations, such as physical injury, pain, and fright, initiate nerve impulses that are relayed to the hypothalamus, stimulating it to secrete CRF. The CRF then promotes secretion of ACTH, causing an increased output of glucocorticoids. The significance of this increased output is not yet completely understood; however, one theory is that in some cases the rapid mobilization of amino acids and fats from cellular storehouses provides not only a source of energy, but also raw materials for the synthesis of substances that are essential to the lives of cells in injured tissues.

3. The **sex hormones** constitute the third group of hormones secreted by the adrenal cortex. Several moderately active male sex hormones (17-ketosteroids) plus minute quantities of female sex hormones are produced by the adrenals in both men and women. Their role in normal body function is uncertain; however, there is some evidence that they are involved in the preadolescent growth spurt and the appearance of axillary and pubic hair.

Hormones of the adrenal medulla

The adrenal medulla produces the catecholamines, epinephrine and norepinephrine. These hormones accumulate in the medulla, usually in a mixture containing 80 to 90 per cent epinephrine, and 10 to 20 per cent norepinephrine. The secretory activity of the adrenal medulla is controlled by the sympathetic division of the ANS. This division is stimulated by pain, hypoglycemia, hypoxia, hypotension, and emotional states.

Epinephrine. This hormone also is known as Adrenalin (a trade name). It induces constriction of arterioles (except those of the skeletal muscles and heart muscle); it increases the rate and the force of the heart beat and raises the blood pressure. It dilates the bronchioles by relaxing the smooth muscle in their walls. It increases the blood sugar by increasing the breakdown of glycogen in the liver. In general, the effects are the same as stimulation of the sympathetic division of the autonomic nervous system, and the function of epinephrine is to reinforce these activities and prolong them. In addition, it stimulates the production of ACTH, which, in turn, stimulates the production of adrenal cortical hormones.

Clinically, epinephrine is used in the treatment of bronchial asthma, urticaria (hives), cardiac failure, circulatory collapse, and with local anesthetics in operations on eye, nose, and throat. It constricts blood vessels and decreases bleeding, causes the blood to coagulate faster, and delays the absorption of the local anesthetic.

Norepinephrine also is known by various trade names (e.g., *Levophed*, *Noradrenalin*). It differs from epinephrine in that it does not affect carbohydrate metabolism in muscle and liver tissue, nor does it cause hyperglycemia. While both hormones produce an elevation of the systolic blood pressure, norepinephrine produces a rise in the diastolic pressure as well. This is because norepinephrine causes vasoconstriction in the skeletal muscles as well as in the skin and the viscera. As a result of this widespread vasoconstrictor effect, there is a marked increase in the peripheral resistance.

In addition to its production by the cells of the adrenal medullae, norepinephrine (sympathin) also is released at the neuroeffector junctions of most sympathetic postganglionic nerve fibers; here it acts as a chemical transmitter substance.

Pathology

Chronic adrenocortical insufficiency, or chronic *hypofunction* of the adrenal cortex, leads to a condition known as Addison's disease. Some of the signs and symptoms that characterize this disease are as follows: (1) marked muscle weakness and easy fatigability; (2) gastrointestinal disturbances, manifested by vomiting, diarrhea, and weight loss; (3) excessive loss of sodium and chloride, and retention of potassium; (4) reduced extracellular fluid volume, with fall in blood pressure and secondary impairment of kidney function; (5) disturbances in carbohydrate metabolism, manifested by decreased gluconeogenesis, low blood sugar, and low liver glycogen; (6) a peculiar pigmentation of the skin (referred to as *bronzing* of the skin); (7) decreased capacity to adapt to even minor changes in the internal or external environments, decreased ability to withstand any form of stress, and acute prostration from even minor infections.

Hyperfunction of the adrenal cortex may involve excessive production of all three groups of hormones, or of just one group. *Cushing's syndrome* is the classic form of hyperactivity; it may be caused by various pituitary tumors that are the source of excess ACTH and lead to hyperplasia of the adrenal cortex. In this condition, the secretion of excess cortisol usually predominates over that of the other cortical hormones. Manifestations of the overall catabolic effect of excess cortisol are marked muscle wasting (especially in the extremities), loss of elasticity and thinning of the skin, a tendency to bruise easily, poor wound healing and osteoporosis. Persons with Cushing's syndrome have an abnormal distribution of fat characterized by a rounding of the face (moon face), deposits of fat in the shoulders (buffalo hump), and a pendulous abdomen; the extremities appear thin. The blood glucose is elevated, probably because of the increased gluconeogenesis; if this persists for many months, the insulin-producing cells in the pancreas may burn out and lead to frank diabetes mellitus. There is slight sodium retention, slight potassium depletion, a mild increase in the extracellular fluid volume, and moderate hypertension.

Adrenogenital syndromes. Excessive secretion of the 17-ketosteroids (male sex hormones) by children results in sexual precocity in boys. In girls, the most obvious sign is masculinization of the external genitalia. In the adult male, an excess of the 17-ketosteroids may not be recognized, inasmuch as they would tend to increase his characteristic maleness. However, hypersecretion in the female may have devastating, masculinizing effects such as baldness, increased growth of facial and body hair, atrophy of the breasts and uterus, enlargement of the clitoris, and increased skeletal muscle mass.

Hyperfunction of the adrenal medulla. Hypersecretion of the medullary hormones accompanies the presence of a tumor known as a *pheochromocytoma.* Two major syndromes are related to this condition—one of intermittent hypertension and one of sustained hypertension.

Intermittent, or paroxysmal, hypertension involves recurrent attacks of extreme, but transient, elevation of the blood pressure. During such attacks the systolic pressure may rise above 300 mm. Hg. The sustained, or chronic, type of hypertension is very similar to malignant hypertension (p. 333). In addition to a consistently high blood pressure, victims of this syndrome have signs of hyperthyroidism (e.g., nervousness and sweating), but their PBI is normal.

ISLET CELLS OF THE PANCREAS (THE ISLETS OF LANGERHANS)

General description

The pancreas is an important accessory organ of the digestive system, and, as such, it was previously described on page 400. At this time we shall concern ourselves with its endocrine function—the secretion of insulin and glucagon.

Irregular clumps of cells, called the islets of Langerhans, constitute the endocrine secretory units of the pancreas. These little islets are scattered throughout the pancreas, and it has been estimated that there are anywhere from 20,000 to 2,000,000 of them.

The islet cells are separated from the surrounding acinar cells by a film of reticular tissue (Fig. 20-9). Each islet has an extensive capillary network to carry off the secretions. Special staining techniques reveal the presence of two cell types in the islets—alpha and beta. These cells are responsible for the secretion of glucagon and insulin.

Hormones of the islet cells

Insulin is secreted by the beta cells of the islets of Langerhans, and its production is controlled primarily by the level of glucose in the blood. As the blood sugar level rises, there is an increased secretion of insulin. With a decline in blood sugar level, insulin secretion declines.

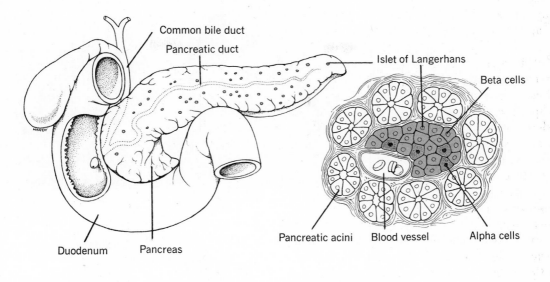

Fig. 20-9. The islets of Langerhans of the pancreas; both alpha and beta cells are shown.

Insulin serves to lower the blood sugar by increasing the utilization of glucose by the tissues, by promoting the storage of glucose as glycogen in the liver and muscles, by promoting the formation of fat from glucose, and by decreasing gluconeogenesis from amino acids. An important basic effect of insulin is its ability to increase the transport of glucose, amino acids, and fatty acids through most of the cell membranes of the body.

The effects of insulin are antagonized by several hormones: epinephrine stimulates the breakdown of glycogen in the liver and produces hyperglycemia; adrenal cortical hormones (glucocorticoids) stimulate gluconeogenesis; and thyroxine stimulates gluconeogenesis.

Glucagon, a hyperglycemic factor, is secreted by the alpha cells of the islets of Langerhans when stimulated by a falling blood sugar. It causes a marked, but temporary, rise in blood sugar level by promoting rapid glycogenolysis in the liver followed by the release of glucose into the blood. Thus it appears that glucagon and insulin constitute a team of hormones, whereby the islet cells help to regulate blood sugar level.

Pathology

Diabetes mellitus is a chronic disorder of carbohydrate metabolism, due to a relative or absolute insulin deficiency. There is a high incidence of this condition throughout the world, and diabetics require medical supervision for the rest of their lives. Diabetes mellitus most frequently develops during and after middle life. It occurs more often in women than in men and is influenced by heredity and obesity.

Many metabolic disturbances occur in diabetes mellitus, but the cardinal symptom is *hyperglycemia* (high blood glucose level), due to the impaired metabolism of carbohydrates. The resulting glycosuria as well as the secondary alterations in protein and fat metabolism lead to polyuria, ketosis, acidosis, dehydration, coma, and death. Abnormal metabolism is described on page 422.

The worst complication in diabetes mellitus involves changes that occur in the walls of the blood vessels. Capillary lesions may produce disastrous results, particularly in the retina of the eye and in the glomeruli of the kidneys. Changes in the walls of the arteries (atherosclerosis and arteriosclerosis) occur more frequently in the diabetic patient and at an earlier age than in the nondiabetic person.

The diabetic person has a marked tendency to develop infections; thus careful and frequent cleansing of the skin is emphasized. Injuries heal

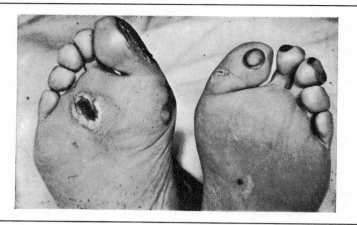

Fig. 20-10. Gangrene of the feet in untreated diabetes mellitus.

slowly and with difficulty, particularly in the lower extremities where circulation frequently is impaired. Ulcers are common, and gangrene may necessitate the amputation of some of the toes. A diabetic ulcer with gangrene of the toes is illustrated in Figure 20-10.

Many obese adult diabetics respond to dietary management and weight reduction alone, but practically all of the younger persons require insulin therapy as well as regulation of the diet. Although insulin is not a cure, it does enable carbohydrate, fat, and protein metabolism to approach the normal pattern.

Insulin must be given by hypodermic injection, since it is destroyed by the digestive enzymes if taken by mouth. There are two principal types of insulin preparations available: soluble or unmodified insulin (rapid acting), and modified preparations (slow acting). Unmodified, *crystalline* insulin acts over a period of six to eight hours. On the other hand, the modified preparations consist of insulin combined with basic proteins and zinc to slow absorption and prolong the effect. For example, *protamine zinc insulin* (PZI) has a period of action of up to thirty hours; *neutral protamine Hagedorn* (NPH) insulin and *Lente* insulin act for about twenty-four hours.

Hyperinsulinism refers to the persistent excessive secretion of insulin by beta cells of the islets of Langerhans. It induces hypoglycemia (low blood glucose level) and is characterized by attacks of weakness, profuse perspiration, and trembling. The attacks occur when meals are delayed, or when unusual exertion is undertaken.

In the early stages, the symptoms are relieved by the taking of food or sugar.

THE GONADS

The cells that produce the gonadal hormones are located in the ovaries and the testes as described in Chapter 21.

The gonadal hormones

Testosterone is produced by the interstitial cells of the testes. The production of this hormone is under the control of ICSH of the anterior lobe of the pituitary gland. Testosterone is responsible for the development of the secondary sexual characteristics of the male and for developing and maintaining the male organs of reproduction.

Estrogen, produced by the cells of the ovarian follicles, frequently is referred to as the growth hormone of the female reproductive system. It is concerned with the development of the secondary sexual characteristics at the age of puberty and with their maintenance during adult life. Estrogen plays an important role in the first half of the menstrual cycle, as it promotes the repair of the endometrium following each menstrual period. It also induces cyclic changes in the breasts and in the vaginal mucosa.

The production of estrogen is controlled by the follicle-stimulating hormone (FSH) of the anterior portion of the pituitary gland; this is described in more detail on page 505.

Progesterone is produced by the cells of the

corpus luteum, under the control of the luteinizing hormone of the anterior lobe of the pituitary gland. Progesterone prepares the uterus for implantation of the fertilized ovum and for the nourishment and the development of the embryo in the early stages of its growth. It supplements the action of estrogen on the mammary glands.

Pathology

Hypofunction of the ovaries. If the ovaries are removed before puberty, or if the anterior lobe of the pituitary gland is underactive, the secondary sexual characteristics fail to develop at the age of puberty. The menstrual cycles never are established. If the ovaries are removed after the age of puberty, the uterus, the vagina, and the external genitalia undergo atrophy.

Hypofunction of the testes. If the testes are removed before puberty, or if the gonadotropic hormones of the anterior lobe of the pituitary gland are lacking, the secondary sexual characteristics fail to develop. The voice remains high-pitched, and there is no growth of hair on the face and the trunk. There is a deposit of fat around the hips and in the pectoral and the gluteal regions. The epiphyses close late; consequently, the person is tall. The lower extremities are disproportionately long in comparison with the trunk. The accessory reproductive organs remain small, and the person is sterile.

If the testes are removed after puberty, there is atrophy of the accessory reproductive organs.

THE PROSTAGLANDINS

Prostaglandins are hormonelike substances that apparently are formed within cell membranes when needed by the cells. The term *prostaglandin* derives from the fact that they were first isolated from human and animal semen, but the name soon proved to be a misnomer as further study showed that they are synthesized in virtually all body organs. Chemically, the postaglandins are 20-carbon carboxylic acids, and they are formed from certain fatty acids.

Functionally, the prostaglandins serve as intracellular regulators whose precise role is unknown. However, they apparently can interact with cyclic AMP, turning it either on or off. The actions mediated by these substances can be quite opposite in nature, depending on the particular prostaglandin. For example, PGE_1 and PGE_2 cause bronchodilatation, whereas PGF_2-alpha causes bronchial constriction. PGE_2 causes peripheral vasodilatation and lowers the blood pressure, but PGF_2-alpha does just the opposite. The prostaglandins affect many other body structures including the gastrointestinal muscle, uterine muscle, the CNS, and blood platelets.

THE THYMUS—AN ENDOCRINE GLAND?

The generally accepted functions of the thymus are discussed on page 327. However, as this book goes to press we wish to mention the exciting work of Dr. A. L. Goldstein and his co-workers, initially at the Albert Einstein College of Medicine and later at the University of Texas. They have progressively improved their ability to extract and fractionate a substance that may be the hormone produced by the thymus; they have named this substance *thymosin* and have devised a test to show that it circulates in the blood of both animals and man. Dr. Goldstein states that there is now a large and rapidly increasing body of evidence from several laboratories that thymosin can act in lieu of an intact thymus to prevent or modify many of the deleterious consequences of neonatal or adult thymectomy on the immune system. He also has collected evidence that links aging to a significant decrease in blood levels of thymosin.

STRESS CONCEPTS

What does the word *stress* mean to you? Is it taking an examination? Living with noise and air pollution? Suffering a broken bone? Experiencing the death of a loved one? Actually, these are not stress in themselves, but rather they serve as *stressors* since they may activate the body's stress mechanisms to some degree—some more than others. Stress itself has been defined as "a dynamic state within the body in response to a demand for adaptation." Thus all of us are experiencing some degree of stress all of the time as we are constantly adapting, or adjusting, to changes in our environment.

Most people think of stress as being a bad thing, but this is not always true. Stress may very well be constructive as well as destructive. Studies of young animals have shown that a certain amount of stress in infancy is necessary for the development of normal adaptive behavior, for their ability to cope with stress when they grow older. In addition, some stressful activities may be the very spice of life (e.g., getting married, participating in competitive sports). However, the activity that serves as an invigorating experience for one individual may be distasteful to another, or may even make him ill. For example, a group of high school students interested in hospital careers were permitted to visit the observation dome of an operating room. Most of them were intensely interested as they watched the preparation for open heart surgery. On the other hand, one student fainted, and another vomited at the first sight of blood. After a brief walk in the fresh air, they returned to the dome, where they were able to adjust to the unfamiliar sights and sounds.

Adaptation to the stresses and strains of life involves a complex interplay between mental and physical activities. First, let us consider three important processes that take place as man interacts with his environment. These are perception, analysis, and reaction.

Perception is an awareness of objects; it is consciousness. Through the activity of our sense organs, the brain is continually receiving information regarding conditions in the environment. This sensory input provides us with background information that is essential for the process of analysis.

Analysis is a separating of the whole into its parts so that we can determine their nature, and their relationships. Heredity and the memory of past experiences are of great importance in this process. Since the brain is continually bombarded with sensory impulses, it is imperative that there be an interpretation of their significance (i.e., whether they are in any way a threat to the person's physical or emotional security. As you might expect, there are many individual differences in interpretation; what appears threatening to one person may not appear so to another. However, when an individual feels threatened, he may react in a variety of ways.

Reaction is the response of the body, or one

of its parts, to stimulation. All of us are constantly reacting either consciously or unconsciously to information received by the CNS. These reactions may be visible to others (overt) or they may be hidden (covert). Also, the way in which one reacts may be influenced by a number of factors including heredity, past experiences, physical condition, mental state, the strength and duration of the stimulus, and others.

In any threatening situation, whether it be physiological or psychological, the hypothalamic action systems are aroused. This means that nerve impulses dispatched from the hypothalamus activate the sympathetic-adrenal medullary mechanism, and the body is prepared for intense muscular activity (p. 225). Also, the hypothalamus secretes CRF (corticotropin releasing factor) which activates the anterior pituitary-adrenocortical mechanism, thus preparing the body to cope with a stressful situation and with potential injury (p. 482). One should keep in mind that these reactions may be elicited not only by the actual existence of danger, but also by threats and symbols of danger experienced in the past. Hopefully, the individual can adapt to stressful situations, but if there is repeated and sustained arousal of the hypothalamic action systems the end result may be bodily disorder or disease.

Stress and disease

Dr. Hans Selye was the pioneer investigator of adaptive reactions to stressful situations. Working with rats, he exposed them to intense cold, intense sound, and physical restraint. Regardless of the stressor used, Selye found that the rats developed a triad of changes: marked hypertrophy of the adrenal cortex, atrophy of lymphatic organs, and deep, bleeding ulcers of the stomach and duodenum. The rats never developed just one lesion without the others. Thus Selye called these responses the *general adaptation syndrome*. He used the term *general* because it was produced by agents having a general effect upon the body, *adaptation* because it stimulated defense, thereby helping the animal to become accustomed to something painful or difficult, and *syndrome* because its manifestations or changes were interdependent.

Selye found that animals go through three distinct stages in their response to stressful situations. First, there is an *alarm reaction*, or *alarm stage*; the hypothalamic action systems are aroused, and the body's defensive forces are called into action. This is followed by the *stage of resistance*, or *adaptation*; since no living creature can be maintained indefinitely in a state of alarm, adaptation is essential for survival. However, with prolonged exposure to severe stress situations, there is loss of the acquired adaptation or resistance that characterizes stage two; this leads to the *stage of exhaustion* and the death of the animal.

In applying this to humans, Selye suggested that most stressors which act upon us produce changes corresponding only to the first and second stages. That is, physical or mental stressors may upset or alarm us, but then we adapt or become accustomed to them. If this were not true we could not live normal lives, we could not carry out all of our activities and resist the many injuries that are to be expected.

Diseases of adaptation

Most diseases have specific causes such as the direct action of bacteria, poisons, or physical injury. On the other hand, Selye proposed that some diseases may be the result of faulty adaptation responses by the body. However, one should keep in mind the following qualifying concepts: (1) "No disease is only a disease of adaptation and nothing else," and (2) "Every disease has some element of adaptation."

Scientists have induced a number of pathologic changes in laboratory animals by subjecting them to various stressors (e.g., intense noise, flashing lights, electric shocks). These animals have developed such diseases as hypertension, gastrointestinal ulcers, nephrosclerosis, and others. The effect of stress on the human body is not quite so clear-cut as it is for lower animals. However, one group of investigators found that sustained hypertension and peptic ulcer were much more prevalent among air traffic controllers at large airports than among a group of airmen, second class, whose role was much less stressful. Other investigators have found that children may fail to grow when they live in an environment that is either physically or psychologically stressful. However, much work remains to be done in the study of adaptive reaction patterns.

SUMMARY

1. **Nervous and endocrine systems are two great communication systems**
   A. Closely coordinated, each depends on other
   B. Neuroendocrinology subdivision of endocrinology
2. **Endocrine glands also called ductless glands; glands of internal secretion**
   A. Secretions called hormones carried in bloodstream
   B. Hormone production regulated in various ways, delicate balance
   C. Cyclic AMP: the second messenger
3. **Pituitary gland (hypophysis cerebri)**
   A. Located at base of brain in sella turcica
   B. Anterior lobe: adenohypophysis
   C. Posterior lobe: neurohypophysis
   D. Hypothalamus controls most activity
      a. Anterior lobe: pituitary portal system; releasing factors
      b. Posterior lobe: direct nervous control
   E. Hormones (see Table 20-1 on p. 491)
   F. Pathology
      a. Anterior lobe
         (1) Hypofunction: dwarfism in child; Simmond's disease in adult
         (2) Hyperfunction: gigantism in child; acromegaly in adult
      b. Posterior lobe
         (1) Hypofunction: diabetes insipidus
         (2) Hyperfunction: not recognized clinically
4. **Pineal gland**
   A. Secretion: melatonin
   B. Biologic clock in regulating gonads (?)
5. **Thyroid gland**
   A. Located below larynx, anterior neck
      a. Composed of follicles containing colloid, thyroglobulin
   B. Hormones: (see Table 20-1 on p. 491)
      a. Stored as part of thyroglobulin molecules
      b. Carried in blood as PBI compounds
   C. Pathology
      a. Iodine deficiency: colloid goiter
      b. Hyperthyroidism: Graves' disease; exophthalmos
      c. Hypothyroidism: cretinism in child; myxedema in adult
      d. Lab tests: PBI, BMR, radioiodine uptake
6. **Parathyroid glands**
   A. Located on posterior surface of thyroid; usually four in number
   B. Hormone: (see Table 20-1 on p. 491)
   C. Pathology
      a. Hypoparathyroidism: tetany

b. Hyperparathyroidism: spontaneous fractures; kidney stones; muscle weakness

7. *Adrenal glands*
A. Located above kidneys; outer cortex, inner medulla
B. Hormones: (see Table 20-1 on p. 491)
C. Pathology
   a. Adrenal cortex
      (1) Hypofunction: Addison's disease
      (2) Hyperfunction: Cushing's syndrome, adrenogenital syndromes
   b. Adrenal medulla
      (1) Hyperfunction: intermittent, or sustained, hypertension

8. *Islet cells of pancreas*
A. Pancreas located in epigastric and left hypochondriac regions
B. Hormones: (see Table 20-1 on p. 491)
C. Pathology
   a. Insulin deficiency: diabetes mellitus
   b. Hyperinsulinism: hypoglycemia, weakness, trembling, sweating

9. *Gonads*
A. Secretory cells in ovaries and testes
B. Hormones: (see Table 20-1 on p. 491)
C. Pathology
   a. Hypofunction before puberty: failure of secondary sex characteristics to develop; sterility
   b. Hypofunction after puberty: atrophy of reproductive organs; sterility

10. *Prostaglandins*
A. Hormonelike substances produced in cells
B. Serve as intracellular regulators of activities

11. *Thymus: endocrine gland? Thymosin?*

12. *Stress concepts*
A. Stress and stressors
B. Perception, analysis, and reaction as man interacts with his environment
C. Stress and disease
      (1) Selye's general adaptation syndrome
      (2) Diseases of adaptation

QUESTIONS FOR REVIEW

1. Compare and contrast the nervous and endocrine communication systems.
2. How do endocrine glands differ from exocrine glands?
3. What is the general function of cyclic AMP? Why is it called a second messenger?
4. How does the hypothalamus control the anterior lobe of the pituitary gland? the posterior lobe?
5. Name and describe the functions of the six major hormones of the anterior pituitary gland.
6. What is the function of antidiuretic hormone? of oxytocin?
7. Distinguish between gigantism and acromegaly?
8. How does diabetes insipidus differ from diabetes mellitus?

9. What are the functions of thyroxine and triiodothyronine?
10. What are some of the signs and symptoms of hyperthyroidism?
11. Compare the functions of thyrocalcitonin and parathyroid hormone. What is tetany?
12. Why is aldosterone called a salt and water hormone? What controls its secretion?
13. Where is cortisol produced? What controls its secretion? What are its functions?
14. Distinguish between epinephrine and norepinephrine in terms of their effect on blood pressure and carbohydrate metabolism.
15. What is Addison's disease?
16. Compare the functions of insulin and glucagon.
17. Why is estrogen referred to as the growth hormone of the female reproductive system?
18. What are prostaglandins? What is their effect on cyclic AMP?
19. Distinguish between stress and stressors.
20. Describe the three processes that take place as man interacts with his environment.
21. Briefly, what is the general adaptation syndrome? What are its three stages?

REFERENCES AND SUPPLEMENTAL READINGS

Beeson, P. B., and McDermott, W.: Cecil-Loeb, Textbook of Medicine, ed. 13. Philadelphia, W. B. Saunders, 1971.

Gardiner, L.: Deprivation dwarfism. Sci. Am., 224:76, Jan. 1971.

Guillemin, R., and Burgus, R.: The hormones of the hypothalamus. Sci. Am., 227:24, Nov. 1972.

Kiely, W. F.: Stress and somatic disease (editorial). JAMA., April 23, 224:521, 1973.

Laros, R. K. Jr., Work, B. A. Jr., and Witting, W. C.: Prostaglandins. AJN, 73:1001, June 1973.

Levey, R. H.: The thymus hormone. Sci. Am., 211:66, July 1964.

Pastan, I.: Cyclic AMP. Sci. Am., 227:97, Aug. 1972.

Pike, J. E.: Prostaglandins. Sci. Am., 225:84, Nov. 1971.

Rassmussen, H., and Pechet, M. M.: Calcitonin. Sci. Am., 223:42, Oct. 1970.

Selye, H.: The Stress of Life. New York, McGraw-Hill, 1956.

——— Stress, it's a G.A.S. Psychology Today, Sept. 1969.

Sutherland, E. W.: On the biological role of cyclic AMP. JAMA, 214:1281, Nov. 16, 1970.

Tepperman, J.: Metabolic and Endocrine Physiology, ed. 2. Chicago, Year Book Publishers, 1968.

Wolf, S., and Goodell, H.: Stress and Disease, ed. 2. Springfield, Ill., Thomas, 1968.

Wurtman, R. J., Axelrod, J., and Kelly, D. E.: The Pineal. New York, Academic Press, 1968.

Table 20-1   Endocrine System in Summary

Endocrine Gland and Hormone	Principal Site of Action	Principal Processes Affected
Pituitary gland		
(a) Anterior lobe		
Growth hormone (Somatotropin)	General	Growth of bones, muscles, and other organs
Thyrotropin	Thyroid	Growth and secretory activity of thyroid gland
Adrenocorticotropin	Adrenal cortex	Growth and secretory activity of adrenal cortex
Follicle-stimulating	Ovaries	Development of follicles and secretion of estrogen
	Testes	Development of seminiferous tubules, spermatogenesis
Luteinizing or interstitial cell stimulating	Ovaries	Ovulation, formation of corpus luteum, secretion of progesterone
	Testes	Secretion of testosterone
Prolactin or lactogenic (luteotropin)	Mammary glands and ovaries	Secretion of milk; maintenance of corpus luteum
Melanocyte-stimulating	Skin	Pigmentation (?)
(b) Posterior lobe Antidiuretic (vasopressin)	Kidney	Reabsorption of water; water balance
	Arterioles	Blood pressure (?)
Oxytocin	Uterus	Contraction
	Breast	Expression of milk
Pineal gland		
Melatonin	Gonads (?)	Sexual maturation (?)
Thyroid gland		
Thyroxine and triiodothyronine	General	Metabolic rate; growth and development; intermediate metabolism
Thyrocalcitonin	Bone	Inhibits bone resorption; lowers blood level of calcium
Parathyroid glands		
Parathormone	Bone, kidney, intestine	Promotes bone resorption; increased absorption of calcium; raises blood calcium level
Adrenal glands		
(a) Cortex		
Mineralocorticoids (e.g. aldosterone)	Kidney	Reabsorpton of sodium; elimination of potassium
Glucocorticoids (e.g. cortisol)	General	Metabolism of carbohydrate, protein, and fat; response to stress; anti-inflammatory
Sex hormones	General (?)	Preadolescent growth spurt (?)
(b) Medulla Epinephrine	Cardiac muscle, smooth muscle, glands	Emergency functions: same as stimulation of sympathetic system
Norepinephrine	Organs innervated by sympathetic system	Chemical transmitter substance; increases peripheral resistance
Islet cells of pancreas		
Insulin	General	Lowers blood sugar; utilization and storage of carbohydrate; decreased gluconeogenesis
Glucagon	Liver	Raises blood sugar; glycogenolysis
Testes		
Testosterone	General	Development of secondary sex characteristics
	Reproductive organs	Development and maintenance; normal function

Table 20-1   (Continued)

Endocrine Gland and Hormone	Principal Site of Action	Principal Processes Affected
Ovaries		
Estrogens	General	Development of secondary sex characteristics
	Mammary glands	Development of duct system
	Reproductive organs	Maturation and normal cyclic function
Progesterone	Mammary glands	Development of secretory tissue
	Uterus	Preparation for implantation; maintenance of pregnancy
Gastrointestinal tract		
Gastrin	Stomach	Production of gastric juice
Enterogastrone	Stomach	Inhibits secretion and motility
Secretin	Liver and pancreas	Production of bile; production of watery pancreatic juice (rich in $NaHCO_3$).
Pancreozymin	Pancreas	Production of pancreatic juice rich in enzymes
Cholecystokinin	Gallbladder	Contraction and emptying

The Reproductive System

". . . male and female created He them."
Genesis 1:27

The ability to reproduce is one of the properties that characterize living matter. Under suitable conditions various forms of life are capable of reproducing themselves from one generation to the next. One-celled organisms reproduce by division of the parent cell into two daughter cells. In higher forms of life, a new individual can be produced only by the union (fertilization) of an ovum of a female and a spermatozoon of a male.

The union of an ovum and a spermatozoon may take place after they have left the body, as is the case in fishes. In birds, the union takes place in the body of the female; then the fertilized ovum leaves the body and develops outside, if kept at the proper temperature.

In mammals, the young are born alive, since the fertilized ovum remains in the uterus of the female, where it grows until ready for independent existence. We find, as we study the development of a new human being, that this is the greatest of the innumerable miracles in human life.

In both men and women, the organs of the reproductive system may be divided into two groups: (1) the gonads (testes and ovaries), which perform the double function of producing germ cells (spermatozoa and ova) and hormones (testosterone, estrogens, and progesterone) and (2) a series of ducts for the transportation of germ cells. In the male, a part of the duct system (penis) is modified for the transfer of germ cells into the body of the female. In the female a part of the duct system (uterus) is modified to support the growth and development of the new individual.

The differentiation of the sexes is a remarkable process. During the first few weeks of intrauterine life there is no structural difference to indicate whether the developing individual is going to be a male or a female. After this, differentiation begins, and, normally, the reproductive organs are completely differentiated, but immature, at birth. At puberty, which is initiated by the hormones of the anterior lobe of the pituitary gland, the organs of reproduction become functionally mature. This occurs between 10 and 14 years in girls and between 14 and 16 years in boys, in temperate climates. Puberty occurs earlier in warmer and later in colder climates.

MALE REPRODUCTIVE SYSTEM

The male reproductive system consists of a pair of male gonads or testes and a system of excretory ducts with their accessory structures. The ducts are the epididymis, the ductus deferens, and the ejaculatory ducts. The accessory structures are the seminal vesicles, the prostate gland, the bulbo-urethral glands, and the penis.

The testes

The two testes are suspended outside of the abdominal cavity in a pouchlike sac called the

scrotum (meaning bag), as shown in Figure 21-1. The scrotum is located between the upper thighs. In early fetal life, the testes lie in the abdominal cavity, near the kidneys. As the fetus grows, the testes move downward through the inguinal canal and usually enter the scrotum a short time before birth.

The descent of the testes into the scrotum is exceedingly important, since the formation of male sex cells (spermatogenesis) can occur only at the scrotal temperature, which is lower than that within the abdominal cavity. Failure of the testes to descend from the abdomen is called cryptorchidism, and the end-result is sterility (in-

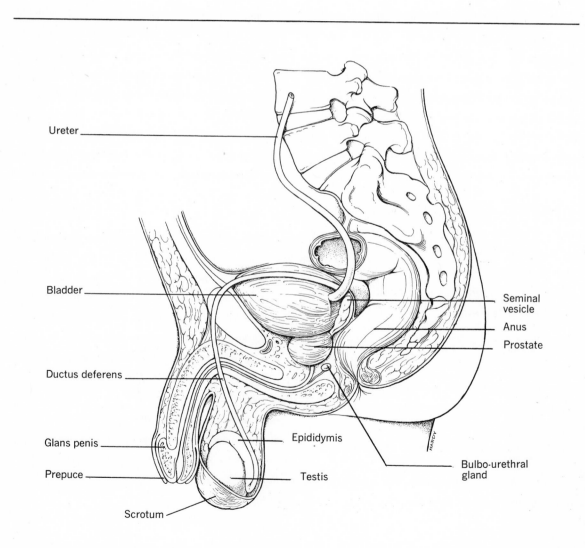

Fig. 21-1.   Organs of the male reproductive system.

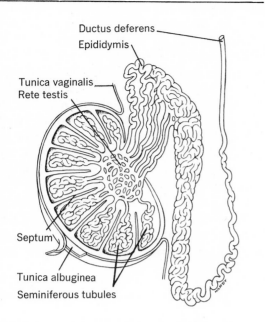

Fig. 21-2. Diagram of structural features of the testis and epididymis.

ability to reproduce).

Each testis is an oval, white body about 1½ inches in length and covered with visceral peritoneum. Beneath this serous membrane is a white fibrous covering that sends partitions into the testis and subdivides it into lobules. Within the lobules are tiny, coiled *seminiferous tubules* (Fig. 21-2) that are concerned with the production of *spermatozoa*, the male sex cells. These tubules have a total length of about one half mile.

Spermatogenesis. The walls of the seminiferous tubules consist of many layers of epithelial cells (the germinal epithelium). As shown in Figure 21-3, the cells farthest from the lumen are called *spermatogonia* (singular: spermatogonium); these are the spermatogenic, or sperm-producing cells. The spermatogonia divide by mitosis (p. 25). Half of the daughter cells remain as spermatogonia, while the other half undergo certain changes before dividing again to produce a new generation of germ cells called *primary spermato-*

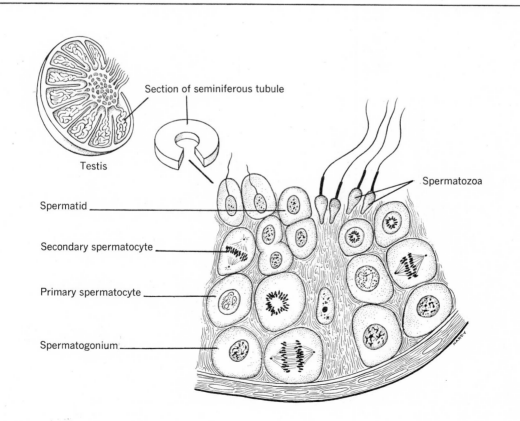

Fig. 21-3. Section of a seminiferous tubule, showing various stages of spermatogenesis.

cytes. After a prolonged prophase, the primary spermatocytes undergo a modified form of mitosis called *reduction division* (meiosis), in which the number of chromosomes is reduced by one half. Although there is a division of the cytoplasm, there is no splitting of chromosomes, and each of the two cells, or *secondary spermatocytes*, that are formed have only 23 chromosomes instead of 46. In one secondary spermatocyte there are 22 regular chromosomes, called *autosomes*, and an "X" sex chromosome; in the other secondary spermatocyte there also are 22 autosomes but the sex chromosome is a "Y."

The two secondary spermatocytes then undergo ordinary mitosis and give rise to four cells called *spermatids.* Two of the spermatids will have an X sex chromosome and two will have a Y sex chromosome, in addition to 22 autosomes. The spermatids then undergo a complex transformation to *spermatozoa*, each developing a head, a middle piece, and a tail piece. The stages of development are shown in Figure 21-4.

When fully formed, spermatozoa enter the lumen of a seminiferous tubule and are moved out of the testis into the duct system of the epididymis (Fig. 21-2). Spermatozoa that have just escaped from the testes are not capable of fertilizing an ovum, but by the time they reach the tail end of the epididymis they have become fully mature and actively motile. By lashing their tails they can move at the rate of 2 or 3 mm. per minute; this makes it possible for them to swim

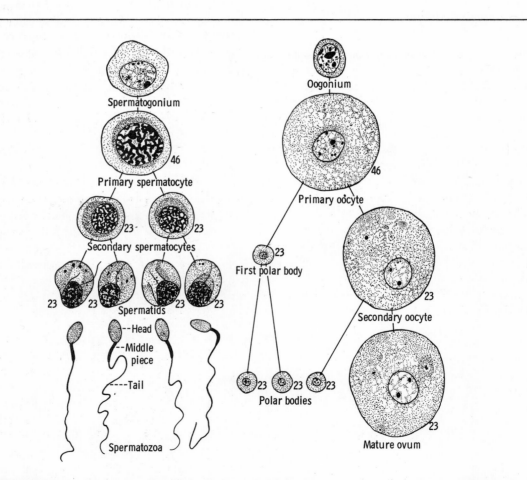

Fig. 21-4. Diagram of gametogenesis. The various stages of spermatogenesis are indicated on the left; one spermatogonium gives rise to four spermatozoa. On the right, oogenesis is indicated; from each oogonium, one mature ovum and three abortive cells are produced. The chromosomes are reduced to one-half the number characteristic for the general body cells of the species. In man, the number in the body cells is 46, and that in the mature spermatozoon and secondary oocyte is 23.

up the female reproductive tract after being deposited in the vagina.

Spermatozoa obtain their energy chiefly from the fermentation of simple sugars; thus they can live under anaerobic conditions. This is important because sperm spend considerable time in relatively oxygen-poor environments such as the lumens of various ducts in the male system, and the vagina, uterus, and uterine tube in the female system. Fructose is thought to be a major metabolic fuel for sperm because it is present in large quantities in the semen.

In the human testes, spermatogenesis begins at puberty (between 14 and 16 years of age) and continues throughout life. It is influenced directly and indirectly by two pituitary gonadotropins: follicle-stimulating hormone (FSH) and interstitial cell-stimulating hormone (ICSH). The germinal epithelium is maintained and stimulated directly by FSH, but ICSH has an indirect effect through its role in stimulating the production of testosterone. Temperature is another factor that affects spermatogenesis; the testes must be maintained at a somewhat lower temperature than that of the body as a whole. Vitamin E is essential for reproductive capacity in lower animals, but its precise function in the human body is not yet known. With advancing age, the seminiferous tubules undergo a slow decline in their ability to produce mature germ cells, but apparently never completely cease to function until death.

Testosterone is the male sex hormone (androgen) that is produced by the testes. It is secreted by little clumps of endocrine cells that lie in the connective tissue stroma between the seminiferous tubules; these cells are known as the *interstitial cells* of the testes, and their secretory activity is controlled by ICSH from the anterior pituitary. Testosterone is essential for the maintenance of a functional reproductive system in the male and for the secondary sexual characteristics (e.g., beard and body structure) that appear at puberty.

The duct system

The duct system serves to transport spermatozoa from each testis to the urethra. It includes the epididymis, the ductus deferens, and the ejaculatory duct.

The **epididymis** is a long, narrow, flattened structure that is attached to the posterior surface of the testis (Fig. 21-1). Efferent ductules from the testis unite to form the greatly coiled ductus epididymis, as illustrated in Figure 21-2. Spermatozoa complete their process of maturation while passing through this part of the duct system, before entering the ductus deferens.

The **ductus deferens,** or *vas deferens,* is a duct that conveys spermatozoa from the epididymis to the ejaculatory duct (Fig. 21-5). As shown in Figure 21-1, the left ductus emerges from the scrotum and enters the abdominal cavity by way of the left inguinal canal. Curving around the urinary bladder, it passes in front of, and medial to, the ureter; the ductus then turns sharply downward to reach the base of the prostate gland between the bladder and the rectum (Fig. 21-6). Each ductus deferens terminates by joining with the duct of the seminal vesicle of its own side to form an ejaculatory duct (Fig. 21-5).

The *spermatic cord* is composed of the testicular arteries, veins, lymphatics, nerves, and the ductus deferens, together with some connective tissue wrappings. Each spermatic cord extends from the testis to the deep inguinal ring of its own side (Fig. 5-26).

The **ejaculatory duct** is formed by the union of a ductus deferens with the duct from the seminal vesicle on the same side (Fig. 21-5). The two ejaculatory ducts then enter the posterior surface of the prostate gland and continue through the substance of this gland for less than an inch before they end in the prostatic portion of the urethra. Thus the male urethra (p. 442) has both excretory and reproductive functions.

Accessory structures

The accessory structures of the duct system of the male include the seminal vesicles, the prostate gland, the bulbo-urethral glands and the penis.

The **seminal vesicles** are elongated, musculo-membranous structures (Fig. 21-6) that lie behind the bladder and in front of the rectum. Each has a mucous membrane lining that is thrown into an extremely large number of folds, thus providing great secretory area and also permitting the vesicle to become distended with stored secretion.

The mucosal cells secrete a thick, yellow, alkaline material that contains large quantities of

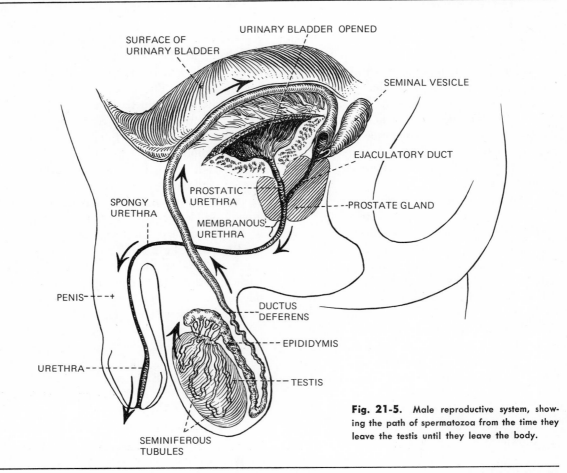

Fig. 21-5. Male reproductive system, showing the path of spermatozoa from the time they leave the testis until they leave the body.

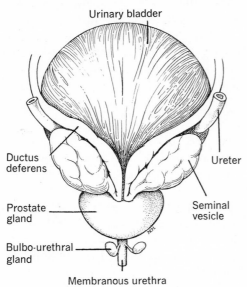

Fig. 21-6. Posterior view of the urinary bladder, the ductus deferens (vas), the seminal vesicles, and the prostate gland.

fructose; this material is emptied into the ejaculatory ducts at the same time that the ductus deferens delivers the spermatozoa. The seminal vesicle secretion serves to provide nourishment for the spermatozoa and to enhance their motility by reducing acidity in their environment. It should be noted that although the duct of each seminal vesicle joins the ductus deferens of its own side, spermatozoa do not enter the seminal vesicle; spermatozoa and the seminal vesicle secretion pass directly into the related ejaculatory duct (Fig. 21-5).

The **prostate gland** is a firm, glandular and muscular organ that is approximately the size of a horse chestnut. Lying below the urinary bladder (Fig. 21-5), it surrounds the proximal, or prostatic, part of the urethra, as well as the two ejaculatory ducts that pass anteriorly to join the urethra.

Beneath a thin, firm covering capsule of connective tissue is a dense layer of smooth muscular tissue; another dense layer of smooth muscular fibers encircles the prostatic urethra. Strong bands of smooth muscular tissue throughout the prostate form a meshwork in which the glandular tissue is embedded. Contraction of the smooth muscle of the prostate provides part of the force needed for expulsion of the semen (ejaculation).

The glandular substance of the prostate consists of a large number of branched, tubular glands whose ducts open separately into the prostatic urethra. The bulk of the prostatic secretion is a thin, milky, alkaline fluid that aids motility of the sperm by helping to maintain an optimum pH.

Benign prostatic hypertrophy is relatively common in older men. In this condition there is an enlargement or overgrowth of the glandular substance. As the prostate enlarges, urination becomes difficult because of compression of the prostatic urethra. When the bladder is not emptied completely, the residual urine leads to distention of the bladder and, in severe cases, to distention of the ureter and kidney (hydroureter and hydronephrosis). Residual urine also favors the occurrence of cystitis. Persons with this condition may become incontinent (i.e., unable to control the emptying of the bladder) and dribble urine. Prostatic hypertrophy usually is treated by surgical removal of the gland.

The **bulbo-urethral glands** (*Cowper's glands*) are small, rounded bodies, yellow in color and about the size of peas. They are inferior to the prostate gland, lying posterior and lateral to the membranous portion of the urethra. Their ducts pass forward to open into the cavernous portion of the urethra. The bulbo-urethral glands secrete an alkaline, mucoid substance that coats the lining of the urethra prior to ejaculation; this probably neutralizes acidity and provides a suitable environment for the spermatozoa.

The **penis** is an external genital organ through which the urethra passes to the exterior of the body (Fig. 21-5). It is composed of three cylindrical masses of erectile (cavernous) tissue that are bound together by elastic connective tissue and covered with skin. Two of these masses, called the *corpora cavernosa,* are located side by side in the dorsal half of the penis. The third mass, called the *corpora cavernosum urethrae,* conducts the urethra in its substance; it lies ventral to the corpora cavernosa. The distal end of the latter cavernous body is expanded into a blunt cone, called the *glans penis*; the glans is covered with a fold of skin called the *prepuce,* or foreskin. Occasionally the prepuce may fit so tightly that it cannot be drawn back over the glans, a condition referred to as *phimosis.* Surgical removal of the prepuce is called *circumcision.*

The substance of the cavernous bodies is a network of connective tissue and smooth muscle, with spaces between. The penis is soft and relaxed when this spongelike network is empty, but when the spaces become filled with blood the penis is firm and erect, a state that is essential for the deposition of spermatozoa in the vaginal canal of the female. The parasympathetic division of the autonomic system is responsible for the dilatation of penile arteries and the resultant state of erection.

Semen

The spermatozoa, plus the secretions of the seminal vesicles, of the prostate gland and of the bulbo-urethral glands, make up the semen, which is ejaculated during the male sexual act. This thick, grayish-white fluid has an average pH of about 7.5.

Semen is ejected by means of contraction of the smooth muscle of the prostate gland and the contraction of the bulbocavernosus muscle which compresses the cavernous portion of the urethra. In the average male, each cubic centimeter of semen contains about 100 million spermatozoa, and the average amount of ejaculate varies between 2 and 4 ml. However, only one spermatozoon can fertilize an oocyte, and the remainder disintegrate.

FEMALE REPRODUCTIVE SYSTEM

The female reproductive system consists of the paired gonads, or ovaries, the paired uterine tubes, a single uterus, and a vagina. The associated structures include the external genitalia and the mammary glands.

The ovaries

Description. The ovaries are small, almond-shaped bodies that are about 1½ inches in length.

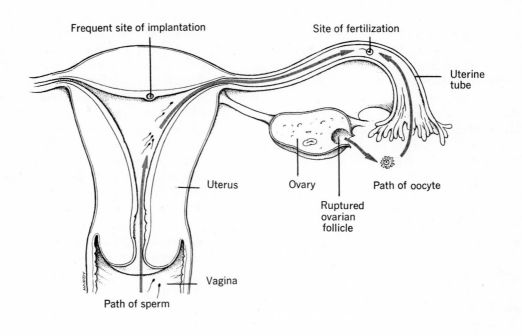

Frequent site of implantation

Site of fertilization

Uterine tube

Uterus

Ovary

Path of oocyte

Ruptured ovarian follicle

Vagina

Path of sperm

Fig. 21-7. Schematic drawing of female reproductive organs, showing path of oocyte from ovary into uterine tube; path of spermatozoa also is shown, as is the usual site of fertilization.

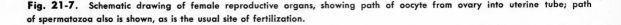

They are located one on either side of the uterus, below the uterine tubes, and attached to the posterior surface of the broad ligament of the uterus. The medial end of each ovary is attached to the uterus by a rounded cord called the *ovarian ligament*. The lateral end of the ovary is in intimate contact with the free end of the uterine tube (Fig. 21-7).

Structurally, each ovary consists of an outer zone, or *cortex*, and an inner zone, or *medulla*. The outermost layer of the cortex consists of the *germinal epithelium*, a single layer of cuboidal cells which covers the free surface of the ovary. Immediately below this surface layer is the connective tissue stroma of the cortex, and it is here that the ovarian follicles are formed. Each *primary follicle* consists of a central germ cell (oogonium) surrounded by a layer of epithelial cells (Fig. 21-8). The medulla is composed of loose connective tissue with numerous blood vessels; there are no follicles in this zone of the ovary.

Oogenesis. At birth several hundred thousand primary follicles are present in the ovary.

However, this number decreases steadily throughout life, as most of them degenerate and die (atresia). During the reproductive years of the average female, only 300 to 400 follicles reach maturity and release their oocytes. This great limitation exists because maturation of follicles does not begin until puberty, and then usually only one follicle and oocyte reach maturity each month.

Under the influence of follicle-stimulating hormone (FSH) from the anterior pituitary, a variable number of primary follicles begin to develop into secondary follicles. By the process of mitosis, the single layer of follicular cells becomes several layers, and the oogonium, which has increased in size, now is referred to as a *primary oocyte*. The follicular cells continue to proliferate by mitosis and fluid begins to accumulate. The oocyte also continues to grow, but not in proportion to the epithelium, which soon makes up the bulk of the follicle as it approaches maturity.

The follicular cells house the oocytes and also secrete *estrogens*, the primary hormones of the

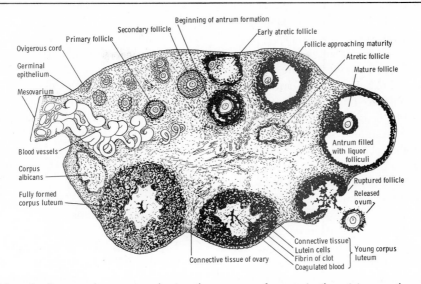

Fig. 21-8. Schematic diagram of an ovary, showing the sequence of events in the origin, growth, and rupture of an ovarian follicle, and the formation and retrogression of a corpus luteum. Atretic follicles are those that show signs of degeneration and death. For proper sequence, follow clockwise around the ovary, starting at the mesovarium. (B. M. Patten: Human Embryology, New York, Blakiston, McGraw-Hill)

ovary. At least six compounds with estrogenic activity have been isolated from the plasma, most abundant being β-estradiol and estrone. Of these, β-estradiol is the more potent; thus it is considered to be the major estrogen secreted by the ovaries. For this reason, the singular form (estrogen) frequently is used instead of the pleural (estrogens).

Estrogen is essential for the growth and development of all female reproductive organs, and then for keeping them in a functional state. The development and maintenance of the secondary female sexual characteristics (e.g., breasts and hair distribution) also depend on adequate estrogen production.

As described in the discussion of the menstrual cycle (p. 505), it appears that both FSH and LH (luteinizing hormone) are necessary for the secretion of estrogen. However, FSH is the predominating hormone in the mixture.

During follicular development, the primary oocyte undergoes a *reduction division* in which the number of chromosomes is reduced by one half (Fig. 21-4). This division gives rise to one *secondary oocyte*, containing 22 autosomes and an X sex chromosome, and a small abortive cell that is called a polocyte (polar body); the polocyte also contains 22 autosomes and an X sex chromo-

some. The secondary oocyte then begins an ordinary mitotic division, but this is not completed unless fertilization occurs (usually within the uterine tube).

Although a variable number of follicles begin this process of development, usually only one reaches maturity each month. At maturity, the follicle is greatly distended with fluid and actually bulges on the surface of the ovary (Fig. 21-8). The rising tide of estrogen produced by the follicular cells now inhibits the secretion of FSH and stimulates a sharp increase in the secretion of LH. In some way not yet completely understood, this predominantly LH mixture provokes ovulation.

Ovulation involves a bursting of the follicular wall and liberation of the secondary oocyte into, or close to, the uterine tube. Rupture of the wall occurs in the area where the follicle bulges on the surface of the ovary and where the overlying tissue is very thin.

Ovulation usually occurs every 28 to 30 days, from puberty to the *climacteric* (end of the reproductive period of life); an exception to this is during pregnancy when follicles do not mature, and ovulation cannot occur. Usually only a single oocyte is discharged, but, if more than one follicle reaches maturity, additional oocytes may be released. After the oocyte is discharged from the follicle, it enters the distal end of the uterine tube where it may be fertilized if spermatozoa are present. If not fertilized within a few days after ovulation, the oocyte disintegrates and disappears.

The Corpus Luteum.   At the site of ovulation the wall of the follicle collapses, and there is a small amount of bleeding into the follicular cavity. The follicular cells enlarge, and a yellowish substance (lutein) accumulates in their cytoplasm. As the entire follicle becomes filled with lutein cells it is called the corpus luteum (meaning *yellow body*); capillaries grow into this mass of cells, forming a fairly typical endocrine structure.

The lutein cells secrete *progesterone*, the secondary ovarian hormone, and estrogen. The chief function of progesterone is to promote the final preparation of the endometrium (lining of the uterus) to receive a fertilized ovum. It also serves to maintain pregnancy by quieting the uterine muscle and by inhibiting ovulation and menstruation. Late in pregnancy, progesterone promotes the development of secretory tissue in the breasts.

Returning to the corpus luteum, we find that two possible courses are open to it (1) if the oocyte is not fertilized, the corpus luteum continues to grow for about 10 to 12 days and then regresses, eventually becoming a small, white, ovarian scar called a *corpus albicans*; or, (2) if the oocyte is fertilized, the corpus luteum becomes very large, forming the corpus luteum of pregnancy. It continues to function for approximately three months, and then slowly degenerates, leaving a white scar on the ovary.

LH is responsible for the formation of the corpus luteum and the secretion of progesterone. However, there is some evidence that luteotropic hormone (LTH) from the anterior pituitary plays a role in maintaining the corpus luteum.

Control of Fertility.   The exact length of time that an oocyte remains viable and capable of being fertilized is uncertain, but apparently it is only a brief period (less than 24 hours). Since most spermatozoa also have a short life span (24 to 48 hours), sexual intercourse usually must occur within a day or two of ovulation if fertilization is to take place. The *rhythm method* of birth control is based on this fact, and intercourse is avoided for three to five days on either side of the calculated date of ovulation. However, this method is of value only if the menstrual cycles occur with great regularity; even then, failures

may occur since not all women ovulate at exactly the same time.

On the other hand an *oral contraceptive agent* ("the pill") may be virtually 100 per cent effective. These medications contain synthetic estrogens or progestins, or both, and when taken regularly they serve to inhibit ovulation. The exact mechanism of oral contraceptive action is uncertain, but it is believed that they act chiefly by inhibiting the hypothalamic secretion of FSH- and LH-releasing factors, thereby blocking pituitary gonadotropic activity.

The uterine tubes

Description.   The uterine tubes are known also as the *fallopian tubes* and *oviducts*. Each tube lies in the upper part of the broad ligament of its own side and is about 4½ inches long. The proximal end penetrates the uterine wall and opens into the uterine cavity. Distally the trumpet-shaped, free end is in intimate contact with the ovary; this extremity of the tube also is called the fimbriated end, due to the presence of fringelike projections or *fimbriae* (Fig. 21-7).

Structure and Function.   The lining of the uterine tube is a mucous membrane that is thrown into longitudinal folds; some of its epithelial cells are ciliated, and others are secretory. The middle coat is composed of circular and longitudinal layers of smooth muscle tissue that are responsible for peristaltic contractions of the tube. The outer coat is serous membrane (visceral peritoneum). An anatomic relationship of considerable clinical significance should be noted at this point. At the open end of the uterine tube the mucosal lining is continuous with the visceral peritoneum. Since the mucosa of the tube also is continuous with that of the uterus and the vagina, it is possible for harmful bacteria to enter the vaginal opening and ultimately cause an infection that could spread to the peritoneal cavity.

Unlike a spermatozoon, the oocyte is incapable of independent movement. Following ovulation, the oocyte enters the uterine tube and is moved along toward the uterus by peristaltic contractions of the tubal wall and by the sweeping action of the ciliated cells. During its passage, one of two courses is possible for the oocyte: (1) it may be fertilized (penetrated by a spermatozoon) in the tube; or (2) if not fertilized, it eventually disintegrates and disappears in the tube or in the uterus. If fertilized, the oocyte completes its mitotic division and becomes a mature *ovum*.

Normally, a fertilized ovum will reach the uterus in four or five days and continue to develop there for the duration of the pregnancy. If for any reason the fertilized ovum fails to reach the uterus, the woman is said to have an *ectopic* pregnancy. For example, an oocyte may remain attached to the ovary and be fertilized there; the result is an *ovarian pregnancy*. Even if the oocyte enters the tube and is fertilized there, its subsequent passage to the uterus may be prevented by adhesions. The result is a *tubal pregnancy*. In ec-

topic pregnancy there is great danger of fatal hemorrhage, because the uterus is the only place that is properly prepared to retain and sustain the growing embryo. Pregnancies that begin to develop elsewhere are a threat to both the new individual and the mother whose function it is to provide a proper home for the newcomer in her uterus.

The uterus

Description. The uterus is a hollow, thick-walled, muscular organ. Its cavity communicates with those of the uterine tubes above and with

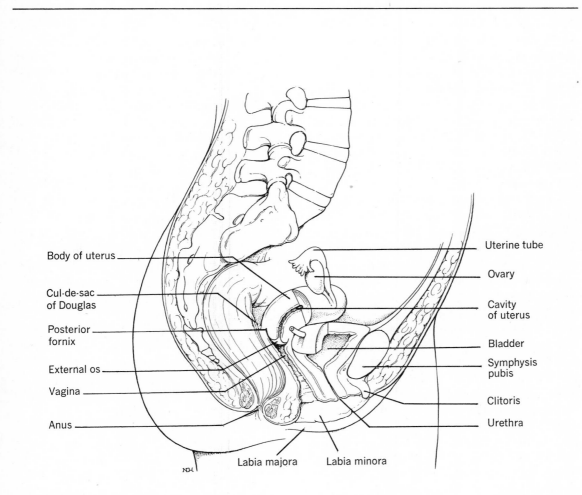

Fig. 21-9. Female reproductive organs as seen in sagittal section.

that of the vagina below. The uterus resembles an inverted pear that is somewhat flattened from front to back; this means that its cavity is slit-like with the anterior and posterior walls close together. The upper portion of the uterus is called the *body,* and the lower constricted portion is called the *cervix.* The portion of the body above the entrance of the uterine tubes is referred to as the *fundus* (Fig. 21-7).

The upper portion of the uterus is free and movable and rests on the upper surface of the urinary bladder (Fig. 21-9). Consequently the position of the uterus changes with the size of the bladder; when the latter is filled, the uterus may be tilted backward (retroversion). The lower portion of the uterus is embedded in the pelvic floor between the bladder and the rectum. The distal end of the cervix projects into the vagina where it terminates in thick, prominent lips that surround the external orifice (os) of the uterus.

Support. The muscles and the fascia of the pelvic diaphragm, particularly the *levator ani* constitute the principal support of the uterus. However, four paired ligaments offer an additional measure of support as they hold the uterus in its proper position in the pelvis. The principal ligaments of the uterus are the broad, cardinal, round, and uterosacral ligaments.

Each *broad ligament* is a transverse sheet of fibrous tissue, covered on both sides with peritoneum, that extends from the lateral border of the uterus across the pelvis to the pelvic wall and floor. It encloses the round and the ovarian ligaments, the uterine tube, the blood vessels, and the nerves of its side. Each *cardinal ligament* is a fibrous sheet of fascia that extends from the cervix and the vagina across the pelvic floor as the base of the broad ligament of its own side. Each *round ligament* is a flattened band of fibrous tissue that passes from the lateral border of the uterus, in front of and below the entrance of the uterine tube of that side, through the broad ligament to the lateral pelvic wall; then it passes through the inguinal canal in a course similar to that of the ductus deferens in the male and merges with the tissue of the external genitalia (labium majus). Each *uterosacral ligament* is a prominent band of fibrous tissue that curves along the lateral pelvic wall of its own side, from the cervix to the sacrum. It is a posterior continuation of the tissue which forms the cardinal ligament.

The Uterine Wall. The wall of the uterus is composed of three main layers: the peritoneum, the myometrium, and the endometrium.

The *peritoneum* or serosal covering of the uterus merges laterally with the peritoneum of the broad ligament. Anteriorly, it is reflected over the urinary bladder. Posteriorly, it forms a deep pouch as it is reflected on to the anterior surface of the rectum; this pouch, called the *cul-de-sac,* or *pouch of Douglas,* is the lowest point in the pelvic cavity (Fig. 21-9).

The *myometrium* consists of smooth muscle. It is very thick and forms the main part of the uterine wall. The size of the uterus is capable of great change; its capacity is about 2 to 5 ml. before pregnancy, and between 5,000 and 7,000 ml. at the height of pregnancy (term). It weighs about 60 Gm. at puberty and about 1,000 Gm. at the end of pregnancy, which is an increase of more than 16 times. The increase in size is due to both hyperplasia (increase in the number of cells) and hypertrophy (increase in size of original cells).

The change in size of the uterus during pregnancy truly is remarkable. It is due in part to hormones, both estrogen and progesterone, and in large measure to the distention by the growing fetus, which is a very effective stimulus to growth. There is also some stretching of the uterine wall, so the elastic and fibrous components of the wall are involved, as well as the smooth muscle. The shortening of the elastic fibers is a large factor in the retraction of the uterus after delivery.

Benign tumors, consisting largely of smooth muscle cells, may develop in the myometrium. These tumors are called *leiomyomas* of the uterus or, more commonly, *fibroids.*

The *endometrium* is a special mucous membrane that lines the uterus. It is composed of two chief layers: a thick, superficial, or *functional,* layer; and a thin, deep, or *basilar,* layer. The functional layer changes greatly during the menstrual cycle and is lost almost completely during menstruation. The basilar layer is not lost during the menstrual flow and remains to regenerate a new functional layer when menstruation ceases.

Blood Supply. Blood is supplied to the uterus by way of branches of the internal iliac and ovarian arteries (Fig. 13-13). After forming a plexus, the veins empty into the tributaries of the internal iliac and the ovarian veins. The blood supply of the superficial, or functional, layer is separate from that of the basilar layer of the endometrium; the former is by way of spiral, or coiled, arteries.

Functions. The initial functions of the uterus are to retain, and to sustain, the new individual during the first 40 weeks of growth and development. The final function of the uterus is to expel the fetus and the placenta (afterbirth) at the end of pregnancy.

The menstrual cycle

In a nonpregnant woman, the superficial layer of the uterine mucosa undergoes cyclic changes that are referred to as *menstrual cycles*. These changes are closely related to changes in the ovaries and involve a complex interplay between pituitary and ovarian hormones. Menstrual cycles recur from puberty to menopause, at intervals of 25 to 35 days, except when pregnancy and lactation intervene. A typical cycle can be described in terms of three broad phases: menstrual, follicular, and luteal.

The Menstrual Phase. The periodic discharge of bloody fluid from the uterine cavity is known as *menstruation*, or the *menses*. This discharge is associated with the shedding of the superficial layer of the endometrium; it consists of epithelial cells, mucus, interstitial fluid, and about 25 to 65 ml. of blood. Usually this blood is liquid, but if the rate of flow is excessive, clots may appear. Menstrual blood probably fails to clot because it first clotted in the interstitial spaces and was liquefied by fibrinolysins before escaping from the uterus.

Although menstruation actually is the terminal event of each cycle, it is the time most easily fixed by any woman; thus, in clinical practice, the first day of the menstrual flow is said to be the first day of the menstrual cycle. This flow may last from one to eight days, but the average duration is four to six days. It is associated with a progressive degeneration of the superficial layer of the endometrium, which is accompanied by patchy areas of bleeding. Fortunately, only one small area at a time is involved; if all areas of the endometrium broke down simultaneously there would be severe hemorrhage.

As patches of bleeding develop, pieces of the superficial endometrium become detached. The uterine glands release their accumulated secretions

and then collapse. The large collection of interstitial fluid that was present also is lost in the menstrual flow. This process spreads gradually until the entire superficial layer of epithelium has disintegrated. The endometrium now is very thin because only the basilar layer of epithelium remains; the latter has a separate blood supply and is left intact throughout the menstrual period.

Ovarian activity during the menstrual phase involves the beginning growth and development of several primary follicles. However, under the influence of pituitary FSH, usually only one follicle is selected for ultimate maturation.

Follicular, or Preovulatory Phase. This phase is associated with a rapidly growing ovarian follicle and the production of estrogen. These activities depend on a mixture of two pituitary gonadotropins: FSH and LH. Early in the phase, FSH is the predominating hormone, but it seems that neither FSH nor LH acting alone can cause the production of estrogen.

For approximately seven to ten days, or from the fourth to thirteenth day of a twenty-eight day cycle, there is a reorganization and proliferation of endometrial cells in response to the stimulating effect of estrogen. Even before the menstrual flow ceases, repair begins in those areas that first underwent destruction. Cells in the basilar layer undergo mitosis and aid in the regeneration of a new superficial layer; the covering epithelium is regenerated first, and then the superficial layer grows until it becomes at least three times as thick as the basilar layer; for this reason, the term *proliferative phase* also may be used in reference to this phase of the menstrual cycle.

As the ovarian follicle continues its growth and secretory activity, the rising estrogen tide causes a sharp shift in the FSH-LH mixture. As FSH secretion is inhibited, LH secretion is greatly increased and ovulation occurs. The follicular phase thus blends into the luteal phase.

Luteal, or Postovulatory Phase. Following ovulation, LH stimulates the development of a corpus luteum and the secretion of progesterone. A considerable quantity of estrogen also is produced by the luteal cells, but progesterone is primarily responsible for the final preparation of the endometrium to receive a fertilized ovum.

This phase of the cycle also is called the *secretory phase* inasmuch as the lumens of the endometrial glands become filled with an abundance of secretions. The superficial endometrium becomes highly vascular, succulent, and rich in glycogen. Thickening of the superficial layer also involves an increase in the amount of interstitial fluid (edema). These changes are maximal about one week after ovulation, corresponding to the expected time of arrival of a fertilized ovum.

If fertilization does not occur, the rising progesterone and estrogen tide serves to inhibit the secretion of LH. As a result, the corpus luteum begins to degenerate, estrogen and progesterone levels decline, and menstruation occurs as the superficial layer of the endometrium is deprived of its hormonal stimulation. On the other hand, if a fertilized ovum is implanted in the uterus, the corpus luteum continues to secrete progesterone for about three months. This is possible because of a hormone (chorionic gonadotropin) produced by the developing embryonic tissues;

this hormone is very similar to LH and serves to maintain the corpus luteum until the embryonic tissues are capable of producing progesterone and estrogen on their own.

The *premenstrual phase*, or late luteal phase, occupies the two or three days immediately preceding menstruation and corresponds to the degeneration of the corpus luteum. At this time there is an infiltration of polymorphonuclear leukocytes and degeneration of the connective tissue stroma of the superficial endometrium begins. Important vascular changes also occur just before the onset of the menstrual flow. Arteries that supply the basilar part of the endometrium are not affected by the decreasing supply of estrogen and progesterone, but the special coiled arteries that supply the superficial layer become markedly constricted. As a result, the overlying endometrium is deprived of blood and has an extremely pale

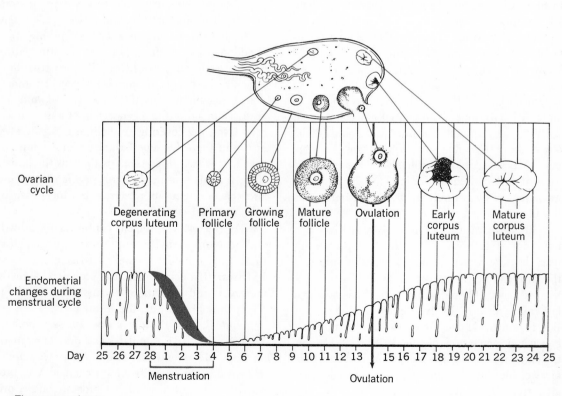

Fig. 21-10. Schematic representation of one ovarian cycle and the corresponding changes in thickness of the endometrium; it is thickest just before the onset of menstruation and thinnest just as it ceases.

appearance. The vasoconstrictive process apparently damages the arteriolar and capillary walls, because when the coiled arteries eventually dilate blood escapes through the walls and accumulates in pools beneath the endometrial surface. This blood soon breaks through the degenerating endometrial stroma, and the menstrual flow begins again. The appearance of the endometrium at various times in the menstrual cycle is shown in Figure 21-10.

Summary. The *menstrual phase* is associated with the shedding of the superficial layer of the endometrium. This has been referred to as a "cleaning of the domicile when the expected tenant failed to arrive."

The *follicular,* or *preovulatory phase,* is associated with the reorganization and proliferation of the endometrium following menstruation. Estrogen is essential for this early development of the uterine mucosa, and it is produced by a rapidly growing ovarian follicle under the influence of a mixture of FSH and LH. However, FSH is the predominating hormone. Eventually the rising estrogen tide inhibits the secretion of FSH and increases the secretion of LH, thus promoting ovulation.

The *luteal,* or *postovulatory phase,* is associated with the formation of a corpus luteum and the production of progesterone and estrogen under the influence of LH. Progesterone is primarily responsible for the final buildup of the endometrium preparing to receive a fertilized ovum. The rising progesterone and estrogen tide shuts off the secretion of LH, and the corpus luteum degenerates. If the oocyte has not been fertilized, the menstrual flow begins again as the hormone level falls. However, if a fertilized ovum is implanted in the uterus, the corpus luteum is maintained by a hormone secreted by the developing embryonic tissue.

Thus ovarian hormones are responsible for the cyclic changes in the uterine mucosa and for those changes that are essential to the maintenance and the development of the embryo, particularly in the early period of pregnancy. The ovaries, in turn, are controlled by the pituitary gonadotropins.

The changes in the uterine mucosa, in relation to the events in the ovaries and the anterior lobe of the pituitary gland, are shown in Table 21-1.

Table 21-1 Correlation of Hormonal Activities with Ovarian and Uterine Changes

Phase	Menstrual	Follicular	Ovulation	Luteal	Premenstrual
DAYS 1 2 3 4 5 6 7 8 9 10 11 12 13 14 15 16 17 18 19 20 21 22 23 24 25 26 27 28 1 2					
Ovary	Degenerating corpus luteum; Beginning follicular development	Growth and maturation of follicle	Ovulation	Active corpus luteum	Degenerating corpus luteum
Estrogen production	Low	Increasing	High	Declining, then a secondary rise	Decreasing
Progesterone production	None	None	Low	Increasing	Decreasing
FSH production	Increasing	High, then declining	Low	Low	Increasing
LH production	Low	Low, then increasing	High	High	Decreasing
Endometrium	Degeneration and shedding of superficial layer. Coiled arteries dilate, then constrict again	Reorganization and proliferation of superficial layer	Continued growth	Active secretion and glandular dilatation; highly vascular; edematous	Vasoconstriction of coiled arteries; beginning degeneration

Puberty and the menarche

Extensive growth and development of ovarian follicles does not occur until the onset of *puberty*, at 10 to 14 years of age. At this time, the secondary sexual characteristics begin to develop. The breasts, uterus, and vagina start to mature; the pubic and axillary hair begin to grow, and there is a general rounding of the female body contour as fat is deposited.

Puberty terminates at *menarche*, the onset of the first menstrual period; the average age at menarche is 13 to 15 years, but it may range from 10 to 18 years. In general, the early cycles are irregular and many of them may be anovulatory (i.e., follicles may fail to rupture).

Puberty is initiated by a marked increase in the secretion of pituitary gonadotropins. However, no one yet knows precisely what triggers pituitary activity. There is some evidence that the pineal gland (p. 477) acts as a biological clock in this respect, but the mechanism has not been determined.

The climacteric and menopause

Between 45 and 50 years of age, a woman may expect to enter the period known as change of life. The critical aspect of this period is the gradual failure of the ovaries to respond to pituitary gonadotropin stimulation. *Climacteric* is a broad term that covers the entire period of gradual ovarian failure, and its highest point is the complete cessation of menstruation, or the *menopause*. Senile ovaries no longer produce ova or secrete hormones, and the period of possible childbearing is over. There is atrophy of the ovaries, uterine tubes, uterus, vagina, external genitalia, and the breasts. During the early years of the menopause there may be various subjective symptoms such as hot flashes and outbursts of sweating due to instability of the vasomotor system. There may be headache, vague muscular pains, and, occasionally, emotional instability. Estrogen therapy frequently is helpful in relieving symptoms of the menopause.

The vagina

The vagina is a collapsible, musculomembranous tube, about 3½ inches long, lying between the bladder and the urethra, anteriorly, and the rectum, posteriorly. As shown in Figure 21-9, it extends downward and forward from the uterus to the external opening in the vestibule.

The upper end of the vagina is attached to the cervix a short distance above the projecting cervical lips, thus forming recessed areas. The deep recess behind the vaginal end of the cervix is called the *posterior fornix*, while the smaller recesses in front and at the sides are called the *anterior* and the *lateral fornices*. At the lower end of the vagina there is a thin fold of mucous membrane called the *hymen*; this fold shows great variation in shape, but usually it forms a circular border around the external vaginal orifice as shown in Figure 21-11. Occasionally the hymen may completely cover the vaginal opening (imperforate hymen), and surgery is required to permit the escape of the menstrual flow.

The wall of the vagina is composed of smooth muscle and fibroelastic connective tissue, lined with mucous membrane, which is thrown into

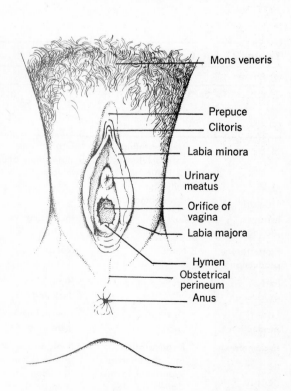

Fig. 21-11. External genitalia of female.

Mons veneris

Prepuce
Clitoris

Labia minora

Urinary meatus

Orifice of vagina

Labia majora

Hymen
Obstetrical perineum
Anus

transverse folds called *rugae*. The epithelium of the mucous membrane is stratified squamous, nonkeratinizing in type.

The vagina serves as the excretory duct of the uterus and provides a sheath for the penis for sperm deposition during sexual intercourse.

The *Papanicolaou test* (Pap smear) is of considerable clinical importance in the early diagnosis of carcinoma of the uterus. This simple, painless test involves the microscopic examination of cells taken from the posterior fornix of the vagina and from the cervix of the uterus; since malignant or cancerous cells have a characteristic appearance, they may be detected very early in the course of the disease and long before symptoms appear.

The external genitalia

The external genital organs (vulva) of the female are illustrated in Figure 21-11.

The *mons pubis*, or *mons veneris*, is the rounded pad of fat in front of the pubic symphysis. It is covered by dense skin and, at puberty, becomes covered with hair.

The *labia majora* are two prominent longitudinal folds of skin and underlying fat that extend backward from the mons pubis toward the anus. The skin of the labia majora contains numerous hair follicles, sweat glands, and sebaceous glands.

The *labia minora* are two small folds of skin that are situated between the labia majora. Anteriorly, the folds meet to form the *prepuce*, a small hood of skin that partially covers the clitoris; the folds then extend backward on either side of the urethral and the vaginal orifices, following much the same course as the labia majora. There are no hair follicles or sweat glands in the labia minora.

The *clitoris* is a small structure that is composed of erectile tissue, corresponding in structure and in origin with the penis in the male. It lies under cover of the prepuce as described previously.

The *vestibule* is the cleft between the labia minora and behind the clitoris. The vagina and the urethra open into this area. The external opening of the urethra is located about 1 inch behind the clitoris and immediately in front of the vaginal orifice. The *greater vestibular glands* (Bartholin's glands) are situated in the floor of

the vestibule, one on either side of the vaginal orifice. These glands secrete a mucoid lubricating fluid, and each drains into a duct that empties into the space between the hymen and the labium minus of its own side.

The *obstetric perineum* (perineal body) is the area between the vagina, anteriorly, and the anal canal, posteriorly (Fig. 21-11). It is composed of a mass of muscular and fibrous tissue. To avoid tearing and undue stretching of these tissues during childbirth, an incision, called *episiotomy*, may be made in the perineum prior to delivery of the infant. The term *anatomic perineum* refers to the area between the pubis, anteriorly, the tip of the coccyx, posteriorly, and the thighs, laterally.

The mammary glands

The mammary glands, or breasts, are functionally related to the reproductive system since they secrete milk for nourishment of the young, but, structurally, they are related to the skin. Lying on the ventral surface of the thorax, each breast extends from the second to the sixth rib, and from the sternum to the anterior border of the axilla of its own side. The size is variable, but each is hemispherical in shape. The posterior surface of the breast is concave as it is molded over the chest wall, primarily in contact with the fascia of the pectoralis major muscle.

Structure. The breast is composed of 15 to 20 lobes, each having its own individual excretory, or *lactiferous duct*. Each lobe consists of lobules of glandular tissue supported by a connective tissue framework that contains a variable amount of adipose tissue. The lobules are drained by intralobular ducts that empty into the main lactiferous ducts.

On the ventral surface of each breast, slightly below center, there is a cylindrical projection or *nipple*. In its rounded tip are 15 to 20 perforations, the openings of the lactiferous ducts. The nipple is surrounded by a pigmented area called the *areola*, which becomes darker during pregnancy. The structure of the breast is shown in Figure 21-12.

Blood is supplied to the mammary glands by way of the internal mammary arteries and is drained into the corresponding veins. Lymphatic vessels drain into the axillary lymph nodes, a fact of considerable clinical importance in the metas-

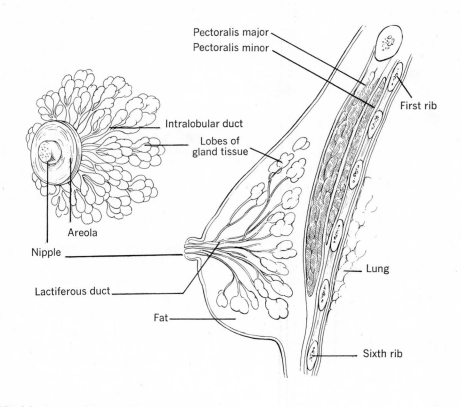

Fig. 21-12. Glandular tissue and ducts of the mammary gland.

tasis (spread) of carcinoma of the breast.

Changes in the breasts during pregnancy and lactation are described on page 517.

SEQUENCE OF EVENTS IN REPRODUCTION

The process of reproduction really begins with the development and the maturation of germ cells of the parents, then progresses through ovulation, fertilization, implantation, development of the ovum through the embryonic and the fetal phases, and, eventually, birth (parturition), or delivery, of the new individual from the uterus to the external world.

The maturation of parental germ cells and the process of ovulation were described earlier in this chapter. Let us now consider the processes of fertilization, implantation, and the development of a new individual during the intrauterine period

of life. For more details on these subjects one should consult a standard textbook on embryology.

Fertilization

Fertilization consists of the penetration of a secondary oocyte by a spermatozoon. Although millions of spermatozoa are deposited in the vagina, only one effects fertilization; the others ultimately disintegrate. Fertilization usually takes place in the distal third of the uterine tube.

Union of the oocyte and spermatozoon restores the diploid number of chromosomes (46) and determines the sex of the offspring, as well as all of its other hereditary complements. Because the oocyte always has an X sex chromosome, it is the spermatozoon (which may carry an X or a Y chromosome) that ultimately determines the sex of the offspring. If an X chromosome-bearing spermatozoon unites with the oocyte, the resulting offspring will be a female (XX). On the other hand, the union of an oocyte and a Y chromosome-bearing spermatozoon will result in a male (XY).

There are a number of disorders associated with the sex chromosomes. These may be the result of the kind of information carried by individual genes on the chromosomes, or they may be the result of abnormalities in the number of chromosomes. Sometimes in the process of spermatogenesis, or oogenesis, the chromosomes tend to stick together so that they do not separate normally. This condition is known as *nondisjunction* and may occur with the autosomes or with the sex chromosomes.

In the most common type of mongolism (*Down's syndrome*, or trisomy 21), for example, during oogenesis, or spermatogenesis, two number 21 chromosomes stick together so that both of them appear in the mature ovum or sperm. If an ovum with two number 21 chromosomes is fertilized by a normal sperm, the resulting zygote will have 47 chromosomes with *three* number 21 chromosomes instead of the normal two. Nondisjunction of chromosome 21 can also occur during spermatogenesis so that the fertilization of a normal ovum by a sperm with two number 21 chromosomes would also result in trisomy 21. The primary manifestation of the extra genetic information carried on chromosome 21 is mental deficiency. However, these individuals also show other typical signs such as a round, full face, a large protruding tongue, hypotonia, other congenital defects (especially heart disorders), and the typical epicanthal folds which make their eyes appear slanted. In fact, this disorder was originally called mongolism because of the appearance of the eyes.

Nondisjunction may also occur with the sex chromosomes. When this occurs during oogenesis, both of the X chromosomes may appear in the mature ovum, or both of them may go into the polar body. Thus the mature ovum may contain XX, or it may have no X chromosome at all. If a mature ovum with XX is fertilized by a sperm with a Y chromosome, the zygote will have 47 chromosomes, 22 pairs of autosomes and XXY. Such a chromosome arrangement results in the condition known as *Klinefelter's syndrome*. This individual has male sex organs, but he is usually sterile and often shows female breast development (gynecomastia). If the ovum with XX is fertilized by a sperm with an X chromosome, the individual also has 47 chromosomes, 22 pairs of autosomes but XXX. This type of individual has been called a super female, but she is usually abnormal both physically and mentally. These super females also tend to be sterile.

On the other hand, if the mature ovum that received neither of the X chromosomes is fertilized by a normal sperm with an X chromosome, the individual would have only 45 chromosomes, 22 pairs of autosomes and XO. This condition is known as *Turner's syndrome* and results in a female of short stature who evidences various skeletal abnormalities; in addition, she has juvenile sex characteristics and the sexual organs of a female, but her ovaries are missing, or, if present, are incapable of producing ova. Although it is theoretically possible to have an individual with a YO chromosome complement, no such individuals have been found; it is believed that this chromosome arrangement is lethal and the zygote never develops.

Other derangements in the number of sex chromosomes have been discovered. In most instances, an abnormal number of chromosomes results in an individual who is abnormal, or, if capable of reproducing, has children who are abnormal. Individuals have been found who show XXY, XXXX, XXXXY, XXYY.

These abnormalities in the number of chromosomes can be detected by an examination of a *karyotype*. Figure 21-13 shows a karyotype of a normal male with an X chromosome in group C and a Y chromosome in Group G. These karyotypes are produced by growing certain cells, usually lymphocytes, in a special medium and stopping the cell division at metaphase. The chromosomes which are visible under the microscope at this stage are then photographed, cut out, and arranged in groups according to their size and form.

Techniques are now available which enable the geneticist to prepare a karyotype of an unborn fetus. Cells in the amniotic fluid are removed, grown, and harvested. The karyotype can then be produced in the same manner as the karyotype made from lymphocytes. Extra chromosomes in the developing fetus can be readily detected by examining this karyotype. The genetic counselor then can advise the prospective parents of the problems that can be expected and may offer termination of the pregnancy if they so desire.

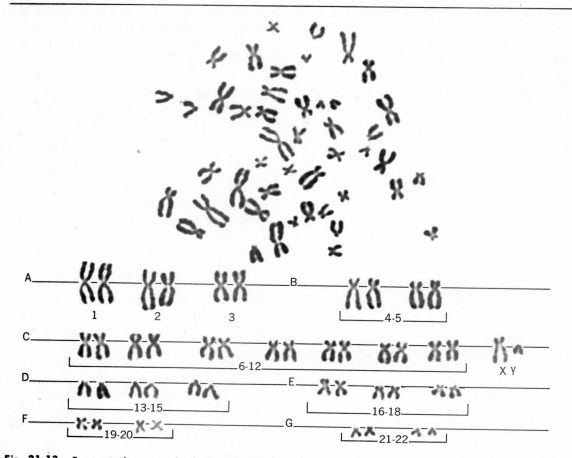

Fig. 21-13. *Top, metaphase spread.* Chromosomes of a normal male photographed during metaphase when they have doubled and are ready to divide. *Bottom, karyotype.* The chromosomes from the metaphase spread, above, arranged in matching pairs according to size and other characteristics. (Courtesy of Kenneth L. Garver, M.D. and Angela M. Ciocio, PhD.)

Other inherited diseases transmitted by the presence of specific genes on the various chromosomes, including the sex chromosomes, cannot be determined by an examination of the karyotype and, therefore, pose a greater problem for the genetic counselor. In many cases, however, the genetic counselor, from his knowledge of the way in which a disease is inherited and from a detailed history of the families involved, can predict the risk of a couple having a child with a particular inherited disease. Many large medical centers now have genetic counseling services to apprise prospective parents of the risk of having a defective child when an inherited disease has appeared in the family of either of the parents.

Implantation

The fertilized ovum (zygote) begins imme-

diately to divide by ordinary mitosis, and passes rapidly through a two-celled, four-celled, eight-celled stage, and so on, until a tiny mulberry-like ball of cells (morula) is formed. The developing mass is moved along through the uterine tube and reaches the cavity of the uterus about three to four days after fertilization. A gradual accumulation of fluid within the morula results in the formation of a cavity, and the dividing cells now differentiate into an outer layer or shell (the trophectoderm or *trophoblast*) and an inner compact mass of cells that is destined to become the embryo (Fig. 21-14). This fluid-filled ball of cells, called a *blastocyst*, adheres to the surface of the endometrium and is ready to implant. That portion of the endometrium to which the blastocyst adheres, and in which it will implant, is called the *decidua basalis*.

The cells in the outer layer (trophoblast) secrete proteolytic enzymes that digest the adjacent endometrial cells and literally eat a hole in the uterine lining. The blastocyst nestles in this hole and buries itself; the superficial wound is closed by a fibrin clot and by a portion of endometrium, the *decidua capsularis*, that is reflected over it. The remainder of the endometrium is referred to as the *decidua parietalis*.

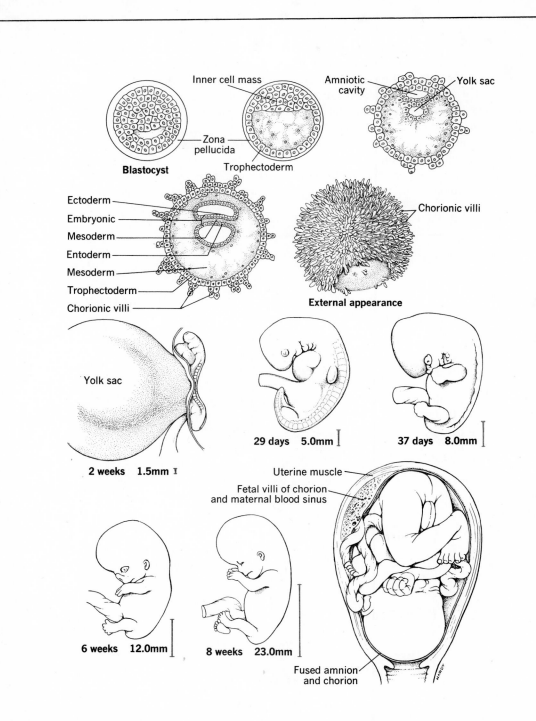

Fig. 21-14. Various stages of development from blastocyst to fetus. Perpendicular lines indicate actual size of embryo in millimeters.

Development of the placenta and membranes

Following implantation, the trophoblast proliferates rapidly, and little fingerlike cords of cells, called villi, invade the surrounding decidua; the resulting disintegrating, liquefying, mucosal components serve as food for the embryo during this early stage of development. As the decidual invasion proceeds, maternal blood vessels are tapped, and a complicated labyrinth of blood-filled lakes, or sinuses, forms in the decidua basalis; this comprises the maternal portion of the placenta.

Meanwhile embryonic blood vessels are developing in the trophoblast, and, as soon as it is vascularized, this structure becomes the *chorion*. Another fetal membrane, the *amnion*, develops around the growing embryo, which is connected with the chorion by the body stalk, as shown in Figure 21-15. The villi in this region enlarge to form the *chorion frondosum* which, in turn, forms the fetal portion of the placenta. Villi on other parts of the chorion cease to grow and undergo almost complete degeneration. Thus denuded of villi, the greater part of the chorion presents a smooth surface and is called the *chorion laeve*. The body stalk ultimately becomes the umbilical cord of the fetus; two umbilical arteries and one umbilical vein convey fetal blood to and from the placenta.

In summary, the placenta consists of maternal and fetal portions: the decidua basalis and the chorion frondosum, respectively (Fig. 21-15). Each part has its own separate circulation, and

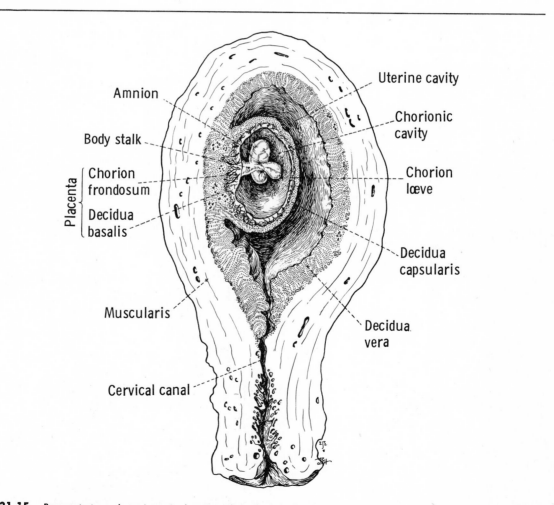

Fig. 21-15. Pregnant uterus shown in sagittal section. The embryo is about one month of age.

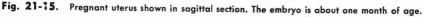

there is no communication between the two (Fig. 21-16). The uterine arteries convey blood to the maternal lakes, or sinuses, and the uterine veins carry blood away. Fetal circulation is illustrated in Figure 13-25 and described on page 318.

The functions of the placenta. This highly vascular organ serves as the respiratory, nutritive and excretory organ of the fetus. Oxygen and nutrients, provided by the mother, diffuse from the maternal blood sinuses through the walls of the villi into the fetal bloodstream. At the same time, waste products produced by the fetus diffuse in the opposite direction—i.e., from the fetal bloodstream into the maternal sinuses—and ultimately are eliminated by the mother.

The intrauterine period

The intrauterine, or *prenatal period*, lasts about 40 weeks, or 10 lunar months (28 days each). It may be subdivided into the period of the ovum, the period of the embryo, and the period of the fetus.

Period of the Ovum. The duration of this period is from 0 (the moment of fertilization) to two weeks. Implantation occurs on the sixth or seventh day, and primitive villi are formed after implantation.

Period of the Embryo. This period extends from the beginning of the third week to the end of the eighth week (second lunar month). The chorionic villi are well developed, both fetal and maternal blood vessels are functioning, and a true placental circulation is established. During this time the rudiments of all the adult main organs are evolved.

Period of the Fetus. The duration of this period is from the beginning of the ninth week to birth, at the end of the 40th week. Few, if any, new major structures are formed, and the development that takes place during this period consists of the growth and maturation of existing structures. The fetus is distinguishable as human at the beginning of this period, and by the end of the 16th week (fourth lunar month) the external genitalia definitely reveal the sex.

By the end of the 24th week (sixth lunar month), the fetus weighs about 600 Gm. The skin is wrinkled, and the first fat deposits are made beneath it. By the end of the 32nd week (eighth lunar month), the skin still is red and wrinkled.

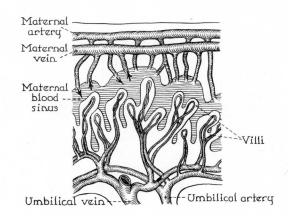

Fig. 21-16. Diagram of placental circulation. Note that maternal and fetal circulations are completely separate.

However, the fetus weighs about 1,700 Gm., cries weakly and moves its extremities quite energetically. A fetus born at this time may survive, with proper care. By the 36th week, the average weight is about 2,500 Gm., and the fetus has a very good chance of survival if born at this time.

Pregnancy reaches full term at the end of the 40th week (tenth lunar month). At this time the fetus is fully developed and has the characteristic features of the newborn infant. Various stages in the development of the embryo and the fetus are shown in Figures 21-14 and 21-17.

Maternal changes

We now return to the mother in whose uterus the above activities have been taking place. From the beginning of pregnancy, the uterus enlarges to keep pace with the growth of the fetus. By the end of the 16th week, the fundus of the uterus is about halfway between the symphysis pubis and the umbilicus. By the end of the 24th week, the fundus has reached the level of the umbilicus. At the end of the 36th week, the fundus has risen to the level of the xiphoid process of the sternum. But at this time, the fetus settles into the pelvis, and the fundus falls a little below the xiphoid process during the last 4 weeks of pregnancy.

Throughout the entire 40 weeks of pregnancy, the mother carries, protects, and nourishes the new life until it is ready for extrauterine existence. By the 20th week she can feel fetal life as the

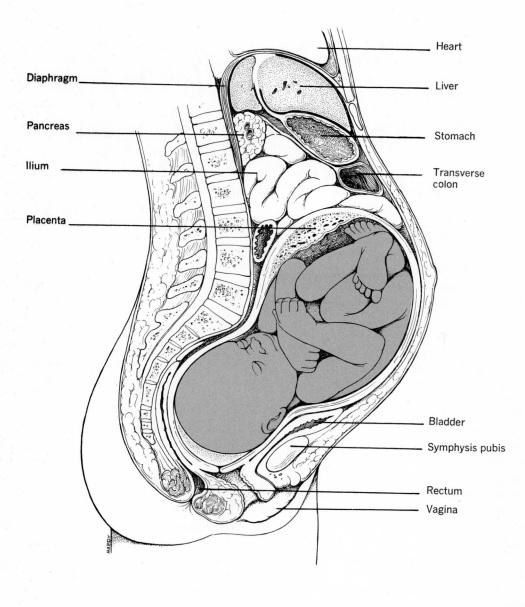

Diaphragm

Pancreas

Ilium

Placenta

Heart

Liver

Stomach

Transverse
colon

Bladder

Symphysis pubis

Rectum

Vagina

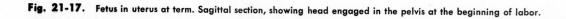

Fig. 21-17.   Fetus in uterus at term. Sagittal section, showing head engaged in the pelvis at the beginning of labor.

fetus moves in her uterus. The fetal heart sounds also can be heard at this time if a stethoscope is placed over the mother's abdominal wall.

During the intrauterine period, the mother provides oxygen and eliminates carbon dioxide for the fetus. In addition, she furnishes food, water, and inorganic salts; inasmuch as developing bones and teeth require large quantities of calcium, the mother must pay particular attention to this item in her diet. The elimination of fetal nitrogenous waste products places an extra load on the mother's kidneys; thus frequent urine analyses and blood pressure determinations are advised during pregnancy to detect any signs of kidney failure or toxicity.

Acute toxemia of pregnancy is characterized by hypertension, edema, and proteinuria. A very severe form of toxemia, referred to as *eclampsia,* is characterized by convulsions; it is one of the most dangerous complications of pregnancy. Hospitalization, bed rest, and sedation are essential in the care of mothers with this condition. Termination of the pregnancy may be necessary to save the mother's life.

Labor, or parturition

At the end of pregnancy, about 40 weeks after the last menstrual period, the uterus begins to undergo rhythmic contractions that ultimately lead to the birth of the infant. The sequence of events involved in this process is referred to as labor or parturition.

The *first stage* of labor consists of *dilatation* of the cervix. With each uterine contraction the amniotic sac of fluid that surrounds the fetus is forced into the cervical canal where it acts as a wedge to dilate the canal. In a typical delivery the amniotic sac ruptures just at the close of this stage, and the fluid escapes.

When the dilatation of the cervical canal is completed, the *second stage* of labor begins. This refers to the *delivery of the infant,* normally head first, through the cervical canal and vagina. It is at the time of the delivery of the head that tears may occur in the pelvic floor. Many obstetricians perform an episiotomy (p. 509) to prevent spontaneous tears; this incision is repaired before the mother leaves the delivery room.

The *third stage* of labor consists of the *delivery of the placenta,* or afterbirth. After this the uterus rapidly decreases in size (involution), primarily due to the contraction of smooth muscle fibers of the myometrium. This tends to close the open spaces of the endometrium or denuded walls of the uterus and prevent bleeding. A fatal hemorrhage can occur if the uterine muscle relaxes at this time.

Cesarean Section. If the diameters of the mother's pelvis are so small that spontaneous delivery seems to be impossible, or if the fetus is in distress, delivery is accomplished by cesarean section. This consists in removing the fetus by the abdominal route. An incision is made through the abdominal and uterine walls; then the fetus is manually removed from the uterus.

Lactation

Estrogens produce growth of the duct system of the mammary glands, and progesterone stimulates development of the secretory tissue. The mammary glands reach their final stage of development at the end of pregnancy. Prolactin, or lactogenic hormone of the anterior lobe of the pituitary gland, stimulates the production of milk in the mammary glands that have been prepared by estrogen and progesterone. The mammary glands secrete milk for the nourishment of the newborn infant. The process is called *lactation.*

Changes begin to take place in the mammary glands as soon as the developing zygote (fertilized ovum) is implanted in the uterus. During the first half of pregnancy the epithelium of the lobes increases rapidly, and secretory portions are formed. During the latter half of pregnancy the cells begin to produce a secretion that is called *colostrum.* This is the fluid obtained from the breasts before the secretion of true milk begins. It may be expressed from the breasts immediately after delivery and is ingested by the newborn during the first two days of postnatal life.

For the maintenance of secretion, removal of milk is absolutely essential. If milk is not removed (e.g., breast-feeding) the swelling of the breasts eventually disappears, milk ceases to be produced, and the mammary glands undergo involution. Under normal conditions, milk production continues for six to nine months after the infant is born.

Extrauterine life

Following parturition, the extrauterine, or postnatal existence, begins. This may be subdivided into various periods or phases.

The *neonatal period,* or period of the newborn, extends from the moment of birth to the end of the fourth week (28th day of extrauterine life).

The *period of infancy* extends from the beginning of the fifth week to the end of the first year.

The *period of childhood* extends from the beginning of the second year to puberty.

The *period of adolescence* extends from puberty to the late 'teens in girls and to the early twenties in boys.

The *period of maturity* extends from the end of the adolescent period to old age, or to the age of senility.

SUMMARY

1. Male

A. Testes
   a. Located in scrotum; descend through inguinal canal
   b. Lobules contain seminiferous tubules; spermatogenesis; directly influenced by FSH, indirectly by ICSH; 22 autosomes, X or Y sex chromosomes
   c. Interstitial cells secrete testosterone; essential for maintenance of sex characteristics and functional system

B. Duct system
   a. Epididymis: receives sperm from testes
   b. Ductus deferens (vas): from epididymis to ejaculatory ducts behind bladder; travels in spermatic cord with blood and lymph vessels and nerves
   c. Ejaculatory duct: formed by union of ductus deferens and duct from seminal vesicle on same side; empties into prostatic urethra

C. Accessory structures
   a. Seminal vesicles: secrete alkaline substance that aids motility of sperm
   b. Prostate gland: surrounds proximal part of urethra; muscular tissue aids in ejaculation; glandular tissue secretes alkaline substance that aids motility of sperm
   c. Bulbo-urethral glands (Cowper's): open into urethra; alkaline secretion neutralizes acidity of urethra
   d. Penis: external genital organ; composed of erectile tissue; distal end (glans) covered by prepuce; tissue spaces become turgid when distended with blood (erection), controlled by sacral autonomics
   e. Semen: consists of sperm and secretions from accessory glands; ejected into vagina by muscular contractions; average 100 million sperm per ml.; average ejaculate 2 to 4 ml.; only one sperm can fertilize ovum

2. Female

A. Ovaries
   a. Located lateral to uterus, attached to posterior surface of broad ligament, inferior to tubes
   b. Ovarian follicles formed in cortex; influenced by FSH, usually only one follicle and oocyte mature each month; 22 autosomes, only X sex chromosome; follicles secrete estrogen; mature follicle bulges on surface of ovary
   c. Ovulation: LH; release of oocyte from follicle; occurs every 28 to 30 days from puberty to menopause except during pregnancy
   d. Corpus luteum forms at site of ovulation; cells secrete progesterone and estrogen; if oocyte fertilized, corpus luteum persists for about three months; if not fertilized, corpus luteum degenerates; controlled by LH

B. Uterine tubes
   a. Lie in upper part of broad ligament; lateral end in contact with ovary; fimbriae at distal end
   b. Receive oocyte from follicle; place where fertilization occurs when sperm present; oocyte moved by peristalsis and ciliated cells; if fertilization does not occur oocyte disintegrates

C. Uterus
   a. Hollow muscular organ in pelvis between bladder and rectum; upper portion called body; lower portion called cervix projects into vagina
   b. Supported by muscles of pelvic floor and by ligaments: broad, cardinal, round, and uterosacral
   c. Posterior fold of peritoneum forms cul-de-sac, or pouch of Douglas; myometrium consists of smooth muscle; endometrium is special mucosal lining composed of functional and basilar layers; functional layer lost during menstruation, regenerated by basilar layer

D. The menstrual cycle
   a. Menstrual phase: flow lasts one to five days; superficial or functional layer of endometrium is lost
   b. Follicular or preovulatory phase: regeneration of superficial layer; follicle grows rapidly during this phase, secretes estrogen essential for preliminary buildup of lining; ovulation occurs at the end of this phase
   c. Luteal phase: progesterone from corpus luteum promotes final buildup of endometrium which becomes thick and engorged with fluid; if oocyte not fertilized, corpus luteum begins to degenerate, and menstrual flow begins again
   d. Correlation of hormonal activities: see Table 21-1 on p. 507.

E. Puberty and menarche
   a. Puberty: secondary sex characteristics develop; reproductive organs mature
   b. Menarche: onset of first menstrual period; end of puberty

F. Climacteric and menopause
   a. Climacteric: period of gradual ovarian failure; atrophy of reproductive organs
   b. Menopause: complete cessation of menstruation; highlight of climacteric

G. Vagina
   a. Collapsible, musculomembranous tube; between bladder and urethra in front and rectum behind;

extends from uterus to vestibule; recesses around vaginal end of cervix called fornices; excretory duct of uterus

    b. Hymen is circular border of mucosa around external vaginal opening.

H. External genitalia: mons pubis, labia majora, labia minora, clitoris, vestibule, perineum

I. Mammary glands

    a. Located on ventral thorax, second to sixth rib, sternum to axilla, overlying pectoralis major muscle

    b. Composed of lobes and lobules of glandular tissue with adipose tissue in framework of connective tissue

    c. Intralobular ducts drain into main lactiferous ducts; milk provides the nourishment for the young

    d. Ducts open onto surface of nipple which is surrounded by pigmented skin called areola

3. *Sequence of events in reproduction*

A. Maturation of germ cells

B. Ovulation

C. Fertilization: penetration of oocyte by sperm

    a. Determination of sex; X and Y chromosomes

    b. Abnormal numbers of chromosomes; Down's, Kleinfelter's, Turner's syndromes.

    c. Karyotype of developing fetus; genetic counseling

D. Implantation embedding of blastocyst in decidua basalis

E. Development of placenta and membranes

    a. Maternal portion: decidua basalis

    b. Fetal portion: chorion frondosum

    c. Fetal membranes: chorion and amnion

F. Intrauterine period lasts forty weeks; period of the ovum; period of the embryo; period of the fetus

G. Maternal changes; enlarging uterus to keep pace with fetus; life felt and fetal heart sounds heard by twentieth week; extra load placed on mother to provide nutrients and remove wastes

H. Labor or parturition

    a. First stage: dilatation of cervix

    b. Second stage: delivery of newborn

    c. Third stage: delivery of placenta

I. Lactation

    a. Estrogens stimulate growth of duct system; progesterone stimulates the secretory tissue

    b. Mammary glands are fully developed at the end of pregnancy; prolactin or lactogenic hormone of pituitary stimulates milk production or lactation

    c. Colostrum is secretion that precedes true milk

    d. Removal of milk essential for continued production

J. Extrauterine life: neonatal period, period of infancy, childhood, adolescence, maturity

QUESTIONS FOR REVIEW

1. What are the male gonads? Where are they formed? What is the significance of their descent into the scrotum?

2. In gametogenesis, what is the significance of reduction division?

3. How do spermatozoa obtain most of their energy? Why is this important?

4. Specifically, where is testosterone produced? What are its functions?

5. List, in proper order, the structures through which spermatozoa must pass to reach an oocyte in the uterine tube.

6. What is the function of the seminal vesicles? the prostate?

7. What is the role of FSH in oogenesis?

9. Specifically, where is estrogen produced and what are some of its functions?

10. What is the source of progesterone and what are its functions?

11. How is it possible for a vaginal infection to result in peritonitis?

12. What are the functions of the uterus?

13. Briefly, describe the three broad phases of a menstrual cycle.

14. Distinguish between the following: menarche, climacteric, and menopause.

15. What is fertilization? What determines the sex of the offspring?

16. What is the meaning of nondisjunction of chromosomes? What is mongolism?

17. What are the functions of the placenta?

18. What are the three stages of labor?

19. Distinguish between estrogen, progesterone, and prolactin as they effect the mammary glands.

REFERENCES AND SUPPLEMENTAL READINGS

Eastman, N. J., and Hellman, L. M.: William's Obstetrics, ed. 13. New York, Appleton-Century-Crofts, 1966.

Gardner, E. J.: Principles of Genetics, ed. 4. New York, John Wiley & Sons, 1972.

Goss, C. M. (ed.): Gray's Anatomy of the Human Body, ed. 28. Philadelphia, Lea & Febiger, 1966.

Guyton, A. C.: Textbook of Medical Physiology, ed. 4. Philadelphia, W. B. Saunders, 1971.

Ham, A. W.: Histology, ed. 6. Philadelphia, J. B. Lippincott, 1969.

Moody, P. A.: Genetics in Man. New York, W. W. Norton & Co., 1967.

Scheinfeld, A.: Heredity in Humans. Philadelphia, J. B. Lippincott, 1972.

Tepperman, J.: Metabolic and Endocrine Physiology, ed. 2. Chicago, Year Book Medical Publishers, 1968.

Glossary

Abdomen (ab-do'men). The part of the body extending from the diaphragm above to the pelvis below.

Abortion (ab-or'shun). The emptying of the uterus prior to the time of fetal viability.

Acapnia (ah-kap'ne-ah). A marked diminution in the amount of carbon dioxide in the blood.

Acetone bodies. Acetoacetic acid, beta-hydroxybutyric acid, and acetone; found in blood and urine in increased amounts whenever too much fat in proportion to carbohydrate is being oxidized. Also called *ketone bodies.*

Achlorhydria (ah-klor-hid're-ah). The absence of hydrochloric acid in the gastric juice.

Acid (as'id). A proton donor.

Acidosis (as-id-o'sis). Diminution in the reserve supply of fixed bases (especially sodium) in the blood.

Acuity (ak-u'it-e). Sharpness, or clearness, especially of vision.

Adaptation (a-dap-ta'shun). Adjustment to a stimulus; also used to denote changes in the retina on exposure to different intensities of light.

Adhesion (ad-he'zyun). Abnormal union of two surfaces as a result of inflammation.

Adolescence (ad-o-les'ens). Period between puberty and adult life.

Adventitia (ad-ven-tish'e-ah). The outermost covering of a structure that does not form an integral part of it.

Albuminuria (al''bu-min-u're-ah). Presence of albumin in the urine.

Alkalosis (al''kah-lo'sis). Increased bicarbonate content of the blood. It may be the result of ingesting large amounts of sodium bicarbonate, prolonged vomiting with loss of hydrochloric acid, or hyperventilation.

Alveolus (al-ve'ol-us). A small cell, or cavity; an air cell, one of the terminal dilatations of the bronchioles in the lungs; an acinus, or terminal lobule of a compound gland.

Ameboid movement (ah-me'boid moov'ment). Movement of a cell by extending from its surface processes of protoplasm (pseudopodia) toward which the rest of the cell flows.

Amenorrhea (am-en-or-e'ah). The absence of the menses.

Amorphous (a-mor'fus). Without definite shape or visible differentiation in structure; not crystalline.

Ampulla (am-pul'ah). A saccular dilatation of a canal.

Amyl nitrite (am'il ni'trīt). A volatile organic compound, which, when inhaled, produces dilatation of the blood vessels; it is used in attacks of angina pectoris.

Anabolism (a-nab'o-lizm). Those reactions in a plant or animal that result in the synthesis of larger molecules from smaller ones.

Analgesia (an-al-je'ze-ah). Loss of sensitivity to pain.

Anaphylactic (an-af-il-ak'tik). Increasing the susceptibility to the action of any foreign protein introduced into the body; decreasing immunity.

Anastomose (an-as'to-mōz). To open one into the other; used in connection with blood vessels, lymphatics, and nerves.

Anesthesia (an-es-the'ze-ah). Loss of sensation.

Aneurysm (an'ur-rizm). A sac formed by the dilatation of the walls of an artery and filled with blood.

Anisocytosis (an-i-so-si-to'sis). A lack of uniformity in the size of red blood corpuscles.

Ankylosis (ang-kil-o'sis). Abnormal immobility and consolidation of a joint due to bony union.

Anoxia (an-ox'e-ah). A diminished supply of oxygen to the body tissues.

Antibody (an'te-bod-ē). A specific substance produced by and in an animal or person as a reaction to the presence of an antigen; they are associated with certain globulin fractions of the plasma proteins.

Antigen (an'te-jen). Any substance which, when introduced into the blood or the tissues, incites the formation of antibodies, or reacts with them.

Antrum (an'trum). A cavity, or chamber, especially one within a bone, such as a sinus; the pyloric end of the stomach.

Aperture (ap'er-chur). An opening, or orifice.

Aphasia (ah-fa'ze-ah). A loss of the faculty of language in any of its forms.

Apoplexy (ap'o-plek-sē). A sudden loss of consciousness, followed by paralysis due to cerebral hemorrhage, or blocking of an artery of the brain by an embolus or a thrombus.

Aqueduct (ak'wē-dukt). A canal for the conduction of a liquid; the cerebral aqueduct of Sylvius connects the third and the fourth ventricles of the brain.

Areola (ar-e'o-lah). Any minute space in a tissue; the pigmented ring around the nipple.

Articulate (ar-tic'u-late). To join together so as to permit motion between parts; enunciation in words and sentences. Divided into joints.

Ascites (as-i'tez). An accumulation of serous fluid within the peritoneal cavity.

Asphyxia (as-fiks'e-ah). Unconsciousness due to interference with the oxygenation of the blood.

Astereognosis (ah-ste''re-og-no'sis). Loss of power to recognize the form of an object by touch.

Asthenia (as-the'ne-ah). Weakness.

Asthma (az'mah). A disease characterized by recurrent attacks of dyspnea, wheezing, and coughing, due to spasmodic contraction of the bronchioles.

Ataxia (ah-taks'e-ah). A loss of the power of muscular coordination.

Atresia (ah-tre'ze-ah). Congenital absence, or pathologic closure, of a normal opening or passage; involution of ovarian follicles.

Atrophy (at'ro-fe). A wasting, or diminution, in the size of a part of the entire body.

Atropine (at'ro-pin). An alkaloid obtained from atropa belladonna; it inhibits the action of the parasympathetic division of the autonomic system.

Auricle (aw'rik-l). The pinna, or flap, of the ear; a small pouch forming the upper portion of each atrium.

Auscultation (aws-kul-ta'shun). The act of listening for sounds within the body; employed as a diagnostic method.

Azygos (az'ig-os). An unpaired anatomic structure; the azygos vein arises from the right ascending lumbar vein and empties into the superior vena cava.

Base (bās). A proton acceptor.

Bolus (bo'lus). A rounded mass of soft consistency.

Bronchiectasis (brong-ke-ek'tas-is). Dilatation of the bronchi or of a bronchus. It is characterized by offensive breath, spasms of coughing, and expectoration of mucopurulent material.

Buffer (buff'er). Any substance that tends to lessen the change in hydrogen ion concentration which otherwise would be produced by adding acids or bases.

Bursa (bur'sah). A closed sac lined with synovial membrane containing fluid, found over an exposed and prominent part, or where a tendon plays over a bone.

Caffeine (kaf'e-in). An alkaloid obtained from the dried leaves of tea or the dried beans of coffee; employed in cardiac disorders.

Calculus (kal'ku-lus). A stone formed in any portion of the body, such as a gallstone.

Calorie (kal'or-e). A unit of heat. A *small calorie* (cal.) is the standard unit and is the amount of heat required to raise 1 Gm. of water from 15° to 16° C. The *large calorie* (Cal.) is used in metabolism and is the amount of heat required to raise 1 Kg. of water from 15° to 16° C.

Canaliculus (kan-al-ik'u-lus). A small canal or channel; in bone, minute channels connect with each lacuna.

Carcinoma (kar-sin-o'mah). A malignant tumor or cancer; a new growth made up of epithelial cells, tending to infiltrate and give rise to metastases.

Caries (ka're-ēz). Molecular decay or death of a bone in which it becomes softened, discolored, and porous. Dental caries, decay of teeth.

Cast (kast). A mold of a tubular structure, such as a bronchial tube or a renal tubule, formed by a plastic exudate.

Castration (kas-tra'shun). Removal of testes or ovaries.

Catabolism (ka-tab'ō-lizm). Those reactions in a plant or animal that result in the degradation, or oxidation, of molecules.

Cataract (kat'ah-rakt). A loss of transparency of the crystalline lens of the eye or of its capsule.

Celiac (se'le-ak). Relating to the abdominal cavity.

Centimeter (sen'tim-e-ter). The hundredth part of a meter; practically, 2/5 inch.

Centriole (sen'tri-ol). Organelle within the centrosome or cell center.

Cerumen (se-ru'men). Ear wax.

Cervix (ser'viks). The neck; any necklike structure; especially the lower cylindrical portion of the uterus.

Chalazion (ka-la'ze-on). A small tumor of the eyelid; formed by the distention of a meibomian gland with secretion.

Cheilosis (ki-lo'sis). A condition characterized by lesions on the lips and at the angles of the mouth.

Chemotherapy (kem''o-ther'a-py). The treatment of disease by administering chemicals that affect the causative organism unfavorably, but do not produce serious toxic effects in the patient.

Chiasma (ki-az'muh). A crossing; specifically, the crossing of the optic nerve fibers from the medial halves of the retinae.

Chromidial (kro-mid'e-al) **substance**. Pertaining to granules of extranuclear chromatin seen in the cytoplasm of a cell.

Chromosome (kro′mo-sōm). A body of chromatin in the cell nucleus that splits longitudinally as the cell divides, one half going to the nucleus of each of the daughter cells; the chromosomes transmit the hereditary characters.

Chyle (kīl). The milky fluid taken up by the lacteals from the food in the intestine after digestion. It consists of lymph and emulsified fat; it passes into the veins by the thoracic duct.

Chyme (kīm). The semifluid mass of partly digested food passed from the stomach into the duodenum.

Ciliary (sil′e-ar-re). Relating to (1) any hairlike process, (2) the eyelashes, (3) certain of the structures of the eyeball.

Cocaine (ko-kān). An alkaloid from coca leaves; it is a local anesthetic, narcotic and, mydriatic.

Coenzyme (kō-en′zīm). A nonprotein substance that is required for activity of an enzyme.

Collateral (kol-at′er-al). Accompanying; running by the side of; not direct; secondary or accessory; a small side branch of an axon.

Colloid (kol′oid). A state of subdivision of matter in which the individual particles are of submicroscopic size and consist either of large molecules, such as proteins, or aggregates of smaller molecules; the particles are not large enough to settle out under the influence of gravity.

Colostomy (ko-los′to-me). Establishment of an artificial anus by an opening into the colon.

Colostrum (ko-los′trum). The first milk secreted at the termination of pregnancy; it contains more lactalbumin and lactoprotein than the later milk.

Coma (ko′mah). A state of profound unconsciousness from which one cannot be roused.

Congenital (kon-jen′it-al). Born with a person; existing at or before birth.

Contralateral (kon″trah-lat′er-al). Situated on, or pertaining to, the opposite side.

Costal (cos′tal). Pertaining to a rib.

Crystalloid (kris′tal-oid). A body that, in solution, can pass through an animal membrane, as distinguished from a colloid that has not this property.

Curare (koo-rah′re). A highly toxic extract that paralyzes muscle; it acts on the motor endplates.

Cutaneous (ku-ta′ne-us). Pertaining to the skin.

Cyanosis (si-an-o′sis). A dark, purplish coloration of the skin and the mucous membrane, due to deficient oxygenation of the blood.

Cystoscopy (sis-tos′ko-pe). The inspection of the interior of the bladder.

Dentition (den-tish′un). Clinically, eruption of the teeth; morphologically the number, shape, and arrangement of the teeth.

Diapedesis (di″ah-pe-de′sis). The passage of blood cells through the unruptured walls of the blood vessels.

Diastole (di-as′to-le). The rhythmic period of relaxation and dilatation of the heart, during which it fills with blood.

Digitalis (dij-it-a′lis). The dried leaves of purple foxglove; it is used in the treatment of certain cardiac disorders.

Diopter (di-op′ter). The unit of refracting power of a lens; noting a lense whose principal focus is at a distance of 1 M.

Diploid (dip′loid). Having two sets of chromosomes, as normally found in the somatic cells of higher organisms.

Diurnal (di-er′nal). Daily.

Ectopic (ek-top′ik). Out of the normal place.

Edema (e-de′mah). An abnormal accumulation of clear, watery fluid in the lymph spaces of the tissues.

Effusion (ef-u′zhun). The escape of fluid from the blood vessels or the lymphatics into the tissues or a cavity.

Electrolyte (e-lek′tro-lite). Any substance that, in solution, conducts an electric current.

Embolism (em′bol-izm). Obstruction, or occlusion, of a vessel by a transported clot, a mass of bacteria, or other foreign material.

Empyema (em-pi-e′mah). The presence of pus in any cavity; when used without qualification, an accumulation of pus in the pleural cavity.

Endoplasmic reticulum (en′do-plas-mic re-tik′u-lum). Complicated system of internal membranes within cytoplasm of cell; ribosomes attached to inner surface.

Enuresis (en-u-re′sis). Involuntary passage of urine after the age of three years.

Enzyme (en′zīm). A protein that catalyzes a biochemical reaction.

Epistaxis (ep-e-staks′is). Nosebleed; hemorrhage from the nose.

Evagination (e-vaj-in-a′shun). A protrusion of some part or organ.

Exophthalmos (ex-of-thal′mus). A protrusion, or prominence, of the eyeball.

Extravasation (ex-trav-as-a′shun). The act of escaping from a vessel into the tissues; said of blood, lymph, or serum.

Extrinsic (ex-trin′sik). Originating outside of the part where it is found or upon which it acts.

Fasciculation (fa-sick″yoo-lay′shun). Localized contraction of muscle fibers, or an incoordinated contraction of skeletal muscle in which the fibers of one motor unit contract.

Fibrinolysin (fi″bri-no-li′sin). Any enzyme that catalyzes the digestion of fibrin.

Fimbria (fim′bre-ah). Fringelike structure; especially the fringelike end of the uterine tube.

Fistula (fis'tu-lah). A pathologic, or abnormal, passage leading from an abscess cavity or a hollow organ to the surface, or from one organ to another.

Flaccid (flak'sid). Relaxed, flabby, soft.

Flatus (flā'tus). Gas or air in the stomach or the intestine; commonly used to denote passage of gas by rectum.

Fovea (fo've-ah). A cup-shaped depression or pit.

Fundus (fun'dus). The bottom of a sac or hollow organ; that farthest removed from the opening.

Gangrene (gan'grēne). A form of necrosis combined with putrefaction; death of the tissue.

Gel (jel). A colloidal system comprising a solid and a liquid phase that exists as a solid or semisolid mass; a jelly or solid or semisolid phase.

Gene (jēn). An ultimate, ultramicroscopic, biologic unit of heredity; self-reproducing; located in a definite position on a particular chromosome.

Gradient (grā'di-ent). An ascending or descending slope. In the body, gradients are determined by the difference in concentration or electric charges across a semipermeable membrane.

Heat of vaporization. The heat energy required to convert 1 Gm. of a liquid into a vapor without a change in the temperature of the substance being vaporized.

Hematuria (hem-at-u're-ah). The presence of blood in the urine.

Hemiplegia (hem-e-ple'je-ah). Paralysis of one side of the body.

Hemorrhoids (hem'or-oidz). Piles; a varicose dilatation of a vein of the superior or inferior hemorrhoidal plexus, causing painful swelling at the anus.

Hemostasis (hēm-o-sta'sis). The arrest of bleeding; the checking of the flow of blood through any part of a vessel.

Heredity (he-red'it-e). The transmission of qualities from parent to offspring.

Hilus (hi'lus). A depression, or pit, at that part of an organ where the vessels and nerves enter or leave.

Homeostasis (ho-mē-o-sta'sis). A tendency to uniformity or stability in the normal body states of the organism (Cannon).

Homologous (ho-mol'o-gus). Corresponding; having similar relations.

Hordeolum (hor-de'o-lum). A sty; an inflammation of a sebaceous gland of the eyelid.

Hyaluronidase (hy″a-lu-ron'i-dase). An enzyme causing breakdown of hyaluronic acid in protective polysaccharide barriers, promoting invasion of cells and tissues by the invading agent; it is a spreading factor.

Hyperplasia (hi″per-plā'ze-ah). The abnormal multiplication, or increase, in the number of normal cells in normal arrangement in a tissue.

Hypertrophy (hi-per'tro-fe). The morbid enlargement, or overgrowth, of an organ or part, due to an increase in size of its constituent cells.

Hypothermia (hi″po-ther'me-ah). Low temperature; especially a state of low body temperature induced for the purpose of decreasing metabolic activities and need for oxygen.

Inflammation (in-flam-a'shun). A series of reactions produced in the tissues by an irritant; it is marked by an afflux of blood with exudation of plasma and leukocytes.

Infundibulum (in-fun-dib'u-lum). A funnel-shaped structure or passage.

Intravascular (in-trah-vas'ku-lar). Within a vessel or vessels.

Intussusception (in″tus-sus-sep'shun). The infolding of one segment of the intestine within another segment.

Invagination (in-vaj-in-a'shun). The pushing of the wall of a cavity into the cavity.

Involution (in-vo-lu'shun). The return of an enlarged organ to normal size; retrograde changes.

Ion (i'on). An electrically charged atom or group of atoms formed by the loss or gain of electrons.

Ipsilateral (ip″si-lat'er-al). Situated on or pertaining to the same side; homolateral.

Isotope (i'so-tōp). An element which has the same atomic number as another, but a different atomic weight. Radioactive isotopes usually refer to elements rendered radioactive by artificial means.

Ketosis (ke-to'sis). The condition marked by excessive production of ketone bodies in the body.

Kilogram (kil'o-gram). 1,000 Gm.; about 2.2 lb. avoirdupois.

Kinesthetic (kin-es-thet'ik). Pertaining to muscle sense, or to the sense by which muscular movement, weight, position are perceived.

Lecithin (les'i-thin). A monoaminomonophosphatide found in animal tissues especially nerve tissue, semen, egg yolk, and in smaller amount in bile and blood.

Leukemia (lu-ke'me-ah). A disease of the blood marked by persistent leukocytosis, associated with changes in the spleen and the bone marrow, or in the lymphatic nodes.

Lumen (lu'men). The space in the interior of a tubular structure such as an artery or the intestine.

Macula (mak′u-lah). A spot.

Malaise (ma-laz′). A feeling of general discomfort or uneasiness; an out-of-sorts feeling, often the first indication of an infection.

Meatus (me-a′tus). A passage, or channel, especially the external opening of a canal.

Menopause (men′o-pawz). Termination of the menstrual cycles.

Metabolism (me-tab′o-lism). The sum of the chemical changes whereby the function of nutrition is effected; it consists of anabolism or the constructive and assimilative changes and catabolism or the destructive and retrograde changes.

Meter (me′ter). A measure of length; 100 cm.; the equivalent of 39.371 inches.

Microgram (mi′kro-gram). One one-millionth of a gram, or 1/1,000 of a milligram.

Micron (mi′kron). One one-millionth of a meter or 1/1,000 of a millimeter; 1/25,000 of an inch.

Millimeter (mil′im-e-ter). One one-thousandth of a meter; about 1/25 inch.

Mitochondria (mit″o-kon′dre-ah). Organelles in the cytoplasm of cells; contain enzymes that make possible the reactions whereby energy is liberated from food and stored temporarily in the chemical bonds of ATP.

Morphine (mor′fin). The chief narcotic principle of opium; used as a hypnotic and analgesic.

Mucus (mu′kus). The viscid watery secretion of the mucous glands; it consists of water, mucin, epithelial cells, leukocytes, and inorganic salts.

Narcosis (nar-co′sis). Stupor, or unconsciousness, produced by some narcotic drug.

Necrosis (ne-kro′sis). Local death of tissue.

Nystagmus (nis-tag″mus). Rhythmic oscillation of the eyeballs, either horizontal, rotary, or vertical.

Orifice (or′if-is). Any aperture or opening.

Ostium (os′te-um). A small opening, especially one of entrance into a hollow organ or canal.

Oxidation (oks-id-a′shun). The combining of food and oxygen in the tissues; chemically, the increase in valence of an element.

Oximeter (ox-im′e-ter). An instrument for measuring the oxygen saturation of hemoglobin in the circulating blood.

Pacchionian bodies (pak-ke-o′ne-an). Small projections of the arachnoid tissue, chiefly into the venous sinuses of the dura mater.

Palpitation (pal-pit-a′shun). Forcible pulsation of the heart perceptible to the individual.

Paralysis (par-al′is-is). A loss of power of voluntary movement in a muscle through injury or disease of its nerve supply.

Parenchyma (par-eng′ki-mah). The essential elements of an organ; the functional elements of an organ, as distinguished from its framework, or stroma.

Parturition (par″tu-rish′un). Giving birth to young.

Pectoral (pek′to-ral). Pertaining to the breast or the chest.

Perimeter (per-im′et-er). An instrument delimiting the field of vision.

Perineum (per-e-ne′um). The portion of the body included in the outlet of the pelvis, extending from the pubic arch to the coccyx and between the ischial tuberosities.

pH. The symbol commonly used in expressing hydrogen ion concentration. It signifies the logarithm of the reciprocal of the hydrogen ion concentration expressed as a power of 10.

Phlebothrombosis (flebo″-throm-bos′is). Thrombosis of a vein without inflammation of its walls.

Physiotherapy (fiz″e-o-ther′ap-e). The use of natural forces, such as heat, light, air, water, and exercise, in the treatment of disease.

Pilocarpine (pi-lo-kar′pin). An alkaloid that stimulates the parasympathetic division of the autonomic nervous system.

Plexus (plex′us). A network, or tangle, of interweaving nerves, veins, or lymphatic vessels.

Potential (po-ten′shal). Possible, but not actual; capable of doing or being, though not yet doing or being.

Pressor (pres′or). Exciting vasoconstrictor activity, producing increased blood pressure; denoting afferent nerves that when stimulated, excite the vasoconstrictor center.

Prolapse (pro′laps). The falling down of an organ or other part, such as the uterus.

Proliferation (pro-lif″er-a′shun). The reproduction, or multiplication, of similar forms, especially of cells.

Psychosomatic (si″ko-so-mat′ik). Pertaining to the mind-body relationship; having bodily symptoms of a psychic, emotional, or mental origin.

Pterygoid (ter′e-goid). Shaped like a wing.

Puberty (pu′ber-te). The age at which the reproductive organs become functionally operative.

Pus (pus). A fluid product of inflammation consisting of a liquid containing leukocytes and the débris of dead cells.

Rachitic (ra-kit′ik). Relating to, or affected by, rickets.

Ramus (ra′mus). A branch; one of the primary divisions of a nerve or a blood vessel; a part of an irregularly shaped bone that forms an angle with the main body.

Receptor (re-sep′tor). Nerve ending that receives a stimulus.

Reflection (re-flek′shun). A turning or bending back.

Refraction (re-frak'shun). The bending of a ray of light as it passes from one medium into another of different density.

Regurgitation (re-gur-gi-ta'shun). The flowing backward of blood through incompletely closed heart valves; the return of small amounts of food from the stomach.

Resorption (re-sorp'shun). The loss of substance through physiologic or pathologic means.

Reticular (re-tik'u-lar). Netlike.

Retinaculum (ret"i-nak'u-lum). A special fascial thickening which holds back an organ or part; helps to retain an organ or tissue in its place.

Retroversion (re-tro-ver'shun). The tipping of an entire organ backward; turning backward without flexion, or bending, of the organ, as of the uterus.

Rh antigen or factor. An agglutinogen, or antigen, first found in the erythrocytes of the rhesus monkey, hence the Rh. Rh positive and Rh negative are terms denoting the presence or absence, respectively, of this antigen.

Ribosomes (rye'bo-somes). Organelles which appear as dots lining the endoplasmic reticulum; they are the protein factories of cells.

Sella turcica (sel'ah tur'sik-ah). A saddlelike depression on the upper surface of the sphenoid bone, in which the hypophysis lies.

Serotonin (sere"o-toe'nin). A compound (5-hydroxytryptamine) found in the bloodstream; it has vasoconstrictive properties. It is possible that serotonin is concerned with some of the functions of the central nervous system.

Sinus (si'nus). A channel for the passage of blood; a hollow in a bone or other tissue; antrum; one of the cavities connecting with the nose; a suppurating cavity.

Sinusoid (si'nus-oid). A blood space in certain organs, as the spleen and the liver.

Spasm (spazm). An involuntary, convulsive, muscular contraction.

Specific heat. The heat energy required to raise the temperature of one gram of a substance one degree centigrade.

Spermatogenesis (sper"mat-o-jen'es-is). The process of formation and development of the spermatozoa.

Sphincter (sfingk'ter). A circular muscle that serves, when in a state of normal contraction, to close one of the orifices of the body.

Squamous (skwa'mus). Scalelike.

Sterility (ster-il'it-e). Infertility, lack of progeny, unproductiveness.

Stricture (strik'tur). A circumscribed narrowing of a tubular structure.

Stroma (stro-mah). The tissue that forms the ground substance, framework, or matrix of an organ, as distinguished from that constituting its functional element, or parenchyma.

Susceptibility (sus-sep-tib-il'it-e). The characteristic that makes one liable to infection by disease.

Syncytium (sin-sit'e-um). A multinucleate mass of protoplasm produced by the merging of cells.

Syndrome (sin'drohm). A group of symptoms and signs, which, when considered together, characterize a disease or lesion.

Synovial (sin-o've-al). Of, or pertaining to, or secreting synovia; synovia is the viscid fluid of a joint or similar cavity.

Systole (sis'to-le). The contraction of the heart muscle.

Tactile (tak'til). Pertaining to the sense of touch.

Tetany (tet'an-e). Intermittent tonic muscular contractions of the extremities.

Thrombophlebitis (throm"bo-fle-bye'tis). The condition in which inflammation of the vein wall has preceded the formation of a thrombus, or intravascular clot.

Thrombosis (throm-boe'sis). The formation of a clot within a vessel during life.

Thrombus (throm'bus). A clot of blood formed within the heart or the blood vessels, due usually to slowing of the circulation of the blood or to alteration of the blood itself or the vessel walls.

Tinnitus (tin'i-tus). A ringing or singing sound in the ears.

Trabecula (tra-bek'u-lah). A septum that extends from an envelope into the enclosed substance, forming an essential part of the stroma of the organ.

Trigone (tri'gōn). Denoting a triangular area, especially of the interior of the bladder between the opening of the ureters and the orifice of the urethra.

Ulcer (ul'ser). An open sore other than a wound; a loss of substance on a cutaneous or mucous surface causing necrosis of the tissue.

Umbilicus (um-bil-i'kus). The navel; the cicatrix (scar) that marks the site of attachment of the umbilical cord.

Urticaria (er-ti-ka're-ah). Nettle-rash; hives; elevated, itching, white patches.

Varicocele (var'ik-o-sēl). Enlargement of the veins of the spermatic cord.

Varicose veins (var'ik-ōs). Unnaturally swollen and tortuous veins.

Vertigo (ver'tig-o). Dizziness, giddiness.

Vesicle (ves'ik-l). A small bladder, or sac, containing liquid.

Vestibular (ves-tib'u-lar). Pertaining to a vestibule; such as the inner ear, larynx, mouth, nose, vagina.

Viscosity (vis-kos'it-e). A condition of more or

less adhesion of the molecules of a fluid to each other so that it flows with difficulty.

Volvulus (vol'vu-lus). A twisting of the intestine, causing obstruction.

COMBINING FORMS AND PREFIXES

These forms, with a prefix or a suffix, or both, are those most commonly used in making medical words. G indicates those from the Greek; L, those from the Latin. Properly Greek forms should be used only with Greek prefixes and suffixes; Latin, with Latin. Often a vowel, usually a, i, or o, is needed for euphony.

A- or **Ab-** (L) *away, lack of:* abnormal, departing from normal.

A- or **An-** (G) *from, without:* asepsis, without infection.

Acr- (G) *an extremity:* acrodermatitis, a dermatitis of the limbs.

Ad- (L) *to, toward, near:* adrenal, near the kidney.

Aden- (G) *gland:* adenitis, inflammation of a gland.

Alg- (G) *pain:* neuralgia, pain extending along nerves.

Ambi- (L) *both:* ambidextrous, referring to both hands.

Ante- (L) *before:* antenatal, occurring, or having been formed, before birth.

Anti- (G) *against:* antiseptic, against or preventing sepsis.

Arth- (G) *joint:* arthritis, inflammation of a joint.

Auto- (G) *self:* autointoxication, poisoning by toxin generated in the body.

Bi- or **Bin-** (L) *two:* binocular, pertaining to both eyes.

Bio- (G) *life:* biopsy, inspection of living organism (or tissue).

Blast- (G) *bud, a growing thing in early stages:* blastocyte, beginning cell not yet differentiated.

Bleph- (G) *eyelids:* blepharitis, inflammation of an eyelid.

Brachi- (G) *arm:* brachialis, muscle for flexing forearm.

Brachy- (G) *short:* brachydactylia, abnormal shortness of fingers and toes.

Brady- (G) *slow:* bradycardia, abnormal slowness of heartbeat.

Bronch- (G) *windpipe:* bronchiectasis, dilation of bronchial tubes.

Bucca- (L) *cheek:* buccally, toward the cheek.

Carcin- (G) *cancer:* carcinogenic, producing cancer.

Cardi- (G) *heart:* cardialgia, pain in the heart.

Cephal- (G) *head:* encephalitis, inflammation of brain.

Cheil- (G) *lip:* cheilitis, inflammation of the lip.

Chole- (G) *bile:* cholecyst, the gallbladder.

Chondr- (G) *cartilage:* chondrectomy, removal of a cartilage.

Circum- (L) *around:* circumocular, around the eyes.

Cleid- (G) *clavicle:* cleidocostal, pertaining to clavicle and ribs.

Colp- (G) *vagina:* colporrhagia, vaginal hemorrhage.

Contra- (L) *against, opposed:* contraindication, indication opposing usually indicated treatment.

Cost- (L) *rib:* intercostal, between the ribs.

Counter- (L) *against:* counterirritation, an irritation to relieve some other irritation (e.g., a liniment).

Crani- (L) *skull:* craniotomy, surgical opening in skull.

Crypt- (G) *hidden:* cryptogenic, of hidden, or unknown, origin.

Cut- (L) *skin:* subcutaneous, under the skin.

Cyst- (G) *sac or bladder:* cystitis, inflammation of the bladder.

Cyto- (G) *cell:* cytology, scientific study of cells; a device for counting and measuring cells.

Dacro- (G) *tear:* dacrocyst, the lacrimal sac.

Derm- or **Dermat-** (G) *skin:* dermatoid, skinlike.

Di- (L) *two:* diphasic, occurring in two stages, or phases.

Dis- (L) *apart:* disarticulation, taking joint apart.

Dys- (G) *pain or difficulty:* dyspepsia, impairment of digestion.

Ecto- (G) *outside:* ectoretina, outermost layer of retina.

Em- or **En-** (G) *in:* encapsulated, enclosed in a capsule.

Encephal- (G) *brain:* encephalitis, inflammation of brain.

End- (G) *within:* endothelium, the layer of cells lining the heart, the blood, and the lymph vessels.

Entero- (G) *intestine:* enteroptosis, falling of intestine.

Epi- (G) *above or upon:* epidermis, outermost layer of skin.

Erythro- (G) *red:* erythrocyte, red blood cell.

Eu- (G) *well:* euphoria, well feeling, feeling of good health.

Ex- or **E-** (L) *out:* excretion, material thrown out of the body or the organ.

Exo- (G) *outside:* exocrine, excreting outwardly (opposite of endocrine).

Extra- (G) *outside:* extramural, situated or occurring outside a wall.

Febri- (L) *fever:* febrile, feverish.

Galacto- (G) *milk:* galactose, a milk-sugar.
Gastr- (G) *stomach:* gastrectomy, excision of the stomach.
Gloss- (G) *tongue:* glossectomy, surgical removal of tongue.
Glyco- (G) *sugar:* glycosuria, sugar in the urine.
Gynec- (G) *woman:* gynecology, science of diseases pertaining to women.

Hem- or **Hemat-** (G) *blood:* hemopoiesis, forming blood.
Hemi- (G) *half:* heminephrectomy, excision of half the kidney.
Hepat- (G) *liver:* hepatitis, inflammation of the liver.
Hetero- (G) *other* (opposite of homo): heterotransplant, using skin from a member of another species.
Hist- (G) *tissue:* histology, science of minute structure and function of tissues.
Homo- (G) *same:* homotransplant, skin grafting by using skin from a member of the same species.
Hydr- (G) *water:* hydrocephalus, abnormal accumulation of fluid in cranium.
Hyper- (G) *above, excess of:* hyperglycemia, excess of sugar in blood.
Hypo- (G) *under, deficiency of:* hypoglycemia, deficiency of sugar in blood.
Hyster- (G) *uterus:* hysterectomy, excision of uterus.

Idio- (G) *self, or separate:* idiopathic, a disease self-originated (of unknown cause).
Im- or **In-** (L) *in:* infiltration, accumulation in tissue of abnormal substances.
Im- or **In-** (L) *not:* immature, not mature.
Infra- (L) *below:* infraorbital, below the orbit.
Inter- (L) *between:* intermuscular, between the muscles.
Intra- (L) *within:* intramuscular, within the muscle.

Kerat- (G) *horn, cornea:* keratitis, inflammation of cornea.

Lact- (L) *milk:* lactation, secretion of milk.
Leuk- (G) *white:* leukocyte, white cell.

Macro- (G) *large:* macroblast, abnormally large red cell.
Mast- (G) *breast:* mastectomy, excision of the breast.
Meg- or **Megal-** (G) *great:* megacolon, abnormally large colon.
Ment- (L) *mind:* dementia, deterioration of the mind.

Mer- (G) *part:* merotomy, division into segments.
Mes- (G) *middle:* mesaortitis, inflammation of middle coat of the aorta.
Meta- (G) *beyond, over, change:* metastasis, change in seat of a disease.
Micro- (G) *small:* microplasia, dwarfism.
My- (G) *muscle:* myoma, tumor made of muscular elements.
Myc- (G) *fungi:* mycology, science and study of fungi.

Necro- (G) *corpse, dead:* necrosis, death of cells adjoining living tissue.
Neo- (G) *new:* neoplasm, any new growth or formation.
Neph- (G) *kidney:* nephrectomy, surgical excision of kidney.
Neuro- (G) *nerve:* neuron, nerve cell.

Odont- (G) *tooth:* odontology, dentistry.
Olig- (G) *little:* oligemia, deficiency in volume of blood.
Oo- (G) *egg:* oocyte, original cell of egg.
Oophor- (G) *ovary:* oophorectomy, removal of an ovary.
Ophthalm- (G) *eye:* ophthalmometer, an instrument for measuring the eye.
Ortho- (G) *straight, normal:* orthograde, walk straight (upright).
Oss- (L) *bone:* osseous, bony.
Oste- (G) *bone:* osteitis, inflammation of a bone.
Ot- (G) *ear:* otorrhea, discharge from ear.

Para- (G) *irregular, around, wrong:* paradenitis, inflammation of tissue in the neighborhood of a gland.
Path- (G) *disease:* pathology, science of disease.
Ped-[1] (G) *children:* pediatrician, child specialist.
Ped-[2] (L) *feet:* pedograph, imprint of the foot.
  [1] Ped—from Greek *pais,* child.
  [2] Ped—from Latin *pes,* foot.
Per- (L) *through, excessively:* percutaneous, through the skin.
Peri- (G) *around, immediately around* (in contradistinction to para): periapical, surrounding apex of root of tooth.
Phil- (G) *love:* hemophilic, fond of blood (as bacteria that grow well in presence of hemoglobin).
Phleb- (G) *vein:* phlebotomy, opening of vein for bloodletting.
Phob- (G) *fear:* hydrophobic, reluctant to associate with water.
Pneum- or **Pneumon-** (G) *lung* (pneum-air): pneumococcus, organism causing lobar pneumonia.
Polio- (G) *gray:* poliomyelitis, inflammation of gray substance of spinal cord.
Poly- (G) *many:* polyarthritis, inflammation of several joints.

Post- (L) *after:* postpartum, after delivery.

Pre- (L) *before:* prenatal, occurring before birth.

Pro- (L and G) *before:* prognosis, forecast as to result of disease.

Proct- (G) *rectum:* proctectomy, surgical removal of rectum.

Pseudo- (G) *false:* pseudoangina, false angina.

Psych- (G) *soul or mind:* psychiatry, treatment of mental disorders.

Py- (G) *pus:* pyorrhea, discharge of pus.

Pyel- (G) *pelvis:* pyelitis, inflammation of pelvis of kidney.

Rach- (G) *spine:* rachicentesis, puncture into vertebral canal.

Ren- (L) *kidney:* adrenal, near the kidney.

Retro- (L) *backward:* retroversion, turned backward (usually, of uterus).

Rhin- (G) *nose:* rhinology, knowledge concerning noses.

Salping- (G) *a tube:* salpingitis, inflammation of tube.

Semi- (L) *half:* semicoma, mild coma.

Septic- (L and G) *poison:* septicemia, poisoned condition of blood.

Somat- (G) *body:* psychosomatic, having bodily symptoms of mental origin.

Sta- (G) *make stand:* stasis, stoppage of flow of fluid.

Sten- (G) *narrow:* stenosis, narrowing of duct or canal.

Sub- (L) *under:* subdiaphragmatic, under the diaphragm.

Super- (L) *above, excessively:* superacute, excessively acute.

Supra- (L) *above, upon:* suprarenal, above or upon the kidney.

Sym- or Syn- (G) *with, together:* symphysis, a growing together.

Tachy- (G) *fast:* tachycardia, fast-beating heart.

Tens- (L) *stretch:* extensor, a muscle extending or stretching a limb.

Therm- (G) *heat:* diathermy, therapeutic production of heat in tissues.

Tox- or Toxic- (G) *poison:* toxemia, poisoned condition of blood.

Trache- (G) *trachea:* tracheitis, inflammation of the trachea.

Trans- (L) *across:* transplant, transfer tissue from one place to another.

Tri- (L and G) *three:* trigastric, having three bellies (muscle).

Trich- (G) *hair:* trichosis, any disease of the hair.

Uni- (L) *one:* unilateral, affecting one side.

Vas- (L) *vessel:* vasoconstrictor, nerve or drug that narrows blood vessel.

Zoo- (G) *animal:* zooblast, an animal cell.

SUFFIXES

-algia (G) *pain:* cardialgia, pain in the heart.

-asis or -osis (G) *affected with:* leukocytosis, excess number of leukocytes.

-asthenia (G) *weakness:* neurasthenia, nervous weakness.

-blast (G) *germ:* myeloblast, bone-marrow cell.

-cele (G) *tumor, hernia:* enterocele, any hernia of intestine.

-cid (L) *cut, kill:* germicidal, destructive to germs.

-clysis (G) *injection:* hypodermoclysis, injection under the skin.

-coccus (G) *round bacterium:* pneumococcus, bacteria of pneumonia.

-cyte (G) *cell:* leukocyte, white cell.

-ectasis (G) *dilation, stretching:* angiectasis, dilatation of a blood vessel.

-ectomy (G) *excision:* adenectomy, excision of adenoids.

-emia (G) *blood:* glycemia, sugar in blood.

-esthesia (G) *(noun) relating to sensation:* anesthesia, absence of feeling.

-ferent (L) *bear, carry:* efferent, carry out to periphery.

-genic (G) *producing:* pyogenic, producing pus.

-iatrics (G) *pertaining to a physician or the practice of healing* (medicine): pediatrics, science of medicine for children.

-itis (G) *inflammation:* tonsillitis, inflammation of tonsils.

-logy (G) *science of:* pathology, science of disease.

-lysis (G) *losing, flowing, dissolution:* autolysis, dissolution of tissue cells.

-malacia (G) *softening:* osteomalacia, softening of bone.

-oma (G) *tumor:* myoma, tumor made up of muscle elements.

-osis (-asis) (G) *being affected with:* atherosis, arteriosclerosis.

-(o)stomy (G) *creation of an opening:* gastrostomy, creation of an artificial gastric fistula.

-(o)tomy (G) *cutting into:* laparotomy, surgical incision into abdomen.

-pathy (G) *disease:* myopathy, disease of a muscle.

-penia (G) *lack of:* leukopenia, lack of white blood cells.

-pexy (G) *to fix:* proctopexy, fixation of rectum by suture.

-phagia (G) *eating:* polyphagia, excessive eating.

-phasia (G) *speech:* aphasia, loss of power of speech.

-phobia (G) *fear:* hydrophobia, fear of water.

-plasty (G) *molding:* gastroplasty, molding or reforming stomach.

-poiesis (G) *making, forming:* hematopoiesis, forming blood.

-pnea (G) *air or breathing:* dyspnea, difficult breathing.

-ptosis (G) *falling:* enteroptosis, falling of intestine.

-rhythmia (G) *rhythm:* arrhythmia, variation from normal rhythm of heart.

-rrhagia (G) *flowing or bursting forth:* otorrhagia, hemorrhage from ear.

-rrhaphy (G) *suture of:* enterorrhaphy, act of sewing up gap in intestine.

-rrhea (G) *discharge:* otorrhea, discharge from ear.

-sthen (ia) (ic) (G) *pertaining to strength:* asthenia, loss of strength.

-taxia or **-taxis** (G) *order, arrangement of:* ataxia, failure of muscular coordination.

-trophia or **-trophy** (G) *nourishment:* atrophy, wasting, or diminution.

-uria (G) *to do with urine:* polyuria, excessive secretion of urine.

Index